BURTON PAUL

Professor of Mechanical Engineering
University of Pennsylvania

KINEMATICS AND DYNAMICS
OF PLANAR MACHINERY

PRENTICE-HALL, INC., *Englewood Cliffs, New Jersey 07632*

Library of Congress Cataloging in Publication Data

Paul, Burton, 1931–
    Kinematics and dynamics of planar machinery.

    Bibliography: p.
    Includes index.
    1. Machinery, Kinematics of. 2. Machinery, Dynamics
of. I. Title.
TJ175.P38   621.8'11   78-25588
ISBN 0-13-516062-6

Editorial production supervision
and interior design by: James M. Chege

Cover design by: Edsal Enterprises

Manufacturing buyer: Gordon Osbourne

Printed in the United States of America

10 9 8 7 6 5 4 3 2

Prentice-Hall International, Inc., *London*
Prentice-Hall of Australia Pty. Limited, *Sydney*
Prentice-Hall of Canada, Ltd., *Toronto*
Prentice-Hall of India Private Limited, *New Delhi*
Prentice-Hall of Japan, Inc., *Tokyo*
Prentice-Hall of Southeast Asia Pte. Ltd., *Singapore*
Whitehall Books Limited, *Wellington, New Zealand*

To *Lois,*

    In loving appreciation for the countless ways in which you helped.

CONTENTS

PREFACE

RAISON D'ÊTRE

This book deals with the kinematics, statics, and kinetics of machines (e.g. reciprocating engines, textile machinery, and linkages generally) which undergo appreciable variations in their effective moments of inertia during normal operation. The textbook literature in this field is characterized by a heavy reliance on graphical methods. This tendency is undoubtedly due to the nonlinear nature of the equations which govern the motion of such systems. It was the unhappy lot of early machine designers to grapple with nonlinear problems at a time when few mathematical tools existed for their solution. Consequently little effort was made to calculate the transient motions (departure from *steady flywheel* conditions), and the stresses in the moving parts could not be calculated with adequate precision. Such an approach leads to a cycle of designing, fabricating, testing, modifying, etc. which often consumes years of development effort at considerable expense.

Fortunately, the solution of nonlinear equations (algebraic or differential) has become a routine matter now that digital computers are widely available. Thus we have, in principal, a way to extricate ourselves from the inadequate and, at best, time-consuming drawing-board approach to kinematics and dynamics. General-purpose computer programs for the dynamic analysis of variable inertia machinery have emerged in recent years, and a brief survey of such programs is given in the final chapter. However, it is decidedly not my intention to merely present a set of recipes (or canned programs) which solve a variety of problems in the subject. The fundamental approach taken in this book is to reframe the pertinent problems of kinematics and dynamics in a form better suited for *numerical* analysis rather than the traditional graphical analysis. The methods expounded may be demonstrated with ordinary electronic calculators, but they are most effectively utilized with the aid of digital computers (or programmable calculators). Ample information is provided for users to generate their own programs at varying levels of sophistication, to utilize the several programs which are fully developed in the text, or to obtain the larger, more sophisticated general-purpose programs.

COVERAGE AND ORGANIZATION OF BOOK

To keep the size of the book within reasonable bounds it was decided to restrict the discussion to *plane* systems. A second major decision was to concentrate on methods of *analysis* as opposed to *direct synthesis* procedures. To some extent, the extreme rapidity of computer methods of analysis provides a capability for synthesis by trial and error. However, many methods of direct kinematic synthesis developed in recent years can be found in references cited.

The text has been arranged into 3 parts. Part 1, *Geometrical Kinematics*, covers classical material in kinematics (e.g., centrodes, cam and gear geometry, four-bar linkage classifications, curvature relations) that would not ordinarily receive much emphasis in the usual prerequisite courses on engineering mechanics. Although I feel that all mechanical engineering students should ultimately be exposed to the many important kinematic ideas considered in Part 1, this material is not essential for subsequent parts of the book, and it may be omitted entirely or in part. A brief treatment of graphical methods for velocity and acceleration analysis is presented as an optional topic because of its value as a check on analytical methods and because so much of the literature in the field presupposes some knowledge of graphical methods. However, any problem which can be solved graphically (and many which cannot) may be solved by the analytic methods of Part 2.

Part 2, *Analytical Kinematics*, represents a break with traditional textbook approaches to kinematics. Starting with single-loop mechanisms, it is shown how algebraic expressions may be systematically derived for the determination of all unknown displacements, velocities, and accelerations. The method used is valid for plane mechanisms with connecting elements of all types and any number of degrees of freedom. It is then shown how the mobility of the most general multiloop mechanism depends on some simple topological characteristics of the linkage network. The essential background on topology needed to understand much of the current literature on mechanisms is included in an appendix. Since topology alone does not determine the mobility of a mechanism in certain singular states (e.g., dead centers, change points, or critical forms), it is necessary to examine classical questions of rank and linear dependence for the fundamental *velocity matrix* associated with any given mechanism. An appendix on matrices and linear algebraic equations provides all the associated mathematical and numerical background that is needed for readers of this book.

The independent loops of the mechanism are exploited to derive appropriate equations for the determination of position, velocity, and acceleration variables for all the driven links of the system. The concept of weak and strong coupling of loops clarifies the difficulties which have been encountered in past attempts to apply graphical methods to so-called *complex mechanisms*.

In part 3, *Analytical Mechanics of Machinery*, it is shown how the fundamental kinematic ideas developed in Part 2 are utilized in a powerful and systematic approach to the *statics* and *dynamics* of all planar machinery. The bridge between kinematics and statics is the *principle of virtual work*, which is carefully developed and illustrated by numerous examples. In building this bridge, we develop and draw freely upon fundamental concepts of *analytical mechanics*, such as generalized coordinates, virtual displacements, workless constraints, and generalized forces. These ideas are blended with the previously developed methods of velocity analysis in order to establish general procedures for the analysis of force equilibrium in arbitrary plane machines.

Having introduced the concepts of generalized coordinates and generalized forces and efficient ways of calculating velocity and centripetal coefficients, we rapidly proceed to establish the generalized equation of motion which governs all mechanical systems with one degree of freedom.

Although an analysis of machines with one degree of freedom covers such important systems as internal combustion engines, punch presses, four-bar mechanisms, and common machine tools, more advanced applications require a more general treatment. It is demonstrated that the appropriate equations of motion may be developed in a form well adapted to computer solution by utilizing Lagrange's form of d'Alembert's principle, i.e., the dynamical generalization of the principle of virtual work.

To fully exploit this formulation of the equations of motion for multifreedom linkages, it is necessary to draw upon the panoply of concepts from analytical kinematics (e.g., velocity and centripetal coefficients) developed and utilized in Part 2 for purely kinematical purposes. This procedure of continuously generalizing and building upon previously developed concepts (which are of great practical utility in their own right) is an example of that "economy of thought" which Ernst Mach has singled out as the major contribution of Lagrangian mechanics.

STUDENT LEVEL AND DESIGN OF COURSES

The level of the book is suitable for undergraduates who have had a traditional course in engineering mechanics. Selected parts of many chapters are suitable for graduate courses in the theory of machines.

The complete book can be covered in one academic year, but the material is arranged in such a way that it is readily adapted to one-semester sequences in either kinematics *or* dynamics or to a compact course in kinematics *and* dynamics of machinery. For example, at the University of Pennsylvania we have only one semester available (at the Junior or Senior level) for the full range of mechanics of machinery. Within that semester I have been able to include excerpts from Part 1 (geometrical kinematics), most of Part 2 (analytical kinematics), and Part 3 (analytical dynamics) through the dynamics of systems with one degree of freedom.

The arrangement of the book affords the utmost flexibility with respect to the student's generation and use of computer programs. In some years, I have covered the range of material just described without assigning any problems that required the use of a computer. In other years, I have devoted about twenty percent of class time, distributed over the semester, to discussion of numerical analysis and computer program utilization. Ample reading material for this purpose is provided in the appendices.

NOTE ON COMPUTER PROGRAMS

The following FORTRAN programs are developed and fully listed in this book:

NEWTR (NEWTon-Raphson algorithm), Appendix **F**
DKINAL (KINematics anALysis), Sec. **10.43**
DYREC (DYnamics of RECiprocating machines), Sec. **12.83**, Appendix **I**
FLYLOOP (FLYwheel design by energy LOOP method), Sec. **12.92**, Appendix **J**

General-purpose programs referred to (but not listed) include

KINMAC (KINematics of MAChines), Sec. **10.43**
DYMAC (DYnamics of MAChines), Sec. **14.40**
STATMAC (STATics of MAChines), Sec. **14.45**
CAMKIN (CAM KINematics), Sec. **14.40**

Source decks, or tapes, for all these programs are available from the author. For information on availability of other programs referred to (e.g., in Sec. **14.30**), contact the individual authors cited.

ACKNOWLEDGMENTS

I would like to thank Pat Burset, Doris Springirth, and Margaret Kirsch for their excellent secretarial assistance. I also appreciate the meticulous care and dedication lavished upon the artwork by Renate Schulz.

Those who collaborated with me in the research upon which much of this book is based include: Drs. Dusan Krajcinovic, George Hud, and Arvind Amin. Dr. Amin also worked out the solution to many of the problems in Chapters 7–12, on the kinematics, statics, and dynamics of machinery, via the computer programs KINMAC, STATMAC, and DYMAC. Mr. Carl Van Dyke collaborated with me in the development of the computer program CAMKIN, which he used to solve many of the problems on cam kinematics in Chapter 4. As the book evolved, various drafts were tested for readability and teachability by the students in my courses MEAM 452 and 518, and by graduate student assistants—Drs. K. P. Singh, G. Hud, W. Woodward, and A. Amin.

Dr. R. S. Berkof was kind enough to read over the material on linkage balancing, and Drs. F. Freudenstein, F.R.E. Crossley, and J. Uicker all encouraged the publisher to accept the manuscript in the face of falling enrollments in engineering colleges. I thank them, along with Mr. M. Melody, and my other editors at Prentice-Hall, for their faith in the book.

I was able to devote a good fraction of my time to work in the subject area of the book due to a research grant from the National Science Foundation. I thank the Foundation and Dr. Clifford Astill for their support over the years.

Finally I would like to express deep appreciation to my wife, Lois, and sons, Jordan and Douglas who, through their patience and understanding, made a major contribution to this book.

BURTON PAUL

University of Pennsylvania
Philadelphia, Pa.

NOTE ON NUMBERING SYSTEM
AND OTHER CONVENTIONS

MAIN SECTIONS AND SUBSECTIONS:

Chapter 1 is divided into *main sections* numbered **1.10**, **1.20**, . . . , **1.50**. *Subsections* are differentiated by the last digit in their identification numbers; e.g., the subsections of Sec. **1.50** are **1.51**, **1.52**, . . . , **1.55**. Similar notation is used in other chapters, with the chapter number preceding the "decimal point."

EQUATIONS, FIGURES AND TABLES

Equations, figures, and *tables* are numbered sequentially, 1, 2, 3, . . . , in each subsection. An expression such as "Eq. (**1.32**-4)" identifies the fourth numbered equation in Sec. **1.32**; *within* that section, the same equation would be cited simply as "Eq. (4)." The same numbering convention applies to figures and tables.

REFERENCES

References are identified by last name of author plus the year of publication in brackets, e.g., Jones [1974]. All references are collected near the end of the book.

APPENDICES

Appendices are labeled **A**, **B**, . . . , etc., and the appropriate letter prefix, e.g., (**A**-10), is used in a cross-reference to a numbered equation or figure appearing in an appendix.

Boldface type is used for words or expressions being defined, to denote three-dimensional vectors (Gibbs' notation), and to denote matrices.

Italic type is used for emphasis.

PART 1

GEOMETRICAL KINEMATICS

"Then my noble friend, geometry will draw the soul towards truth and create the spirit of philosophy, and raise up that which is now, unhappily, allowed to fall down."

—Plato, The Republic*

---

*Coolidge [1963, p. 26].

Drawing upon the Greek word for motion (*kinema*), Ampère introduced the name *cinématique* for the science ". . . in which movements are considered in themselves . . . in solid bodies all about us, and especially in the assemblages called machines."* The anglicized equivalent, **kinematics**, has since been broadened to include all aspects of motion, for both solid and fluid bodies, which can be discussed independently of the concept of *force*. In fact, Kelvin and Tait [1879] devoted 218 pages of their monumental *Treatise on Natural Philosophy* to kinematics in this broader sense. It is of some interest to note the following classification of mechanics as outlined in their preface:

> "We adopt the suggestion of Ampère, and use the term *Kinematics* for the purely geometrical science of motion in the abstract. Keeping in view the properties of language, and following the example of the most logical writers, we employ the term *Dynamics* in its true sense as the science which treats of the action of *force*, whether it maintains relative rest, or produces acceleration of relative motion. The two corresponding divisions of Dynamics are thus conveniently entitled *Statics* and *Kinetics*."

In this book, we follow this general classification of the subject, with the discussion of kinematics further divided into two parts—geometrical and analytical. Since kinematics is the *geometry of motion* and Euclid's basic axioms on congruence are fundamentally kinematic in nature, it is redundant to speak of *geometrical kinematics*. The tautology could be avoided by use of the term *synthetic kinematics*. Here the word *synthetic* is used in the sense of geometricians, such as Coolidge [1963, p. viii], who defines *synthetic geometry* as ". . . the geometry with practically no algebraic substructure, where we consider figures directly and not through equations." However, to avoid the danger of confusing the idea of *synthetic kinematics* with that of *kinematic synthesis* (defined in Sec. **5.12**), we adopt the title "Geometrical Kinematics" for Part 1.

The main purpose of this part is to provide an outline† of the most important developments in the field of kinematics of machinery which arose prior to the widespread introduction of digital computers. Much of this material is background matter which is not strictly needed in order to follow the analytical treatment of Part 2. But

---

*Quoted by Ferguson [1962] from Ampère's two-volume treatise on the classification of all human knowledge, published in 1838.
†For an in-depth treatment of geometrical kinematics, see the book by Dijksman [1976].

it should provide a perspective which allows one to better appreciate the advantages of the newer analytic methods and (particularly in Chapter 1) to better appreciate the *lore* of mechanisms and machinery.

Chapter 2 introduces a specialized tool (complex vectors) which forms a bridge between the synthetic and algebraic treatment of plane geometry. This material may be omitted on first reading, but it does allow for such a streamlined, elegant, and penetrating study of many aspects of plane kinematics that serious students will want to consider it sooner or later—especially if they intend to read the current periodical literature. Chapter 3, on gears, and Chapter 4, on cams, give the minimum background material, on these ubiquitous machine components, that every mechanical engineer should be exposed to. However, there is no essential requirement that these chapters be covered prior to subsequent chapters. The one important result, from Chapter 3, which is used explicitly in the subsequent analytical work is the *fundamental equation of epicyclic gear trains*, Eq. (3.32-3), which can be developed when needed.

Although the computational methods of Parts 2 and 3 make no use of the concept of centrodes, this concept is essential to a satisfactory understanding of kinematics and plays a prominent role in the older literature. It is therefore recommended that all students of kinematics be exposed at some point to the earlier parts of Chapter 5 (especially Secs. **5.11, 12, 13**, and **15**). The latter parts of Chapter 5 contain supplementary material on trochoidal motions and curvature theory (the Euler-Savary equation) that will be of interest to more advanced readers.

The graphical methods presented in Chapter 6 are the subject of a major proportion of all traditional books on kinematics of machinery. These methods have been superseded in many respects by the analytical and computer methods of Part 2, and they may be omitted in a first reading. Nevertheless, engineers who work in this field will wish to learn something about these methods in order to read the older literature and that decreasing fraction of the newer literature which deals with graphical kinematics. Graphical methods are also useful for spot-checking computer output; this possibility should be greatly appreciated by experienced computer users, who know how often a trustworthy program can be induced to produce nonsensical output due to subtle input errors.

MECHANISM TERMINOLOGY
AND CHARACTERISTICS

1.10 INTRODUCTION

Typically, the design of a new machine begins when there is a perceived need for a mechanical device with certain (often vaguely defined) characteristics. The degree of need may be determined by a formal or informal market analysis and economic study.

To fulfill the identified need, a **conceptual** or **inventive phase** of the design process is required to establish the general nature and type of machine. It may well be that the art of invention is not communicable to any great degree.[1] But all of us, whether endowed with great or little talent for invention, can profit through familiarity with the major features of existing machines. Many of the most important mechanisms in use today will be studied in this book. However, it is highly recommended that the prospective inventor peruse the many existing qualitative descriptions, surveys, and catalogues of existing machines. References to such source materials are given in Sec. **1.55**.

Having arrived at a concept of the general form of the device, one usually makes a preliminary geometric layout, which is subjected to a **kinematic analysis** in order to determine whether the displacements, velocities, and accelerations are suitable for the intended purpose. Closely related to the process of analysis is the process of **kinematic**

---

[1]"At present, questions of this kind can only be solved by that species of intuition which long familiarity with a subject usually confers upon experienced persons, but which they are totally unable to communicate to others. When the mind of a mechanician is occupied with the contrivance of a machine, he must wait until, in the midst of his meditations, some happy combination presents itself to his mind which may answer his purpose."—R. Willis [1870].

synthesis,[2] which is a systematic attempt to compute the major kinematic design variables (link lengths, fixed angles, pivot locations, etc.). In this book, we shall concentrate on the problem of *analysis*, as opposed to *direct synthesis*. Sufficiently general and automatic analysis procedures, such as those to be developed in Part 2, make it practical to reach a satisfactory design by repeated iterations of an "analyze-modify" procedure. Direct procedures for kinematic synthesis may be found in a number of publications cited in Sec. **5.12**.

For low-speed machinery, the kinematic analysis may be followed directly by a **static analysis** of the forces which the links and bearings must withstand. On the basis of these forces, the link cross sections and bearing dimensions may be properly designed by suitable methods of stress analysis and lubrication theory.

For high-speed machinery, or systems with several degrees of freedom, it is frequently necessary to make a complete **dynamic analysis** of the system, because of the pronounced effects of inertia forces. Considerations of statics and dynamics will be taken up in Part 3.

We begin our study of kinematics of machinery by introducing certain fundamental terms and concepts.

1.20 TERMINOLOGY OF MECHANISM THEORY

The evolution of machines and mechanisms is interwoven with the development of man's social and economic systems. Thus we find an extensive discussion of mechanical devices, and their impact on society, in the work of the noted economic historian A. B. Usher [1929]. A beautifully succinct historical outline of the science of kinematics is given by Hartenberg and Denavit [1964]; a longer, more general, discussion of machinery, from the time of Watt, is provided by Ferguson [1962]; and the role of machinery in the history of mechanical engineering is well described by Burstall [1965]. All writers, on both the historical and technical aspects of mechanisms, owe a great debt to the penetrating work of Franz Reuleaux [1876].

1.21 Definitions of Machine

Until the time of Reuleaux it was usually accepted that there exists a small number of *simple machines*, which, acting in combination to form so-called *compound machines*, could produce the most general form of mechanical device. However, as pointed out by Reuleaux [1876, p. 275], previous writers could not even agree on the number of simple machines, much less their form. Hero of Alexandria (ca. 50 A.D.) had described five *simple machines* or *mechanical powers* (wheel and axle, lever, pulley, wedge, and screw) used for lifting weights. According to Usher [1929, p. 120], the Arabic version of Hero's book was called "the book on the raising of heavy weights." Upon translation into Greek, the title became "Baroulkos" ("the elevator") or "Mechanics." This last term, says Usher, literally meant "lifting of heavy weights" in Hero's time. Over the years the word *mechanics* took on a more general meaning, but most writers on mechanics clung to the idea that *all* machines could be compounded

---

[2]The determination of design dimensions is often called **size synthesis**, as opposed to **type synthesis**, which denotes the *conceptual* or *inventive phase* of the design cycle.

from the simple machines. The number of such *simple machines* wandered up considerably (in order to include the inclined plane, toothed wheel, cord, etc.) and down to as few as two (since the wheel, lever, and pulley have similar static characteristics, and the wedge, inclined plane, and screw have similar static characteristics).

Although the concept of mechanical powers had some merit when restricted to weight lifting machines, it has no great significance in modern times and has been virtually ignored in contemporary serious works on mechanics.

Most definitions of the term *machine* to be found in current technical literature are variations on the following definition[3] of Reuleaux: "A machine is a combination of resistant bodies so arranged that by their means the mechanical forces of nature can be compelled to do work accompanied by certain determinate motions." It should also be noted that in current technical usage the term *machine* connotes a device which transmits significant levels of force, as in an automobile engine. When the force transmitted is small and the principal function of the device is to transmit or modify motion, as in a clock, we usually refer to it as a mechanism. A more precise definition of the term *mechanism* will now be given.

1.22   *Kinematic Chains and Mechanisms*

The individual rigid bodies which collectively form a machine are said to be **members** or **links**. Links may consist of nonrigid bodies, such as cables or fluid columns, which momentarily serve the same function as rigid bodies and are sometimes referred to as **resistant bodies**. The links are interconnected in pairs at points of contact called **joints**. That part of a link's surface which contacts another link is called a **pair element**. The combination of two such elements constitutes a **kinematic pair**. Note the difference between a *pair* and a *joint*. A joint connecting two links constitutes a **simple pair**. **Double pairs**, **triple pairs**, or **multiple pairs**, in general, occur at joints where three, four, or more links are connected. Simple and double pairs are illustrated at joints C and D, respectively, in Fig. 2(b). A joint which connects $N + 1$ members is said to have **multiplicity** N.

An assemblage of interconnected links is called a **kinematic chain**. When one link of a kinematic chain is held fixed, the chain is said to form a **mechanism**, provided that any of its links may move. The **fixed link** is also called the **ground link** or **frame**. Different mechanisms may be formed from the same chain by fixing different links; all mechanisms formed from the same chain are said to be **inversions** of one another. When all motion is prohibited by the fixing of one link, the chain is referred to as a **structure**.

Most mechanisms consist of **closed chains** wherein each link is connected to at least two other links of the system. Links containing two, three, or four joints are called **binary**, **ternary**, or **quaternary links**. A chain which is not closed, such as a double pendulum, is said to be **open**. An open chain can contain links with only one joint, which are properly called **unitary links**, although they are sometimes referred to by the overworked adjective **singular**. A **closed mechanism** is formed from a closed chain, and an **open mechanism** from an open chain.

---

[3]See p. 35 in Reuleaux [1876]. Note that *24* other definitions, used by earlier writers, are listed on pp. 583–588 of this reference.

Any variable angle or length, associated with a given system, is said to be a **position variable** or **configuration variable**. The minimum number of position variables needed to fully define the configuration of a system is called the **degree of freedom (DOF)** of the system. The DOF of a *mechanism* is often called its **mobility**. Figure 1 shows various closed kinematic chains, each with N **hinged** joints and N links ($N = 3, 4, 5$) including the ground link (1); such systems are called **N-bar chains**.

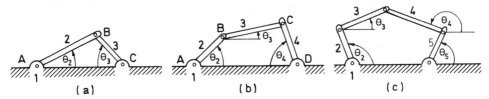

Figure 1.22-1. (a) 3-bar chain or 3 hinged truss (0 D.O.F.); (b) 4-bar mechanism (1 D.O.F.); (c) 5-bar mechanism (2 D.O.F.).

For the three-bar chain of Fig. 1(a) we see that the system is rigid, and therefore it constitutes a *structure* (called a **three-bar truss**). In other words, the angles θ_2 and θ_3 are fixed, and the system has zero DOF. For the four-bar mechanism of Fig. 1(b) we see that if angle θ_2 is specified, the other configuration variables, θ_3 and θ_4, may be calculated[4] (or measured from a drawing); therefore the system has one DOF. Similarly, if we fix angles θ_2 and θ_3 in Fig. 1(c), the remaining angle variables are determined, and the five-bar mechanism is seen to have 2 DOF.

The closed three-bar truss contains the least number of members of any N-bar hinged polygonal chain. Accordingly, it represents the fundamental *building block* for the generation of rigid structures. For example, if we add the two links 4 and 5 to the three-bar truss 1, 2, 3 of Fig. 2(a), the new system is a rigid structure. The process may be repeated by adding bars two at a time (e.g., bars 6 and 7) to form a triangulated rigid structure of just about any shape desired.

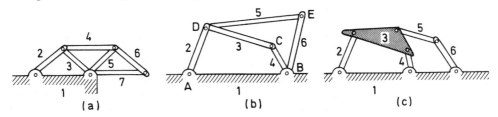

Figure 1.22-2. (a) Rigid truss; (b) A six-bar mechanism with one D.O.F.; (c) A six-link mechanism with one D.O.F.

Similarly, the four-bar mechanism contains the least number of members of any closed N-bar mechanism with at least one DOF. Therefore, mechanisms of great complexity can be built up by adding links, two at a time, to a given four-bar mechanism, as illustrated in Figs. 2(b) and 2(c). A combination of two links pinned together (such as links 5 and 6) is called a **dyad**. We see from Fig. 2 that the addition of a dyad to a mechanism results in a new mechanism without changing the original DOF. By

---

[4]Details of the calculation may be found in Sec. **1.35.**

induction, it may be established that the DOF for mechanisms formed from N links and P_s *simple pairs* is

$$F = 3(N - 1) - 2P_s \qquad (1)$$

This result is a special case of more general mobility criteria discussed in greater depth in Sec. **8.20**.

Reuleaux [1876, p. 47] has defined a *mechanism* as a *closed constrained* kinematic chain with one link fixed. He introduced the term *constrained* to describe a mechanism with one DOF. This usage is contrary to the older and still current tradition of mechanics wherein a **constraint** is any condition which reduces the DOF of a system but not necessarily to a single DOF. Henceforth we shall use the expression *constraint* only in the broader sense of mechanics.

Note that our definition of a mechanism includes open chains and those with several degrees of freedom; thus, unlike Reuleaux, we would classify a pendulum and the five-bar linkage of Fig. 1(c) as *mechanisms*.

1.23   Kinematic Pairs

If all particles of a given system undergo motion in parallel planes, we say that the system experiences **planar motion**. In this book we shall consider only **planar mechanisms**, in which all links undergo planar motion. For planar mechanisms, only four types of joints or kinematic pairs arise; they are called

1. **Hinge** (also called **turning, revolute**, or **pinned**) **pairs**
2. **Sliding** (also called **prismatic**) **pairs**
3. **Rolling** (or **gear**) **pairs**
4. **Cam pairs**

All of the joints shown in Figs. **1.22**-1 and **1.22**-2 are hinge pairs.

If the link 4 of the four-bar mechanism shown in Fig. **1.22**-1(b) is very long, it may be desirable to use the more compact, *kinematically equivalent*[5] arrangement shown in Fig. 1(a). When the radius R approaches infinity, the connection between

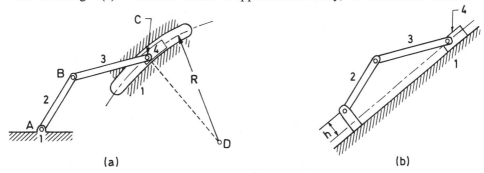

Figure 1.23-1. (a) Kinematic equivalent of 4-bar mechanism; (b) Slider-crank mechanism with offset h.

---

[5]The concept of *kinematic equivalence* is discussed more fully in Sec. **1.54**.

members 4 and 1 is called a **sliding** or **prismatic pair**. The mechanism shown in Fig. 1(b) is called an **offset slider-crank mechanism,** and the distance h is called the **offset.**

When two friction wheels roll about fixed centers A and B, as shown in Fig. 2, and no slip occurs at points of contact, the kinematic pair is said to be a **rolling pair**. If teeth are cut into the wheels, to provide a driving force independent of friction, the wheels become **gears,** and the equivalent rolling circles are called the **pitch circles** of the gears, which touch at the **pitch point** P; thus the name **gear pair** is equivalent to **rolling pair.**

It is possible for *noncircular* **pitch curves** to roll together without slipping. When meshing teeth are cut into the pitch curves we speak of the members as **noncircular gears.** Unless indicated otherwise, we shall henceforth assume that the gears we deal with have circular pitch curves.

Figure 3 shows two members with curved outlines which touch at a common point C, where sliding action occurs at the so-called **cam pair**. The distinction between the cam pair and the prismatic pair is that the former has a line contact and the latter has surface contact. Reuleaux classified joints with surface contact as **lower pairs** and those with point or line contact as **higher pairs**. Other writers have redefined the terms so that a lower pair refers to the case where the paired members have one DOF relative to each other. For simplicity, we accept Reuleaux's definition whereby hinges and sliding pairs are classified as lower pairs and gear and cam pairs are considered higher pairs.

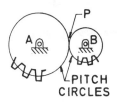

Figure 1.23-2. Friction wheels or gears. No slip occurs at pitch point P, where pitch circles contact.

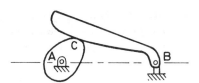

Figure 1.23-3. Cams. Sliding occurs at contact point C.

It has been suggested by Hartenberg and Denavit [1964] that a mechanism which contains only lower pairs be called a **linkage**. This is a useful distinction, which we shall observe. However, the reader should be warned that the term *linkage* has been used widely as a synonym for *mechanism* and in other specialized ways; e.g., by Beggs [1955, p. 403].

Frequently we shall use stylized drawings called **kinematic skeletons,** as in Fig. 4, to represent the essential kinematic features of a mechanism. The following conventions should be noted: The joints are lettered, the links are numbered, and hinges are shown as small circles. Sliding pairs are illustrated in Fig. 4 at M and A. The small triangle on the slider at A indicates that it is welded to link 2.

Binary links are shown as straight lines (e.g., 5) or bent lines (e.g., 9). Ternary and higher-order links may be shown as shaded polygons (e.g., $CFDE$) or by a continuous line (e.g., HG) which carries a hinge (e.g., B) in its interior. When a link (such as 11)

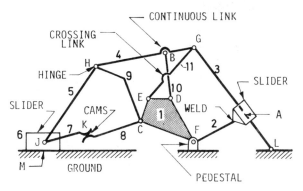

Figure 1.23-4. Conventions used in drawing kinematic skeletons.

crosses over another link (such as 10) a small break is made in the bottom link. Sometimes the crossing is emphasized by a small bend in the upper link, over the crossing point. A fixed link is indicated by cross-hatching, and fixed pivots are shown on pedestals (as at *F*) or embedded in the fixed link (as at *L*). Cams are indicated by small contacting arcs (as at *K*).

Additional conventions will be introduced as needed; e.g., in Fig. **3.33**-1 for gear trains.

1.24 Problems

Define the following terms:

1. kinematic analysis
2. kinematic synthesis
3. static analysis
4. dynamic analysis
5. resistant body
6. kinematic pair
7. elements of a pair
8. kinematic chain
9. mechanism
10. ground link
11. frame
12. kinematic inversion
13. structure
14. closed mechanism
15. open mechanism
16. binary link
17. ternary link

18. quaternary link
19. unitary link
20. position variable
21. displacement variable
22. degree of freedom
23. mobility
24. simple pair
25. pair multiplicity
26. hinge pair
27. turning pair
28. revolute pair
29. pinned pair
30. sliding pair
31. prismatic pair
32. rolling pair
33. gear pair
34. cam pair

35.	kinematic equivalence		**41.**	rolling pair
36.	slider-crank mechanism		**42.**	lower pair
37.	offset slider-crank mechanism		**43.**	higher pair
38.	four-bar linkage		**44.**	linkage
39.	pitch circle		**45.**	dyad
40.	pitch point		**46.**	*N*-bar chain

1.30 KINEMATIC ANALYSIS OF THE FOUR-BAR MECHANISM

We have seen that the four-bar mechanism is the simplest closed kinematic chain of hinged links with a single degree of freedom (after fixing one link) and that more complex mechanisms can be built up by using one four-bar mechanism to drive one or more others. Because of this property, and because of the wide variety of motions[6] which can be generated directly by four-bar mechanisms, they are often found at the heart of machines and subsystems such as punch presses, film transports, quick returns, analog computers, and function generators. The study of the four-bar linkage is well justified not only because of its many direct applications but also because most of the basic problems encountered in more general linkages show up in a simpler and more understandable way in the four-bar linkage.

1.31 Summary of Basic Types

There are many schemes to classify the various types of four-bar mechanisms. We shall follow a modified version[7] of a classification scheme introduced by Grashof [1883], but first it will be necessary to introduce some basic nomenclature.

In a four-bar chain, we shall refer to the *line segment between hinges* on a given link as a **bar**[8] and shall let

$$s = \text{length of shortest bar}$$

$$l = \text{length of longest bar}$$

$$p, q = \text{lengths of intermediate bars}$$

We shall also use $s, l, p,$ and q as the names of the corresponding bars or links. When a link of the chain is fixed it is called the **frame**, or **fixed link**. The opposite link is called the **coupler link**, and the links which are hinged to the frame are called **side links**. We shall frequently use the following symbols (e.g., as in Fig. **1.32**-1):

$$c = \text{length of coupler bar}$$

$$f = \text{length of fixed bar}$$

$$a, b = \text{length of side bars}$$

A link which is free to rotate through 360° with respect to a second link will be

---

[6]To appreciate the variety of curves traced out by a point on the coupler bar of a four-bar linkage one should peruse the 7300 **coupler curves** which occupy 730 large pages (11 × 17 in.) of the book by Hrones and Nelson [1951]. An interesting historical account of the application of coupler curves is given by Nolle [1974a, 1974b, 1975].
[7]See Appendix A.
[8]Note: We distinguish between a *bar*, which is a line segment, and a *link*, which is an extended plane.

said to **revolve** relative to the second link. A side link which revolves relative to the frame is called a **crank**. Any link which does not revolve is called a **rocker**.

If it is possible for all four bars to become simultaneously aligned, such a state is called a **change point** and the linkage is said to be a **change-point mechanism**.

Grashof's theorem states that *a four-bar mechanism has at least one revolving link if*

$$s + l \leq p + q \qquad (1)$$

and all three links will rock if

$$s + l > p + q \qquad (2)$$

Inequality (1) is **Grashof's criterion**, and mechanisms which satisfy it are called **Grashof mechanisms**. Mechanisms which violate Grashof's criterion are called **non-Grashofian**.

It may be shown (see Appendix **A**) that all four-bar mechanisms fall into one of the five categories listed in Table 1 and briefly described below:

TABLE 1.31-1 *Classification of Four-Bar Mechanisms*

Case	$l + s \lesseqqgtr p + q$	Shortest Bar	Type
1	<	Frame	Double-crank
2	<	Side	Rocker-crank
3	<	Coupler	Double rocker
4	=	Any	Change point
5	>	Any	Triple-rocker

1. In a **double-crank**, also known as a **drag-link** mechanism, both side links revolve. See Fig. 1(a).

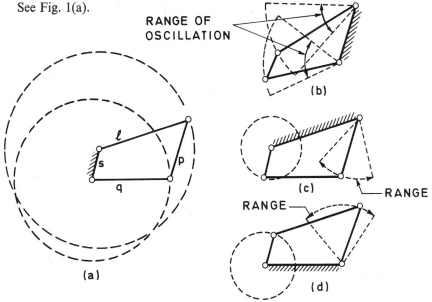

Figure 1.31-1. All possible inversions of a Grashof chain ($s + l < p + q$): (a) Drag-link; (b) Double-rocker; (c) Rocker-crank; (d) Rocker-crank.

2. In a **rocker-crank** the shorter side link revolves and the other one rocks (i.e., oscillates). See Figs. 1(c) and (d).
3. In this inversion, both side links rock and the coupler revolves. This *revolving-coupler double rocker* may be designated more simply as a **double-rocker**, provided we note the distinction between it and case 5. See Fig. 1(b).
4. A **change-point mechanism** is a limiting case which separates Grashof and non-Grashofian mechanisms. Its special characteristics will be discussed in Sec. **1.33**.
5. In a **triple-rocker** all three links rock; Grashof called this mechanism a "double-rocker"—the same name he used for case 3, despite the important differences between the two cases.

Figure 1 shows all possible inversions of a Grashof chain. For a non-Grashofian chain all inversions are triple-rockers.

For an interesting discussion of practical mechanisms (e.g., lawn sprinklers, automobile-hood-raising linkages, forklift mechanisms, etc.) which utilize the various types of four-bar linkages, see Hall [1966, pp. 1–11].

1.32 Dead-Point Configurations

When a side link, such as *a* in Fig. 1, becomes aligned with the coupler *c*, it can only be compressed or extended by the coupler. In this configuration, a torque applied to the other side link *b* cannot induce rotation in link *a*, which is therefore said to be at a **dead point** (sometimes called a **toggle point**).

If *a* is a crank, we see, from Fig. 1, that it can become aligned with *c* in full *extension*[9] along the line AD_1C_1 or in *flexion* with AD_2 folded over D_2C_2. We shall denote the angle ABC by ϕ and BAD by θ and shall use subscripts 1 or 2 to denote the extended or flexed state of links *a* and *c*. In the extended state, point C cannot move clockwise along the circular arc CC_1 without stretching or compressing the theoretically rigid line AC_1. Therefore, link *b* cannot move into the "forbidden zone" below BC_1, and ϕ_1 must be at one of its two extreme positions; in other words, link *b* is at an **extremum**. A second extremum of link *b* occurs with $\phi = \phi_2$. Note that the *extreme positions of a side link occur simultaneously with the dead points of the opposite link.*

Although *b* is temporarily immovable at its extremum points, *a* can effectively serve as the *driving* link with finite angular velocity ω_a. If ω_b denotes the angular velocity of link *b*, we can conclude that at the dead points of link *a*, $\omega_a/\omega_b = \infty$. In the neighborhood of the dead point, the velocity ratio ω_a/ω_b can be huge; i.e., a mechanism can function as a *velocity* (or *displacement*) *magnifier* near its dead points.

To find the four critical angles ($\phi_1, \theta_1, \phi_2, \theta_2$), consider triangle ABC and introduce the notation

$$e \equiv AC, \qquad \phi^* = \angle ABC, \qquad \theta^* = \angle BAC \qquad (1)$$

---

[9]The words **extension** and **flexion** are used in their anatomical sense. For example, a fully outstretched arm is in extension, but a complete bend of the elbow joint puts it into a state of flexion.

where $0 \leq \phi*$ and $\theta* \leq \pi$. From the law of cosines we find that

$$\phi* = \arccos\left(\frac{f^2 + b^2 - e^2}{2fb}\right) \tag{2}$$

$$\theta* = \arccos\left(\frac{f^2 + e^2 - b^2}{2fe}\right) \tag{3}$$

For the rocker-crank shown in Fig. 1, it is apparent that in state 1 (extension of a and c)

$$e \equiv AC_1 = c + a, \quad \phi_1 = \phi*, \quad \theta_1 = \theta* \tag{4}$$

whereas for state 2 (flexion of a and c)

$$e \equiv AC_2 = c - a, \quad \phi_2 = \phi*, \quad \theta_2 = \theta* + \pi \tag{5a}$$

For the case of a revolving coupler mechanism (Fig. 2), Eqs. (1)–(4) remain valid but Eqs. (5a) should be replaced by

$$e \equiv e_2 = a - c, \quad \phi_2 = \phi*, \quad \theta_2 = \theta* \tag{5b}$$

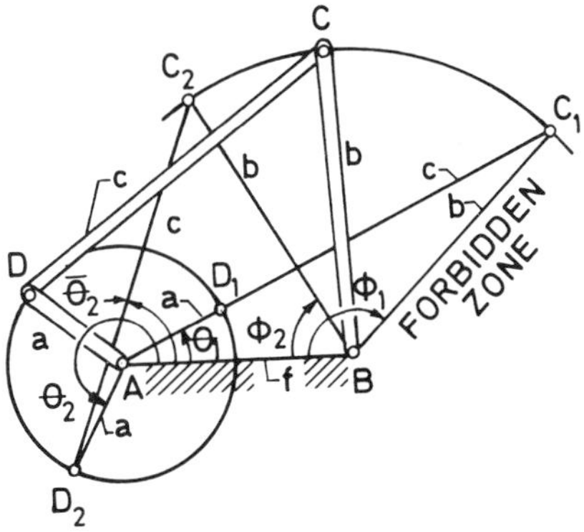

Figure 1.32-1. Dead points of a rocker-crank $\phi_{max} = \phi_1$, $\phi_{min} = \phi_2$.

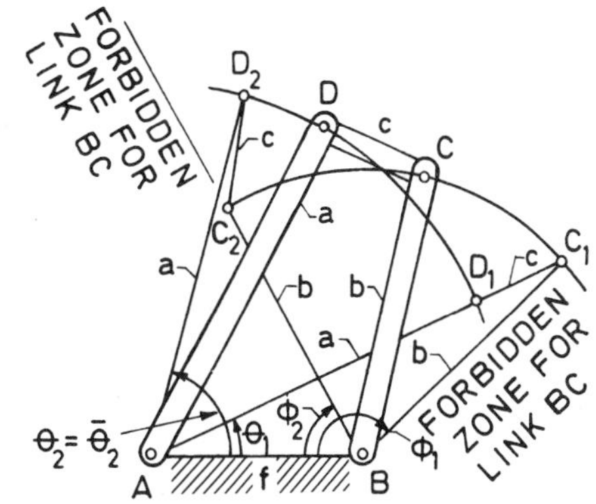

Figure 1.32-2. Dead points of link b of a revolving-coupler double rocker.

To find the extremum points of rocker-link a of Fig. 2 it is only necessary to interchange a and b and to interchange symbols ϕ and θ everywhere in Eqs. (1)–(4) and in Eq. (5b). Then subscript 1 (2) represents the state where b and c are aligned in extension (flexion).

Since double-crank mechanisms cannot (by definition) have any extremum points, Eqs. (1)–(5) cover all possible Grashof mechanisms.[10] For non-Grashofian linkages (triple-rockers), the extremum angles (ϕ_1, ϕ_2) of the follower bar b and the associated dead points (θ_1, θ_2) of the input bar a are given in Table 1. The required angles $\phi*$ and $\theta*$ are

TABLE 1.32-1 *Extremum Angles (ϕ) and Associated Dead Points (θ) for Triple-Rockers*

Case	Longest Bar Is	e	ϕ_1	θ_1	ϕ_2	θ_2
1	Frame, f	$a + c$	$\phi*$	$\theta*$	$-\phi*$	$-\theta*$
2	Coupler, c	$c - a$	$2\pi - \phi*$	$\pi - \theta*$	$\phi*$	$\pi + \theta*$
3	Follower, b	$a + c$	$\phi*$	$\theta*$	$-\phi*$	$2\pi - \theta*$
4	Input, a	$a - c$	$2\pi - \phi*$	$-\theta*$	$\phi*$	$\theta*$

[10]Some minor reinterpretation of symbols may be necessary for linkages that are assembled in modes opposite to those shown in Figs. 1 and 2. See the discussion of mode of assembly in Sec. **1.35**.

to be found from Eqs. (1)–(3), using the value of e indicated in the table. Verification of these results is left as an exercise (Problems **1.37**-22, 23, and 24).

A rocker-crank mechanism can serve as a **quick-return mechanism**. To see this, note that the rocker tip C in Fig. 1 moves over path C_1CC_2 during the time that the crank tip D moves over path D_1DD_2. If the crank rotates counterclockwise at uniform speed ω, the time for this "forward motion" is

$$t_f = \frac{\theta_2 - \theta_1}{\omega} \tag{6a}$$

The "return motion" (from C_2 to C_1) of the rocker occurs in the interval

$$t_r = \frac{2\pi - (\theta_2 - \theta_1)}{\omega} \tag{6b}$$

The **time ratio** (TR) is therefore given by

$$\text{TR} \equiv \frac{t_f}{t_r} = \frac{\theta_2 - \theta_1}{2\pi - (\theta_2 - \theta_1)} \tag{6c}$$

where θ_1 and θ_2 are found from Eqs. (4) and (5a).

1.33   Change-Point Configurations

A type of singularity known as a *change point* can arise when the equality form of Grashof's criterion, Eq. (**1.31**-1), is satisfied; i.e., $s + 1 = p + q$. By definition, *a change-point configuration occurs when all of the bars of the mechanism become collinear.*

Perhaps the most familiar type of *change-point mechanism* is the **parallelogram linkage**, formed by the horizontal coupling rod on locomotive wheels, as shown in Fig. 1(a) by $ABPQ$. When the rod moves into position $P'Q'$ (aligned with the frame AB), the motion becomes kinematically indeterminate. To see this, note that when point Q' on the forward wheel rotates to position Q'' the corresponding point P'' may remain on the parallelogram linkage as in Fig. 1(a), or else the rear wheel may change direction at the change point, so that the link $Q''P''$ "crosses" the frame as shown in Fig. 1(b). The linkage shown in Fig. 1(b) has been called the **antiparallel**

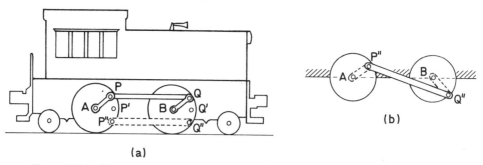

(a)

(b)

Figure 1.33-1. Change points in parallelogram linkage: (a) Gas turbine locomotive, circa 1954. [After R. T. Sawyer (1972)]; (b) Antiparallel crank (butterfly, bow-tie) linkage.

crank linkage by Reuleaux [1876, p. 290], although the shorter terms **butterfly linkage** and **bow-tie linkage** are more descriptive. Whether the mechanism goes into the parallelogram or butterfly arrangement depends on how it is guided through the change point. For example, if the wheels are rotating at a reasonable speed, the *flywheel effect* (i.e., inertia torque) could carry it through in the parallelogram mode. On the other hand, a positive kinematic drive can be assured by adding a second connecting rod (perhaps on the other side of the locomotive) whose change point does not coincide with that of the first rod (see Fig. 2).

When a side rod broke on an old-time locomotive, the engineers knew that the opposite rod had to be demounted before the locomotive was to be restarted. Otherwise the good rod might start in the butterfly mode, and either the rod or one of the pins would break. Similar ideas were used to keep one crank of a multicylinder steam engine from starting in the wrong direction if one piston happened to be at dead center upon starting. (See Problems **1.37**-9 and -10 for further examples.)

Another interesting mechanism with change points is the **Galloway mechanism**[11] shown in Fig. 3. This so-called **isosceles** or **kite linkage** has one crank length equal to

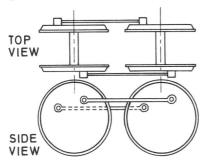
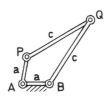

Figure 1.33-2. Guiding a linkage through change-point configuration.

Figure 1.33-3. Galloway mechanism.

the frame length and the other equal to the coupler. The reader should verify that if $a < c$ and if link AP revolves continuously, link AP executes two revolutions for every one of link BQ.

It may be shown (Paul [1979], Dijksman [1976]) that a linkage can be a movable change-point mechanism if, and only if,

$$s + l = p + q \qquad (1)$$

1.34 Transmission Angle

If AD is the input link in Fig. 1, the force applied to the output link BC is transmitted through the coupler link DC. For sufficiently slow motions (negligible inertia forces) the force in the coupler rod is pure tension or compression (negligible bending action) and is directed along DC. For a given force in the coupler rod, the torque communicated to the output bar (about point B) is a maximum when the angle μ between coupler bar DC and output bar BC is 90°. When μ, the **transmission angle**, deviates significantly from 90°, the torque on the output bar falls off and may not be

[11]Described by Reuleaux [1876, p. 197].

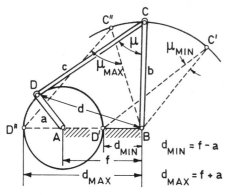

$d_{MIN} = f - a$

$d_{MAX} = f + a$

Figure 1.34-1. Transmission angle μ.

sufficient to overcome the friction in the system. For this reason, it is desirable that the **deviation angle** $\Delta \equiv |90° - \mu|$ not be too great. There is no definite upper limit that can be placed upon Δ, because in practice the existence of inertia forces may eliminate the undesirable force relationships that prevail under static conditions. Nevertheless, it has been suggested[12] that the criterion

$$\Delta_{max} \equiv |90° - \mu|_{max} < 50°$$

be observed, where

$$\Delta_{max} = \text{maximum of } |90° - \mu_{max}| \text{ or } |90° - \mu_{min}|$$

From the law of cosines applied to triangle BCD in Fig. 1, it follows that

$$\mu = \text{arc cos} \frac{c^2 + b^2 - d^2}{2bc}$$

where d is the diagonal DB of the quadrilateral $ABCD$. It is evident from the figure that

$$\mu = \mu_{max} \qquad \text{when } d = d_{max}$$
$$\mu = \mu_{min} \qquad \text{when } d = d_{min}$$

Further, we see that when the driver is a crank (as shown)

$$d_{max} = f + a$$
$$d_{min} = |f - a|$$

On the other hand, when the driver is a rocker, we know from the discussion of Sec. **1.32** that its dead points occur when the follower link becomes aligned (in extension or flexion) with the coupler. Thus $\mu_{max} = 180°$ and $\mu_{min} = 0$; at the two dead points of a rocking driver.

Hall [1966, p. 42] shows how one may design a rocker-crank with minimum possible deviation angle Δ when the time ratio and range of rocker motion are specified.

1.35 Displacement Analysis

Figure 1 shows a typical four-bar linkage with frame f, coupler c, and side links a and b. If the linkage has at least one crank,[13] we shall let AD represent a crank;

---

[12]Based on recommendations of Alt, according to Hartenberg and Denavit [1964, pp. 47, 48].
[13]This will be the case if the linkage is a Grashof linkage with either a or f as the shortest member, i.e., a rocker-crank or double-crank.

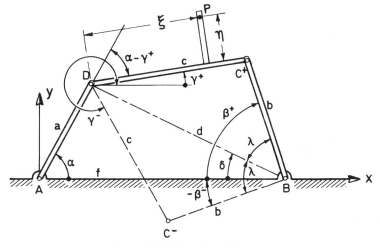

Figure 1.35-1. Four-bar linkage.

otherwise AD represents either side link. In any case, we shall consider AD as an input link and BC as output. The angles which the vectors $\overrightarrow{AD}$, $\overrightarrow{BC}$, $\overrightarrow{DC}$ make with the frame axis $\overrightarrow{AB}$ are denoted by α, β, γ, measured positive as shown. For a given angle α, the linkage may be assembled in either of two different ways, as shown by the sequences ABC^+D and ABC^-D. The link BC lies on opposite sides of the diagonal BD in the two possible mechanisms. Note that $\overrightarrow{BC^-}$ lags behind $\overrightarrow{BD}$ by an angle λ, whereas $\overrightarrow{BC^+}$ leads $\overrightarrow{BD}$ by angle λ. Therefore we shall refer to the configuration ABC^-D as the **lagging form** and to ABC^+D as the **leading form** of the linkage. The linkage can switch from one form to the other only if point C crosses the diagonal BD. Note that when point C reaches the diagonal, the link BC is in a dead center position (which may or may not coincide with a change point). The angle between $\overrightarrow{BA}$ and $\overrightarrow{BD}$ is denoted by δ, measured positive as shown.

To decide upon the proper *form* of mechanism, let us first consider the case where a is a crank and no change points occur (i.e., $s + l < p + q$ and $a = s$). For these *crank mechanisms*, link a revolves continuously without extremum points; i.e., links b and c never become aligned.[14] Therefore point C never reaches the diagonal DB, and we conclude that *crank mechanisms without change points retain the leading or lagging form in which they were originally assembled.*

If $s + l = p + q$, change points are encountered where links b and c become aligned with a and c, and point C will be on the diagonal BD. Whether or not C crosses the diagonal (resulting in a change of form) is not determined by kinematics alone. In other words, *at change points, the form of linkage is kinematically indeterminate and must be specified by auxiliary guiding devices or dynamic considerations.*

Finally, consider the case where a is a rocker. Then a is capable of driving until it reaches one of its extremum points; there b and c necessarily line up, and C reaches the diagonal BD. No further motion is possible in this state if a is the driver. However, if b becomes the driver, the motion of point C is well defined. That is, the mechanism changes form if b *continues* to rotate in the same direction; if b reverses direction as

---

[14]Recall from Sec. **1.32** that extremum points of a side link occur if and only if the opposite link becomes aligned with the coupler (in extension or flexion).

soon as a reaches its dead point, the mechanism *retains* its form. In any case, the motion remains well defined with b driving and continues so until b reaches an extremum point (if any), whereupon a may be treated as a driver again. In short, we can say that *between extremum points of a rocking side link, a mechanism retains its form.*

Having decided upon the form to be used for any given phase of the linkage, we may calculate angle β for a given α as follows: First, find the projections (X, Y) of BD along and perpendicular to the frame axis BA; i.e.,

$$X \equiv f - a \cos \alpha, \qquad Y \equiv a \sin \alpha \tag{1}$$

Then find[15]

$$\delta = \arctan_2 (Y, X) \tag{2}$$

$$d = (X^2 + Y^2)^{1/2} \tag{3}$$

$$\lambda = \cos^{-1} \left(\frac{d^2 + b^2 - c^2}{2bd} \right) \tag{4}$$

$$\beta^+ = \delta + \lambda \qquad \text{(leading form)} \tag{5a}$$

$$\beta^- = \delta - \lambda \qquad \text{(lagging form)} \tag{5b}$$

Note that λ may be restricted to the range $(0, 180°)$ without loss of generality, whereas all other angles are unambiguously defined within the range $(0, 360°)$.

If b is the driving link, interchange a with b and α with β in Eqs. (1)–(5).

After having decided on the appropriate value of β, γ is given uniquely by the following expressions for the projections of DC perpendicular and parallel to line AB, respectively:

$$c \sin \gamma = b \sin \beta - a \sin \alpha \tag{6}$$

$$c \cos \gamma = f - b \cos \beta - a \cos \alpha \tag{7}$$

$$\gamma = \arctan_2 (\sin \gamma, \cos \gamma) \tag{8}$$

If ξ and η are given coordinates of a point P fixed to the coupler, we can find the curve traced by point P as angle α is varied. From Fig. 1, we see that the x and y coordinates of the **coupler-point curve** are given by

$$x_P = a \cos \alpha + \xi \cos \gamma - \eta \sin \gamma \tag{9}$$

$$y_P = a \sin \alpha + \xi \sin \gamma + \eta \cos \gamma \tag{10}$$

1.36 *Velocity and Acceleration Analysis*[16]

If each term in Eqs. (**1.35-6**) and (**1.35-7**) is differentiated with respect to time one finds[17]

$$b\dot{\beta} \cos \beta - c\dot{\gamma} \cos \gamma = a\dot{\alpha} \cos \alpha \tag{1}$$

$$b\dot{\beta} \sin \beta + c\dot{\gamma} \sin \gamma = -a\dot{\alpha} \sin \alpha \tag{2}$$

---

[15]See Appendix C for a definition of the function $\arctan_2 (Y, X)$. Equation (4) follows from the law of cosines applied to triangle BCD.

[16]The method used in this section is a special case of a general method of analytical kinematics, described in Sec. **7.10**.

[17]The derivative of a variable with respect to time is denoted by a dot above the variable.

Solving Eqs. (1) and (2), one finds the velocities in the form

$$\dot{\beta} = \dot{\alpha}\frac{a}{b}\frac{\cos\alpha\sin\gamma - \sin\alpha\cos\gamma}{\cos\beta\sin\gamma + \sin\beta\cos\gamma} = \dot{\alpha}\frac{a}{b}\frac{\sin(\gamma-\alpha)}{\sin(\gamma+\beta)} \tag{3}$$

$$\dot{\gamma} = \dot{\alpha}\frac{a}{c}\frac{-\cos\beta\sin\alpha - \sin\beta\cos\alpha}{\cos\beta\sin\gamma + \sin\beta\cos\gamma} = -\dot{\alpha}\frac{a}{c}\frac{\sin(\beta+\alpha)}{\sin(\gamma+\beta)} \tag{4}$$

At change points, Eqs. (3) and (4) are indeterminate forms of the type 0/0.

For brevity we introduce the notation

$$\dot{\beta} = k_{\beta\alpha}\dot{\alpha} \tag{5}$$

$$\dot{\gamma} = k_{\gamma\alpha}\dot{\alpha} \tag{6}$$

where the **influence coefficients** (or **velocity ratios**) are defined by

$$k_{\beta\alpha} \equiv \frac{\dot{\beta}}{\dot{\alpha}} = \frac{a}{b}\frac{\sin(\gamma-\alpha)}{\sin(\gamma+\beta)} \tag{7}$$

$$k_{\gamma\alpha} \equiv \frac{\dot{\gamma}}{\dot{\alpha}} = -\frac{a}{c}\frac{\sin(\beta+\alpha)}{\sin(\gamma+\beta)} \tag{8}$$

Upon differentiation of Eqs. (5) and (6) we find that the accelerations are given by

$$\ddot{\beta} = k_{\beta\alpha}\ddot{\alpha} + k'_{\beta\alpha}\dot{\alpha}^2 \tag{9}$$

$$\ddot{\gamma} = k_{\gamma\alpha}\ddot{\alpha} + k'_{\gamma\alpha}\dot{\alpha}^2 \tag{10}$$

where the **centripetal coefficients** are functions of position only and are given (see Problem **1.37**-18) by

$$k'_{\beta\alpha} \equiv k_{\beta\alpha}[(k_{\gamma\alpha} - 1)\cot(\gamma-\alpha) - (k_{\gamma\alpha} + k_{\beta\alpha})\cot(\gamma+\beta)] \tag{11}$$

$$k'_{\gamma\alpha} \equiv k_{\gamma\alpha}[(k_{\beta\alpha} + 1)\cot(\beta+\alpha) - (k_{\gamma\alpha} + k_{\beta\alpha})\cot(\gamma+\beta)] \tag{12}$$

An alternative way of finding accelerations is described in Problem **1.37**-19.

1.37 Problems

1. Define the following terms:
   a. side link
   b. coupler link
   c. crank
   d. rocker
   e. rocker-crank mechanism
   f. double-crank mechanism
   g. double-rocker mechanism
   h. drag-link mechanism
   i. Grashof's theorem
   j. dead-point configuration
   k. toggle-point configuration
   l. change-point configuration
   m. parallelogram linkege
   n. antiparallel crank linkage

   o. butterfly linkage
   p. bow-tie linkage
   q. Galloway mechanism
   r. isosceles linkage
   s. kite linkage
   t. leading form of four-bar mechanism
   u. lagging form of four-bar mechanism
   v. influence coefficient
   w. velocity ratio
   x. centripetal coefficient
   y. coupler-point curve
   z_1. Grashof's criterion
   z_2. Grashof mechanism
   z_3. Non-Grashofian mechanism

2. (Exercises in the manipulation of inequalities) Given the inequalities $a < b$, $c < d$, and $e > f$, which of the following statements are necessarily true? (Give proofs of correct statements.)

$$\text{addition:} \quad a + c < b + d$$

$$\text{subtraction:} \quad a - c < b - d$$

$$\text{multiplication:} \quad ma < mb \quad \text{(with } m > 0)$$

$$ma < mb \quad \text{(with } m < 0)$$

$$ma > mb \quad \text{(with } m < 0)$$

$$\text{subtraction:} \quad a - e < b - f$$

3. Show that four links cannot be assembled into a four-bar chain if the length of any bar exceeds the sum of the lengths of the remaining bars. What can you say about the chain if the longest length is exactly equal to the sum of the other lengths?

4. Show that a mechanism which satisfies Grashof's criterion, inequality (**1.31**-1), will never have one bar which equals or exceeds in length the sum of the other bar lengths.

5. Prove that any side of a triangle exceeds in length the absolute value of the difference in length of the remaining two sides.

6. If AD in Fig. 1 is driven at constant speed ω_1, will the link AG oscillate or will it turn continuously through complete revolutions? The proportions of the mechanism are as shown, but the scale is arbitrary.

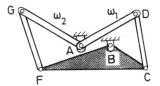

Figure 1.37-1 (Prob. 6).

7. Show that the shortest bar of a four-bar mechanism cannot exceed the mean of the intermediate bar lengths.

8. A rocker-crank mechanism is to be designed with the following dimensions (see Fig. **1.34**-1): $a = 8$ cm, $b = 12$ cm, $f = 15$ cm. Find the coupler length c which makes the maximum deviation angle Δ_{max} as small as possible. *Hint:* plot Δ_{max} versus c for several values of c which satisfy Grashof's criterion for rocker cranks.

9. Can the parallelogram mechanism $ABCD$ shown in Fig. 2 switch into an antiparallel crank mechanism? If so, demonstrate by a sketch. If not, why not?

10. Can the parallelogram mechanism $ABCD$ shown in Fig. 3 switch over to an antiparallel crank mechanism? If so, demonstrate by a sketch. If not, why not?

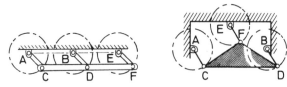

Figure 1.37-2. (Prob. 9).   **Figure 1.37-3** (Prob. 10).

11. Is the Galloway mechanism of Fig. **1.33**-3 a double-rocker, rocker-crank, or double-crank? Prove it.

12. Demonstrate that link *AP* of Fig. **1.33**-3 can make two revolutions for each revolution of link *BQ*.

13. Is the behavior described in Problem 12 consistent with the statement "A four-bar linkage has one degree of freedom; therefore the location of point *P*, in Fig. **1.33**-3, is uniquely defined by the location of point *Q*"?

14. Figure 4 shows the top view of a locomotive wheel set with the connecting rods and sliders of separate cylinders of a steam engine. Draw a side view showing how the rods may be hinged to the wheels in such a way that the wheels are guided unambiguously through the dead points of the engine.

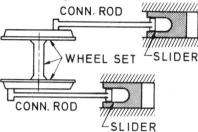

Figure **1.37**-4 (Prob. 14).

15. In a four-bar mechanism, *a* and *b* represent the side bar lengths, *c* is the coupler length, and *f* is the fixed link length.

   a. Show that change-point configurations are possible only if

   $$\pm a \pm b \pm c = f$$

   b. Sketch each of the eight possible combinations which can satisfy the above equation.

16. In Problem 15, show that one of the eight "possible" configurations will not be possible for a closed chain, that four others constitute locked mechanisms, and that the remaining three cases are covered by the condition $s + l = p + q$.

17. Show that the "kite" linkage of Fig. **1.33**-3 can collapse into a singular configuration which behaves exactly like a single rigid bar (or pendulum) pinned to the frame.

18. Verify Eqs. (**1.36**-9)–(**1.36**-12). Note that if $f = f(\alpha, \beta, \gamma)$,

$$\frac{df}{d\alpha} = \frac{\partial f}{\partial \alpha} + \frac{\partial f}{\partial \beta}\frac{d\beta}{d\alpha} + \frac{\partial f}{\partial \gamma}\frac{d\gamma}{d\alpha}$$

   where $d\beta/d\alpha = \dot{\beta}/\dot{\alpha} = k_{\beta\alpha}$ and similarly $d\gamma/d\alpha = k_{\gamma\alpha}$.

19. Show that differentiation of Eqs. (**1.36**-1) and (**1.36**-2) gives

$$\begin{bmatrix} C_\beta & -C_\gamma \\ S_\beta & S_\gamma \end{bmatrix} \begin{bmatrix} b\ddot{\beta} \\ c\ddot{\gamma} \end{bmatrix} = a\ddot{\alpha} \begin{bmatrix} C_\alpha \\ -S_\alpha \end{bmatrix} + \begin{bmatrix} R_1 \\ R_2 \end{bmatrix}$$

   where all velocity terms have been grouped together in the form

$$\begin{bmatrix} R_1 \\ R_2 \end{bmatrix} = \begin{bmatrix} -a\dot{\alpha}^2 S_\alpha + b\dot{\beta}^2 S_\beta - c\dot{\gamma}^2 S_\gamma \\ -a\dot{\alpha}^2 C_\alpha - b\dot{\beta}^2 C_\beta - c\dot{\gamma}^2 C_\gamma \end{bmatrix}$$

and $S_\alpha \equiv \sin \alpha$, $C_\beta \equiv \cos \beta$, etc. Then show that the angular accelerations $\ddot{\beta}$ and $\ddot{\gamma}$ are given by

$$Db\ddot{\beta} = a\ddot{\alpha}(C_\alpha S_\gamma - S_\alpha C_\gamma) + R_1 S_\gamma + R_2 C_\gamma$$
$$Dc\ddot{\gamma} = a\ddot{\alpha}(-C_\beta S_\alpha - S_\beta C_\alpha) + C_\beta R_2 - S_\beta R_1$$

where

$$D \equiv S_\beta C_\gamma + C_\beta S_\gamma = \sin(\gamma + \beta)$$

20. Redraw Fig. 1.32-2 to show link a at each of its two extreme positions. From your sketches, derive expressions for the extreme values of angle BAD and the associated values of angle ABC. Verify that your results follow from Eqs. (1.32-1)–(1.32-4) and (1.32-5b) after interchanging a and b and interchanging θ and ϕ, with suitable change of subscript.

21. Figure 5 shows a four-bar mechanism used in an oscillating lawn sprinkler. Link AD is rotated continuously by a hydraulic motor, and link CBJ contains a means (not shown) which allows bar b and the angle α between BC and the water jet axis BJ to be adjusted to the following settings:

	1. Full	2. Left	3. Right	4. Center
b (in.)	0.875	1.50	1.50	2.125
α (deg)	105	140	88	130

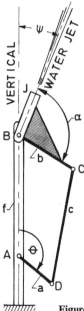

Figure 1.37-5 (Prob. 21): Oscillating lawn sprinkler.

The other links are fixed at $a = 0.625$ in., $c = 1.625$ in., and $f = 1.625$ in. For the setting(s) designated by your instructor, find the minimum and maximum elevation angles ψ of the water jet, measured from the vertical, and find the crank angle θ corresponding to each extreme value of ψ calculated.

22. Verify Table **1.32**-1 for all triple-rockers where s is adjacent to l, p is adjacent to l, and $p < q$.

23. The same as Problem 22 but $p > q$.

24. Verify Table **1.32**-1 for all triple-rockers where s is opposite l.

25. For the four-bar linkage of Fig. **1.35**-1, use Eqs. (**1.35**-1)–(**1.35**-10) to write a program for a digital computer, or programmable calculator, which finds
   a. output angles β and γ versus input angle;
   b. the path of a point P fixed to the coupler link DC.
   Use the program to plot several coupler curves for different locations of point P on a given mechanism. Repeat for several different mechanism proportions.

1.40   THE SLIDER-CRANK CHAIN

When one turning pair of a four-bar chain is replaced by a sliding pair, the chain becomes a slider-crank, as we have seen in Fig. **1.23**-1. We shall now examine the various inversions of this kinematic chain.

1.41   *Ordinary Slider-Crank Mechanism*

When an element of the sliding pair is fixed, we obtain the so-called **ordinary slider-crank mechanism**, which is commonly seen in reciprocating engines and compressors.

Figure 1 shows an ordinary slider-crank mechanism with the usual nomenclature. The distance e between the fixed pivot O and the straight line path of point B is called the **offset**. In most reciprocating engines, the offset is zero. However, offsets are utilized in mechanisms where a *quick-return* action is desired.

The link OA is called the **crank**, the floating link AB is called the **connecting rod**, and the sliding link is called a **piston, crosshead, plunger, slider, ram**, etc., depending on the application.

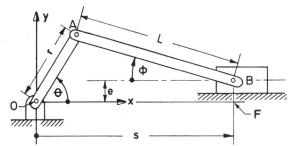

Figure 1.41-1. Ordinary slider-crank mechanism with offset.

To find relationships between the variables, θ, ϕ, and s, we observe that the closure of polygon $OABFO$ requires the satisfaction of the two "loop equations"

$$r \cos \theta + L \cos \phi - s = 0$$
$$r \sin \theta - L \sin \phi - e = 0 \tag{1}$$

For a given crank angle θ, Eqs. (1) can be solved explicitly for the connecting rod angle ϕ and the piston displacement s in the form

$$\phi = \arcsin\left(\frac{r}{L}\sin\theta - \frac{e}{L}\right) \tag{2}$$

$$s = r\cos\theta + L\cos\phi \tag{3}$$

From Eq. (2), we see that if the crank revolves completely, the connecting rod angle ϕ achieves its minimum (maximum) at $\theta = -90°$ $(\theta = 90°)$; i.e.,

$$\phi_{min} = \arcsin\frac{-r - e}{L} \tag{4a}$$

$$\phi_{max} = \arcsin\frac{r - e}{L} \tag{4b}$$

To find the conditions under which the crank can execute complete revolutions, it suffices to consider $e \geq 0$, since negative values of e represent a reflection of the mechanism of Fig. 1 about the x axis. The angle ϕ_{min}, defined by Eq. (4a), cannot be reached unless

$$r + e \leq L \tag{5a}$$

However, if inequality (5a) holds, so too must the inequality

$$|r - e| \leq r + e \leq L \tag{5b}$$

This inequality guarantees that the value ϕ_{max}, defined by Eq. (4b), can be reached. Thus, condition (5a) guarantees that the crank can rotate continuously,[18] as is necessary for engines and pumps.

If the mechanism is assembled with the slider on the positive half of the x axis and if condition (5a) is satisfied, the **principal value** of the *arc sin* function is to be used in Eq. (2) et seq.; i.e.,

$$-\frac{\pi}{2} \leq \phi \leq \frac{\pi}{2}$$

When initially assembled with the slider on the negative half of the x axis, the correct branch is

$$\frac{\pi}{2} \leq \phi \leq \frac{3\pi}{2}$$

As long as condition (5a) is satisfied, the mechanism always remains in that branch in which it was originally assembled.

It is left as an exercise (Problem **1.43**-5) to show that

1. When condition (5a) is satisfied, the extreme positions of the slider are given by

$$s_{max} \equiv s_1 = [(L + r)^2 - e^2]^{1/2} \tag{6a}$$

$$s_{min} \equiv s_2 = [(L - r)^2 - e^2]^{1/2} \tag{6b}$$

---

[18]See Problem **1.43**-11 for a proof by Grashof's criterion.

2. The crank angles corresponding to these extreme positions are

$$\theta_1 = \arc\sin\frac{e}{L+r} \tag{6c}$$

$$\theta_2 = \pi + \arc\sin\frac{e}{L-r} \tag{6d}$$

3. The piston moves

$$\begin{Bmatrix} \text{forward } (\dot{x} < 0) \\ \text{backward } (\dot{x} > 0) \end{Bmatrix} \quad \text{when} \quad \begin{Bmatrix} \theta_1 < \theta < \theta_2 \\ \theta_2 < \theta < 2\pi + \theta_1 \end{Bmatrix}$$

Hence, if $\dot{\theta}$ is constant, the ratio of time required for the forward stroke (t_f) to that for the return stroke (t_r) is given by the *time ratio*

$$\text{TR} \equiv \frac{t_f}{t_r} \equiv \frac{\theta_2 - \theta_1}{2\pi - (\theta_2 - \theta_1)} \tag{6e}$$

This equation characterizes the *quick-return action* of the offset slider-crank mechanism.

When condition (5a) is violated, it may readily be seen from a sketch[19] that there is a forbidden zone for the crank angle; i.e., θ is confined to the range

$$-\theta^* \leq \theta \leq \pi + \theta^* \tag{7}$$

where

$$\theta^* = \arc\sin\frac{L-e}{r} \tag{8}$$

The "velocity equations" resulting from the differentiation of Eqs. (1) are

$$\dot{\phi} = \frac{d\phi}{d\theta}\frac{d\theta}{dt} = k_\phi \dot{\theta}$$

$$\dot{s} = \frac{ds}{d\theta}\frac{d\theta}{dt} = k_s \dot{\theta} \tag{9}$$

where

$$k_\phi \equiv \frac{d\phi}{d\theta} = \frac{r}{L}\frac{\cos\theta}{\cos\phi}$$

$$k_s \equiv \frac{ds}{d\theta} = -(r\sin\theta + Lk_\phi\sin\phi) \tag{10}$$

From Eqs. (9) we find accelerations

$$\ddot{\phi} = k_\phi \ddot{\theta} + k_\phi' \dot{\theta}^2$$
$$\ddot{s} = k_s \ddot{\theta} + k_s' \dot{\theta}^2$$

---

[19]Problem **1.43-2.**

where

$$k'_\phi \equiv \frac{dk_\phi}{d\theta} = -\frac{r}{L}\frac{\sin\theta}{\cos\phi} + k^2_\phi \tan\phi$$

$$k'_s \equiv \frac{dk_s}{d\theta} = -(r\cos\theta + Lk'_\phi \sin\phi + Lk^2_\phi \cos\phi)$$

(11)

In reciprocating engines, all position variables are periodic functions of the crank angle θ and may be expressed as Fourier series in θ. For later reference, we indicate here how these series may be developed for the case of nonoffset engines (i.e., $e = 0$).[20]
Setting $e = 0$ in the second of Eqs. (1) gives

$$\sin\phi = \frac{r}{L}\sin\theta = \lambda \sin\theta$$

(12)

where

$$\lambda = \frac{r}{L}$$

(13)

Hence

$$\cos\phi = (1 - \sin^2\phi)^{1/2} = (1 - \lambda^2 \sin^2\theta)^{1/2}$$

(14)

Introducing the binomial expansion

$$(a + b)^n = a^n + \binom{n}{1}a^{n-1}b + \binom{n}{2}a^{n-2}b^2 + \binom{n}{3}a^{n-3}b^3 + \cdots$$

(15a)

where

$$\binom{n}{1} = \frac{n}{1!}, \quad \binom{n}{2} = \frac{n(n-1)}{2!}, \quad \binom{n}{3} = \frac{n(n-1)(n-2)}{3!}, \quad \cdots$$

(15b)

we may express Eq. (14) in the form

$$\cos\phi = 1 - \tfrac{1}{2}(\lambda^2 \sin^2\theta) - \tfrac{1}{8}(\lambda^4 \sin^4\theta) - \tfrac{1}{16}(\lambda^6 \sin^6\theta) + \cdots$$

(16)

Now we make use of a group of trigonometric identities which are given immediately below, without proof. Their derivation is left as an exercise in the use of the complex variable notation introduced in Chapter 2 (see Problems **2.50**-41, 42, 43, and 44).

$$\sin^2\theta = \tfrac{1}{2}(-\cos 2\theta + 1)$$

(17a)

$$\sin^3\theta = \tfrac{1}{4}(-\sin 3\theta + 3\sin\theta)$$

(17b)

$$\sin^4\theta = \tfrac{1}{8}(\cos 4\theta - 4\cos 2\theta + 3)$$

(17c)

$$\sin^5\theta = \tfrac{1}{16}(\sin 5\theta - 5\sin 3\theta + 10\sin\theta)$$

(17d)

$$\sin^6\theta = \tfrac{1}{32}(-\cos 6\theta + 6\cos 4\theta - 15\cos 2\theta + 10)$$

(17e)

$$\cos^2\theta = \tfrac{1}{2}(\cos 2\theta + 1)$$

(18a)

$$\cos^3\theta = \tfrac{1}{4}(\cos 3\theta + 3\cos\theta)$$

(18b)

$$\cos^4\theta = \tfrac{1}{8}(\cos 4\theta + 4\cos 2\theta + 3)$$

(18c)

$$\cos^5\theta = \tfrac{1}{16}(\cos 5\theta + 5\cos 3\theta + 10\cos\theta)$$

(18d)

$$\cos^6\theta = \tfrac{1}{32}(\cos 6\theta + 6\cos 4\theta + 15\cos 2\theta + 10)$$

(18e)

---

[20]For engines with $e \neq 0$, consult Biezeno and Grammel [1954, pp. 35, 36] and Mewes [1968]. Listings of the Fourier coefficients are also given by Taylor [1968, Vol. II, Chap. 8].

Upon substituting from Eqs. (17) into Eq. (16), we find

$$\cos \phi = \lambda(A_0 + \tfrac{1}{4}A_2 \cos 2\theta - \tfrac{1}{16}A_4 \cos 4\theta + \tfrac{1}{36}A_6 \cos 6\theta + \cdots) \tag{19}$$

where

$$A_0 = \frac{1}{\lambda} - \frac{1}{4}\lambda - \frac{3}{64}\lambda^3 - \frac{5}{256}\lambda^5 - \cdots \tag{20a}$$

$$A_2 = \lambda + \frac{1}{4}\lambda^3 + \frac{15}{128}\lambda^5 + \cdots \tag{20b}$$

$$A_4 = \frac{1}{4}\lambda^3 + \frac{3}{16}\lambda^5 + \cdots \tag{20c}$$

$$A_6 = \frac{9}{128}\lambda^5 + \cdots \tag{20d}$$

Since λ is usually in the range of $\tfrac{1}{6}$ to $\tfrac{2}{5}$ for internal combustion engines, it is usually unnecessary to retain more than a few terms in Eqs. (20) or to carry more than the first two or three terms in Eq. (19).

Upon using Eq. (19) in Eq. (3), the piston displacement is found to be given by

$$\frac{s}{r} = \cos \theta + \frac{1}{\lambda} \cos \phi \tag{21}$$

$$= A_0 + \cos \theta + \tfrac{1}{4}A_2 \cos 2\theta - \tfrac{1}{16}A_4 \cos 4\theta + \tfrac{1}{36}A_6 \cos 6\theta + \cdots$$

The corresponding velocities and accelerations are given by

$$\frac{\dot{s}}{r} = \dot{\theta}\left(-\sin \theta - \frac{A_2}{2}\sin 2\theta + \frac{A_4}{4}\sin 4\theta - \frac{A_6}{6}\sin 6\theta + \cdots\right), \tag{22}$$

$$\frac{\ddot{s}}{r} = \dot{\theta}^2(-\cos \theta - A_2 \cos 2\theta + A_4 \cos 4\theta - A_6 \cos 6\theta + \cdots)$$

$$+ \ddot{\theta}\left(-\sin \theta - \frac{A_2}{2}\sin 2\theta + \frac{A_4}{4}\sin 4\theta - \frac{A_6}{6}\sin 6\theta + \cdots\right) \tag{23}$$

The angular velocity of the connecting rod is found by differentiating Eq. (16) to obtain

$$\dot{\phi}\sin\phi = \dot{\theta}\left(\lambda^2 \sin \theta \cos \theta + \frac{\lambda^4}{2}\sin^3 \theta \cos \theta + \frac{3}{8}\lambda^6 \sin^5 \theta \cos \theta + \cdots\right) \tag{24}$$

If $\sin \phi$ is eliminated via Eq. (12) and use is made of the identities (17) and (18), it follows that

$$\dot{\phi} = \lambda\dot{\theta}\left(C_1 \cos \theta - \frac{C_3}{3}\cos 3\theta + \frac{C_5}{5}\cos 5\theta + \cdots\right) \tag{25}$$

and

$$\ddot{\phi} = -\lambda\dot{\theta}^2(C_1 \sin \theta - C_3 \sin 3\theta + C_5 \sin 5\theta + \cdots)$$

$$+ \lambda\ddot{\theta}\left(C_1 \cos \theta - \frac{C_3}{3}\cos 3\theta + \frac{C_5}{5}\cos 5\theta + \cdots\right) \tag{26}$$

where

$$C_1 \equiv 1 + \tfrac{1}{8}\lambda^2 + \tfrac{3}{64}\lambda^4 + \cdots \tag{27a}$$

$$C_3 \equiv \tfrac{3}{8}\lambda^2 + \tfrac{27}{128}\lambda^4 + \cdots \tag{27b}$$

$$C_5 \equiv \tfrac{15}{128}\lambda^4 + \cdots \tag{27c}$$

For most engines, it is sufficiently accurate to consider only the first two harmonics in expression (21) for the piston displacement and to use only the leading terms in Eqs. (20), thus giving the approximations

$$s \doteq L + r\left(\cos\theta + \frac{\lambda}{4}\cos 2\theta\right) \tag{28a}$$

$$\dot{s} \doteq -r\dot{\theta}\left(\sin\theta + \frac{\lambda}{2}\sin 2\theta\right) \tag{28b}$$

$$\ddot{s} \doteq -r\dot{\theta}^2(\cos\theta + \lambda\cos 2\theta)$$
$$\qquad - r\ddot{\theta}\left(\sin\theta + \frac{\lambda}{2}\sin 2\theta\right) \tag{28c}$$

where the terms involving θ and 2θ are referred to, respectively, as **primary** and **secondary** components.

1.42   Inversions of the Slider-Crank Mechanism

Figure 1 shows the four possible inversions of the slider-crank chain. Figure 1(a) shows the ordinary slider-crank, and Fig. 1(b) shows the **Whitworth quick-return mechanism** which results from fixing the crank in Fig. 1(a). The dashed links show the usual arrangement for rapidly returning a workpiece or cutting tool after a cutting stroke. This same inversion has been used[21] for aircraft engines where link 1, contain-

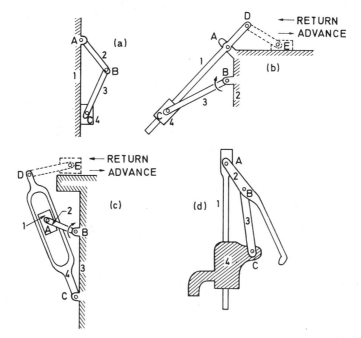

Figure 1.42-1. Inversions of slider-crank mechanism: (a) Ordinary slider-crank; (b) Whitworth quick-return; (c) Crank-shaper (ordinary quick-return); (d) Hand pump.

---

[21]See Hartenberg and Denavit [1964, pp. 56–58].

ing the cylinder and engine housing, was fastened directly to the rotating propeller shaft. When the connecting rod 3 is fixed, one obtains the **crank-shaper** (or **ordinary quick-return**) **mechanism** shown by the solid lines in Fig. 1(c). This inversion has been used[21] in the past for oscillating steam engines. The last inversion is illustrated by the **hand pump** of Fig. 1(d).

1.43 Problems

1. Show, by means of a sketch, that it would be impossible to assemble a slider-crank mechanism for which $e - r > L$ (symbols defined in Fig. **1.41**-1).

2. Show with the aid of sketches that θ is confined to the range defined by inequality (**1.41**-7) when $r + e > L$. Consider the four possible cases: $(r > e, e < L)$, $(r > e, e > L)$, $(r < e, e < L)$, $(r < e, e > L)$.

3. By integration of Eq. (**1.41**-25), find the Fourier series for connecting rod angle ϕ up to term(s) involving 5θ.

4. Show that terms of type $A_n \cos n\theta$ and $B_n \sin n\theta$ $(n = 1, 2, 3, \dots)$ appear in the Fourier series for s/r calculated from Eq. (**1.41**-21) if $e \neq 0$. Find A_1, A_2, B_1, and B_2 in terms of the "crank ratio" $\lambda \equiv r/L$ and the offset ratio $\beta \equiv e/L$.

5. a. Sketch the slider-crank mechanism (of Fig. **1.41**-1) when the crank and connecting rod are fully extended (OAB is a straight line of length $L + r$); also sketch the flexed position (where $OB = L - r$). Use this sketch to verify Eqs. (**1.41**-6a)–(**1.41**-6e).
   b. Find the time ratio TR defined by Eq. (**1.41**-6e) for selected pairs of r/L and e/r from the following table:

	Case					
	1	*2*	*3*	*4*	*5*	*6*
r/L	1/2	1/2	1/2	3/4	3/4	3/4
e/r	1/4	1/2	3/4	0.1	0.2	0.3

6. If link 3 of Fig. **1.42**-1(b) rotates with constant speed, find the ratio of the time required for the advance stroke to that for the return stroke of the workpiece. Use the following dimensions (units arbitrary): $AB = 3$, $BC = 10$, $AD = 6$, $DE = 24$.

7. If link 2 of Fig. **1.42**-1(c) rotates with constant speed, find the ratio of the time required for the advance stroke to that for the return stroke of the workpiece. Use the following dimensions (units arbitrary): $AB = 1.3$, $BC = 2.6$, $CD = 4.5$, $DE = 3.3$.

8. Show that the hand pump of Fig. **1.42**-1(d) may be correctly classified as an ordinary slider-crank mechanism with offset.

9. Show that, except for proportions, the mechanisms comprising loop $ABCA$ are the same in Figs. **1.42**-1(b) and (c). Describe the difference in placement of point D in the two cases.

10. Show that, except for proportions, the mechanisms are kinematically identical in Figs. **1.42**-1(a) and (d). Describe the difference, in the nature of the sliding pair elements, in the two cases.

11. Let *OC* be the fixed link of a four-bar mechanism with hinges at *O, A, B,* and *C*. The link lengths are $OA = r$, $AB = L$, $OC = p$, and $CB = p + e$.

    a. Show that as $p \to \infty$, the four-bar mechanism approaches a slider-crank mechanism with crank *OA*, connecting rod *AB*, and offset *e* (as defined in Fig. **1.41**-1).

    b. Use Grashof's criterion to show that if $r < L$, a slider-crank mechanism can have a completely revolving crank if and only if $e + r \leq L$.

1.50 DESCRIPTIONS OF SOME COMMON LINKAGES

1.51 Scotch Yoke and Its Inversions

We have seen that the four-bar mechanism is the fundamental building block of linkage design and have examined all of its possible inversions in Sec. **1.31**. When one hinge of the four-bar mechanism is replaced by a sliding pair, we obtain one of the four inversions of the slider-crank mechanism discussed in Sec. **1.42**. We now inquire as to the possible linkages which result when two hinges in a four-bar mechanism are replaced by sliding pairs.

Figure 1(a) shows a linkage, with four members, two hinges, and two sliding pairs, which is called a **Scotch yoke**. It is readily verified that the link 4 reciprocates in the fixed track with displacement $a \cos \theta$, where *a* is the length of the crank 2. Thus, the Scotch yoke is an exact sinusoid generator, unlike the slider-crank linkage which only approximates sinusoidal motion for extremely long connecting rods.

If member 2, which contains the two hinge pairs, is fixed, one obtains the mecha-

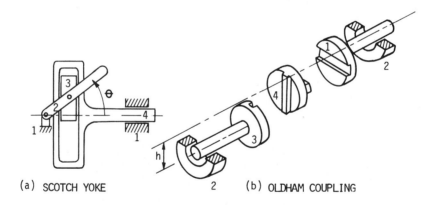

(a) SCOTCH YOKE   2   (b) OLDHAM COUPLING

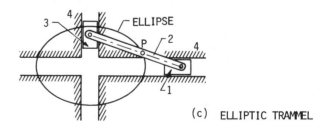

(c) ELLIPTIC TRAMMEL

Figure 1.51-1. Inversions of the Scotch yoke.

nism known as the **Oldham coupling**, shown in exploded view in Fig. 1(b). This mechanism is used to connect two rotating shafts with parallel axes which are separated by a fixed distance h. Although it might appear that the Oldham coupling is a space mechanism, it is readily seen that all particles of the system move in planes which are perpendicular to either shaft axis, thus fulfilling the definition of planar motion.

When link 4 (which contains two sliding pairs) is fixed, the linkage, shown in Fig. 1(c), becomes an **elliptic trammel**. This name stems from the readily verified fact that any point P on member 2 traces out an ellipse.

The mechanism which results when three hinges of a four-bar linkage are replaced by sliders has not been particularly useful, although it does have some interesting theoretical aspects. For example, the remaining hinge is theoretically *passive* and could in principle be welded fast to create *a movable three-bar linkage*. However, this linkage is *overconstrained* and will tend to lock, in practice, due to the friction in its joints brought about by large joint forces associated with small thermal or elastic distortions.

1.52   Pantograph

The **pantograph** is a device which is used to copy any plane figure in a reduced or enlarged scale. It is used in mapmaking, metal cutting, and the production of metal type molds, among many other applications.

Figure 1 shows two typical arrangements where C is the tracing point and P is the drawing point (or vice versa). The paths traced by P and C will be geometrically similar if the radius vector OP is always collinear with radius vector OC and the ratio OP/OC remains constant throughout the motion. For the systems shown, the similarity of triangles OPD and OCE guarantees the required conditions. Note that this linkage is not *closed* and not *constrained* in the sense of Reuleaux (i.e., it has two DOF rather than one). Thus it does not satisfy *Reuleaux's definition of a mechanism* on two counts, but it does satisfy our *more general definition of a mechanism*, i.e., a kinematic chain with one fixed link.

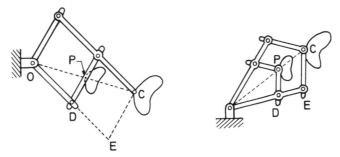

Figure 1.52-1. Pantographs.

1.53   Geneva Wheel

The **Geneva wheel**, shown in Fig. 1, is an example of a device which converts continuous motion into intermittent motion.

The pin P mounted on the driver wheel (1) engages slot EA in the Geneva wheel

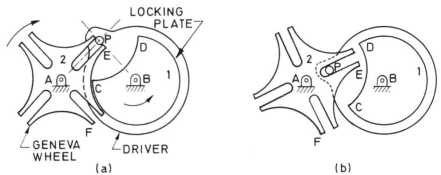

Figure 1.53-1. Geneva wheel: (a) Upon entry of pin; (b) During rotation of the Geneva wheel.

(2) when the angle BPA is 90°. As the driver rotates counterclockwise it will drive the Geneva wheel until slot AE occupies the position formerly held by slot AF in Fig. 1(a), whereupon the pin escapes from the slot and the Geneva wheel is no longer driven. To keep the wheel from rotating when the pin is disengaged, the locking plate is mounted integrally on the wheel as shown. The arc CD must be cut out of the locking plate to provide clearance for the tip E of the wheel, which would otherwise cut into the locking plate when the wheel rotates as illustrated in Fig. 1(b). A Geneva wheel with N slots would be turned through $1/N$ revolutions while the driving pin is engaged with a slot; then the wheel would **dwell** as the driving wheel completed its revolution, and the process would repeat indefinitely.

If one slot of the Geneva wheel is filled in, the driver will make a limited number of turns before it strikes the filled slot and is brought to a halt. Such an arrangement is called a **Geneva stop** and is used to prevent overwinding of watches or other instruments. A detailed discussion of Geneva wheel design is given by Bickford [1972].

1.54 Equivalent Mechanisms

The **kinematic state** of a particle is defined by its position, velocity, and acceleration. In a mechanism with one DOF, we may designate one link as the input link and specify the kinematic state of one point on it. Such a point is called an **input point** or **driving point**. For single-DOF mechanisms, the kinematic state of all other points (**output points** or **driven points**) will be uniquely determined by that of the input point.

If two mechanisms have the same kinematic states at their respective input and output points, the mechanisms are said to be **kinematically equivalent**.

If, for example, point B is the driving point for all of the mechanisms shown in Fig. 1 and point C is the driven point, all the mechanisms in Fig. 1 will have the same input-output relations with respect to position, velocity, and acceleration at all times. They are therefore **permanently kinematically equivalent**.

Although they perform the same kinematic functions, there are important design differences among the various mechanisms shown. Most noticeably, the reduction of contact area due to the replacement of the slider by a roller will result in very much higher contact stresses ("Hertz" stress) at the contact zone between roller and frame. The replacement of the roller by the cam will result in increased wear due to the sliding action at the contact zone between cam and frame.

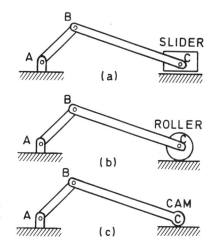

Figure 1.54-1. Slider-crank linkage and equivalent mechanisms with higher pairing.

Next, consider the case of two cams rotating about fixed centers O_1 and O_2. If, as in Fig. 2, both cams have circular profiles with centers at C_1 and C_2 respectively, it is apparent that the distance $C_1 C_2$ will remain constant throughout the entire range of motion. Since distances $O_1 C_1$ and $O_2 C_2$ will also remain constant during the motion, the four-bar linkage, determined by points O_1, C_1, C_2, O_2, and shown in Fig. 2, will give exactly the same relationship between angles θ_1 and θ_2 as will the original cam system, *at all times*. In other words, the relation between $\theta_1(t)$ and $\theta_2(t)$ at any time t is the same for both mechanisms shown in Fig. 2. Therefore the velocities $\dot{\theta}_1$ and $\dot{\theta}_2$, as well as the accelerations $\ddot{\theta}_1$ and $\ddot{\theta}_2$, will be identical for both mechanisms at all times, and the two mechanisms are *permanently kinematically equivalent*.

If either of the two cam profiles are noncircular, as in Fig. 3, we may attempt to create an instantaneously equivalent four-bar mechanism by first replacing the cam profiles by their respective **circles of curvature**[22] through the contact point. The centers of curvature C_1 and C_2 define an equivalent four-bar mechanism $O_1 C_1 C_2 O_2$ for the substitute constant curvature cam set. Since the distance $C_1 C_2$ will change when the actual cams rotate through any finite angle, the equivalent mechanism shown in Fig. 3 can be, at best, an **instantaneously equivalent mechanism**.

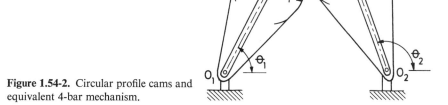

Figure 1.54-2. Circular profile cams and equivalent 4-bar mechanism.

---

[22]A unique circle may be passed through any three neighboring points P_1, P_2, P_3 on a given curve. As the three points converge toward a common point P, the circle just mentioned will converge to a limiting circle called the **circle of curvature** at P. The radius and center of the limiting circle are called the **radius of curvature** and **center of curvature** respectively. Other properties of the circle of curvature are discussed in Sec. **2.30.**

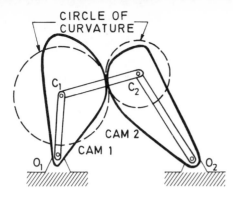

Figure 1.54-3. Noncircular cam profiles replaced by circles of curvature, and equivalent four-bar linkage.

To show that it is indeed an equivalent mechanism, we shall now demonstrate that the kinematic state of any point on the output link O_2C_2 of the four-bar mechanism is identical with that of the same point on cam 2 when cam 1 and input link O_1C_1 have identical kinematic states.

Figure 4 shows segments of two cams at three neighboring instants of time t, t', and t'', separated by the interval Δt. Point A on the driving cam is in contact with point a on the driven cam at time t. Similarly B and C on the driving cam contact b and c on the driven cam at times t' and t'', respectively. If the position of the driving cam is a specified function of time, the locations of points A, B, C and a, b, c are uniquely fixed by the shapes of the cam profiles and the three selected times t, t', t''. For clarity and generality, frame 3 has been allowed to move in Fig. 4.

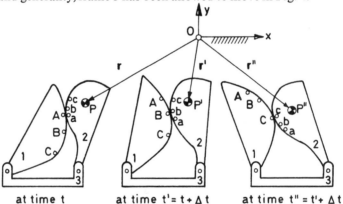

Figure 1.54-4. Showing cam positions at times t, t', and t'', and path of point P in driven cam.

Let $\mathbf{r}$ be the radius vector, at time t, from a fixed origin O to an arbitrary point P fixed in the driven cam. At the successive times t' and t'', P will occupy points P' and P'', defined by the corresponding radius vectors $\mathbf{r}'$ and $\mathbf{r}''$.

The velocity $\mathbf{v}_P$ and acceleration $\mathbf{a}_P$ of point P are given, respectively, by the defining relations

$$\mathbf{v}_P = \lim_{\Delta t \to 0} \left(\frac{\mathbf{r}' - \mathbf{r}}{\Delta t} \right)$$

$$\mathbf{a}_P = \lim_{\Delta t \to 0} \frac{1}{\Delta t} (\mathbf{v}'_P - \mathbf{v}_P) = \lim_{\Delta t \to 0} \frac{1}{\Delta t} \left[\frac{1}{\Delta t} (\mathbf{r}'' - \mathbf{r}') - \frac{1}{\Delta t} (\mathbf{r}' - \mathbf{r}) \right]$$

Thus we see that the velocity and acceleration at P depend only on the position vectors of the three consecutive points P, P', P''. These points, in turn, are uniquely determined by the position of points A, B, C and a, b, c. The shape of the cams outside the range of the arcs $\overset{\frown}{ABC}$ and $\overset{\frown}{abc}$ has no bearing on the acceleration (or velocity) of point P at time t. In the limit (as Δt vanishes), points B and C converge upon A, points b and c converge upon a, and the arcs $\overset{\frown}{ABC}$ and $\overset{\frown}{abc}$ approach arcs on the circles of curvature for the two cams at A and a, respectively. Hence, for sufficiently small time increments, the given cam profiles may be replaced by their circles of curvature, and the corresponding velocity and acceleration of any point in the driven cam will approach that of the original cam system as Δt vanishes. In short, *cam profiles may be instantaneously replaced by their circles of curvature without influencing the velocities or accelerations in the system.*

This result justifies the construction shown in Fig. 3 and verifies that the four-bar linkage shown there is kinematically equivalent to the original cam system at the instant shown.

Note that higher derivatives[23] such as $d\mathbf{a}_P/dt$ will depend on *four* or more consecutive points on the cam profiles and will not, in general, be correctly predicted when the cam profiles are replaced by their circles of curvature.

1.55 *Compilations of Mechanisms*

We have described a number of widely used mechanisms, and others will be mentioned subsequently (e.g., see Sec. **5.22**). However, it is not our intention to present a survey of the vast number of ingenious mechanical devices which have been invented. Those readers who will sooner or later become involved in the conceptual design of mechanisms will want to explore in greater depth the rich heritage left by previous generations of inventors. To aid in this activity we shall record here a variety of sources of such information.

Good historical guides to the subject are given by Ferguson [1962] and Hartenberg and Denavit [1964], who list many of the older catalogues of mechanisms. For example, a number of "sketchbooks" exist such as Barber [1934],[24] Clark [1943], Jones [1936–1967], Moskalenko [1964], and Grafstein and Schwarz [1971], which show hundreds of mechanisms for achieving a wide variety of motions. Some of the traditional textbooks on kinematics of machinery, such as Schwamb et al. [1947], contain many useful illustrations and descriptions. Many college libraries are well stocked with such books (from the era 1870–1950), and they are well worth perusing. The often-quoted classics of Reuleaux [1876] and Willis [1870] fall into this category.

Among the more recent textbooks, Beggs [1955] is notable for a lengthy picture catalogue of mechanisms, Hain [1967] for an exhaustive bibliography of 2364 items on applied kinematics, and Artobolevskii [1975] for an "atlas" of 2000 mechanisms.

Periodicals such as *Machine Design, Product Engineering, Design News,* and *Mechanical Engineering* frequently contain articles on mechanisms. Many such arti-

---

[23]The rate of change of acceleration, sometimes called **jerk**, is sometimes cited as a factor in motion sickness and ride quality of passenger vehicles.

[24]Barber, in his preface, expresses the hope that his work will be of service to those ". . . engaged in the head-splitting, exhausting work of scheming and devising machinery, than which I can conceive of no headwork more wearing and anxious."

cles, drawn largely from *Product Engineering*, have been compiled in book form by Chironis [1965, 1966].

Numerous examples of special-purpose mechanisms may be found in books such as Bickford [1972] on mechanisms for intermittent motion, Molian [1968b] on cam-linkages, Chironis [1967] on gears and gear linkages, and Artobolevskii [1964] and Svoboda [1948] on plane curve generators.

A large variety of intricate **valve gear mechanisms** has been designed to regulate the flow of steam into reciprocating engines. Interest in these mechanisms has waned with the decline of the steam engine itself. Although the growing interest in steam-powered automotive vehicles may revive interest in valve gear mechanisms, for the present they serve mainly as examples in books on kinematics. Sinclair [1907, Chap. 26] gives a 50-page discussion of their evolution; collections of valve gear diagrams may be found in Klein [1917, p. 112-a], Wittenbauer [1923, pp. 510 et seq.], and Beggs [1955, p. 210].

A good encyclopedia-type article covering the whole field of kinematics is that of Freudenstein and Sandor [1964], which also contains an excellent bibliography.

The items mentioned above and the references contained therein are a rich source of ideas for conceptual design of mechanisms. However, the prospective inventor should not overlook the wide variety of mechanisms which lies exposed (or nearly so) in many ordinary devices such as garage door hinges and locks, auto-hood hinge linkages, desk chair seat adjustments, lawn sprinklers, thrill rides, etc.

Finally, it must be mentioned that the patent literature is a vast storehouse of information for those who have the patience to wade through the surrounding legal jargon, which often seems designed more to confuse and intimidate possible competitors than to disclose the essence of an invention.

2

COMPLEX VECTORS IN KINEMATICS

When dealing with vectors in three-dimensional problems, the usual notation[1] involving $\mathbf{i}, \mathbf{j}, \mathbf{k}, \mathbf{A} \cdot \mathbf{B}, \mathbf{A} \times \mathbf{B}$, etc., is most useful. However, systems of coplanar vectors can often be treated more conveniently by means of the complex number notation which we shall now review. When we wish to emphasize the vector character of complex numbers we shall refer to them as **complex vectors**.

A comprehensive treatment of advanced plane geometry, by means of complex vectors, has been given by Zwikker [1963]. The application of complex vectors to problems of kinematics is discussed in this chapter and in Secs. **5.30** and **5.40**. This material is considered, from a similar point of view, by Dijksman [1976].

2.10 REVIEW OF COMPLEX VARIABLE NOTATION

The vector connecting two points, such as O and P in Fig. 1, may be represented by $\overrightarrow{OP}$ or by the complex number Z, where

$$\overrightarrow{OP} = Z = x + iy \tag{1}$$

and

$$i^2 = -1 \tag{2}$$

We note that x and y are called the **real** and **imaginary** parts of Z and are written symbolically in the form

$$x = \operatorname{Re} Z, \qquad y = \operatorname{Im} Z \tag{3}$$

---

[1]Due to J. Willard Gibbs (1839–1903).

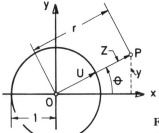

Figure 2.10-1. Representation of complex vector $Z = x + iy$.

Equation (1) gives the **rectangular form** of Z, which can also be expressed in the **polar form**

$$Z = r(\cos \theta + i \sin \theta) \tag{4}$$

In Eq. (4), r is said to be the **magnitude** of Z, denoted by $|Z|$, and θ is called the **argument** of Z, denoted by $\arg(Z)$. Since the magnitude of Z is given by

$$r = (x^2 + y^2)^{1/2} \tag{5}$$

the term in parentheses in Eq. (4) plays the role of a **unit vector** U which points in the direction of $\overrightarrow{OP}$; i.e.,

$$\cos \theta + i \sin \theta \equiv U \tag{6}$$

It may be shown by expansions in Taylor series, or otherwise (Appendix **B**), that

$$e^{i\theta} = \cos \theta + i \sin \theta \tag{7}$$

where $e\,(2.718281\ldots)$ is the base of the natural logarithm system. Equation (7), known as **Euler's formula**, expresses the unit vector U in **exponential form**. Thus the exponential form of Z is given by Eqs. (4) and (7) as

$$Z = re^{i\theta} \tag{8}$$

Complex numbers are assumed to obey all the formal rules of real algebra (associative, distributive, commutative, etc.). Hence, addition (or subtraction) of two complex vectors Z_1 and Z_2 is given simply by addition (or subtraction) of their respective components. That is, if we define

$$Z_1 = x_1 + iy_1 = r_1 e^{i\theta_1} \tag{9}$$
$$Z_2 = x_2 + iy_2 = r_2 e^{i\theta_2} \tag{10}$$

we can write

$$Z_1 \pm Z_2 = (x_1 \pm x_2) + i(y_1 \pm y_2) \tag{11}$$

If we represent the product of Z_1 and Z_2 by $Z_3 = r_3 e^{i\theta_3}$,

$$Z_3 = r_3 e^{i\theta_3} = r_1 e^{i\theta_1} r_2 e^{i\theta_2} = r_1 r_2 e^{i(\theta_1 + \theta_2)} \tag{12}$$

Thus we see that the magnitude of the product of two complex numbers equals the product of the magnitudes of the factors, and the argument of the product equals the sum of the arguments of the factors. From Eq. (12) we see that

$$Z_1 = \frac{Z_3}{Z_2} = \frac{r_3 e^{i\theta_3}}{r_2 e^{i\theta_2}} = \frac{r_3}{r_2} e^{i(\theta_3 - \theta_2)} \tag{13}$$

Hence, the quotient of two complex numbers has magnitude equal to the quotient of their magnitudes and argument equal to the difference of their arguments. From Eq. (12) we see that multiplication of a complex vector Z_1 by a unit vector of the form $e^{i\alpha}$ has the net effect of rotating the vector Z_1 through the angle α (counterclockwise if α is positive).

IMPORTANT

It is frequently useful to introduce the **complex conjugate**

$$\bar{Z} \equiv x - iy = re^{-i\theta} \tag{14}$$

ROTATION ABOUT REAL AXIS.

associated with the complex quantity $Z = x + iy = re^{i\theta}$. The following useful operations involving complex conjugation are readily confirmed:

$$(\overline{A \pm B}) = \bar{A} \pm \bar{B} \tag{15}$$

$$(\overline{AB}) = \bar{A}\bar{B} \tag{16}$$

$$|Z| = (Z\bar{Z})^{1/2} \tag{17}$$

$$\mathrm{Re}\, Z = \frac{Z + \bar{Z}}{2} \tag{18}$$

$$\mathrm{Im}\, Z = i\frac{\bar{Z} - Z}{2} \tag{19}$$

Given two vectors $\mathbf{a} = a_x\mathbf{i} + a_y\mathbf{j}$ and $\mathbf{b} = b_x\mathbf{i} + b_y\mathbf{j}$, we often wish to find the **scalar product**

$$\mathbf{a} \cdot \mathbf{b} \equiv a_x b_x + a_y b_y \tag{20}$$

and the **vector product**

$$\mathbf{a} \times \mathbf{b} = (a_x b_y - b_x a_y)\mathbf{k} \equiv (\mathbf{a} \times \mathbf{b})_z \mathbf{k} \tag{21}$$

where $\mathbf{i}$ and $\mathbf{j}$ are unit vectors in the direction of the x and y axes, respectively, and $\mathbf{k} \equiv \mathbf{i} \times \mathbf{j}$ is a unit vector perpendicular to the x, y plane.

It is readily proved (see Problems **2.50**-16 and 17) that the scalar and vector products may be expressed in the following compact forms:

$$\mathbf{a} \cdot \mathbf{b} = \mathrm{Re}\,(A\bar{B}) = \mathrm{Re}\,(\bar{A}B) \tag{22}$$

$$(\mathbf{a} \times \mathbf{b})_z = \mathrm{Im}\,(\bar{A}B) = -\mathrm{Im}\,(A\bar{B}) \tag{23}$$

where

$$A \equiv a_x + ia_y, \qquad B \equiv b_x + ib_y \tag{24}$$

are complex numbers.

Examples of how complex vectors can simplify geometric analyses follow.

EXAMPLE 1

The **involute,** of a circle of radius a, is traced out by the tip P of a string which is held taut as it is unwrapped from the circle. Show that the equation of the involute is given by

$$Z = ae^{i\theta}(1 - i\theta) \tag{25}$$

where θ is a real variable.

Proof:

Before unwrapping the string, its tip is at point A on the circle of center O (Fig. 2).

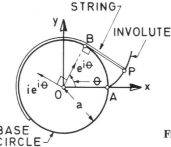

Figure 2.10-2. Involute of circle.

After the string has been unwraped over the arc $\widehat{AB}$ of the circle, the tip P will be at a distance $BP = \widehat{AB} = a\theta$ from the tangent point B.

The radius vector to P is

$$Z \equiv \overrightarrow{OP} \equiv \overrightarrow{OB} + \overrightarrow{BP} \tag{26}$$

where

$$\overrightarrow{OB} = ae^{i\theta} \tag{27}$$

To find $\overrightarrow{BP}$, we note that it has magnitude $a\theta$ and that it points 90° clockwise of $\overrightarrow{OB}$; but $ie^{i\theta}$ points 90° counterclockwise of $\overrightarrow{OB}$; hence $\overrightarrow{BP}$ has the direction of $-ie^{i\theta}$. That is,

$$\overrightarrow{BP} = a\theta(-ie^{i\theta}) \tag{28}$$

Upon substituting Eqs. (27) and (28) into Eq. (26) we find

$$Z = ae^{i\theta}(1 - i\theta) \tag{29}$$

as the equation of the involute. Equation (29) gives the location of point P for any specified angle θ. The real and imaginery parts of Z are

$$x = a(\cos\theta + \theta\sin\theta)$$
$$y = a(\sin\theta - \theta\cos\theta) \tag{30}$$

These are the parametric equations of the involute.

EXAMPLE 2

The power of complex number notation in kinematics is illustrated by the following proof of the **Roberts-Chebyshev theorem,**[2] which states that *three different four-bar linkages will trace identical coupler curves.*

Figure 3 shows a four-bar linkage $ABCD$ with point P rigidly attached to the coupler link DC. The path traced out by the *coupler point* P is called a *coupler curve*. It

---

[2]Due to S. Roberts (1876) and P. L. Chebyshev (1878). Historical remarks are given by Hartenberg and Denavit [1964, pp. 168–179], who also discuss limiting cases, e.g., when P lies on the line DC, or the case of the slider-crank mechanism.

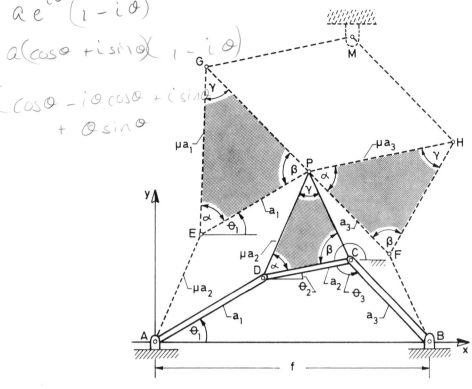

handwritten in margin:
$a e^{i\theta}(1 - i\theta)$

$a(\cos\theta + i\sin\theta)(1 - i\theta)$

$a(\cos\theta - i\theta\cos\theta + i\sin\theta + \theta\sin\theta)$

Figure 2.10-3. Cognate linkages, and the Roberts-Chebyshev theorem.

will be shown that there exist two other four-bar linkages (*AEGM* and *BFHM*) which trace out the same coupler curve. To find these so-called **cognate mechanisms** we add links *PE* and *EA* to the given linkage such that *ADPE* forms a parallelogram. Then we add links *PF* and *FB* such that *PFBC* is a parallelogram. Next, triangles *EPG* and *PFH*, which are both *similar* to the given triangle *DCP*, are fixed to links *EP* and *PF*, respectively (note the order of the angles α, β, γ in each triangle). Finally links *GM* and *HM* are added to complete the parallelogram *GPHM*.

It may be seen intuitively (and checked by the methods of Sec. **8.20**) that the single degree of freedom of the system is not altered by the formation of the three parallelogram loops. Therefore we would expect that, as point *P* traces out a coupler curve determined by the given four-bar *ABCD*, point *M* traces out some well-defined curve. However, it will be shown that *point M remains stationary*. Therefore, links *AEGM* constitute a four-bar mechanism, and links *BFHM* also constitute a four-bar mechanism. The point *P* may be viewed as a coupler point of either of these mechanisms or of the originally given mechanism *ABCD*.

To show that point *M* remains stationary, we shall show that the complex vector

$$Z_M = \overrightarrow{AE} + \overrightarrow{EG} + \overrightarrow{GM} \tag{31}$$

does not vary as point *P* moves about. We now introduce the following notation:

$$\overrightarrow{AD} = a_1 e^{i\theta_1}$$
$$\overrightarrow{DC} = a_2 e^{i\theta_2} \tag{32}$$
$$\overrightarrow{CB} = a_3 e^{i\theta_3}$$

From the similarity of the three shaded triangles,

$$\mu = \frac{DP}{DC} = \frac{EG}{EP} = \frac{PH}{PF}$$

The last-noted relationship can be expressed in the form

$$
\begin{aligned}
DP &= \mu DC = \mu a_2 \\
EG &= \mu EP = \mu a_1 \\
PH &= \mu PF = \mu a_3
\end{aligned}
\tag{33}
$$

It therefore follows that the desired complex vectors are given by

$$
\begin{aligned}
\overrightarrow{AE} &= \overrightarrow{DP} = \mu a_2 e^{i(\theta_2 + \alpha)} \\
\overrightarrow{EG} &= \mu a_1 e^{i(\theta_1 + \alpha)} \\
\overrightarrow{GM} &= \overrightarrow{PH} = \mu a_3 e^{i(\theta_3 + \alpha)}
\end{aligned}
\tag{34}
$$

where the angles of each vector, *with respect to the x axis*, have been used as arguments of the corresponding complex numbers.

Upon substituting Eqs. (34) into Eq. (31), we find that

$$Z_M = \mu e^{i\alpha}(a_2 e^{i\theta_2} + a_1 e^{i\theta_1} + a_3 e^{i\theta_3}) \tag{35}$$

However, the quantity in parentheses in Eq. (35) is the vector sum

$$\overrightarrow{DC} + \overrightarrow{AD} + \overrightarrow{CB} \equiv \overrightarrow{AB} \tag{36}$$

But $\overrightarrow{AB}$ is a constant vector whose complex representation is equal to the real number f, and we see that

$$Z_M = \mu f e^{i\alpha} = \text{constant} \tag{37}$$

Thus, point M has been shown to remain stationary, and the theorem is proved.

2.20 POSITION, VELOCITY, AND ACCELERATION OF A PARTICLE

The position of a particle is specified by the complex number, or position vector,

$$Z = x + iy = re^{i\theta} \tag{1}$$

where (x, y) are Cartesian coordinates and r, θ are polar coordinates of the particle. To find the complex vector V which represents the velocity of the particle we merely differentiate Eq. (1) with respect to time t and find

$$V = \dot{Z} = \dot{x} + i\dot{y} \quad \text{(Cartesian form)} \tag{2}$$

or

$$V = (\dot{r} + ir\dot{\theta})e^{i\theta} \quad \text{(polar form)} \tag{3}$$

If we introduce the unit vectors

$$U = e^{i\theta}, \qquad iU = ie^{i\theta} \tag{4}$$

which point in the radial and circumferential directions, as shown in Fig. 1, we see that Eq. (3) can also be expressed in the polar form

$$V = \dot{r}U + r\dot{\theta}(iU) \tag{5}$$

From Eq. (2) we see that the acceleration vector A is given in Cartesian form by

$$A = \ddot{x} + i\ddot{y} = \ddot{Z} = \dot{V} \tag{6}$$

Hence the polar form is found, from differentiation of Eq. (3), to be

$$A = \underbrace{(\ddot{r} - r\dot{\theta}^2)}_{a_r}\underbrace{e^{i\theta}}_{U} + \underbrace{(r\ddot{\theta} + 2\dot{r}\dot{\theta})}_{a_\theta}\underbrace{ie^{i\theta}}_{iU} \tag{7}$$

Equation (7) clearly shows the two components which constitute the **radial acceleration** (a_r) and the **transverse acceleration** (a_θ) in polar coordinates.

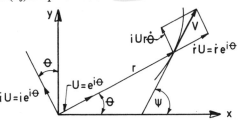

Figure 2.20-1. Illustrating complex velocity vector **V**.

EXAMPLE 1

Water discharges outward with constant speed v_R relative to the vanes of a centrifugal pump impeller. A typical vane is a circular arc of radius a and center C located at a distance a from the axis O of the impeller. The vanes are welded to the impeller wheel, which rotates with constant angular speed ω. Find the absolute velocity and acceleration of a particle at a point P located by the angles ϕ and θ as shown in Fig. 2. The

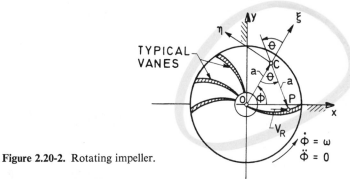

Figure 2.20-2. Rotating impeller.

position vector of particle P is given by

$$Z = \overrightarrow{OC} + \overrightarrow{CP} \tag{8}$$

$$= ae^{i\phi} - ae^{i(\phi+\theta)} \tag{9}$$

The velocity is therefore

$$V \equiv \dot{Z} = ai\dot{\phi}e^{i\phi} - ai(\dot{\phi} + \dot{\theta})e^{i(\phi+\theta)} \tag{10}$$

Recognizing that

$$\dot{\phi} = \omega, \qquad a\dot{\theta} = v_R \qquad (11)$$

we see that the velocity can be expressed as

$$V = i[a\omega e^{i\phi} - (a\omega + v_R)e^{i(\phi+\theta)}] \qquad (12)$$

The acceleration vector is therefore

$$A \equiv \dot{V} = i^2 a\omega\dot{\phi}e^{i\phi} - i(a\omega + v_R)i(\dot{\phi} + \dot{\theta})e^{i(\phi+\theta)}$$

$$A = \left[-a\omega^2 + a\left(\omega + \frac{v_R}{a}\right)^2 e^{i\theta}\right]e^{i\phi} \qquad (13)$$

If desired, we can find the components of acceleration along various directions. For example, the components (a_ξ, a_η) along the ξ, η axes which are rotating with the wheel are readily found by noting that $e^{i\phi} = U_\xi$ and $ie^{i\phi} \equiv U_\eta$ are the unit vectors parallel to ξ and η, respectively. Therefore we may use Euler's formula

$$e^{i\theta} = \cos\theta + i\sin\theta \qquad (14)$$

to express Eq. (13) in the form

$$A \equiv a_\xi U_\xi + a_\eta U_\eta = a_\xi e^{i\phi} + a_\eta ie^{i\phi} \qquad (15)$$

where

$$a_\xi \equiv -a\omega^2 + a\left(\omega + \frac{v_R}{a}\right)^2 \cos\theta \qquad (16)$$

$$a_\eta \equiv a\left(\omega + \frac{v_R}{a}\right)^2 \sin\theta \qquad (17)$$

If we desire the components of acceleration (a_x, a_y) along the fixed x and y axes, we merely find the real and imaginary parts of A, after substituting $\cos\phi + i\sin\phi$ for $e^{i\phi}$ in Eq. (15); i.e.,

$$a_x \equiv \mathrm{Re}\, A = a_\xi \cos\phi - a_\eta \sin\phi \qquad (18)$$

$$a_y \equiv \mathrm{Im}\, A = a_\xi \sin\phi + a_\eta \cos\phi \qquad (19)$$

EXAMPLE 2. Whirl-Producing Forces

If the net force acting on a particle of mass m can be expressed as a complex vector

$$F = F_x + iF_y = F(Z, \dot{Z}) \qquad (20)$$

which is a function of position Z and velocity $\dot{Z}$, the acceleration can be expressed[3] in the form

$$F(Z, \dot{Z}) = m\ddot{Z} \qquad (21)$$

Equation (21) is an ordinary differential equation in the complex variable Z.

For example, suppose that a particle of mass m is acted on by a force of magnitude kmv (where k is constant and v is particle speed) directed 90° counterclockwise of the

[3]Equation (21) expresses Newton's second law, which is all that we need to borrow from the subject of *dynamics* at this point; the remainder of the problem involves only *kinematics*.

velocity vector. Find the path of the particle if it starts from the origin with initial speed u directed along the x axis.

Solution:

The given force is $F = kmi\dot{Z}$; therefore Eq. (21) yields

$$\ddot{Z} = \frac{1}{m}F(Z, \dot{Z}) = ik\dot{Z} \tag{22}$$

Integrating once, we find

$$\dot{Z} - \dot{Z}_0 = ik(Z - Z_0) \tag{23}$$

Note that initially

$$\dot{Z}(0) \equiv \dot{Z}_0 = u$$

and

$$Z(0) \equiv Z_0 = 0$$

Hence

$$\frac{dZ}{dt} = \dot{Z} = u + ikZ = ik\left(Z - \frac{iu}{k}\right)$$

This "separable" differential equation may be expressed as

$$\frac{d(Z - iu/k)}{Z - iu/k} = ik\,dt$$

which may be integrated to yield

$$Z - \frac{iu}{k} = Ce^{ikt}$$

The constant of integration C is found by noting that $Z(0) = 0$; hence

$$Z - \frac{iu}{k} = -\frac{iu}{k}e^{ikt} \tag{24}$$

Equation (24) tells us that the particle describes a circle of radius u/k, counterclockwise about a center located u/k above the origin, as shown in Fig. 3. In real terms, Eq. (24)

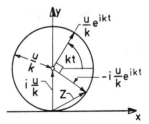

Figure 2.20-3. Particle path, example 2.

gives the path coordinates:

$$x = \text{Re } Z = \frac{u}{k}\sin kt$$

$$y = \text{Im } Z = \frac{u}{k}(1 - \cos kt)$$

The result just found helps to explain the cause of "whirling motions" in fluid film journal bearings. A hydrodynamic analysis of lubricant flow shows that when a journal (i.e., a shaft) moves radially into its surrounding fluid film, the fluid exerts a force component which acts at right angles to the direction of motion. Equation (24) indicates that the shaft will undergo a circular "orbiting" or "whirling" motion. This means that a momentary disturbance which knocks the shaft off its equilibrium position will tend to produce an undesirable whirling of the shaft center, unless sufficient fluid damping exists to control the whirl amplitude. Whirl-producing forces (normal to velocity) result from a variety of mechanisms, including (1) internal "solid" damping[4] in the shaft material or at the interface of a hub and shaft; (2) dry friction on an unlubricated shaft, thrust pad, guide, or shroud[5]; and (3) fluid trapped in a hollow rotor.[6]

2.30 INTRINSIC COORDINATES AND PATH CURVATURE

It is frequently convenient to work with a set of moving coordinate axes oriented along the tangent and normal to the path of a moving particle. Such axes define an **intrinsic coordinate** system. Geometric and kinematic properties of such a system will now be described.

2.31 *Velocity and Acceleration*

To express the velocity in terms of *intrinsic coordinates* we note from Fig. 1 that as the particle moves from position P to P', in time Δt, it sweeps out an arc Δs, and its position vector changes from Z to $Z + \Delta Z$.

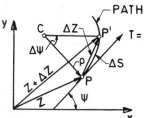

Figure 2.31-1. Illustrating complex tangent vector **T**.

By definition, the velocity vector is

$$V = \dot{Z} = \lim_{\Delta s \to 0} \frac{\Delta Z}{\Delta s} \frac{\Delta s}{\Delta t} = \frac{dZ}{ds} \dot{s} \tag{1}$$

From Fig. 1 we see that

$$\frac{dZ}{ds} = \lim_{\Delta s \to 0} \frac{\Delta Z}{\Delta s} \equiv T \equiv e^{i\psi} \tag{2}$$

where T is the unit tangent vector inclined at an angle ψ to a fixed reference axis (usually chosen as the x axis). Accordingly, Eq. (1) can be written in the form

$$V = \dot{s}T = \dot{s}e^{i\psi} \tag{3}$$

---

[4]Den Hartog [1956, p. 295].
[5]Den Hartog [1956, p. 292].
[6]Ehrich [1967].

The angle ψ may be found from the expressions[7]

$$\psi = \arg \dot{Z} = \arg(\dot{x} + i\dot{y}) = \arctan_2 (\dot{y}, \dot{x}) \tag{4}$$

The intrinsic form of the *acceleration* vector is found, from differentiation of Eq. (3), in the form

$$A = \dot{V} = \ddot{s}T + \dot{s}\dot{T} = \ddot{s}e^{i\psi} + \dot{s}\dot{\psi}ie^{i\psi} \tag{5}$$

Let us now define the **unit normal vector**

$$N \equiv iT = ie^{i\psi} \tag{6}$$

which obviously has unit length and points 90° counterclockwise of T. With this notation, the acceleration becomes

$$A = \ddot{s}T + \dot{s}\dot{\psi}N \tag{7}$$

We now express the angular velocity $\dot{\psi}$ of the unit tangent vector in the form

$$\dot{\psi} = \frac{d\psi}{ds}\frac{ds}{dt} = \kappa\dot{s} \tag{8}$$

where

$$\kappa \equiv \frac{d\psi}{ds}$$

defines the **curvature** of the path at point P.

Accordingly, the acceleration is given, via Eq. (7), in the form

$$A = \ddot{s}T + \kappa\dot{s}^2 N \tag{9}$$

The quantity

$$\rho \equiv \left|\frac{1}{\kappa}\right| = \left|\frac{ds}{d\psi}\right| \tag{10}$$

is called the **radius of curvature** of the path at point P. From the definition (10) we see that ρ is the limiting ratio

$$\rho = \lim_{\Delta s \to 0} \left|\frac{\Delta s}{\Delta \psi}\right| \tag{11}$$

which, as shown in Fig. 1, is the *limit of the radius of the circle that is tangent to the path at P and which passes through a neighboring point P' on the path.* As P' approaches P, the circle just described approaches a limiting circle called the **circle of curvature** or **osculating circle**, the center of which is called the **center of curvature**.

Note that ρ is positive by definition, and Eq. (9) can be expressed in the form

$$A = \ddot{s}T \pm \left(\frac{\dot{s}^2}{\rho}\right)N \tag{12}$$

---

[7]Recall that the function $\arctan_2 (Y, X)$ is defined in Appendix **C**.

where the $+$ or $-$ sign is to be used accordingly as $\dot{\psi}$ is positive or negative. The first term in Eq. (12) is called the **tangential acceleration**, and the second term is called the **centripetal acceleration** because it always points toward the center of curvature.

2.32 Curvature

Recall that the direction of the tangent vector T, at a point P, is that of the limiting line PQ passing through *two* neighboring points P, Q on the curve. Therefore, the circle of curvature just defined may be described as the limit of the circle through *three* neighboring points $(P, Q, \text{ and } R)$ on a curve, as Q and R converge on P; the three points are shown in Fig. 1. This "three-point definition" of the circle of curvature is seen to be fully equivalent to the definition which follows Eq. (**2.31**-11).

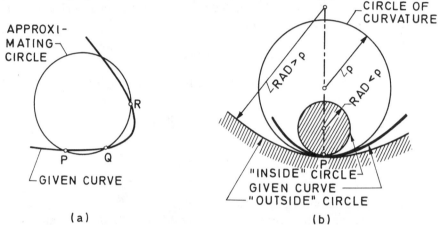

APPROXI-MATING CIRCLE

GIVEN CURVE

CIRCLE OF CURVATURE

"INSIDE" CIRCLE
GIVEN CURVE
"OUTSIDE" CIRCLE

(a) (b)

Figure 2.32-1. (a) Circle through three consecutive points on a curve crosses the curve; (b) Circle of curvature separates "inside" circles from "outside" circles.

This result is sometimes expressed by the following statement: **The circle of curvature is the limit of the circles through three ultimately coincident points on a curve.** Figure 1(a) shows that it is necessary for a circle to cross the given curve at Q if it is to pass through three neighboring points. In the limit, as Q and R converge upon P, we arrive at the geometric theorem that a curve crosses its circle of curvature at the point of tangency.[8] This theorem is illustrated in Fig. 1(b), which also suggests the verifiable conclusion that of all the tangent circles (with centers on the concave side of the curve) at a point of the curve, those which are smaller than the circle of curvature lie "inside" the curve in the neighborhood of the contact point, and those which are larger lie "outside" of the curve in the neighborhood of the contact point.

To express the curvature of the path in terms of Cartesian components of velocity and acceleration, we may use Eq. (**2.31**-4):

$$\psi = \arctan_2 (\dot{y}, \dot{x}) \tag{1}$$

Then, with the aid of Eq. (**2.31**-10) we see that

---

[8]This theorem is violated only at isolated points of the curve such as the vertices of an ellipse or more generally wherever the curve is cut by an axis of symmetry. Of course the theorem is violated if the given curve is itself a circle, because every normal to a circle is an axis of symmetry.

$$\frac{1}{\rho} = |\kappa| = \left|\frac{d\psi}{ds}\right| = \left|\frac{d\psi/dt}{ds/dt}\right| = \left|\frac{\dot{\psi}}{\dot{s}}\right| = \left|\frac{1}{\dot{s}}\frac{d}{dt}\arctan\left(\frac{\dot{y}}{\dot{x}}\right)\right| \qquad (2)$$

By using the rule for differentiation of the arc tangent function and by recalling that $\dot{s} = |V| = (\dot{x}^2 + \dot{y}^2)^{1/2}$, we arrive at the result

$$\frac{1}{\rho} = |\kappa| = \frac{|(1 + \dot{y}^2/\dot{x}^2)^{-1}(\dot{x}\ddot{y} - \dot{y}\ddot{x})/\dot{x}^2|}{(\dot{x}^2 + \dot{y}^2)^{1/2}} = \frac{|\dot{x}\ddot{y} - \dot{y}\ddot{x}|}{(\dot{x}^2 + \dot{y}^2)^{3/2}} \qquad (3)$$

where Eq. (C-10) has been utilized. Equation (3) gives us the means for calculating the curvature of a path when the x and y coordinates are prescribed functions of time.

The following useful results may be proved as exercises:

1. If a curve is specified by giving y as a function of x in the form

$$y = f(x)$$

then Eq. (3) implies (letting $t \equiv x$) that

$$\frac{1}{\rho} = |\kappa| = \left|\frac{d^2f/dx^2}{[1 + (df/dx)^2]^{3/2}}\right| \qquad (4)$$

2. If a curve is specified, in polar coordinates, by the relationship

$$r = f(\theta)$$

then

$$\frac{1}{\rho} = |\kappa| = \left|\frac{2(df/d\theta)^2 - f(d^2f/d\theta^2) + f^2}{[f^2 + (df/d\theta)^2]^{3/2}}\right| \qquad (5)$$

2.33 Tangent and Normal Vectors

When the Cartesian coordinates of a point on a curve are given functions of a parameter t, the unit tangent and normal vectors may be found from Eqs. (2.31-4) and (2.31-6). Frequently, however, the radius r of a point is a given function of the polar angle ϕ of the form

$$r = F(\phi) \qquad (1)$$

This polar representation is especially convenient to describe the contour of a disk cam, as illustrated in Fig. 1, where ϕ is measured from a fixed x axis. We seek to find expressions for the unit tangent vector T, the unit outward normal vector N_o, and the angle v between the radius vector and N_o.

Let T represent a tangent vector described by a point P which traces out the given profile in the direction assumed for increasing arc length s. The complex radius vector $\overrightarrow{OP}$ is given by

$$Z = re^{i\phi} \equiv F(\phi)e^{i\phi} \qquad (2)$$

If an observer moving forward with P sees the cam material (or the "interior" of an arbitrary closed curve) on his left, as in Fig. 1, the outward normal N_o will point to

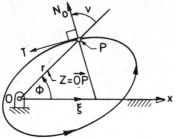

Figure 2.33-1. Tangent and outward normal vector for closed curve.

his right, and the unit tangent vector can be represented by

$$T = \frac{dZ}{ds} = \frac{dZ}{d\phi}\frac{d\phi}{ds} = \left(\frac{dF}{d\phi} + iF\right)e^{i\phi}\frac{d\phi}{ds} \tag{3}$$

where $dZ/d\phi$ has been found from Eq. (2).

From Fig. 1 we see that the outward unit normal is given by $N_o = -iT$; hence

$$N_o = \left(-i\frac{dF}{d\phi} + F\right)e^{i\phi}\frac{d\phi}{ds} = e^{i\phi}\left(F\frac{d\phi}{ds} - i\frac{dF}{d\phi}\frac{d\phi}{ds}\right) \tag{4}$$

Thus the angle which N_o makes with the x axis is

$$\arg N_o = \phi + \arctan_2\left(-\frac{d\phi}{ds}\frac{dF}{d\phi}, \frac{d\phi}{ds}F\right) \tag{5}$$

If we introduce the **sign indicator** σ defined as

$$\sigma = \begin{Bmatrix} 1 \\ -1 \end{Bmatrix} \quad \text{if} \quad \frac{d\phi}{ds} \begin{Bmatrix} \geq 0 \\ < 0 \end{Bmatrix} \tag{6}$$

we see that

$$\arg N_o = \phi + \arctan_2\left(-\sigma\frac{dF}{d\phi}, \sigma F\right) \tag{7}$$

From Fig. 1 we also see that

$$\arg N_o = \phi + \nu \tag{8}$$

Upon comparing Eqs. (7) and (8) we may conclude that

$$\nu = \arctan_2\left(-\sigma\frac{dF}{d\phi}, \sigma F\right) \tag{9}$$

2.40 MOTION RELATIVE TO A MOVING FRAME

2.41 *Direct Method*

Suppose that a point P momentarily occupies position coordinates ξ, η in a reference frame M which is itself moving as a rigid body with respect to a fixed frame F, as shown in Fig. 1. The motion of the moving frame is completely defined by the posi-

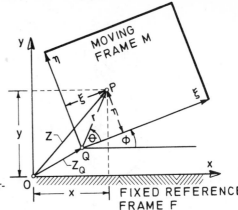

Figure 2.41-1. Fixed and moving reference frames.

FIXED REFERENCE FRAME F

tion vector Z_Q of a point Q fixed in M and by the angle ϕ between the moving ξ axis and the x axis fixed in frame F.

An observer stationed in the moving frame would express the complex position vector ζ of the particle P by

$$\zeta = \xi + i\eta \tag{1}$$

We call ζ the **position vector relative to the moving frame,** or the **relative position vector** for brevity.

An observer stationed in the fixed frame would represent the absolute[9] position vector $\overrightarrow{OP}$ as

$$Z = \overrightarrow{OP} = \overrightarrow{OQ} + \overrightarrow{QP} \tag{2}$$

In complex vector notation[10] we may write

$$\overrightarrow{OQ} = Z_Q$$
$$\overrightarrow{QP} = \xi e^{i\phi} + \eta i e^{i\phi} = \zeta e^{i\phi} \tag{3}$$

Thus the absolute position vector is

$$Z = Z_Q + \zeta e^{i\phi} \tag{4}$$

To solve typical problems in particle kinematics, with the aid of complex vectors, it is only necessary to write down an expression for the complex position vector Z in the form of Eq. (4). Then, by two successive differentiations, one finds the velocity $\dot{Z}$ and the acceleration $\ddot{Z}$ in a completely systematic way.

EXAMPLE 1

A rigid rod (Fig. 2) slides over a fixed circle of radius a. The distance of the rod's tip P, measured from the contact point Q, is denoted by s, and the rod makes an angle ϕ with the fixed x axis. Find the acceleration components (a_x, a_y) of point P with respect to the fixed frame (x, y). Also find the components (a_ξ, a_η) of the absolute acceleration in the direction of the moving axes (ξ, η).

---

[9]The word **absolute** denotes *relative to the fixed frame*.

[10]Capital letters will denote complex vectors whose real and imaginary parts are measured along the fixed axes x and y, respectively.

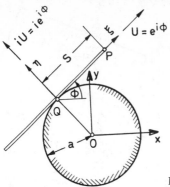

Figure 2.41-2. Rod slides on fixed circle.

Solution:

The complex position vector of point P is

$$Z = \overrightarrow{OQ} + \overrightarrow{QP} = ae^{i(\phi+\pi/2)} + se^{i\phi}$$
$$= (s + ia)e^{i\phi} \tag{5}$$

velocity, $\quad V = \dot{Z} = \dot{s}e^{i\phi} + i\dot{\phi}(s + ia)e^{i\phi} = (\dot{s} - a\dot{\phi})e^{i\phi} + s\dot{\phi}ie^{i\phi} \tag{6}$

acceleration, $\quad A = \dot{V} = (\ddot{s} - a\ddot{\phi} - s\dot{\phi}^2)e^{i\phi} + (2\dot{s}\dot{\phi} - a\dot{\phi}^2 + s\ddot{\phi})ie^{i\phi} \tag{7}$

We may, if desired, find the components of acceleration in various directions. For example, the unit vectors along the axes of ξ and η are given, respectively, by

$$U \equiv e^{i\phi} \quad \text{and} \quad iU \equiv ie^{i\phi} \tag{8}$$

Thus we recognize that the parentheses in the equation for A enclose the desired components of acceleration in the direction of ξ and η, respectively; i.e.,

$$a_\xi = \ddot{s} - a\ddot{\phi} - s\dot{\phi}^2 \tag{9}$$
$$a_\eta = s\ddot{\phi} + 2\dot{s}\dot{\phi} - a\dot{\phi}^2 \tag{10}$$

If we wish to find the components of acceleration in the directions of x and y, we merely have to find the real and imaginary parts of A. This is readily done by recalling that $e^{i\phi} = \cos\phi + i\sin\phi$ and rewriting Eq. (7) in the form

$$A = a_\xi(\cos\phi + i\sin\phi) + a_\eta(i\cos\phi - \sin\phi) \tag{11}$$

Hence

$$a_x = \text{Re } A = a_\xi \cos\phi - a_\eta \sin\phi \tag{12}$$
$$a_y = \text{Im } A = a_\xi \sin\phi + a_\eta \cos\phi \tag{13}$$

2.42 Coriolis' Theorem

A famous theorem on relative motion is named after G. G. de Coriolis who discovered it (in somewhat veiled terms) in his investigation of the theory of water wheels in 1832.[11] Upon differentiating Eq. (**2.41**-4) we find the velocity in the form

$$V = \dot{Z} = \dot{Z}_Q + i\dot{\phi}\zeta e^{i\phi} + \dot{\zeta}e^{i\phi} \tag{1}$$

_____

[11]Dugas [1955, p. 374].

The last term in Eq. (1) is due strictly to the motion of point P in the moving frame and is accordingly called the **relative velocity**,

$$V_{\text{rel}} \equiv \dot{\zeta} e^{i\phi} \tag{2}$$

To interpret the meaning of the other terms in Eq. (1) let us assume for the moment that point P is permanently glued to the moving frame M (i.e., $\dot{\zeta} = 0$). Such a point will be referred to as a **base point**, and its velocity is given by

$$V_{\text{base}} \equiv \dot{Z}_Q + i\zeta\dot{\phi} e^{i\phi} \tag{3}$$

This **base velocity** is also referred to in the literature by such names as **transport velocity**, **constraint velocity**, **drag velocity**, **frame velocity**, **vehicular velocity**, *schleppgeschwindigkeit*, etc.

In short, we see that the absolute velocity consists of two terms:

$$V = V_{\text{base}} + V_{\text{rel}} \tag{4}$$

To find the absolute acceleration A, we differentiate Eq. (1) and obtain

$$A = \dot{V} = \ddot{Z}_Q + i\zeta(\ddot{\phi} + i\dot{\phi}^2)e^{i\phi} + 2i\dot{\zeta}\dot{\phi} e^{i\phi} + \ddot{\zeta} e^{i\phi} \tag{5}$$

Equation (5) expresses **Coriolis' theorem:**

$$A = A_{\text{base}} + A_{\text{Cor}} + A_{\text{rel}} \tag{6}$$

where

$$A_{\text{base}} \equiv \ddot{Z}_Q + (i\zeta\ddot{\phi} - \zeta\dot{\phi}^2)e^{i\phi} \tag{7}$$

is the acceleration of the base point (previously defined) and is called **base acceleration**.

$$A_{\text{rel}} \equiv \ddot{\zeta} e^{i\phi} \tag{8}$$

is the **relative acceleration**, and

$$A_{\text{Cor}} \equiv i2\dot{\phi}\dot{\zeta} e^{i\phi} = i2\dot{\phi} V_{\text{rel}} \tag{9}$$

is the **Coriolis acceleration**.

In summary, *the absolute acceleration of a point on a moving reference frame is the vector sum of base acceleration, relative acceleration, and Coriolis' acceleration.*

Note that if $\dot{\phi}$ is positive the Coriolis acceleration clearly points 90° counterclockwise from the vector V_{rel}, as shown in Fig. 1(a). If $\dot{\phi}$ is negative, A_{Cor} is directed 90° clockwise from V_{rel}, as shown in Fig. 1(b). A simple rule, worth remembering, is that *the direction of A_{Cor} is found by rotating the vector V_{rel} 90° in the direction of $\dot{\phi}$.*

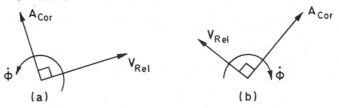

Figure 2.42-1. Coriolis' acceleration vector is turned 90° from relative velocity vector, in sense of angular velocity.

Problems involving moving reference frames may always be solved with the aid of Coriolis' theorem (6). However, a great advantage of the complex vector notation is that it allows us to solve such problems directly, without reference to this theorem, as was illustrated in Sec. **2.41**.

2.50 PROBLEMS

1. A and B in Fig. 1 are two points fixed in a lamina. Point P lies on the line joining A and B, at a directed distance μAB from A. Show that the velocity and acceleration of P are given by

$$\dot{Z} = (1 - \mu)\dot{Z}_A + \mu\dot{Z}_B$$
$$\ddot{Z} = (1 - \mu)\ddot{Z}_A + \mu\ddot{Z}_B$$

where Z, Z_A, Z_B are position vectors of points P, A, B, respectively.

2. Points A and B in Fig. 2 are fixed in a lamina, and P is a point fixed in the lamina at a distance $\mu|\overrightarrow{AB}|$ from A, with angle $BAP = \beta$. Show that the acceleration of P is given by

   a. $\ddot{Z} = (1 - U)\ddot{Z}_A + U\ddot{Z}_B$, where $U \equiv \mu e^{i\beta}$.

   b. $\begin{Bmatrix} \ddot{x} \\ \ddot{y} \end{Bmatrix} = \begin{Bmatrix} \ddot{x}_A \\ \ddot{y}_A \end{Bmatrix} + \mu \begin{bmatrix} \cos\beta & -\sin\beta \\ \sin\beta & \cos\beta \end{bmatrix} \begin{Bmatrix} \ddot{x}_B - \ddot{x}_A \\ \ddot{y}_B - \ddot{y}_A \end{Bmatrix}$, where x and y are arbitrarily oriented rectangular axes.

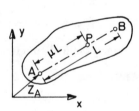
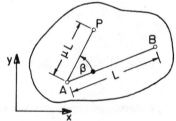

Figure 2.50-1 (Prob. 1)   **Figure 2.50-2** (Prob. 2).

3. For the double pendulum of Fig. 3, show that the acceleration components of point P are given by

$$\ddot{x} = -a\dot{\theta}^2 \cos\theta + l\dot{\phi}^2 \cos(\beta + \phi) + a\ddot{\theta} \sin\theta + l\ddot{\phi} \sin(\beta + \phi)$$
$$\ddot{y} = -a\dot{\theta}^2 \sin\theta - l\dot{\phi}^2 \sin(\beta + \phi) + a\ddot{\theta} \cos\theta + l\ddot{\phi} \cos(\beta + \phi)$$

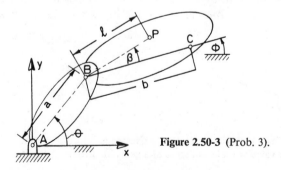

Figure 2.50-3 (Prob. 3).

4. A particle P moves in a plane with constant angular speed ω about a fixed point O. If the rate of increase in its acceleration is parallel to PO,

a. Prove that

$$\frac{d^2r}{dt^2} = \frac{1}{3}r\omega^2$$

b. Find r as a function of t if, at time zero, $r = r_0$ and $dr/dt = \dot{r}_0$.

5. P is a point on the tangent which touches a circle of radius a at point Q. QP is of length r and makes an angle θ with a fixed tangent. Show that the components of acceleration of P, along and normal to QP, are, respectively,

$$\ddot{r} - r\dot{\theta}^2 + a\ddot{\theta} \quad \text{and} \quad r\ddot{\theta} + 2\dot{r}\dot{\theta} + a\dot{\theta}^2$$

6. Point Q moves uniformly on a fixed circle of radius a with angular speed ω about center O. P moves in the same plane, describing relative to Q a circle with center Q and radius b. The angular velocity of ray QP is $-\omega$. Express the acceleration of point P in terms of vector $\overrightarrow{OP}$.

7. A small pin P moves in a fixed circular groove of radius r (Fig. 4). A slotted rod OA passes over the pin and rotates, about the fixed point O on the circumference of the groove, with a constant angular velocity ω. Find the absolute acceleration of the pin.

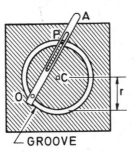

Figure 2.50-4 (Prob. 7).

8. Prove that the speed of a particle is constant if the acceleration vector is always perpendicular to the velocity vector.

9. If we erect a Cartesian coordinate system with origin at a point P of a curve, the ξ axis in the tangent direction and the η axis toward the center of curvature (Fig. 5), show that the curvature at P, as defined by Eq. (2.32-4), is equivalent to

$$\kappa = \lim_{\xi \to 0} \left(\frac{2\eta}{\xi^2}\right)$$

(Hint: Use Taylor series expansion.)

10. Newton defined the circle of curvature at a point P of a curve as the limit of a circle which is tangent to the curve at P and passes through a neighboring point Q. If the ξ axis is directed along the tangent and the η axis toward the concave side of the curve as shown in Fig. 6, prove that the radius of the tangent circle through P and Q is given by

$R_1 = \frac{1}{2}[\eta + (\xi^2/\eta)]$; therefore the radius of curvature is

$$R = \lim_{\xi \to 0} R_1 = \lim_{\xi \to 0} \frac{\xi^2}{2\eta}$$

Note: Comparison of this result with that of **Problem 9** shows that Newton's definition is consistent with the more commonly used definition

$$\frac{1}{R} = \frac{\eta''}{[1 + (\eta')^2]^{3/2}}$$

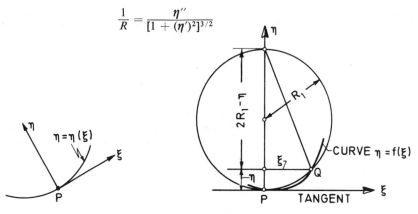

Figure 2.50-5 (Prob. 9).   Figure 2.50-6 (Prob. 10).

11. If, in Fig. **2.20-2**, the angular acceleration is given by $\ddot{\phi} = \alpha$, find an expression for the complex acceleration vector A of the water particle P. Also find the components of acceleration along the axes ξ and η.

12. Water discharges outward along the vanes of a centrifugal pump impeller with a constant speed V_r relative to the vanes (Fig. 7). The impeller rotates at speed ω in a clockwise direction. Find the acceleration components (a_x, a_y) of a particle of water P as it leaves the impeller if $V_r = 120$ ft/sec and $\omega = 1600$ rpm? Use the dimensions shown, and assume $\dot{\omega} = 0$.

13. The side AB of a carpenter's square ABP is rolled over a circle of center O and radius a, as illustrated in Fig. 8. Show that if $BP \equiv a$, the point P describes an Archimedian spiral; i.e., $|OP|$ is proportional to the angle through which ray OP turns.

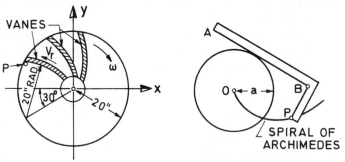

Figure 2.50-7 (Prob. 12).   Figure 2.50-8 (Prob. 13).

14. A cam follower F translates along a tangent to the base circle of an involute cam profile (Fig. 9). Show that the "rise" y of the follower is, at all times, proportional to the angle θ through which the cam has turned.

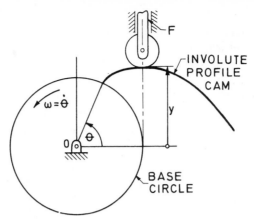

Figure 2.50-9 (Prob. 14).

15. The path of a particle is described by coordinates x, y. Show that the angular velocity $\dot{\psi}$ of the unit tangent is given by $\dot{\psi} = (\dot{x}\ddot{y} - \dot{y}\ddot{x})/(\dot{x}^2 + \dot{y}^2)$.

16. Show that the scalar product

$$\mathbf{a} \cdot \mathbf{b} = \text{Re}(A\bar{B})$$

where

$$A = a_x + ia_y, \qquad \bar{B} = b_x - ib_y$$

and (a_x, a_y), (b_x, b_y) are the rectangular resolutes of $\mathbf{a}$ and $\mathbf{b}$ in the x and y directions, respectively.

17. If $A = a_x + ia_y$ and $B = b_x + ib_y$, show that the z component of the cross product $\mathbf{a} \times \mathbf{b}$ is given by $\text{Im}(\bar{A}B)$, where $(a_x, a_y, 0)$ and $(b_x, b_y, 0)$ are the rectangular resolutes of $\mathbf{a}$ and $\mathbf{b}$, respectively, along the x, y, z axes.

18. Show that an arbitrary vector $V_3 = |V_3| e^{i\theta_3}$ can be expressed in the form

$$V_3 = C_1 V_1 + C_2 V_2$$

where

$$V_1 = |V_1| e^{i\theta_1} \quad \text{and} \quad V_2 = |V_2| e^{i\theta_2}$$

are two arbitrary (but assumed known) nonparallel vectors and C_1, C_2 are real constants. Show that

$$C_1 = \frac{|V_3| \sin (\theta_3 - \theta_2)}{|V_1| \sin (\theta_1 - \theta_2)} = \frac{\text{Im} (V_3/V_2)}{\text{Im} (V_1/V_2)}$$

$$C_2 = \frac{|V_3| \sin (\theta_3 - \theta_1)}{|V_2| \sin (\theta_2 - \theta_1)} = \frac{\text{Im} (V_3/V_1)}{\text{Im} (V_2/V_1)}$$

19. Use Eq. (**2.10**-22) to prove that the components of acceleration, tangent and normal, to the path of a moving particle are given, respectively, by

$$a_T = \frac{\dot{x}\ddot{x} + \dot{y}\ddot{y}}{(\dot{x}^2 + \dot{y}^2)^{1/2}}$$

$$a_N = \pm \frac{\dot{x}\ddot{y} + \dot{y}\ddot{x}}{(\dot{x}^2 + \dot{y}^2)^{1/2}}$$

20. Verify that path curvature is given by $\kappa = |\mathrm{Im}(A\bar{V})|/|V|^3$, where V and A are complex vectors of velocity and acceleration of a moving particle.

21. A particle moves along a spiral path given in polar coordinates (r, ϕ) by $r = ae^{-\beta\phi}$. Show that the radius of curvature at any position $Z = re^{i\phi}$ is given by $\rho = |BZ/\mathrm{Re}\, B|$, where $B \equiv 1 + i\beta$. Then show that $\rho = a(1 + \beta^2)^{1/2}e^{-\beta\phi}$. (Hint: Assume that $\dot{\phi} = 1$. Why is this justifiable? Then use the result of Problem 20.)

22. Find the acceleration of points Q and N if the straight edge MQ of body 2 rolls without slipping over the fixed circle of radius a in Fig. 10.

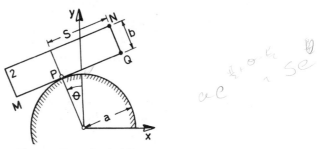

Figure 2.50-10 (Prob. 22).

23. The motion of a particle in the x, y plane is governed by the differential equation $\ddot{Z} - p^2Z = 0$, where p is a real positive constant and $Z = x + iy$.
    a. Find the complex position vector Z at an arbitrary time t; i.e., solve the differential equation for $Z(t)$ using arbitrary initial conditions.
    b. Sketch the path of the particle, and indicate the direction of motion along the path.

24. a. Solve the differential equation $\ddot{Z} + p^2Z = 0$, where p is a real positive constant and initial conditions are given by $Z(0) \equiv Z_0$, $\dot{Z}(0) \equiv \dot{Z}_0$.
    b. Sketch the solution curve $Z(t)$. (Hint: See Problem 32.)

25. Show that the tangent of the curve described by $Z = Z(t)$ makes an angle ψ with the real axis, where $\psi = (1/2i) \ln (\dot{Z}/\bar{\dot{Z}})$.

26. Show that the curvature at an arbitrary point of the curve described by $Z = Z(t)$ can be expressed in the form

$$\kappa \equiv \frac{d\psi}{ds} = \frac{1}{2i} \frac{\ddot{Z}\bar{\dot{Z}} - \dot{Z}\bar{\ddot{Z}}}{(\dot{Z}\bar{\dot{Z}})^{3/2}}$$

27. Show that the curvature at a point of the curve $Z = Z(t)$ is given by

$$\kappa = \frac{\mathrm{Im}(\bar{\dot{Z}}\ddot{Z})}{|\dot{Z}|^3}$$

28. Show that the curvature at a point of the curve $Z = Z(s)$ is given by

$$\kappa = \frac{d\psi}{ds} = \mathrm{Im}\left(\frac{d\bar{Z}}{ds}\frac{d^2Z}{ds^2}\right)$$

where s is arc length measured from an arbitrary point on the curve.

29. A curve is specified in polar coordinates (r, θ) by giving the radius in the form $r = f(\theta)$.
    a. Verify that the unit tangent vector T is given by $T = e^{i\psi}$, where

$$\psi = \arctan_2 [(f\dot{\theta} \cos \theta + \dot{f} \sin \theta), (-f\dot{\theta} \sin \theta + \dot{f} \cos \theta)].$$

b. Verify that the angle ν between the radius vector and a normal to the curve is given by $\nu = -\arctan_2 (\dot{f}, -f\dot{\theta})$.

30. Three points A, B, C are located by complex vectors Z_A, Z_B, Z_C, respectively. Show that the centroid of triangle ABC is given by $Z = \frac{1}{3}(Z_A + Z_B + Z_C)$.

31. A rod OQ, of length q, rotates with constant angular velocity ω about a fixed point O. A rod QP of length p is hinged to the first rod at Q and rotates with constant absolute angular velocity $-\omega$. Show that point P describes an ellipse with semidiameters $q + p$ and $|q - p|$.

32. a. Show that the complex vector $Z \equiv x + iy = Ae^{i\omega t} + Be^{-i\omega t}$ traces out an ellipse as time t increases, where A and B are complex constants and ω is a real constant.
    b. Show that the semidiameters of the ellipse are given by $|A| + |B|$ and $||A| - |B||$ along the major and minor axes, respectively.
    c. Show that the major axis of the ellipse is inclined to the x axis by the angle $\psi = \frac{1}{2}(\arg A + \arg B)$. (Hint: Rotate the axes of the ellipse in Problem 31 through the angle ψ and replace ωt by $\omega t + \phi_0$, where ϕ_0 is an arbitrary constant.)

33. An ellipse is generated by the sum of two rotating vectors $\overrightarrow{OM} = Ae^{i\omega t}$ and $\overrightarrow{ON} = Be^{-i\omega t}$ such that

$$\overrightarrow{OP} = Z = Ae^{i\omega t} + Be^{-i\omega t}$$

Show that the normal to the ellipse at P is parallel to the vector $\overrightarrow{NM}$ and the tangent is perpendicular to $\overrightarrow{NM}$.

34. Show that the motion described by the equation $Z = Ae^{i\omega t} + Be^{-i\omega t}$, when $|A| = |B|$, is a simple harmonic motion along a straight segment of length $4|A|$, inclined at an angle $\frac{1}{2}(\arg A + \arg B)$ to the real axis.

35. The sum of the squares of the distances, from two fixed points A and B, to a coplanar moving point P remains constant. Prove that point P describes a circle whose center is the midpoint of segment AB.

36. Figure 11 shows a rectangle $ABCD$ where side AB has length L. The points A and B are constrained to move along fixed straight lines OA and OB which make an angle α with each other. The point P, fixed to the plane $ABCD$, has Cartesian coordinates ξ and η in the moving plane. Prove that the point P traces out an ellipse in the fixed coordinate frame x, y. (Hint: Use the result of Problem 32.)

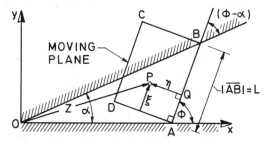

Figure 2.50-11. General elliptic trammel.

37. A particle of mass m is acted on by a force of magnitude kmv, where k is constant, m is particle mass, and v is instantaneous velocity. If the force always makes a fixed angle α with the direction of velocity v, show that

a. The path of the particle is a spiral.

b. The particle spirals outward if $|\alpha| < 90°$ and inward to a fixed point if $90° < \alpha < 270°$. Find the ultimate position of the particle in the latter case.

38. Verify Eq. (2.32-4).

39. Verify Eq. (2.32-5).

40. Find the rectangular coordinates for the path of an arbitrary point on the cross-link of an Oldham coupling [link 4 in Fig. 1.51-1(b)].

41. Show that

$$\cos\theta = \frac{1}{2}(e^{i\theta} + e^{-i\theta}), \qquad \sin\theta = \frac{1}{2i}(e^{i\theta} - e^{-i\theta})$$

and use this information to derive Eqs. (1.41-18).

42. Starting from Eqs. (1.41-14) and (1.41-15), show that the general form corresponding to Eqs. (1.41-18) is

$$\cos^n\theta = \frac{1}{2^{n-1}}\left[\cos n\theta + \binom{n}{1}\cos(n-2)\theta + \binom{n}{2}\cos(n-4)\theta + \cdots \right.$$
$$\left. + K\binom{n}{I}\cos(n-2I)\theta\right]$$

where I = integer part of $(n/2)$ and

$$K = \begin{cases} 1 & \text{if } n \text{ is odd} \\ \frac{1}{2} & \text{if } n \text{ is even} \end{cases}$$

43. Starting as in Problem 41, derive Eqs. (1.41-17).

44. Starting as in Problem 43, show that the general form corresponding to Eqs. (1.41-17) is as follows:

a. When n is *odd*,

$$\sin^n\theta = \frac{i^{n-1}}{2^{n-1}}\left[\sin n\theta - \binom{n}{1}\sin(n-2)\theta\right.$$
$$\left. + \binom{n}{2}\sin(n-4)\theta - \cdots\right]$$

[brackets contain $(n+1)/2$ terms with alternating signs; $i^2 \equiv -1$).

b. When n is *even*,

$$\sin^n\theta = \frac{i^n}{2^{n-1}}\left[\cos n\theta - \binom{n}{1}\cos(n-2)\theta\right.$$
$$\left. + \binom{n}{2}\cos(n-4)\theta - \cdots + (-1)^{n/2}\frac{1}{2}\binom{n}{n/2}\right]$$

45. Starting as in Problem 41, show that
a. $\cos m\theta \cos n\theta = \frac{1}{2}[\cos(m+n)\theta + \cos(n-m)\theta]$.
b. $\cos m\theta \sin n\theta = \frac{1}{2}[\sin(m+n)\theta + \sin(n-m)\theta]$.
c. $\sin m\theta \sin n\theta = \frac{1}{2}[\cos(m-n)\theta - \cos(m+n)\theta]$.

3

GEARS

3.10 INTRODUCTION

There are numerous works available which discuss the history of gear technology. However, recent historical studies of Derek de Solla Price [1974, 1975] indicate that many previously accepted interpretations of the early history of gearing are untrustworthy. Therefore, we shall give a recapitulation of the essential historical background on gears.

It has been known from ancient times that wheels in contact may transmit rolling motion by virtue of the friction developed at their interface. Since the available frictional force is limited by nature, it was an inevitable forward step to cut teeth into the contacting surfaces, thereby emulating that microscopic roughness which was thought to be the cause of passive dry friction. The toothed wheels, or **gears,** so formed are now among the most commonly used machine components.

The earliest[1] known description of friction wheels appears in the *Mechanical Problems* of Aristotle (384–322 B.C.). Formerly, it was believed that these wheels might have had teeth cut into their rims, but Drachmann [1963] has shown that there is no explicit reference to such teeth in the pertinent text. The oldest known unequivocal reference to toothed wheels (i.e., *gears*) was written by the Roman architect Vitruvius (ca. 25 B.C.), who described a rack-and-pinion[2] arrangement used in the water clock of Ktesibios[3] (Ctebius) in about 280 B.C. Whether this arrangement was actually used is unknown because nothing like it has been

---

[1]For the early history of gear technology see Price [1974, 1975] and de Camp [1974].

[2]A **pinion** is either a small gear or the smaller of two mating gears. A **rack** is a plane surface, with teeth cut into it; it may also be considered as part of a gear of infinite diameter.

[3]Ktesibios, apparently a prolific inventor, has been called the "Edison of Ptolmaic Alexandria" by de Camp [1974].

found in later water clocks. It seems more certain that the combination of *worm*[4] and *worm-gear* was used in the war engines of Archimedes, in about 250 B.C., and thereby became quite well known. For example, Hero of Alexandria (ca. 60 A.D.) described the application of the wormgear to instruments such as the hodometer (also known as odometer, cyclometer, or taximeter), shown in Fig. 1. A single tooth, attached to the hub E of a carriage wheel, advances the *crown gear ABCD* by one tooth spacing for each revolution of the carriage wheel. The dial pointers, driven by the worm-wheel train, might be the first known example of an automatic recording device.

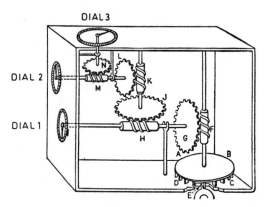

Figure 3.10-1. The hodometer described by Hero of Alexandria. From Cohen and Drabkin [1948, p. 343] (Courtesy, Harvard University Press).

Vitruvius also described a grinding mill, in which a horizontal shaft, extending from a vertical water wheel, carries a *crown gear* which meshes with a **lantern gear** (consisting of vertical wooden pegs sandwiched between, and distributed around the periphery of, two horizontal discs) having a vertical axis. An illustration of this early example of *bevel gears* (gears with nonparallel shafts) may be seen in Usher [1929, p. 169]. This "Vitruvian mill" is noteworthy as the first known example where gears were used to transmit appreciable amounts of power.

The earliest known literature on trains of gears with parallel shafts (i.e., *spur gears*) is Hero's description of a lifting device, similar to that of Fig. **3.34**-16.

The rudimentary gear systems described above would appear to represent the state of the art up to the first century A.D. For the following 1000 years, there is no existing record of progress in gear technology. We know that Islamic astronomical computing instruments with ingenious gear trains were produced around 1000–1200 A.D. Shortly afterwards, sophisticated astronomical clocks were produced in Europe by Richard of Wallingford (ca. 1330) and Giovanni de' Dondi[5] (ca. 1348–1364). The de' Dondi clock had at least seven dials, a large number of elliptical as well as circular gears, linkwork, an automated calendar, and indicators for planetary motions. Thereafter, the production of elaborate and sophisticated clockwork mechanisms accelerated rapidly, especially for use in cathedrals.

From the above synopsis, it would appear that progress in gear technology flourished briefly for 250 years after the time of Archimedes and stagnated for a millenium afterwards. Then—in one magnificent burst of invention—the most ingenious and complex geared systems appeared *ab initio*. To explain this apparent discontinuity in the evolution of the gear art we must jump ahead to our own century.

In the year 1900, a group of Greek spongefishers stumbled upon the wreck of a great

---

[4]Gear terminology printed in *italic* will be defined in later sections of this chapter.

[5]See Bedini and Maddison [1966] for a detailed description of the de'Dondi clock as well as a chronology of related clockwork.

ship, under 180 ft of water, off the Aegean island of *Antikythera*.[6] Among the many treasures recovered (bronze and marble statues, glass vessels, Hellenistic and Roman pottery, and commercial amphoras) was a formless, lime-encrusted lump which went unnoticed for nearly 8 months after completion of the salvage operation. Under the corrosion deposits of centuries, this lump contained fragments of a wooden case which, in the process of drying out, split open to reveal the existence of badly corroded pieces of bronze that faintly resembled fragments of toothed wheels.

Initially the "mechanism" was identified as an astrolabe,[7] but a controversy soon developed over this interpretation. There the situation rested until 1953 when Derek de Solla Price began his studies of the object. By 1959 he was ready to conjecture that it was an "Ancient Greek computer." By June 1973, Price was satisfied that the Antikythera mechanism was a sun and moon computing mechanism which was made around 87 B.C. More importantly, he established that the device was a sophisticated instrument with more than 30 gears, including spur gears, crown gears, compound gears, planetary trains, and the first known example of the *differential gear* (a device for finding the sum or difference of two input shaft rotations).

Perhaps the greatest significance of Price's work lies in his realization that the sophistication of this device exceeds by far anything that has been described in the Hellenistic literature. Therefore, if the date of 87 B.C. is approximately correct, he has shown that the level of gear technology had, by then, reached heights unappreciated by previous historians, and the imagined 1000-year hiatus in gear technology, prior to the appearance of the great cathedral mechanisms, never did occur. Thus, in Price's words, ". . . this singular artifact, the oldest existing relic of scientific technology, and the only complicated mechanical device we have from antiquity, quite changes our ideas about the Greeks and makes visible a more continuous historical evolution of one of the most important main lines that lead to our modern civilization."[8]

Prior to Price's investigations, the first use of a *differential* gear was thought by Needham [1965] to have occurred in the Chinese south-pointing chariot (ca. 120–250 A.D.). This supposed forerunner of modern navigational systems is believed to have been made somewhat as shown in the reconstructed model of Fig. 2.

If the chariot wheels do not slip and they are separated by exactly one diameter ($L \equiv D$), it can be verified, by methods of gear train analysis developed below, that the man's extended arm, shown in Fig. 2, will continue to point in the direction it is initially set regardless of the motions of the chariot. The slightest deviation from the assumptions mentioned would destroy the accuracy required for a trip of significant length.

Numerous other examples of gear trains, used from 400 A.D. onward, may be found in a compact monograph *The Evolution of the Gear Art* by Dudley [1969], in the historical works of Usher [1929] and Burstall [1965], and in the brief article of Davison [1963].

Although the use of trains of toothed wheels can be traced back to antiquity, the proper form of *tooth profile* required for gears to turn smoothly and with constant speed ratios was not discovered until the seventeenth century. According to Reuleaux [1876, p. 154], Desargues (1593–1662) was the first to solve this problem, by the use of so-called *cycloidal* teeth, although the honor is often attributed to the astronomer Roemer (1664–1710). The modern form of gear tooth profile, the *involute*, was recommended by de La Hire (a student of Desargues) in 1694, and later mathematical studies of mating gear teeth were made by Camus

---

[6]See Weinberg et al. [1965] and Price [1974].

[7]The astrolabe is a mathematically ingenious but mechanically simple device which is essentially a circular star map that can be rotated to show the position of the stars at various times. It might have been invented as early as the second century B.C. by Hipparchos (Price [1975, p. 38]).

[8]Price [1974, p. 13].

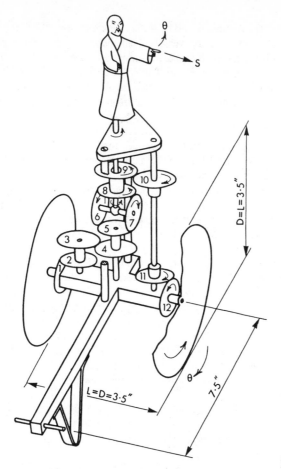

Wheels 1,2; 5,6,7,8;11,12:- Bevel gears,1:1 ratio
Wheels 3,4;9,10:- Spur gears,1:1 ratio

Figure 3.10-2. Model of Chinese South-pointing chariot. From Burstall [1968] (Courtesy, Edward Arnold, Ltd.).

(1733) and Euler (1754). It was Camus who stated the *fundamental law of gearing*, discussed in Sec. **3.41,** and described practical methods for the design and manufacture of cycloidal teeth for *spur* and bevel gears. A chronology of the evolution of gear tooth profiles is given by Dudley [1969, p. 35], who also covers the development of manufacturing methods from 1540 onward. A more extensive treatment of manufacturing methods is given by Woodbury [1959].

The net effect of properly designed gear teeth is to produce a motion equivalent to the rolling of friction wheels. Thus the detailed form of tooth profiles does not enter directly into a discussion of the kinematics of gear trains. Accordingly, we shall first study ways of analyzing the motions of various types of gear sets (e.g., *spur, bevel, worm*) and gear trains (*simple, compound, epicyclic, differential*). Although we are focusing our attention on problems of *planar* motion, we shall investigate gear trains containing nonparallel shafts, because of the importance of such configurations; fortunately, this limited departure from strictly planar motion does not introduce serious complications.

After we have shown how to analyze gear trains of arbitrary arrangement, we shall briefly discuss a few important topics related to the design and manufacture of suitable tooth profiles. This is an immense subject with a correspondingly huge literature. It is our intention to present only the minimum amount of information required by designers of general machinery—information which will enable one to select appropriate gears from catalogues and to communicate effectively with representatives of the gear manufacturing industry.

More detailed information[9] will be found in specialized works such as Merritt [1971], Dudley [1954, 1962a], Buckingham [1963], and Chironis [1967]. It will also prove rewarding to examine some of the older kinematics texts, such as Willis [1870], Albert and Rogers [1938], and Schwamb et al. [1947], which contain a more extensive treatment of the subject than is given herein.

The material on tooth profiles is independent of that on gear trains, and the remainder of this book is essentially independent of this chapter, except for an occasional cross-reference to some simple formulas on gear train ratios.

3.20 COMMON TYPES OF GEAR SETS

In this section we shall describe the most commonly used types of rolling surfaces and their associated gear sets.

Spur Gears

If two parallel circular[10] cylinders (as in Fig. 1) are pressed together, it is possible to transmit rotary motion between them by virtue of the friction developed at the line of contact. The rolling surfaces are called **pitch** surfaces, and their intersections with a plane normal to their axes are called **pitch circles**. For sufficiently low contact forces, the cylinders will roll without relative slip, but slip will develop when the applied torques cannot be balanced by the maximum attainable friction forces on the pitch surfaces. A more reliable kinematic drive is obtained if teeth are cut into the cylinders, thereby converting them from **friction wheels** to **gears**.

For example, suppose that a series of uniformly spaced lines (called **tooth traces**) are drawn, parallel to the cylinder axes, on both pitch surfaces as shown in Fig. 1(a). If material is removed from every other strip formed by the tooth traces, leaving a set

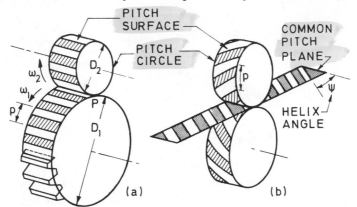

Figure 3.20-1. Rolling parallel cylinders. Development of: (a) Spur gears; (b) Helical gears.

[9]A bibliography of 814 items on planetary gearing is given by Levai [1969].
[10]Pure rolling contact of noncircular cylinders is possible if the rolling surfaces satisfy a criterion established in Sec. **5.15**. The current chapter is restricted to circular pitch cylinders and circular gears. Information on noncircular gears will be found in Chironis [1967] and Olsson [1953].

of teeth in the alternate strips (shown shaded in Fig. 1), the gears so formed are called **spur gears**. The profiles of the gear teeth, in planes normal to the tooth traces, must be carefully chosen to ensure smooth rolling action of the pitch cylinders. The design of suitable tooth profiles will be discussed in Secs. **3.40-3.45**.

Imagine that a sheet of paper has been inserted between the gears, as in Fig. 1(b), with a pattern of uniformly spaced parallel bands inked onto both sides of the strip as shown. If the strip is wrapped tautly about each cylinder, in turn, the inked bands will be transferred to the pitch surfaces of both gears and establish a suitable pattern for teeth to be cut in both gears. If the bands are straight and parallel (the usual case), their boundaries (the tooth traces) form a set of parallel helixes on the pitch surfaces, and the gears are called **helical gears** (sometimes called helical *spur* gears).

When the "strip of paper" shown in Fig. 1(b) is flat, it defines the **common pitch plane** of the rolling cylinders. If teeth are cut into the common pitch plane along the tooth traces, they form a **rack** which will mesh with the gears of the set. A rack may also be thought of as a gear of infinite diameter.

Figure 2 illustrates the physical appearance of straight and helical spur gears and

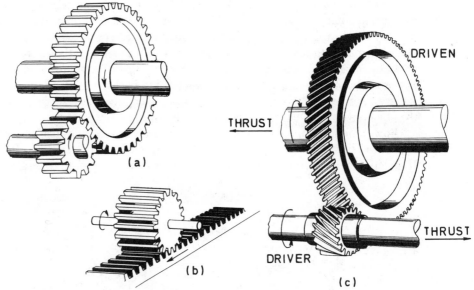

Figure 3.20-2. Spur gear forms: (a) Straight spur gears; (b) Rack and pinion; (c) Helical (spur) gears (Courtesy, Mobil Oil Corp.).

the limiting case (a rack) of the former. From simple considerations of statics it may be seen that the helical gears of Fig. 2(c) produce axial forces, or thrusts, on one another in the directions indicated. These thrusts will have to be absorbed by suitable thrust bearings, which can be quite large and expensive if a large amount of power is to be transmitted through the gear set. The necessity for such thrust bearings can be eliminated by cutting helical gears, of opposite helix angle, on the same gear blank as shown in Fig. 3. Often these **double-helical** or **herringbone** gears will have a circumferential groove between the two sets of teeth to help circulate the lubricating oil, or else the two sets of teeth are staggered to improve oil flow.

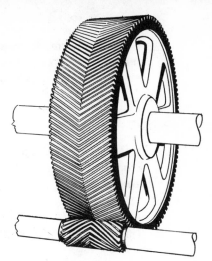

Figure 3.20-3. Double helical or herring-bone gears transmit no axial thrust to bearings (Courtesy, Mobil Oil Corp.).

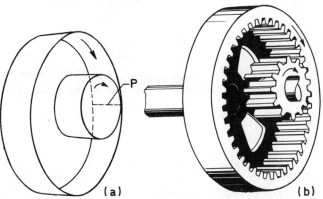

(a) (b)

Figure 3.20-4. Gears making internal contact: (a) Pitch surfaces; (b) Pinion and annular gear (Courtesy, Mobil Oil Corp.).

When one pitch surface surrounds another, as in Fig. 4, the corresponding gears are said to make **internal contact**, and the enclosing gear is called a **ring gear, annular gear,** or **internal gear**.

The distance p (see Fig. 1) spanned along the pitch circle, by one tooth and its neighboring space, is called the **pitch** or **circular pitch**. For a gear with pitch circle diameter D, and N equally spaced teeth, it follows that

$$\pi D = Np \tag{1}$$

The pitch circles make contact at the **pitch point** P where surface particles in both wheels must move at a common speed v to fulfill the postulated condition of no slip. Accordingly, the angular speeds (ω_1, ω_2) of the wheels are related by the expressions

$$\frac{\omega_1 D_1}{2} = \frac{\omega_2 D_2}{2} = v \tag{2}$$

where D_1, D_2 are the pitch circle diameters, also known as **pitch diameters**.

To ensure smooth meshing, it is essential that both gears of a set have the same number of teeth per inch of circumference. It has become traditional to refer to the number of teeth per inch of pitch diameter as the **diametral pitch** P_d, i.e.,

$$P_d = \frac{N}{D} \tag{3}$$

Most stock gears are manufactured with diametral pitches having integer values. From Eqs. (1) and (3) it follows that circular pitch and diametral pitch are related by the expression

$$pP_d = \pi \tag{4}$$

The *speed* ratio of two mating gears follows from Eqs. (2) and (3) in the form

$$\left|\frac{\omega_1}{\omega_2}\right| = \frac{D_2}{D_1} = \frac{N_2}{N_1} \tag{5}$$

Equation (5) is valid for the speed ratio of both internally and externally mating gears. By observing that the pitch point P has the same velocity on both gears of a set, we see from Fig. 1 that externally contacting gears rotate in opposite directions. Similarly we see from Fig. 4 that internally contacting gears rotate in the same direction.

Crossed-Helical Gears

If the axes of the friction wheels are not parallel, it is possible to ink a suitable set of tooth traces on both pitch surfaces, as shown in Fig. 5(a), where the common pitch plane (marked with parallel tooth traces inclined to both cylinder axes) is alternately wrapped around each of the pitch surfaces. The gear teeth cut between these inclined tooth traces form crossed-helical gears. Because the pitch surfaces touch at a point (rather than along a line, as in Fig. 1), these gears have a tendency to wear rapidly and are seldom used to transmit significant loads.

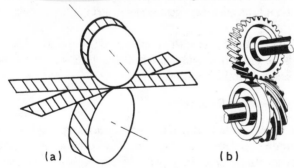

(a) (b)

Figure 3.20-5. Crossed helical gears: (a) Pitch cylinders with non-parallel axes. Development of tooth traces; (b) Physical appearance of crossed-helical gears (Courtesy, Mobil Oil Corp.).

Bevel Gears

When tooth traces are drawn along the generators of rolling cones, as in Fig. 6(a), they form **straight bevel gears**. If the tooth traces are curved, as in Fig. 6(b), they form

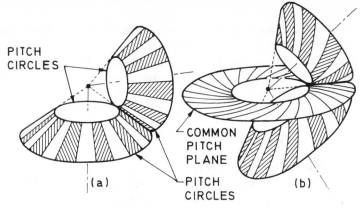

PITCH
CIRCLES

COMMON
PITCH
PLANE

(a)

PITCH
CIRCLES

(b)

Figure 3.20-6. Rolling cones.

spiral bevel gears.[11] The common pitch plane shown in Fig. 6(b) contains the curved tooth traces which define the spiral teeth. Teeth cut into the tooth traces on the common pitch plane would form a **crown gear**. A crown gear may also be viewed as a bevel gear with a pitch cone having a vertex angle of 180°; it is analogous to the rack which mates with a spur gear. Bevel gears with orthogonal axes are shown in Fig. 7.

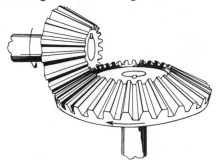

Figure 3.20-7. Straight bevel gears
(Courtesy, Mobil Oil Corp.).

---

[11]Straight lines inclined to the cone generators cannot lie on the cone surface, but can form hyperboloids of revolution, which are the pitch surfaces of so-called **hyperboloidal** or **skew bevel gears**. The axes of these gears do not intersect, and they are not now used very much (Merritt [1971, p. 33]). The "hypoid" gears manufactured by the Gleason Works resemble skew bevel gears in that the gear axes are offset, and resemble spiral bevel gears by virtue of their curved tooth traces. (See Fig. 8.)

(a)   (b)

Figure 3.20-8. Spiral bevel and hypoid gears: (a) Spiral bevel gears (curved tooth traces; shafts intersect); (b) Hypoid gears (Curved tooth traces; shafts do not intersect) (Courtesy, Deere and Company).

Any pair of touching circles on the conical pitch surfaces can be referred to as *pitch circles* for the bevel gears, although it is common practice to refer to the largest diameter when using the definition of diametral pitch ($P_d = N/D$). When the pitch circles of a pair of bevel gears are of equal size and meet at right angles, the gears are said to be **miter gears**.

It is readily established that Eq. (5) is valid for the speed ratio of bevel gears.

Worm and Wormgear Sets

A **worm and wormgear set** (Fig. 9) is very much like a pair of crossed-helical gears where one of the gears (the **worm**) has a relatively small diameter. The axes of the two gears are usually (but not necessarily) perpendicular. In practice, the larger gear (the **wormgear**) is usually modified from a helical gear and given a concave form in its axial cross section, as shown in Fig. 9(a), for improved strength. Such a configuration is referred to as a **single-enveloping wormgear**. Still greater strength will be developed by also making the worm's outer surface concave in its axial cross section (e.g., an hourglass form), rather than straight as shown in Fig. 9(b); such a mutually enveloping configuration is called a **double-enveloping worm set**.

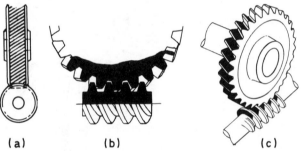

(a)   (b)   (c)

Figure 3.20-9. Single enveloping wormgear and worm: (a) Cross-section through wormgear axis; (b) Cross-section through worm axis; (c) Pictorial view (Courtesy, Mobil Oil Corp.).

A worm resembles a screw, and its teeth are often called **threads**. There may be a single thread winding around a worm, as shown schematically in Fig. 10(a), or several parallel threads, as in Fig. 10(b), which shows two threads. Many more threads are frequently used in practice, as illustrated in Fig. 10(c). Depending on the number of threads, the worms are called **single-threaded**, **double-threaded**, etc.; they are also referred to as **single-worms**, **double-worms**, etc.

Regardless of the number of threads in a worm, the **axial pitch** p, which is the distance between corresponding points on adjacent teeth in an axial cross section, must equal the circular pitch p of the wormgear, as shown in Fig. 10. Therefore, when a single-threaded worm rotates through one revolution, it will rotate the pitch circle of the worm gear through the pitch distance p, as indicated in Fig. 10(a). A double-threaded worm, such as in Fig. 10(b), will rotate the pitch circle of the worm gear through twice the pitch distance, $2p$, for each revolution of the worm. The circumferential displacement L of the pitch circle, per revolution of the worm, is called the

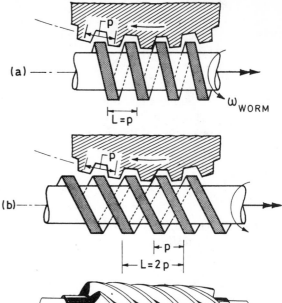

Figure 3.20-10. Right-handed worms:
(a) Single-thread, schematic; (b) Double-
thread, schematic; (c) Eight threads, **(c)**
pictorial (Courtesy, Mobil Oil Co.).

lead and is given by

$$L = N_w p \qquad (6)$$

where N_w is the number of threads on the worm.

If the wormgear has N_g teeth, its pitch circle has circumference pN_g; hence the fraction of a revolution that it turns through, during one revolution of the worm, is $L/(pN_g)$. This fraction is identical with the ratio of the angular speed of the gear (ω_g) to that of the worm (ω_w). By virtue of Eq. (6), we find

$$\frac{\omega_g}{\omega_w} = \frac{L}{pN_g} = \frac{N_w}{N_g} \qquad (7)$$

Note that the velocity ratio is proportional to the tooth number ratio, as in Eq. (5), but is not proportional to the diameter ratio of the pitch circles. This remark applies to all crossed-helical gears.

The direction of gear rotation depends on the inclination of the worm threads. A thread is called **right-handed** if, as in Fig. 10, the visible threads inclines downward to the right when the worm axis is horizontal; otherwise it is called **left-handed**. If a right-handed screw is grasped in the right hand and rotated in the direction of the coiled fingers, it will advance, in the direction of the thumb, into a stationary nut.

Thus if the worm in Fig. 10 were rotated in the direction indicated, it would advance in the direction of the double-headed arrow[12] *relative to a stationary nut*; i.e.,

---

[12]Henceforth a double-headed arrow will represent a rotation in the direction of a right-handed screw advancing in the direction of the arrowhead. This convention, called the **right-hand rule**, is also used to depict the sense of a torque.

the nut would move *relative to the worm* in the opposite direction. Therefore, the gear, which plays the role of the nut, moves to the left *relative to the worm*, as indicated in Fig. 10.

Screws

Suppose that the worm gear, of Fig. 10, is *fixed* so that it behaves like a stationary nut and that the worm (or screw) is allowed to advance axially into the nut. For each revolution of the screw, it will advance axially through the lead L. Therefore, if the screw rotates through an angle $\Delta\theta$, it will simultaneously advance axially through Δx, where

$$\Delta x = \frac{\Delta\theta}{2\pi}L \tag{8}$$

Thus the screw's axial speed $v_{ax} = \Delta x/\Delta t$ is related to its angular speed $\omega = \Delta\theta/\Delta t$ by the relation

$$v_{ax} = \omega\frac{L}{2\pi} \tag{9}$$

The direction of advance is given by the right-hand rule for a right-handed thread and by the corresponding **left-hand rule** for left-handed threads.

A **differential screw** consists of two nested coaxial screws with threads of the same hand. For example, Fig. 11 shows two right-handed screws with leads L_1 and L_2,

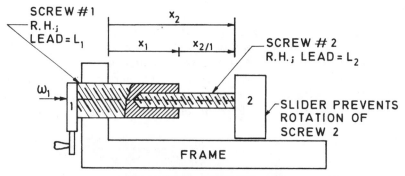

Figure 3.20-11. Differential screw.

respectively. If screw 1 is rotated in the direction indicated through one revolution, it will advance axially through distance $\Delta x_1 = L_1$ relative to the frame. At the same time, screw 2, which is restrained against rotation, moves a distance L_2 *into* body 1; i.e., distance $x_{2/1}$ decreases, and its increment may be written as $\Delta x_{2/1} = -L_2$. The net motion (positive to the right) of body 2 is therefore

$$\Delta x_2 = \Delta(x_1 + x_{2/1}) = \Delta x_1 + \Delta x_{2/1} = L_1 - L_2 \tag{10}$$

This device enables very small axial motions to be produced by large rotary motions. For example, if $L_1 = \frac{1}{16}$ in. and $L_2 = \frac{1}{20}$ in., $\Delta x_2 = \frac{1}{80}$ in./revolution of the screw. Exactly the same net result would be achieved if the left end of screw 2 were welded to screw 1 and its right end were threaded into the slider at the right end.

It may be easily verified that when the two screws have threads of opposite *hand*, the axial displacement per revolution is $\Delta x_2 = L_1 + L_2$. Such **compound screws** have relatively few applications compared to the differential screw.

For all of the gear sets discussed above, the speed ratio of a pair is given by

$$\left|\frac{\omega_1}{\omega_2}\right| = \frac{N_2}{N_1} \tag{11}$$

where N represents the number of teeth in a gear or the number of threads in a worm. The relative direction of rotation of the mating gears can best be found in individual cases by noting whether threads are right- or left-handed and whether gears make internal or external contact. The procedure will be illustrated for the various classes of gear trains discussed in succeeding sections.

3.21 Problems

Define the following terms, using sketches where appropriate:

1.	pitch circle	14.	spiral bevel gear
2.	pitch point	15.	crown gear
3.	tooth trace	16.	hypoid gear
4.	spur gear	17.	miter gears
5.	helical gear	18.	worm
6.	rack	19.	wormgear
7.	herringbone gear	20.	single-threaded
8.	annular gear	21.	double-threaded
9.	internal gear	22.	lead (of worm)
10.	circular pitch	23.	left-handed thread
11.	pitch diameter	24.	right-handed thread
12.	diametral pitch	25.	differential screw
13.	straight bevel gear		

3.30 GEAR ARRANGEMENTS

Gears may be assembled in an extremely wide variety of arrangements. We shall restrict our discussion to methods of *analysis* for the most common and useful arrangements. Further details and discussions of direct design (synthesis) may be found in Dudley [1962b], Merritt [1971], Chironis [1967], and Tao [1967].

3.31 Gears on Fixed Axes

a. Simple Trains

When each shaft in a gear train rotates about a fixed axis and carries only a single gear, as shown in Fig. 1, the system is called a **simple gear train**.

It is convenient to think of one end gear (e.g., 1) as the **driver** or **first gear** in a

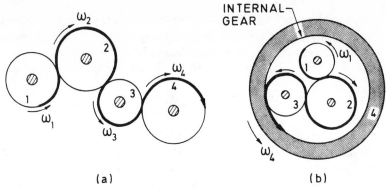

Figure 3.31-1. Simple gear train. All gears are on fixed axes. Heavy lines show pitch line direction: (a) $R = \dfrac{\omega_1}{\omega_4} = -\dfrac{N_4}{N_1}$; (b) $R = \dfrac{N_4}{N_1}$.

simple train and the gear at the other end (e.g., 4) as the **driven gear** or **last gear**. The ratio of their angular velocities

$$R = \frac{\omega_{\text{driver}}}{\omega_{\text{driven}}} \tag{1}$$

is called the **train ratio** or **train value**.[13] Its magnitude $|R|$ may readily be found by successive applications of Eq. **3.20**-11). For the example of Fig. 1(a) we find

$$|R| = \left|\frac{\omega_1}{\omega_4}\right| \equiv \left|\frac{\omega_1}{\omega_2}\frac{\omega_2}{\omega_3}\frac{\omega_3}{\omega_4}\right| = \left(\frac{N_2}{N_1}\right)\left(\frac{N_3}{N_2}\right)\left(\frac{N_4}{N_3}\right) = \frac{N_4}{N_1} \tag{2}$$

For the train of Fig. 1(b), similar reasoning shows that $|R| = N_4/N_1$.

Note that the **speed ratio** $|R|$ is determined by the number of teeth in the first and last gears of the train. The only function of the **intermediate gears** or **idler gears** (2 and 3) is to provide the desired spacing between the first and last shafts and to determine the relative direction of rotation. In general, for the train ratio of a simple train we can write

$$R \equiv \frac{\omega_F}{\omega_L} = \pm\frac{N_L}{N_F} \tag{3}$$

where F and L designate arbitrarily chosen "first" and "last" gears in the train. The proper sign to be used in Eq. (3) can most simply be found by noting the direction of the moving pitch line (heavy lines in Fig. 1) as it winds, like a belt, around successive gears.

b. Compound Trains

Frequently, several gears are keyed to the same shaft, as in Fig. 2, thereby forming a **compound train**. In Fig. 2, the number of teeth are indicated in parentheses. Note that two gears (with n_2 teeth and N_2 teeth) are keyed to the same shaft. An assemblage of gears keyed to a given shaft is called a **compound gear** or **cluster gear**.

---

[13]Some writers define the inverse of this ratio as train value.

$$\frac{\omega_1}{\omega_4} = \left(\frac{N_2}{n_1}\right)\left(\frac{N_3}{n_2}\right)\left(\frac{-N_4}{n_3}\right)$$

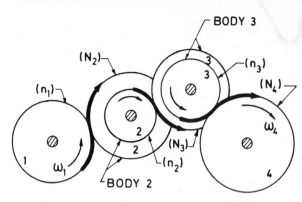

Figure 3.31-2. Compound train (Numbers of teeth are shown in parentheses).

To find the train value, let us designate gear 1 as *driver* and gear 4 as *driven*. Gear 1 has n_1 teeth and meshes with a gear having N_2 teeth on body 2. Also fixed to body 2 is a gear (with n_2 teeth) that drives a gear (with N_3 teeth) on body 3. Body 3 has a gear (of n_3 teeth) which drives the follower gear (of N_4 teeth). The lowercase letters (n_1, n_2, n_3) connote *driving* gears, and the capital letters (N_2, N_3, N_4) connote *driven* gears. The train value $R \equiv \omega_1/\omega_4$ is given by

$$R = \frac{\omega_1}{\omega_2}\frac{\omega_2}{\omega_3}\frac{\omega_3}{\omega_4} = \left(-\frac{N_2}{n_1}\right)\left(-\frac{N_3}{n_2}\right)\left(-\frac{N_4}{n_3}\right) = -\frac{N_2 N_3 N_4}{n_1 n_2 n_3} \qquad (4)$$

In general, we may express the train ratio for any compound train in the form

$$R = \frac{\omega_{\text{driver}}}{\omega_{\text{driven}}} = \pm \frac{\text{product of teeth in driven gears}}{\text{product of teeth in driving gears}} \qquad (5)$$

The reader should verify that Eq. (5) reduces to the simpler expression (3) for the case of *simple trains* such as in Fig. 1.

c. Reverted Trains

When the input and output gears of a compound train are coaxial, as Fig. 3, the configuration is called a **reverted train**.

In a reverted gear train, the number of teeth in the two gear pairs must be chosen

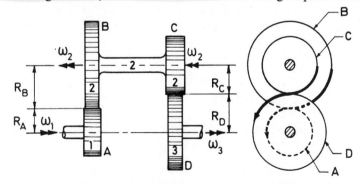

Figure 3.31-3. Reverted gear train. Numbers 1, 2, 3 denote bodies. Letters A, B, C, D denote gears.

so that the common centerline distance is maintained. For example, in Fig. 3,

$$R_A + R_B = R_C + R_D \qquad (6)$$

where the letter R denotes a pitch circle radius. Recalling, from Eq. (**3.20**-3), that the diametral pitch $P_d = N/2R$ must be the same for mating gears, we see that Eq. (6) takes the form[14]

$$\frac{N_A + N_B}{P_B} = \frac{N_C + N_D}{P_C} \qquad (7)$$

where P_B and P_C are the diametral pitches of gears B and C, respectively, and N_i is the number of teeth in gear i ($i = A, B, C,$ or D).

The foregoing ideas may be illustrated by the following examples.

EXAMPLE 1

Figure 4 shows a differential screw press driven by the left-handed triple-threaded worm A. It is desired to find the displacement of platen 5 due to one revolution of the worm in the direction indicated (counterclockwise.)

The first step is to determine the direction of rotation for worm gear 2, and that of

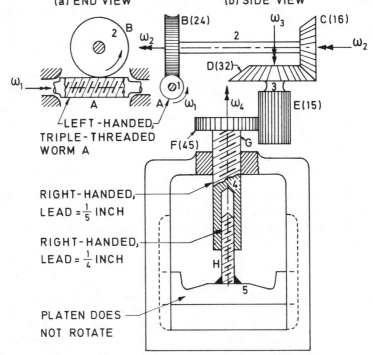

Figure 3.31-4. Example 1. Differential screw press (Number of teeth in parentheses).

---

[14]Equation (7) is valid when the gears are mounted on *standard centers*; i.e., the pitch circle diameters are exactly N/P. However, it is a great advantage of the involute gear tooth form that the centerline spacing can be varied within moderate limits without changing the speed ratio (see Sec.**3.41**). Therefore, it is possible to use *nonstandard* or *extended center* distances which violate Eq. (7) but still satisfy Eq. (6).

each subsequent shaft, by mentally traversing the pitch line from gear B to gears C, D, E, and F. The reader should verify that the directions shown are compatible with the prescribed input direction.

Next we determine the train ratio $R \equiv \omega_1/\omega_4$ for the compound train of gears $ABCDEF$. Toward this end, we note that the gears may be classified as follows:

$$\text{driving gears:} \quad A, C, E$$
$$\text{driven gears:} \quad B, D, F$$

Hence, by Eq. (5), the train value is[15]

$$R \equiv \frac{\omega_1}{\omega_4} = \frac{N_B N_D N_F}{N_A N_C N_E} = \frac{(24)(32)(45)}{(3)(16)(15)} = 48$$

Note that the triple-threaded worm is treated as a "gear" with three teeth in accordance with Eq. (3.20-7).

Now we note from Eq. (3.20-10) that the differential-screw combination (G, H) advances body 5 through a distance

$$\Delta x_5 = L_G - L_H = \tfrac{1}{5} - \tfrac{1}{4} = -\tfrac{1}{20} \text{ in.}$$

for each revolution of screw G in the positive sense of advance for a right-hand screw; that is, a positive value of Δx_5 would indicate motion of the platen *upward* for the direction of ω_4 shown. The negative value of Δx_5 indicates that the platen moves downward $\tfrac{1}{20}$ in. per revolution of body 4.

Since body 4 rotates through $\tfrac{1}{48}$ revolution for each revolution of worm 1, the downward motion of platen 5, corresponding to one revolution of the worm, is

$$\Delta x_5 = (\tfrac{1}{20})(\tfrac{1}{48}) = \tfrac{1}{960} \text{ in. (per revolution of worm)}$$

EXAMPLE 2. Typical Standard Transmission for Automobile

Internal combustion engines, unlike steam engines and some electric motors, can only operate efficiently over a rather narrow range of speeds. They therefore require a gear train, or other form of "transmission," to step down the speed from engine shaft to the (usually lower[16]) speeds desired at the driving axles. In addition, there is no convenient way of reversing the direction of rotation of an internal combustion engine without the use of gears. The "standard" or "manual" three-speed gearbox (or "clash box" as it was accurately called in the old days) typically consists of an arrangement similar to that of Fig. 5.

The input shaft 1 (from the engine) has a gear A fixed to it, which meshes at all times with a gear E that is fixed to a **countershaft** 3. Also fixed to this countershaft are the gears F, G, and H, which together with E constitute a four-gear *cluster* (or *compound gear*). The output shaft (2) is splined and has two gears, (B) and (C), which slide axially on it and rotate with it.

By moving a gear-shaft lever, the operator can move C to the left (or right), engaging it with G or I. Gear I is an idler which rotates on its own shaft (4) and serves merely

---

[15]The sign of R is considered positive for the directions of ω_1 and ω_4 shown in the figure. Some such sign convention must be established for all problems involving nonparallel shafts.

[16]If the speed is stepped up, the transmission is said to be in the *overdrive* mode.

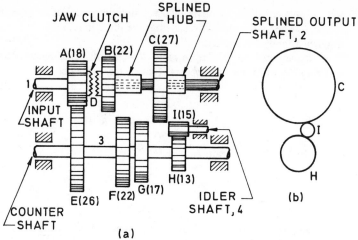

Figure 3.31-5. Standard "three speed" transmission (Note: Symbols used for splined shaft and hub, and jaw clutch. Number of teeth per gear shown in parentheses.): (a) Schematic of gear arrangement; (b) Partial end view showing staggered idler.

to reverse the direction of the output shaft, as will be seen. Likewise, gear *B* can be shifted to the right to engage *F*, or to the left to engage the **jaw clutch**[17] (or **dog clutch**) which causes shafts 1 and 2 to rotate together as a unit. Thus three forward "speeds" and a reverse "speed" are possible, as summarized in Table 1. It is left as a problem (Problem **3.34**-2) to verify the train values (gear reduction ratios) listed for the system shown in the figure.

TABLE 3.31-1 *Standard Transmission*

"Speed"	Shift Action	Mesh Path	Train Value (ω_1/ω_2)
1st	*C* to *G*	1*AEGC*2	2.294
2nd	*B* to *F*	1*AEFB*2	1.444
3rd	*D* engaged	1*D*2	1
Reverse	*C* to *I*	1*AEHIC*2	−3.000

It should be noted that the train values calculated do not represent the final speed ratio seen at the rear (i.e., powered) wheels, since the output shaft from the transmission usually serves as input to a final gear train (the *rear end* or *differential;* see Fig. **3.33**-5), which gives a further stage of reduction with a typical train value of about 3.4. For such a rear end, the *overall train value*, from engine to rear axles, in first gear would be $(2.294)(3.4) = 7.8.$

---

[17]We represent a jaw clutch schematically by a set of serrations and can think of it physically as two "hole saws" in mesh. Refinements can be made on jaw clutches to reduce the clashing effect associated with them (see Tao [1967]). However, the clashing effect of the jaw clutch, and sliding gears, is removed in modern **synchromesh transmissions** by devices called **synchronizers**, which first bring freewheeling gears up to the speed of a splined rotating shaft by means of friction cones and *then* lock them to the shaft. For a description of synchronizer action, see Deere [1969] or Ramous [1972].

An **epicyclic train** is one in which the center of a gear P (the **planet** gear) moves in a circle about the center of another gear S (the **sun** gear) while the angular velocities of both gears (with respect to a third *fixed* body) maintain a fixed ratio.[18] The center distance is maintained constant by a rotating **arm** or **carrier** C as shown in Fig. 1. There may or may not be intermediate (idler) gears between the sun and planet gears, and any of the gears in the train might be annular gears.

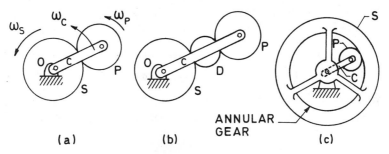

Figure 3.32-1. Epicyclic gear trains.

In practice, the hub of an annular gear would be connected to the gear by a set of spokes [as in Fig. 1(c)] which are ordinarily omitted from schematic diagrams. If the carrier is fixed, the system becomes a *simple* or *compound* train, with one degree of freedom. If the carrier is free to move, the resulting epicyclic train acquires a second degree of freedom. Thus, an epicyclic train is seen to have *two degrees of freedom*, or *two inputs*.

To analyze the motion of epicyclic trains, it is necessary to introduce a lemma on relative angular velocity. Figure 2 shows two bodies, 1 and 2, which are undergoing planar motion relative to a fixed plane (x, y). Lines AB and CD, scribed in bodies 1 and 2, make respective angles θ_1 and θ_2 with the fixed x axis, and CD makes the angle $\theta_{2/1}$ with line AB. All angles are measured positive counterclockwise. From the figure

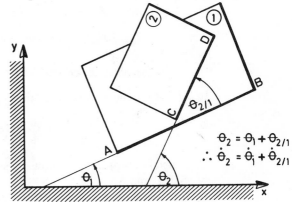

$$\theta_2 = \theta_1 + \theta_{2/1}$$
$$\therefore \dot{\theta}_2 = \dot{\theta}_1 + \dot{\theta}_{2/1}$$

Figure 3.32-2. Relative angular motion of planes.

---

[18]The nomenclature is based on an obvious analogy with the solar system. For a discussion of epicyclic *curves* in relation to the solar system, see Sec. **5.14.**

we see that

$$\theta_{2/1} = \theta_2 - \theta_1$$

Hence the rates of change of these angles are related by the expression

$$\omega_{21} \equiv \dot{\theta}_{2/1} = \dot{\theta}_2 - \dot{\theta}_1 \equiv \omega_2 - \omega_1 \tag{1}$$

$\dot{\theta}_1$ and $\dot{\theta}_2$ are the *absolute*[19] angular velocities of bodies 1 and 2, and $\dot{\theta}_{2/1}$ is the angular velocity of body 2 *relative* to body 1. In short, Eq. (1) states that the *relative angular velocity of two bodies is the difference of their absolute angular velocities.*

Applying this lemma to the epicyclic gear trains of Fig. 1, we arrive at the result

$$\omega_{S/C} = \omega_S - \omega_C$$
$$\omega_{P/C} = \omega_P - \omega_C \tag{2}$$

where $\omega_{S/C}$ and $\omega_{P/C}$ are the velocities of gears S and P, respectively, as seen by an observer on carrier C. This same observer would calculate the train ratio

$$\bar{R}_{SP} \equiv \frac{\omega_{S/C}}{\omega_{P/C}} = \frac{\omega_S - \omega_C}{\omega_P - \omega_C} \tag{3}$$

for the fixed-axes gear train that results when the carrier is fixed. We have discussed in Sec. **3.31** how the train ratio may be computed for any such train; therefore, we may assume henceforth that the **fixed-carrier train ratio** $\bar{R}_{SP}$ can be calculated without difficulty, and Eq. (3) thus becomes the **fundamental equation of epicyclic gear trains**. It clearly shows that when the velocities of any two of the three members P, S, or C are given, the third velocity is easily calculated. Note the use of an overbar to distinguish the *fixed-carrier* train ratio $\bar{R}_{SP}$ from the ordinary velocity ratio

$$R_{SP} = \omega_S/\omega_P.$$

When any one of the members P, S, or C is fixed, the system has a single remaining degree of freedom, and Eq. (3) gives the following results:

$$\text{gear } P \text{ fixed } (\omega_P = 0): \quad \omega_S = \omega_C(1 - \bar{R}_{SP}) \tag{4a}$$

$$\text{gear } S \text{ fixed } (\omega_S = 0): \quad \omega_P = \omega_C\left(1 - \frac{1}{\bar{R}_{SP}}\right) = \omega_C(1 - \bar{R}_{PS}) \tag{4b}$$

$$\text{carrier } C \text{ fixed } (\omega_C = 0): \quad \omega_S = \omega_P\bar{R}_{SP} \tag{4c}$$

In Eq. (4b) we have introduced the definition

$$\bar{R}_{PS} \equiv \frac{\omega_{P/C}}{\omega_{S/C}} = \frac{1}{\bar{R}_{SP}} \tag{5}$$

Of course Eq. (4b) should follow, by symmetry, from Eq. (4a) upon interchanging the symbols P and S.

It is readily demonstrated from the discussion in Sec. **3.31** that the fixed-carrier train values $\bar{R}_{SP}$ for the three cases shown in Fig. 1 are

---

[19]The word *absolute* means "relative to the assumed fixed plane."

$$\text{case (a):} \quad \bar{R}_{SP} = -\frac{N_P}{N_S}$$

$$\text{case (b):} \quad \bar{R}_{SP} = +\frac{N_P}{N_S}$$

$$\text{case (c):} \quad \bar{R}_{SP} = +\frac{N_P}{N_S}$$

3.33 Planetary Trains

Although the terms *planetary train* and *epicyclic train* are frequently used inter-changeably, it is convenient to define[20] a ***planetary train*** as one in which two or more *independent coaxial gears are meshed with a number of similar, equally spaced gears or gear assemblies* (***planets***) *mounted on intermediate shafts, which in turn are fixed to a member called a* ***carrier***.

The **principal members** are the carrier and the coaxial gears, known as **sun gears**. Ordinarily, external torques are applied only to the principal members. A **simple planetary train** contains only *two sun gears*, whereas a **compound planetary train** contains *more* than two sun gears. Note that a *compound planetary train* may or may not contain compound gears (or cluster gears) as defined in Sec. **3.31.** Also note that since a planetary train contains a carrier, it will contain simple epicyclic trains as sub-assemblies.

In the examples which follow, two types of diagrams are used—**schematic** and **skeleton**—with symbols as defined in Fig. 1.

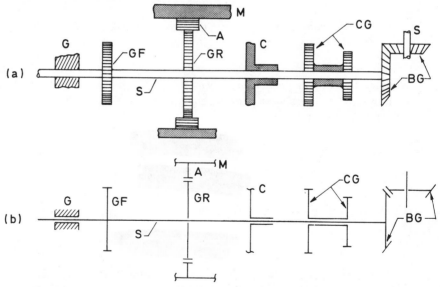

Figure 3.33-1. Symbols used in: (a) Schematic diagrams; (b) Skeleton diagrams: S = Shaft, G = ground (fixed link); GF = Gear fixed to shaft; GR = Gear rotating on shaft; A = Annular gear fixed to member M; C = Carrier or rotating case; CG = Compound (or "cluster") gears; BG = Bevel gears; M = Member.

---

[20]Following Merritt [1971] and Glover [1969], who should be consulted for more extensive treatments.

Figure 2 shows a simple planetary train in which one of the coaxial gears A is an annular gear mounted on a housing which is free to rotate. The other coaxial gear is designated by S, and the entire assemblage is referred to as a **sun-planet-annulus train**. Note that although three planet gears are present, two of them are kinematically redundant (they are there primarily to share the load and give better balance) and play no role in the kinematic analysis. In fact only one planet gear P is generally shown in schematic and skeleton diagrams. If desired, one may simplify the diagrams still further by omitting all matter below the centerline of the sun gears.

Any two of the members (S, A, C, P) may serve as independent inputs. Usually one member (not P) is given the special input of zero velocity; i.e., it is *grounded*, and either S, A, or C is driven as the input member.

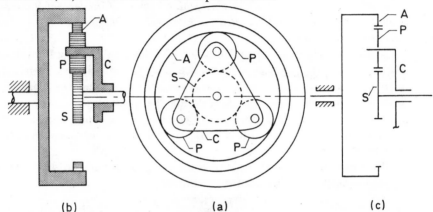

(b) (a) (c)

Figure 3.33-2. Simple planetary train: (a) Viewed parallel to shaft axes; (b) Schematic; (c) Skeleton.

The velocity ratios for *any* planetary train (no matter how complex) may always be found by using the *fundamental equation for epicyclic trains*, Eq. (**3.32**-3), at every point where two gears mesh. For example, in Fig. 2, the meshing pairs are (S, P) and (P, A); thus we can write

$$\frac{\omega_S - \omega_C}{\omega_P - \omega_C} = -\frac{N_P}{N_S} \tag{1}$$

$$\frac{\omega_P - \omega_C}{\omega_A - \omega_C} = +\frac{N_A}{N_P} \tag{2}$$

Upon multiplying these two equations we find

$$\frac{\omega_S - \omega_C}{\omega_A - \omega_C} = -\frac{N_A}{N_S} \tag{3}$$

Equation (3) could have been derived more directly by using the idea of the *fixed-carrier train ratio* discussed in Sec. **3.32**; that is,

$$\frac{\omega_{S/C}}{\omega_{A/C}} = \frac{\omega_S - \omega_C}{\omega_A - \omega_C} = \bar{R}_{SA} \tag{4a}$$

or

$$\omega_S = \bar{R}_{SA}\omega_A + (1 - \bar{R}_{SA})\omega_C \tag{4b}$$

where $\bar{R}_{SA} = -N_A/N_S$ is the fixed-carrier train value for the case of Fig. 2. Note that ω_P and N_P do not appear in Eqs. (4) since P is an idler which serves primarily to influence the sign of $\bar{R}_{SA}$.

The sun-planet-annulus train of Fig. 2 can provide a fairly large speed ratio in a compact space and has many direct applications; for examples, see Problems **3.24**-17, 18, 24, 25, and 26.

More generally, the reasoning behind Eqs. (4a) and (4b) is valid for any planetary system in which S and A are coaxial gears associated with carrier C, and $\bar{R}_{SA}$ is the appropriate *fixed-carrier train value*. Using the notation $\bar{R}_{AS} \equiv 1/\bar{R}_{SA}$, it is readily shown that Eq. (4b) can be written in the more symmetrical form

$$\omega_C = \frac{\omega_S}{1 - \bar{R}_{SA}} + \frac{\omega_A}{1 - \bar{R}_{AS}} \qquad\qquad \left(R_{SA} = \frac{-N_A}{N_S} \right) \qquad (5)$$

which governs all simple planetary trains.

When all principal members are free to move, the simple planetary train is sometimes called a **simple differential**. In a **symmetrical differential**, $\bar{R}_{SA} = \bar{R}_{AS} = -1$; hence

$$\omega_C = \tfrac{1}{2}(\omega_S + \omega_A) \qquad (6)$$

That is, *the symmetrical differential is an averaging or summing device*; a common example of such a device will be described below in connection with Fig. 5 (also see Fig. **3.34**-1).

By fixing any of the principal members, it is possible to obtain the following variety of speed ratios:

$$
\begin{aligned}
A \text{ fixed:} \quad & \omega_S = \omega_C(1 - \bar{R}_{SA}) \\[4pt]
S \text{ fixed:} \quad & \omega_A = \omega_C(1 - \bar{R}_{AS}) = \omega_C\left(1 - \frac{1}{\bar{R}_{SA}}\right) \\[4pt]
C \text{ fixed:} \quad & \omega_S = \omega_A \bar{R}_{SA}
\end{aligned}
\qquad (7)
$$

There is no need to memorize Eqs. (7). Rather it is suggested that the analysis of planetary trains always start with the *fundamental equation* (4a). The particular value of fixed-carrier ratio $\bar{R}_{SA}$ to be used depends in a simple way on details of gear arrangement. The determination of $\bar{R}_{SA}$ for two specific gear trains will now be illustrated.

EXAMPLE 1. Determination of Fixed-Carrier Train Values

To calculate the appropriate values of $\bar{R}_{SA}$ needed for the systems of Figs. 3 and 4, the reader should verify that, with the carrier fixed, the directions of rotation are as indicated in the figures; then it may be shown that the fixed-carrier train ratios are

$$\bar{R}_{SA} = \frac{N_P N_A}{N_S N_Q} \qquad \text{(for Fig. 3)} \qquad (8)$$

$$\bar{R}_{SA} = -\frac{N_A}{N_S} \qquad \text{(for Fig. 4)} \qquad (9)$$

Note that the gear A is not necessarily an annulus (as it is in Fig. 2) but that it *is* always one of the two coaxial gears in a simple planetary train.

In Fig. 3, the two planets (P and Q) are compounded (i.e., rotate together as a unit), but since there are only two sun gears, the system is still classified as a *simple planetary*

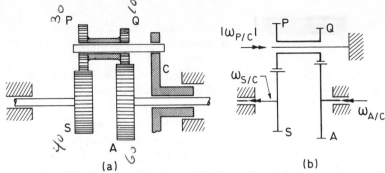

Figure 3.33-3. Simple planetary train with compound planets: (a) Schematic; (b) Skeleton of fixed-carrier train showing directions of rotation with carrier fixed.

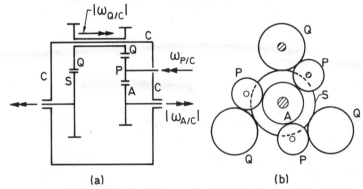

Figure 3.33-4. Simple planetary train with intermediate (idler) planets: (a) Skeleton showing directions of rotation in fixed-carrier train; (b) End view.

train.[21] In Fig. 4(a) the gear Q is shown as a compound gear (distorted) in order to clearly show the proper meshing relationships.

Equations (7) give the appropriate velocity relations when any one of the three principal members (A, S, or C) are fixed. It may be easily shown (Problem **3.34**-5) that if the planet P (or indeed any idler on the carrier) is fixed to the frame[22] the entire mechanism becomes locked to the frame.

It is also possible (e.g., by means of mechanical, hydraulic, or electric clutches) to lock any two of the moving links (A, S, C, P) together; then it may be seen (Problem **3.34**-7) that *all* the moving links rotate together as a rigid body.

EXAMPLE 2. Bevel Gear Differential

Figure 5 shows a simple planetary train with bevel gears. Since the motions are not all coplanar, we must adopt a suitable ad hoc sign convention. We shall therefore assume that the coaxial members S, A, and C have positive velocities ω_S, ω_A, and ω_C when their respective rotation vectors point to the left. The angular velocity of planet

---

[21]Although some writers, e.g., Dudley [1962b], would call this arrangement a *compound epicyclic train.*
[22]If P is not completely fixed to the frame but we merely enforce the condition that $\omega_P = 0$ (a difficult feat to execute mechanically), the mechanism does not lock and retains a single degree of freedom (see Problem **3.34**-6).

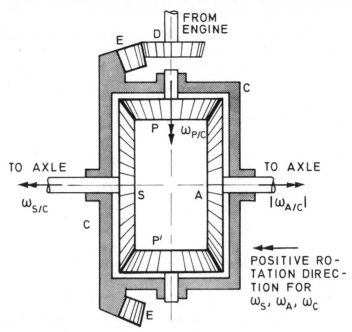

Figure 3.33-5. Bevel gear differential train (S, A, P, C). Fixed-carrier train rotation directions shown by arrows. Bevel gears D, E are not part of differential train.

P is the vector sum of ω_C and $\omega_{P/C}$, where the latter is assumed positive when its rotation vector is directed toward the common apex of the pitch cones, as indicated in the figure. If we imagine the carrier C to be fixed, the gears will rotate in the directions shown and the fixed-carrier train value is clearly

$$\bar{R}_{SA} = \frac{\omega_{S/C}}{\omega_{A/C}} = -\frac{N_A}{N_S} \tag{10}$$

This is the value of $\bar{R}_{SA}$ to be used in Eq. (5). Note that for the special case when $N_A = N_S$ Eq. (5) predicts that

$$\omega_C = \tfrac{1}{2}(\omega_S + \omega_A) \tag{11}$$

In other words, the bevel planetary train with equal sun gears constitutes a *summing mechanism*[23] and has been used as such in analog computers. If one defines the positive direction of rotation for sun gear A opposite to that used above and denotes its angular velocity by $\omega'_A \equiv -\omega_A$, then Eq. (11) reads

$$\omega_C = \tfrac{1}{2}(\omega_S - \omega'_A) \tag{12}$$

and the device may be used to find the difference of inputs to the coaxial gears S and A. For this reason, it is known as a **bevel gear differential**. Such mechanisms are available in stock sizes for use in analog and control devices.

A common use of the bevel gear differential is to distribute power to the rear end of an automobile. In this application, the carrier C is driven by the engine drive shaft (through a set of bevel gears D, E), and the shafts of A and S are each keyed to a rear

---

[23]If gears S and A are not identical, the carrier speed will be a weighted mean of ω_A and ω_S (see Problem 3.34-22).

wheel axle. This arrangement allows the rear wheels to turn at different speeds, as is required when the automobile moves in a curve. Note from Eq. (11) that the mean speed of the two rear wheels is always equal to the speed of the carrier.

Differentials can be made with spur gearing rather than bevel gearing, as illustrated in Problem 3.34-8.

In general, we may analyze *all* planetary trains (simple or compound) in either of two ways:

1. Apply the fundamental equation of epicyclic trains to every pair of meshing gears in the train.
2. Apply the fundamental equation once for each epicyclic train in the assembly, ignoring the motion of idlers.

The second method is more direct, but the first method will serve as a quick and reliable check of results. In addition, the first method will provide the angular velocities of any idler gears of interest. Each of these methods will be illustrated in the following examples of compound planetary trains.

EXAMPLE 3. High Ratio Speed Reducer

In Fig. 6, the epicyclic train (A, E, B, C) drives the epicyclic train (F, G, C) through the common carrier C and the cluster gear EF.

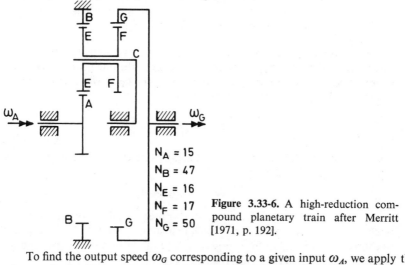

$N_A = 15$
$N_B = 47$
$N_E = 16$
$N_F = 17$
$N_G = 50$

Figure 3.33-6. A high-reduction compound planetary train after Merritt [1971, p. 192].

To find the output speed ω_G corresponding to a given input ω_A, we apply the fundamental epicyclic equation to the three meshing pairs $(A, E), (E, B), (F, G)$, and get

$$\frac{\omega_A - \omega_C}{\omega_E - \omega_C} = \bar{R}_{AE} = -\frac{N_E}{N_A} \tag{i}$$

$$\frac{\omega_E - \omega_C}{\omega_B - \omega_C} = \bar{R}_{EB} = \frac{N_B}{N_E} \tag{ii}$$

$$\frac{\omega_F - \omega_C}{\omega_G - \omega_C} = \bar{R}_{FG} = \frac{N_G}{N_F} \tag{iii}$$

Upon noting that $\omega_B = 0$ and $\omega_F \equiv \omega_E$, the three given equations suffice to find the three unknowns ω_C, ω_E, and ω_G. From Eqs. (i) and (ii) (or by inspection)

$$\frac{\omega_A - \omega_C}{\omega_B - \omega_C} = -\frac{N_B}{N_A}$$

Hence

$$\omega_C = \frac{\omega_A}{1 + (N_B/N_A)} \qquad \text{(iv)}$$

Then, from Eq. (ii) we find

$$\omega_E = \left(1 - \frac{N_B}{N_E}\right)\omega_C = \frac{1 - (N_B/N_E)}{1 + (N_B/N_A)}\omega_A \qquad \text{(v)}$$

Upon substituting Eqs. (iv) and (v) into Eq. (iii) (with $\omega_F \equiv \omega_E$), we find the desired output speed:

$$\omega_G = \frac{1 - (N_F/N_G)(N_B/N_E)}{1 + (N_B/N_A)}\omega_A \qquad \text{(vi)}$$

For the numbers of teeth shown in Fig. 6, the overall gear ratio is

$$\frac{\omega_A}{\omega_G} = \frac{1 + \frac{47}{15}}{1 - (\frac{17}{50})(\frac{47}{16})} = 3306.666\ldots,$$

In this example, the direct application of the fundamental epicyclic equation to each pair of meshing gears proves to be efficient.

EXAMPLE 4. Compound Planetary Auto Transmission

Planetary trains lend themselves nicely to automatic transmissions in automobiles. By locking one or more members of such a system, a wide variety of speed ratios can be achieved (see Problem 3.34-26). Figure 7 is a skeleton of a three-speed planetary transmission in the intermediate speed position, where coaxial gear F has been fixed by means of a brake. The input gear is A, and the output gear is E. Note that the presence of three coaxial gears (F, A, E) makes this system a compound planetary train.

There are four meshing gear pairs (FD, DE, AB, BD) which can provide the four independent epicyclic equations needed to find the four unknown speeds ω_B, ω_C, ω_D, ω_E. However, it proves to be convenient in this problem to take advantage of the fact that train (A, B, D, F) is an epicyclic train built on carrier C, which drives the epicyclic train (F, D, E) built on the same carrier.

Accordingly, we can use the *shortcut method* and write the fixed carrier train value as

$$\frac{\omega_A - \omega_C}{\omega_F - \omega_C} = \bar{R}_{AF} = -\frac{N_F}{N_A} \qquad \text{(i)}$$

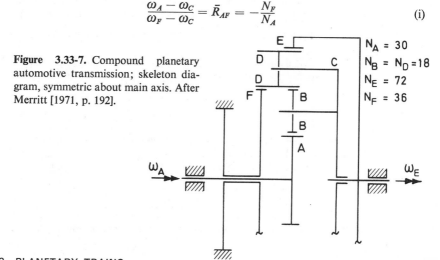

Figure 3.33-7. Compound planetary automotive transmission; skeleton diagram, symmetric about main axis. After Merritt [1971, p. 192].

$N_A = 30$
$N_B = N_D = 18$
$N_E = 72$
$N_F = 36$

Upon recognizing that $\omega_F = 0$ in this problem, Eq. (i) gives the carrier velocity:

$$\omega_C = \frac{\omega_A}{1 + (N_F/N_A)} \qquad \text{(ii)}$$

Next we find the fixed-carrier train value:

$$\frac{\omega_E - \omega_C}{\omega_F - \omega_C} = \bar{R}_{EF} = -\frac{N_F}{N_E} \qquad \text{(iii)}$$

This (upon setting $\omega_F = 0$) leads to

$$\omega_E = \left(1 + \frac{N_F}{N_E}\right)\omega_C = \frac{1 + (N_F/N_E)}{1 + (N_F/N_A)}\omega_A \qquad \text{(iv)}$$

For the numerical values shown in Fig. 7 the overall speed ratio is $\omega_A/\omega_E = 1.467\ldots.$

3.34 Problems

1. Define the following terms:

    a. simple train
    b. train ratio (train value)
    c. idler (intermediate) gear
    d. compound gear
    e. compound train (not planetary)
    f. cluster gear
    g. reverted train
    h. synchronizer
    i. epicyclic train

    j. carrier (or arm)
    k. fundamental equation of epicyclic gear trains
    l. fixed-carrier train ratio
    m. planetary train
    n. simple planetary train
    o. compound planetary train
    p. sun-planet-annulus train
    q. bevel gear differential

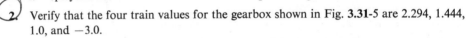

2. Verify that the four train values for the gearbox shown in Fig. 3.31-5 are 2.294, 1.444, 1.0, and -3.0.

3. Find the train values in first, second, third, and reverse gears for the transmission of Fig. 3.31-5 if the gears have the following numbers of teeth: $N_A = 15$, $N_B = 20$, $N_C = 27$, $N_E = 27$, $N_F = 22$, $N_G = 15$, $N_H = 12$, $N_I = 17$.

4. Verify Eqs. (3.33-8) and (3.33-9) for the cases shown in Figs. 3.33-3 and 3.33-4, respectively.

5. Show that a simple planetary train (e.g., Figs. 3.33-2, 3, 4, and 5) becomes completely immobile if a planet gear (or any idler on the carrier) is completely fixed to the frame.

6. Given a simple planetary train consisting of sun gears (S and A), carrier (C), and planet (P),

    a. Show that if the planet P is forced to have zero angular speed ($\omega_P = 0$), the velocity ratio for the two sun gears (S and A) is

    $$\frac{\omega_S}{\omega_A} = \frac{1 - \bar{R}_{SP}}{1 - \bar{R}_{AP}}$$

    where $\bar{R}_{SP}$ and $\bar{R}_{AP}$ are fixed-carrier values of ω_S/ω_P and ω_A/ω_P observed when carrier C is held fixed and gear P is rotated.

    b. For the system of Fig. 3.33-2, show that this expression reduces to

    $$\frac{\omega_S}{\omega_A} = \frac{N_A(N_A + N_P)}{N_S(N_A - N_P)}$$

    c. Find an expression for ω_A/ω_C, in terms of tooth numbers, when $\omega_P = 0$.

7. Show that there are six possible ways of locking together any two of the four moving parts (S, A, C, P) of a simple planetary train (e.g., Fig. **3.32**-2) and that all four parts move together as a rigid body in each of the six cases.

8. For the symmetrical *spur gear differential* of Fig. 1, show that $\omega_3 = \frac{1}{2}(\omega_1 + \omega_2)$. Note that $N_1 = N_2$ and $N_4 = N_5$ and that the gears make contact in the sequence 1-4-5-2.

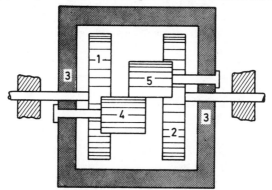

Figure 3.34-1. Cross-section through Spur Gear Differential, Prob. 8.

9. Let the numbers of teeth in Fig. **3.33**-3 be $N_S = 40$, $N_P = 30$, $N_A = 60$, $N_Q = 10$. Find the train ratios (ω_S/ω_C, ω_S/ω_A, ω_C/ω_A) that result upon fixing member A, C, or S, respectively.

10. a. Draw a skeleton diagram for the system of Fig. 2.
   b. Find an algebraic expression for ω_6 in terms of ω_1, ω_5 and the tooth numbers N_1, N_2, N_3, N_4, N_5.
   c. What is the numerical value of ω_6 if $\omega_1 = 8$ rpm and $\omega_5 = 1$ rpm?

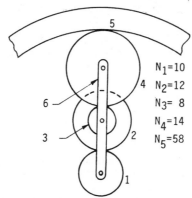

$N_1 = 10$
$N_2 = 12$
$N_3 = 8$
$N_4 = 14$
$N_5 = 58$

Figure 3.34-2 (Prob. 10) (Note: Gears 2 and 3 are keyed together).

11. In Fig. 2, let gear 5 be fixed. Find ω_1/ω_6 in terms of N_1, N_2, ..., N_5. What is the numerical value of ω_1/ω_6?

12. In Fig. 2, let gear 1 be fixed. Find ω_5/ω_6 in terms of N_1, N_2, ..., N_5. What is the numerical value of ω_5/ω_6?

13. In Fig. 2, let arm 6 be fixed. Find ω_5/ω_1.

14. If $\omega_1 = 1$ rpm in Fig. 2, what must be the value of ω_5 such that the arm 6 remains stationary?

15. In Fig. 3, B and F form a compound gear.
    a. Find an algebraic expression for the ratio ω_C/ω_D if gear A is held fixed.
    b. What is the numerical value of ω_D/ω_C if the tooth numbers are $N_A = 99$, $N_B = 100$, $N_F = 101$, $N_D = 100$?
    c. Sketch a three-speed transmission, based on Fig. 3, which will produce $\omega_C/\omega_D = -100, 0, 100$. (Hint: Use a sliding cluster gear on a spline.)

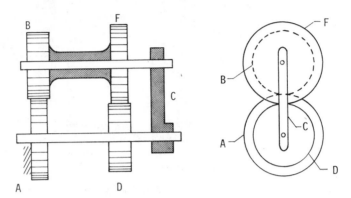

Figure 3.34-3 (Probs. 15, 16).

16. Is it possible to assemble the gears with tooth numbers as given in Problem 15b? If so, explain what precautions are necessary, and give the pitch diameter and diametral pitch of each gear in a suitably meshing set.

17. Figure 4 shows the skeleton of a planetary speed reduction system for driving an airplane propeller. The input from the engine shaft turns the annular gear A, and the propeller shaft is fixed to the carrier C, which holds the pinions E. The sun gear B is fixed to the aircraft frame. If the engine shaft speed is 2700 rpm, find the propeller shaft speed using the tooth numbers shown in parentheses.

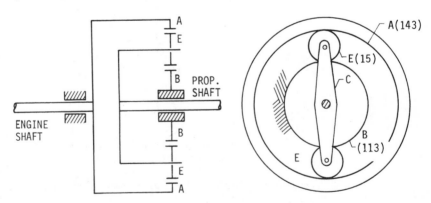

Figure 3.34-4. Planetary reduction gear (Prob. 17).

18. Figure 5 shows a planetary speed reducer of large ratio with input to pinion 1 and output from the carrier 7.

a. Find an algebraic expression for the reduction ratio ω_1/ω_7. What is the signed value of ω_1/ω_7 if the tooth numbers are as shown in parentheses in Fig. 5?
b. Can the gears be assembled on standard centers (i.e., with pitch diameters of N/P_d) with the given numbers of teeth? If the number of teeth in gear 2 is changed so that the gears can be assembled with standard center distances, what does the reduction ratio become, and what happens to the sense of rotation of the output shaft?

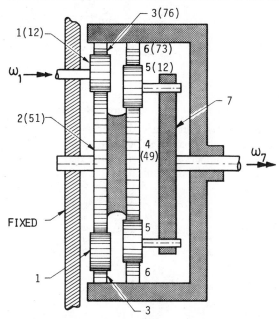

Figure 3.34-5 (Prob. 18) After Spotts [1967].

19. Find the reduction ratio ω_5/ω_4 for Fig. 6 if $N_1 = 60$, $N_2 = 61$, $N_3 = 60$, $N_4 = 61$.

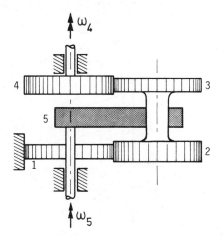

Figure 3.34-6 (Probs. 19, 20).

20. Repeat Problem 19 using the tooth numbers $N_1 = 61$, $N_2 = 60$, $N_3 = 61$, $N_4 = 60$.

21. Find the ratio ω_9/ω_2 in Fig. 7.

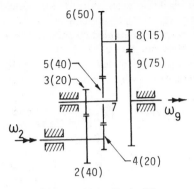

Figure 3.34-7. (Prob. 21).

22. Figure 8 shows an unbalanced bevel gear differential. Show that the carrier velocity is a weighted average of the velocities of gears 1 and 2; i.e., express ω_4 in terms of ω_1, ω_2 and tooth numbers N_1, N_2, N_3. Note that the balanced differential (Fig. **3.33**-5) is a special case of this mechanism.

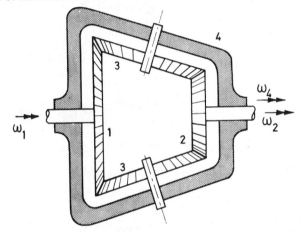

Figure 3.34-8 (Prob. 22).

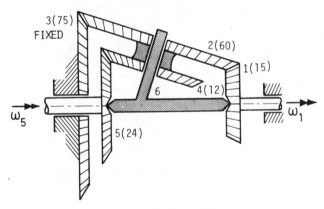

Figure 3.34-9 (Prob. 23).

23. Figure 9 shows a nest of bevel gear differentials known as **Humpage's gear train.** Note that 2 and 4 form a cluster gear and that 3 is fixed.
   a. Find an algebraic expression for ω_5/ω_1 in terms of tooth numbers $N_1, \ldots, N_5$.
   b. Evaluate ω_5/ω_1 for the tooth numbers given in Fig. 9.

24. Figure 10 shows an epicyclic hoist in which the carrier 4 is integral with a sprocket wheel that carries the load chain. The sun gear 1 is keyed to a sprocket wheel which carries the hand chain.
   a. Find the velocity ratio ω_1/ω_4.
   b. For the direction of ω_1 shown, does the load (which is on the segment of the load chain behind the sprocket wheel 4) rise or fall?
   c. What is the ideal mechanical advantage (ratio of chain speeds) of the hoist if the sprocket wheels have a ratio of $R/r = 2$?

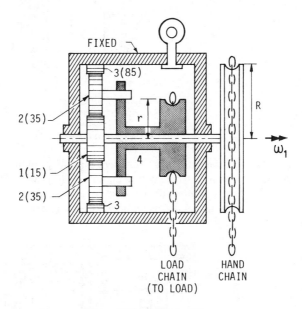

Figure 3.34-10 (Prob. 24).

25. Figure 11 shows a *two-speed* planetary transmission which produces *high*, *low*, and *neutral* speeds by engaging or disengaging a suitable clutch and/or brake. When the clutch is engaged, sun gear 1 is locked to the carrier 4; when the brake is engaged, annular gear 3 is locked to the frame. Fill in the gear ratio ω_4/ω_1 in the following table (for high and low drive) if the tooth numbers are $N_1 = 39$, $N_2 = 27$, $N_3 = 99$. Discuss possible values of ω_4/ω_1 for neutral drive.

Drive	Clutch	Brake	ω_4/ω_1
High	Engaged	Disengaged	
Low	Disengaged	Engaged	
Neutral	Disengaged	Disengaged	

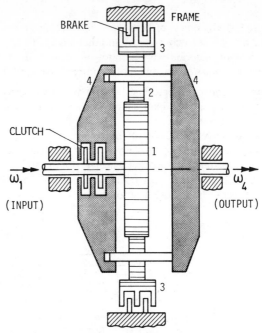

Figure 3.34-11 (Prob. 25).

26. Figure 12 is a schematic of a *three-speed* planetary automotive transmission which consists of two epicyclic trains with a common sun gear S. The input shaft turns annular gear A, and the output shaft is turned by annular gear a. The different speeds are obtained by engaging and disengaging brakes and clutches not shown.[24] In all speeds, the

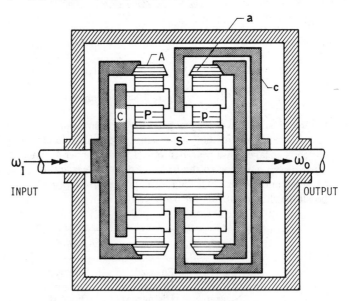

Figure 3.34-12 (Prob. 26).

---

[24]For arrangements of clutches and brakes, see Murray [1974, p. 123].

carrier C is locked to the annular gear a; i.e., $\omega_a = \omega_C = \omega_o$, where ω_o is output velocity. If ω_I is input speed and the clutches and brakes impose the conditions shown in the second column of the table below, verify that the reduction ratio (ω_I/ω_o) is as shown in the third column. Calculate the reduction ratios corresponding to the following tooth numbers: $N_A = N_a = 49$, $N_P = N_p = 13$, $N_S = 23$.

Speed	Conditions Imposed	Reduction Ratio (ω_I/ω_o)
Low (1st)	$\omega_A = \omega_I$ $\omega_c = 0$	$2 + N_S/N_A$
Med. (2nd)	$\omega_A = \omega_I$ $\omega_S = 0$	$1 + N_S/N_A$
high (3rd)	$\omega_A = \omega_I$ $\omega_S = \omega_I$	1
Reverse	$\omega_S = \omega_I$ $\omega_c = 0$	$-N_A/N_S$

27. An electrical device sits on top of the rotating table 7 in Fig. 13. Electric power is supplied from a stationary base by means of a flexible cable, which would become increas-

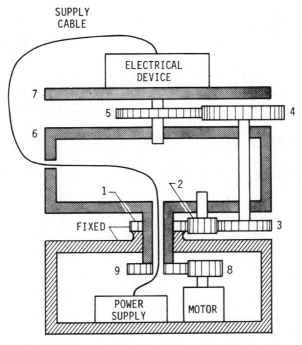

Figure 3.34-13 (Prob. 27). After D. A. Adams, U.S. Patent No. 3,586,413.

ingly twisted, and ultimately broken, unless special means are provided to keep it untwisted. It has been shown by D. A. Adams (see Stong [1975]) that the cable will revert to its original (untwisted) state after every two revolutions of the table 7, provided that it is guided through an auxiliary carrier 6 which rotates at *half* the angular velocity of the table 7. The planetary train shown will provide the ratio $\omega_7/\omega_6 = 2$ provided that the number of teeth in the gears is properly chosen.

a. Express the train ratio ω_7/ω_6 in terms of the numbers of teeth $N_1, N_2, N_3 \ldots$.

b. If the motor rotates gear 8 at 100 rpm, table 7 is to rotate at $\omega_7 = 300$ rpm, and $N_9 = 20$, find a set of tooth numbers $N_1, N_2, N_3, N_4, N_5, N_8$ such that $\omega_7/\omega_6 = 2$.. Assume standard center distances between gears.

28. Show that the pointing arm of the south-pointing chariot (Fig. 3.10-2) always points in the same direction, relative to ground, regardless of how the chariot is steered. Assume that the wheels roll without slip over the ground. Comment on the practicality of this device.

29. Figure 3.10-1 shows Hero's description of the hodometer (or taximeter). The wheel E, fixed to the axle of a carriage wheel, has a single tooth which engages crown wheel $ABCD$. In Hero's example the crown wheel has eight teeth, the worm wheels each have 30 teeth, and the carriage wheel has a circumference of 10 cubits.[25] If the carriage is moving to the left in the figure, find

a. The distance traveled (in cubits) corresponding to one revolution of each indicating dial.

b. The direction of rotation (clockwise or counterclockwise) of each dial as seen from outside the instrument housing. Indicate which worms are right-handed and which are left-handed.

30. Figure 14 shows how the carrier output ω_c of a sun-planet-annulus train (S, P, A) can be varied while the main input velocity ω_s to the sun gear is held constant. The adjustment of output speed is accomplished by providing controlled rotation to the annular

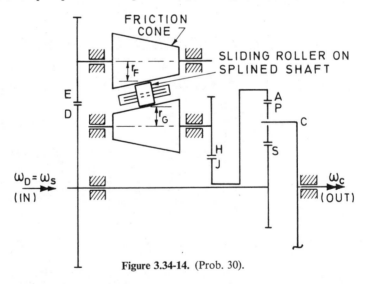

Figure 3.34-14. (Prob. 30).

---

[25]From the Latin *cubitum* ("elbow"); the cubit represents the length of a man's forearm. Its average value in Roman times has been estimated to be 17.4 in. (see Smith [1958, Vol. 2, p. 641]).

gear A through the gears H and J. The latter are driven by two friction cones, with a movable roller between them that provides effective pitch radii of r_F and r_G for the friction drive. The roller may be shifted (on a splined shaft) parallel to the cone generators, while the system is in motion, giving *continuously variable* speed control. The rollers are driven from the main input shaft through the gear pair DE.

    a.   Express the train ratio ω_C/ω_S in terms of tooth numbers and the roller cone ratio $\epsilon = r_F/r_G$.

    b.   Suppose that $N_D = N_E$, $N_H = N_J$. Show that $\omega_C/\omega_S = R_0 + \epsilon(1 - R_0)$, where R_0 is the train ratio (ω_C/ω_S) of a sun-planet-annulus train with fixed annulus and $\epsilon = r_F/r_G$.

31. When balanced bevel gear differentials are coupled together as in Fig. 15, two inputs $(\omega_1$ and $\omega_9)$ can be used to produce two outputs $(\omega_3$ and $\omega_4)$. If $N_3 = N_5$, $N_4 = N_7$, and $N_1/N_2 = N_9/N_8$, show that one output is proportional to the sum of the inputs and that the other output is proportional to the difference of the inputs.

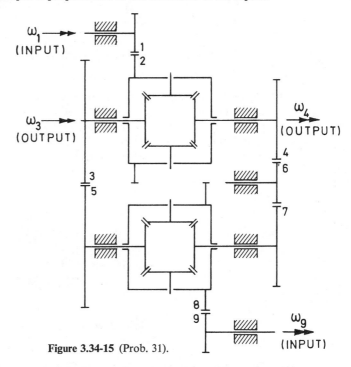

Figure 3.34-15 (Prob. 31).

32. Figure 16 shows a lifting device described by Hero of Alexandria. It is an early example of modern speed reduction units. Assume that the worm which drives wheel XW is single-threaded. Compound gears are made up of gears XW and UF, gears ST and PR, and gears JO and MN. The pitch diameters of the gears are $D_{MN} = D_{PR} = D_{UF} = 1$ in., $D_{HG} = D_{JO} = D_{ST} = 5$ in., $D_{XW} = \frac{8}{3}$ in. The crank handle Q moves in a circle of diameter 8 in., the diameter of the axle EZ is 1 in., and the gear XW has 50 teeth.

    a.   Find the speed reduction ratio (ω_Y/ω_Z) of shafts Y and Z.

    b.   Find the speed of the suspended weight if the pitch circle of gear XW moves at a linear speed of $V_X = 1$ in./sec.

c. Find the speed V_Z of the suspended weight when handle Q moves at a speed of $V_Q = 25$ in./sec.

d. With the left-handed thread on the worm as shown, should handle Q be rotated clockwise or counterclockwise in order to raise the weight?

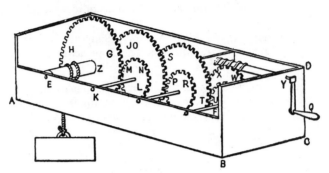

Figure 3.34-16. A hoisting device described by Hero of Alexandria. From Cohen and Drabkin [1948, p. 229]. (Courtesy, McGraw-Hill Book Co.).

33. Figure 17 shows three epicyclic trains with a common *fixed* sun gear (1), a common carrier (6), and a common intermediate gear (2). For the numbers of teeth shown in parentheses, find the velocity ratios (ω_3/ω_6), (ω_4/ω_6), (ω_5/ω_6). Why do you think this train has been dubbed a "paradox?" Explain why it is, or is not, possible for all the gears to have the same diametral pitch.

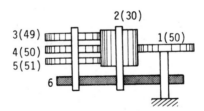

Figure 3.34-17 (Prob. 33) Ferguson's paradox.

34. Because of a patent dispute, James Watt chose not to use the ordinary slider-crank mechanism in his first rotative engine. Instead, he utilized the *sun and planet* arrangement of Fig. 18, in which the connecting rod AP is rigidly fixed to a planet gear P and the sun gear S is keyed to the driven shaft. In Watt's design both gears had the same number of teeth.

a. Supposing that angle θ is negligibly small, show that the engine shaft rotates at twice the speed of the carrier arm SP.

b. If θ is not always zero but varies cyclically with the rotation of arm SP, show that the engine shaft makes two revolutions for every one of arm SP.

c. Show that by varying the number of teeth in gears S and P it is possible to vary the number of shaft revolutions per cycle of beam oscillation.

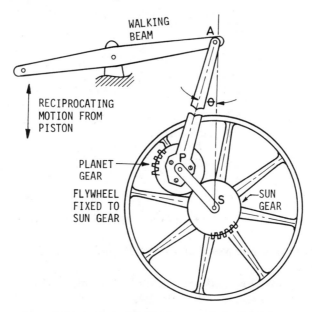

Figure 3.34-18 (Prob. 34) Watt's sun and planet gearing.

3.40 GEAR TOOTH PROFILES

We now wish to discuss the significance of gear tooth profiles. A fuller treatment of this vast subject will be found in the references cited at the end of Sec. **3.10** and in the publications of the American Gear Manufacturers' Association (AGMA), which continuously issues and updates standards that cover almost every imaginable aspect of gear technology. Of particular interest in this chapter is their standard 112.04,[26] "Standard Gear Nomenclature. . ."; the main contents of this publication will also be found in Scott [1962].

Figure 1 shows the principal terms used to describe spur gear teeth, regardless of their specific form. The terms *pitch circle* and *circular pitch* have already been defined (in Sec. **3.20**) but are included for convenience in the following list of definitions associated with Fig. 1:

> **transverse plane:** a plane normal to gear axis.
> **tooth profile:** intersection of curved surface of tooth and the transverse plane.
> **pitch circles:** the rims of imaginary friction wheels having the same velocity ratio

---

[26]Nomenclature is now fairly stable, but all AGMA standards are subject to periodic revision. For a comparison of British and American nomenclatures (which differ primarily in symbols rather than fundamental definitions), see Merritt [1971].

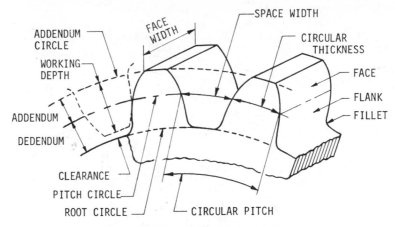

Figure 3.40-1. Gear tooth nomenclature.

as a given gear pair. For a rack, the pitch circle becomes a straight line, the **pitch line**.

pitch point: point of contact of pitch circles.

pitch diameter: diameter of pitch circle.

circular pitch: distance along pitch circle between corresponding points on adjacent teeth.

diametral pitch: number of teeth divided by pitch diameter (in inches).

module: pitch diameter divided by number of teeth. The diameter is usually specified in inches or millimeters; in the former case the module is the inverse of diametral pitch.

addendum circle: the circle[27] which passes through the tooth crests.

addendum: distance between pitch circle and addendum circle.

root circle: circle tangent to bottoms of tooth spaces.

dedendum: distance between pitch circle and root circle.

clearance: dedendum of a gear minus addendum of mating gear.

working depth: sum of addenda of mating gears.

whole depth: addendum plus dedendum.

circular thickness: length of arc between two sides of the same tooth, measured along the pitch circle.

space width: length of arc between closest sides of two adjacent teeth, measured along the pitch circle.

backlash: space width minus circular thickness.

face: the surface of the tooth between the pitch circle and addendum circle.

flank: the surface of the tooth between the pitch circle and the root circle, excluding the fillet region which joins the tooth profile to the root circle.

The definitions given apply to any spur gear, regardless of the specific tooth profile used, and may be suitable generalized for helical and bevel gears.

---

[27]Unless otherwise specified all circles and distances referred to lie in a *transverse plane*.

If a driving gear rotates at a *constant* velocity (ω_1), it is highly desirable that the angular velocity (ω_2) of the driven gear also be constant. Otherwise, the angular acceleration ($\dot{\omega}_2$) will result in fluctuating inertia moments that could vibrate the machine and its supporting structure and cause rapid wear at tooth contact and bearing surfaces.

It is therefore important to understand the geometric requirements on the profiles of gear teeth which ensure that the speed ratio (ω_2/ω_1) of mating gears will be constant.

Figure 1 shows parts of two gears[28] (1 and 2) which contact at point C and rotate

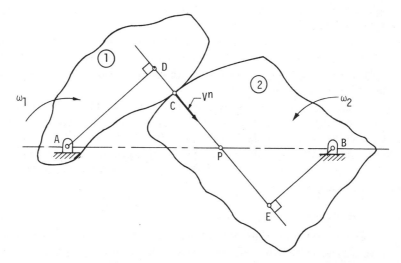

Figure 3.41-1. Bodies in sliding contact.

about fixed centers (A and B). The **common normal** to both tooth profiles at C passes through the *line of centers* (AB) at a point P. The velocity of that point on gear 1 which momentarily coincides with the moving contact point (C) will be designated as $\mathbf{V}_{C1}$. This vector can be resolved into components (V_1^n and V_1^t) along and perpendicular to the common normal (CP). Similarly, the velocity ($\mathbf{V}_{C2}$) of point C on gear 2 can be resolved into components (V_2^n and V_2^t) along and perpendicular to the common normal. As long as the teeth maintain contact and do not interpenetrate, V_1^n and V_2^n must be *equal*, and we designate their common value by V^n; i.e.,

$$V_1^n = V^n = V_2^n \tag{1}$$

To find V^n, we consider the velocity $\mathbf{V}_D$ of point D (in gear 1), which is the foot of the perpendicular from A to the common normal. The component (V_D^n) of $\mathbf{V}_D$ along the line DC must equal V^n; otherwise the rigid line DC would have to stretch or con-

---

[28]We are thinking in terms of gears, but the discussion of this section is valid for any two bodies making contact at a moving point C; i.e., the results derived are true for *cam pairs* in general.

tract.[29] But, by construction, $V_D^n = |\mathbf{V}_D| = \omega_1 AD$; i.e.,

$$V_D^n = \omega_1 AD = V^n \tag{2}$$

Similarly,

$$V_E^n = \omega_2 BE = V^n \tag{3}$$

where E (in gear 2) is the foot of the perpendicular from B to the common normal. The velocity ratio is therefore given by

$$\frac{\omega_2}{\omega_1} = \frac{AD}{BE} = \frac{AP}{PB} \tag{4}$$

where the rightmost term in Eq. (4) follows from the similarity of the right triangles PAD and PBE.

The ratio AP/PB will be constant if, and only if, the point P remains fixed on the line of centers AB. In short, **the velocity ratio of mating gears will be constant if, and only if, the common normal to the tooth profiles, at the point of contact, intersects the line of centers at a fixed point.**

This important statement is called the **fundamental law of gearing**.[30] Pairs of mating tooth profiles which satisfy this law are said to be **conjugate** to one another. Starting with a given tooth profile, it is possible (Buckingham [1963]) to find the profile of the conjugate tooth, but we shall not pursue this geometrical problem here. Instead, we shall concentrate on the only two forms of teeth that have ever been widely used, viz. *cycloidal* and *involute* teeth.

Involute Teeth

The overwhelming majority of gears used today have tooth profiles which are the involutes of circles, hereafter called simply **involute profiles**. It will be recalled that this curve is defined as the locus of a point attached to a taut string which is unwrapped from the circumference of a **base circle**, as illustrated by the curve BC in Fig. 2(a).

To show that any involute tooth profile will be conjugate to any other involute profile we shall consider the involute from a slightly different point of view. Figure 2(b) shows a string Tt wrapped around two pulleys which are pivoted at fixed points, A and a. As the string moves along the path Tt (tangent to the pulleys) and drives the pulleys without slip, a point C on the string will trace out a curve BC on a plate fixed to the lower pulley. To an observer located on the lower pulley, the point C appears to be traced by a point C on a taut string TC which unwinds from a circle of radius TA. Therefore, the curve BC traced on the plate is an involute of the base circle of radius TA. If the plate were cut along the curve BC, it would form an involute tooth attached to the lower pulley (or base circle). Similarly, if a plate were attached to the upper pulley, the point C would trace out an involute bC of the base circle of radius ta. From Fig. 2(a) we see that the tracing point C always moves perpendicular[31] to the

---

[29]It will prove useful to remember that all points lying on any straight line in a *rigid* body must have equal velocity components in the direction of the line.

[30]An alternative derivation of this rule is given in Sec. **5.15**.

[31]If point C had a velocity component along TC, the string would have to stretch or buckle.

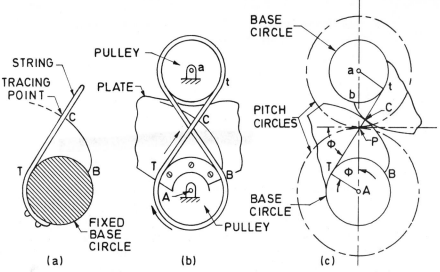

Figure 3.41-2. Involute teeth: (a) Involute generated by point on a taut string unwinding from a fixed base circle; (b) Involute drawn on a plate fixed to a pulley; (b) Tooth profiles cut out of plate.

line TC; hence the two profiles just described have a common tangent at C, as shown in Fig. 2(c).

Furthermore, the common normal at C is the straight line TCt which intersects the line of centers Aa at a fixed point P. Thus, the fundamental law of gearing is satisfied by involute profiles.

The pitch point P is the contact point of the imaginary friction circles, or pitch circles, in which the teeth are presumed to be cut. These pitch circles have radii PA and Pa.

The following nomenclature is useful in discussing involute gears, where lowercase symbols are associated with the smaller gear (pinion) and uppercase symbols with the larger gear:

base circle radii: $R_B = AT$, $r_b = at$.
pitch circle radii: $R = AP$, $r = aP$.
number of teeth: N, n.
pressure angle: $\phi =$ angle between tangent (Tt) to both base circles and the common tangent to both pitch circles.[32]
center distance: $c = Aa$.

It should be carefully noted that the involutes are uniquely defined at all points outside of their base circles, except for a rigid body rotation about the base circles, which provides the freedom to space the teeth evenly around the gear.

On the other hand, the pressure angle and the pitch circles are not *intrinsic* characteristics of an individual gear but are defined only after the center distance (Aa) has been fixed for two gears in mesh.

---

[32]Strictly speaking, this defines the **pressure angle of the two gears in mesh**, which should be distinguished from the **involute pressure angle**, $\angle TAC$, for a generic point C on the gear profile. It is common practice to use the term *pressure angle* in the former sense.

It may readily be verified from Fig. 2(c) that the pressure angle ϕ is defined, for given base circles and center distance, by

$$\cos \phi = \frac{AT}{AP} = \frac{at}{ap} = \frac{AT + at}{AP + Pa} = \frac{R_B + r_b}{c} \tag{5}$$

and the pitch radii are given by

$$R = AP = AT \sec \phi = R_B \sec \phi$$
$$r = aP = at \sec \phi = r_b \sec \phi \tag{6}$$

From these equations, or Fig. 2(c), we see that the main effect of altering the center distance is to change the pressure angle and the pitch radii, but *the conjugate action of the teeth is not influenced by adjustments of the center distance*. This remarkable property of involute teeth enables one to relax tolerances on center distances, to make adjustments for backlash and wear, and to permit the same physical gear to be used in situations where different pitch diameters may be desired (as in Problems **3.34**-16 and 18).

Finally, we note from Eqs. (6) that the speed ratio

$$\frac{N}{n} = \frac{R}{r} = \frac{R_B}{r_b} \tag{7}$$

is independent of the pressure angle and therefore independent of the center distance.

These properties, along with ease of manufacturing, have won the field for involute gears over that of their only close competitor, *cycloidal* gears. Although the latter are still used, in modified form (Dean [1962]), for clockwork gearing, they are mainly of historical interest. For this reason, we devote a few paragraphs to show that cycloidal gears exhibit conjugate tooth action.

Cycloidal Teeth

In the seventeenth century, it was discovered by Desargues that certain curves traced out by points on rolling circles[33] make conjugate tooth profiles. These teeth are commonly called **cycloidal teeth** but are more properly called **trochoidal teeth**.

The face of a cyloidal tooth is generated by rolling a **describing circle** (of diameter D_1) over the *outside* of the pitch circle (p_1) of a given gear, as shown in Fig. 3(a).

As the describing circle rolls from position C to C', the describing point E (fixed on the describing circle) traces out the epicycloid EE' which forms the face of the tooth. If a second describing circle (of diameter D_2) rolls *inside* the pitch circle, a describing point H will trace out the hypocycloid EH, which forms the flank of the tooth. The tooth itself is included between the profile just described and its mirror image about a radius of the pitch circle.

In a similar way, the face (flank) of a mating gear 2 is formed by rolling a describing circle of diameter D_2 (D_1) on the outside (inside) of pitch circle p_2, as in Fig. 3(b).

---

[33]A point fixed on a circle which rolls on the outside (inside) of another circle describes a curve known as an **epicycloid (hypocycloid)**. Such curves are members of the more general family of *trochoids*, described in Sec. **5.20** et seq.

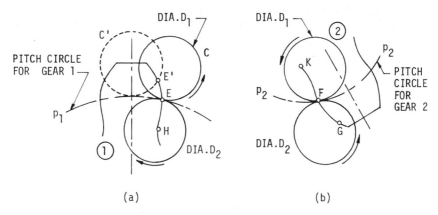

(a)   (b)

Figure 3.41-3. Formation of cycloidal teeth by rolling of describing circles on pitch circles.

To show that the teeth just described do indeed satisfy the fundamental law of gearing, consider Fig. 4, which shows two pitch circles (p_1 and p_2) rolling together with common pitch-line speed V_p. That is, the corresponding gears (1 and 2) are rotating about fixed centers not shown in the figure.

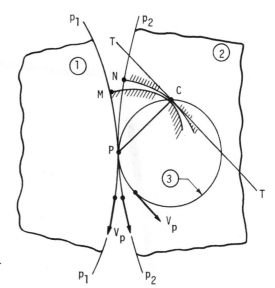

Figure 3.41-4. Conjugate action of cycloidal teeth.

To visualize the tooth action, assume that a describing circle (3) simultaneously rolls over the outside of circle p_1 and over the inside of circle p_2. This will be possible if the circumferential speed of the describing circle has the same value (V_p) as that of the other two circles. Under these conditions, the mutual contact point of all three circles remains fixed at the pitch point (P) of the two pitch circles. To an observer sitting on gear 1, it will appear that a point C on the describing circle is moving at right angles to the radius (CP) from the *instantaneous center*[34] of rotation (P) as it

---

[34]The concept of instantaneous center is described in detail in Sec. **5.13.**

traces out the epicycloid *CM*. That is, the normal at *C*, to the curve *CM*, passes through the pitch point *P*. Similarly, an observer fixed in body 2 would see point *C* trace out the hypocycloid *CN*, with the normal at point *C* passing through point *P*. In short, the common normal to the mating tooth surfaces passes through the pitch point *P* at all times. The fundamental law is satisfied, and the teeth are conjugate, as was to be proved.

Note that no restriction has been placed upon the diameters D_1 and D_2 of the describing circles, exeept for the implied restriction that D_2 (D_1) must be smaller than the pitch diameter of gear 1 (2) in order to roll "inside" the pitch circle. For interchangeability it is desirable that $D_1 = D_2$.

This is as far as we shall pursue the theory of cycloidal gearing. For further information on the subject, consult the older literature, e.g., Albert and Rogers [1938].

3.42   *Properties of Involute Teeth*

From the fundamental diagram, Fig. **3.41**-2, we see that the point of contact *C* between mating involute teeth traverses the straight line *Tt* which is tangent to both base circles. This line is therefore called the **path of contact,** or **line of action,** and its extremities (*T* and *t*) are called **interference points**, for reasons which will appear shortly.

Figure 1 shows a pair of mating teeth, moving from left to right, which make their

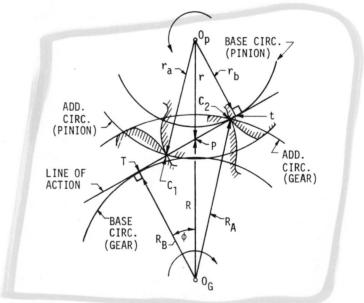

Figure 3.42-1. Tooth action of involute gears. Tt is line of action. Contact begins at C_1 and ends at C_2.

first contact at a point C_1 and break contact at a point C_2; both points, of course, must be on the path of contact *Tt*. We refer to the lower member as the *gear* and to the upper as the *pinion*. The first point on the pinion tooth to make contact is the outer tip, i.e., the point on the pinion addendum circle. Thus, C_1 is at the intersection of the pinion addendum circle and the line of action. Similarly, C_2 is located at the intersection of the gear addendum circle and the line of action. The distance C_1C_2 is called the **length of action** and is usually denoted by the symbol Z.

As the contact point moves from C_1 to C_2, it is apparent, from Fig. **3.41**-2, that both base circles rotate through the arc $C_1C_2 = Z$. The number of teeth included by the arc Z on the base circle of the pinion is

$$m_c \equiv \frac{Z}{2\pi r_b} n = \frac{Z}{p_b} \tag{1}$$

where r_b is the radius of the pinion base circle, n is the number of teeth on the pinion, and $p_b = 2\pi r_b/n$ is the **base pitch** of the pinion. The quantity m_c represents the *average number of teeth in contact* as the gears rotate and is called the **contact ratio**.

By substituting the word *gear* for *pinion* in the above paragraph it is apparent that we could have substituted the base pitch ($P_b = 2\pi R_B/N$) of the gear for p_b in Eq. (1). Indeed, it follows from Eq. (**3.41**-7) that $P_B = p_b$; i.e., the base pitch is identical for gear and pinion.

To calculate the contact length Z, we note from Fig. 1 that

$$\begin{aligned} Z = C_1C_2 &= C_1t + TC_2 - Tt \\ &= C_1t + TC_2 - (TP + Pt) \end{aligned} \tag{2}$$

If we utilize the notation

$$(R, R_B, R_A): \quad \text{pitch, base, and addendum radii for gear}$$
$$(r, r_b, r_a): \quad \text{pitch, base, and addendum radii for pinion}$$
$$\phi: \quad \text{pressure angle}$$

it follows from Eq. (2), and Fig. 1, that

$$Z = (r_a^2 - r_b^2)^{1/2} + (R_A^2 - R_B^2)^{1/2} - (R + r) \sin \phi \tag{3}$$

If desired, the base radii can be expressed in terms of the pitch radii via Eqs. (**3.41**-6).

For smooth action, it is necessary that the contact ratio exceed 1—otherwise there will be intervals where no teeth are in contact. The AGMA recommends a minimum contact ratio of 1.2 (Dean [1962]); however, the minimum recommended value of 1.40 has been quoted (without attribution) in numerous textbooks (e.g., Billings [1943]) for many years. Recently, it has been claimed (Kohl [1976]) that a contact ratio of 2 is desirable to reduce noise in gears. However, such a large ratio is difficult to achieve with standard gears, as shown by the charts of contact ratio in Peterson [1958].

In Fig. 1, the extremities (C_1 and C_2) of the actual path of contact both fall within the line segment Tt. In this case, contact occurs only on the involute portion of the teeth (outside the base circles), and there is no need to consider what shape must be given to the portion of the teeth that lies below the base circle.

For conjugate tooth action, it is essential that contact take place on the involutes only, but the inactive portion of the teeth below the base circle can be given any convenient form such as a radial line tangent to the involute at the base circle—which is the usual design.

However, it is a disadvantage of the involute profile (unlike the cycloidal profile) that situations can arise where the teeth make contact on a noninvolute surface, thereby destroying the desired conjugate action. This phenomenon, known as **interference**, is seen most clearly when a rack meshes with a pinion.

It will be recalled that a rack is the limiting case of a gear of infinite radius. The tooth profile, for a rack which is conjugate to an involute pinion tooth, is a straight line. To see this, we note from Fig. **3.41**-2(a) that the radius of curvature of the involute BC at point C is $\rho_G = TC$; similarly, the radius of curvature of the mating involute is $\rho_p = Ct$. The sum of these two radii is therefore

$$\rho_G + \rho_p = AT \tan \phi + at \tan \phi = (R_B + r_b) \tan \phi \qquad (4)$$

where $AT = R_B$ and $at = r_b$ are the base radii of the two involutes. Now, with ϕ constant, imagine that the upper gear is held fixed (ρ_p, r_b, both constant), while the lower gear is gradually increased in size. In the limit, as $R_B \longrightarrow \infty$, the lower gear becomes a rack, and Eq. (4) shows that the *radius* of curvature ρ_G of the rack becomes infinite. In other words, the *curvature* ($1/\rho_G$) of the rack profile vanishes; that is, *the tooth profile of the rack, conjugate to an involute, is a straight line normal to the line of action tP*. Furthermore, the interference point T moves to infinity as $AT \longrightarrow \infty$. Such a rack is shown in Fig. 2 in mesh with a pinion.

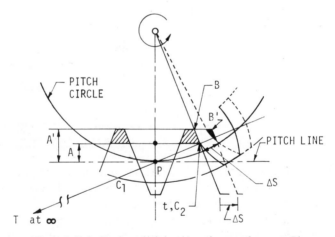

Figure 3.42-2. Rack and pinion; interference shown at B'.

Consider the case where the rack addendum is A and the extreme contact point C_2 coincides with the interference point t on the line of action. Based on the discussion of Fig. 1, we see that as the rack in Fig. 2 moves from left to right, contact begins at C_1 (the intersection of the pinion addendum circle with the line of action) and continues until separation occurs at t (the intersection of the rack addendum line with the line of action).

Now suppose that the rack addendum is increased to A', thereby adding the crosshatched material to the rack tooth. This additional material does not influence the motion of the system as the contact point moves from C_1 to t. However, in the latter position, the additional material on the rack tooth lies flush against the straight radial flank (tB) of the pinion tooth. If the rack then advances a further distance Δs, the rack tooth tip (B) moves to B'. If pure rolling of the pitch surfaces is to occur, the pinion must rotate through an arc Δs on its pitch circle, bringing the pinion tooth into the dashed position. However, with the pinion tooth in its new position, the rack tooth tip (B') would have interpenetrated the pinion tooth to the extent shown by the blackened area just under point B'.

In practice, the teeth shown would bend, abrade, and interfere with the pinion teeth to the left (not shown) which are trying to maintain a pure rolling motion. This is the phenomenon called **involute interference**. With some methods of tooth manufacture (Sec. **3.44**), a cutting tool which simulates the rack would dig out a scallop on the pinion tooth in the neighborhood of point B'. This **undercutting** weakens the pinion tooth and destroys part of the involute near B'.

To understand the phenomena of interference and undercutting from a different point of view, consider Fig. 3, where the rack is assumed to be rolling over a stationary pinion with angular velocity ω as shown.

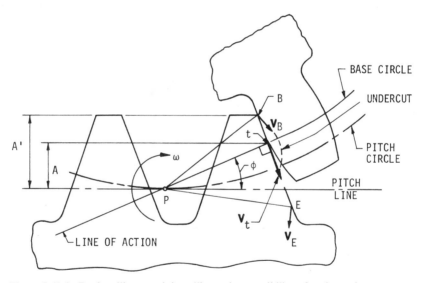

Figure 3.42-3. Rack rolling on pinion; illustrating possibility of undercutting.

The figure represents the same condition as in Fig. 2; i.e., the contact point has just reached the interference point t, so that the portion of the rack above point t lies flush alongside the radial flank of the pinion tooth. If the rack were to roll further, the point B at the rack tooth tip would have a velocity vector $\mathbf{V}_B$. Because the rack is, at this instant, rolling about the pitch point P, the velocity vector $\mathbf{V}_B$ points at right angles to the radius PB.[35] From Fig. 3, we see that $\mathbf{V}_B$ has a component normal to the pinion surface at B, and the postulated rolling motion could not occur with perfectly

---

[35]In Sec. **5.13** it will be verified that the velocities of points in the rack are the same as they would be if the rack were hinged at the *instantaneous center* of rotation P.

rigid teeth; i.e., *interference* occurs. On the other hand, if the rack tooth shown is actually a cutting tool, it will scoop out part of the pinion tooth enclosed by the dashed curve[36]; i.e., a rack-shaped cutting tool can create an *undercut* on the pinion tooth.

Since the rack profile is normal to the line of action, the point t on the rack tooth, which is momentarily on the line of action, will have a velocity $\mathbf{V}_t$ directed along the common tangent to the tooth surfaces; i.e., point t on the rack profile has no component pointing into the pinion. Similarly, the velocity $\mathbf{V}_E$ of any point on the rack profile beneath t will have a component pointing *into* the rack profile and away from the pinion profile. Therefore, to avoid interference and undercutting, it is merely necessary to keep the rack addendum at or below the dimension A shown in Figs. 2 and 3. That is, *to avoid interference the extremities of contact* (C_1, C_2) *must lie inside the segment of the line of action bounded by the interference points* (t, T).

This statement also applies to a pinion meshing with a gear of finite pitch radius. In addition, we can state that if a pinion meshes with a rack, without interference, it will mesh, without interference, with any gear of diameter larger than its own, provided that the addendum of the gear is equal to or less than that of the rack. This can be inferred from Fig. 2, since the addendum line (corresponding to A) will be replaced by an addendum circle which curves downward and intersects the line of action (Pt) in a point C_2 which lies to the left of the interference point t.

To find the maximum addenda for a pair of gears that operate without interference, it is only necessary that the addendum circles of the gear and pinion pass through the respective interference points $(t$ and $T)$. This condition is illustrated in Fig. 4, from which we see that the contact length Tt is given by

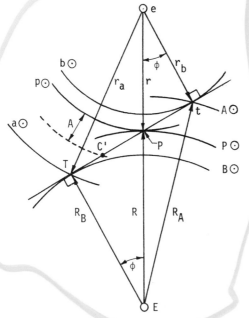

Figure 3.42-4. Maximum addenda to avoid interference: $A\odot, P\odot, B\odot$ are addendum, pitch, and base circles of gear; $a\odot, p\odot, b\odot$ are addendum, pitch, and base circles of pinion.

---

[36]Equations for this *trochoidal* curve are given by Buckingham [1963, pp. 49–51]. These equations verify that the trochoid extends beyond the base circle of the pinion, as shown in Fig. 3. That is, undercutting removes part of the pinion's involute profile, resulting in reduction of contact ratio.

$$Tt = EP \sin \phi + eP \sin \phi = c \sin \phi \qquad (5)$$

where

$$c = EP + eP = R + r \qquad (6)$$

is the *center distance Ee*.

From the figure, we also see that the critical addendum *radii* for gear and pinion are

$$R_A = [R_B^2 + (Tt)^2]^{1/2} = [R_B^2 + (c \sin \phi)^2]^{1/2} \qquad (7a)$$
$$r_a = [r_b^2 + (Tt)^2]^{1/2} = [r_b^2 + (c \sin \phi)^2]^{1/2} \qquad (7b)$$

Addendum radii smaller than these values will permit operation without interference. These equations provide a useful check for interference, as illustrated in the following example.

EXAMPLE 1

A gear with 24 teeth and a pinion with 18 teeth operate at a pressure angle of 20° and diametral pitch of 5. Find the maximum addendum radii to avoid interference.
Solution:
From the definition of diametral pitch, we find the pitch radii:

$$r = \frac{n}{2P_d} = \frac{18}{(2)(5)} = 1.8 \text{ in.}$$

$$R = \frac{N}{2P_d} = \frac{24}{(2)(5)} = 2.4 \text{ in.}$$

The corresponding base radii are

$$r_b = r \cos \phi = 1.8 \cos 20° = 1.6914 \text{ in.}$$
$$R_B = R \cos \phi = 2.4 \cos 20° = 2.2553 \text{ in.}$$

The center distance is $c = r + R = 4.2$ in.
The required maximum addendum radii are given by Eqs. (7) as

$$\max R_A = [(2.2553)^2 + (4.2 \sin 20°)^2]^{1/2} = 2.6739 \text{ in.}$$
$$\max r_A = [(1.6914)^2 + (4.2 \sin 20°)^2]^{1/2} = 2.2191 \text{ in.}$$

The maximum addenda of gear and pinion are therefore given by

$$\max A = \max R_A - R = 2.6739 - 2.4 = 0.2739 \text{ in.}$$
$$\max a = \max r_a - r = 2.2191 - 1.8 = 0.4191 \text{ in.}$$

We shall see below that for standard *full-depth* teeth the addendum is $1/P_d$, which in this case is 0.2 in. Since this is less than max A (and max a), no interference will occur.

3.43 Standard Tooth Forms

The basic references for this subject are the various AGMA standards.[37] However, the pertinent information in these standards has been summarized and interpreted by Dean [1962].

---

[37]Available from AGMA, 1330 Massachusetts Ave., N.W. Washington, D.C. 20005.

For our purposes, the most important standardized dimensions are the addendum and dedendum, which are given below in terms of the diametral pitch P_d for the most important systems now in use. Some obsolete systems are included because machines with such gears still exist. We are considering only **equal-addendum gearing**. Other systems exist where gear and pinion have different addenda. Table 1 summarizes the pertinent information.

TABLE 3.43-1 *Standardized Tooth Proportions of Spur Gears*

Item	Coarse Pitch ($P_d < 20$) Full Depth		Fine Pitch ($P_d \geq 20$) Full Depth	Obsolete	Obsolete Fellows 20° Stub
Pressure angle	20°	25°	20°	$14\frac{1}{2}$°	20°
Addendum	$\dfrac{1}{P_d}$	$\dfrac{1}{P_d}$	$\dfrac{1}{P_d}$	$\dfrac{1}{P_d}$	$\dfrac{0.8}{P_d}$
Dedendum (min.)	$\dfrac{1.250}{P_d}$	$\dfrac{1.250}{P_d}$	$\dfrac{1.200}{P} + 0.002$ in.	$\dfrac{1.157}{P_d}$	$\dfrac{1}{P_d}$
Minimum teeth					
In pinion*	18	12	18	32	14
In equal gears†	13	9	13	23	10
AGMA standard	201.02	201.02	207.05	201.02*A*	—

*To avoid interference in rack and pinion. See Eq. (7).
†To avoid interference in two equal gears. See Eq. (8).

In all these systems, the tooth thickness and tooth space, on the pitch circle, are equal to $\frac{1}{2}\pi D/N = \frac{1}{2}\pi/P_d$.

If the number of teeth in the *pinion* exceeds the number listed in the table, no interference will occur with *gears* of any size, including a rack. To see how these numbers were developed, we first note, from Fig. **3.42**-4, that when the pinion is given the same addendum (A) as the gear, the pinion addendum circle (shown dashed in the figure) intercepts the line of action at a point C' well within the segment Tt of the line of action. Therefore when interference occurs with equal addendum teeth, it is the gear tooth tip which interferes with, or undercuts, the pinion tooth. Accordingly, no interference will take place if the gear addendum radius is less than the maximum given by Eq. (**3.42**-7a); i.e., the largest gear which will mesh properly with a given pinion has addendum radius R_A given by

$$R_A^2 = R_B^2 + (R + r)^2 \sin^2 \phi \qquad (1)$$

If we use the fact that $R_B = R \cos \phi$, Eq. (1) can be expressed as

$$\left(\frac{R_A}{R}\right)^2 = \cos^2 \phi + \left(1 + \frac{r}{R}\right)^2 \sin^2 \phi = \cos^2 \phi + \left(1 + \frac{n}{N}\right)^2 \sin^2 \phi \qquad (2)$$

We now note from Table 1 that the addendum for standard gears is of the form

$$A = \frac{k}{P_d} = \frac{k2R}{N} \qquad (3)$$

where $k = 0.8$ for the 20° stub system and 1.0 for all other systems discussed. Hence

$$\frac{R_A}{R} = \frac{R + A}{R} = 1 + \frac{2k}{N} \tag{4}$$

and Eq. (2) can be expressed in the form

$$\left(1 + \frac{2k}{N}\right)^2 = \cos^2 \phi + \left(1 + \frac{n}{N}\right)^2 \sin^2 \phi \tag{5}$$

Solving for N, we find

$$N = \frac{4k^2 - (n \sin \phi)^2}{2n \sin^2 \phi - 4k} \tag{6}$$

This represents the maximum size gear that will mesh properly with a given pinion. When N, predicted by Eq. (6), is not an integer, the integer part of N should be used.

To find the smallest pinion that will mesh properly with a rack, we note that $N \rightarrow \infty$ in Eq. (6) when n approaches the limiting value

$$n = \frac{2k}{\sin^2 \phi} \tag{7}$$

If n is made smaller than this value, the interference point t (see Fig. 3.42-2) will move toward the pitch point, and interference will result. Equation (7) was used to find the fourth line of Table 1.

If n is set equal to N in Eq. (6), we arrive at the following condition for determining the minimum number of teeth in each of two *equal* gears that mesh without interference:

$$3N^2 \sin^2 \phi - 4kN - 4k^2 = 0 \tag{8}$$

Equation (8) was used to find the fifth line in Table 1.

3.44 Gear Manufacturing Methods

In this section we intend to briefly describe the fundamental kinematic bases of the most important methods for machining accurate involute gears. The historical background of this topic is given by Woodbury [1959], and a short illustrated survey of manufacturing processes is given by Merritt [1971]. For more detailed information, see the specialized literature, e.g., Dudley [1954, 1962a], Candee [1961], and Michalec [1966].

Gears may be *cast* if a sufficient quantity is needed to justify the individual molds required for each form and if the inherent inaccuracies are acceptable.

A *milling cutter* of the exact shape and size for a given gear tooth can cut one tooth space to the desired depth. Then the gear blank is indexed (rotated through $360°/N$), and each of the N desired teeth is cut in succession. Since the shape of the profile is dependent on the base circle diameter, different milling cutters are needed for gears of different diameters. In practice, the same cutter is used over a range of tooth numbers, but accuracy suffers at the ends of the ranges.

Today, accurate gears are usually manufactured by so-called *generation methods*, the principles of which will now be described.

Rack-Based Generating Methods

The methods of *gear hobbing, gear grinding,* and *generation with rack-shaped cutters* all depend on the following **generating-rack principle:** *When the pitch line of a rack rolls over the* *pitch circle of a pinion, the (straight-line) rack tooth profiles envelop a series of involute tooth profiles symmetrically spaced around the pitch circle.*

This principle is illustrated in Fig. 1, which shows the successive positions occupied by the rack teeth as the rack pitch line rolls over the pinion pitch circle. We can imagine that the "rack" has been replaced by a cardboard template and that a cardboard disc or **blank** rolls over the pitch line drawn on the rack template. If the rack outline is traced on the blank after every two or three degrees of rotation, the successive rack outlines will all be tangent to a set of curves, on the blank, which define regularly spaced "teeth" around the pitch circle. These "teeth" are the *envelopes*[38] of the straight-line rack tooth profiles. That these envelopes are actually *involutes* of the pinion base circle will be proved at the end of this section.

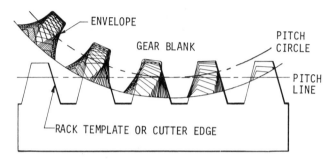

Figure 3.44-1. As the blank rolls over the rack, the rack tooth profiles envelop a set of involute tooth profiles in the blank.

Rack-Form Cutters

To generate the pinion teeth in *metal*, a hardened steel *cutter*, having the shape of a rack, is reciprocated normal to the plane of a stationary gear blank and is fed in radially to a desired depth, thereby cutting the straight rack profile into the blank. The cutter is then withdrawn normal to the plane of the blank; the blank is rotated through a small angle, and its center is moved a corresponding distance parallel to the pitch line of the rack (simulating rolling action). Then the cutter moves normal to the plane of the blank, cutting a small chip. The process is repeated until the blank has advanced through one circular pitch, whereupon the blank is returned without rotation to its starting point and the process repeated.

A variation of this procedure is to keep the center of the blank stationary while the blank is indexed angularly and the rack cutter is translated parallel to its pitch line by the amount needed to simulate pure rolling action.

Gear Hobbing

To avoid the need for long rack cutters, a cutting tool called a **hob** has been developed. To visualize a hob, consider a screw thread such that a cross section through the screw axis has the form of a rack. If a number of axial gashes are made on the outer surface of the thread, a series of cutting edges will be formed, as illustrated in Fig. 2. This hardened steel cutting tool is called a hob.

The hob is slowly fed along the axis of the gear blank, while the hob and blank rotate about their axes by amounts which simulate rolling of the rack on the pinion. To cut a spur

---

[38]A curve which is tangent to each member of a *family* of curves is called the **envelope** of the family, and the curves of the family are said to **envelop** the envelope.

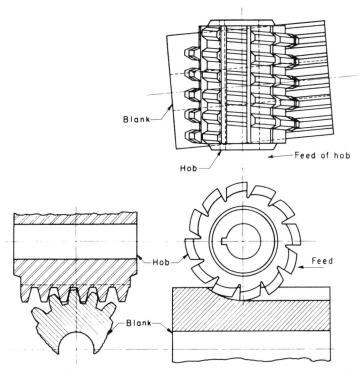

Figure 3.44-2. Illustrating hobbing of a spur gear. From De Garmo [1969] (Courtesy, Macmillan Publishing Co., Inc.).

gear, the axis of the hob must be inclined to that of the blank, as shown, to compensate for the helix angle of the "screw thread" on the hob. By increasing the inclination still more, a helical gear will be formed.

A great advantage of the hobbing process (as with any rack-based generating method) is that a single hob (or generating rack) will cut all gears of the same diametral pitch and pressure angle, *regardless of the number of teeth*. In addition, the depth of the generated tooth can be controlled by adjusting the center distance while cutting, and the standard pitch diameter of the cut tooth can be adjusted by altering the ratio of rack travel (along its pitch line) to the gear blank rotation.

Gear Grinding

After a cut gear has been heat-treated it will often have inaccuracies due to metallurgical changes. Since it is difficult to machine extremely hard parts, it is necessary to remove the irregularities by abrasive grinding. One way of doing so is to let the *flat* side of a grinding wheel serve as the side of a rack tooth. In all other major respects the process is similar to that used with a rack-form cutter.

Gear-Shaper Process

It is shown at the end of this section that just as a rack-form cutter can generate all mating gears of the same diametral pitch, so too can a gear-form cutter. Figure 3 shows the Fellows method of generating spur gears with such a cutter.

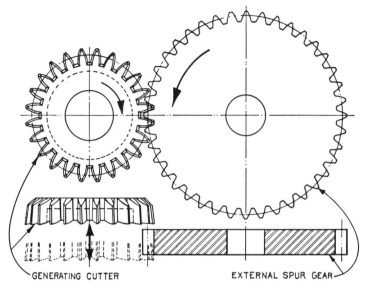

Figure 3.44-3. Spur gear generation by a gear-form cutter (Courtesy, Fellows Corp., Subsidiary of Emhart Industries).

The cutter is reciprocated parallel to its axis, and both cutter and blank are rotated, between cutting strokes, so that the pitch circles on each rotate through the same arc. The development of the cut tooth is illustrated in Fig. 4.

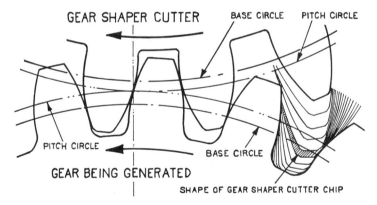

Figure 3.44-4. Diagram illustrating the generating action of the Fellows gear shaper cutter. Showing how the involute portion, flank, and fillet of the teeth are produced (Courtesy, Fellows Corp., Subsidiary of Emhart Industries).

Properties of Generated Gears

The properties of generated gears (as opposed to cast or milled gears) are determined by the equivalent generating rack, the relative spacing of rack and blank, and their relative speed during the cutting process.

A rack is kinematically defined by its profile slope (pressure angle ϕ) and by the distance between successive teeth (pitch). When this rack is used to cut a gear blank, with pitch-line speed V_p, the angular speed ω of the blank determines the **nominal pitch diameter (cutting pitch diameter)** by the relationship

$$\frac{D}{2} = \frac{V_p}{\omega}$$

The **nominal diametral pitch** of such a gear, with N teeth, is therefore

$$P_d = \frac{N}{D} = \frac{\omega N}{2V_p}$$

Thus, when one buys a gear that is classified as "8 pitch, 20°, 24 teeth," it means that it will mesh with any gear cut from a standard rack having 20° pressure angle, and a tooth spacing of

$$\frac{\pi D}{N} = \frac{\pi}{P_d} = \frac{\pi}{8} = 0.3927 \text{ in.}$$

If this gear is meshed with an "8 pitch, 20°, 18 teeth" gear at the **standard center distance** (or **cutting center distance**)

$$c = \frac{1}{2}(D_1 + D_2) = \frac{1}{2}\frac{N_1 + N_2}{P_d} = \frac{1}{2}\frac{24 + 18}{8} = 2.625 \text{ in.}$$

the **operating pressure angle** will equal the **standard pressure angle** (or **cutting pressure angle**) of 20°. If the **operating center distance** is made greater than the standard value of 2.625in., the operating pressure angle will increase; the *actual* (or *operating*) pressure angle can be calculated from Eq. (**3.41-5**).

Proof of the Generating-Rack Principle

In our discussion of Fig. 1, we assumed that the curve enveloped by a rack tooth profile is an involute. To justify this statement, we first recall from Sec. **3.42** that the rack tooth profile conjugate to an involute is a straight line. In Fig. 5(a), such a straight-line rack tooth is shown in contact with an involute tooth attached to a base circle of fixed center O. As the rack's pitch line translates parallel to itself, and the pitch circle rotates about its center O, the contact point C travels along the line tP, which passes through the pitch point P and is tangent to the base circle at t.

Two successive positions of the moving teeth are shown in Fig. 5(a), where contact occurs at points C_1 and C_2, respectively, along the path of contact. An observer fixed in the gear would note that the rack tooth profile C_1G_1 is tangent to the gear tooth profile when contact occurs at point C_1. Similarly, he would observe that the rack tooth profile is tangent to the gear's involute profile for any other position (e.g., C_2) of the contact point. Figure 5(b) shows how two successive positions of the rack tooth profile (C_1G_1 and C_2G_2) are positioned *relative to the gear*. In short, the straight-line rack tooth profile always remains tangent to the involute gear tooth profile as the rack rolls over the gear—in other words, the rack tooth profile *envelops* the involute gear tooth profile as the rack rolls relative to the gear.

This proves the **generating-rack principle,** which can be restated in gear terminology as follows: *A straight line, attached at an angle of* $90° - \phi$ *to the pitch line of a rack, which rolls over a pinion of pitch radius r, envelops an involute of a concentric base circle of radius* r_b $= r \cos \phi$.

As a corollary of the *generating-rack principle*, we may prove the **generating-gear principle,** which states the following: *When the pitch circle of a gear rolls over the pitch circle of a second*

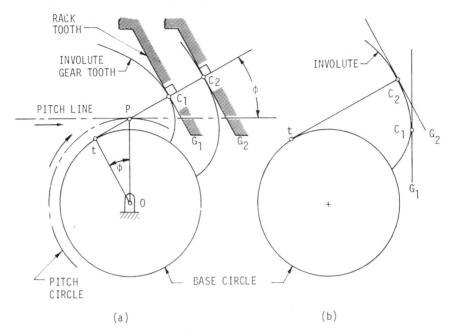

Figure 3.44-5. (a) Rack translates parallel to its pitch line; gear rotates about center O; rack tooth profile shown at successive positions (C_1G_1, C_2G_2); (b) Successive positions of rack tooth shown relative to gear.

gear, the tooth profiles of the first gear envelop a series of conjugate tooth profiles on the second gear.

To prove this statement, consider Fig. 6, which shows two gears in mesh. Interposed between the teeth is a straight-line rack profile. If the lower half of the rack profile is solid (as indicated on the left-hand side), the rack represents a cutter which could form the teeth of the upper gear. Similarly, if the upper half of the rack profile is solid (as indicated on the right),

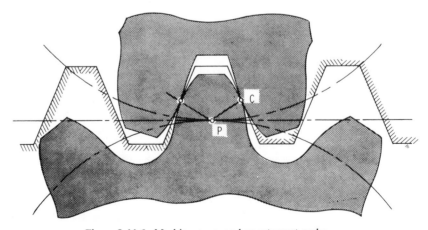

Figure 3.44-6. Meshing gears and counterpart racks.

the corresponding rack cutter would form the lower gear teeth. These two physically different racks, formed from the selfsame profile, are called **counterpart racks.** Any gears generated by such counterpart racks will be conjugate, because the counterpart rack profiles are kinematically indistinguishable.

Now imagine that the rack profile translates to the right with a speed V_p, thereby causing both gears to rotate about their fixed centers, while both pitch circles roll on each other with peripheral speed V_p. This leaves the pitch point fixed in space, and the path of contact for the rack and any conjugate tooth must (by the fundamental law of gearing) be the normal PC to the rack profile. As the contact point C moves out along the line PC, it traces tooth profiles on both rolling pitch circles (or gear blanks), but both tooth profiles are normal to PC at the instantaneous contact point (by definition of contact). That is, *both are tangent to the rack at the moving contact point and therefore are tangent to each other.*

From the generating-rack principle (previously proved), we know that the rack profile envelops the lower tooth profile. But at each point where the rack profile touches the lower tooth profile, the upper tooth profile touches the rack. That is, the upper tooth profile is tangent to the lower tooth profile at all positions of C. In short, *the upper tooth profile envelops the lower tooth profile,* and the envelope (lower tooth profile) is identical with the envelope formed by the rack. Furthermore, since the pitch-line speeds of both gears are identical with that of the rack, the teeth formed by envelopment from a gear are conjugate. This completes the proof of the generating-gear principle.

This principle is valid for any kind of generating gear; it is not restricted to involute gear profiles. In the special case of involute teeth, we have seen that all involutes are conjugate, regardless of the base circle they are formed from. Furthermore, all involute teeth generated from the same straight-sided rack will have the same pitch and will therefore intermesh properly.

3.45 PROBLEMS

Note: Unless otherwise noted, all given data refer to standard gears, operating at standard centers; ϕ = pressure angle; P_d = diametral pitch.

1. Define the following terms:

a.	pitch point	h.	clearance
b.	pitch diameter	i.	circular thickness
c.	circular pitch	j.	space width
d.	diametral pitch	k.	backlash
e.	module	l.	face
f.	addendum	m.	flank
g.	dedendum		

2. Define the following terms:

a.	fundamental law of gearing	k.	path of contact
b.	conjugate tooth profiles	l.	line of action
c.	involute of circle	m.	interference points
d.	base circle of involute	n.	length of action
e.	pressure angle	o.	base pitch
f.	center distance	p.	contact ratio
g.	trochoid	q.	involute interference
h.	epicycloid and hypocycloid	r.	undercutting
i.	describing circle (for trochoids)	s.	AGMA
j.	cycloidal teeth		

3. Define the following terms:
   a. generating rack
   b. envelope
   c. rack-form cutter
   d. gear-hobbing process
   e. gear-shaper process
   f. cutting pitch diameter
   g. cutting pressure angle
   h. operating pitch diameter
   i. operating pressure angle
   j. operating center distance
   k. generating-rack principle
   l. generating-gear principle
   m. counterpart racks

4. To get a feeling for the size of gear teeth with different diametral pitches, draw standard racks (three teeth will suffice) having a 20° pressure angle and the diametral pitches indicated. You may omit the fillet radius specified by AGMA, but use the AGMA standards for addendum and dedendum. Diametral pitches are
   a. 2.5, 3, 3.5
   b. 4, 5, 6
   c. 7, 8, 10
   d. 12, 14, 16.

5. In Fig. 3.41-2(c), denote the angle TAC by ψ, the angle BAC by θ, the radius AC by r, and the base circle radius by r_b. Show that the parametric equations of the involute can be expressed in the polar coordinate form

$$r = r_b \sec \psi$$
$$\theta = \tan \psi - \psi$$

The polar angle θ is sometimes called the **involute function**: $\text{inv}(\psi) \equiv \tan \psi - \psi$.

6. Using the results of Problem 5, show that the thickness t of an involute tooth, at any radius r, can be expressed in terms of its thickness t_0 at some known radius r_0 by the expression

$$t = 2r \left(\frac{t_0}{2r_0} + \text{inv} \, \psi_0 - \text{inv} \, \psi \right)$$

where $\psi = \sec^{-1}(r/r_b)$. and $\psi_0 = \sec^{-1}(r_0/r_b)$.

7. Find the distance between centers of two external gears of diametral pitch P_D if the gears have N_1 and N_2 teeth, respectively. Assume no backlash. Use the numerical values $P_d = 8$, $N_1 = 22$, $N_2 = 36$.

8. Find the distance between centers of two external gears (without backlash) if the circular pitch p is known and the number of teeth in the gears is given as N_1 and N_2, respectively. For numerical values, use $p = 2$ in., $N_1 = 5$, $N_2 = 10$. Would you expect gear manufacturers to have these gears "in stock?"

9. Solve Problem 7 if the gears mesh internally.

10. Solve Problem 8 if the gears mesh internally.

11. Show that each of the following quantities is uniquely determined if the diametral pitch P_d and pressure angle ϕ are specified:
    a. number of teeth per inch of pitch line.
    b. thickness of teeth along the pitch line (no backlash).
    c. ratio of involute base circle diameter to pitch circle diameter.

12. Given a standard gear with $\phi = 14.5°$, $P_d = 4$, and pitch diameter $= 8$ in., find (a) number of teeth, (b) addendum, (c) clearance, (d) root circle diameter, (e) addendum circle diameter, (f) base circle diameter.

13. Repeat Problem 12 for $\phi = 20°$.

14. Repeat Problem 12, but change P_d to 5 and the pitch diameter to 4 in.

15. Given that the center distance between two gears is 10 in., their speeds are 400 and 600 rpm, and $P_d = 4$, find the pitch diameters, tooth numbers, and circular pitch.

16. Given that a gear with 30 teeth, $P_d = 3$, and $\phi = 14.5°$ meshes with a second gear having a 15-in. pitch diameter, find (a) number of teeth on second gear, (b) center distance, (c) circular pitch, (d) standard addendum, (e) speed ratio (f) base circle diameter of first gear.

17. For standardized equal addendum gears, show that the contact ratio given by Eqs. (**3.42**-1) and (**3.42**-2) can be expressed in terms of pressure angle ϕ, tooth numbers (n and N), and the coefficient k as defined by Eq. (**3.43**-3). That is:

$$\pi m_c = \left[\left(\frac{n}{2} + k\right)^2 \sec^2 \phi - \left(\frac{n}{2}\right)^2\right]^{1/2} + \left[\left(\frac{N}{2} + k\right)^2 \sec^2 \phi - \left(\frac{N}{2}\right)^2\right]^{1/2} - \frac{N + n}{2} \tan \phi$$

18. Show that the contact ratio m_c for a gear of n teeth is given by the expression below for a rack and pinion with standardized equal addenda ($a = k/P_d$) and pressure angle ϕ.

$$m_c = \frac{[(n + 2k)^2 - (n \cos \phi)^2]^{1/2} - n \sin \phi + 2k/\sin \phi}{2\pi \cos \phi}$$

19. Using the result of Problem 17 find the contact ratio for the smallest pair of equal gears that mesh without interference. Assume standard full-depth, coarse pitch, 20° pressure angle teeth as described in Table **3.43**-1.

20. Using the result of Problem 18, find the contact ratio for a rack and the smallest pinion it will mesh with, without interference. Assume standard full-depth, coarse pitch, 20° pressure angle teeth as described in Table **3.43**-1.

21. Repeat Problem 19 for a pressure angle of 14.5°.

22. Repeat Problem 20 for a pressure angle of 14.5°.

23. Will standard, full-depth, 20° pressure angle gears, of diametral pitch 5, mesh without interference, at standard centers, if the number of teeth for gear and pinion are 24 and 16?

24. What is the contact ratio for the gears of Problem 23?

25. Will the gears of Problem 23 mesh without interference if the pinion is changed to one with 14 teeth?

26. A gear set consists of 24- and 14-tooth full-depth, 20° pressure angle gears of diametral pitch 5. At what center distance can these gears operate without interference? What is the contact ratio under these conditions?

4

CAMS

4.10   INTRODUCTION

In our discussion of kinematic pairs in Sec. **1.23**, we defined a cam pair as a combination of two bodies which contact at a point (or line, in three dimensions). The individual bodies making such contact are called **cams**. Sometimes the word *cam* is used only for the *driving* member, and the word **follower** is used for the *driven* member of the pair. Thus, the contacting teeth of gears form cam pairs, and we may think of conjugate gears as cams where the cam and follower have a constant speed ratio.

Cams come in a variety of two- and three-dimensional designs (cylinder cams, globoidal cams, etc.), but we shall restrict our discussion to the two-dimensional case, where the cam is essentially a closed curve cut from a plate—this accounts for the commonly used names **plate cam** or **disk cam** for these machine elements. Such cams are also referred to as **radial** cams by some writers.

Descriptions of two- and three-dimensional cams and extensive discussions of the kinematics, dynamics, design, and manufacture of cams of all types will be found in the books of Rothbart [1956], Jensen [1965], Molian [1968b], Neklutin [1969], Koster [1974], Tesar and Matthew [1976], Chakraborty and Dhande [1977]; and the survey paper by Chen [1977] contains a useful list of references.

In this chapter, it is our intention to briefly illustrate how cams can be designed to generate virtually any follower motion desired, by either graphical or analytical methods. From this discussion, it will become clear that the great popularity of cams is due, in large measure, to the straightforward design procedures associated with them. In other words, cams are well suited to the *synthesis* of mechanisms. This situation contrasts with the case of linkages—which are relatively easy to *analyze* (once they are designed) but are more difficult to synthesize for a specified motion program. The *analysis* of linkages containing cams with *given* profiles will be deferred until Chapter 7, where we discuss analysis for mechanisms in general.

Most of the standard terminology listed below is illustrated in Fig. 1 or in subsequent figures of this chapter:

> **knife edge follower:** the follower comes to a point where it contacts the cam. This is not a practical design because of rapid wear at the knife edge.
> **roller follower:** the follower is a circular disc free to rotate about its center. This is the most commonly used type (Fig. 1).
> **flat-faced follower:** the follower surface is plane at the point of contact.
> **trace point:** the point of a knife edge follower, the center of a roller follower, or a reference point on a flat-faced follower (Fig. 1).
> **translating follower:** the trace point translates relative to the frame in which both cam and follower move (Fig. 1).
> **pivoted** or **oscillating follower:** the trace point rotates about a fixed point on the frame (Figs. **4.33**-1 and **4.35**-2).
> **radial follower:** the trace point translates along a straight line through the cam's center of rotation (Fig. 1).
> **offset follower:** the trace point translates along a straight line located at a fixed distance (the offset) from the cam's center of rotation (Fig. **4.32**-1).
> **cam profile:** the working surface of the cam, in contact with the follower (Fig. 1).

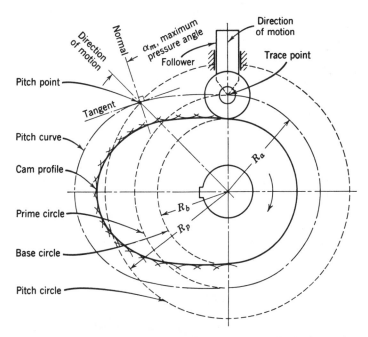

Figure 4.10-1. Cam nomenclature. A radial translating roller is shown. A knife edge follower is indicated by broken lines converging on the trace point. From Rothbart [1956] (Courtesy, John Wiley & Sons, Inc.).

base circle: smallest circle, centered on cam axis, that touches cam profile (Fig. 1).
pitch curve: path of trace point (Fig. 1).
prime circle: smallest circle, centered on cam axis, that touches pitch curve (Fig. 1).
pressure angle: angle between the normal to the pitch curve and the instantaneous direction of trace point motion (Fig. 1).
pitch point: point on pitch curve where pressure angle is maximum (Fig. 1).
pitch circle: circle, through pitch point, centered on cam axis (Fig. 1).

If the pressure angle is zero, the force exerted by the cam on the follower is in the direction of follower motion, and no side thrust is produced on the follower. This is desirable because side thrust increases the friction on a translating follower and results in higher stresses in the system. The side thrust will increase with increasing pressure angle; therefore it is good practice to keep the pressure angle as low as possible. Rothbart [1956, p. 64] recommends limiting the pressure angle to about 30°; however, he states that, with light loads and low-friction bearings (e.g., ball bushings), pressure angles as high as 47.5° have proved satisfactory. The design variable which influences the pressure angle most significantly is the base circle radius. For a given total displacement of the follower, the maximum pressure angle can always be reduced by increasing the size of the base circle. For an extremely large base circle radius, the cam profile (in Fig. 1) will approach a circular disk, and the pressure angle will approach zero. The disadvantage of a large base circle is that the size of the cam becomes too great for many applications.

In general, one should choose the largest base circle that space permits. Charts for finding the minimum base circle consistent with a specified pressure angle have been developed[1] for specific types of followers and specific types of motion. However, in all cases, a simple trial and error process using two or three different trial base circles will enable one to choose a suitable base circle without the use of charts.

4.20 DISPLACEMENT DIAGRAMS

The function of a cam is to cause a specified follower displacement for a given position of the cam. We shall designate the cam rotation angle by θ and the follower displacement (linear or angular) by y. The cam must be so designed that the **displacement function**

$$y = f(\theta) \tag{1}$$

is satisfied. A graph of y versus θ is called a **displacement diagram**.

From the displacement function, we see that the velocity and acceleration are given by

$$\dot{y} = \frac{dy}{d\theta}\frac{d\theta}{dt} = \dot{\theta}y' \tag{2a}$$

$$\ddot{y} = \ddot{\theta}y' + \dot{\theta}\frac{dy'}{d\theta}\frac{d\theta}{dt} = \ddot{\theta}y' + \dot{\theta}^2 y'' \tag{2b}$$

---

[1] See Mabie and Ocvirk [1975, pp. 74–76] and Jensen [1965, Chap. 5].

where we have introduced the abbreviations

$$\frac{d(\ )}{dt} = (\dot{\ }), \qquad \frac{d(\ )}{d\theta} = (\ )' \tag{3}$$

Usually, the cam runs at a constant angular speed ω, so that the follower velocity and acceleration are simply

$$\dot{y} = \omega y', \qquad \ddot{y} = \omega^2 y'' \qquad (\dot{\theta} = \omega, \ddot{\theta} = 0) \tag{4}$$

We shall therefore follow the usual practice and consider the cam angle θ as the independent variable.

During any given increment $\Delta\theta$ of cam rotation, the follower is said to **rise, dwell,** or **return** accordingly as the corresponding displacement Δy is *positive, zero,* or *negative.* Periods of rise, dwell, or return may occur in any sequence during a complete revolution of the cam.

Let y_1, y_2 be the displacements corresponding to cam angles θ_1, θ_2. The total displacement

$$h \equiv (y_2 - y_1) \tag{5a}$$

during the angular interval

$$\beta \equiv (\theta_2 - \theta_1) \tag{5b}$$

is called the **lift** or **fall** accordingly as h is positive or negative; we assume that $\theta_2 > \theta_1$ so that β is never negative.

For slow-moving systems, the detailed nature of the displacement function is not important, and any convenient curves can be used as displacement diagrams, provided that they give appropriate rise, return, and dwell periods.

However, when cam speeds are high, the acceleration $\omega^2 y''$ will also be high unless $|y''|$ is sufficiently small. Large accelerations result in large inertia forces, with accompanying large stresses, vibrations, and wear. Therefore, a principal objective of cam design is to keep the accelerations at reasonably low levels. We shall now show how the form of the displacement diagram is influenced by such considerations.

Figure 1 shows a typical displacement diagram having phases of rise (AB), dwell (BC), and return (CD), defined by the four given **control points** A, B, C, D. The simplest curve which passes through these points is the piecewise straight line, shown solid. However, this straight-line diagram (or piecewise constant velocity diagram) has discontinuities in slope, at the control points, which correspond to infinite accelerations ($y'' = \infty$) at those points. To ameliorate this condition, it is desirable to use a

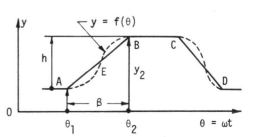

Figure 4.20-1. Displacement diagram for dwell-rise return cycle.

curve, such as that shown dotted in Fig. 1. This curve has continuous slope everywhere —thereby assuring that $\ddot{y} = \omega^2 y''$ is *finite* everywhere—although y'' may undergo jump discontinuities at the control points. Where y'' is discontinuous, the quantity $\dddot{y} = \omega^3 y'''$ (called **jerk** or **pulse**) is infinite.

For high-quality machinery, it is essential that accelerations remain finite, even if not continuous. For even better results, it is desirable to have the acceleration continuous as well (i.e., finite jerk). However, the manufacturing accuracies required to attain continuous acceleration can become excessively severe.

In any case, several forms of displacement diagram with finite acceleration and some with finite jerk have proved effective. We shall discuss only the following types of such diagrams, which should suffice for most purposes:

1. Simple harmonic
2. Cycloidal
3. Piecewise constant acceleration

Detailed discussions of a much wider variety of displacement curves will be found in Rothbart [1956], Jensen [1965], and Molian [1968b].

Case 1. Simple Harmonic Motion (SHM)

Perhaps the simplest "S"-shaped curve of the type shown in Fig. 1 is the sinusoid, given by

$$y = y_1 + \frac{h}{2}\left[1 - \cos\frac{\pi(\theta - \theta_1)}{\beta}\right] \tag{6a}$$

$$y' = \frac{\pi h}{2\beta}\sin\frac{\pi(\theta - \theta_1)}{\beta} \tag{6b}$$

$$y'' = \frac{\pi^2 h}{2\beta^2}\cos\frac{\pi(\theta - \theta_1)}{\beta} \tag{6c}$$

where h and β are defined by Eqs. (5). It is easily verified that these equations represent a *rise* or a *return* accordingly as h is *positive* or *negative*.

From Eqs. (6-a, b, c), and Eq. (4), the maximum magnitude of velocity and acceleration are seen to be

$$|\dot{y}|_{\max} = |\omega y'|_{\max} = \frac{\pi}{2}\left|\frac{\omega h}{\beta}\right| = 1.571\left|\frac{\omega h}{\beta}\right| \tag{7a}$$

$$|\ddot{y}|_{\max} = |\omega^2 y''|_{\max} = \frac{\pi^2}{2}\left|\frac{\omega^2 h}{\beta^2}\right| = 4.935\left|\frac{\omega^2 h}{\beta^2}\right| \tag{7b}$$

EXAMPLE 1

(a) Give the displacement equation for the displacement diagram having the characteristics specified in the first four columns of the following table. (b) Find the follower displacement, velocity, and acceleration when the cam angle is $\theta = 145°$, $\omega = 100$ rpm, and $\dot{\omega} = 0$.

It is advisable to use radian measure in all *analytical* work, since differentiation and integration are thereby expedited. The required functions, found by repeated application of Eq. (6a), are listed in the last column of the table given below.

Domain $(\theta_1 \longrightarrow \theta_2)$	$\Delta\theta°$ (β rad)	h (in.) = $y_2 - y_1$	Motion	Displacement y
$0° \longrightarrow 90°$	$90°$ ($\pi/2$)	2	SHM	$1 - \cos 2\theta$
$90° \longrightarrow 120°$	$30°$ ($\pi/6$)	0	Dwell	2
$120° \longrightarrow 180°$	$60°$ ($\pi/3$)	-0.5	SHM	$2 - \frac{1}{4}[1 - \cos 3(\theta - \frac{2}{3}\pi)]$
$180° \longrightarrow 270°$	$90°$ ($\pi/2$)	0	Dwell	1.5
$270° \longrightarrow 360°$	$90°$ ($\pi/2$)	-1.5	SHM	$1.5 - \frac{3}{4}[1 - \cos 2(\theta - \frac{3}{2}\pi)]$

For $\theta = 145°$, the third line of the table should be used. The required displacement and its derivatives are therefore

$$y = 2 - 0.25[1 - \cos 3(\theta - \tfrac{2}{3}\pi)] = 1.815 \text{ in.}$$
$$\dot{y} = \omega y' = -\omega(3)(0.25) \sin 3(\theta - \tfrac{2}{3}\pi) = -7.586 \text{ in./sec}$$
$$\ddot{y} = \omega^2 y'' = -\omega^2(9)(0.25) \cos 3(\theta - \tfrac{2}{3}\pi) = -63.84 \text{ in./sec}^2$$

where $\omega = 100(2\pi/60) = 10.47$ rad/sec, $3(\theta - \tfrac{2}{3}\pi) = 75°$, and use has been made of Eqs. (4).

Case 2. Cycloidal Displacement

The so-called **cycloidal**[2] **displacement** function is a modification of a sinusoid which gives zero acceleration at the end points A and B in Fig. 1. It is defined by the relations

$$y = y_1 + \frac{h}{2\pi}\left[\frac{2\pi(\theta - \theta_1)}{\beta} - \sin\frac{2\pi(\theta - \theta_1)}{\beta}\right] \tag{8a}$$

$$y' = \frac{h}{\beta}\left[1 - \cos\frac{2\pi(\theta - \theta_1)}{\beta}\right] \tag{8b}$$

$$y'' = \frac{2\pi h}{\beta^2} \sin\frac{2\pi(\theta - \theta_1)}{\beta} \tag{8c}$$

where a positive (negative) value of h denotes a rise (return).

Note that the maximum magnitudes of velocity and acceleration for this motion are given via Eqs. (4) and (8), by

$$|\dot{y}|_{\max} = |\omega y'|_{\max} = 2\left|\frac{\omega h}{\beta}\right| \tag{9a}$$

$$|\ddot{y}|_{\max} = |\omega^2 y''|_{\max} = 2\pi \frac{|\omega^2 h|}{\beta^2} \tag{9b}$$

These values exceed those given by Eqs. (7) for the simple harmonic motion. This is the price we pay for insisting on a continuous acceleration (finite jerk) at the ends of the interval.

EXAMPLE 2

Repeat Example 1(a) but change the initial rise and final return to a cycloidal motion rather than SHM. The problem statement and solution are tabulated below:

---

[2]The displacement y is usually called *cycloidal* because it represents the horizontal distance traversed by a point on a circle of radius ($h/2\pi$) as the circle rolls on a horizontal plane through the angle $\phi = 2\pi(\theta - \theta_1)/\beta$ [cf Eq. **5.21**-1].

Domain (θ_1, θ_2)	$\Delta\theta°$ (β rad)	$h =$ (in.)	Motion	Displacement y
$0° \longrightarrow 90°$	$90°$ ($\pi/2$)	2	Cycloidal	$(1/\pi)(4\theta - \sin 4\theta)$
$90° \longrightarrow 120°$	$30°$ ($\pi/6$)	0	Dwell	2
$120° \longrightarrow 180°$	$60°$ ($\pi/3$)	-0.5	SHM	$2 - \frac{1}{4}[1 - \cos 3(\theta - \frac{2}{3}\pi)]$
$180° \longrightarrow 270°$	$90°$ ($\pi/2$)	0	Dwell	1.5
$270° \longrightarrow 360°$	$90°$ ($\pi/2$)	-1.5	Cycloidal	$1.5 - (0.75/\pi)[4(\theta - \frac{3}{2}\pi)$ $- \sin 4(\theta - \frac{3}{2}\pi)]$

Case 3. Piecewise Constant Acceleration (Parabolic Motion)

If the "acceleration" y'' has the same constant magnitude A but opposite signs on each side of the transition point E in Fig. 1, we can write

$$y = y_1 + \tfrac{1}{2}A(\theta - \theta_1)^2; \qquad (0 \le \theta - \theta_1 \le \tfrac{1}{2}\beta) \tag{10a}$$

$$y = y_2 - \tfrac{1}{2}A(\theta_2 - \theta)^2; \qquad (\tfrac{1}{2}\beta \le \theta - \theta_1 \le \beta) \tag{10b}$$

where the second equation represents a reflection of the lower half of the curve about point E. To have $y - y_1 = h/2$ when $\theta - \theta_1 = \beta/2$, it is necessary that

$$A \equiv 4h/\beta^2 \tag{11}$$

The displacement function and its derivatives may therefore be expressed in the form

$0 \le \theta - \theta_1 \le \beta/2$	$\beta/2 < \theta - \theta_1 \le \beta$	
$y = y_1 + 2h\left(\dfrac{\theta - \theta_1}{\beta}\right)^2$	$y = y_1 + h - 2h\left(\dfrac{\theta_2 - \theta}{\beta}\right)^2$	
$y' = \dfrac{4h}{\beta}\left(\dfrac{\theta - \theta_1}{\beta}\right)$	$y' = \dfrac{4h}{\beta}\left(\dfrac{\theta_2 - \theta}{\beta}\right)$	(12)
$y'' = \dfrac{4h}{\beta^2}$	$y'' = \dfrac{-4h}{\beta^2}$	

These equations represent a lift or return accordingly as h is positive or negative. This type of displacement is often referred to as a **constant acceleration** function or **balanced parabolic motion**, rather than by its more accurate name *piecewise constant acceleration function*.

The peak magnitudes of velocity and acceleration are therefore given via Eqs. (4) and (12) by

$$|\dot{y}|_{\max} = |\omega y'|_{\max} = 4\left|\frac{\omega h}{\beta}\right| \tag{13a}$$

$$|\ddot{y}|_{\max} = |\omega^2 y''|_{\max} = 4\left|\frac{\omega^2 h}{\beta^2}\right| \tag{13b}$$

Let us define the *nondimensional velocity* y^* and *nondimensional acceleration* y^{**} by the expressions

$$y^* \equiv \frac{\beta \dot{y}}{\omega |h|} = \frac{\beta y'}{|h|} \qquad (14a)$$

$$y^{**} \equiv \frac{\beta^2 \ddot{y}}{\omega^2 |h|} = \frac{\beta^2 y''}{|h|} \qquad (14b)$$

We can now compare the peak nondimensionalized velocities and accelerations for the three cases investigated:

	SHM	Cycloidal	Parabolic
$\|y^*\|_{\max}$:	1.571	2	2
$\|y^{**}\|_{\max}$:	4.935	6.283	4

Although the parabolic motion gives rise to the smallest peak acceleration, it has the disadvantage of creating an infinite jerk at its midpoint.

EXAMPLE 3

Repeat Example 2 but change the cycloidal motions to parabolic motions. The following table shows the equations to be used:

Domain (θ_1, θ_2)	$\Delta\theta°$ (β rad)	$h =$ (in.)	Motion	Displacement y
$0° \longrightarrow 45°$ $\atop 45° \longrightarrow 90°$	$90°$ $(\pi/2)$	2	Parabolic	$\{16(\theta/\pi)^2$ $\atop 2 - 4(1 - 2\theta/\pi)^2$
$90° \longrightarrow 120°$	$30°$ $(\pi/6)$	0	Dwell	2
$120° \longrightarrow 180°$	$60°$ $(\pi/3)$	-0.5	SHM	$2 - 0.25[1 - \cos 3(\theta - \tfrac{2}{3}\pi)]$
$180° \longrightarrow 270°$	$90°$ $(\pi/2)$	0	Dwell	1.5
$270° \longrightarrow 315°$ $\atop 315° \longrightarrow 360°$	$90°$ $(\pi/2)$	-1.5	Parabolic	$\{1.5 - 3(2\theta/\pi - 3)^2$ $\atop 3(4 - 2\theta/\pi)^2$

Unbalanced Parabolic Motion

If it is desired to have constant (but different) accelerations, on either side of the *transition point E* (see Fig. 1), the displacement function and its derivatives are given by

$\theta_1 \leq \theta \leq \theta_E$	$\theta_E \leq \theta \leq \theta_2$
$y = y_1 + \tfrac{1}{2}A_1(\theta - \theta_1)^2$	$y = y_2 - \tfrac{1}{2}A_2(\theta_2 - \theta)^2$
$y' = A_1(\theta - \theta_1)$	$y' = A_2(\theta_2 - \theta)$
$y'' = A_1$	$y'' = -A_2$

(15)

where A_1 and A_2 are two constants, to be determined.[3]

The location θ_E of the transition point E is found by noting that y' must be continuous at $\theta = \theta_E$; hence

$$A_1(\theta_E - \theta_1) = A_2(\theta_2 - \theta_E) \qquad (16)$$

---

[3] From Eq. (4), the A_i are related to the actual accelerations by $\ddot{y} = \omega^2 A_1$ if $\theta \leq \theta_E$; $\ddot{y} = \omega^2 A_2$ if $\theta \geq \theta_E$.

from which it follows that

$$\theta_E = \frac{\theta_1 + r\theta_2}{1 + r} \tag{17a}$$

or

$$\frac{\theta_E - \theta_1}{\theta_2 - \theta_1} = \frac{r}{1 + r} \tag{17b}$$

where

$$r \equiv \frac{A_2}{A_1} \tag{18}$$

From the first two lines of Eqs. (15), it is readily shown that the velocity-displacement relation is

$$(y')^2 = 2A_1(y - y_1) \quad \text{for} \quad \theta \le \theta_E$$
$$(y')^2 = 2A_2(y_2 - y) \quad \text{for} \quad \theta \ge \theta_E \tag{19}$$

Hence, at the transition point where $y \equiv y_E$, it follows from Eqs. (19) that

$$y_E = \frac{y_1 + ry_2}{1 + r} \tag{20a}$$

or

$$\frac{y_E - y_1}{y_2 - y_1} = \frac{r}{1 + r} \tag{20b}$$

When Eqs. (17b) and (20b) are substituted into the first of Eqs. (15), we find

$$A_1 = \frac{2(1 + r)}{r} \frac{y_2 - y_1}{(\theta_2 - \theta_1)^2} = \frac{2(1 + r)}{r} \frac{h}{\beta^2} \tag{21a}$$

hence

$$A_2 = rA_1 = 2(1 + r)\frac{h}{\beta^2} \tag{21b}$$

Thus, if we know r, or θ_E, or y_E, we can calculate A_1 and A_2, and the required displacement function follows from Eqs. (15).

These results are illustrated in the following example.

EXAMPLE 4

Find the piecewise constant acceleration curve $y = y(\theta)$ corresponding to a total lift of 2 in. in 180° of cam rotation with the given acceleration ratio $A_1/A_2 = 3$. Assume that $y_1 = \theta_1 = 0$.

SOLUTION:
$h = 2$, $\beta = \pi$, $r = A_2/A_1 = \frac{1}{3}$, $y_1 = 0$, $\theta_1 = 0$, $\theta_2 = \pi$, $y_2 = 2$ in..

From Eq. (17a) we find $\theta_E = \dfrac{0 + \frac{1}{3}\pi}{1 + \frac{1}{3}} = \dfrac{\pi}{4}$ rad.

From Eqs. (21a) and (21b) we find $A_1 = \dfrac{2(1 + \frac{1}{3})}{\frac{1}{3}}\dfrac{2}{\pi^2} = \dfrac{16}{\pi^2}$, $A_2 = \frac{1}{3}A_1 = \dfrac{16}{3\pi^2}$

Finally, the displacement function is given by Eqs. (15) as

$$y = \frac{8}{\pi^2}\theta^2 \qquad \text{for } 0 \le \theta \le \frac{\pi}{4}$$

$$y = 2 - \frac{8}{3\pi^2}(\pi - \theta)^2 \qquad \text{for } \frac{\pi}{4} \le \theta \le \pi$$

4.30 GRAPHICAL METHODS OF CAM DESIGN

A few examples should suffice to show how plate cams may be designed by graphical methods. In all these examples, we consider an inversion of the cam-follower-frame mechanism in which the cam is considered the fixed member. The follower is then located in its proper position relative to the cam, corresponding to a number of different cam rotation angles. The cam profile is the envelope of follower profiles, as the follower moves relative to the cam.

The examples which follow represent typical problems in graphical cam design. Variants of the methods used will be found in the literature cited in Sec. **4.10** as well as in most books on kinematics of machinery. The construction methods are illustrated by well-executed drawings taken, with permission, from Maxwell [1960].

Graphical methods have inherent accuracy limitations and should therefore be used only for relatively low-speed cams. The analytical methods of Sec. **4.40** et seq. are better suited to higher-speed applications.

4.31 *Translating Radial Roller Follower*

Figure 1(a) shows a cam and follower which are supposed to generate the given displacement curve of Fig. 1(b). To find the required cam profile, we mark on the abscissa of the displacement diagram a number (N) of uniformly spaced points (such as those labeled 0, 1, 2, . . . , 12) which are located at a constant angular interval of $\Delta\theta$ (30° in the example). The

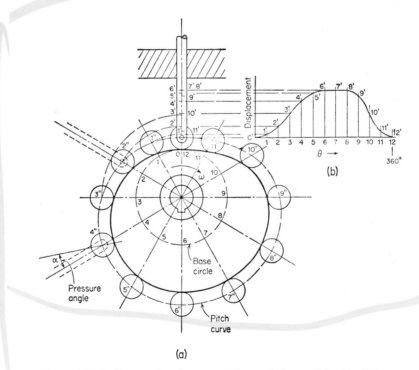

Figure 4.31-1. Construction for cam with translating radial roller follower. From Maxwell [1960] (Courtesy, Prentice-Hall, Inc.).

corresponding displacements (0', 1', 2', ..., 12') are marked on the translation axis of the follower.

We want the *trace point* (center of the roller) to occupy the positions 1', 2', 3, ... when the cam has rotated clockwise through the angles $\Delta\theta$, $2\Delta\theta$, $3\Delta\theta$, ..., 2π. Toward this end, we next select a convenient radius for the roller follower and a convenient radius for the base circle. The roller radius should be large enough to keep the contact stresses within acceptable limits,[4] and the base circle should be as large as space considerations permit. Wide latitude exists in both of these choices. The roller profile and base circle are drawn in the position of minimum lift, thereby fixing the center of the base circle. Next the base circle is divided into equal sectors of central angle $\Delta\theta$ by rays marked 1, 2, ..., N.

When the cam has rotated clockwise through angle $\Delta\theta$, the follower axis will have rotated counterclockwise relative to the cam by angle $\Delta\theta$ and will coincide with the ray marked 1 on the base circle. The distance of the trace point from the cam center should be the same as the distance of point 1'. Accordingly, this distance is laid off along ray 1, and the corresponding position of the trace point is labeled 1''. Similarly, the distance of point 2' from the cam center is laid off on ray 2, to define the relative position 2'', of the trace point when the cam has rotated through angle $2\Delta\theta$.

The process is repeated for each ray, thereby defining the locus of the trace points (1'', 2'', 3'', ...) relative to the cam, i.e., defining the desired *pitch curve*. The points 1'', 2'', 3'', ... represent the centers of the roller profile circles corresponding to the cam angles $\Delta\theta$, $2\Delta\theta$, $3\Delta\theta$, ..., and the corresponding positions of the roller relative to the cam are shown by the small circles in Fig. 1(a). Since the roller is tangent to the cam profile at all times, the cam profile is the envelope of the family of roller profile circles. That is, the *cam profile is found by drawing a smooth curve which is tangent to each of the roller circles drawn as in Fig. 1(a).*

The pressure angle α is measured between the normal to the pitch curve and the direction of follower motion, which in this example is the direction of the cam radius at any point of contact. A typical pressure angle is shown at position 4'' in Fig. 1. The maximum value of pressure angle α_{max} should be located for any given profile. If it exceeds 30° (see the discussion in Sec. 4.10), a larger base circle should be assumed, and the enlarged cam profile should be found by the method just described. By plotting α_{max} versus base circle radius for two or three cases, the desired base circle is quickly found. For this trial procedure it is unnecessary to find the complete cam contour. Only that portion of the cam in the neighborhood of α_{max} is needed; fortunately, the location of α_{max} will not change drastically as the base circle radius is gradually increased.

4.32 *Offset Translating Roller Follower*

Figure 1 is similar to Fig. **4.31**-1, except that the follower axis is located at a fixed distance (the offset) from the cam center.

The procedure for finding the pitch curve of the cam is very similar to that used for the case of zero offset. The main difference is that the successive positions of the follower axis (relative to the cam) are all drawn tangent to an **offset circle**. The offset circle is centered on the cam rotation axis and has a radius equal to the given offset. Each successive tangent drawn should make the uniform angle $\Delta\theta$ with its predecessor, and the distance along the tangent from the offset circle to a typical point (e.g., 2'') on the pitch curve must equal the corresponding distance (e.g., 02') measured along the actual (fixed) follower axis from point 0 on the offset circle.

---

[4]Problems of stress analysis are beyond the scope of this book, but see Rothbart [1956, Chap. 4] for a discussion of the topic.

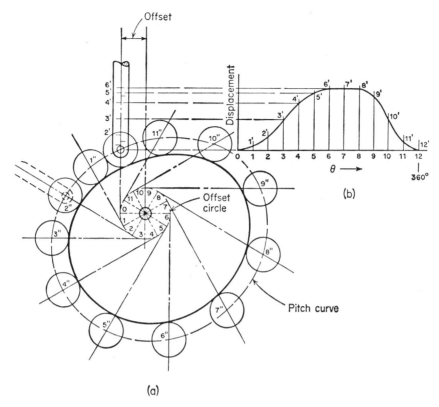

Figure 4.32-1. Construction for offset translating roller follower. From Maxwell [1960] (Courtesy, Prentice-Hall, Inc.).

Once the pitch curve has been established, the cam profile may be found as the envelope of successive roller profiles, exactly as in Fig. **4.31**-1.

If the maximum pressure angle is too great, the minimum distance between the cam center and the roller center should be increased, and the construction should be repeated in the neighborhood of α_{max}. The process should be repeated often enough to bring α_{max} down to the desired level.

4.33 *Oscillating Roller Follower*

Figure 1(a) shows an *oscillating* or *pivoted* roller follower where the roller center (trace point) traces out a circular arc centered on the fixed pivot of an oscillating arm which carries the roller axis.

An important advantage of an oscillating follower over a translating follower is the absence of sliding friction. This permits the use of much greater pressure angles for oscillating follower designs.

The ordinate of the displacement diagram in Fig. 1(b) represents arc length along the path of the trace point. These ordinates are marked on the trace point arc as points $0'$, $1'$, $2'$, $3'$, . . ., which define the roller position, in fixed space, for the cam angles 30°, 60°, 90°, ...

Let the reference line $Q0$ be the position of a given cam radius when $\theta = 0$. The points a, b, c, . . . are marked on the reference line at radial distances $Q1'$, $Q2'$, $Q3'$, . . . from the fixed point Q.

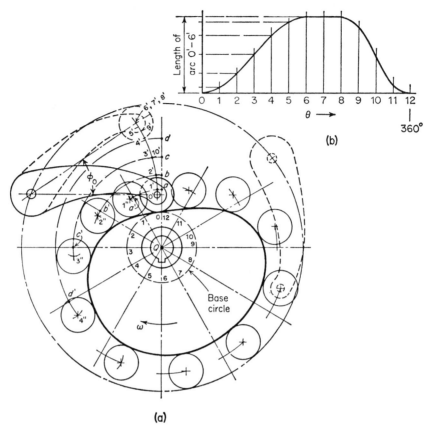

Figure 4.33-1. Construction for oscillating roller follower. From Maxwell [1960] (Courtesy, Prentice-Hall, Inc.).

To locate the positions ($1''$, $2''$, $3''$, . . .) of the trace point relative to the cam, the rays $Q1$, $Q2$, $Q3$, . . . are drawn from the cam center Q at angular intervals of 30°. These rays represent the position of the moving reference line which initially occupied position $Q0$. The point $4''$, for example, lies at the distance $Q4'$ from Q, i.e., on an arc of center Q and radius $Q4'$. The distance of $4''$ from the ray $Q4$ is the same as the distance of $4'$ from d; hence the chordal distance $d'4'' = d4'$ is laid off from the ray $Q4$ on the circle of center Q, which passes through d and $4'$. This uniquely locates point $4''$.

The construction is repeated to find the locus of trace points $1''$, $2''$, $3''$, . . . which define the pitch curve. The cam profile is the envelope of the family of circles centered on the pitch curve with radius equal to the roller radius.

4.34 Roller Diameter Limitations

The roller diameter should be large enough to prevent excessive contact stresses on the cam surface. However, if the roller radius exceeds the radius of curvature of the pitch curve at any point of contact, the theoretical cam profile is a self-intersecting curve, which is totally unsuitable for practical use. To see this, consider Fig. 1, which shows a given pitch curve and the associated cam profile (envelope of roller circles) required for three different choices of roller radius R. In all cases the envelope of the moving roller is drawn in the neighborhood of a point where the pitch curve is externally convex and has radius of curvature ρ_P.

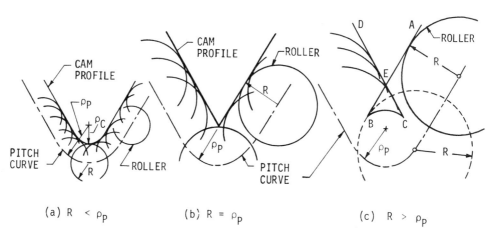

(a) $R < \rho_P$ (b) $R = \rho_P$ (c) $R > \rho_P$

Figure 4.34-1. The same pitch curve with different rollers: (a) $\rho_C = \rho_P - R \geq 0$; (b) $\rho_C = \rho_P - R = 0$; (c) Theoretical cam profile is self-intersecting.

In Fig. 1(a), the roller radius R is less than the pitch curve radius ρ_P, and the cam profile (envelope of the roller profiles) is unambiguously defined. It may be seen that the cam profile is a **parallel curve** to the pitch curve, and it has a local radius of curvature given by

$$\rho_C = \rho_P - R \tag{1}$$

Hence, the cam curvature is finite and positive for the case illustrated in Fig. 1(a).

Figure 1(b) shows the special case where $R = \rho_P$, and the radius of cam curvature is zero, according to Eq. (1). That is, the curvature of the cam is infinite when $R = \rho_P$, and the cam is required to have a sharp point. Naturally, this condition produces very high stresses and is to be avoided. We therefore strive *to make the roller radius smaller than the minimum radius of curvature of the cam.*

If the roller radius exceeds the minimum radius of curvature of the cam, as in Fig. 1(c), the envelope of the roller circles is a curve $ABCD$, which intersects itself at E. In other words, proper cam action for the segment AB of the theoretical cam profile will entail removal of the material BCE, which is necessary for proper action along the segment CD. The removal of part of the useful profile is called **undercutting** (just as in the case of gear generation). If the region BCE is removed in the manufacturing process, the roller will not be fully constrained when it passes through the undercut region, as shown by the dashed position of the roller in Fig. 1(c).

For externally *convex* pitch curves, undercutting is prevented by making $R < \rho_P$. If necessary, ρ_P can be increased by enlarging the base circle. For externally *concave* pitch curves, no undercutting can occur with roller followers. A simple sketch will show that externally concave cam surfaces cannot be used with flat-faced followers.

4.35 Flat-Faced Followers

The design of a flat-faced translating follower is illustrated in Fig. 1.

The trace point is the intersection of the flat face and the axis of the follower stem. The successive positions of the trace point, in space, are indicated by $1'$, $2'$, $3'$, . . . , corresponding to cam angles of $\Delta\theta$, $2\Delta\theta$, $3\Delta\theta$, ... After selecting a base circle, the rays numbered 1, 2, 3, . . . are laid out from the center of the base circle at an angular interval of $\Delta\theta$ (30°, in the figure). When the cam has rotated through angle $\Delta\theta$, the trace point will occupy position $1''$ relative to the cam. Point $1''$ lies on ray 1, at a distance from the cam center equal to that of point $1'$. Similarly, point $2''$ lies on ray 2, at the same distance from the cam center as point $2'$. Through the points $1''$, $2''$, . . . , so constructed, a series of straight lines is drawn normal to the rays 1, 2, ... These straight lines represent the instantaneous positions of the follower's flat face, relative to the cam, when $\theta = \Delta\theta$, $2\Delta\theta$, ... The cam profile is always tangent to the follower face and may be drawn as the smooth curve which is everywhere tangent to the family of straight lines representing the successive positions of the follower face.

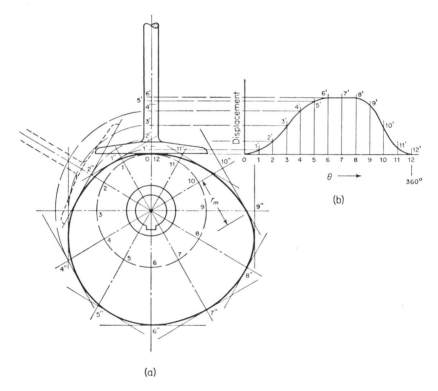

Figure 4.35-1. Construction for translating flat-faced follower. From Maxwell [1960] (Courtesy, Prentice-Hall, Inc.).

The flat face usually has a circular boundary whose radius must be at least as great as the maximum distance r_m from any contact point to the stem axis. This distance, shown in Fig. 1(a), is easily picked off the cam drawing. However, it can also be found by noting that the follower velocity is given by

$$\frac{dy}{dt} = r \frac{d\theta}{dt}$$

where r is the instantaneous distance of the contact point from the follower stem axis. Therefore $r = dy/d\theta$, and the absolute maximum value of r is

$$r_m = \left| \frac{dy}{d\theta} \right|_{\max}$$

which value may be found at the point of absolute maximum slope on the displacement diagram of Fig. 1(b) (at approximately $\theta = 300°$ in this example).

It is worth noting that the pressure angle is always zero for a flat-faced follower, and the only tendency to jamming of the slider in its guide is due to the moment of the normal force acting on moment arm r and to the frictional force on the face of the follower. Therefore, the size of the cam is not restricted by pressure angle considerations (as it is for a translating roller follower), and the use of a flat-faced follower permits relatively compact cam designs.

When a flat-faced follower is pivoted about a fixed point Q_0, the construction for the cam profile is as shown in Fig. 2.

The trace point is located on the flat face at an arbitrary radius r_1 from the point Q_0, and the successive trace point positions $0'$, $1'$, $2'$, $\ldots$ are transferred from the displacement diagram to the fixed arc of center Q_0 and radius r_1.

A base circle, center O, is chosen, and a set of rays $0, 1, 2, 3, \ldots$ are drawn on the base circle at equal angular intervals $\Delta\theta$ (e.g., 30°). The ray numbered 0 coincides with the fixed line OQ_0. The flat face of the follower is always tangent to the cam and tangent to a small circle of known radius r_t, centered on the point Q_0. We shall call this small circle the *hub circle*.

When the cam has rotated through angle $\Delta\theta$ clockwise, the pivot Q_0 will have rotated, relative to the cam, in the opposite direction to the point Q_1 which lies on the circle of radius OQ_0, at the intersection with ray 1. To find the position $1''$ on the trace point relative to the cam, we recall that the trace point is at distance r_1 from point Q_1 and at distance $O1'$ from O. Therefore $1''$ is located at the intersection of the arc of radius r_1, centered at Q_1, and the arc of radius $O1'$ centered at O. We now recall that the flat face of the follower passes through the trace point and is always tangent to the hub circle. Accordingly, the position of the flat follower face, relative to the cam, is the straight line through point $1''$ which is tangent to the hub circle of radius r_t centered at point Q_1. Similarly, point $2''$ is at the intersection of an arc of radius r_1 (centered at Q_2) and an arc of radius $O2'$ (centered at O), and similarly for points $3''$, $4''$, $\ldots$. The successive positions of the flat follower face, relative to the cam, are represented by the lines through points $2''$, $3''$, $\ldots$ which are tangent to hub circles centered at points Q_2, Q_3, $\ldots$, respectively. The cam profile is the smooth curve which is tangent to all members of the family of straight lines representing the follower face.

The minimum amount of straight bearing surface required on the follower is found by noting the maximum distance of the contact point from the points $1''$, $2''$, $\ldots$. These distances might extend on either side of the points $1''$, $2''$, $\ldots$.

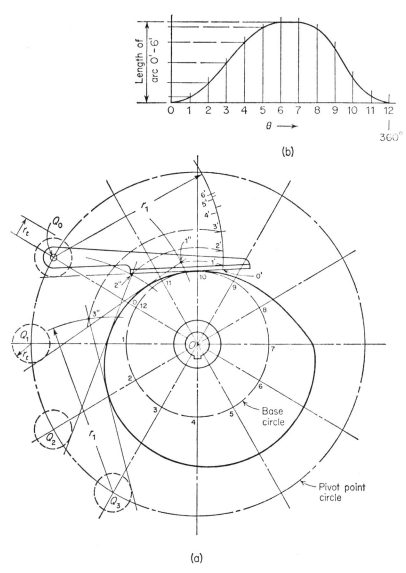

(b)

(a)

Figure 4.35-2. Construction for pivoted flat-faced follower. From Maxwell [1960] (Courtesy, Prentice-Hall, Inc.).

4.40 ANALYTICAL METHODS OF CAM DESIGN

To *design* a cam, we must know the distances and angles required to locate the grinding or cutting surfaces of the forming tools. It is also important to know the radius of curvature and the pressure angle at the point of contact in order to evaluate contact stresses. We may also wish to know the equations of the cam profile, although this information is usually not essential for manufacturing purposes.

There are many ad hoc analytical solutions for specific types of cam-follower systems,[5] but that of Raven [1959] gives the radius of curvature of the cam in a particularly direct and attractive form. This will be illustrated for the four cam-follower systems which are most likely to arise in practice.

A computer program called CAMKIN (CAM KINematics), based upon the analysis of Secs. **4.41–4.44**, is available[6] for the synthesis of the four major types of cam-follower systems with arbitrarily prescribed displacement functions.

4.41 Flat-Faced Translating Follower

Figure 1 shows the system to be investigated, with the contact point at C and the corresponding center of curvature at A. O is the center of cam rotation, and B is the initial point of contact on the follower. The initial contact point on the cam will move to position B' after the cam has rotated *clockwise* through the angle θ.

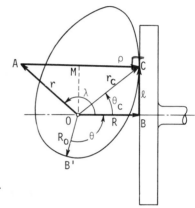

Figure 4.41-1. Flat-faced translating follower.

We are given the follower displacement function

$$R = R_0 + f(\theta) \quad \text{(Given)} \tag{1}$$

where R_0 is the base circle radius (OB'), and we seek to find the following quantities: radius of curvature ρ, location (r, λ) of the center of curvature, contact distance $l = BC$, and the polar coordinates (r_C, ϕ) of the cam profile, where $\phi \equiv \angle B'OC = \theta + \theta_C$.

We begin by noting that all points of the cam located on the line AC (such as points C and M) must be moving along line AC with the same velocity[7]; i.e., $V_C = V_M = \omega OM = \omega l$, where $\omega = d\theta/dt$. Since AC is normal to the follower surface, the follower velocity (dR/dt) equals V_C. Therefore

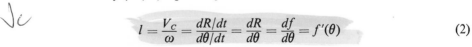

$$l = \frac{V_C}{\omega} = \frac{dR/dt}{d\theta/dt} = \frac{dR}{d\theta} = \frac{df}{d\theta} = f'(\theta) \tag{2}$$

where f is the function given by Eq. (1).

---

[5]See Rothbart [1956] and Molian [1968b].
[6]See Van Dyke and Paul [1979], or contact author.
[7]Recall the footnote relating to Eq. **(3.41-2)**. In Fig. 1, OM is perpendicular to AC.

The quantity l is an important design variable, since the flat follower must be at least as long as

$$l_{max} - l_{min} = \left(\frac{df}{d\theta}\right)_{max} - \left(\frac{df}{d\theta}\right)_{min} \tag{3}$$

The extreme values of $df/d\theta$ may be easily obtained from the slope of the displacement diagram.

From Fig. 1 it may be seen that the angle θ_C is given by

$$\theta_C = \arctan_2 [l, R] = \arctan_2 [f', R_0 + f] \tag{4}$$

where we have introduced the notation

$$(\ )' \equiv \frac{d(\ )}{d\theta} \tag{5}$$

The polar angle, $\phi \equiv \angle B'OC$, of the contact point and its radius vector r_C are given, respectively, by

$$\phi \equiv \theta + \theta_C = \theta + \arctan_2 [f', R_0 + f] \tag{6}$$
$$r_C = [l^2 + R^2]^{1/2} = [(f')^2 + (R_0 + f)^2]^{1/2} \tag{7}$$

These are the parametric equations of the cam contour.

The radius of curvature ρ at any point of the contour can be found from the parametric equations by means of Eq. (**2.32**-5), expressed in the form

$$\frac{1}{\rho} = \left| \frac{2(dr_C/d\phi)^2 - r_C(d^2 r_C/d\phi^2) + r_C^2}{[r_C^2 + (dr_C/d\phi)^2]^{3/2}} \right| \tag{8}$$

Because r_C has been found as a function of θ rather than a function of ϕ, we cannot use Eq. (8), as it stands, but must transform it by use of the identities

$$\frac{d(\ )}{d\phi} = \frac{d\theta}{d\phi} \frac{d(\ )}{d\theta} = \frac{1}{\phi'}(\ )' \tag{9a}$$

$$\frac{d^2(\ )}{d\phi^2} = \frac{1}{\phi'} \frac{d}{d\theta} \frac{d(\ )}{d\phi} = \frac{1}{\phi'} \frac{d}{d\theta} \left[\frac{1}{\phi'}(\ )' \right]$$

$$= \frac{\phi'(\ )'' - (\ )'\phi''}{(\phi')^3} \tag{9b}$$

When Eqs. (9) are substituted into Eq. (8) we find the immediately applicable expression

$$\frac{1}{\rho} = \left| \frac{2(r_C')^2\phi' - r_C(\phi' r_C'' - r_C'\phi'') + r_C^2(\phi')^3}{[(r_C\phi')^2 + (r_C')^2]^{3/2}} \right| \tag{10}$$

This is a general expression which is valid for any cam. However, its use requires a substantial amount of algebraic manipulation. Raven [1959] has described the following alternative method for finding ρ. From Fig. 1, the radius of curvature at C is seen to be

$$p = R - r \cos \lambda \qquad (11)$$

where

$$r = OA, \qquad \lambda = \angle BOA \qquad (12)$$

It was shown in Sec. **1.54**-3 that displacements, velocities, and accelerations are correctly predicted if an arbitrary cam is instantaneously replaced by a circular cam with the same radius of curvature and center of curvature. If we imagine the given cam to be replaced by its *equivalent circular cam*, both p and r may be treated as constants. Therefore differentiation of Eq. (11) results in

$$-r(\sin \lambda)\frac{d\lambda}{d\theta} = \frac{dR}{d\theta} = f'(\theta) \qquad (13)$$

where Eq. (1) has been used.

We now observe that the line OA rotates, *with the equivalent cam*, at the angular velocity $\dot{\lambda}$ (positive *counterclockwise*). But the cam, by definition, rotates with angular speed $\dot{\theta}$ *clockwise*. Hence

$$\dot{\lambda} = -\dot{\theta} \quad \text{or} \quad \frac{d\lambda}{d\theta} = -1 \qquad (14)$$

and Eq. (13) may be expressed as

$$r \sin \lambda = f'(\theta) \qquad (15)$$

Differentiating Eq. (15), we find

$$r(\cos \lambda)\lambda' = -r \cos \lambda = f''(\theta) \qquad (16)$$

Substituting this last result into Eq. (11), we find the desired *radius of curvature*:

$$p = R + f''(\theta) = R_0 + f(\theta) + f''(\theta) \qquad (17)$$

To avoid high stresses, the radius of curvature must not be too small. Of course, the limiting case of vanishing p corresponds to sharp points (cusps) on the cam, where infinite stresses occur. To avoid cusps or concavities, it is necessary that $p > 0$.

The expression (17) can help us to decide upon the desired radius R_0 of the base circle. For example, if, on the basis of a contact stress analysis, we know that p must exceed a critical value p_{crit}, it follows from Eq. (17) that

$$R_0 + f(\theta) + f''(\theta) \geq p_{\text{crit}} \qquad (18)$$

Since this inequality must hold for all values of θ, it follows that

$$R_0 \geq p_{\text{crit}} - [f(\theta) + f''(\theta)]_{\min} \qquad (19)$$

where the minimum value of $f(\theta) + f''(\theta)$ can be found for any given displacement function $f(\theta)$.

Finally, the *coordinates of A* can be found (if desired) from

$$r = [(\rho - R)^2 + l^2]^{1/2} \tag{20}$$

$$\lambda = \arctan_2 (l, R - \rho) \tag{21}$$

4.42   Translating Offset Roller Follower

In Fig. 1, the center P (trace point) of the roller, of radius r_r, is offset a known distance e from the center O of cam rotation. Point B is the projection of P on the line through O parallel to the direction of follower translation. Initially, point C_o on the cam's base circle was in contact with the roller, point P_o on the cam's prime circle was at the initial trace point P, and point B_o on the cam was on the fixed reference line OB. The cam is shown in Fig. 1 after it has rotated *clockwise* through the angle θ from its initial position. The common center of curvature for point C on the cam profile and point P on the pitch curve is denoted by A. The pressure angle is denoted by α, measured positive anticlockwise from ray AP.

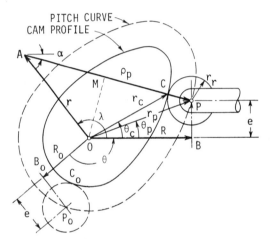

Figure 4.42-1. Translating roller follower.

We are given the displacement function

$$R = R_o + f(\theta) \tag{1}$$

where R_o, the minimum value of R, can be expressed in terms of the prime circle radius, $R_{po} \equiv OP_o$, by

$$OB_o \equiv R_o = [R_{po}^2 - e^2]^{1/2} \tag{2}$$

We seek to find the following quantities, for arbitrary values of θ: polar coordinates $(r_P, \theta + \theta_P)$ of the pitch curve, pressure angle α, radii of curvature ρ_P (for pitch curve) and ρ_C (for cam profile), polar coordinates $(r_C, \theta + \theta_C)$ of the cam profile, and polar coordinates $(r, \theta + \lambda)$ of the center of curvature.

From Fig. 1 we see that $\theta_P = \arctan_2 (e, R)$, where R is given by Eq. (1). Hence the *parametric polar equations of the pitch curve* are given simply by

$$\phi_P \equiv \theta + \theta_P = \theta + \arctan_2 (e, R) \tag{3}$$

$$r_P = (R^2 + e^2)^{1/2} \tag{4}$$

To find the pressure angle α, we observe that V_C' the velocity component parallel to AC, of point C, must equal the velocity V_M of point M; i.e.

$$V_C' = V_M = OM\,\omega \tag{5}$$

where OM is the distance from O to line AC, and $\omega = d\theta/dt$. It may be seen from Fig. 1 that

$$OM = OB \sin \alpha + BP \cos \alpha = R \sin \alpha + e \cos \alpha \tag{6}$$

Since the velocity component of P, in the direction CP, must equal V_C, and P is constrained to move parallel to OB, with resultant velocity $V_P = \dot{R}$, it follows that

$$V_P \cos \alpha = \dot{R} \cos \alpha = V_C' \tag{7}$$

Combining this result with Eqs. (5) and (6), we find

$$\frac{dR}{dt} \cos \alpha = (R \sin \alpha + e \cos \alpha) \frac{d\theta}{dt} \tag{8}$$

Upon making use of Eq. (1), we can solve Eq. (8) for

$$\tan \alpha = \frac{1}{R}\left(\frac{dR}{d\theta} - e\right) \quad \text{or} \quad \alpha = \tan^{-1}\left[\frac{f'(\theta) - e}{R_0 + f(\theta)}\right] \tag{9}$$

which gives the *pressure angle* within the range[8]

$$-\frac{\pi}{2} \le \alpha \le \frac{\pi}{2} \tag{10}$$

With the pressure angle known, we can express the orthogonal projections of $\overrightarrow{CP}$ (parallel to $\overrightarrow{OB}$ and to $\overrightarrow{BP}$, respectively) as

$$CP \cos \alpha = r_r \cos \alpha, \qquad -CP \sin \alpha = -r_r \sin \alpha \tag{11}$$

Hence, the *polar coordinates of the cam profile* are given by

$$r_C = [(R - r_r \cos \alpha)^2 + (e + r_r \sin \alpha)^2]^{1/2} \tag{12}$$

$$\phi_C \equiv \theta + \theta_C = \theta + \arctan_2 [(e + r_r \sin \alpha), (R - r_r \cos \alpha)] \tag{13}$$

To find ρ_P by Raven's method, we note from the loop $OAPBO$ in Fig. 1 that

$$\rho_P \cos \alpha = R - r \cos \lambda \tag{14}$$

$$\rho_P \sin \alpha = r \sin \lambda - e \tag{15}$$

---

[8]We may accept inequality (10) as part of the definition of pressure angle, which has hitherto been defined within an addend of $\pm\pi$.

To solve for ρ_P we shall need an independent expression for $r \cos \lambda$. This is found by differentiating Eq. (15) with respect to θ. When performing the differentiation, we may treat ρ_P and r as constants and use the relation $d\lambda/d\theta = -1$, for the reasons given in Sec. **4.41**. Thus we find

$$\rho_P(\cos \alpha)\alpha' = r \cos \lambda(-1) \tag{16}$$

Upon substituting $r \cos \lambda$ into Eq. (14), we find

$$\rho_P = \frac{R \sec \alpha}{1 - \alpha'} \tag{17}$$

The only unknown in Eq. (17) is α', which may be found by differentiating Eq. (9); i.e.,

$$\alpha' = \frac{f'' - f' \tan \alpha}{R \sec^2 \alpha} \tag{18}$$

Since α is given by Eq. (9), Eqs. (17) and (18) uniquely define ρ_P.
By using the identity

$$\sec^2 \alpha = 1 + \tan^2 \alpha = 1 + \frac{(f' - e)^2}{R^2} \tag{19}$$

we may combine Eqs. (17) and (18) to yield the compact expression

$$\frac{\rho_P}{R} = \frac{(1 + \tan^2 \alpha)^{3/2}}{1 + \tan \alpha[(e/R) + 2 \tan \alpha] - (f''/R)} \tag{20}$$

which simplifies to the following expression for the pitch curve's *radius of curvature*

$$\rho_P = \frac{[R^2 + (e - f')^2]^{3/2}}{R^2 + (e - f')[2(e - f') - e] - Rf''} \tag{21}$$

After ρ_P has been calculated, the coordinates of the center of curvature A may be found from Eqs. (14) and (15) in the form

$$r = [(\rho_P \cos \alpha - R)^2 + (e + \rho_P \sin \alpha)^2]^{1/2} \tag{22}$$

$$\lambda = \arctan_2 [(e + \rho_P \sin \alpha), (R - \rho_P \cos \alpha)] \tag{23}$$

Finally, the *radius of curvature of the cam profile* is given by Eq. (**4.34**-1) as

$$\rho_C \equiv AC = \rho_P - r_r \tag{24}$$

4.43 *Pivoted Flat-Faced Follower*

We wish to find the form of a disc cam, such as that in Fig. 1, which will produce the given follower displacement function:

$$\psi = \psi_0 + f(\theta) \tag{1}$$

The distances a, b, and e are all given.

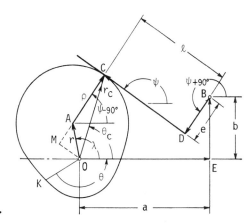

Figure 4.43-1.

As usual, A denotes the center of curvature of the cam for the contact point C. To find the distance l, defined by DC on the follower, we note from Fig. 1 that all points on line AC (e.g., M, A, C) have the same velocity component V_n in the direction AC, i.e.,

$$V_n = \dot{\theta}\, OM \tag{2}$$

where M is the projection of fixed point O on CA extended. This velocity is transmitted to the follower at point C, so that

$$V_n = -l\dot{\psi} \tag{3}$$

Hence

$$l\dot{\psi} = -OM\,\dot{\theta} \quad \text{or} \quad l\frac{d\psi}{d\theta} = -OM \tag{4}$$

By projecting the segments of length l, b, and a on the line OM it may be shown that

$$OM = l + b\sin\psi + a\cos\psi \tag{5}$$

Upon substituting this value into Eq. (4) we find

$$l = -\frac{b\sin\psi + a\cos\psi}{1 + \psi'} \tag{6}$$

where, as usual, we let

$$(\ )' \equiv \frac{d(\ )}{d\theta} \tag{7}$$

To find the Cartesian coordinates (x, y) of point C along the horizontal and vertical axes of the figure, we project the segments of length e and l onto these axes and find

$$x = a - e\sin\psi + l\cos\psi \tag{8a}$$
$$y = b + e\cos\psi + l\sin\psi \tag{8b}$$

Thus, the radius $r_c \equiv OC$ is given by

$$r_c = (x^2 + y^2)^{1/2} \tag{9}$$

and the angle EOC is

$$\theta_c = \arctan_2 (y, x) \tag{10a}$$

The angle ϕ between the fixed reference line (OK) on the cam, from which θ is measured, and the cam radius OC is

$$\phi \equiv \theta + \theta_c \tag{10b}$$

At this stage, we could find the radius of curvature by use of Eq. (**4.41**-10). Alternatively, we can use the following version of the method of Raven [1959]: We first express the fact that the loop $OACDBEO$ is closed, i.e., that the sum of the projections of $\overrightarrow{OA}$, $\overrightarrow{AC}$, etc., on the x and y axes must vanish. This gives the two equations

$$a - e \sin \psi + l \cos \psi - \rho \sin \psi - r \cos \lambda = 0 \tag{11}$$
$$b + e \cos \psi + l \sin \psi + \rho \cos \psi - r \sin \lambda = 0 \tag{12}$$

We now differentiate Eq. (12) with respect to θ (treating r and ρ as constants and setting $d\lambda/d\theta = -1$, as explained in Sec. **4.41**), thereby finding

$$-e\psi' \sin \psi + l' \sin \psi + l\psi' \cos \psi - \rho\psi' \sin \psi = -r \cos \lambda \tag{13}$$

To express $r \cos \lambda$ in terms of ψ and its derivatives we must evaluate l' by differentiating Eq. (6); i.e.,

$$l' = \frac{\psi'(a \sin \psi - b \cos \psi) - l\psi''}{1 + \psi'} \tag{14}$$

To find ρ, we calculate l from Eq. (6), l' from Eq. (14), $r \cos \lambda$ from Eq. (13), and ρ from Eq. (11). If the indicated algebraic operations are carried out (details omitted), one finds the desired *radius of curvature*

$$\rho = \frac{(a \sin \psi - b \cos \psi)(1 + 2\psi') - l\psi''}{(1 + \psi')^2} - e \tag{15}$$

where $\psi' = f'(\theta)$ and $\psi'' = f''(\theta)$ are known from Eq. (1).

With ρ known, one may find $r \cos \lambda$ and $r \sin \lambda$, if desired, from Eqs. (11) and (12) and then find

$$r = [(r \cos \lambda)^2 + (r \sin \lambda)^2]^{1/2} \tag{16}$$
$$\lambda = \arctan_2 [r \sin \lambda, r \cos \lambda] \tag{17}$$

4.44   Oscillating Roller Follower

In Fig. 1, O is the cam pivot, B is the follower pivot, and θ is the rotation angle of reference line OK, on the cam, measured clockwise from the x-axis. The follower angle ψ is specified by the given relation

$$\psi = \psi_0 + f(\theta) \tag{1}$$

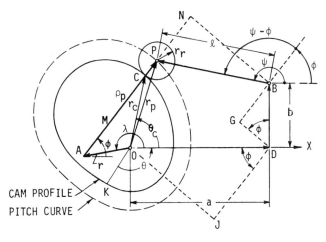

Figure 4.44-1. Oscillating roller follower.

where ψ_0 is the value of ψ when reference point K on the cam lies on the x axis. The distances a, b, l and r_r are all given.

The radius vector r_P of the pitch curve is seen to have orthogonal components

$$x_P = a + l\cos\psi$$
$$y_P = b + l\sin\psi \tag{2}$$

in the direction of x and perpendicular thereto. Thus, the polar *coordinates* (r_P, ϕ_P) of the trace point P are clearly

$$r_P = (x_P^2 + y_P^2)^{1/2}$$
$$\phi_P \equiv \theta_P + \theta = \arctan_2(y_P, x_P) + \theta \tag{3}$$

To find the angle ϕ, we note that all points of the cam (e.g., M and C) on the line AC have the same velocity component V_C^n along the direction of $\overrightarrow{AC}$, and

$$V_C^n = OM\,\dot{\theta} \tag{4}$$

where M is the projection of O on AC, A is the center of curvature of the pitch curve at trace point P, and C is the point of contact between cam and roller. Similarly, the velocity component of C in the direction of $\overrightarrow{CP}$ is given by

$$V_C^n = -BN\,\dot{\psi} \tag{5}$$

where N is the projection of pivot B on line $\overrightarrow{CP}$ extended.

From Eqs. (4) and (5) it follows that

$$OM = -BN\frac{d\psi}{d\theta} \equiv -BN\psi' \tag{6}$$

From the figure, we see that

$$BN = l\sin(\psi - \phi) = l(\sin\psi\cos\phi - \cos\psi\sin\phi) \tag{7}$$

and

$$OM = JM - JO = (DG + BN) - JO$$
$$OM = b \cos \phi + BN - a \sin \phi \tag{8}$$

Upon substitution of Eqs. (7) and (8) into Eq. (6), one finds

$$\tan \phi = \frac{D}{C} \tag{9}$$

where

$$D \equiv b + l(1 + \psi') \sin \psi$$
$$C \equiv a + l(1 + \psi') \cos \psi \tag{10}$$

With ϕ known we can find the coordinates of C (*cam profile*) in the *Cartesian form*,

$$x_C = a + l \cos \psi - r_r \cos \phi$$
$$y_C = b + l \sin \psi - r_r \sin \phi \tag{11}$$

or in the *polar form*,

$$r_C = (x_C^2 + y_C^2)^{1/2}$$
$$\phi_C \equiv \theta + \theta_C = \theta + \arctan_2 (y_C, x_C) \tag{12}$$

To find the radius of curvature of the pitch curve, we note, from the projections of loop $OACPBDO$ on and perpendicular to the x axis, that

$$r \cos \lambda + \rho_P \cos \phi - l \cos \psi - a = 0 \tag{13}$$
$$r \sin \lambda + \rho_P \sin \phi - l \sin \psi - b = 0 \tag{14}$$

Equation (13) will give us $\rho \cos \phi$ if we can find $r \cos \lambda$. The latter term can be found, by differentiating Eq. (14), to be

$$r \cos \lambda = \rho_P \phi' \cos \phi - l\psi' \cos \psi \tag{15}$$

where we have utilized the fact that $\lambda' = -1$ and have treated ρ_P and r as constants during the differentiation, for the reasons given in connection with Eqs. (**4.41**-13) and (**4.41**-14).

By a similar process of differentiation, we find from Eq. (9) that

$$\phi' = \frac{CD' - C'D}{C^2 + D^2} \tag{16}$$

where use has been made of the identity

$$\sec^2 \phi = 1 + \tan^2 \phi = \frac{C^2 + D^2}{C^2} \tag{17}$$

Upon substitution of Eqs. (14) and (15) into Eq. (13), one finds

$$\rho_P = \frac{l(1 + \psi') \cos \psi + a}{(1 + \phi') \cos \phi} \tag{18}$$

where all terms on the right are now known.

If desired, one may utilize Eqs. (16) and (17) to express Eq. (18) in the form

$$\rho_P = \frac{(C^2 + D^2)^{3/2}}{C^2 + D^2 + CD' - C'D} \tag{19}$$

or (after further algebraic manipulations) in the form

$$\rho_P = \frac{(C^2 + D^2)^{3/2}}{E} \tag{20}$$

where

$$E \equiv (C^2 + D^2)(1 + \psi') - (aC + bD)\psi'$$
$$+ (a \sin \psi - b \cos \psi) l \psi''(\theta) \tag{21}$$

With ρ_P known, we may use Eq. (**4.34**-1) to find the *radius of curvature of the cam profile* ρ_C; i.e.,

$$\rho_C = \rho_P - r_r \tag{22}$$

At this point, Eqs. (13) and (14) may be solved for r and λ, if desired.

4.50 PROBLEMS

Note: The following notation is used in Problems 10–44 of this section. The types of follower displacement to be used will be designated by the following symbols:

S: simple harmonic, Eqs. (**4.20**-6)

C: cycloidal, Eqs. (**4.20**-8)

P: balanced parabolic, Eqs. (**4.20**-12)

P_r: unbalanced parabolic, Eqs. (**4.20**-15), with acceleration ratio $r = A_2/A_1$

D: dwell

A portion of the displacement diagram with total rise h during cam rotation angle β will be denoted by $T(\beta, h)$, where T represents any of the five motion *types* (S, C, P, D, or P_r). Unless indicated otherwise, β will be given in degrees and h in inches. For example, the motion specified in Example 3 of Sec. **4.20** is represented by $P(90, 2)$, $D(30, 0)$, $S(60, -0.5)$, $D(90, 0)$, $P(90, -1.5)$.

The following motion schedules will be referred to in the problems below:
(a) $P(150, 1)$, $D(60, 0)$, $P(90, -1)$, $D(60, 0)$
(b) $P(150, 1)$, $D(60, 0)$, $S(90, -1)$, $D(60, 0)$
(c) $S(150, 1)$, $D(60, 0)$, $C(90, -1)$, $D(60, 0)$
(d) $C(90, 1.5)$, $C(90, -0.5)$, $D(90, 0)$, $C(90, -1)$

(e) $S(90, 1.5)$, $C(90, -0.5)$, $D(90, 0)$, $P(90, -1)$

(f) $S(120, 1)$, $D(30, 0)$, $S(180, -1)$, $D(30, 0)$

(g) $S(60, 0.75)$, $D(30, 0)$, $C(180, -0.75)$, $D(90, 0)$

(h) $C(150, 1.25)$, $D(30, 0)$, $C(180, -1.25)$

(i) $P(150, 1.5)$, $D(30, 0)$, $P(180, -1.5)$

(j) $P_{0.3}(150, 1.5)$, $D(30, 0)$, $P_3(180, -1.5)$

1. Define the following terms related to cams:
   a. disk cam
   b. plate cam
   c. trace point
   d. translating follower
   e. pivoted (oscillating) follower
   f. radial follower
   g. offset follower
   h. base circle
   i. prime circle
   j. pitch curve
   k. pressure angle
   l. dwell, rise, return
   m. jerk
   n. simple harmonic displacement
   o. cycloidal displacement
   p. parabolic displacement
   q. undercutting

2. a. Derive expressions, analogous to Eqs. (**4.20**-6), for the displacement, velocity, and acceleration of a follower which moves with *constant velocity* through a total lift h while the cam rotates through angle β.
   b. Find the maximum absolute values of nondimensional velocity (y^*) and acceleration (y^{**}) as defined by Eqs. (**4.20**-14).
   c. Describe the advantages and disadvantages of this displacement function.

3. Rework Example 1 of Sec. **4.20**, with the SHM displacement function replaced by a function that gives constant velocity during a period of rise or return.

4. What factors must be considered when deciding upon the base circle diameter and roller follower diameter?

5. a. Why is it desirable to have a small pressure angle when using a translating roller follower?
   b. What design changes can be made in order to increase the pressure angle?
   c. What can you say about the significance of the pressure angle when using a translating flat-faced follower?
   d. What can you say about the significance of the pressure angle when using a pivoted follower?

6. If a cam is rotating at a uniform angular velocity of 200 rpm and the follower is moving as in Example 1 of Sec. **4.20**, write expressions for the displacement, velocity, and acceleration of the follower as a function of time. Plot the diagrams of displacement, velocity, and acceleration versus displacement.

7. Repeat Problem 6 using Example 2 of Sec. **4.20**.

8. Repeat Problem 6 using Example 3 of Sec. **4.20**.

9. Repeat Problem 6 using Example 4 of Sec. **4.20**.

For Problems 10–14: Find the maximum magnitude of velocity and acceleration for the motion schedule indicated. Assume a constant cam speed of 300 rpm.

10. Schedule (a).

11. Schedule (b).

12. Schedule (h).

13. Schedule (i).

14. Schedule (j).

For Problems 15–29: Use a *graphical* method to determine the cam profile for the conditions specified. Use the minimum number of points needed to sketch in the major features of the cam. Use a radius of 1 in. wherever an initial choice of base radius or prime radius is required. If cusps occur, increase the size of the cam until the cusps disappear. From your drawing, estimate the minimum radius of curvature of the cam profile and the maximum pressure angle. The direction of cam rotation is indicated by CW (clockwise) or ACW (anticlockwise). All roller followers are to have diameters of 0.5 in.

15. Translating radial roller follower; schedule (c); CW.

16. Translating radial roller follower; schedule (d); ACW.

17. Translating radial roller follower; schedule (e); CW.

18. Translating roller follower with offset of 0.25 in. to left; schedule (c); CW.

19. Translating roller follower with offset of 0.25 in. to the left; schedule (d); ACW.

20. Translating roller follower with offset of 0.25 in. to left; schedule (e); CW.

21. Translating flat-faced follower; schedule (f); CW.

22. Translating flat-faced follower; schedule (g); ACW.

23. Translating flat-faced follower; schedule (h); CW.

24. Oscillating roller follower with dimensions (refer to Fig. **4.44**-1) $a = 4$ in., $b = 0.75$ in., $l = 3.5$ in., $r_r = 0.25$ in. Use motion schedule (f) with the indicated lift laid out along the circular arc described by the roller center. Cam rotation is CW.

25. Repeat Problem 24 with motion schedule (d).

26. Repeat Problem 24 with motion schedule (e).

27. Oscillating flat-faced follower with dimensions (refer to Fig. **4.43**-1) $a = 4$ in., $b = 1$, $e = 0.375$ in. Use motion schedule (f) with the indicated lift laid out along an arc of radius 3.5 in. centered at the follower pivot. Cam rotation is CW.

28. Repeat Problem 27 with motion schedule (h).

29. Repeat Problem 27 with motion schedule (c).

30–44. Repeat Problems 15–29 but use analytical rather than graphical methods. "Hand" calculations need to be made only for a few typical values of cam rotation angle θ. If a programmable calculator (or digital computer) is used, give results for increments in θ of 10° (or 1°) over a full revolution. For problem N ($30 \le N \le 44$), use the conditions specified in problem $N - 15$.

45. The displacement function of a flat-faced translating follower (see Fig. **4.41**-1) is given by $R = R_o + e \sin \theta$, where R_o and e are constants.
   a. Show that the cam profile is circular.
   b. What is the radius of the circular cam?
   c. How far from the center of the circular profile is the center of rotation located?

46. Use Eqs. (**4.41**-1)–(**4.41**-10) to find an expression for the curvature of a cam with a flat-faced translating follower if the displacement function is given by $R = R_o + e \sin \theta$, where R_o and e are constants (see Fig. **4.41**-1).

47. The polar coordinates (r, ϕ) of a cam profile are given in the parametric form $r = \theta$, $\phi = (1/a) \ln \theta$, where a is a constant and θ is the cam rotation angle. Use Eq. (**4.41**-10) to show that the radius of curvature of the cam is $\rho = r(1 + a^2)^{1/2} = \theta(1 + a^2)^{1/2}$. Verify this result by substituting the expression $r = e^{a\phi}$ into Eq. (**4.41**-8).

48. Derive Eq. (**4.43**-15) from Eqs. (**4.43**-6), (**4.43**-11), (**4.43**-13), and (**4.43**-14).

49. Verify that Eq. (**4.44**-20) follows from Eq. (**4.44**-19) and prior equations.

5

MOTION OF A LAMINA

5.10 FINITE MOTIONS IN A PLANE

A rigid body [Fig. 1(a)] can move in such a way that all its particles trace plane curves which are parallel to a fixed reference plane. The motion of the body is then fully defined by the motion of any cross section which is parallel to the reference plane. Alternatively, the motion is defined by a thin slice (called a **lamina**) of the three-dimensional body [Fig. 1(b)] which is parallel to the reference plane. The plane in which the lamina moves will be referred to henceforth as the **plane of motion**. The position of the lamina with respect to the fixed plane of motion is defined by three coordinates which may be chosen in a variety of ways. For example, we may locate a reference point Q fixed in the lamina and a reference line QM, where M is an arbitrary point

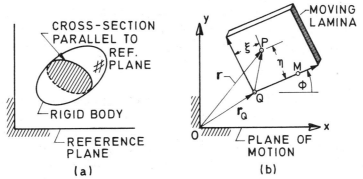

Figure 5.10-1. Plane motion of a rigid body: (a) Cross-section of rigid body; (b) Moving lamina.

which is fixed in the lamina. The Cartesian coordinates (x_Q, y_Q) of point Q (with respect to the fixed plane of motion) and the angle ϕ of the reference line constitute three degrees of freedom which uniquely define the configuration of the lamina, as shown in Fig. 1(b) (reflections about the reference line are prohibited). It is evident that the position of any point $P(\xi, \eta)$ in a lamina is fixed by the position of any single line segment (e.g., QM) which is fixed in the lamina.

5.11   Any Planar Motion of a Lamina Is Equivalent to a Pure Rotation

To prove this basic theorem it is only necessary to show that any given line AB in the lamina can be moved to an arbitrary position $A'B'$ by a pure rotation about a point C. To show this, consider Fig. 1 where the length AB in the original position equals the distance $A'B'$ in the final position, as is required for a *rigid line*. It will now

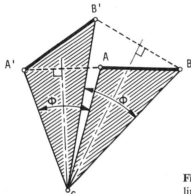

Figure 5.11-1. Finite motion of a rigid line.

be proved that the center C is located at the intersections of the midnormals (i.e., perpendicular bisectors) of AA' and BB'. The construction used ensures that $CA = CA'$ and $CB = CB'$. Because of the rigidity requirement, $AB = A'B'$; hence triangles ABC and $A'B'C'$ have three matching sides and are congruent. Thus we see that the segment AB can be superposed upon its desired final position $A'B'$ by a rigid body rotation about the point C through the angle $\phi = \angle ACA'$. Note that point C, which is called the **pole** of the rotation, moves to infinity when $A'B'$ is parallel to AB. In other words, *we may consider a translational motion to be a rotation about an infinitely remote pole*. In all subsequent discussions, we shall assume that the moving lamina is sufficiently large (infinite, if necessary) to include the center of rotation.

5.12   Linkages for Two and Three Positions of a Link

The theorem just proved may be used to advantage in some simple problems of linkage design. For example, suppose that we wish to move a link of a mechanism (the link is defined by its two end points at A and B) from an initial position A_1B_1 to a second specified position A_2B_2 as shown in Fig. 1. To do so, let us locate a point C_1 anywhere on the midnormal to A_1A_2 and another point C_2 anywhere on the midnormal to B_1B_2. With C_1 and C_2 as fixed pivots we may construct the four-bar mechanism $C_1A_1B_1C_2$ shown in Fig. 1, which will give the desired motion.

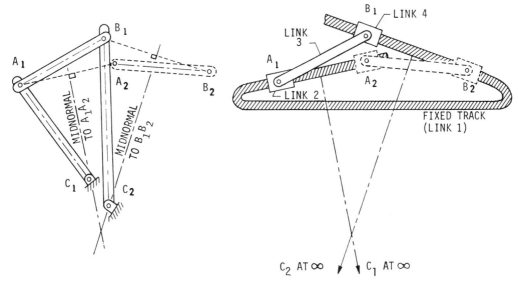

Figure 5.12-1. Mechanisms for two positions of a link: (a) Nonunique 4-bar mechanism; (b) Unique 4-link sliding mechanism.

The mechanism shown in Fig. 1(a) is not unique because C_1 and C_2 can lie anywhere on their respective midnormals. If, however, we agree to locate C_1 and C_2 infinitely far from A_1 and B_1 (but still on the correct midnormals), we shall in essence constrain point A_1 to move along a straight path from A_1 to A_2, and similarly for point B. This may be achieved mechanically by the four-link mechanism consisting of a link AB connected between two sliders which translate along tracks A_1A_2 and B_1B_2, respectively, which are part of a fixed frame, as shown in Fig. 1(b). This linkage is uniquely defined.

If we seek to move a bar through the *three* positions A_1B_1, A_2B_2, A_3B_3, as shown in Fig. 2, we can do so by means of the four-bar mechanism shown. In Fig. 2, the fixed

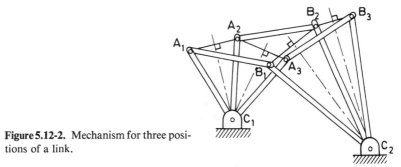

Figure 5.12-2. Mechanism for three positions of a link.

center C_1 is at the intersection of the midnormals to A_1A_2 and A_2A_3, and the fixed center C_2 is at the intersection of the midnormals to B_1B_2 and B_2B_3.

It is possible to design a four-bar mechanism which will move a link into *four* or *five* specified positions (see Freudenstein and Sandor [1961], [1964]). However, the details of the procedure would draw us deeply into the branch of mechanical design called **kinematic synthesis**. For further information on this large field of literature the reader may consult Sandor [1959], Hartenberg and Denavit [1964], Beyer [1963], Tao

[1964], Hall [1966], Hain [1967], Harrisberger [1961], Soni [1974], or Suh and Radcliffe [1978].

5.13 Continuous Motion of a Lamina (Centrodes)

Suppose that a reference line AB, which is scribed on a moving lamina, is known to occupy positions A_1B_1, A_2B_2, etc., at successive times t_1, t_2, etc., as shown in Fig. 1 (which may be thought of as a multiple-exposure photograph). From the known positions A_iB_i we may locate the successive poles $C_1, C_2, \cdots$ by use of the *midnormal intersection* construction described in Sec. **5.11**. The moving lamina rotates about the fixed points C_1, C_2, etc., through angles ϕ_1, ϕ_2, etc. The very same motion may be achieved by scribing a polygon with vertices C_1', C_2', etc., on the moving lamina at time t_1 and then rolling the polygon C_1', $C_2', \ldots$ on the fixed polygon $C_1, C_2, \ldots$. The rolling polygon C_1', C_2' may be formed by the construction shown in Fig. 2, where the angles α_1', α_2', etc., are given by

$$\alpha_i' = \alpha_i + \phi_i \tag{1}$$

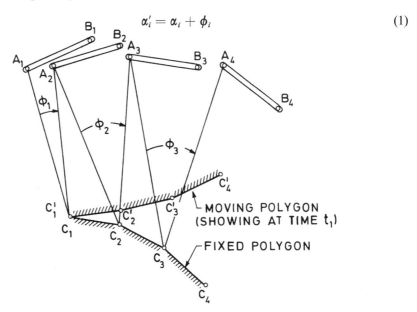

Figure 5.13-1. Successive rotations of a lamina equivalent to rolling of polygons.

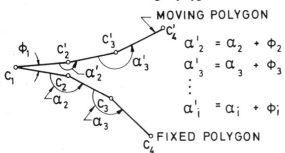

$$\alpha_2' = \alpha_2 + \phi_2$$
$$\alpha_3' = \alpha_3 + \phi_3$$
$$\vdots$$
$$\alpha_i' = \alpha_i + \phi_i$$

Figure 5.13-2. Showing how the moving polygon may be constructed from the fixed polygon.

and the side lengths are given by

$$C_1'C_2' = C_1C_2 \ldots \tag{2}$$

In short, we see that a lamina may be moved through any given sequence of positions by rolling a polygon which is attached to the lamina on a second polygon which is fixed in space. In the limit, as the distance between successive positions becomes infinitesimal, the two polygons become smooth curves, as shown in Fig. 3. The limiting curves formed by the moving polygon and fixed polygon are referred to as the **moving centrode** and **fixed centrode**, respectively.

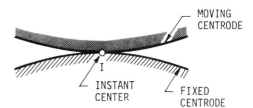

Figure 5.13-3. Plane motion of a lamina is equivalent to rolling of moving centrode on fixed centrode.

The idea of reducing all motion to the common status of rolling curves has an undeniable elegance, which prompted the following burst of exuberance from Reuleaux [1876, p. 84]:

> ... Just as the old philosopher compared the constant gradual alteration of things to a flowing, and condensed it into the sentence: "Everything flows"; so we may express the numberless motions in that wonderful production of the human brain which we call a machine in one word, "Everything rolls. ..."
> ... the machine becomes instinct with a life of its own through the rolling geometric forms everywhere connected with it. ... In the midst of the distracting noise of their material representatives they carry on their noiseless life-work of rolling. They are as it were the soul of the machine. ...

The point of tangency I of the two centrodes is called the **instantaneous center** or **instant center** (abbreviated as I.C.).[1]

We shall now consider some examples of constrained motions and the associated centrodes.

EXAMPLE 1. Locating the I.C. for Guided Laminas

The velocity $\mathbf{V}_A$ of a point A in a rigid lamina must be perpendicular to the ray AI from A to the I.C. Therefore if we know the direction of the velocity at two points, A and B, of a lamina, we can locate the I.C. at the intersection of the normals to $\mathbf{V}_A$ and $\mathbf{V}_B$ passing through points A and B, respectively. Several examples of this construction are given in Fig. 4.

In cases (a)–(c) of Fig. 4 the directions of motion of points A and B are specified by the constraints in an obvious manner. In Fig. 4(d), the velocity of point B (fixed on the

---

[1]Some authors refer to the I.C. as the **pole** and to the centrodes as **polodes**. Kennedy (Reuleaux [1876]) used the spelling "centroid," which of course has a different meaning today. The terms **polhode** and **herpolhode** are sometimes used for the moving and fixed centrodes.

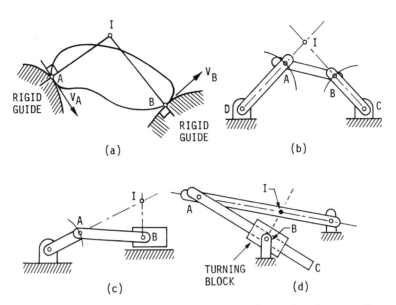

Figure 5.13-4. Location of I.C. at the intersection of normals to known paths: (a) Body sliding on two rigid guides; (b) Four-bar linkage; (c) Slider-crank mechanism; (d) Turning-block mechanism.

rod AC) must be along the line AC because the pin on the turning block prohibits motion of the block at B and thereby prohibits motion normal to AC of point B on the rod AC.

The locus of point I in Fig. 4 traces out the fixed centrode which may be graphically constructed pointwise. The moving centrode may then be found by the construction of Fig. 2.

EXAMPLE 2. Centrodes of Trammels

When the end A of a rod of length L slides along a line (the x axis) and the end B slides along an orthogonal line (the y axis), the instant center I must be located such that IA is perpendicular to the velocity of point A and IB is perpendicular to the velocity of point B. Thus I is located as shown in Fig. 5. It is left as a problem to show that the

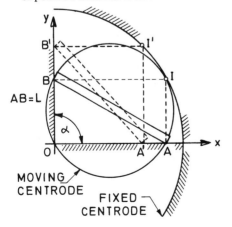

Figure 5.13-5. Centrodes of the elliptic trammel.

moving centrode is a circle of diameter L and the fixed centrode is a circle of diameter $2L$.

Even when the angle α between the two straight guides is not a right angle, the centrodes are circles (Problem **5.16**-10). The mechanism is known as the *trammel of Archimedes* or as an **elliptic trammel**, because any point attached to rod AB traces out an ellipse (see Problem **5.16**-13).

EXAMPLE 3. Elliptical Gears

Figure 6 shows a four-bar linkage in which the fixed bar DC and the opposite bar BA have the same length b, whereas the side bars, DA and CB, have equal length a, with $b < a$. Just as in Fig. 4(b), the I.C. of the link AB is at I, the intersection of the two side bars. Because of the symmetrical *butterfly* configuration, it may be proved that

$$IC = IA \tag{3}$$

Hence it follows that

$$DI + IC = DI + IA = DA = a \tag{4}$$

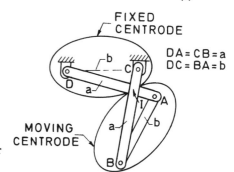

Figure 5.13-6. Elliptic centrodes for a four-bar mechanism.

Equation (4) states that the *sum of the distances from I to the fixed points D and C is constant.* Therefore we see that the locus of I is an ellipse with foci at D and C. In other words, the fixed centrode for member AB is an ellipse. Now we note that

$$BI + IA = BI + IC = BC = a \tag{5}$$

That is, the sum of the distances from I to points B and A is constant. Therefore, to an observer fixed on AB, the locus of I will be an ellipse with foci B and A. In other words, the moving centrode is an ellipse congruent with the fixed centrode.

Thus the motion of AB may be produced by the rolling of an ellipse over an identical ellipse. If teeth are cut into the ellipses in order to ensure lack of slippage at the contact point, we have the case of elliptical gears.

It may be verified (see Problem **5.16**-12) that Eqs. (4) and (5) are valid only if $b < a$. If $b > a$, the centrodes are hyperbolas rather than ellipses.

5.14 Equations for Centrodes

In many cases (e.g., Fig. **5.13**-6) the nature of the centrodes may be determined most conveniently from ad hoc geometric arguments. If, however, the geometric approach is not self-evident, it is desirable to use an analytic procedure, which we shall now develop.

Figure 1 shows a moving plane M with a reference point Q through which embedded coordinate axes (ξ, η) pass. An arbitrary point P fixed in M is defined by the complex constant $\zeta = \xi + i\eta$, and the position vector of P, referred to the fixed

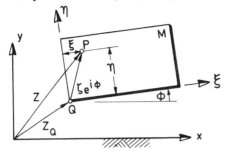

Figure 5.14-1. Point P is fixed to moving plane.

coordinates (x, y), is given by

$$Z = x + iy = Z_Q + \overrightarrow{QP} = Z_Q + \zeta e^{i\phi} \tag{1}$$

The velocity of P is therefore given by

$$V = \dot{Z}_Q + i\dot{\phi}\zeta e^{i\phi} \tag{2}$$

The I.C. is that point I for which $V = 0$, and from Eq. (2) we see that its position, $\zeta_I = \xi_I + i\eta_I$, is given by

$$\dot{Z}_Q + i\dot{\phi}\zeta_I e^{i\phi} = 0 \tag{3}$$

which may be solved for ζ_I in the form

$$\zeta_I = -\frac{\dot{Z}_Q}{i\dot{\phi}e^{i\phi}} = \frac{i(dZ_Q/dt)}{d\phi/dt}e^{-i\phi} = i\frac{dZ_Q}{d\phi}e^{-i\phi} \tag{4}$$

The rectangular components of ζ_I in the moving plane are

$$\xi_I = \operatorname{Re}\zeta_I$$
$$\eta_I = \operatorname{Im}\zeta_I \tag{5}$$

(Note: See Problems 5.16-16 and 18 for expressions in real terms.)

If the relationship between Z_Q and ϕ is known, $dZ_Q/d\phi$ may be found for any value of ϕ, and Eqs. (4) or (5) constitute the parametric equations for the path of I in M. That is, Eqs. (4) or (5) describe the moving centrode.

When Eq. (4) is substituted into Eq. (1) we find the position vector of I in the form

$$Z_I = Z_Q + \zeta_I e^{i\phi} = Z_Q + i\frac{dZ_Q}{d\phi} \tag{6}$$

whence

$$x_I = \operatorname{Re}Z_I$$
$$y_I = \operatorname{Im}Z_I \tag{6a}$$

(Note: See Problem 5.16-17 for expressions in real terms.)

Equations (6) or (6a) describe the path of I in the fixed plane; that is, they describe the fixed centrode.

EXAMPLE 1

The center Q of a disk falls vertically downward with constant acceleration a, and the angular velocity ω of the disk remains constant. Find expressions for and sketch the fixed and moving centrodes.

SOLUTION:

At time t, the center Q of the falling disk will be at a point on the y axis at a distance $at^2/2$ below the horizontal x axis, as shown in Fig. 2(a). Therefore, when the disk has

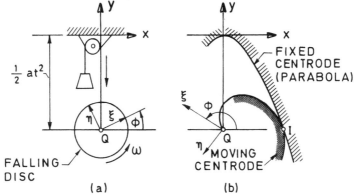

Figure 5.14-2. (a) Sketch of falling disk; (b) Moving centrode (spiral of Archimedes) rolls on fixed centrode (parabola).

rotated through the angle

$$\phi = \omega t \tag{7}$$

its center will be at point

$$Z_Q = -\frac{iat^2}{2} \tag{8}$$

Accordingly,

$$\frac{dZ_Q}{d\phi} = \frac{\dot{Z}_Q}{\dot{\phi}} = -\frac{iat}{\omega} \tag{9}$$

and the fixed centrode is given, via Eq. (6), by

$$Z_I = Z_Q + i\frac{dZ_Q}{d\phi} = -\frac{iat^2}{2} + \frac{at}{\omega} \tag{10}$$

Thus the parametric equations of the fixed centrode are given by

$$x = \operatorname{Re} Z_I = \frac{at}{\omega} \tag{11}$$

$$y = \operatorname{Im} Z_I = -\frac{at^2}{2} \tag{12}$$

or

$$y = -\frac{a}{2}\left(\frac{\omega x}{a}\right)^2 = \frac{-\omega^2 x^2}{2a} \tag{13}$$

which is clearly the equation of a *parabola*, as sketched in Fig. 2(b).

The moving centrode is given, by Eq. (4), in the form

$$\zeta_I = i\frac{dZ_Q}{d\phi}e^{-i\phi} = \frac{at}{\omega}e^{-i\omega t} = \frac{a}{\omega^2}\phi e^{-i\phi} \tag{14}$$

Equation (14) represents a spiral, winding clockwise about the center Q of the disk, with a radius vector whose length is proportional to its angle, as sketched in Fig. 2(b). Such a curve is called an **Archimedian spiral.**

In summary, the moving centrode is a spiral of Archimedes which rolls on a fixed parabolic centrode.

EXAMPLE 2. Epicyclics and Trochoids

Suppose that point C of a bar PC (of length h) is hinged to a bar CO (of length l) which is fixed at O, as shown in Fig. 3. This *dyad* has two degrees of freedom defined by

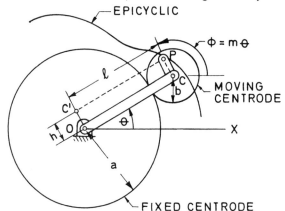

Figure 5.14-3. Centrodes of epicyclic (case of $m > 1$ shown).

the angles ϕ and θ which are measured from a fixed line OX. If the angle ϕ is a fixed multiple of θ given by

$$\phi = m\theta \tag{15}$$

the point P will trace out a curve known as an **epicyclic.**[2]

We wish to find the fixed and moving centrodes for the motions of link CP. Note that point C in Fig. 3 plays the role of Q in Fig. 1; hence we can write

$$Z_Q = Z_C = \overrightarrow{OC} = le^{i\theta} \tag{16}$$

$$\frac{dZ_Q}{d\phi} = \frac{dZ_Q}{d\theta}\frac{d\theta}{d\phi} = (ile^{i\theta})\left(\frac{1}{m}\right) \tag{17}$$

where use has been made of Eq. (15).

Therefore the moving centrode is given by Eq. (4) in the form

$$\zeta_I = i\left(\frac{dZ_Q}{d\phi}\right)e^{-i\phi} = \frac{-le^{i\theta}}{m}e^{-im\theta} = -\frac{l}{m}e^{i\theta(1-m)} \tag{18}$$

---

[2]According to Ptolemy (ca. 140 A.D.) the planets move in a circle (called the **epicycle**) about a center C, which itself moves in a circle (called the **deferent**) about a center O fixed in the stationary earth. The path of P is therefore an *epicyclic.* Ptolemy is to astronomy somewhat as Euclid is to geometry, i.e., the compiler, organizer, and preserver of the known body of knowledge in his science. He acknowledged that much of the material presented in his great work *The Almagest* is due to Hipparchus, who lived three centuries earlier. Hipparchus in turn drew upon the work of Appollonius (who lived a century before him) in introducing epicyclic motions into astronomy (Berry [1898]).

Since $e^{i\theta(1-m)}$ is a unit vector, *the moving centrode is a circle*, centered on C, of radius

$$b = \frac{l}{|m|} \tag{19}$$

The fixed centrode is given by Eq. (6) in the form

$$Z_I = Z_Q + i\frac{dZ_Q}{d\phi} = le^{i\theta} - \frac{le^{i\theta}}{m} = \frac{l}{m}(m-1)e^{i\theta} \tag{20}$$

which represents *a circle*, centered on O, of radius

$$a = \left|\frac{m-1}{m}\right| l = |m-1|\,b \tag{21}$$

Thus the epicyclic curve can equally well be described by a point P attached to a circle of radius b, which rolls over a fixed circle of radius a. Curves generated by rolling circles in this manner are called **trochoids**. In other words, *epicyclics are identical with trochoids*. When $m > 0$, as in Fig. 3, the epicyclic is said to be **direct**; when $m < 0$ the epicyclic is said to be **retrograde**.

Various types of trochoids associated with the two types of epicyclics are described in Sec. **5.21.** For now, we merely remark that the epicyclic traced by the *dyad OCP* could equally well have been generated by the dyad $OC'P$, shown in Fig. 3, where $OCPC'$ is a parallelogram. In other words, *the same epicyclic can be generated by two distinct linkages*. The centrodes of link $C'P$ are circles of radii a' and b', where

$$a' = |1-m|\,h \tag{22a}$$
$$b' = |m|\,h \tag{22b}$$

These circles determine the *same* trochoid through P as did the circles of radii a and b. This dual way of generating the same trochoid has been called the **double-generation theorem**.

5.15 Relative Rotation of Three Planes (*Kennedy-Aronhold Theorem*)

Suppose that a link 2 is hinged to another link 1 at a point I_{21} as shown in Fig. 1. If 1 is a fixed link, point I_{21} is the I.C. of link 2. But if link 1 is moving, point I_{21} is a point fixed in link 2 which is at rest *relative* to link 1 and is therefore called the **relative I.C.** of links 1 and 2. The relative I.C. of any two links (m and n) will be designated as I_{mn} or I_{nm}.

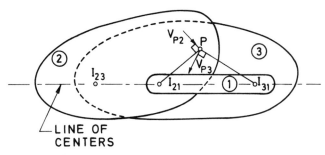

Figure 5.15-1. For proof of theorem of three centers.

Suppose that another link, 3, is hinged to link 1 at the point I_{31}. We have seen in Sec **5.13** that there must be a point on link 3 which is instantaneously at rest relative to link 2; that point is therefore the relative I.C. of links 2 and 3 and is denoted by I_{23} (or I_{32}). We now show that point I_{23} must lie on the **line of centers** $I_{21}I_{31}$.

Measuring velocities relative to link 1, we note that a point P of link 2 must have a velocity $\mathbf{V}_{P2}$ which is perpendicular to ray $I_{21}P$. The overlapping point on link 3 must have a velocity $\mathbf{V}_{P3}$ perpendicular to ray $I_{31}P$. The two velocities $\mathbf{V}_{P2}$ and $\mathbf{V}_{P3}$ will therefore not be in the same direction unless rays $I_{21}P$ and $I_{31}P$ are collinear; but these rays can be collinear only for those points (P^*) which lie on the line of centers $I_{21}I_{31}$. Therefore all points P^* for which $\mathbf{V}_{P2}$ and $\mathbf{V}_{P3}$ are parallel lie on this line of centers. Since I_{23} is defined as the point where $\mathbf{V}_{P2} = \mathbf{V}_{P3}$, it belongs to the subset of points P^*. This completes the proof that I_{23}, I_{31}, and I_{21} are collinear.

The above demonstration may be summed up in the **theorem of three centers**: *The relative instant centers associated with three moving planes lie on a straight line.* The theorem was discovered independently by Kennedy in 1886 and by Aronhold in 1872. As an example, consider the four-bar chain of Fig. 2. The hinges are the instantaneous

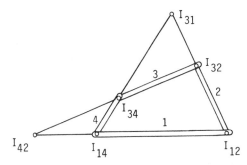

Figure 5.15-2. Showing all relative instant centers for a four-bar chain.

centers of neighboring links. The relative I.C. for two opposite links, such as 3 and 1, can be found from the theorem of three centers. Applying the theorem to links 3, 4, and 1, we see that I_{31} must lie along the line joining I_{34} and I_{14}. Applying the theorem to links 3, 2, and 1, we see that I_{31} must lie along the line joining I_{32} and I_{12}. The desired center therefore lies at the intersection of two known lines. Similar reasoning locates point I_{42}.

Location of all[3] the I.C.'s for more complicated kinematic chains requires the use of special techniques. If one plans to use relative I.C.'s extensively for graphical kinematic analysis, further discussion is needed. However, we shall develop alternative methods of analysis which do not require such a discussion.

Corollaries on Sliding and Rolling Contact

Five important corollaries of the *theorem of three centers* will now be derived. In every case we shall consider two bodies (or cams) which touch at a point C and are pinned to fixed centers I_{13} and I_{23} as shown in Fig. 3. The line joining I_{13} and I_{23} is called the **line of centers**. Denote the common normal to the touching surfaces, at the point of contact, by Cn and

---

[3]The number of I.C.'s for n links equals the number of combinations of n objects taken two at a time; i.e., $n(n-1)/2$.

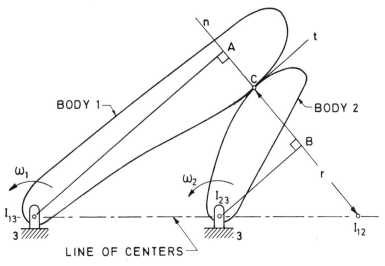

Figure 5.15-3. Bodies in contact.

the common tangent by Ct. Now the particle in body 1 at point C must slide along the tangent Ct, relative to body 2. Therefore, the relative I.C. of bodies 1 and 2 must be on the ray normal to Ct; i.e., I_{12} must lie on the normal Cn. But, according to the theorem of three centers, I_{12} must lie on the line of centers. We may therefore state the following:

Corollary 1: *The relative I.C. of two cams lies on the intersection of the line of centers and the common normal.*

If ω_1 and ω_2 represent the angular velocities of bodies 1 and 2, in Fig. 3, the angular velocity of link 1 relative to link 2 is $\omega_{1/2} \equiv \omega_1 - \omega_2$, and the velocity of the point C on body 1, relative to body 2 (velocity of sliding), is

$$v_s \equiv r(\omega_1 - \omega_2)$$

where r is the distance CI_{12}.

If the sliding velocity vanishes, the bodies are said to be in a state of **pure rolling contact**. A sufficient condition[4] for the sliding speed to vanish is that r vanishes. Hence we may state the following:

Corollary 2: *Cams are in pure rolling contact when the contact point lies on the line of centers.*

The point C on body 1 must have the same component of velocity v^n along the common normal as the point it touches in body 2. Otherwise, the bodies would interpenetrate or separate. Furthermore, all points such as A on the common normal must have the same component of velocity v^n along the normal.[5] If we choose A such that AI_{13} is normal to CA, it follows that A moves in the direction of An with speed

$$v_A = \omega_1 AI_{13} = v^n \tag{1}$$

Similarly, if B is chosen on the common normal such that $I_{23}B$ is normal to BC, B moves in

---

[4]If $r \neq 0$, v_s may vanish when $\omega_1 = \omega_2$. However, it is easily shown that $\omega_1 = \omega_2$ only when distance $AI_{13} = BI_{23}$. This in turn implies that I_{13} and I_{23} coincide and that both members revolve as a single rigid body about a common pivot.

[5]For any two points A, B on a rigid body the components of velocity along the line AB must be equal; otherwise the line AB would have to stretch or contract.

the direction of Bn with speed

$$v_B = \omega_2 BI_{23} = v^n \tag{2}$$

From the last two equations it follows that

$$\frac{\omega_1}{\omega_2} = \frac{BI_{23}}{AI_{13}} \tag{3}$$

However, from the two similar right triangles shown in Fig. 3, we see that

$$\frac{BI_{23}}{AI_{13}} = \frac{I_{23}I_{12}}{I_{13}I_{12}} \tag{4}$$

Therefore the angular velocity ratio may be expressed as

$$\frac{\omega_1}{\omega_2} = \frac{I_{23}I_{12}}{I_{13}I_{12}} \tag{5}$$

Equation (5) expresses the following:

Corollary 3: *The angular velocities of two cams vary inversely as the distances between their fixed pivots and the intercept of the common normal with the line of centers.*

Combining this result with Corollary 2, we find the following:

Corollary 4: *In pure rolling contact, the angular velocities of two bodies vary inversely as the radii from their fixed pivots to the contact point.*

Finally we note from Eq. (5) that the angular velocity ratio of two cams remains constant only if their relative instant center I_{12} remains at a fixed point. This may be expressed as follows:

Corollary 5: *The velocity ratio of two cams remains constant if the common normal cuts the line of centers at a fixed point.*

Corollary 5 is the **fundamental law of gearing,** which was derived by another method in Sec. **3.41**.

Willis [1870, p. 76] describes a clever application of Corollaries 2 and 4 to the design of a lever which is actuated by direct contact of a second lever such that there exists a prescribed velocity ratio with minimum wear on the contacting surfaces (see Fig. 4).

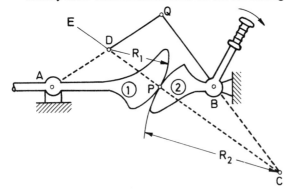

Figure 5.15-4. Design of a "constant" velocity ratio lever actuation system for minimum wear.

Let A and B be the fixed pivots and P the contact point, chosen such that the ratio BP/AP equals the desired speed ratio $[\omega_1/\omega_2]$. Select an arbitrary point Q such that $QP \perp AB$ and draw an arbitrary inclined line EP through P. Locate C at the intersection of QB and EP and locate D at the intersection of QA and EP. Form the required cam surfaces by striking arcs of radius CP and DP as shown.

By Corollaries 2 and 4, the cams will be in pure rolling motion, with the proper velocity ratio, for the position shown. However, the construction ensures that the dashed lines AD, DC, CB form an *equivalent* four-bar mechanism (see Fig. **1.54**-2). Furthermore, the I.C. of coupler bar DC is at point Q [see Fig. **5.13**-4(b)]. Therefore, point P will start to move $\perp$ to QP, i.e., along the line of centers. Thus, the construction guarantees that the criterion of pure rolling contact (Corollary 2) will be approximately satisfied for moderate departures from the initial state. Hence, the cams are always in an approximate state of pure rolling, and wear due to sliding is minimized.

Cams which undergo pure relative rolling are called *contour cams* by Molian [1968b], who discusses aspects of their design.

5.16 Problems

1. If Z_Q is a complex vector representing the known point Q in Fig. **5.10**-1, show that the lamina could be moved into the position shown by a rotation through an angle ϕ about a point C located at position $\zeta_C \equiv \xi_C + i\eta_C = Z_Q/(1 - e^{i\phi})$ in the moving lamina. Assume that initially the (ξ, η) axes coincide with the fixed (x, y) axes.

2. Suppose that the moving (ξ, η) axes of the lamina in Fig. **5.10**-1 initially coincided with fixed (x, y) axes. Show that the lamina could move into the arbitrary position depicted by means of a rotation through an angle ϕ about a pole C located at coordinates (ξ_C, η_C) in the moving lamina, where

$$\xi_C = \frac{1}{2}\left[x_Q - \frac{y_Q \sin\phi}{1 - \cos\phi}\right]$$

$$\eta_C = \frac{1}{2}\left[y_Q + \frac{x_Q \sin\phi}{1 - \cos\phi}\right]$$

and (x_Q, y_Q) represent the known final coordinates of point Q.

3. If (x_A, y_A) are the Cartesian coordinates of point A in Fig. **5.11**-1 and $(x_{A'}, y_{A'})$ are those of point A', show that the pole of the rotation shown is located at the Cartesian coordinates

$$x_C = \frac{1}{2}\left[(x_A + x_{A'}) - (y_{A'} - y_A)\cot\frac{\phi}{2}\right]$$

$$y_C = \frac{1}{2}\left[(y_A + y_{A'}) + (x_{A'} - x_A)\cot\frac{\phi}{2}\right]$$

4. A point Q on a moving lamina M moves to position Q', and the lamina rotates about Q' through the angle ϕ, as shown in Fig. 1, where W represents the magnitude of $\overrightarrow{QQ'}$ and ψ its angle. Using complex vectors, show that a point P fixed in the moving plane with constant coordinates r, θ moves to a new position P' where the vector displacement $\Delta Z = \overrightarrow{PP'} = We^{i\psi} + re^{i\theta}(e^{i\phi} - 1)$. Then show that there is a *fixed point* or pole C in the lamina which does not move at all. Show that the coordinates of the fixed point in the lamina are given by

$$r = r_C = W\Big/\left(2\sin\frac{\phi}{2}\right)$$

$$\theta = \theta_C = \frac{\pi}{2} - \frac{\phi}{2} + \psi$$

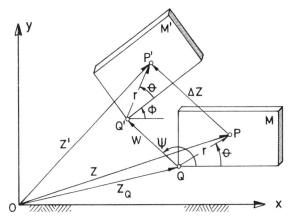

Figure 5.16-1 (Prob. 4)

5. A map which was lying flat on a table slides randomly over the table to another position.
    a. Show that there is one point on the map which occupies the same position on the table before and after the map was moved. Assume that the map is sufficiently large so that the point in question remains on the map.
    b. If the origin of a Cartesian coordinate system on the map is translated through the vector displacement (u, v) and the map is rotated through an angle ϕ, find the location (in x, y coordinates) of that point in the map which remains fixed on the table.

6. A *half-scale* photograph of a drawing is randomly placed (right side up) over a full-scale drawing on a flat table.
    a. Show that one point on the original drawing occupies the same position on the table as its image on the photograph.
    b. If the full-scale drawing has coordinate axes x, y and the half-scale drawing has coordinate axes ξ, η, find the (x, y) coordinates of the common point if the origin of the ξ, η system is shifted through displacements (u, v) and then the reduced drawing is rotated through an angle ϕ about its displaced origin.

7. Show directly from the geometry of Fig. **5.11**-1 that the center of rotation C is given by the complex vector (measured from an arbitrary origin)

$$Z_C = Z_A + \frac{Z_{A'} - Z_A}{2}\left(i + \cot\frac{\phi}{2}\right) \equiv Z_A + \frac{Z_{A'} - Z_A}{1 - e^{i\phi}}$$

8. A rigid rod AB moves into a new coplanar position $A'B'$ (Fig. 2). Let us denote by A, A',

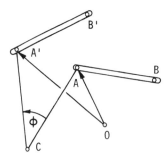

Figure 5.16-2 (Prob. 8).

B, B' the complex vectors $\overrightarrow{OA}, \overrightarrow{OA'}, \overrightarrow{OB}, \overrightarrow{OB'}$, respectively, where O is an arbitrary origin in the plane of motion. Show that the specified motion can be accomplished by a rigid body rotation through an angle ϕ about a center C, where $C = (A'B - B'A)/(B - B' - A + A')$ and $e^{i\phi} = (A' - B')/(A - B)$.

9. Draw a portion of the fixed centrode for the body AB shown in Figs. **5.13**-4(a)–(d) by locating three positions of I in the neighborhood of that shown. Then use the construction of Fig. **5.13**-2 to draw a portion of the moving centrode. Use any convenient scale.

10. Show that the fixed and moving centrodes for the problem shown in Fig. **5.13**-5 are circles with a diameter ratio of $2:1$ for any value of angle α.

11. Based on the concept of an elliptic trammel (Fig. **5.13**-5), sketch out a design for a drafting instrument used to draw ellipses of any specified shape.

12. Figure 3 shows a four-bar mechanism with link DC fixed; $AD = BC$, and $AB = CD$. Show that the fixed and moving centrodes of the link AB are

$$\left.\begin{matrix} \text{hyperbolas} \\ \text{ellipses} \end{matrix}\right\} \quad \text{if} \quad \left\{\begin{matrix} CD > AD \\ CD < AD \end{matrix}\right\}$$

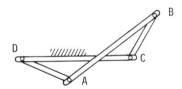

Figure 5.16-3 (Prob. 12).

13. The ends A and B of a rigid rod of length L slide along the x and y axes, respectively (Fig. 4). A point P is fixed on the rod at a distance μL measured from B to A. Show that

   a. Point P describes an ellipse; express the semidiameters of the ellipse in terms of μ and L.

   b. The sum of the semidiameters is independent of where P is located along the rod.

   c. The ellipse will be tangent to AB at P if $\mu = \sin^2 \theta$.

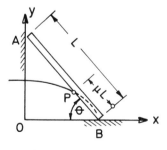

Figure 5.16-4 (Prob. 13).

14. Given an ellipse with major and minor semidiameters a and b, respectively, verify the following construction for the normal to the ellipse at an arbitrary point P (Fig. 5). Pass a line MN of length $a + b$ through P such that M lies on the major axis and N lies

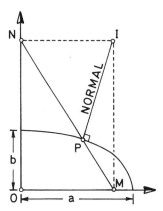

Figure 5.16-5 (Prob. 14).

on the minor axis of the ellipse. The normal is PI, where I is the fourth vertex of the rectangle $MONI$.

15. Show that there is a point J in the moving plane of Fig. **5.14**-1 for which the the acceleration vanishes and that the path of point J is described by the equation $\zeta_J = \ddot{Z}_Q e^{-i\phi} / (\dot{\phi}^2 - i\ddot{\phi})$.

16. If the velocity of point Q in Fig. **5.14**-1 is known to have components $v_{Q\xi}$ and $v_{Q\eta}$ along the axes of ξ and η, respectively, show that the moving centrode is described by the equations $\xi_I = -v_{Q\eta}/\dot{\phi}$ and $\eta_I = v_{Q\xi}/\dot{\phi}$. Suggestion: (a) Prove directly from geometry or (b) start from Eq. (**5.14**-5).

17. If the velocity of point Q in Fig. **5.14**-1 has components $\dot{x}_Q$ and $\dot{y}_Q$ along the axes of x and y, respectively, show that the fixed centrode is described by the equations $x = x_Q - \dot{y}_Q/\dot{\phi}$ and $y = y_Q + \dot{x}_Q/\dot{\phi}$. Suggestion: (a) Prove directly from geometry or (b) start from Eqs. (**5.14**-6a).

18. Show that the moving centrode of the lamina shown in Fig. **5.14**-1 is described by the equations

$$\xi_I = \frac{dx_Q}{d\phi} \sin \phi - \frac{dy_Q}{d\phi} \cos \phi$$

$$\eta_I = \frac{dx_Q}{d\phi} \cos \phi + \frac{dy_Q}{d\phi} \sin \phi$$

Suggestion: (a) Use results of Problem 16 or (b) start from Eq. (**5.14**-5).

19. The top R of a swinging rod is hinged to a tiny roller which moves along a horizontal track in the plane of motion (Fig. 6). Because no horizontal force acts on the rod, the mass center C of the rod must move in a vertical line. C is located a distance a from R along the rod. Describe
   a. The path of a point P on the rod at a distance L from R.
   b. The fixed and moving centrodes of the rod.

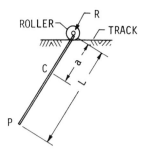

Figure 5.16-6 (Prob. 19).

20. The involute of a circle of radius a slides along the fixed orthogonal axes (x, y) (Fig. 7). Show that the moving centrode is a circle of radius $a\sqrt{2}$ and that the fixed centrode is a straight line of slope 1 and y-intercept $(2 - \pi/2)a$. (The involute of a circle is defined in Example 1 of Sec. **2.10**.)

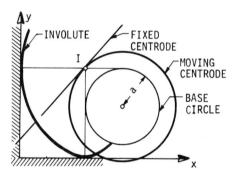

Figure 5.16-7 (Prob. 20).

21. A point Q fixed in a moving lamina describes the *equiangular spiral*

$$Z_Q = ae^{(\theta_1 \cot \alpha + i\theta_1)}$$

where a and α are constants and θ_1 is a variable. The moving lamina rotates about point Q with angular velocity $\dot{\theta}_1/n$. Show that both centrodes of the motion are *equiangular spirals* [i.e., curves described in polar coordinates (r, θ) by an equation of the form $r = ce^{k\theta}$] where c and k are constants.

22. In Fig. 8, the rod AB touches the fixed circle of radius a and center O while B slides

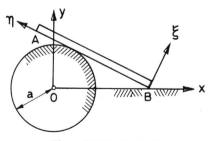

Figure 5.16-8 (Prob. 22).

along the line OB. Show that the fixed and moving centrodes are described by the two following equations: $x_I^2(x_I^2 - a^2) = a^2 y_I^2$; $\eta_I^2 = a\xi_I$.

23. Find equations for the fixed and moving centrodes of the connecting rod BC of the slider-crank mechanism shown in Fig. 9.

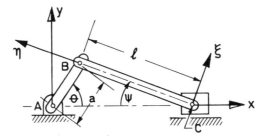

Figure 5.16-9 (Prob. 23).

24. If the ratio a/l is small enough so that we may use the approximations $\sin \psi \doteq \psi$ and $\cos \psi \doteq 1$ for the slider-crank mechanism of Problem 23, show that the equations of the fixed and moving centrodes of rod BC are given approximately by $(x_I - l)^2(x_I^2 + y_I^2)$ $= a^2 x_I^2$ and $l^2 \eta_I^2(l^2 + \xi_I^2) = a^2 \eta_I^4$.

25. If the weight W rises vertically, find the fixed and moving centrodes of pulleys A and B, which have radii a and b, respectively (Fig. 10).

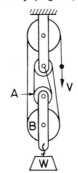

Figure 5.16-10 (Prob. 25).

26. Find the fixed and moving centrodes for the connecting rod BC of the slider-crank mechanism shown in Fig. 11 with $BC = AB$.

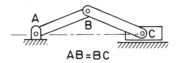

AB = BC

Figure 5.16-11 (Prob. 26).

27. The rod BD always passes through the fixed point C in the *swinging block* inversion of the slider-crank mechanism shown in Fig. 12. Find the fixed and moving centrodes for the rod BD.

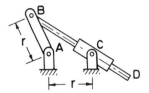

Figure 5.16-12 Prob. 27).

28. Figure 13 shows schematically how a front wheel is mounted to the frame of an auto-mobile chassis. We may treat the wheel and link *CD* as one rigid link. Automobile designers frequently assume that the tire is pinned to the ground at *E* and seek the center of rotation (*roll center*) for the motion of the chassis frame (car body) relative to the ground. If the frame tilts, show that the center of rotation sought is at *G* when the geometry of the system is symmetrical about the centerline *GH* of the frame.

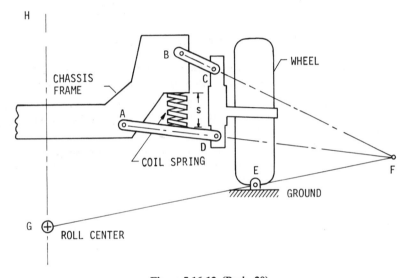

Figure 5.16-13 (Prob. 28).

29. Find the roll center in Problem 28 when the spring, on one side only, deflects to half of the length *s* shown in Fig. 13. Use dimensions equal to or proportional to those shown in the figure.

30. A *sliding rod suspension* for the front wheel of an automobile consists of a rod *BE* which is fixed to the axle *DF* and which slides in a block 5 that is hinged to the chassis frame at *C*, as shown in Fig. 14. Assuming that the tire (link 4) is pivoted at the ground contact point *G*, find the relative I.C. for rotation of the frame (link 2) with respect to the ground (link 1). The geometry is symmetric about the centerline of the frame.

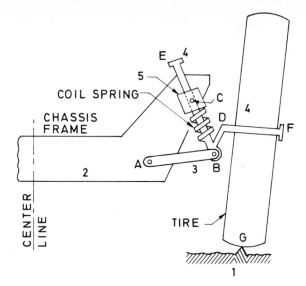

Figure 5.16-14 (Prob. 30).

31. A disc rotates with constant angular velocity, while its center moves along a straight line with constant acceleration. Show that the fixed centrode is a parabola.

32. A lamina rotates with constant angular velocity ω, while a given point of the lamina moves along a straight line with constant speed v. Find expressions for and sketch the fixed and moving centrodes.

33. One end of a rod slides along the y axis, while another point of the rod slides over a fixed point of the x axis at a distance b from the origin (Fig. 15). Find expressions for the fixed and moving centrodes. Sketch the centrodes.

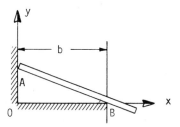

Figure 5.16-15 (Prob. 33).

34. Three bars, AB, BC, CD form a triple pendulum $ABCD$, where A is a fixed pivot. The absolute angular velocities of bars AB, BC, and CD are ω, 2ω, and 3ω, respectively. Show that the instantaneous center of bar CD is always at the centroid of triangle ABC. [Hint: If points A, B, C are located at Z_A, Z_B, Z_C, respectively, the centroid of triangle ABC is located at $\frac{1}{3}(Z_A + Z_B + Z_C)$.]

5.20 PROPERTIES AND APPLICATIONS OF ROLLING CIRCLES

As we have seen, the instantaneous velocity distribution of a moving lamina is governed by the rolling of one curve (moving centrode) on another (fixed centrode). Because the position, velocity, and acceleration are correctly related[6] if we replace the centrodes, at any instant, by their circles of curvature, we are motivated to study the behavior of rolling circles. Finite motions of rolling circles can be mechanized by friction wheels or gears.

Displacement Velocity

The point of contact I between rolling circles is the instantaneous center (I.C.) for the motion of one circle m (the *moving centrode*) relative to the other circle f (the *fixed centrode*). When I is considered as a point attached to either of the centrodes, it has zero velocity by definition, but the I.C. does move over the two centrodes. To visualize this motion, we may imagine that I is a fixed point on a *carrier* link OM which is hinged to the center O of the fixed circle and to the center M of the moving circle, as shown in Fig. 1, where ρ_f and ρ_m represent the radii of the fixed and moving circles.

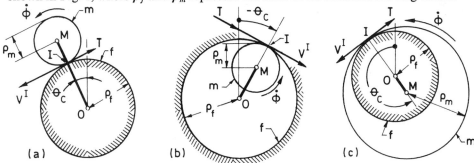

Figure 5.20-1. Moving circle m rolls over fixed circle f. Displacement velocity V^I shown for case of $\dot{\phi} > 0$.

The velocity of point I on the carrier link may be represented as a complex vector V^I which is tangent to the centrodes at I and *points in the counterclockwise direction around the fixed centrode*. V^I is called the **displacement velocity** or **pole velocity**. To find V^I, we note from Fig. 1(a) that for the case of external contact

$$V^I = -\rho_f \dot{\theta}_c T \qquad (1)$$

where $\dot{\theta}_c$ is the angular velocity (*positive counterclockwise*) of the carrier arm OM and T is a unit vector pointing *clockwise* along the fixed centrode at I. Equation (1) is also valid for the case of internal contact—whether m is inside f [Fig. 1(b)] or f is inside m [Fig. 1(c)]. Note that in the case of Fig. 1(b) V^I will point clockwise when $\dot{\theta}_c$ is negative (clockwise).

To express V^I in terms of the angular velocity $\dot{\phi}$ of the moving centrode, we utilize the fundamental equation of epicyclic gearing, Eq. (**3.32**-3), which, for the case of

---

[6]See Sec. **1.54**.

Fig. 1, can be expressed as

$$\frac{\dot{\phi} - \dot{\theta}_c}{0 - \dot{\theta}_c} = \mp \frac{\rho_f}{\rho_m} \quad \text{or} \quad \dot{\theta}_c = \frac{\rho_m \dot{\phi}}{\rho_m \pm \rho_f} \tag{2}$$

where the upper or lower sign is to be used accordingly as contact is external [Fig. 1(a)] or internal [Figs. 1(b) and (c)]. Combining Eqs. (1) and (2), we see that the displacement velocity is given by

$$V^I = -\frac{\rho_f \rho_m}{\rho_m \pm \rho_f} \dot{\phi} T = -D\dot{\phi} T \tag{3}$$

where

$$D \equiv \frac{\rho_f \rho_m}{\rho_m \pm \rho_f} \tag{4}$$

is a signed quantity [negative only for the case of Fig. 1(b)] which plays an important role in the study of path curvatures (Sec. **5.42**).

5.21 *Geometry of Rolling Circles*

Curves traced out by points of circles which roll upon other circles or straight lines have been studied by mathematicians, astronomers, and natural scientists, including Galileo, Descartes, Huygens, the Bernoullis, and Newton. These early scientists were concerned with properties of isochronous (equal period) pendula, tautochrones and brachistochrones (curves along which particles descend in equal times or in minimum times), and the proper forms for toothed wheels. The geometric properties of rolling circles were studied assiduously by all mathematical students of earlier generations.[7] Our main interest here is merely to record the terminology used for the various curves and to point out some of their features useful in the study of mechanisms. Any curve generated by rolling circles can be produced mechanically by means of epicyclic gearing.

Nomenclature: The path of a point attached to the plane of the rolling circle is called a **trochoid**. Trochoids are called **epitrochoids** if the rolling circle is outside of the fixed circle, **hypotrochoids** if inside, and **cycloids** if the fixed circle degenerates to a straight line. The trochoids are classified as **prolate, ordinary**, or **curtate** accordingly as the tracing point lies inside, on, or outside of the rolling circumference. Ordinary trochoids are also referred to as **epicycloids** or **hypocycloids**. This nomenclature is summarized in Table 1. Some writers, e.g., A. B. Kennedy (see his notes on Reuleaux[8]), use the name **peritrochoid** for those epitrochoids which arise when the rolling circle encloses the fixed circle.[9]

Figures 1, 2, and 3 show typical trochoids of the various types. Parametric equa-

---

[7]Maxwell [1849], Proctor [1878], Edwards [1892], Yates [1959].
[8]Reuleaux [1876, p. 592].
[9]The distinction between peritrochoids and other epitrochoids is blurred by the fact that any peritrochoid is identical with some epitrochoid as a consequence of the so-called **double-generation theorem** referred to in Example 2 of Sec. **5.14**.

Table 5.21-1 *Nomenclature for Trochoids*

		All Trochoids	
Location of Tracing Point	*Circle Rolls on Straight Line, Cycloids*	*Rolling Circle Outside Fixed Circle, Epitrochoids*	*Rolling Circle Inside Fixed Circle, Hypotrochoids*
On circle	Ordinary cycloid	Epicycloid	Hypocycloid
Outside circle	Curtate cycloid	Curtate epitrochoid	Curtate hypotrochoid
Inside circle	Prolate cycloid	Prolate epitrochoid	Prolate hypotrochoid

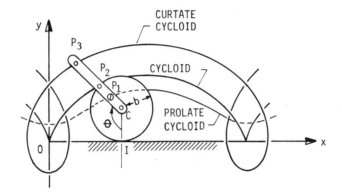

Figure 5.21-1. Cycloids.

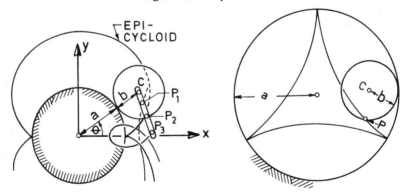

Figure 5.21-2. Epitrochoids.

Figure 5.21-3. Hypocycloid of three cusps $b = h = \frac{1}{3}a$

tions for the different classes of curves are as follows, where a is the radius of the fixed circle, b is the radius of the moving circle, C is its center, and $h = CP$ is the radius, on the moving circle, of the tracing point P:

$$\text{cycloids:} \qquad x = b\theta - h \sin \theta \qquad (1)$$

$$y = b - h \cos \theta \qquad (2)$$

$$epitrochoids: \quad x = (a + b) \cos \theta - h \cos \left(\frac{a + b}{b} \theta \right) \qquad (3)$$

$$y = (a + b) \sin \theta - h \sin \left(\frac{a + b}{b} \theta \right) \qquad (4)$$

$$hypotrochoids: \quad x = (a - b) \cos \theta + h \cos \left(\frac{a - b}{b} \theta \right) \qquad (5)$$

$$y = (a - b) \sin \theta - h \sin \left(\frac{a - b}{b} \theta \right) \qquad (6)$$

It may be observed from Eqs. (3) and (4), or from Fig. 2, that if the ratio $(b/a) \rightarrow \infty$ the rolling circle approaches a straight line, and the *epicycloid* traced by a point on the rolling line becomes identical to the *involute of a circle* as described in Fig. **2.10**-2.

The use of epi- and hypocycloids for the generation of gear teeth was discussed in Sec. **3.41**. Other applications of rolling circles will now be described.

5.22   *Mechanical Applications of Trochoids*[10]

a. Straight-Line Generators

The special case of a hypocyloid with $a = 2b$ is often referred to as the case of **Cardan circles**. Let the moving circle m (of center C) contact the fixed circle f at point P initially, as shown in Fig. 1. When the moving circle has moved to position m' (with center C'), the point P will have moved to a new position P', and the contact point will have moved to Q. Note that the diameter of m', which passes through Q, also passes through the center G of the fixed circle; therefore $QC'G$ is a straight line. By definition of pure rolling $\overset{\frown}{QP} = s = \overset{\frown}{QP'}$. Therefore in circle m',

$$\angle QGP' = \frac{1}{2} \angle QC'P' = \frac{1}{2} \frac{s}{b}$$

In circle f,

$$\angle QGP = \frac{\overset{\frown}{QP}}{2b} = \frac{s}{2b}$$

Thus we see that $\angle QGP = \angle QGP'$, or, in other words, P' is collinear with G and P. This proves that *P travels along a straight line which is a diameter of the fixed circle.* By symmetry, all hypocycloids of the moving circle trace out diameters of the fixed circle.

A straight-line mechanism (attributed[11] to both Lahire and White) can be based on this geometric principle, as shown in Fig. 2. The device, for which White was given a medal by Napoleon, found limited use in printing presses and steam engines. In modern times it has been used as an analog instrument to convert an angular input θ into a linear output proportional to $\sin \theta$.

---

[10]Additional applications of cycloids, e.g., to newspaper folding, paper carton folding, speed reducers, and straight-line generators, are described by Pollitt [1960] and Jensen [1966].

[11]Ferguson [1962]. The same device was utilized by Murray in 1805 to avoid an infringement of Watt's patents; see Clark [1963].

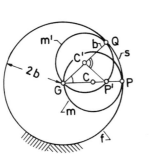

Figure 5.22-1. Straight line generation by Cardan circles.

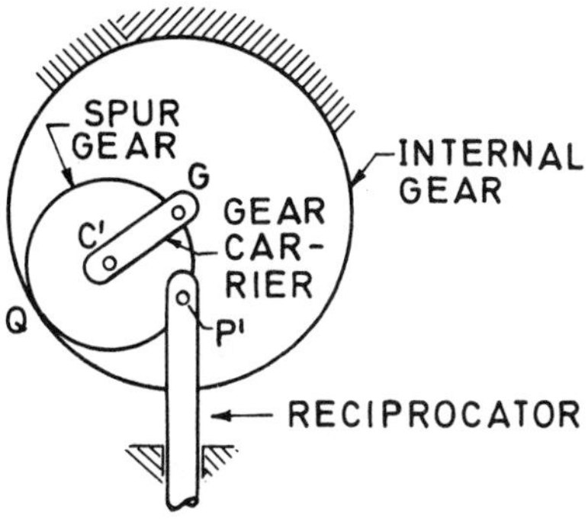

Figure 5.22-2. Lahire's (or White's) straight line mechanism.

It is easily shown (Problem **5.23**-11) that any point on the moving Cardan circle traces out an ellipse; thus the gear set shown in Fig. 2 can be used as the basis of an ellipsograph.

b. Properties of a Cycloid; the Isochronous Pendulum

The preeminent development in the history of accurate time measurement was the invention of the pendulum clock by Christian Huygens in 1656. The theory behind this breakthrough in the art of horology was described in Huygen's treatise *Horologium Oscillatorium*[12] in 1673. This ingenious work contains many original contributions to kinematics and dynamics, including (1) discovery of the theory of evolutes, (2) proof that the evolute of a cycloid is an identical cycloid, (3) the theory of the pendulum, and (4) a method for correcting the *circular error* in the period of the ordinary pendulum by the use of cycloidal cheeks. To better understand these matters it is necessary to derive a few properties of the cycloid P_0PC [Fig. 3(a)], traced out by a point P attached to a generating circle of radius b and center M as it rolls along a fixed line Q_0I.

The point Q diametrically opposite P on the generating circle traces out a congruent cycloid Q_0Q whose cusps lie midway between those of the given cycloid. After the generating circle has rolled through angle θ, the initial I.C. at Q_0 will have moved to I, the initial contact point Q_0 (on the moving circle) will have moved to Q such that $Q_0I = \widehat{QI} = b\theta$, and the coordinates of point P on the given cycloid will be given by

$$\xi = Q_0I + MP \sin \theta = b\theta + b \sin \theta \tag{1}$$

$$\eta = KM - MP \cos \theta = b - b \cos \theta \tag{2}$$

where point K is diametrically opposite I on the moving circle. Since P moves perpendicularly to the ray IP through the I.C., the path tangent at P is normal to IP and makes the angle $\psi = \theta/2$ with the ξ axis. As the generating circle rolls through an

---

[12]". . . a work of the highest genius which has influenced every science through its mastery of the principles of dynamics. It is second in scientific importance only to the *Principia*, which is in some respects based on it" (Singer [1959, p. 302]).

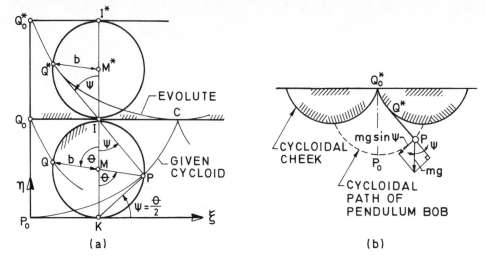

Figure 5.22-3. Properties of the cycloid: (a) Generation of cycloid and its evolute; (b) Huygen's cycloidal pendulum.

incremental angle $d\theta$, the point P describes an increment of arc ds given by

$$ds = IP \, d\theta = KI \cos \psi \, d\theta = 2b \cos \psi \, d(2\psi) \tag{3}$$

Equation (3) may be integrated to give the arc length $\widehat{P_0 P}$ in the form

$$s = 4b \sin \psi = 4b \sin \frac{\theta}{2} \tag{4}$$

An alternative useful expression for s follows from the observation that

$$\eta = KP \sin \psi = 2b \sin \psi \sin \psi \tag{5}$$

which, with the help of Eq. (4) can be expressed as

$$\eta = \frac{s^2}{8b} \tag{6}$$

From Eq. (4) or (6) it follows that the arc distance $\widehat{P_0 C}$ is $4b$ and that the arc length between cusps on a cycloid is $8b$.

From Eq. (4) we see that the radius of curvature ρ is given by

$$\rho \equiv \frac{ds}{d\psi} = 4b \cos \psi = 2IP \tag{7}$$

since $IP = 2b \cos \psi$. Further, we note that the center of curvature Q^* lies along the normal PI and must therefore be located at a distance $IQ^* = PI$ along the extension of ray PI.

We now construct a circle of radius b and center M^* which passes through Q^* and I and just touches the original generating circle at I. Since triangle IM^*Q^* is congruent to triangle IMP, it follows that $\overrightarrow{M^*Q^*}$ is equal and parallel to $\overrightarrow{PM}$ and hence also to $\overrightarrow{MQ}$. In short, point Q^* lies on the parallelogram QQ^*M^*M and always remains at a fixed distance $2b$ vertically above point Q. Hence, as point Q traces out the cycloid Q_0Q, point Q^* traces out the identical cycloid $Q_0^*Q^*C$, which is displaced vertically upward from the former cycloid by a distance $2b$.

The curve traced out by the center of curvature Q^* of the given curve is, by definition, the **evolute** of the given cycloid, and the given cycloid is the **involute** of the evolute.

Huygens reasoned that since the involute cycloid P_0PC could be traced by a taut string $Q_0^*Q^*P$, tied to point Q_0^*, and wrapped around the evolute arc $\overset{\frown}{Q_0^*Q^*}$, a mass m attached to a string Q_0^*P (of length $4b$) would describe the cycloidal path P_0P if the string were surrounded symmetrically by two rigid cycloidal cheeks as shown in Fig. 3(b). The advantage of this scheme over that of the ordinary pendulum (without cycloidal cheeks) is that the cycloidal pendulum will oscillate with a constant period τ independently of the amplitude of motion. To show this, we note from Fig. 3(b) that the component of gravity force mg acting along the path tangent direction is $mg \sin \psi$ in the direction of negative s, where s is arc length $\overset{\frown}{P_0P}$. Hence Newton's second law requires that

$$m\ddot{s} = -mg \sin \psi \tag{8}$$

However, $\sin \psi = s/4b$, according to Eq. (4); therefore Eq. (8) assumes the form

$$\ddot{s} + \left(\frac{g}{4b}\right)s = 0 \tag{9}$$

The solution to Eq. (9) is

$$s = s_m \sin \left(\sqrt{\frac{g}{4b}}\, t\right) \tag{10}$$

which represents an oscillation of period $\tau = 2\pi\sqrt{4b/g}$ *independently of the amplitude of motion* s_m; i.e., the pendulum is **isochronous**. It is readily verified that in the absence of the cycloidal cheeks the period of a pendulum depends on the amplitude of motion, although this dependence is not strong for small motions of the pendulum. As a matter of practice, cycloidal cheeks are not widely used in pendulum clocks. It was discovered by Bessel and others that the effect of the cheeks could be approximated by merely attaching the upper end of the pendulum to a small elastic leaf spring.[13]

c. Roots Blower

In Fig. **3.41**-3 it was shown how "describing" circles form cycloidal gear teeth as they roll over a fixed "pitch" circle. If the describing circles are one-fourth the size of the pitch circle, they will describe a two-lobed gear of the form shown in Fig. 4. Such a gear, when meshed with an identical gear and housed in a suitable enclosure, as shown, can be used to pump a fluid. The arrangement is called a **Roots**[14] **blower**.

Note that the common normal passes through the line of centers for the configuration shown in Fig. 4. This means that the gears cannot transmit torque in the position

---

[13]Sommerfeld [1964, p. 96].

[14]However, an Englih patent for such a device was granted in 1858, nine years prior to Roots's disclosure. See Reuleaux [1876, pp. 413, 609].

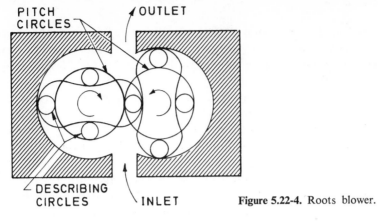

PITCH
CIRCLES

OUTLET

DESCRIBING
CIRCLES

INLET

Figure 5.22-4. Roots blower.

shown. This fact is of little consequence, as the two rotors are usually driven through an external set of spur gears, not shown.

A number of other ingeniously contrived trochoidal rotors may be utilized in a manner similar to Roots. Several of these are described by Reuleaux [1876] in his chapter on chamber-wheel trains.

d. Rotary Internal Combustion Engine

The **Wankel engine** consists of an annular gear which rotates around a fixed spur gear and carries with it a three-lobed rotor ABC, as shown in Fig. 5. The points A,

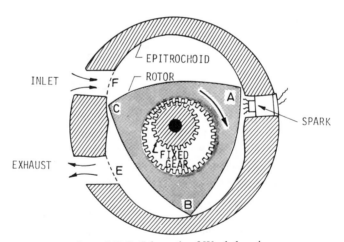

Figure 5.22-5. Schematic of Wankel engine.

B, C are all equidistant from the center of the rotor and trace out a peritrochoid. The peritrochoid is divided into three compartments by *apex* seals on the rotor at points A, B, C. In the phase shown, a spark has just ignited the fuel mixture in the compartment bounded by arc AB; the net gas force will drive the rotor clockwise. Simultaneously, exhaust gases are being pushed out of port E by the surface CB, and a fresh charge is being drawn in at port F. Shortly after the phase shown, rotor tip C will close off inlet port F, and the gases enclosed by arc CA will be compressed. The cycle (intake, compression, expansion, exhaust) is repeated for the gases in each of the three pockets during each revolution of the rotor. The net effect is to produce continuous rotation, with only one moving part.[15]

5.23 Problems

1. Derive Eqs. (5.21-1) and (5.21-2).

2. Derive Eqs. (5.21-3) and (5.21-4).

3. Derive Eqs. (5.21-5) and (5.21-6).

4. Show that epi- and hypocycloids are closed curves only if the ratio b/a (b = radius of rolling circle, a that of fixed circle) is a rational number. If b/a is a rational number which can be written as a fraction n/d, show that the associated curve has d cusps and that it closes up after the moving circle has made n circuits about the fixed circle.

5. If θ is the angle of rotation of the gear carrier arm in Fig. 5.22-2, show that the reciprocator gives a displacement proportional to $\cos\theta$ and is therefore a suitable cosine generator for an analog computer. Where would you locate a second reciprocator to give an output proportional to $\sin\theta$?

6. Find the differential equation for the large motion of an ordinary circular pendulum of length $4b$. Show that for small motions of the pendulum the period approaches that of the cycloidal pendulum shown in Fig. 5.22-3(b).

7. A curve is formed by a point P on the circumference of a circle of radius b and center C which rolls within a fixed circle of radius $a = nb$ and center O. Let θ denote the angle between the x axis and the line of centers OC of the two circles. Let the variable point C be defined by the variable unit vector $U = e^{i\theta}$.
   a. Show that the path of P (a hypocycloid) is described by the complex vector equation $Z = b[(n-1)U + U_0^n U^{1-n}]$, where U_0 is a constant unit vector.
   b. Show that U_0 points from O toward a cusp of the hypocycloid.
   c. If n is an integer, it represents the number of cusps in the hypocyloid; prove this.

8. If the rolling circle of Problem 7 is outside of the fixed circle, show that
   a. The path of P (an epicycloid) is described by $Z = b[(n+1)U - U_0^{-n}U^{n+1}]$, where U_0 is a constant unit vector
   b. U_0 points from O toward a cusp of the epicycloid.

9. Show that the results of Problem 7 are formally valid for the case of the pericycloid if $n < 1$.

10. From the results of Problem 7, show that the hypocycloid of two cusps is a straight line.

11. Show that a point on the moving circle of Fig. 5.22-2 traces an ellipse.

5.30 VELOCITY AND ACCELERATION RELATIONSHIPS

We shall now develop some useful relationships between the velocities and accelerations of groups of points which are fixed to a common moving lamina. Unless otherwise indicated, complex vector notation will be used in the remainder of this chapter.

---

[15]For further information, see Norbye [1971] and Ansdale [1968].

Suppose that a number of points $B, C, D, \ldots$ are permanently marked on a moving lamina, as shown in Fig. 1(a). The points are fully specified by known vectors $\zeta_B = \xi_B + i\eta_B, \ldots$, which are fixed in the moving plane. As the point Q (fixed in the

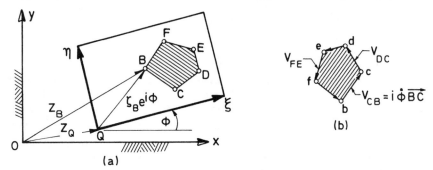

Figure 5.31-1. (a) Polygon BCDEFB; (b) Velocity image of polygon BCDEFB.

lamina) translates through Z_Q and the lamina rotates through an angle ϕ, the vectors $\zeta_B, \zeta_C, \ldots$ rotate with the lamina and become vectors $\zeta_B e^{i\phi}, \zeta_C e^{i\phi}, \ldots$ referred to the fixed plane. From Fig. 1(a), we see that the vector OB can be expressed as

$$Z_B = Z_Q + \zeta_B e^{i\phi} \tag{1}$$

The absolute velocity of point B is therefore given by

$$V_B = \dot{Z}_B = \dot{Z}_Q + i\dot{\phi}\zeta_B e^{i\phi} \tag{2}$$

The velocity difference between any two points, such as C and B, is found from Eq. (2) to be given by

$$\begin{aligned} V_C - V_B &= (\dot{Z}_Q + i\dot{\phi}\zeta_C e^{i\phi}) - (\dot{Z}_Q + i\dot{\phi}\zeta_B e^{i\phi}) \\ &= i\dot{\phi}(\zeta_C e^{i\phi} - \zeta_B e^{i\phi}) = i\dot{\phi}(\overrightarrow{QC} - \overrightarrow{QB}) = i\dot{\phi}\overrightarrow{BC} \end{aligned} \tag{3}$$

Equation (3) tells us that the **velocity difference** between the points B and C has magnitude

$$|V_C - V_B| = |\overrightarrow{BC}||\dot{\phi}| \tag{4}$$

and is directed *perpendicular to the line joining B and C*. Thus, in visualizing the velocity difference $V_C - V_B$ we may imagine that a link BC is momentarily pinned at B and is rotating about B with angular speed $\dot{\phi}$. Henceforth we shall refer to $(V_C - V_B)$ as the **velocity of C about B** and shall denote it by V_{CB}.

From Eq. (3) we can write

$$V_{CB} = V_C - V_B = \dot{\phi}i\overrightarrow{BC} \tag{5}$$

and similarly

$$V_{DC} = \dot{\phi}i\overrightarrow{CD}$$
$$V_{ED} = \dot{\phi}i\overrightarrow{DE}$$

$$
\begin{array}{cc}
\cdot & \cdot \\
\cdot & \cdot \\
\cdot & \cdot
\end{array}
$$

In short, the *velocity difference vector corresponding to any two points attached to a moving plane is found by rotating the relative position vector of the two points through 90° counterclockwise and multiplying by the scalar $\dot{\phi}$*. Note that if $\dot{\phi}$ is negative, the net effect is to rotate through 90° clockwise. Thus, if the polygon $B, C, D, \ldots$ shown in Fig. 1(a) is rotated bodily through 90° and magnified by the factor $\dot{\phi}$, it will form the similar polygon $b, c, d, \ldots$. The latter polygon is called the **velocity image**[16] of the space polygon $BCD \ldots$ and will prove useful in the graphical analysis of mechanisms. Note that the circuits $b, c, d, \ldots$ and $B, C, D, \ldots$ both run in the same direction, e.g., counterclockwise in Fig. 1.

For future reference, we note from Eq. (5) that the real and imaginary components of V_{CB} give the velocity components of point C about point B in the form

$$
\begin{aligned}
\dot{x}_{CB} &\equiv \dot{x}_C - \dot{x}_B = -\dot{\phi}y_{CB} \\
\dot{y}_{CB} &\equiv \dot{y}_C - \dot{y}_B = \dot{\phi}x_{CB}
\end{aligned}
\tag{6}
$$

where

$$
\begin{aligned}
x_{CB} &\equiv x_C - x_B = \xi_{CB}\cos\phi - \eta_{CB}\sin\phi \\
y_{CB} &\equiv y_C - y_B = \xi_{CB}\sin\phi + \eta_{CB}\cos\phi
\end{aligned}
\tag{7}
$$

are the components of the vector $\overrightarrow{BC}$ along the fixed coordinate axes and

$$
\xi_{CB} \equiv \xi_C - \xi_B, \qquad \eta_{CB} \equiv \eta_C - \eta_B
\tag{8}
$$

For later use, we shall now find the speed (v_C) of a point C in terms of the speeds (v_A, v_B) of two points (A, B) where all three points (A, B, C) are fixed on a straight line in a moving link (Fig. 2).

If the *oriented*[17] distances AC and CB are represented by a and b, positive in the sense of AB, and U is the unit vector along the direction of AB, the velocities of points A and B are

$$
\begin{aligned}
V_A &= V_C - i\dot{\phi}AC = V_C - i\dot{\phi}aU \\
V_B &= V_C + i\dot{\phi}CB = V_C + i\dot{\phi}bU
\end{aligned}
\tag{9}
$$

---

[16]Also called **Mehmke's diagram**, from a paper of R. Mehmke in *Civ. Ing.*, in 1883.

[17]Given three points A, C, B on a straight line, the line may be said to have a *positive orientation* in the direction from A to B. Then the **oriented** (or **directed**) line segment AB is considered positive, and $BA = -AB$ is considered negative. If AC has the same orientation as AB, then $AC = -CA$ is considered positive too. It is easily verified that the relationship

$$AC + CB = AB$$

is correct for any ordering of the three points on the line, and for any assumed positive orientation of the line. The **magnitude** or **length** of an oriented segment such as CA, is denoted by $|CA| = |AC|$.

Hence

$$v_A^2 = V_A \bar{V}_A = (V_C - i\dot{\phi}aU)(\bar{V}_C + i\dot{\phi}a\bar{U})$$
$$= v_C^2 + a^2\dot{\phi}^2 + i\dot{\phi}a(V_C\bar{U} - U\bar{V}_C)$$
$$v_B^2 = V_B \bar{V}_B = (V_C + i\dot{\phi}bU)(\bar{V}_C - i\dot{\phi}b\bar{U})$$
$$= v_C^2 + b^2\dot{\phi}^2 - i\dot{\phi}b(V_C\bar{U} - U\bar{V}_C) \tag{10}$$

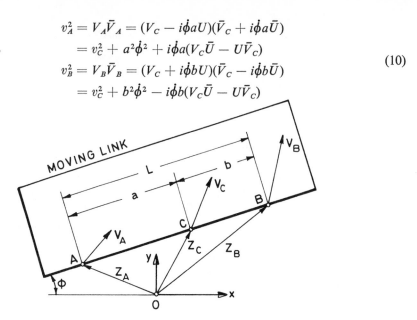

Figure 5.31-2. Collinear points A, C, B on a moving link.

Upon multiplying the first of these equations by b, the second by a, and adding, we find

$$bv_A^2 + av_B^2 = (a + b)v_C^2 + ab(a + b)\dot{\phi}^2 \tag{11}$$

But since $a + b = L$, it follows that

$$v_C^2 = \frac{b}{L}v_A^2 + \frac{a}{L}v_B^2 - ab\dot{\phi}^2 \tag{12}$$

This compact result will be used in Sec. **12.31** to simplify the calculation of the kinetic energy of floating links.

A Note on Terminology

The velocity difference $V_C - V_B$ which will be designated by V_{CB} is sometimes said to be the velocity of C *relative* to B. Actually, a velocity is a vector quantity which must be defined relative to a specified coordinate frame, not with respect to a point. Therefore it is rather loose usage to speak of a velocity as "relative to a point." To illustrate the confusion which can arise if the word *relative* is misused, consider the following example.

An observer A sits on a horizontal disc which is rotating at speed ω about a vertical axis through a fixed point O (Fig. 3). A second observer B stands vertically over A on a stationary platform. Both observers train telescopes at a flagpole P fixed to the rotating disc at a horizontal distance r from the axis of rotation. Both observers are located at distance a from the axis.

Considering motion in the horizontal plane, we see that A and B are momentarily coincident points, but the velocity differences V_{PA} and V_{PB} are quite different since

$$V_{PA} = V_P - V_A = r\omega - a\omega$$

and

$$V_{PB} = V_P - V_B = r\omega$$

This shows that the velocity difference of two points is not fully specified unless we know the reference frames to which the points are attached.

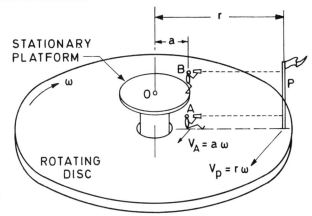

Figure 5.31-3. Illustrating that "velocity relative to a point" is meaningless.

Furthermore, we should not call V_{PA} the velocity of P "relative to" A because point P never passes out of the field of view of observer A's telescope; insofar as Mr. A is concerned, point P is stationary relative to him. By the same token, P passes from left to right past the viewing field of observer B, provided that Mr. B remains perfectly stationary and does not move his telescope relative to the stationary platform. Therefore we should speak of V_{PB} as the velocity of P relative to the platform on which B stands or, more briefly, as the velocity of P *about* B. The common practice of calling V_{PA} the velocity of P "relative" to A should be avoided.

5.32   Acceleration Polygon

Upon differentiation across Eq. (**5.31**-2) we find that the absolute acceleration of B is given by

$$A_B = \dot{V}_B = \ddot{Z}_Q + \zeta_B(i\ddot{\phi} - \dot{\phi}^2)e^{i\phi} \qquad (1)$$

Similarly, the absolute acceleration of point C is given by

$$A_C = \ddot{Z}_Q + \zeta_C(i\ddot{\phi} - \dot{\phi}^2)e^{i\phi} \qquad (2)$$

Therefore the acceleration difference $A_C - A_B$ is given by

$$A_{CB} \equiv A_C - A_B = (i\ddot{\phi} - \dot{\phi}^2)(\zeta_C - \zeta_B)e^{i\phi} \qquad (3)$$

We call A_{CB} the **acceleration of C about B**. Note from Fig. **5.31**-1 that

$$(\zeta_C - \zeta_B)e^{i\phi} = \overrightarrow{BC} \qquad (4)$$

It now proves convenient to define the complex number (see Fig. 1)

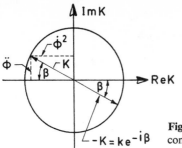

Figure 5.32-1. Illustrating definition of complex number K.

$$K \equiv -\ddot{\phi}^2 + i\ddot{\phi} \equiv -ke^{-i\beta} = ke^{i(\pi-\beta)} \tag{5}$$

where

$$k = |K| = \sqrt{\ddot{\phi}^2 + \dot{\phi}^4} = \sqrt{\alpha^2 + \omega^4} \tag{6}$$

$$\tan \beta = \frac{\ddot{\phi}}{\dot{\phi}^2} \equiv \frac{\alpha}{\omega^2} \qquad \left(-\frac{\pi}{2} \le \beta \le \frac{\pi}{2}\right)^{18} \tag{7}$$

In Eqs. (6) and (7) we have introduced the conventional nomenclature

$$\omega \equiv \dot{\phi}, \qquad \alpha \equiv \ddot{\phi} \tag{8}$$

From Eqs. (6) and (7) or directly from Fig. 1 we see that

$$k = \frac{\dot{\phi}^2}{\cos \beta} \tag{9}$$

Equation (3) states that

$$A_{CB} = K \overrightarrow{BC} \tag{10}$$

In other terms, the acceleration difference A_{CB} is found by stretching the position vector BC in the ratio $k{:}1$, reversing its sign, and rotating the vector so obtained through the acute angle $-\beta$ (β is measured positive counterclockwise, as usual). The construction is illustrated in Fig. 2(a). An alternative construction for A_{CB} is shown in Fig. 2(b), where the components of A_{CB} along and perpendicular to BC are denoted by

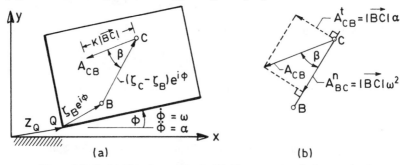

(a)   (b)

Figure 5.32-2. (a) Showing angle β; (b) Components of A_{CB} resolved along CB (A_{CB}^n) and perpendicular to CB (A_{CB}^t).

---

[18] Why can we be sure that $|\beta| \le \pi/2$?

A_{CB}^n and A_{CB}^t, respectively. It follows directly from Eq. (3) or from the geometry of Fig. 2(b) that

$$A_{CB}^n = |\overrightarrow{BC}|\,\omega^2 \tag{11}$$

$$A_{CB}^t = |\overrightarrow{BC}|\,\alpha \tag{12}$$

The **centripetal component** A_{CB}^n is always directed from C to B, and the **transverse component** A_{CB}^t points in the sense that α would tend to rotate the line BC if the latter were pinned at B.

By analogy with Eq. (10) we may write

$$A_{DC} = K\,\overrightarrow{CD}, \qquad A_{ED} = K\,\overrightarrow{DE}, \ldots \tag{13}$$

Therefore we can form an *acceleration image* polygon $b', c', d', \ldots$ whose sides represent the acceleration differences $A_{CB}, A_{DC}, \ldots$ as follows:

1. Elongate each leg of polygon $BCD \ldots$ by the factor k.
2. Rotate the polygon so formed through $180° - \beta$. The construction is illustrated in Fig. 3. The acceleration image may be used to find the absolute acceleration of any point in a link once the acceleration is known for any single point in the link. If the circuit DCB is clockwise, so too is $d'c'b'$.

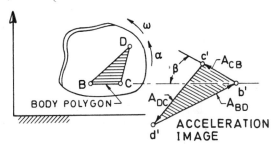

Figure 5.32-3. Body polygon and acceleration image.

For future reference, we note from Eqs. (3) and (4) that

$$A_{CB} = (i\ddot{\phi} - \dot{\phi}^2)\overrightarrow{BC} = (i\ddot{\phi} - \dot{\phi}^2)(x_{CB} + iy_{CB}) \tag{14}$$

where we have used the notation of Eq. **(5.31-7)** for x_{CB} and y_{CB}—the components of $\overrightarrow{BC}$ on the fixed coordinate axes. The real and imaginary parts of A_{CB} give the acceleration components $\ddot{x}_{CB}, \ddot{y}_{CB}$ in the form

$$
\begin{aligned}
\ddot{x}_{CB} &\equiv \ddot{x}_C - \ddot{x}_B = -y_{CB}\ddot{\phi} - x_{CB}\dot{\phi}^2 \\
\ddot{y}_{CB} &\equiv \ddot{y}_C - \ddot{x}_B = x_{CB}\ddot{\phi} - y_{CB}\dot{\phi}^2
\end{aligned}
\tag{15}
$$

EXAMPLE 1

Let us use the foregoing results to calculate the acceleration of a point on a wheel which rolls over the surface of a fixed wheel. In Fig. 4, the moving wheel (m) has radius ρ_m and center Q; the fixed wheel (f) has radius ρ_f and center O. We choose the fixed

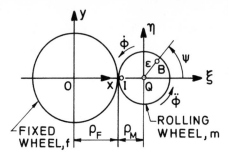

Figure 5.32-4. One wheel rolling on another.

axes (x, y) such that x coincides with the line OQ at the instant of interest, and the moving axes (ξ, η) are chosen to pass through Q and to be parallel to (x, y) at the moment of interest (hence $\phi = 0$). An arbitrary point B on the moving wheel has polar coordinates (ϵ, ψ); hence

$$\zeta_B = \epsilon e^{i\psi}$$

Since the circles touch at the instant center I, the instantaneous *speed* of point Q is clearly

$$\dot{s}_Q = \rho_m \dot{\phi} \tag{16}$$

Since point Q travels about point O on a circle of radius $\rho_m + \rho_f$, it follows that the acceleration of Q is given by

$$\ddot{Z}_Q = \frac{-|\dot{s}_Q|^2}{OQ} + \ddot{s}_Q i$$

or

$$\ddot{Z}_Q = -\frac{\rho_m^2}{\rho_m + \rho_f}\dot{\phi}^2 + \rho_m\ddot{\phi}i \tag{17}$$

where use has been made of Eq. (16).

From Eq. (1) we see that the acceleration of point B is

$$A_B = \ddot{Z}_Q + \zeta_B(i\ddot{\phi} - \dot{\phi}^2) \tag{18}$$

Hence the desired acceleration is

$$A_B = -\left(\frac{\rho_m^2}{\rho_m + \rho_f} + \zeta_B\right)\dot{\phi}^2 + i(\rho_m + \zeta_B)\ddot{\phi} \tag{19}$$

Now let us consider the acceleration of that point which momentarily coincides with the contact point I; for this point $\zeta_B = -\rho_m$; hence

$$A_B = A_I = \frac{\rho_m\rho_f}{\rho_m + \rho_f}\dot{\phi}^2 \tag{20}$$

If we introduce the definition

$$D \equiv \frac{\rho_m\rho_f}{\rho_m \pm \rho_f} \tag{21}$$

as defined by Eq. (**5.20-4**), where the upper (lower) sign is used for external (internal)

CHAPTER 5 MOTION OF A LAMINA

192

contact, Eq. (20) can be expressed in the form

$$A_I = D\dot{\phi}^2 \tag{22}$$

It may be verified (Problems **5.33**-9 and 10) that Eq. (22) is also valid when the wheels contact externally.

5.33 Problems

1. Show that vector velocity difference $\mathbf{v}_Q - \mathbf{v}_P$ between two points P and Q fixed in a moving rigid body must be normal to the vector PQ because of the inextensibility of the line segment PQ.[19]

2. Prove **Mehmke's theorem**: "The heads of the velocity vectors of points in a rigid lamina form a figure which is geometrically similar to the array of points if the tails of the velocity vectors all are bound to a common origin."

3. If ω and α are the angular velocity and acceleration, respectively, of the moving frame shown in Fig. 1 and if the velocity and acceleration of point Q in the moving frame are

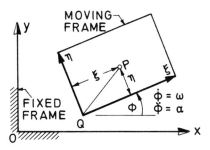

Figure 5.33-1 (Prob. 3).

denoted by

$$\mathbf{v}_Q = v_{Q\xi}\hat{\boldsymbol{\xi}} + v_{Q\eta}\hat{\boldsymbol{\eta}}$$
$$\mathbf{a}_Q = a_{Q\xi}\hat{\boldsymbol{\xi}} + a_{Q\eta}\hat{\boldsymbol{\eta}}$$

where $\hat{\boldsymbol{\xi}}$ and $\hat{\boldsymbol{\eta}}$ are unit vectors in the moving frame, show that

a. The velocity and acceleration components of any point P fixed in the moving lamina are given by

$$\mathbf{v} = \mathbf{v}_Q - \eta\omega\hat{\boldsymbol{\xi}} + \xi\omega\hat{\boldsymbol{\eta}}$$
$$\mathbf{a} = \mathbf{a}_Q - (\eta\alpha + \xi\omega^2)\hat{\boldsymbol{\xi}} + (\xi\alpha - \eta\omega^2)\hat{\boldsymbol{\eta}}$$

b. The coordinates of the instant center are given by

$$\xi = -\frac{v_{Q\eta}}{\omega}$$

$$\eta = \frac{v_{Q\xi}}{\omega}$$

---

[19]Note that this result applies to three-dimensional motions, not just to planar motion.

c. The coordinates of the *instantaneous center of acceleration* (point of vanishing acceleration) are given by

$$\xi = \frac{\omega^2 a_{Q\xi} - \alpha a_{Q\eta}}{\omega^4 + \alpha^2}$$

$$\eta = \frac{\omega^2 a_{Q\eta} + \alpha a_{Q\xi}}{\omega^4 + \alpha^2}$$

4. The angular velocity and acceleration of the moving frame, shown in Fig. 1, are ω and α, respectively. The position, velocity, and acceleration of the point Q (fixed in the moving frame) are given by

$$\mathbf{r}_Q = x_Q \hat{\mathbf{x}} + y_Q \hat{\mathbf{y}}$$

$$\mathbf{v}_Q = \dot{x}_Q \hat{\mathbf{x}} + \dot{y}_Q \hat{\mathbf{y}}$$

$$\mathbf{a}_Q = \ddot{x}_Q \hat{\mathbf{x}} + \ddot{y}_Q \hat{\mathbf{y}}$$

where $\hat{\mathbf{x}}$ and $\hat{\mathbf{y}}$ are fixed unit vectors in the direction of the axes x and y. Show that

a. The velocity and acceleration of any point P (fixed in the moving frame) are given by

$$\mathbf{v} = \mathbf{v}_Q - (y - y_Q)\omega \hat{\mathbf{x}} + (x - x_Q)\omega \hat{\mathbf{y}}$$

$$\mathbf{a} = \mathbf{a}_Q + [-(y - y_Q)\alpha - (x - x_Q)\omega^2]\hat{\mathbf{x}} + [(x - x_Q)\alpha - (y - y_Q)\omega^2]\hat{\mathbf{y}}$$

b. The coordinates of the instant center are given by

$$x = x_Q - \dot{y}_Q/\omega$$

$$y = y_Q + \dot{x}_Q/\omega$$

c. The coordinates of the instantaneous center of acceleration (point of vanishing acceleration) are given by

$$x = x_Q + (\omega^2 \ddot{x}_Q - \alpha \ddot{y}_Q)/(\omega^4 + \alpha^2)$$

$$y = y_Q + (\omega^2 \ddot{y}_Q + \alpha \ddot{x}_Q)/(\omega^4 + \alpha^2)$$

5. A wheel of radius 5 in. is rolling along a fixed wheel of 10-in. radius as shown in Fig. 2. If the angular velocity and acceleration of the rolling wheel are $\omega = 10$ rad/sec and $\alpha = 4$ rad/sec², find the acceleration of points P and Q located as shown.

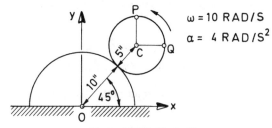

Figure 5.33-2 (Prob. 5).

6. A railroad wheel, with the dimensions shown in Fig. 3, rolls without slipping on a straight track. The car is accelerating *uniformly* from rest to a top speed of 70 mph in 30 sec. Find the acceleration of points P and Q on the rim of the flange when the car has achieved a speed of 20 mph.

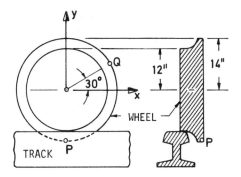

Figure 5.33-3 (Prob. 6).

7. A wheel of radius a rolls with constant angular speed ω over a circular depression (*low spot* or *pothole*) of radius R. Find the acceleration of point P at the top of the wheel and point Q, located as shown in Fig. 4, when the wheel passes over the bottom of the depression. Assume $a = 12$ in., $R = 36$ in., $\beta = 30°$, and the speed of the center of the wheel is $V = 60$ mi./hr.

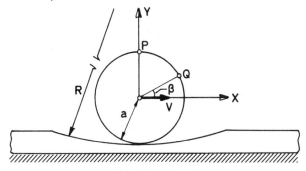

Figure 5.33-4 (Prob. 7).

8. Repeat Problem 7 but assume that the circular depression is replaced by a *circular hump;* i.e., the fixed surface is convex.

9. Verify Eq. (**5.32**-22) for the case where a moving wheel of radius ρ_m rolls *inside* a fixed ring of radius $\rho_f > \rho_m$.

10. Verify Eq. (**5.32**-22) for the case where a rolling ring of radius ρ_m surrounds a fixed wheel of radius $\rho_f < \rho_m$.

11. Derive Eq. (**5.32**-19) by expressing the complex vector Z_B in terms of ϕ and then differentiating Z_B twice.

12. Three distinct points (A, B, C) are represented by the vectors $\mathbf{r}_A, \mathbf{r}_B, \mathbf{r}_C$ from a common origin O in three-dimensional space.
   a. Show that A, B, C are collinear if, and only if, $\mathbf{r}_C = \mu \mathbf{r}_A + (1 - \mu)\mathbf{r}_B$, where μ is a scalar.
   b. Show that the points lie in the order

$$(A, B, C) \quad \text{if} \quad \mu < 0$$
$$(A, C, B) \quad \text{if} \quad 0 < \mu < 1$$
$$(C, A, B) \quad \text{if} \quad 1 < \mu$$

13. Using the results of Problem 12, show that
    a. The velocities v_A, v_B, v_C of points A, B, C are related by the expression $v_C^2 = \mu^2 v_A^2 + (1 - \mu)^2 v_B^2 + 2\mu(1 - \mu)\mathbf{v}_A \cdot \mathbf{v}_B$, where $v_i^2 \equiv \mathbf{v}_i \cdot \mathbf{v}_i$ ($i = A$, B, or C).
    b. The above expression for v_C^2 reduces to Eq. (**5.31**-12) for plane motion.

14. If $\boldsymbol{\omega}$ is the angular velocity vector for a rigid body in three dimensions and $\mathbf{r}_{AB}$ is the vector $\overrightarrow{AB}$ between two points (A, B) fixed in the body, show that the velocities v_A, v_B of the points (A, B) are related by $2\mathbf{v}_B \cdot \mathbf{v}_A = v_A^2 + v_B^2 - \boldsymbol{\omega} \times \mathbf{r}_{AB}$.

15. Use the results of Problems 13 and 14 to show that
    a. $v_C^2 = \mu v_A^2 + (1 - \mu)v_B^2 + \mu(1 - \mu)(\boldsymbol{\omega} \times \mathbf{r}_{AB})^2$.
    b. The last result reduces to Eq. (**5.31**-12) for planar motion.

5.40 LOCAL PROPERTIES OF LAMINA MOTION

We now wish to establish some results concerning velocity, acceleration, and path curvature of points fixed in a moving lamina. These are *local properties* in the sense that they are uniquely defined by the instantaneous position, angular velocity, and angular acceleration of the lamina.

Throughout Secs. **5.41** and **5.42** we shall use complex vector notation. In Sec. **5.42** we shall also use the concept of **oriented line segments** described in footnote 17 of Sec. **5.31**.

5.41 Instantaneous Centers of Velocity and Acceleration

If P is a point fixed in the moving lamina shown in Fig. 1, its absolute position and velocity vectors Z and $\dot{Z}$ are given by

$$Z = R + \zeta e^{i\phi} \tag{1}$$

$$\dot{Z} = \dot{R} + \zeta \dot{\phi} i e^{i\phi} \tag{2}$$

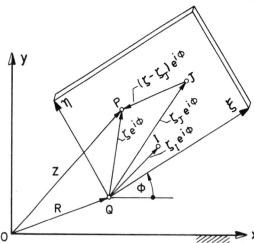

Figure 5.41-1. Showing notation associated with moving lamina.

where $\dot{R}$ is the velocity of the moving origin Q and $\dot{\phi}$ is the angular velocity of the lamina. If there is indeed a point I which is instantaneously at rest, its velocity $\dot{Z}_I = 0$, and its position ζ_I in the moving lamina is readily found by setting $\dot{Z} = 0$ in Eq. (2), which leads to the result

$$\zeta_I = -\frac{\dot{R}}{i\dot{\phi}e^{i\phi}} = \frac{\dot{R}}{\dot{\phi}}ie^{-i\phi} \tag{3}$$

Equation (3) states that a well-defined I.C. does exist, unless $\dot{\phi} = 0$, in which case we have (at least momentarily) a pure translation with the I.C. located at infinity.

Upon elimination of $\dot{R}$ between Eqs. (2) and (3) we see that the velocity of any point can be expressed in the form

$$V = \dot{Z} = \dot{\phi}(\zeta - \zeta_I)ie^{i\phi} = \dot{\phi}i\overrightarrow{\text{IP}} \tag{4}$$

Equation (4) states that the velocity of point P is proportional to its distance from the I.C. and points at right angles to the vector IP.

At this point, it is natural to ask whether there exists a point in the lamina where the acceleration instantaneously vanishes. To answer this question we differentiate Eq. (2) and get

$$\begin{aligned} \ddot{Z} &= \ddot{R} - (\dot{\phi}^2 - i\ddot{\phi})\zeta e^{i\phi} \\ &= \ddot{R} + K\zeta e^{i\phi} \end{aligned} \tag{5}$$

where we have utilized the complex number

$$K = -(\dot{\phi}^2 - i\ddot{\phi}) \equiv ke^{i(\pi - \beta)} = -ke^{-i\beta} \tag{6}$$

where k and $\pi - \beta$ are the magnitude and argument of K. If there is a point J where the acceleration vanishes, it may be found by setting $\ddot{Z} = 0$, which leads to

$$\zeta_J = \frac{-\ddot{R}e^{-i\phi}}{K} = \frac{\ddot{R}e^{i(\beta-\phi)}}{k} \tag{7}$$

Thus, a so-called **center of acceleration** exists unless k vanishes (i.e., $\dot{\phi}$ and $\ddot{\phi}$ are simultaneously zero) in which case all points in the body are momentarily moving along parallel straight paths.

Upon elimination of $\ddot{R}$ between Eqs. (5) and (7) we see that

$$A = \ddot{Z} = K(\zeta - \zeta_J)e^{i\phi} = K\overrightarrow{\text{JP}} \tag{8}$$

since $(\zeta - \zeta_J)e^{i\phi}$ represents the vector leading from J to P, as may be seen in Fig. 1. The vector $K\overrightarrow{\text{JP}}$ is readily found from the construction shown in Fig. **5.32**-2 if we interpret point B in that figure as the acceleration center J. This construction has been used in Fig. 2.

In Fig. 2 we see that the vector $K\overrightarrow{\text{JP}}$ has magnitude kd and points in the direction β (clockwise) from $\overrightarrow{\text{PJ}}$, where $d = |\overrightarrow{\text{PJ}}|$. Figure 2 also illustrates that all points (such as P', P'', etc.) which lie on the ray JP have parallel acceleration vectors which are proportional in magnitude to the distances (d', d'', etc.) from J to P', P'', etc.

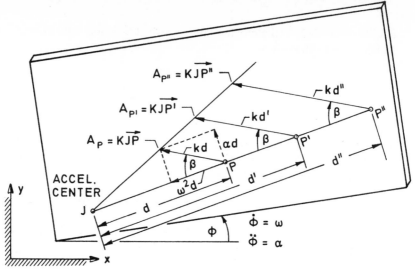

Figure 5.41-2. Acceleration of points on a ray through the acceleration center J.

In summary, we see that the location of the I.C. and the acceleration center can be found via Eqs. (3) and (7). Then the absolute velocity and acceleration of any point P are readily found through Eqs. (4) and (8) or through simple geometric constructions, as indicated in Fig. 2.

Another useful expression for the acceleration follows from Eq. (4) expressed in the form

$$V = \dot{Z} = i\dot{\phi}(Z - Z_I) \tag{9}$$

where Z_I is the location vector $\overrightarrow{OI}$ of the instantaneous center. By differentiating Eq. (9), we find the acceleration

$$\ddot{Z} = i\ddot{\phi}(Z - Z_I) + i\dot{\phi}\dot{Z} - i\dot{\phi}\frac{d}{dt}Z_I \tag{10}$$

We now note from Eq. (**5.20**-3) that the displacement velocity of the I.C. is given by

$$\frac{dZ_I}{dt} \equiv V^I = -D\dot{\phi}T \tag{11}$$

where D is defined by Eq. (**5.20**-4) and T is a unit vector tangent to the centrodes at I and pointing *clockwise* along the fixed centrode. The direction of T is called the **pole tangent**, and that of $N \equiv iT$ is called the **pole normal**.

Upon substituting Eqs. (9) and (11) into Eq. (10) we find

$$\ddot{Z} = i\ddot{\phi}(Z - Z_I) - \dot{\phi}^2(Z - Z_I) + D\dot{\phi}^2 iT \tag{12a}$$

or

$$\ddot{Z} = K(Z - Z_I) + D\dot{\phi}^2 iT \tag{12b}$$

where K is defined by Eq. (6).

The acceleration $\ddot{Z}_I$ of that particle on the moving lamina which momentarily coincides with the I.C. is found by setting $Z \equiv Z_I$ in Eq. (12); i.e.,

$$\ddot{Z}_I = D\dot{\phi}^2 iT = -i\dot{\phi}V^I \tag{13}$$

This shows that the particle located at I has an acceleration which is always normal to the pole tangent T.

5.42 Curvature Relationships

As shown in Fig. 1, the path of a moving point P has center of curvature C located somewhere on the line IP through the instant center I. Equation (5.41-12a) shows that

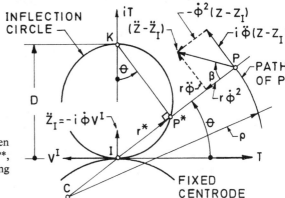

Figure 5.42-1. Relationship between points I, K, P, P^*, C; $r = IP$, $r^* = IP^*$, $\rho = CP$. All oriented line segments along IP are positive in sense of IP.

the acceleration of point P is the sum of the following three vectors:

$$i\dot{\phi}(Z - Z_I) = i\dot{\phi}\overrightarrow{IP} \tag{1a}$$

$$-\dot{\phi}^2(Z - Z_I) \doteq \dot{\phi}^2\overrightarrow{PI} \tag{1b}$$

$$iTD\dot{\phi}^2 = -i\dot{\phi}V^I \tag{1c}$$

These three vectors are shown in proper orientation in Fig. 1, where $r = |IP|$ and angle θ is measured positive counterclockwise between T and $\overrightarrow{IP}$ [recall the definition of T following Eq. (5.41-11)].

The component of acceleration along the path normal IP is given by

$$a_n = \frac{-|V_P|^2}{\rho} = \frac{-(r\dot{\phi})^2}{\rho} = \frac{-(r\dot{\phi})^2}{CP} \tag{2}$$

where ρ is the radius of curvature, positive when CP has the same sense as IP.

Alternatively, we can express a_n as the sum of the projections of the three vectors given by Eqs. (1a)–(1c) on the directed line IP; i.e.,

$$a_n = -r\dot{\phi}^2 + D\dot{\phi}^2 \sin \theta \tag{3}$$

Upon comparing Eqs. (2) and (3) we see that the path curvature is

$$\frac{1}{CP} = \frac{1}{\rho} = \frac{r - D \sin \theta}{r^2} \tag{4}$$

From Eq. (4) we see that the path curvature vanishes for all points P^* for which the radius r^* ($\equiv IP^*$) satisfies the relationship

$$r^* = D \sin \theta \tag{5}$$

Any point lying on the locus of Eq. (5) is a point of vanishing curvature, i.e., an *inflection point* of the path. These points are of some importance in mechanism design because the particle paths are nearly straight lines in the neighborhood of such points.

Equation (5) is clearly the equation of a circle with diameter IK (of magnitude D) lying in the direction of the vector $iTD\dot{\phi}^2$, as shown in Fig. 1 (where it is assumed[20] that $D > 0$). Thus, the locus of all inflection points in the particle path is a circle, called the **inflection circle** or the circle of **de la Hire**.[21] The point K, diametrically opposite I on the inflection circle, is called the **inflection pole**. Note that the inflection pole K, like the instantaneous center I (and the acceleration center J), is an *invariant point* determined by the gross motion of the lamina, whereas P^* is different for each choice of θ.[22]

Equation (4) can be written, with the help of Eq. (5), in the form

$$CP = \rho = \frac{r^2}{r - D \sin \theta} = \frac{(IP)^2}{IP - IP^*} = \frac{(IP)^2}{P^*P} \tag{6}$$

Note that the direction of IP determines the positive direction for all other *oriented segments*, such as IP^*, P^*P, and CP, all of which are shown positive in Fig. 1.

Equation (6) is one form of the **Euler-Savary equation**, which is useful in kinematic synthesis.[23] It relates four points on the ray IP, namely points I, P, P^*, and C, and allows us to find any of the four points from a knowledge of the other three. To see this more clearly, let us note from Fig. 1 that

$$\rho = CP = IP - IC \tag{7}$$

Recall that we consider IC to be an *oriented line segment* which is positive when the vector $\overrightarrow{IC}$ is in the direction of $\overrightarrow{IP}$ (as in Fig. 1). It then follows from Eq. (7) that ρ will be positive whenever C and I lie on the same side of P and will be negative otherwise. In other terms, *ρ is positive when the path of P is concave toward the I.C.* We may note, with the aid of Eq. (6), that the path of any point outside (inside) the inflection circle is concave (convex) toward I, as shown in Fig. 2.

Upon substitution of Eq. (7) into Eq. (6) we find the algebraic relationship

$$(IP - IC)(IP - IP^*) = (IP)^2$$

Hence

$$-(IP)(IP^*) - (IC)(IP) + (IC)(IP^*) = 0$$

---

[20]When D is negative, the inflection circle is the reflection, about the pole tangent, of that shown in Fig. 1, and IP^* has the opposite sign to that shown in Fig. 1.

[21]See Hartenberg and Denavit [1964] for reference to a publication of de la Hire in 1706.

[22]For a generalized discussion of invariants of motion, see Bottema [1961].

[23]Historical background on Euler's work of 1765 and Savary's in the decade 1831–1841 is given on p. 204 of Hartenberg and Denavit [1964]. Additional information about Savary's work (which was never formally published) may be found in the discussion of the paper by de Jonge [1943, pp. 680–681].

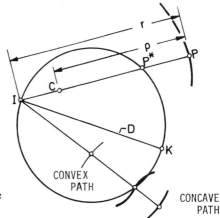

Figure 5.42-2. Shows paths which are concave and convex towards *I*.

"Dividing" this last equation through by the common denominator $(IP)(IP^*)(IC)$, we find a **second form of the Euler-Savary equation:**

$$\frac{1}{IC} = \frac{1}{IP} - \frac{1}{IP^*} \tag{8}$$

where it is to be emphasized that *IC* and *IP** are to be interpreted as *oriented line segments* which are positive in the sense of *IP*.

EXAMPLE 1. Elliptic Trammel

A rod maintains contact with a vertical wall at one end *E* and with a horizontal floor at the other end *F*. Find the curvature of the path of an arbitrary point *P* on the rod for the orientation of the rod shown in Fig. 3.

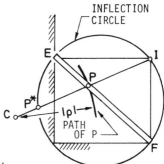

Figure 5.42-3. Example 1.

SOLUTION:

Because their paths are straight lines, points *E* and *F* are both inflection points. Thus the inflection circle passes through *E*, *F*, and *I*, the instant center, which is located as shown. The radius of curvature of point *P* is given by Eq. (6) as

$$\rho = CP = \frac{(IP)^2}{P^*P} \tag{9}$$

where P^* is the known point where ray IP intersects the inflection circle. Since P^*P has a negative sense in Fig. 3 (recall that IP determines the positive sense), Eq. (9) predicts a negative value of ρ, which means that the directed segment CP defined by Eq. (9) is negative and must point in the opposite sense from IP, as is illustrated in Fig. 3. The magnitude of $|\rho| = |CP|$ is readily found from Eq. (9).

EXAMPLE 2. Four-Bar Linkage

Find the curvature of the path of a point P attached to the *coupler* link AB of the four-bar linkage shown in Fig. 4.

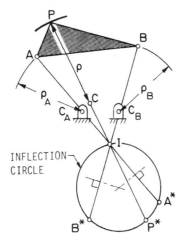

Figure 5.42-4. Example 2.

PROCEDURE:

As in Fig. **5.13**-4(b), the I.C. of the coupler bar is located at the intersection of the side bars. Because A rotates about the permanent center C_A, its radius of curvature is $\rho_A = C_A A$. Hence Eq. (8) locates the inflection point A^* on the ray AI at distance IA^*, given by

$$\frac{1}{IA^*} = \frac{1}{IA} - \frac{1}{IC_A}$$

Note that IA^* will be negative. This means that A^* is located on the side of I opposite to A. Similarly, the point B^* is located by

$$\frac{1}{IB^*} = \frac{1}{IB} - \frac{1}{IC_A}$$

The three points I, B^*, and A^* define the inflection circle which is drawn as shown. The ray PI intersects the inflection circle at P^*.

The center of curvature C of the path of P is located from Eq. (8) in the form

$$\frac{1}{IC} = \frac{1}{IP} - \frac{1}{IP^*} = \frac{1}{IP} + \frac{1}{P^*I}$$

Since P^*I points in the positive sense (in the same direction as IP), we see that IC is positive. Also, since $1/IC > 1/IP$, we see that $IC < IP$, and thus C must lie between I

and P as shown. The radius of curvature is of course $\rho = CP$. The same conclusion could have been reached by using Eq. (6) in the form $\rho = (IP)^2/(P^*P)$.

EXAMPLE 3. Rocking Boulder[24]

Figure 5 shows a boulder resting upon a stationary surface. The contact surfaces are approximated as parallel cylinders in the neighborhood of contact. The center of gravity G is vertically over the contact line when the boulder is in equilibrium. It may be shown (Problem **5.43**-1) that the configuration is stable only if G falls within the inflection circle defined by the local curvatures of the contacting rock surfaces.[25]

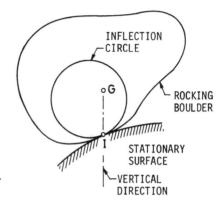

Figure 5.42-5. The problem of the "rocking boulder."

Other applications of the Euler-Savary equation are given in Sec. **5.43** and in Hall [1958, 1966], Hartenberg and Denavit [1964], and Rosenauer and Willis [1953].

5.43 Problems

1. Prove that the boulder in Fig. **5.42**-5 is stable for small oscillations if the center of gravity G lies within the inflection circle. Start from the fact that in the equilibrium state point G is at its minimum height.

2. A rectangular block is set on top of a rough circular cylinder of radius a. If, in this position, the mass center C of the block is located at a vertical distance b above the topmost point T of the cylinder and at a horizontal distance e from T, as shown in Fig. 1(a),

Figure 5.43-1 (Prob. 2).

---

[24]Kelvin and Tait [1879, p. 111].
[25]It is shown in Part 2 that the equilibrium configuration will be stable if the path of G is concave upward.

a. Show that the block will be in equilibrium under gravity loading if it rolls over the block through an angle θ_e, as shown in Fig. 1(b), where

$$\theta_e = \frac{e}{a} + \frac{b}{a} \tan \theta_e$$

Assume that sufficient friction exists so that the block does not slide over the cylinder surface.

b. Show that the equilibrium position θ_e is stable if $b/a < \cos^2 \theta_e$. Note that stable equilibrium occurs when C is at its minimum vertical height.

c. Find θ_e for $b/a = 0.2$ and $e/a = 0.1$. Is this a stable equilibrium position?

3. Given the instant center I, the center of acceleration J, and the angle $\beta = \tan^{-1}(\alpha/\omega^2)$ in Fig. 2 [where $-(\pi/2) < \beta < (\pi/2)$], show that

a. If angle $IPJ = (\pi/2) + \beta$, the acceleration of point P is normal to the ray IP; hence P is an inflection point.

b. The locus of all points P which subtend the same angle from a given line segment is a circle; hence the locus of all inflection points is a circle.

c. The diameter IK of the circle which is the locus of inflection points is given by $D = |IJ|/\cos \beta$.

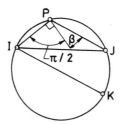

Figure 5.43-2 (Prob. 3).

4. Find the inflection circle and the center of curvature of the point P on the connecting rod of the slider-crank mechanism in the position shown in Fig. 3. Draw the circle of curvature through point P.

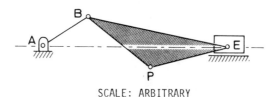

SCALE: ARBITRARY

Figure 5.43-3 (Prob. 4).

5. If $\gamma = 45°$ and $\psi = 30°$, in Fig. 4, find the inflection circle and the radius of curvature of point P, which is rigidly attached to the rod AB, whose ends slide along the straight lines OA and OB. A semigraphical procedure is suggested.

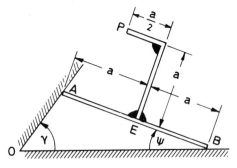

Figure 5.43-4 (Prob. 5).

6. The four-bar mechanism shown in Fig. 5 is to be redesigned with the fixed pivots C_A and C_B relocated at given points C_P and C_Q. Relocate the pivots A and B on the coupler to new points P and Q such that *all* points on the coupler link instantaneously move along the same tangent and have the same curvature as in the original design for the given position of the coupler link.

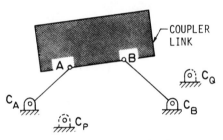

Figure 5.43-5 (Prob. 6).

7. Find three points P, Q, and R on the coupler plane (attached to coupler rod AB) which trace paths of zero curvature which are parallel to the given unit vectors, $\hat{t}_1$, $\hat{t}_2$, $\hat{t}_3$, respectively, at the instant shown in Fig. 6.

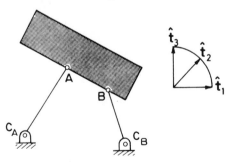

Figure 5.43-6 (Prob. 7).

8. Use an Euler-Savary equation to locate the center of curvature and to find the radius of curvature of point G on the connecting rod of the slider-crank mechanism shown in Fig. 7.

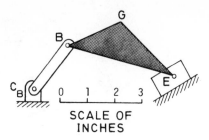

SCALE OF
INCHES

Figure 5.43-7 (Prob. 8).

9. Use an Euler-Savary equation to verify that the radius of curvature at a point P on a cycloid is given by $2IP$, where I is the instantaneous center of the generating circle (see Fig. **5.22**-3).

10. Use an Euler-Savary equation to find an expression for the curvature at an arbitrary point of an epicycloid.

11. Use an Euler-Savary equation to find an expression for the curvature at an arbitrary point of a hypocycloid.

6

GRAPHICAL KINEMATICS

6.10 GRAPHICAL ANALYSIS OF VELOCITIES IN MECHANISMS

The governing kinematic equations for typical linkages are highly nonlinear in the displacement variables. Therefore—prior to the advent of electronic computers—only graphical methods were practical for finding the positions of moving links. Once seated before a drawing board it was natural for machine designers to seek graphical methods for the determination of velocities and accelerations as well as displacements. In fact, a wide variety of graphical techniques was developed, and a large part of linkage analysis consisted of the classification of problems into categories where one or another known graphic technique could be applied. For illustrations of such an approach, see Rosenauer and Willis [1953] and Hain [1967].

Our goal is to give examples of *one* graphical method which works well for four-bar mechanisms, slider-crank chains, and most of the simpler mechanisms encountered in practice.[1] This method is based on the idea of the velocity polygon (Sec. 5.31) and could be described exclusively in terms of complex vectors; however, in this chapter we shall, unless otherwise noted, use the following conventional notation for graphical solutions:

$$\mathbf{V}_P = \text{velocity vector of point } P$$

$$V_P = |\mathbf{V}_P| = \text{magnitude of } \mathbf{V}_p$$

$$\overrightarrow{AB} = \text{vector of directed line segment } AB$$

$$AB = |\overrightarrow{AB}|$$

---

[1] In Part 2 we shall present a uniform analytic method which will apply to all problems.

6.11 Linkages with Lower Pairs (*Pin or Slider Joints*)

EXAMPLE 1. Four-Bar Mechanism

Figure 1(a) shows a four-bar mechanism where the velocity $\mathbf{V}_B$ of point B is known. To find the velocity of point C, we use the vector identity

$$\mathbf{V}_C = \mathbf{V}_B + (\mathbf{V}_C - \mathbf{V}_B) = \mathbf{V}_B + \mathbf{V}_{CB}$$

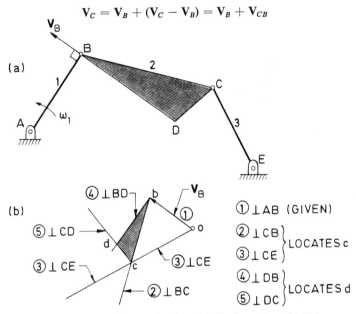

Figure 6.11-1. Four-bar mechanism: (a) Configuration; (b) Velocity polygon.

where $\mathbf{V}_{CB} \equiv \mathbf{V}_C - \mathbf{V}_B$. It is readily shown that $\mathbf{V}_{CB}$ *must be perpendicular to line BC* (e.g., see Sec. **5.31**), and we speak of $\mathbf{V}_{CB}$ as the **velocity of C about B**. To find the magnitude of $\mathbf{V}_{CB}$ we proceed as follows:

1. From an arbitrary origin o lay off the line ob [marked ① in Fig. 1(b)] representing the known vector $\mathbf{V}_B$ to a suitable scale. Point b is called the **velocity image** of the physical point B.
2. Draw a line ② through b, perpendicular to BC. Because the velocity difference $\mathbf{V}_{CB} = \mathbf{V}_C - \mathbf{V}_B$ is perpendicular to BC, the tip of vector $\mathbf{V}_{CB} \equiv \overrightarrow{bc}$ must lie along line ②, although point c is not yet known.
3. Because EC rotates about fixed center E, the velocity $\mathbf{V}_C \equiv \overrightarrow{oc}$ must be $\perp$ to EC[2]; therefore c must line on line ③, which is drawn through $o \perp$ to EC. We have now located c at the intersection of lines ② and ③; hence $\mathbf{V}_{CB}$ is now given by $\overrightarrow{bc}$, and

$$\mathbf{V}_C = \mathbf{V}_B + \mathbf{V}_{CB} = \overrightarrow{ob} + \overrightarrow{bc} = \overrightarrow{oc}$$

To find the velocity of other points of interest, such as D, we observe that

$$\mathbf{V}_D = \mathbf{V}_B + (\mathbf{V}_D - \mathbf{V}_B) \equiv \mathbf{V}_B + \mathbf{V}_{DB}$$

---

[2]The symbols $\perp$ and $\parallel$ denote *perpendicular* and *parallel*, respectively.

and

$$\mathbf{V}_D = \mathbf{V}_C + (\mathbf{V}_D - \mathbf{V}_C) \equiv \mathbf{V}_C + \mathbf{V}_{DC}$$

These two expressions for $\mathbf{V}_D$ suggest the next two steps in the graphical solution.

4. $\mathbf{V}_{DB}$ is $\perp$ to $\overrightarrow{BD}$; hence d must lie on line ④ drawn through b, $\perp$ to $\overrightarrow{BD}$.

5. $\mathbf{V}_{DC}$ is $\perp$ to $\overrightarrow{CD}$; hence d must lie on line ⑤ drawn through c, $\perp$ to $\overrightarrow{CD}$. Thus d, the *velocity image* of point D, is located at the intersection of lines ④ and ⑤. Note that

$$\mathbf{V}_D = \mathbf{V}_B + \mathbf{V}_{DB} = \overrightarrow{ob} + \overrightarrow{bd} = \overrightarrow{od}$$

The velocity of any other point on link BCD can be found in a similar way—or by noting that bcd is similar to BCD and that any point in BCD is similarly situated in the *velocity polygon bcd*. To find the angular velocity of link BC, we recall, from Eq. (5.31-3), that the angular velocity ω_2 of line segment BC is given by the *complex vector* relation

$$V_{CB} = V_C - V_B = \omega_2 i \overrightarrow{BC}$$

where V_C and V_B represent complex vectors. Thus the magnitude of ω_2 is given by

$$|\omega_2| = \left|\frac{\mathbf{V}_{CB}}{BC}\right| = \left|\frac{\overrightarrow{bc}}{BC}\right|$$

and its direction is that which is required to turn BC into alignment with bc, i.e., *clockwise* in this example.

EXAMPLE 2. Slider-Crank Linkage

Given the angular velocity $\omega = 2$ rad/sec for link AB of the slider-crank linkage shown in Fig. 2, we begin the velocity construction by drawing the velocity $\mathbf{V}_B$ of magnitude ωAB and direction $\perp$ to $\overrightarrow{AB}$ as shown in the figure by ①. To find $\mathbf{V}_C$, we note that $\overrightarrow{oc}$ must be $\parallel$ to $\overrightarrow{AC}$; hence c must lie on a line ② drawn $\parallel$ to $\overrightarrow{AC}$. Since $\mathbf{V}_C = \mathbf{V}_B + \mathbf{V}_{CB}$, where $\mathbf{V}_{CB}$ is $\perp$ to $\overrightarrow{CB}$, we draw a line ③ through b, $\perp$ to $\overrightarrow{CB}$. We know that c must lie somewhere on line ③, but it must also lie on ②; hence c is located at the intersection of ② and ③, and $\mathbf{V}_C = \overrightarrow{oc}$ is now known. To find $\mathbf{V}_D$ we observe that $\mathbf{V}_D = \mathbf{V}_B + \mathbf{V}_{DB}$ and $\mathbf{V}_D = \mathbf{V}_C + \mathbf{V}_{DC}$, where $\mathbf{V}_{DB}$ is $\perp$ to $\overrightarrow{DB}$ and $\mathbf{V}_{DC}$ is $\perp$ to $\overrightarrow{CD}$. Therefore d is located at the intersection of ④ ($\perp \overrightarrow{DB}$) and ⑤ ($\perp \overrightarrow{CD}$). The final solu-

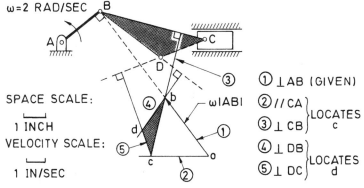

Figure 6.11-2. Slider-crank mechanism. Numbers show the sequence for drawing construction lines.

tion is seen to be

$$\mathbf{V}_C = \overrightarrow{oc} \quad (3.3 \text{ in./sec})$$

$$\mathbf{V}_D = \overrightarrow{od} \quad (4.2 \text{ in./sec})$$

The angular velocity of link BC is denoted by ω_{BC}, and

$$|\omega_{BC}| = \left|\frac{\mathbf{V}_C - \mathbf{V}_B}{CB}\right| = \frac{|\overrightarrow{bc}|}{CB} = \frac{3.3}{4.4} = 0.75 \text{ rad/sec (clockwise)}$$

The direction of ω_{BC} is clockwise because the velocity image bcd is turned 90° clockwise from BCD. Alternatively, the velocity difference $\mathbf{V}_C - \mathbf{V}_B = \overrightarrow{bc}$ indicates that point C moves clockwise *about* point B.

When numerical answers are required, it is advisable to always indicate the scale of the mechanism drawing (space scale) and the scale of the velocity diagram (velocity scale) in a manner similar to that shown in Fig. 2.

EXAMPLE 3. Series of Simple Sublinkages

If, as in Fig. 3, the crank AB of a four-bar mechanism is used to convert a circular

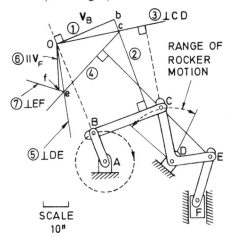

Figure 6.11-3. Four-bar and slider-crank mechanisms in tandem.

motion into the oscillatory motion of the link DC (the rocker), the rocker may be used to drive a slider-crank mechanism DEF. The net effect is to create a reciprocating rectilinear motion of slider F. To find the velocity of F one may use the construction of Fig. 1 to find $\mathbf{V}_E$ and then use the construction of Fig. 2 to find $\mathbf{V}_F$. The velocity of any other points in the mechanism can be read off the resulting velocity polygons which were built up in the sequence indicated by the circled numbers. This approach is characteristic for mechanisms which consist of a series of simpler mechanisms, each of which provides sufficient information for the solution of its neighboring sublinkage. Unfortunately, not all mechanisms are so simply arranged; therefore the method is not universally applicable.

EXAMPLE 4. Complex Mechanism

The velocity of point B in the six-bar mechanism of Fig. 4 (Stephenson linkage) is given (at say 8.5 in./sec). We are to find the velocity of all lettered points. Since there is no grounded four-bar chain in this mechanism, we cannot use previously studied methods; a subterfuge is necessary. We note [see Fig. **5.13-4(b)**] that point I, the rela-

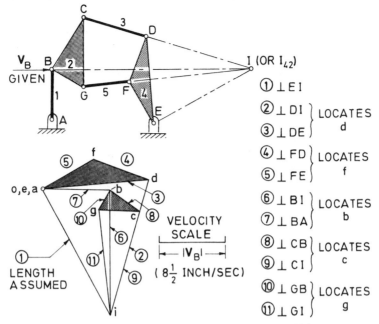

Figure 6.11-4. Velocity construction for Stephenson linkage.

tive I.C. of links 2 and 4, may be viewed as a point on link 2 or on link 4 with a common velocity. Viewed as a point on link 4, we see that I must have a velocity $\perp$ to ray EI since E is a fixed center for link 4. Therefore we assume an arbitrary[3] length for $\mathbf{V}_I$ and draw the vector as line ① in the velocity polygon. The velocities of points D and F are then readily found from the construction lines ②, ③, ④, ⑤ in the standard fashion for points on a given rigid link. Because I and B are both on link 2, we may find the velocity of point B at the intersection of lines ⑥ ($\perp \overrightarrow{BI}$) and ⑦ ($\perp \overrightarrow{BA}$). Knowing the velocities of points B and I on link 2, we locate c at the intersection of ⑧ ($\perp \overrightarrow{CB}$) and ⑨ ($\perp \overrightarrow{CI}$). Finally g is located at the intersection of ⑩ ($\perp \overrightarrow{GB}$) and ⑪ ($\perp \overrightarrow{GI}$).

The velocity image is now complete, but because oi was drawn with an arbitrary length, the scale of the diagram was not known to begin with. However, we are given the length $|\mathbf{V}_B|$ (e.g., 8.5 in./sec); hence we know that length $|\overrightarrow{ob}| = |\mathbf{V}_B|$, and the scale is therefore fixed as shown.

Although we managed to solve Example 4 by a judicious starting procedure, we cannot hope to always be so canny, and many so-called *complex mechanisms* will elude our powers of solution by means of a direct graphical approach. Methods have been devised[4] using *auxiliary points* (i.e., nonobvious points) which sometimes work and sometimes don't. A method based on superposition has been presented[5] which can in principle work for any complex plane linkage; however, this *method of influence coefficients* requires graphical solutions for two or more sublinkages together with the

---

[3]Note that merely changing the scale on a velocity diagram has the effect of multiplying all velocities in a mechanism by a common factor. Therefore we may always choose the driving point in a mechanism (with one degree of freedom) to suit our convenience. Later the velocity scale may be adjusted to give any point a prescribed velocity.

[4]Hall and Ault [1943].

[5]Modrey [1959].

solution of one or more linear algebraic equations. The basis for this method will best be seen when we discuss universally applicable algebraic and numerical methods in Part 2.

6.12 Linkages with Higher Pairs (Equivalent Linkages)

EXAMPLE 1. Circular Profile Cams

Figure 1(a) shows a three-link mechanism with links 1 and 2 making sliding contact at a point P. Point P is called P_1 in link 1 and P_2 in link 2. Link 1 (2) is pinned at A (B) and has a *circular contact profile* of radius R_1 (R_2) with center at C_1 (C_2). We are given the angular velocity ω_2 or the velocity V_{C_2}, and we seek the angular velocity ω_1 or V_{C_1}.

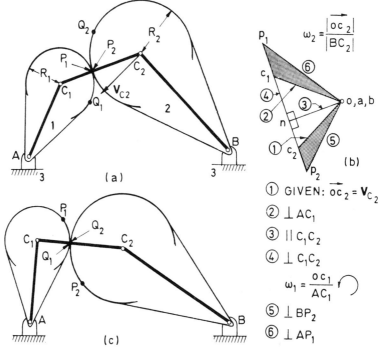

$$\omega_2 = \frac{|\overrightarrow{oc_2}|}{|BC_2|}$$

① GIVEN: $\overrightarrow{oc_2} = \mathbf{V}_{C_2}$

② $\perp AC_1$

③ $\parallel C_1C_2$

④ $\perp C_1C_2$

$$\omega_1 = \frac{oc_1}{AC_1}$$

⑤ $\perp BP_2$

⑥ $\perp AP_1$

Figure 6.12-1. (a) Initial configuration; (b) Velocity polygon; (c) Later configuration.

We begin the velocity polygon with the known vector $\overrightarrow{oc_2}$. Next, we draw line ② in the known direction of V_{C_1}, i.e., $\perp$ to $\overrightarrow{AC_1}$. Next, line ③ is drawn $\parallel$ to $\overrightarrow{C_1C_2}$. We note that the velocity component of V_{C_2} along the common normal C_2C_1 is given by $\overrightarrow{on}$, where n is the projection of c_2 on line ③. Since all points along the normal C_1C_2 must have the same velocity component along C_1C_2 (otherwise C_1P_2 and C_2P_2 would stretch or contract), n is also the projection of c_1 on line ③; therefore c_1 is at the join of line ② and ④ (the extension of c_2n). We now have $V_{C_1} = \overrightarrow{oc_1}$ as required. To find the velocity of P_2, we locate p_2 on the image polygon of BC_2P_2 in the usual way, and similarly for p_1 (the velocity image of P_1).

The velocity with which P_2 slides past P_1 is given by

$$\mathbf{V}^{\text{slide}} = \mathbf{V}_{P_2} - \mathbf{V}_{P_1} = \overrightarrow{op_2} - \overrightarrow{op_1} = \overrightarrow{p_1p_2}$$

Note that a four-bar linkage AC_1C_2B would predict exactly the same velocity ratio ω_1/ω_2. Therefore the four-bar is said to be an *equivalent mechanism*.[6] Because the contact surfaces have circular profiles, the very same *four-bar AC_1C_2B* continues to predict the correct ratio ω_1/ω_2 after both links undergo a finite change of position, as shown in Fig. 1(c). Therefore the four-bar AC_1C_2B is a *permanently equivalent mechanism*.

The above formulation is not necessarily the quickest way to find ω_1/ω_2 (for example, see Problems **6.14**-1 and 2); however, it is useful to illustrate the idea of an equivalent mechanism, created by the introduction of a link along the common normal to sliding surfaces, with pins at the centers of curvature.

EXAMPLE 2. Variable Curvature Contact Surfaces

If the contact surfaces in Fig. 1 were not circular but had variable curvature, the construction of Fig. 1(b) would still be valid if we used the centers of curvature for C_1 and C_2. However, the distance C_1C_2 would not remain constant with time. Therefore a new equivalent mechanism would be required at each instant, and we should therefore speak of *instantaneous equivalent mechanisms* when surfaces of variable curvature are involved.

As shown in Sec. **1.54**, we can rest assured that the *instantaneous equivalent mechanism for acceleration is the same as that for velocity*.

6.13   *Linkages with Rolling Elements or Gears*

EXAMPLE 1. Cam, Roller, and Swinging Follower

The cam (1) drives the roller (2), which is carried by the follower (3) in Fig. 1. If ω_1 is given, we can draw the velocity image op (line ①) of the contact point P in link 1. Since the velocity of P is the same in links 2 and 1 (why?), the velocity difference $\mathbf{V}_{CP} = \mathbf{V}_C - \mathbf{V}_P$ is in the direction of ②, $\perp PC$. But since C is a point in link 3, its velocity $\overrightarrow{oc}$ must be that of ③, $\perp CB$; hence c is located at the crossing of ② and ③. The angular velocities of links 2 and 3 are given by

$$\omega_2 = -\left|\frac{\overrightarrow{pc}}{PC}\right| \quad \text{(clockwise)}$$

$$\omega_3 = -\left|\frac{\overrightarrow{oc}}{BC}\right| \quad \text{(clockwise)}$$

where a minus sign is used to denote a clockwise angular velocity.

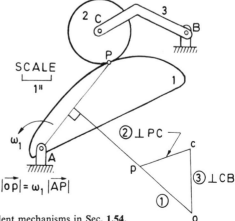

Figure 6.13-1. Cam, roller and swinging follower.

---

[6]This is consistent with the discussion of equivalent mechanisms in Sec. 1.54.

EXAMPLE 2. Geared Four-bar

Figure 2 shows a four-bar linkage $ABCD$ where the crank (1) is driven at known speed ω_1 and carries a gear fixed rigidly to it. The gears (5 and 6) pivot freely about pins C and D. The relative I.C. of gears 1 and 5 is denoted by E, and that of gears 5 and 6 is denoted by F. The velocity of point C is found by the standard procedure for a four-bar (see Fig. **6.11**-1) utilizing construction lines ① ($ob = \omega_1 AB$); ②, $\perp BC$; and ③, $\perp CD$. The image of point E on link 1 is found from ④ ($\mathbf{V}_{EB} \perp BE$) and ⑤ ($\mathbf{V}_E \perp AE$). Since $\mathbf{V}_E = \overrightarrow{oe}$ is the same whether we consider E a point in link 1 or link 5, we now know $\mathbf{V}_E$ and $\mathbf{V}_C$ in link 5. Hence we can find $\mathbf{V}_F$ in link 5 at the crossing of lines ⑥ ($\perp EF$) and ⑦ ($\perp CF$).

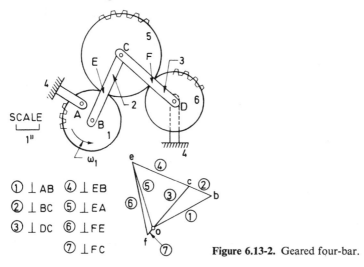

① $\perp$ AB ④ $\perp$ EB
② $\perp$ BC ⑤ $\perp$ EA
③ $\perp$ DC ⑥ $\perp$ FE
⑦ $\perp$ FC

Figure 6.13-2. Geared four-bar.

The angular velocities of gears 5 and 6 are

$$\omega_5 = -\left|\frac{\overrightarrow{ce}}{CE}\right| \quad \left(\text{or} \ -\left|\frac{\overrightarrow{cf}}{CF}\right|\right) \qquad \text{(clockwise)}$$

$$\omega_6 = \left|\frac{\overrightarrow{of}}{DF}\right| \qquad \text{(counterclockwise)}$$

Again it may be observed that the correct instantaneous velocities would be obtained by substituting a pin joint for the rolling contacts at E and F. However, incorrect accelerations would be predicted by this substitute mechanism. A correct substitute mechanism for acceleration analysis will be described in Sec. **6.23**.

6.14 Problems

1. The link 1 (2) in Fig. 1 has radius of curvature R_1 (R_2) at the point of contact, and C_1 (C_2) is the corresponding center of curvature. Show that

$$\frac{\omega_1}{\omega_2} = \frac{b}{a}$$

where a (b) is the perpendicular distance from A (B) to line $C_1 C_2$.

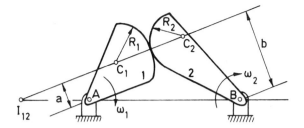

Figure 6.14-1 (Prob. 1).

2. Show that I_{12}, the relative I.C. of link 1 with respect to link 2, occurs at the intersection of lines C_2C_1 and AB in Fig. 1. Then show that

$$\frac{\omega_1}{\omega_2} = \frac{BI_{12}}{AI_{12}}$$

where AI_{12} and BI_{12} are directed line segments.

3. Suppose that the contact surface of link 1 in Fig. **6.12**-1(a) is perfectly straight at point P_1. In other words, link 1 is a flat-faced cam follower with $P_1C_1 = \infty$. Link 2 is exactly as drawn. Assume the figure is full scale.
   a. Draw the mechanism and the velocity image diagram for $\omega_2 = 1$ rad/sec.
   b. Calculate ω_1.
   c. For the same input (ω_2), would this mechanism produce the same output (ω_1) as the mechanism in Fig. **6.12**-1(a)?
   d. Draw an equivalent mechanism, with lower pairs only, for the mechanism with a flat-faced follower.

4. Do parts a and b of Problem **6.24**-9.

5. Do parts a and b of Problem **6.24**-10.

6. Do parts a and b of Problem **6.24**-11.

7. Find the extreme positions (up and down) of the piston in Fig. **6.11**-3 as the crank AB rotates through a complete revolution.

8. Determine, by graphical methods, the velocity of point G, in Fig. 2, if link AB rotates

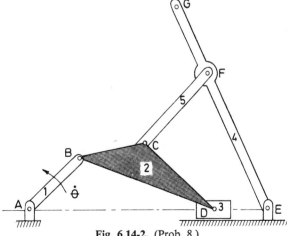

Fig. 6.14-2. (Prob. 8.)

counterclockwise at 1 rad/sec. Indicate the sequence of construction by numbers such as ①, ②, ③, ... *Indicate your velocity scale.* Suggestion: Use 1 in. = 1 in./sec. Use the following lengths (in.) and angles: $AB = 2$, $AE = 7$, $BD = 4$, $BC = 2$, $CF = 2.5$, $EF = 4$, $EG = 6$, $\angle DBC = 30°$, $\angle DAB = 45°$.

6.20 GRAPHICAL ANALYSIS OF ACCELERATION IN MECHANISMS

We shall show how the acceleration polygon referred to in Sec. **5.32** can be used to analyze certain classes of mechanisms. Just as in the case of velocity analysis, graphical methods will either fail completely or become exceedingly complicated for sufficiently complex mechanisms. Therefore we shall restrict our discussion of graphical methods to simple mechanisms. In later sections we shall develop an algebraic and numerical method that is applicable for all cases.

6.21 *Notation and Review of Relative Motion*

First we must establish some notational conventions. Figure 1 shows a general link with angular velocity ω and angular acceleration α (both are positive if counterclockwise). The points P and Q are two fixed points in the link whose absolute[7]

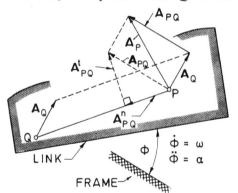

Figure 6.21-1. Points P and Q fixed in a moving link.

accelerations are denoted by $\mathbf{A}_P$ and $\mathbf{A}_Q$. The acceleration difference vector is

$$\mathbf{A}_{PQ} = \mathbf{A}_P - \mathbf{A}_Q \tag{1}$$

and, as stated in Sec. **5.32**, is correctly called the acceleration of point P *about* point Q.

The vector $\mathbf{A}_{PQ}$ can be resolved into components $\mathbf{A}_{PQ}^n$ and $\mathbf{A}_{PQ}^t$ which are, respectively, in the direction of $\overrightarrow{PQ}$ and $\perp$ to $\overrightarrow{PQ}$ (in the direction that α tends to rotate the tip of vector $\overrightarrow{QP}$), as shown in Fig. 1. Thus we have

$$\mathbf{A}_{PQ} = \mathbf{A}_{PQ}^n + \mathbf{A}_{PQ}^t \tag{2}$$

It was shown in Sec. **5.32** [Eqs. (**5.32**-13) and (**5.32**-14)] that

---

[7]Recall that *absolute acceleration* is the acceleration relative to an agreed-upon *fixed* or *ground* link.

$$A_{PQ}^n = |A_{PQ}^n| = \omega^2 PQ \qquad \text{(centripetal)} \qquad (3)$$

$$A_{PQ}^t = |A_{PQ}^t| = \alpha PQ \qquad \text{(transverse)} \qquad (4)$$

where PQ denotes the length of $\overrightarrow{PQ}$.

This notation will be used frequently and is summarized below for convenient recall:

$\mathbf{A}_P$ = acceleration of point P with respect to ground

$\mathbf{A}_{PQ} = \mathbf{A}_P - \mathbf{A}_Q = \mathbf{A}_{PQ}^n + \mathbf{A}_{PQ}^t$ = acceleration of P *about* Q

$\mathbf{A}_{PQ}^n$ = centripetal component of $\mathbf{A}_{PQ}$ (in direction of $\overrightarrow{PQ}$)

$\mathbf{A}_{PQ}^t$ = component of $\mathbf{A}_{PQ} \perp$ to PQ in sense of complex vector $i\alpha\overrightarrow{QP}$

$A_{PQ}^n = |A_{PQ}^n| = \omega^2 PQ$

$A_{PQ}^t = |A_{PQ}^t| = |\alpha| PQ$

ω = angular velocity of link (positive counterclockwise)

α = angular acceleration of link (positive counterclockwise)

Problems which involve links that slide upon rotating guides will require the use of Corioli's acceleration in graphical constructions. For such problems some additional notation is needed.

Figure 2 shows a link 2 rotating at speed ω_2 relative to the ground link. Link 3 is constrained to move on link 2 in such a way that a point P_3 (in 3) traces out a path Γ in link 2 with arc length s_{P3} (measured from a point O_s fixed in Γ), radius of curvature R, unit normal $\hat{n}$ (pointing toward the center of curvature of Γ at P_3), and unit tangent $\hat{t}$ (positive in direction of increasing s_{P3}). The point in 2, momentarily coincident with P_3, will be denoted as P_2.

According to Coriolis' theorem (see Sec. **2.42**), the acceleration, relative to ground, of point P_3 is given by[8]

$$\mathbf{A}_{P3} = \mathbf{A}_{P3}^{\text{BASE}} + \mathbf{A}_{P3}^{\text{COR}} + \mathbf{A}_{P3}^{\text{REL}} \qquad (5)$$

The terms of Eq. (5) will now be examined individually. The base acceleration is, by

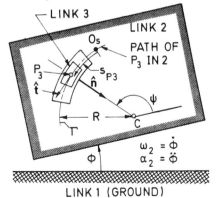

Figure 6.21-2. Link 3 constrained to move in rotating link 2.

----

[8]We write A_{P3} rather than A_{P_3} because subscripted subscripts are a nuisance in writing, reading, and typesetting. This practice will be followed except on rare occasions.

definition, the acceleration of the point in plane 2 which is currently coincident with P; it will be denoted by

$$\mathbf{A}_{P3}^{\text{BASE}} = \mathbf{A}_{P2} \tag{6}$$

The Coriolis acceleration has been shown to be

$$\mathbf{A}_{P3}^{\text{COR}} = \pm 2\omega_2 \, \dot{s}_{P3} \hat{\mathbf{n}} \tag{7}$$

(where the $+$ sign is used when $\hat{\mathbf{t}}$ lags $\hat{\mathbf{n}}$ by $90°$ as shown in Fig. 2). We may also express the velocity of P_3 relative to link 2 as

$$\dot{s}_{P3}\hat{\mathbf{t}} = \mathbf{V}_{P3} - \mathbf{V}_{P2} \tag{8}$$

where $\mathbf{V}_{P2}$ is the velocity of the point in plane 2 which is coincident with P. If it is ever necessary to explicitly indicate that the velocity is relative to link 2, we may write[9] $\dot{s}_{P3/2}$ for $\dot{s}_{P3}$. The relative acceleration consists of tangential and normal components:

$$\mathbf{A}_{P3}^{\text{REL}} = \mathbf{A}_{P3/2} = \mathbf{A}_{P3/2}^{t} + \mathbf{A}_{P3/2}^{n} = A_{P3/2}^{t}\hat{\mathbf{t}} + A_{P3/2}^{n}\hat{\mathbf{n}} \tag{9}$$

where

$$A_{P3/2}^{t} = \ddot{s}_{P3} \tag{10}$$

$$A_{P3/2}^{n} = \frac{(\dot{s}_{P3})^2}{R} \tag{11}$$

If it ever becomes necessary to emphasize that the *relative acceleration* is with respect to link 2, we may write $\ddot{s}_{P3/2}$ for $\ddot{s}_{P3}$. Finally we introduce the abbreviations

$$\mathbf{V}_{P3,P2} = \mathbf{V}_{P3} - \mathbf{V}_{P2} = V_{P3/2} \tag{12}$$

$$\mathbf{A}_{P3,P2} = \mathbf{A}_{P3} - \mathbf{A}_{P2} \tag{13}$$

It follows from Eq. (8) that the velocity difference $\mathbf{V}_{P3,P2}$ of the *coincident* points at P is indeed the *relative* velocity $(\mathbf{V}_{P3/2})$ of $P3$ with respect to link 2. However, it should be carefully noted that the acceleration difference $A_{P3,P2}$ is *not* the acceleration of P_3 *relative* to P_2; in fact $A_{P3,P2}$ is seen from Eq. (5) to be given by

$$\mathbf{A}_{P3,P2} = \mathbf{A}_{P3} - \mathbf{A}_{P2} = \mathbf{A}_{P3}^{\text{COR}} + \mathbf{A}_{P3/2}^{\text{REL}} \tag{14}$$

These results should be contrasted with the case of two *discrete* points A and B fixed to the same link, where the velocity difference $\mathbf{V}_{AB}$ is the velocity of point A *about* point B, and $\mathbf{A}_{AB}$ is the acceleration of A *about* B.

The notation introduced above is summarized below for future convenience:

P_3 = point on link 3

P_2 = point on link 2, momentarily coincident with P_3

---

[9] A slash ("/") will always indicate motion relative to the frame of reference identified by the symbol following the slash.

$\mathbf{V}_{P3}$ = velocity of P_3 relative to ground link

$\mathbf{A}_{P3}$ = acceleration of P_3 relative to ground link

$\mathbf{V}_{P3,P2} = \mathbf{V}_{P3} - \mathbf{V}_{P2}$ = velocity of P_3 relative to link 2 $(= \mathbf{V}_{P3/2})$

$\mathbf{A}_{P3,P2} = \mathbf{A}_{P3} - \mathbf{A}_{P2}$ (*not relative* acceleration)

$\dot{s}_{P3} = \dot{s}_{P3/2}$ = tangential component of velocity of P_3 relative to link 2

$\ddot{s}_{P3} = \ddot{s}_{P3/2}$ = tangential component of acceleration of P_3 relative to link 2

$\mathbf{A}_{P3/2}^n = [(\dot{s}_{P3})^2/R]\hat{\mathbf{n}}$ = normal acceleration of P_3 relative to link 2

$\mathbf{A}_{P3/2}^t = \ddot{s}_{P3}\hat{\mathbf{t}}$ = tangential acceleration of P_3 relative to link 2

$\mathbf{A}_{P3}^{\text{BASE}} = \mathbf{A}_{P2}$

$\mathbf{A}_{P3}^{\text{COR}} = \pm 2\omega_2 \dot{s}_{P3}\hat{\mathbf{n}}$

$\mathbf{A}_{P3}^{\text{REL}} = \mathbf{A}_{P3/2}^n + \mathbf{A}_{P3/2}^t$

$\mathbf{A}_{P3} = \mathbf{A}_{P3}^{\text{BASE}} + \mathbf{A}_{P3}^{\text{COR}} + \mathbf{A}_{P3}^{\text{REL}}$

6.22 Mechanisms with Lower Pairs

EXAMPLE 1. Four-Bar Linkage

Figure 1 shows a four-bar linkage *ABCE*. It is desired to find the acceleration of points *C* and *D* and the angular accelerations α_2 and α_3 of links 2 and 3. In addition to the configuration and dimensional data given, we are told that the angular velocity and acceleration of link 1 are $\omega_1 = \frac{3}{4}$ rad/sec and $\alpha_1 = \frac{5}{8}$ rad/sec².

The first step is to construct the velocity image, exactly as in Fig. **6.11**-1 [reproduced in Fig. 1(b)], where line *ob* represents $|\mathbf{V}_B| = \omega_1 AB = (\frac{3}{4})(4) = 3$ in./sec.

To start construction of the acceleration image [Fig. 1(c)], we note that acceleration $\mathbf{A}_B$ of point *B*, with respect to ground, can be resolved into a tangential component $\mathbf{A}_{BA}^t \perp$ to *AB* and a centripetal or normal component $\mathbf{A}_{BA}^n$. Hence we can draw the two known components: $A_{BA}^t = (\frac{5}{8})(4) = 2.5$ in./sec², and $A_{BA}^n = V_B^2/AB = 2.25$ in./sec², as shown by construction lines ①' and ②' which fix $\overrightarrow{o'b'} = \mathbf{A}_B$.

To find $\mathbf{A}_C$ we note that

$$\mathbf{A}_C = \mathbf{A}_B + \mathbf{A}_{CB} = \overrightarrow{o'b'} + \mathbf{A}_{CB}^n + \mathbf{A}_{CB}^t$$

The normal component A_{CB}^n has known magnitude $V_{CB}^2/CB = 2.52$ and direction $\overrightarrow{CB}$, as shown by line ③'. The direction of $\mathbf{A}_{CB}^t$ is known ($\perp CB$) as shown by line ④', but its magnitude is not known at this point. To locate point *c'* along line ④' we note that $\mathbf{A}_C = \mathbf{A}_{CE}^n + \mathbf{A}_{CE}^t$, where $A_{CE}^n = V_C^2/CE$, as shown by line ⑤'. Although we do not know the magnitude of A_{CE}^t, we do know its direction ($\perp CE$), as shown by line ⑥'. Since *c'* must lie along line ④' *and* along ⑥', is it uniquely defined at the intersection of the two lines. The desired acceleration $\mathbf{A}_C = \overrightarrow{o'c'}$.

To find the acceleration image of any point, such as *D* on link 2, we merely draw the triangle *b'c'd'* which is geometrically similar to triangle *BCD*, being careful to note that the circuit *b'c'd'* must have the same direction (i.e., clockwise) as circuit *BCD* (see Sec. **5.32**).

The magnitude of angular acceleration of link 2 is given [Eq. (**5.32**-12)] by

$$|\alpha_2| = \frac{A_{CB}^t}{BC}$$

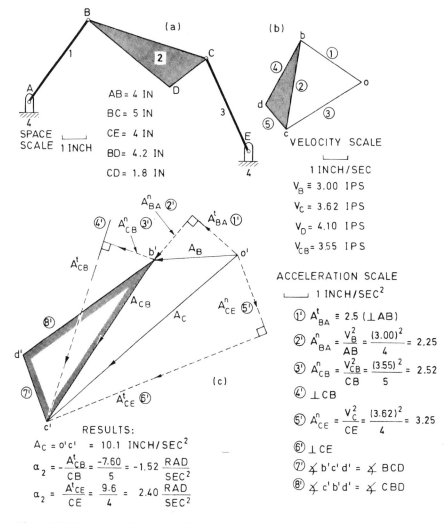

SPACE
SCALE └─┘ 1 INCH

AB = 4 IN
BC = 5 IN
CE = 4 IN
BD = 4.2 IN
CD = 1.8 IN

VELOCITY SCALE

└──┘
1 INCH/SEC

$V_B \equiv 3.00$ IPS

$V_C = 3.62$ IPS

$V_D = 4.10$ IPS

$V_{CB} = 3.55$ IPS

ACCELERATION SCALE

└──┘ 1 INCH/SEC^2

① $A^t_{BA} \equiv 2.5$ ($\perp$ AB)

② $A^n_{BA} = \dfrac{V^2_B}{AB} = \dfrac{(3.00)^2}{4} = 2.25$

③ $A^n_{CB} = \dfrac{V^2_{CB}}{CB} = \dfrac{(3.55)^2}{5} = 2.52$

④ $\perp$ CB

⑤ $A^n_{CE} = \dfrac{V^2_C}{CE} = \dfrac{(3.62)^2}{4} = 3.25$

⑥ $\perp$ CE

⑦ ∡ b'c'd' = ∡ BCD

⑧ ∡ c'b'd' = ∡ CBD

RESULTS:

$A_C = o'c' = 10.1$ INCH/SEC^2

$\alpha_2 = -\dfrac{A^t_{CB}}{CB} = \dfrac{-7.60}{5} = -1.52 \dfrac{\text{RAD}}{\text{SEC}^2}$

$\alpha_2 = \dfrac{A^t_{CE}}{CE} = \dfrac{9.6}{4} = 2.40 \dfrac{\text{RAD}}{\text{SEC}^2}$

Figure 6.22-1. Four-bar linkage: (a) Instantaneous configuration; (b) Velocity image; (c) Acceleration image.

and the sign of α_2 is negative because the sense of A^t_{CB} in Fig. 1(c) is such as to rotate point C clockwise around point B. According to the scale of the diagram, $A^t_{CB} = 7.60$ in./sec²; hence

$$\alpha_2 = -\frac{7.60}{5} = -1.52 \text{ rad/sec}^2 \quad \text{(clockwise)}$$

Similarly,

$$\alpha_3 = \frac{A^t_{CE}}{CE} = \frac{9.6}{4} = 2.40 \text{ rad/sec}^2 \quad \text{(counterclockwise)}$$

EXAMPLE 2. Quick-Return Mechanism

Figure 2 shows a quick-return mechanism ABP which is a well-known inversion of the slider-crank linkage. The *constant* angular velocity of crank AP is given as $\omega_2 = -100$ rpm (10.5 rad/sec), and we seek the angular acceleration α_4 of link BC.

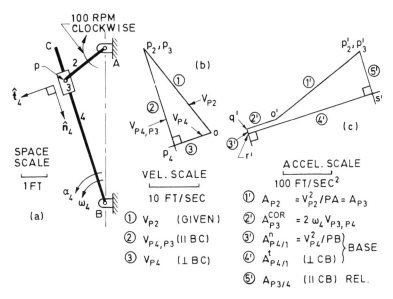

Figure 6.22-2. Acceleration analysis of quick-return mechanism: (a) Configuration; (b) Velocity polygon; (c) Acceleration polygon.

The velocity is constructed graphically from the vector equation

$$\mathbf{V}_{P4} = \mathbf{V}_{P3} + \mathbf{V}_{P4,P3} \tag{1}$$

where $V_{P3} = V_{P2} = |\omega_2 AP| = (10.5)(2) = 21$ ft/sec.

From the velocity image we can scale off the magnitudes

$$V_{P4} = op_4 = 7.5 \text{ ft/sec}$$

$$V_{P3,P4} = p_3 p_4 = 20 \text{ ft/sec}$$

The angular velocity of link 4 can now be calculated in the form

$$\omega_4 = \frac{op_4}{BP} = \frac{7.5}{5} = 1.50 \text{ rad/sec} \qquad \text{(counterclockwise)}$$

The acceleration analysis is based on the fact that we can express the acceleration of P_3 in two different ways. First, if we think of P as a point on both link 3 and link 2, we may write

$$\mathbf{A}_{P3} = \mathbf{A}_{P2} = \mathbf{A}^n_{P2/1} + \mathbf{A}^t_{P2/1} \tag{2}$$

where the fixed link has been designated as body 1. The magnitude $A^n_{P2/1} = V^2_{P2}/PA$ $= (21)^2/2 = 220.5$ ft/sec^2, and $A^t_{P2/1} = 0$; hence Eq. (2) serves to locate p'_2 and p'_3, the same point, as shown by line ① in Fig. 2(c).

Next, we consider P_3 to be a moving point on the link BC which is rotating at angular velocity ω_4. Coriolis' theorem [Eq. (**6.21**-5)] therefore gives us a second expression for A_{P3}:

$$\mathbf{A}_{P3} = \mathbf{A}^{COR}_{P3} + \mathbf{A}^{BASE}_{P3} + \mathbf{A}^{REL}_{P3/4} \tag{3}$$

To find A_{P3}^{COR} we need the relative velocity of P_3 on link 4:

$$V_{P3}^{REL} = V_{P3/4} = V_{P3,P4} = \overrightarrow{p_4 p_3}$$

Therefore

$$A_{P3}^{COR} = 2\omega_4 V_{P3,P4} \hat{t}_4 \tag{4}$$

$$A_{P3}^{BASE} = A_{P4} = A_{P4/1}^n + A_{P4/1}^t \tag{5}$$

$$A_{P3}^{REL} = A_{P3/4} \qquad (\| CB) \tag{6}$$

The reader should verify from the previously calculated quantities ω_4 and $V_{P3,P4}$ $(= \overrightarrow{p_4 p_3})$ that the direction of the Coriolis term is given correctly by Eq. (4). Unit vectors $\hat{t}_4$ ($\perp CB$) and $\hat{n}_4$ ($\| CB$) are defined as shown in Fig. 2(a).

The known numerical values to be used in Eqs. (4) and (5) are

$$2\omega_4 V_{P3,P4} = (2)(1.5)(20) = 60 \text{ ft/sec}^2$$

$$A_{P4/1}^n = \frac{V_{P4}^2}{PB} = \frac{(7.5)^2}{5} = 11.25 \text{ ft/sec}^2$$

Since A_{P3}^{COR} is completely known, it may be laid off as line ② $(\overrightarrow{o'q'})$ in the acceleration image. The term $A_{P4/1}^n$ in Eq. (5) is known in direction and magnitude and is laid off as line ③ $(\overrightarrow{q'r'})$. $A_{P4/1}^t$ is unknown at this point, but its (unoriented) direction ④ ($\perp CB$) is known.

Finally, we note that the magnitude of $A_{P3/4}$ is unknown, but its direction is known to be that of $\hat{n}_4$; further we know [from Eq. (3)] that $A_{P3/4}$ must pass through a point s' on line ④ such that

$$\overrightarrow{o'p_3'} = \overrightarrow{o'q'} + \overrightarrow{q'r'} + \overrightarrow{r's'} + A_{P3/4} \tag{7}$$

Therefore s' is the foot of the perpendicular dropped from p_3' on line ④. We now see that $\overrightarrow{r's'}$ points opposite to $\hat{t}_4$, and its magnitude, scaled from the drawing, is 267 ft/sec^2; therefore

$$A_{P4/1}^t = \overrightarrow{r's'} = -267 \,\hat{t} \text{ ft/sec}^2$$

We see that the *tangential* acceleration of P_4 ($A_{P4/1}^t = r's'$) would produce a *clockwise* rotation of link 4; hence

$$\alpha_4 = -\frac{A_{P4/1}^t}{PB} = -\frac{267}{5} = -53.4 \text{ rad/sec}^2$$

In summary, we note that the acceleration analysis hinges upon the fact that the acceleration of a sliding point on a rotating link can be expressed in two different ways—which provide two different vector paths to the same point on the acceleration diagram. This is the key to the graphical analysis of mechanisms with rotating sliding links. For *complex mechanisms*, this "key" may not open the door, and more ingenious graphical methods will be required (see Rosenauer and Willis [1953] or Hirschhorn [1962]). On the other hand, a perfectly general algebraic-numeric method will be described in Part 2.

6.23 Mechanisms with Higher Pairs

EXAMPLE 1. Cams and Followers

For curved cam followers, such as in Fig. 1(a), one may analyze the equivalent four-bar linkage AC_2C_3B, which was shown, in Sec. **1.54**, to produce the correct velocity and acceleration of the output link for the instant shown. The acceleration analysis for the four-bar linkage follows the pattern of Example 1 of Sec. **6.22**.

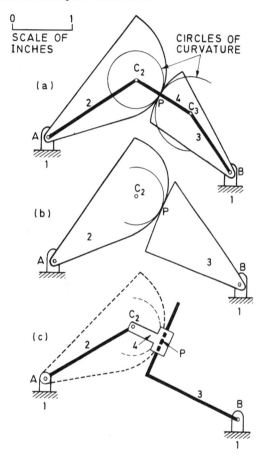

Figure 6.23-1. (a) Curved cams; equivalent linkage AC_2C_3B; (b) Flat-faced follower cam; point C_3 at ∞; (c) Equivalent linkage for (b).

If one cam has a flat face, as in Fig. 1(b), its center of curvature moves to ∞, and the equivalent four-bar linkage is not a useful model for acceleration analysis. However, it may readily be verified that the equivalent linkage of Fig. 1(c) will produce the same velocities as the original mechanism in three infinitesimally close configurations. Therefore, Fig. 1(c) shows a substitute linkage which will produce the same accelerations in links 2 and 3 as in Fig. 1(b). This equivalent mechanism is kinematically similar to that of Fig. **6.22**-2, which was fully analyzed in Example 2 of the previous section.

EXAMPLE 2. Mechanism with Rolling Elements

Figure 2(a) shows a linkage $BDEFG$, which consists of three moving links 1, 2, 3 and a ground link 4. Link 3 has the form of a roller of radius ρ_M which contacts a fixed

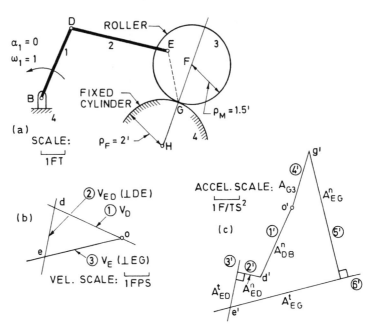

Figure 6.23-2. Acceleration analysis of mechanism with rolling element: (a) Space diagram; (b) Velocity polygon; (c) Acceleration polygon.

cylinder, of radius ρ_F, at point G. Assuming pure rolling of 3 on 4, find the acceleration of point E and the angular accelerations of links 2 and 3 when link 1 rotates uniformly at an angular velocity of $\omega_1 = 1$ rad/sec.

SOLUTION:

It is readily verified that the velocity polygon is as shown in Fig. 2(b); hence the angular velocity of the roller may be found to be

$$\omega_3 = \frac{oe}{GE} = \frac{3.2}{2} = 1.6 \text{ rad/sec}$$

To find the acceleration of point E we use the relationship

$$\mathbf{A}_E = \mathbf{A}_D + \mathbf{A}_{ED} = \mathbf{A}_D + \mathbf{A}_{ED}^n + \mathbf{A}_{ED}^t$$

where

①: $A_D = A_{DB}^n = \omega_1^2 BD = 1^2(3) = 3$ ft/sec²   ($\parallel DB$)

②: $A_{ED}^n = \dfrac{(de)^2}{DE} = \dfrac{(2.0)^2}{4.0} = 1$ ft/sec²   ($\parallel ED$)

③: $\mathbf{A}_{ED}^t$ ($\perp DE$)

The known vectors $\mathbf{A}_{DB}^n$ and $\mathbf{A}_{ED}^n$ are laid out as shown by lines ① and ② in Fig. 2(c). Only the direction of ③ ($\perp DE$) is known at this point, so we can infer that e' lies somewhere on line ③.

To proceed further we note that $\mathbf{A}_E$ can also be expressed in the form

$$\mathbf{A}_E = \mathbf{A}_{G3} + \mathbf{A}_{EG} = \mathbf{A}_{G3} + \mathbf{A}_{EG}^n + \mathbf{A}_{EG}^t$$

where G is thought of as a point of link 3.

From Eq. (**5.32**-20) we note that A_{G3} points from G to F and has magnitude given by

$$\text{④: } \quad A_{G3} = \omega_3^2 \frac{\rho_M \rho_F}{\rho_M + \rho_F} = \frac{(1.6)^2(1.5)(2.0)}{1.5 + 2.0} = 2.19 \text{ ft/sec}^2 \qquad (\|\overrightarrow{GF})$$

Next we note

$$\text{⑤: } \quad A_{EG}^n = \frac{(oe)^2}{EG} = \frac{(3.2)^2}{2} = 5.12 \text{ ft/sec}^2 \qquad (\|\overrightarrow{EG})$$

The two vectors $\mathbf{A}_{G3}$ and $\mathbf{A}_{EG}^n$ are now completely known and are laid off as lines ④ and ⑤. Finally we note

$$\text{⑥: } \quad A_{EG}^t \qquad (\perp EG)$$

and lay out line ⑥ accordingly. Since e' must lie on line ③ and on line ⑥, we have thereby located e' at the intersection of the two lines.

The desired acceleration

$$\mathbf{A}_E = \overrightarrow{o'e'}$$

and has magnitude $A_E = o'e' = 4.70 \text{ ft/sec}^2$. The required angular accelerations are seen to be

$$\alpha_3 = \frac{A_{EG}^t}{EG} = \frac{4.55}{2.00} = 2.28 \text{ rad/sec}^2 \qquad \text{(counterclockwise)}$$

$$\alpha_2 = -\frac{A_{ED}^t}{ED} = -\frac{1.45}{4.00} = -0.36 \text{ rad/sec}^2 \qquad \text{(clockwise)}$$

where A_{EG}^t and A_{ED}^t have been scaled from the acceleration polygon.

6.24 Problems

1. Explain the difference between the acceleration difference $\mathbf{A}_{PQ}$ defined by Eq. (**6.21**-1) and the acceleration difference $\mathbf{A}_{P3,P2}$ defined by Eq. (**6.21**-13). Is it ever correct to speak of either of the two vectors as a relative acceleration? Explain.

2. With the point P_3 defined by intrinsic coordinates R and ψ, as shown in Fig. **6.21**-2, show that the acceleration difference $\mathbf{A}_{P3,P2}$ is

$$\mathbf{A}_{P3,P2} = R(\dot\psi^2 + 2\omega_2 \dot\psi)\hat{\mathbf{n}} + R\ddot\psi \hat{\mathbf{t}}$$

3. Rod AB is pinned to link 2 at point A. Point P_3 on a link 3, sliding along the rod, has polar coordinates r, θ with respect to link 2, as shown in Fig. 1. If the link 2 has absolute angular velocity ω, show that the acceleration difference $\mathbf{A}_{P3,P2}$ is given by

$$\mathbf{A}_{P3,P2} = (\ddot r - r\dot\theta^2 - 2\omega r\dot\theta)\hat{\mathbf{r}} + (r\ddot\theta + 2\omega\dot r + 2\dot\theta\dot r)\hat{\boldsymbol{\theta}}$$

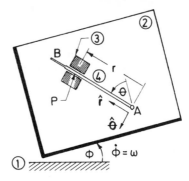

Figure 6.24-1 (Prob. 3).

4. Redraw the acceleration image of Fig. **6.22**-1 to scale. Show that the acceleration of point D could be found in two alternative ways from the vector relations

$$A_D = A_B + A_{DB}^n + A_{DB}^t$$

and

$$A_D = A_C + A_{DC}^n + A_{DC}^t$$

Make the appropriate constructions and compare your result to that of the figure.

5. Find the inflection circle for link 3 rolling on link 4 in Fig. **6.23**-2. Then use the Euler-Savary equation to find the center of curvature for point E. Use the information obtained to calculate the acceleration of E and the angular accelerations of links 2 and 3 when link 1 rotates at uniform angular velocity $\omega_1 = 1$ rad/sec.

6. Find the angular accelerations of links 2 and 3 in Fig. **6.23**-2 when link 1 rotates with uniform angular velocity $\omega_1 = 10$ rad/sec.

7. Find the acceleration of the piston F in Fig. **6.11**-3 when the link AB rotates clockwise at a uniform angular speed of 10 rad/sec.

8. Find the angular accelerations for links 2, 3, 5, and 6 of the mechanism shown in Fig. **6.13**-2 if link 1 is driven at a uniform angular velocity of $\omega_1 = 10$ rad/sec.

9. Using graphical methods, for Fig. 2, find
   a. The velocity of point E.
   b. The angular velocity of link FEC.

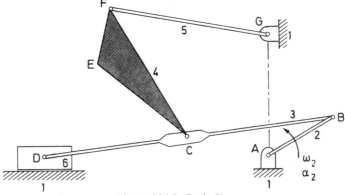

Figure 6.24-2 (Prob. 9).

c. The acceleration of point E.

d. The angular acceleration of link FEC.

Assume that $\omega_2 = 3$ rad/sec, $\dot{\omega}_2 = 0$. Note: Make explanatory comments, and show the sequence of construction by circled numbers, e.g., ①', ②', $\cdots$. Use the following dimensions (ft): $AB = 2$, $BC = CD = 3.8$, $CF = 3.75$, $FG = 4.25$, $GA = 3$, $FE = 1.4$; $\angle BAG = 60°$, $\angle EFC = 45°$.

10. Repeat Problem 9 but assume that $\omega_3 = 10$ rad/sec, $\dot{\omega}_2 = 0$.

11. Repeat Problem 9 but assume that $\omega_4 = 10$ rad/sec, $\dot{\omega}_4 = 0$. Note: It will be necessary to use a trial and error procedure or some variant of a superposition technique to find accelerations.

12. Find the angular velocity and acceleration of link 3 in Fig. **6.23**-1(a) when link 2 is driven (clockwise) with $\omega_2 = -10$ rad/sec, $\dot{\omega}_2 = 0$.

13. Find the angular velocity and acceleration of link 3 in Fig. **6.23**-1(b) when link 2 is driven (clockwise) with $\omega_2 = -10$ rad/sec, $\dot{\omega}_2 = 0$.

PART 2

ANALYTICAL KINEMATICS

Maclaurin had dealt in lines and figures—those characters, as Galileo has finely said, in which the great book of the Universe is written. Lagrange, on the contrary, pictured the Universe as an equally rhythmical theme of numbers and equations; . . . .

—H. W. Turnbull [1951]

Because kinematics is the *geometry* of motion, classical geometrical demonstrations have played a dominant role in its development. However, it will prove to be advantageous to convert the essentially geometric content of kinematics into an algebraic or analytical mode. This approach will introduce all the familiar benefits of algebraic geometry, but more important, it enables us to quantify the subject and thereby permits extensive use of digital calculation procedures. In addition, we shall find that *analytical kinematics* forms a natural and necessary prelude to *analytical dynamics*—the subject of Part 3.

Unlike the great inventor of analytical mechanics, we shall not go to the extreme of banishing geometrical diagrams.* On the contrary, we shall use them freely to avoid tedious verbal descriptions.

To illustrate the basic ideas of analytical kinematics in a simple, direct, and practical manner, we begin, in Chapter 7, with a discussion of mechanisms having a single closed loop. In Chapter 8 we consider general principles. Following that, we develop a uniform and systematic technique for deriving the governing kinematic equations of any plane mechanism.

It will be seen that displacements in mechanisms are governed by nonlinear algebraic equations which may not be solvable in closed algebraic form. However, even when it is not possible to find *explicit algebraic expressions* for the displacements in a particular mechanism, we can always find *numerical values* for them. It will also be shown that velocities and accelerations are governed by linear algebraic equations which enable us to solve explicitly for all velocities and accelerations once the displacements are all known.

It will be recalled that the graphical methods, described in Chapter 6, break down for so-called *complex* mechanisms (which may in fact look disarmingly simple on casual examination). The analytical techniques to be described are completely effective for all plane mechanisms.

Much of the material in this part is based on the concept of independent kinematic loops, as described by Paul [1960] and Paul and Krajcinovic [1970a, 1970b]. A related approach is described by Raven [1958].

---

*The following statement appears in the preface of Lagrange's *Mecanique Analytique* (1788): "No diagrams will be found in this work. The methods that I explain in it require neither constructions nor geometrical or mechanical arguments, but only the algebraic operations inherent to a regular uniform process. Those who love Analysis will, with joy, see mechanics become a new branch of it and will be grateful to me for thus having extended its field" (Dugas [1955, p. 333]).

SINGLE-LOOP MECHANISMS

7.10 THE FOUR-BAR MECHANISM

The simplest closed-loop mechanisms are those in which all the links lie on a single closed polygon, as illustrated in Fig. 1. It is convenient to define a positive vector sense for each rotating link, e.g., from A to B, and to define the angle of the link in the trigonometrically positive sense (counterclockwise) from a fixed x axis. With this convention, link CD in Fig. 1(a) has an angle θ_3 which is momentarily in the fourth quadrant. Link BC, in Fig. 1(b), has a variable length r_2, and its angle is the same as that of the slider (link 3) which it carries; i.e., $\theta_3 = \theta_2$.

Since the vector sum of a closed vector polygon is zero, we may write for Fig. 1(a) [or Fig. 1(d)]

$$\overrightarrow{AB} + \overrightarrow{BC} + \overrightarrow{CD} + \overrightarrow{DA} = 0 \tag{1}$$

A similar **vector loop equation** may be written for all the mechanisms in Fig. 1. We shall now illustrate by specific examples how the loop equations enable us to solve for position, velocity, acceleration, or any other kinematic variable.

The vector equation (1) is equivalent to the two **position loop equations**[1]

$$a_1 \cos \theta_1 + a_2 \cos \theta_2 + a_3 \cos \theta_3 - f = 0$$
$$a_1 \sin \theta_1 + a_2 \sin \theta_2 + a_3 \sin \theta_3 = 0 \tag{2}$$

Equations (2) merely state that the sum of the x (and of the y) projections of the vec-

---

[1]Sometimes referred to as **displacement loop equations**.

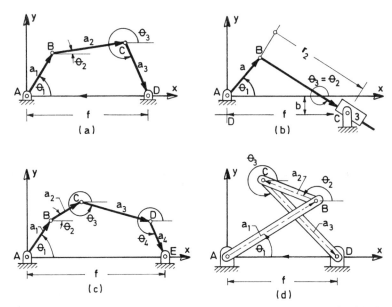

Figure 7.10-1. Single loop mechanisms: (a) 4-bar (1 D.O.F.); (b) rotating block inversion of slider crank (1 D.O.F.); (c) 5-bar (2 D.O.F.); (d) crossed 4-bar (1 D.O.F.).

tors in a closed vector polygon vanishes; this fact is of course familiar to all students of statics.

Let us assume, for the moment, that link 3 is the driving link. Therefore θ_3 is known, and Eqs. (2) provide two equations in the unknown variables θ_1 and θ_2 which may be solved for if the mechanism can indeed function for the given numerical data. It is possible, for example, to specify link lengths which prohibit closure (e.g., if $a_1 + a_2 + a_3 < f$) or to specify a prohibited angle for θ_3 (see Sec. **1.32**). As a matter of fact, it has been shown in Sec. **1.35** that Eqs. (2) can be solved for explicit algebraic expressions for θ_1 and θ_2 as functions of θ_3. Here we only wish to point out that the two loop equations provide sufficient information to define the configuration if one *input* (or *driving*) variable is specified. Thus the system has a *single degree of freedom* (1 DOF).

Let us suppose that for a given value of θ_3 ($\theta_3 \equiv q$) we have somehow (either algebraically, numerically, or graphically) found the corresponding values of θ_1 and θ_2. The position variables are now known, and we seek to find the velocities of all links when the driving velocity $\dot{\theta}_3$ is given some numerical value denoted by $\dot{q}$. By differentiating across Eqs. (2) we find the **velocity loop equations**

$$a_1\dot{\theta}_1 \sin \theta_1 + a_2\dot{\theta}_2 \sin \theta_2 + a_3\dot{\theta}_3 \sin \theta_3 = 0$$
$$a_1\dot{\theta}_1 \cos \theta_1 + a_2\dot{\theta}_2 \cos \theta_2 + a_3\dot{\theta}_3 \cos \theta_3 = 0 \tag{3}$$

Since $\dot{\theta}_3 = \dot{q}$ is known, Eqs. (3) represent two linear equations in two unknowns, which we can write in the matrix form

$$\begin{bmatrix} a_1 S_1 & a_2 S_2 \\ a_1 C_1 & a_2 C_2 \end{bmatrix} \begin{bmatrix} \dot{\theta}_1 \\ \dot{\theta}_2 \end{bmatrix} = \begin{bmatrix} -a_3 \dot{q} S_3 \\ -a_3 \dot{q} C_3 \end{bmatrix} \tag{4}$$

where we have introduced the abbreviations

$$S_i \equiv \sin \theta_i, \qquad C_i \equiv \cos \theta_i \tag{5}$$

Equations (4) may be readily solved[2] to give

$$\begin{aligned} \dot{\theta}_1 &= \frac{-a_3 a_2 \dot{q}(S_3 C_2 - C_3 S_2)}{a_1 a_2 (S_1 C_2 - C_1 S_2)} = \dot{q} \frac{a_3}{a_1} \frac{\sin (\theta_2 - \theta_3)}{\sin (\theta_1 - \theta_2)} \\ \dot{\theta}_2 &= \frac{-a_1 a_3 \dot{q}(S_1 C_3 - C_1 S_3)}{a_1 a_2 (S_1 C_2 - C_1 S_2)} = \dot{q} \frac{a_3}{a_2} \frac{\sin (\theta_3 - \theta_1)}{\sin (\theta_1 - \theta_2)} \end{aligned} \tag{6}$$

Sometimes it is convenient to express velocities in terms of the **influence coefficients** or **velocity ratios** k_{13} and k_{23}, defined by

$$k_{13} = \frac{\dot{\theta}_1}{\dot{\theta}_3}, \qquad k_{23} = \frac{\dot{\theta}_2}{\dot{\theta}_3} \tag{7a}$$

If it is understood that link 3 is the driving link, we may drop the second subscript in Eq. (7a) and simply write

$$\dot{\theta}_1 = k_1 \dot{q}, \qquad \dot{\theta}_2 = k_2 \dot{q} \tag{7b}$$

Until further notice, we shall assume that link 3 is the driving link and use the single-subscript notation.

From Eqs. (6) it is seen that

$$\begin{aligned} k_1 &\equiv \frac{\dot{\theta}_1}{\dot{\theta}_3} \equiv \frac{a_3}{a_1} \frac{\sin (\theta_2 - \theta_3)}{\sin (\theta_1 - \theta_2)} \\ k_2 &\equiv \frac{\dot{\theta}_2}{\dot{\theta}_3} \equiv \frac{a_3}{a_2} \frac{\sin (\theta_3 - \theta_1)}{\sin (\theta_1 - \theta_2)} \end{aligned} \tag{8}$$

Thus we see that it is a relatively simple matter to calculate velocities once the position variables are known.

To find accelerations, we differentiate the velocity loop equations (3) and obtain the **acceleration loop equations**

$$\begin{aligned} a_1 S_1 \ddot{\theta}_1 + a_2 S_2 \ddot{\theta}_2 &= -a_3 S_3 \ddot{\theta}_3 - \sum_i a_i C_i \dot{\theta}_i^2 \equiv R_1 \\ a_1 C_1 \ddot{\theta}_1 + a_2 C_2 \ddot{\theta}_2 &= -a_3 C_3 \ddot{\theta}_3 + \sum_i a_i S_i \dot{\theta}_i^2 \equiv R_2 \end{aligned} \tag{9}$$

If $\ddot{\theta}_3$ has a given numerical value $\ddot{q}$, the right-hand sides of Eqs. (9) (denoted by R_1 and R_2) are known, provided that we have previously solved for the position and

---

[2]Cramer's rule is especially convenient for solving sets of two linear equations in algebraic terms (see Appendix E.7).

velocity variables. Therefore, we find the accelerations by solving Eqs. (9) in the form

$$\ddot{\theta}_1 = \frac{R_1 a_2 C_2 - R_2 a_2 S_2}{a_1 a_2 (S_1 C_2 - C_1 S_2)} = \frac{R_1 C_2 - R_2 S_2}{a_1 \sin(\theta_1 - \theta_2)}$$

$$\ddot{\theta}_2 = \frac{a_1 S_1 R_2 - a_1 C_1 R_1}{a_1 a_2 (S_1 C_2 - C_1 S_2)} = \frac{-R_1 C_1 + R_2 S_1}{a_2 \sin(\theta_1 - \theta_2)}$$

(10)

7.11 Points of Interest

It is often desired to find the position, velocity, or acceleration of a point P attached to a specified link in a mechanism. For example, the point P on the coupler link of the four-bar linkage shown in Fig. 1 is identified by the fixed coordinates ξ_2 and η_2 measured in a coordinate system which is embedded in link 2. Similar coordinate systems (ξ_1, η_1) and (ξ_3, η_3) may be attached to links 1 and 3 in order to identify points fixed to those links (see Fig. 2).

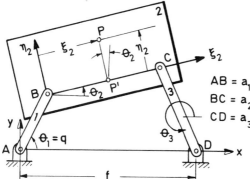

AB = a_1
BC = a_2
CD = a_3

Figure 7.11-1. Four-bar linkage.

The vector $\overrightarrow{AP}$ has coordinates (x_P, y_P) in the fixed coordinate system x, y. Upon denoting by P' the projection of P on line BC, we may write the **partial-loop**[3] vector equation

$$\overrightarrow{AP} = \overrightarrow{AB} + \overrightarrow{BP'} + \overrightarrow{P'P}$$

(1)

Equation (1) represents the two scalar equations

$$x_P = a_1 \cos\theta_1 + \xi_2 \cos\theta_2 - \eta_2 \sin\theta_2$$
$$y_P = a_1 \sin\theta_1 + \xi_2 \sin\theta_2 + \eta_2 \cos\theta_2$$

(2)

Upon differentiating Eqs. (2), we find the **partial-loop velocity** equations

$$\dot{x}_P = -a_1 S_1 \dot{\theta}_1 - (\xi_2 S_2 + \eta_2 C_2)\dot{\theta}_2$$
$$\dot{y}_P = a_1 C_1 \dot{\theta}_1 + (\xi_2 C_2 - \eta_2 S_2)\dot{\theta}_2$$

(3)

If $\theta_1, \theta_2, \dot{\theta}_1$, and $\dot{\theta}_2$ are all known from a previous analysis of the complete loop $ABCDA$, Eqs. (3) uniquely specify the desired velocity components. The acceleration

---

[3]The term *partial loop* is used to distinguish between Eq. (1) and equations such as Eqs. (7.10-1) which come from consideration of a *complete loop*.

components of the coupler point are readily found, by differentiating Eqs. (3), to be

$$\ddot{x}_P = -a_1(S_1\ddot{\theta}_1 + C_1\dot{\theta}_1^2) - (\xi_2 S_2 + \eta_2 C_2)\ddot{\theta}_2 - (\xi_2 C_2 - \eta_2 S_2)\dot{\theta}_2^2$$
$$\ddot{y}_P = a_1(C_1\ddot{\theta}_1 - S_1\dot{\theta}_1^2) + (\xi_2 C_2 - \eta_2 S_2)\ddot{\theta}_2 - (\xi_2 S_2 + \eta_2 C_2)\dot{\theta}_2^2 \tag{4}$$

The velocities and accelerations of points on the side links are found by analogous procedures, utilizing embedded coordinate axes (ξ, η), as illustrated in Fig. 2 (see Problem **7.60**-1).

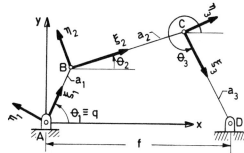

Figure 7.11-2. Local coordinate axes for a four-bar linkage.

7.12 *Application of Complex Vectors (Lamy's Theorem)*[4]

We now wish to show how the foregoing results may be obtained utilizing complex vectors.

The vector loop equation (**7.10**-1) may be expressed as

$$a_1 e^{i\theta_1} + a_2 e^{i\theta_2} + a_3 e^{i\theta_3} - f = 0 \tag{1}$$

Differentiating each term of Eq. (1) results in the velocity loop equation

$$i[a_1\dot{\theta}_1 e^{i\theta_1} + a_2\dot{\theta}_2 e^{i\theta_2} + a_3\dot{\theta}_3 e^{i\theta_3}] = 0 \tag{2}$$

Recognizing that $e^{i\theta_1}$, $e^{i\theta_2}$, and $e^{i\theta_3}$ are unit vectors in the directions of $\overrightarrow{AB}$, $\overrightarrow{BC}$, and $\overrightarrow{CD}$ (Fig. **7.11**-2), we can express the bracketed part of Eq. (2) in the form

$$g_1 U_1 + g_2 U_2 + g_3 U_3 = 0 \tag{3}$$

where

$$g_k \equiv a_k\dot{\theta}_k, \qquad U_k \equiv e^{i\theta_k} \qquad (k = 1, 2, 3) \tag{4}$$

Note that the vector equation (3) contains two scalar unknowns, g_1 and g_2. All the unit vectors U_k are completely known, as is the scalar $g_3 \equiv a_3\dot{\theta}_3$. We shall find that such equations occur repeatedly in subsequent work, and it is worthwhile to find a general solution of the vector equation. This is readily done with the aid of Fig. 1, which represents Eq. (3) as a closed vector triangle where $\overrightarrow{AB} \equiv g_1 U_1$, $\overrightarrow{BC} \equiv g_2 U_2$, $\overrightarrow{CA} \equiv g_3 U_3$ and g_1, g_2, g_3 represent the side lengths. By applying the law of sines twice to triangle *ABC*, we find

$$\frac{g_1}{g_3} = \frac{\sin(\theta_2 - \theta_3)}{\sin(\theta_1 - \theta_2)} \tag{5}$$

$$\frac{g_2}{g_3} = \frac{\sin(\theta_3 - \theta_1)}{\sin(\theta_1 - \theta_2)} \tag{6}$$

[4]This section is not essential for continuity.

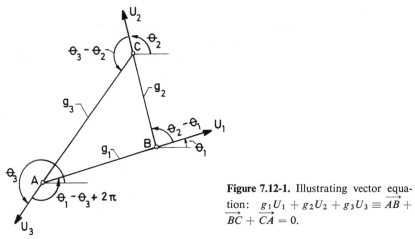

Figure 7.12-1. Illustrating vector equation: $g_1U_1 + g_2U_2 + g_3U_3 \equiv \overrightarrow{AB} + \overrightarrow{BC} + \overrightarrow{CA} = 0.$

Equations (5) and (6) represent the general solution to the vector equation $g_1U_1 + g_2U_2 + g_3U_3 = 0$. For this particular case, where $g_k = a_k\dot{\theta}_k$, Eqs. (5) and (6) duplicate the velocity ratios given in Eqs. (7.10-8). This illustrates how one obtains explicit expressions for velocity ratios directly from the velocity loop equations in complex form.

Equations (5) and (6), which will be referred to as **Lamy's theorem**,[5] may be easily recalled by noting that the subscripts in any given line follow a **cyclic permutation** of digits: (1, 2, 3), (2, 3, 1), or (3, 1, 2).

An alternative use of Eq. (2) is as a guide to the graphical construction of the velocity polygon *abc* shown in Fig. **6.11**-1 (see Problem **7.60**-25).

It is left as an exercise (Problem **7.60**-26) to show that the acceleration loop equation

$$a_1\ddot{\theta}_1(iU_1) + a_2\ddot{\theta}_2(iU_2) + a_3\ddot{\theta}_3(iU_3) - a_1\dot{\theta}_1^2U_1 - a_2\dot{\theta}_2^2U_2 - a_3\dot{\theta}_3^2U_3 = 0 \qquad (7)$$

obtained by differentiating Eq. (2) represents a set of instructions for the graphical construction of the acceleration polygon shown in Fig. **6.22**-1.

Alternatively, we may cast Eq. (7) into the form

$$h_1V_1 + h_2V_2 + h_3V_3 = 0 \qquad (8)$$

where

$$h_1V_1 \equiv a_1\ddot{\theta}_1iU_1 - \sum_{i=1}^{3} a_i\dot{\theta}_i^2U_i \equiv h_1e^{i\phi_1} \qquad (9)$$

$$h_2 \equiv a_2\ddot{\theta}_2, \qquad h_3 \equiv a_3\ddot{\theta}_3 \qquad (10)$$

$$V_2 = iU_2 \equiv e^{i\phi_2}, \qquad V_3 \equiv iU_3 \equiv e^{i\phi_3} \qquad (11)$$

Note that the arguments of the unit vectors V_i have been denoted by ϕ_i.

If $\dot{\theta}_1$ and $\ddot{\theta}_1$ are given and $\dot{\theta}_2, \dot{\theta}_3$ have been obtained from the previous velocity analysis, we can apply Lamy's theorem to Eq. (8) and solve for the unknown accelerations $\ddot{\theta}_2$ and $\ddot{\theta}_3$ in the form

$$\frac{a_2\ddot{\theta}_2}{h_1} = \frac{h_2}{h_1} = \frac{\sin(\phi_3 - \phi_1)}{\sin(\phi_2 - \phi_3)}, \qquad \frac{a_3\ddot{\theta}_3}{h_1} = \frac{h_3}{h_1} = \frac{\sin(\phi_1 - \phi_2)}{\sin(\phi_2 - \phi_3)} \qquad (12)$$

In evaluating Eqs. (12), we may find h_1 and ϕ_1 from Eq. (9), and we may note from Eqs. (11) that ϕ_2 and ϕ_3 are merely $\theta_2 + 90°$ and $\theta_3 + 90°$, respectively.

---

[5]Bernard Lamy (1645–1715) introduced Eqs. (5) and (6) to describe the equilibrium of three force vectors in his *Traite de Mecanique*, 1679 (see Lamb [1928, p. 15]).

7.20 MECHANISMS WITH SLIDERS

One example should suffice to illustrate the procedure for mechanisms with sliders. The displacement loop equations for the linkage of Fig. 1 are

$$a \cos \theta + r \cos \phi - d = 0 \tag{1}$$

$$a \sin \theta - r \sin \phi \quad\quad = 0 \tag{2}$$

If θ is considered the input variable, the solution for displacements is simply

$$\phi = \arctan_2 (r \sin \phi, r \cos \phi) = \arctan_2 (a \sin \theta, d - a \cos \theta) \tag{3}$$

$$r = \frac{a \sin \theta}{\sin \phi} \tag{4}$$

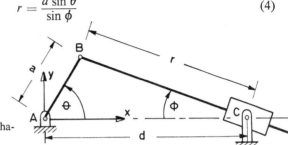

Figure 7.20-1. Oscillating engine mechanism.

Differentiating Eqs. (1) and (2) with respect to time, we find the velocity loop equations

$$-a\dot{\theta} \sin \theta + \dot{r} \cos \phi - r\dot{\phi} \sin \phi = 0 \tag{5}$$

$$a\dot{\theta} \cos \theta - \dot{r} \sin \phi - r\dot{\phi} \cos \phi = 0 \tag{6}$$

In matrix form Eqs. (5) and (6) are

$$\begin{bmatrix} \cos \phi & -\sin \phi \\ -\sin \phi & -\cos \phi \end{bmatrix} \begin{bmatrix} \dot{r} \\ r\dot{\phi} \end{bmatrix} = a\dot{\theta} \begin{bmatrix} \sin \theta \\ -\cos \theta \end{bmatrix} \tag{7}$$

Equations (7) are easily solved by Cramer's rule to give

$$\dot{r} = \frac{-(\sin \theta \cos \phi + \cos \theta \sin \phi)}{-(\cos^2 \phi + \sin^2 \phi)} a\dot{\theta} = a\dot{\theta} \sin (\theta + \phi) \tag{8}$$

$$r\dot{\phi} = \frac{-\cos \phi \cos \theta + \sin \phi \sin \theta}{-1} a\dot{\theta} = a\dot{\theta} \cos (\theta + \phi) \tag{9}$$

The accelerations $\ddot{r}$ and $\ddot{\phi}$ may be found by differentiating Eqs. (8) and (9).

7.30 MECHANISMS WITH GEARS OR ROLLING PAIRS

When gears or rolling pairs are present, the loop equations must be supplemented by **auxiliary equations** which express the no-slip condition at the contact point. Two examples will illustrate the procedure.

EXAMPLE 1

In Fig. 1, a wheel 3 of radius R_3 rolls on track 4 and is pinned at D to the dyad DBA.

The loop $ABDCPA$ gives the two position equations

$$a_1 \cos \theta_1 + a_2 \cos \theta_2 + a_3 \cos \theta_3 - s = 0 \qquad (1)$$

$$a_1 \sin \theta_1 + a_2 \sin \theta_2 + a_3 \sin \theta_3 - R_3 = 0 \qquad (2)$$

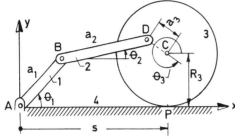

Figure 7.30-1. Mechanisms with rolling element.

Note that Eqs. (1) and (2) contain the four variables $\theta_1, \theta_2, \theta_3$, and s. Initially (at time $t \equiv t_0$) we may assemble the mechanism by arbitrarily choosing two variables, e.g., $\theta_1 \equiv \theta_{1,0}$ and $s \equiv s_0$; then Eqs. (1) and (2) may be solved (using the same analysis as for the case of a four-bar mechanism) to find the corresponding initial values $\theta_2 \equiv \theta_{2,0}$ and $\theta_3 \equiv \theta_{3,0}$.

After the mechanism has been assembled, the wheel rolls without slip at point P, satisfying the condition

$$s - s_0 = -R_3(\theta_3 - \theta_{3,0}) \qquad (3)$$

Thus we see that for all subsequent time, the three equations (1), (2), and (3) may be used to find θ_2, θ_3, and s for any specified value of θ_1. In other words, the system has only *one* DOF despite the fact that both $\theta_{1,0}$ and s_0 must be specified to define the *initial* configuration.

The velocity equations, found by differentiation of Eqs. (1) and (2), are

$$S_1 a_1 \dot{\theta}_1 + S_2 a_2 \dot{\theta}_2 + S_3 a_3 \dot{\theta}_3 + \dot{s} = 0 \qquad (4)$$

$$C_1 a_1 \dot{\theta}_1 + C_2 a_2 \dot{\theta}_2 + C_3 a_3 \dot{\theta}_3 \quad\quad = 0 \qquad (5)$$

where $S_i = \sin \theta_i$ and $C_i = \cos \theta_i$.

If one velocity (e.g., $\dot{\theta}_1$) is given, Eqs. (4) and (5) contain three unknowns and hence are not sufficient to solve the problem. A third equation is provided by differentiating Eq. (3) to give

$$\dot{s} = -R_3 \dot{\theta}_3 \qquad (6)$$

Equation (6) is the desired third equation, which can be substituted into Eqs. (4) and (5) to yield

$$\begin{bmatrix} S_2 & (S_3 - R_3/a_3) \\ C_2 & C_3 \end{bmatrix} \begin{bmatrix} a_2\dot{\theta}_2 \\ a_3\dot{\theta}_3 \end{bmatrix} = a_1\dot{\theta}_1 \begin{bmatrix} -S_1 \\ -C_1 \end{bmatrix} \qquad (7)$$

The solution of Eqs. (7) is

$$\dot{\theta}_2 = \frac{a_1}{a_2} \frac{(-R_3/a_3) \cos \theta_1 - \sin (\theta_1 - \theta_3)}{(R_3/a_3) \cos \theta_2 + \sin (\theta_2 - \theta_3)} \dot{\theta}_1 \qquad (8)$$

$$\dot{\theta}_3 = \frac{a_1}{a_3} \frac{\sin(\theta_1 - \theta_3)}{(R_3/a_3)\cos\theta_2 + \sin(\theta_2 - \theta_3)} \dot{\theta}_1 \qquad (9)$$

Having found $\dot{\theta}_3$, we can use Eq. (6) to find the third unknown, $\dot{s}$.

Thus we see that the "new twist" required when rolling elements are present is to add a sufficient number of *auxiliary equations* to those obtained from the standard loop traversals. Here, the auxiliary equation expresses the no-slip condition (3) or (6).

Note that accelerations are straightforwardly found by differentiating Eqs. (4), (5), and (6) and solving the linear system so obtained for $\ddot{\theta}_2$, $\ddot{\theta}_3$, and $\ddot{s}$.

EXAMPLE 2

In Fig. 2, a gear with pitch radius R_2 and center C is rigidly fixed to link 2 and meshes with gear 4, which has pitch radius R_4 and fixed center D. We consider link 1 as driving link.

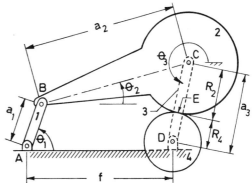

Figure 7.30-2. Mechanism with gear pairs.

The gears may be held in contact by a physical link DC or by means of *force closure* (e.g., gravity or spring forces). In the latter case, we draw the imaginary link DC (shown dashed in Fig. 2) on the kinematic skeleton and treat it henceforth as a real link.[6] Therefore, the motions of links 1, 2, 3 are precisely those of the four-bar mechanism $ABCD$, and the corresponding velocity ratios are given, via Eqs. (**7.10**-8), by

$$\frac{\dot{\theta}_2}{\dot{\theta}_1} = k_{21} \equiv \frac{a_1 \sin(\theta_3 - \theta_1)}{a_2 \sin(\theta_2 - \theta_3)} \qquad (10)$$

$$\frac{\dot{\theta}_3}{\dot{\theta}_1} = k_{31} \equiv \frac{a_1 \sin(\theta_1 - \theta_2)}{a_3 \sin(\theta_2 - \theta_3)} \qquad (11)$$

To find the velocity $\dot{\theta}_4$, we note that links 2, 3, and 4 form an epicyclic gear train. Since the velocities of gears 2 and 4 relative to arm 3 are $\dot{\theta}_2 - \dot{\theta}_3$ and $\dot{\theta}_4 - \dot{\theta}_3$, respectively, the relative velocity ratio for the gear pair is obviously[7]

$$\frac{\dot{\theta}_4 - \dot{\theta}_3}{\dot{\theta}_2 - \dot{\theta}_3} = -\frac{R_2}{R_4} \qquad (12)$$

Hence

$$R_4\dot{\theta}_4 = (R_2 + R_4)\dot{\theta}_3 - R_2\dot{\theta}_2 \qquad (13)$$

---

[6]The necessity of adding such carrier arms to gear pairs is discussed in greater depth in Sec. **8.21**.
[7]Recall the fundamental equation of epicyclic gear trains, Eq. (**3.32**-3).

$$\frac{\dot{\theta}_4}{\dot{\theta}_1} = \left(1 + \frac{R_2}{R_4}\right)\frac{\dot{\theta}_3}{\dot{\theta}_1} - \frac{R_2}{R_4}\frac{\dot{\theta}_2}{\dot{\theta}_1} \tag{14}$$

Introducing the velocity coefficients defined by Eqs. (10) and (11), we find

$$\frac{\dot{\theta}_4}{\dot{\theta}_1} = \left(1 + \frac{R_2}{R_4}\right)k_{31} - \frac{R_2}{R_4}k_{21} \tag{15}$$

If desired, the ratio R_2/R_4 can be replaced by N_2/N_4, where N_2 and N_4 represent the number of teeth in gears 2 and 4. Therefore

$$k_{41} \equiv \frac{\dot{\theta}_4}{\dot{\theta}_1} = \left(1 + \frac{N_2}{N_4}\right)k_{31} - \frac{N_2}{N_4}k_{21} \tag{16}$$

In short, we see that the presence of gears in a mechanism requires that the ordinary velocity loop equations be supplemented by an *auxiliary equation* of the form

$$\frac{\dot{\theta}_{\mathrm{I}} - \dot{\theta}_{\mathrm{III}}}{\dot{\theta}_{\mathrm{II}} - \dot{\theta}_{\mathrm{III}}} = \mp\frac{R_{\mathrm{II}}}{R_{\mathrm{I}}} = \mp\frac{N_{\mathrm{II}}}{N_{\mathrm{I}}} \tag{17a}$$

or

$$\dot{\theta}_{\mathrm{I}} - \dot{\theta}_{\mathrm{III}} = \mp\frac{N_{\mathrm{II}}}{N_{\mathrm{I}}}(\dot{\theta}_{\mathrm{II}} - \dot{\theta}_{\mathrm{III}}) \tag{17b}$$

for each[8] gear pair. In Eqs. (17), I and II represent meshing gears, and III represents the carrier arm. The negative sign is to be used when the gears contact externally (relative velocities have opposite signs) and the positive sign is used when the gears contact internally (relative velocities have the same sign).

To find accelerations we proceed exactly as usual for the nongeared system and then append auxiliary equations of the type

$$\ddot{\theta}_{\mathrm{I}} - \ddot{\theta}_{\mathrm{III}} = \mp\frac{N_{\mathrm{II}}}{N_{\mathrm{I}}}(\ddot{\theta}_{\mathrm{II}} - \ddot{\theta}_{\mathrm{III}}) \tag{18}$$

for each gear pair.

The angular position of gear 4 (in Fig. 2) is determined from integration of Eq. (13) in the form

$$\theta_4 - \theta_{4,0} = \left(1 + \frac{R_2}{R_4}\right)(\theta_3 - \theta_{3,0}) - \frac{R_2}{R_4}(\theta_2 - \theta_{2,0}) \tag{19}$$

where $\theta_{4,0}$ is an arbitrary constant, determined by the position of a mark scribed on gear 4 at a time t_0 when $\theta_2 = \theta_{2,0}$ and $\theta_3 = \theta_{3,0}$. Of course, θ_2 and θ_3 must be found from the displacement analysis of the four-bar mechanism $ABCD$.

Discussions on design aspects of gear linkages are given by Freudenstein and Primrose [1963], Dimarogonas et al. [1971], Oleksa and Tesar [1971], Mohan Rao and Sandor [1971], Buchsbaum and Freudenstein [1970], and Sandor et al. [1970].

---

[8]As pointed out in Secs. 3.32 and 3.33, Eqs. (17) may be applied directly to a train of many gears if $\mp(R_{\mathrm{II}}/R_{\mathrm{I}})$ is replaced by the so-called fixed-carrier train ratio.

7.40 CAM-LINKAGES

A linkage containing cam pairs (a **cam-linkage**) can always be analyzed by reduction to an equivalent mechanism with only lower pairs as described in Sec. **1.54**. This is a useful approach for cams with circular profiles, since the same equivalent mechanism will serve for all positions of the system (see Fig. **1.54**-2). However, for noncircular cam profiles, the equivalent mechanism changes its proportions as the configuration varies (see Fig. **1.54**-3). Therefore, an alternative approach is desirable for the analysis of cams with noncircular profiles. Both methods of analysis will now be illustrated.

7.41 *Circular Disk Cam*

Figure 1(a) shows a circular disk cam with a flat-faced oscillating follower. It is to be emphasized that we are focusing attention on the case where the radius of curvature ρ is constant and the center of curvature C remains at a fixed distance b from the pivot point A of the cam. If desired, we may replace the cam mechanism by the kinematically equivalent slider-crank inversion of Fig. 1(b). The following analysis applies equally well to Fig. 1(a) and (b).

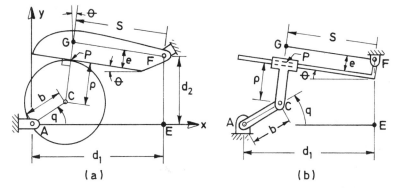

Figure 7.41-1. (a) Cam with flat-faced oscillating follower; (b) Equivalent mechanism with lower pairs only.

The displacement vector loop equation is readily seen to be

$$\overrightarrow{AE} + \overrightarrow{EF} + \overrightarrow{FG} + \overrightarrow{GP} + \overrightarrow{PC} + \overrightarrow{CA} = 0 \tag{1}$$

Resolving all vectors into x and y components, we find the independent equations

$$d_1 - s \cos \theta - (e + \rho) \sin \theta - b \cos q = 0 \tag{2}$$

$$d_2 + s \sin \theta - (e + \rho) \cos \theta - b \sin q = 0 \tag{3}$$

It will be shown presently that we can solve these equations explicitly for θ and s as functions of the input angle q. Even if we could not solve in algebraic terms, we could easily measure θ and s from a scale drawing; accordingly we assume that all configuration variables are known and proceed to find the unknown velocities $\dot{\theta}$ and $\dot{s}$ in terms of the driving velocity $\dot{q}$.

Upon differentiating Eqs. (2) and (3) we find

$$\begin{bmatrix} -\cos\theta & [s\sin\theta - (e+p)\cos\theta] \\ \sin\theta & [s\cos\theta + (e+p)\sin\theta] \end{bmatrix} \begin{bmatrix} \dot{s} \\ \dot{\theta} \end{bmatrix} = b\dot{q}\begin{bmatrix} -\sin q \\ \cos q \end{bmatrix} \tag{4}$$

Equations (4) are readily solved by Cramer's rule to yield

$$\dot{s} = [\sin(q+\theta) - \frac{e+p}{s}\cos(q+\theta)]b\dot{q} \tag{5}$$

$$\dot{\theta} = \frac{b}{s}\dot{q}\cos(q+\theta) \tag{6}$$

Accelerations may be found either by differentiating Eqs. (5) and (6) or else by differentiating Eq. (4) and solving the resulting linear equations for $\ddot{\theta}$ and $\ddot{s}$.

We now return to the somewhat tangential issue of solving explicitly for position variables. Multiplying Eqs. (2) and (3) by $\sin\theta$ and $\cos\theta$, respectively, and adding, we find

$$A\sin\theta + B\cos\theta = (e+p) \tag{7}$$

where

$$\begin{aligned} A &\equiv d_1 - b\cos q \\ B &\equiv d_2 - b\sin q \end{aligned} \tag{8}$$

A standard technique for solving an equation of the form (7) is to recast it in the form

$$\cos\beta\sin\theta + \sin\beta\cos\theta = \frac{e+p}{C} \tag{9}$$

where by definition

$$C = (A^2 + B^2)^{1/2} \tag{10}$$

$$\left.\begin{aligned} \cos\beta &= \frac{A}{C} \\ \sin\beta &= \frac{B}{C} \end{aligned}\right\} \quad \text{or} \quad \beta = \arctan_2(B, A) \tag{11}$$

The left-hand side of Eq. (9) can be expressed as $\sin(\theta + \beta)$; hence

$$\theta = \arcsin\left(\frac{e+p}{C}\right) - \beta \tag{12}$$

For the proportions shown in Fig. 1, the principal value of the arc sin function $[-(\pi/2) \le \theta \le \pi/2]$ should be used.

For a given input angle q, we may evaluate θ from Eq. (12) and then use either Eq. (2) or (3) to solve for s. Alternatively, we can eliminate $e + p$ between Eqs. (2) and (3) to find

$$s = b\sin(q-\theta) + d_1\sin\theta - d_2\cos\theta \tag{13}$$

Had the cam shown in Fig. 1 been noncircular, the fundamental loop equations would still be valid. However, the radius of curvature ρ and the distance b to the center of curvature would not be constants but would depend on the location of the contact point P along the cam profile. The approach just illustrated would lead to cumbersome algebraic expressions. Therefore an alternative approach for variable curvature cams will be described next.

7.42 Noncircular Cam Profiles

For noncircular cams it is convenient to express the cam profile in polar coordinates with the radius r given as a function of the polar angle ϕ in the form

$$r = F(\phi) \tag{1}$$

It will be recalled that the angle v between the radius vector and the outward normal vector at any point on the cam profile is given by Eq. (2.33-9) as

$$v = \arctan_2\left(-\sigma\frac{dF}{d\phi}, \sigma F\right) \tag{2}$$

where σ, the sign indicator, is defined by

$$\sigma = \left\{\begin{matrix} 1 \\ -1 \end{matrix}\right\} \quad \text{if} \quad \frac{d\phi}{ds}\left\{\begin{matrix} \geq 0 \\ < 0 \end{matrix}\right\} \tag{3}$$

and s represents oriented arc length on the cam profile.

The analysis of noncircular cam profiles will be illustrated by examples, showing how the usual loop equations are to be supplemented by auxiliary equations.

EXAMPLE 1

First we consider the case of an oscillating flat-faced follower as shown in Fig. 1. This problem is the same as that depicted in Sec. 7.41-1 except that the radius of curva-

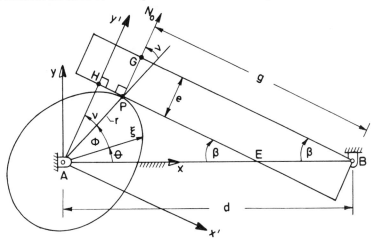

Figure 7.42-1. Cam with variable curvature and oscillating flat-faced follower.

ture and the center of curvature both vary. The driving angle θ is measured between the fixed x axis and an axis ξ engraved in the cam. The radius r is given by a known function $F(\phi)$ of the polar angle ϕ measured counterclockwise from the ξ axis.

We could work with the equivalent mechanism of Fig. 7.41-1, but it is more direct to proceed by writing the displacement loop equations (for loop $AHPGBA$) with respect to the moving axes (x', y'), where the y' axis is normal to the flat follower face:

$$r \sin \nu + g - d \cos \beta = 0 \tag{4}$$

$$r \cos \nu + e - d \sin \beta = 0 \tag{5}$$

Had we referred the loop equations to the fixed axes (x, y), we would have formed two equations that could be algebraically reduced to the above simple form. Equations (4) and (5) may be rewritten in the form

$$\sin \beta = \frac{e + F(\phi) \cos \nu}{d} \tag{6}$$

$$g = d \cos \beta - F(\phi) \sin \nu \tag{7}$$

Although the loop equations are hopelessly transcendental, if we think of θ as the independent angle, they become uncoupled if we utilize ϕ as the independent variable. For a fixed value of ϕ, Eq. (2) defines ν; then Eq. (6) defines[9] β, and Eq. (7) defines g. Finally we note from the triangle APE that $\theta + \phi + \nu = (\pi/2) - \beta$; hence

$$\theta = \frac{\pi}{2} - \nu - \beta - \phi \tag{8}$$

In short, all displacement variables are now known functions of the single variable ϕ.

To find the velocity ratios for the mechanism, Eqs. (6) and (7) are differentiated to give

$$\dot{\beta} d \cos \beta = \dot{\phi} F'(\phi) \cos \nu - \dot{\nu} F(\phi) \sin \nu \tag{9}$$

$$\dot{g} = -\dot{\beta} d \sin \beta - \dot{\phi} F'(\phi) \sin \nu - \dot{\nu} F(\phi) \cos \nu \tag{10}$$

From Eq. (2) we find that

$$\tan \nu = -F'(\phi)/F(\phi) \tag{11}$$

which, upon differentiation, gives

$$\dot{\nu} \sec^2 \nu = \dot{\phi}[-FF'' - (F')^2]/F^2 \tag{12}$$

From Eq. (12) we find the velocity ratio

$$\dot{\nu}/\dot{\phi} \equiv k_{\nu\phi} = -\cos^2 \nu \, [FF'' + (F')^2]/F^2 \tag{13}$$

Now Eq. 9 may be written in the form

$$\dot{\beta}/\dot{\phi} \equiv k_{\beta\phi} = (F' \cos \nu - k_{\nu\phi} F \sin \nu)/(d \cos \beta) \tag{14}$$

From Eq. (10) we find

---

[9]For the case illustrated in Fig. 1, β is limited to the range $-90° \leq \beta \leq 90°$.

$$g/\dot{\phi} \equiv k_{g\phi} = -(k_{\beta\phi}d\sin\beta + F'\sin v + k_{v\phi}F\cos v) \tag{15}$$

Finally, from Eq. (8) we find

$$\dot{\theta}/\dot{\phi} \equiv k_{\theta\phi} = -(\dot{v} + \dot{\beta} + \dot{\phi})/\dot{\phi} = -(k_{v\phi} + k_{\beta\phi} + 1) \tag{16}$$

Equations (13)–(16) give all of the instantaneous velocities in terms of any selected driving velocity. For example, if $\dot{\theta}$ is the driving velocity, we find

$$\dot{\phi} = \frac{\dot{\theta}}{k_{\theta\phi}}, \qquad \dot{\beta} = \frac{\dot{\theta}k_{\beta\phi}}{k_{\theta\phi}}, \qquad \text{etc.} \tag{17}$$

Accelerations may be found, if desired, by suitable differentiation; for example,

$$\ddot{\theta} = k_{\theta\phi}\ddot{\phi} + \frac{dk_{\theta\phi}}{d\phi}\dot{\phi}^2 \tag{18}$$

EXAMPLE 2

Consider the noncircular cam with a translating roller follower, shown in Fig. 2, where a roller of radius a is centered at C.[10] The cam position variable is θ, and the

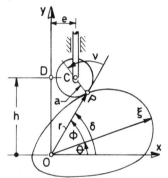

Figure 7.42-2. Noncircular disk cam with translating roller follower.

contact point P makes an angle ϕ with the axis ξ embedded in the cam. From the loop $OPCDO$ in Fig. 2 we see that

$$r\cos\delta + a\cos(\delta + v) - e = 0 \tag{19a}$$
$$r\sin\delta + a\sin(\delta + v) - h = 0 \tag{19b}$$

where

$$\delta \equiv \theta + \phi \tag{20}$$

Equation (19a) may be expanded to give

$$(r + a\cos v)\cos\delta - (a\sin v)\sin\delta = e \tag{21}$$

Following a pattern similar to that of Eqs. (7.41-7)–(7.41-11), we introduce the nomenclature

$$C = [(r + a\cos v)^2 + (a\sin v)^2]^{1/2} = (r^2 + a^2 + 2ar\cos v)^{1/2} \tag{22}$$

---

[10]This example has been treated in a different fashion by Townes and Blackketter [1971].

$$\cos \beta = \frac{r + a \cos v}{C} \tag{23}$$

$$\sin \beta = \frac{a \sin v}{C} \tag{24}$$

Then Eq. (21) can be expressed as

$$\cos \beta \cos \delta - \sin \beta \sin \delta = \frac{e}{C} \tag{25}$$

or

$$\cos (\beta + \delta) = \frac{e}{C} \tag{25a}$$

$$\theta + \phi \equiv \delta = \arccos \frac{e}{C} - \beta \tag{26}$$

It may be shown that the required branch of the arc cos function will make $\delta + \beta$ fall in the range $0°$ to $90°$ (see Problem 7.60-30).

If we treat ϕ as the independent variable, we see that all terms on the right-hand side of Eq. (26) are known for a given cam profile and a given value of ϕ; hence, θ may be calculated from Eq. (26). Finally, h may be calculated from Eq. (19b). Thus, all displacement quantities are known as functions of ϕ.

To find $\dot{\delta} = \dot{\theta} + \dot{\phi}$, we differentiate Eq. (19a) with respect to time and find

$$\frac{\dot{\delta}}{\dot{\phi}} \equiv k_{\delta\phi} = \frac{F'(\phi) \cos \delta - ak_{v\phi} \sin (\delta + v)}{r \sin \delta + a \sin (\delta + v)} \tag{27}$$

where $k_{v\phi}$ is defined as in Eq. (13). Since $\dot{\delta} = \dot{\theta} + \dot{\phi}$, it follows that

$$\frac{\dot{\theta}}{\dot{\phi}} = \frac{\dot{\delta}}{\dot{\phi}} - 1 = k_{\delta\phi} - 1 \tag{28}$$

From Eq. (19b) we find the follower pivot velocity:

$$\frac{\dot{h}}{\dot{\phi}} \equiv k_{h\phi} = F'(\phi) \sin \delta + k_{\delta\phi} r \cos \delta + (k_{\delta\phi} + k_{v\phi})a \cos (\delta + \phi) \tag{29}$$

Note that the analysis used is completely independent of whether the circular follower is free to rotate about point C or is frozen at point C. If the follower rolls without slipping on the cam, its angular velocity may be found by introducing another auxiliary equation as described in the following short example.

EXAMPLE 3

Figure 3 illustrates the case where a roller follower 4 is used with a cam 1.

The motion of the follower 3 is the same whether the roller is free to pivot about C or is locked to the follower. In the latter case we have to deal with an ordinary sliding cam, which is readily analyzed via the instantaneous equivalent four-bar linkage $ABCD$, where B is the center of curvature of the cam profile at P. If the roller is free to turn about C, we merely add the auxiliary equation [see Eq. (7.30-17)]

$$\frac{\dot{\theta}_4 - \dot{\theta}_2}{\dot{\theta}_1 - \dot{\theta}_2} = -\frac{R_1}{R_4}$$

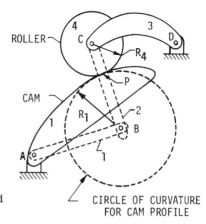

Figure 7.42-3. Mechanism with cam and roller or noncircular gears.

CIRCLE OF CURVATURE FOR CAM PROFILE

Hence

$$\dot{\theta}_4 = \left(1 + \frac{R_1}{R_4}\right)\dot{\theta}_2 - \frac{R_1}{R_4}\dot{\theta}_1 = \left[\left(1 + \frac{R_1}{R_4}\right)k_{21} - \frac{R_1}{R_4}\right]\dot{\theta}_1$$

Note that for a variable curvature cam R_1 will vary with position. If the roller has variable curvature, then R_4 will also vary with position. The analysis just given is valid if the profiles of members 1 and 4 represent the pitch lines of noncircular gears.

7.43 Fundamental Auxiliary Equation for Cam Pairs

In the cam linkage examples discussed so far, there has been a single noncircular cam hinged to a fixed pivot. For all such cases, we were able to find explicit algebraic solutions to the displacement equations. However, for the more general case of contact between cams of arbitrary profile, we shall see that a numerical solution of the displacement equations will usually be required.

Figure 1 shows two disc cams of arbitrary profile making contact at point P. The

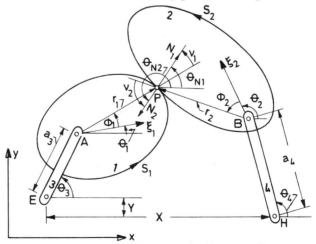

Figure 7.43-1. Floating cams of arbitrary profile.

cams are hinged at points A and B to links 3 and 4, which may be fixed or moving.[11] Axes ξ_1 and ξ_2 are engraved in the cams and make angles θ_1 and θ_2 with a fixed x axis. The radii of points on the cam profile are given, respectively, by known functions

$$r_1 = F_1(\phi_1)$$
$$r_2 = F_2(\phi_2) \tag{1}$$

of the polar angles ϕ_1 and ϕ_2 measured, about poles A and B, with respect to the embedded axes ξ_1 and ξ_2. The angles v_1 and v_2, measured from the outward normals N_1 and N_2, are given according to Eq. (7.42-2) by

$$v_i = \arctan_2 \left[-\sigma_i \frac{dF_i}{d\phi_i}, \sigma_i F_i \right] \qquad (i = 1, 2) \tag{2}$$

From Fig. 1 it may be noted that the angle θ_{N1} between the fixed x axis and the outward normal of cam 1 (at point P) is given by

$$\theta_{N1} = \theta_1 + \phi_1 + v_1 \tag{3a}$$

Similarly,

$$\theta_{N2} = \theta_2 + \phi_2 + v_2 \tag{3b}$$

Since the outward normals N_1 and N_2 point in opposite directions, it follows that

$$\theta_{N2} - \theta_{N1} = \pi \pm 2n\pi \tag{4}$$

where n is zero or an integer. Usually the value of n is immaterial, but if it is needed, it can be found from the initial state. Thereafter θ_{N1} and θ_{N2} will change continuously and n remains fixed. In the case of Fig. 1, $n = 0$. Upon substitution of Eqs. (3) into Eq. (4) we find the **fundamental auxiliary equation for cam pairs**:

$$(\theta_2 - \theta_1) + (\phi_2 - \phi_1) + (v_2 - v_1) = \pi \pm 2n\pi \tag{5}$$

Since v_i is a well-defined function of ϕ_i, Eq. (5) can be written in the form

$$\theta_2 - \theta_1 = G_1(\phi_1) - G_2(\phi_2) + \pi \pm 2n\pi \tag{6}$$

where

$$G_i \equiv \phi_i + v_i \tag{7}$$

The displacement loop equations, corresponding to loop $EAPBHE$ in Fig. 1, are

$$a_3 \cos \theta_3 + F_1(\phi_1) \cos (\theta_1 + \phi_1) - F_2(\phi_2) \cos (\theta_2 + \phi_2) - a_4 \cos \theta_4 - X = 0 \tag{8}$$
$$a_3 \sin \theta_3 + F_1(\phi_1) \sin (\theta_1 + \phi_1) - F_2(\phi_2) \sin (\theta_2 + \phi_2) - a_4 \sin \theta_4 + Y = 0 \tag{9}$$

where X and Y are the projections of line EH on the fixed axes and use has been made of the definitions $r_i \equiv F_i(\phi_i)$.

---

[11]A member which is not paired directly to the ground link is said to *float*. Figure 1 shows two *floating cams*.

Let us assume for the moment that links 3 and 4 are fixed so that $a_3, a_4, \theta_3, \theta_4, X$, and Y are all known constants. For this case the three equations (6), (8), and (9) involve the four variables $\theta_1, \phi_1, \theta_2, \phi_2$. Hence, if any of the four configuration variables are specified (e.g., θ_1), we have a set of three equations which (barring singular states[12]) are necessary and sufficient to solve for the remaining three variables. For the general case of arbitrary cam profiles, the three equations will have to be solved numerically. In special cases (e.g., one cam has a flat face) an explicit algebraic solution may be possible.

When links 3 and 4 are not fixed but constitute part of a kinematic chain not shown in Fig. 1, the three displacement equations found above will have to be supplemented with the loop equations corresponding to the complete kinematic chain. One auxiliary equation of type (6) may be written for each cam pair in the system.

7.50 MULTIFREEDOM SINGLE-LOOP SYSTEMS

Figure 7.10-1(c) is an example of a single-loop mechanism with more than one DOF. Its displacement loop equations are

$$\sum_{i=1}^{4} a_i C_i = f$$

$$\sum_{i=1}^{4} a_i S_i = 0 \tag{1}$$

where C_i and S_i represent $\cos \theta_i$ and $\sin \theta_i$. Accordingly, the velocity equations are of the form

$$\sum_{i=1}^{4} a_i S_i \dot{\theta}_i = 0$$

$$\sum_{i=1}^{4} a_i C_i \dot{\theta}_i = 0 \tag{2}$$

If any two of the four angular velocities $(\dot{\theta}_1, \dot{\theta}_2, \dot{\theta}_3, \dot{\theta}_4)$ are given, the two equations (2) enable us to find the others. For example, if $\dot{\theta}_1$ and $\dot{\theta}_3$ are known, they should be transferred to the right-hand sides of Eqs. (2) to give

$$\begin{bmatrix} S_2 & S_4 \\ C_2 & C_4 \end{bmatrix} \begin{bmatrix} a_2 \dot{\theta}_2 \\ a_4 \dot{\theta}_4 \end{bmatrix} = \begin{bmatrix} -(a_1 S_1 \dot{\theta}_1 + a_3 S_3 \dot{\theta}_3) \\ -(a_1 C_1 \dot{\theta}_1 + a_3 C_3 \dot{\theta}_3) \end{bmatrix} \equiv \begin{bmatrix} R_1 \\ R_2 \end{bmatrix} \tag{3}$$

Hence the unknown velocities are given by

$$a_2 \dot{\theta}_2 = \frac{R_1 C_4 - R_2 S_4}{S_2 C_4 - C_2 S_4}$$

$$a_4 \dot{\theta}_4 = \frac{S_2 R_2 - C_2 R_1}{S_2 C_4 - C_2 S_4} \tag{4}$$

---

[12]Singular states are discussed in Sec. 8.23.

If the angular accelerations of any two links are known, we may solve for the accelerations of the remaining links by differentiating Eqs. (2) with respect to time and solving the two resulting linear equations. We may then proceed to solve for the velocities and accelerations of points of interest, on any of the links, by setting up partial loop equations analogous to Eqs. (**7.11**-3) and (**7.11**-4).

Problems involving several degrees of freedom frequently arise in the analysis of open-loop subsystems. For example, suppose that we are given the velocity V_A of a point A on one link and the velocity V_B of a point B on a second link, and we are asked to find the velocity V_C of a *floating hinge* C when the two links are pinned together, as illustrated in Fig. 1, where complex vectors are indicated by capital letters with subscripts.

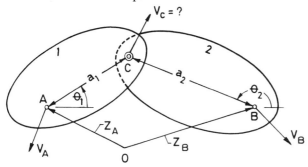

Figure 7.50-1. Floating hinge problem.

With an arbitrary origin at point O, the displacement loop equation for loop $ACBOA$ is

$$a_1 e^{i\theta_1} - a_2 e^{i\theta_2} - Z_B + Z_A = 0 \tag{5}$$

The corresponding velocity loop equation is

$$ia_1\dot\theta_1 e^{i\theta_1} - ia_2\dot\theta_2 e^{i\theta_2} - \dot Z_B + \dot Z_A = 0$$

or

$$a_1\dot\theta_1 U_1 - a_2\dot\theta_2 U_2 + g_3 U_3 = 0 \tag{6}$$

where

$$U_1 = e^{i\theta_1}, \quad U_2 = e^{i\theta_2}, \quad \text{and} \quad g_3 U_3 = \frac{-\dot Z_B + \dot Z_A}{i} = i(V_B - V_A) \tag{7}$$

In the last equation, $g_3 \equiv |V_B - V_A|$; and θ_3, the argument of unit vector U_3, is given by

$$\theta_3 \equiv \arg i(V_B - V_A) = \arg (V_B - V_A) + \frac{\pi}{2} \tag{8}$$

From Lamy's theorem [Eqs. (**7.12**-5) and (**7.12**-6)], the solution of Eq. (6) is given by

$$a_1\dot\theta_1 = g_3 \frac{\sin (\theta_2 - \theta_3)}{\sin (\theta_1 - \theta_2)}, \quad a_2\dot\theta_2 = -g_3 \frac{\sin (\theta_3 - \theta_1)}{\sin (\theta_1 - \theta_2)} \tag{9}$$

Finally, we see that the position of joint C is given by

$$Z_C = Z_A + a_1 e^{i\theta_1}$$

Hence the desired velocity is given by

$$V_C = \dot{Z}_C = \dot{Z}_A + ia_1\dot{\theta}_1 e^{i\theta_1} = V_A + ie^{i\theta_1}g_3 \frac{\sin(\theta_2 - \theta_3)}{\sin(\theta_1 - \theta_2)} \qquad (10)$$

where all quantities on the right-hand side are now known.

For the analogous problem of a *floating link*, see Problem **10.60**-14.

For systems with several DOF, the presence of several kinematic loops does not significantly complicate the analysis. Accordingly, we shall adopt a more general point of view for such systems. However, it is desirable to first discuss general criteria for identifying the DOF of such systems; this is done in Chapter 8.

7.60 Problems

1. Derive expressions analogous to Eqs. (**7.11**-3) and (**7.11**-4) for velocities and accelerations of points on the side links of the four-bar linkage of Fig. **7.11**-2.

2. Find the velocity ratios $k_{21} \equiv \dot{r}_2/\dot{\theta}_1$, $k_{31} \equiv \dot{\theta}_3/\dot{\theta}_1$ for the mechanism of Fig. **7.10**-1(b).

3. Derive an expression for the angular velocity of link CD in Fig. **7.10**-1(c) in terms of given velocities $\dot{\theta}_1$ and $\dot{\theta}_2$.

4. Derive Eqs. (**7.11**-2)–(**7.11**-4) by writing the partial-loop displacement equation in complex form and differentiating.

5. Form the complex displacement and velocity loop equations corresponding to Fig. **7.20**-1, and use Lamy's theorem to verify Eqs. (**7.20**-8) and (**7.20**-9).

6. With r, $\dot{r}$, and $\ddot{r}$ given, for the oscillating engine of Fig. **7.20**-1, find explicit algebraic solutions for (a) θ and ϕ, (b) $\dot{\theta}$ and $\dot{\phi}$, and (c) $\ddot{\theta}$ and $\ddot{\phi}$.

7. Deduce the accelerations $\ddot{r}$ and $\ddot{\phi}$ for the mechanism of Fig. **7.20**-1 using a complex loop equation and Lamy's theorem.

8. Verify Eqs. (**7.41**-5) and (**7.41**-6) by applying Lamy's theorem to the complex velocity loop equation associated with Fig. **7.41**-1.

9. The disk in Fig. 1 is pressed against a *frictionless* fixed plane surface which passes through point A as shown. The angular velocity of the disk is given by $\dot{\theta}_2 = \omega$. For an arbitrary configuration, find expressions for the angular velocity of link 1 and the velocity of point C.

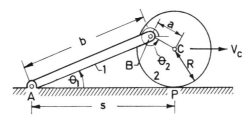

Figure 7.60-1 (Prob. 9).

10. For the pump mechanism shown in Fig. 2, find general expressions for the velocity of the plunger C when the *horizontal* velocity component of point H on the pump handle is given as $\dot{x}_H = u$.

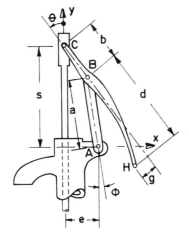

Figure 7.60-2 (Prob. 10).

11. A marine steering gear (Rapson's slide) is actuated by a push rod AB which controls the tiller angle θ (Fig. 3). Find an expression for $\dot{\theta}$ and $\ddot{\theta}$ when the velocity of point A is V_A.

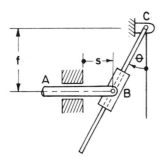

Figure 7.60-3 (Prob. 11).

12. Link BP, of length a, has a fixed pivot at B and drives a pin P in a circular slot, of constant radius R and center C, situated in a curved link 1 which has a fixed pivot at A (Fig. 4). The orientation of the slotted link is measured by ϕ, the angle between AB and the tangent to the slot arc, at A.

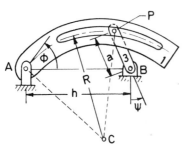

Figure 7.60-4 (Prob. 12).

a. Derive an expression for the angular velocity ratio $\dot{\phi}/\dot{\psi}$ in terms of position variables. *HINT:* consider a four-bar mechanism built on loop $ABPCA$.
b. Derive an expression for the angular acceleration $\ddot{\phi}$ as a function of $\dot{\psi}, \ddot{\psi}$, and position variables.

13. In Fig. 5, tooth numbers are shown in parentheses.
    a. Find expressions for the velocity ratios $k_{51} = \dot{\theta}_5/\dot{\theta}_1$ and $k_{61} = \dot{\theta}_6/\dot{\theta}_1$ for the geared linkage shown.
    b. Find expressions for $\ddot{\theta}_5$ and $\ddot{\theta}_6$ in terms of $\ddot{\theta}_1$.

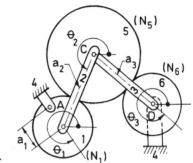

Figure 7.60-5 (Prob. 13).

14. Disc 3 rolls without slip over the fixed disk 4 in Fig. 6.
    a. Express $\dot{\theta}_3$ in terms of $\dot{\theta}_1$.
    b. Express $\ddot{\theta}_3$ in terms of $\dot{\theta}_1$ and $\ddot{\theta}_1$.

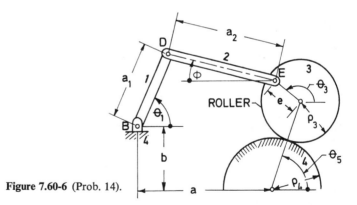

Figure 7.60-6 (Prob. 14).

15. In the quick-return mechanism of Fig. 7, link 2 rotates with known velocity $\dot{\theta}_2$ and acceleration $\ddot{\theta}_2$.
    a. Find expressions for $\dot{r}_3, \dot{\theta}_4$.
    b. Find expressions for $\ddot{r}_3, \ddot{\theta}_4$.
    c. Show that the loop equation $\overrightarrow{AP} + \overrightarrow{PB} + \overrightarrow{BA} = 0$, expressed in complex form, leads to a set of instructions for the graphical construction of velocities and accelerations shown in Fig. **6.22**-2.

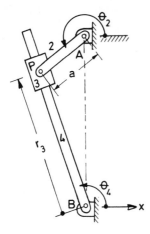

Figure 7.60-7 (Prob. 15).

16. In Fig. 8, the straight link *BPC* slides over the fixed circle with center *D* and radius
    DP = *b*. Show that the complex velocity loop equation is of the form

$$g_1 U_1 + g_2 U_2 + g_3 U_3 = 0 \tag{i}$$

where U_i are the unit vectors shown and g_i are real variables.
   a. From the vector triangle shown, deduce that

$$r\dot{\phi} = a\dot{\theta} \cos(\phi + \theta) \tag{ii}$$

$$\dot{r} = \left[\sin(\phi + \theta) + \frac{b}{r} \cos(\phi + \theta)\right] a\dot{\theta} \tag{iii}$$

   b. Solve Eq. (i) by Lamy's theorem, and verify Eqs. (ii) and (iii).

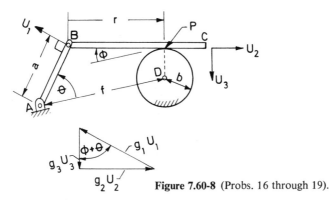

Figure 7.60-8 (Probs. 16 through 19).

17. In Fig. 8, link *BC* slides over the fixed circle centered at *D*, and θ, $\dot{\theta}$, $\ddot{\theta}$ are given. Find
    expressions for (a) *r* and ϕ, (b) $\dot{r}$ and $\dot{\phi}$, and (c) $\ddot{r}$ and $\ddot{\phi}$.

18. Let the link *BC* in Problem 17 represent a toothed rack which meshes with a gear, of
    pitch radius *b*, which is free to turn about a pivot at *D*. Find $\dot{\phi}$, $\dot{r}$, and the angular veloc-
    ity of the gear for a given value of $\dot{\theta}$.

19. Express the acceleration components $(\ddot{x}_P, \ddot{y}_P)$ of point P in link BC of Problem 16 in terms of $\ddot{\theta}$ and $\dot{\theta}$.

20. In Fig. 9, the slotted rod BC bears against a fixed pin at P and slides along rod AD.
    a. Express the velocity components of point C in terms of $\dot{\theta}$.
    b. Express the acceleration components of point C in terms of $\ddot{\theta}$ and $\dot{\theta}$.

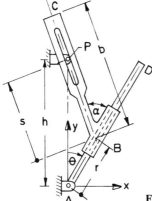

Figure 7.60-9 (Prob. 20).

21. In Fig. 10, a wheel of radius b rolls without slipping along the x axis. A connecting rod of length a is pinned at B on the periphery of the wheel and to a slider at A which moves with *constant* speed $\dot{s}$ along a wall inclined at angle β to the x axis.
    a. Find $\dot{\theta}$, $\dot{\phi}$.
    b. Find $\ddot{\theta}$, $\ddot{\phi}$.

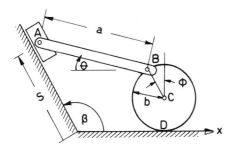

Figure 7.60-10 (Probs. 21 and 22).

22. In Problem 21, ϕ and s are initially set at ϕ_0 and s_0. Find the equations which determine ϕ and θ for all subsequently specified values of s.

23. Verify Eqs. (**7.12-5**) and (**7.12-6**) by simultaneously solving Eq. (**7.12-3**) and its complex conjugate equation: $g_1\bar{U}_1 + g_2\bar{U}_2 + g_3\bar{U}_3 = 0$.

24. Verify Eqs. (**7.12-5**) and (**7.12-6**) by simultaneously solving for the real and imaginary parts of Eq. (**7.12-3**).

25. Show that Eq. (**7.12-2**) prescribes the velocity polygon of Fig. **6.11**-1.

26. Show that Eq. (**7.12-7**) prescribes the acceleration polygon of Fig. **6.22**-1.

27. *Slot and Crank Coupling.*[13] Body 1 (Fig. 11) consists of a shaft AA' rigidly connected to a disc which has a slot along a diameter through A. Pin P slides in the slot and is hinged to a link 2 which is welded at B to a shaft BB'. Bearings at A' and B' keep the shafts AA' and BB' parallel and at a fixed distance e. Fixed orthogonal axes x, y lie in the plane of the disc with origin at A and with y lying in the plane formed by the two shafts. Let θ_1 denote the angle between the x axis and the slot axis, let $\omega_1 \equiv \dot{\theta}_1$ denote the angular velocity of the shaft 1, and let ω_2 denote the angular velocity of shaft 2.
   a. Find the ratio ω_2/ω_1 for an arbitrary angle θ_1.
   b. Find the acceleration of the driven shaft 2 as a function of θ_1, ω_1 and $\alpha_1 \equiv \ddot{\theta}_1$.
   c. Show that if $\omega_1 =$ constant, the driven shaft has an approximate dwell ($\omega_2 \doteq 0$) over a reasonable part of the cycle if $0 < 1 - (e/a) \ll 1$ (e.g., $e/a = 0.95$).

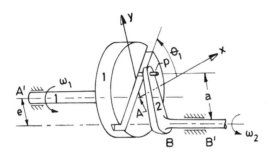

Figure 7.60-11 (Prob. 27).

28. For the disc cam with profile specified by $r = F(\phi)$ and a flat-faced translating follower, find the cam angle θ and the rise h for an arbitrary value of the angle ϕ. Also find expressions for the velocity ratios $\dot{\phi}/\dot{\theta}$, $\dot{h}/\dot{\theta}$. Note that (ξ, η) are coordinates embedded in the cam. Refer to Fig. 12.

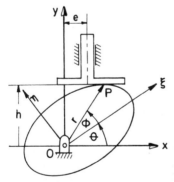

Figure 7.60-12 (Prob. 28).

29. The disc cam shown in Fig. 13 acts on an oscillating flat-faced follower. The profile of the cam is specified by $R = F(\phi)$. Coordinates (ξ, η) are embedded in the cam, with the ξ axis making an angle ψ with the fixed x axis. ρ is the radius of curvature, and C is the center of curvature, associated with the contact point P.
   a. Use the vector loop $ACPGFEA$ to establish a necessary and sufficient set of displace-

---

[13]Neklutin [1969].

ment equations which can be used to solve for all position variables as a function of either ψ or ϕ.

b. Comment upon the feasibility of solving the displacement equations in explicit algebraic form when ψ is given—what if ϕ is given?

c. Establish the velocity loop equations and solve for the ratio $\dot{\psi}/\dot{\phi}$.

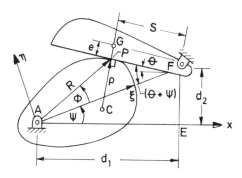

Figure 7.60-13 (Prob. 29).

30. Show that the quantities C and β defined by Eqs. (7.42-22)–(7.42-24) have the following geometrical significance: C is the distance OC, and β is the angle POC in Fig. 7.42-2. Verify that $\delta + \beta$ falls in the range between 0° and 90° for the case shown in Fig. 7.42-2.

31. Express the term $dk_{\theta\phi}/d\phi$, which appears in Eq. (7.42-18), in terms of position variables and their derivatives with respect to ϕ.

32. Find an expression for the accelerations $\ddot{h}$ (for the cam system of Fig. 7.42-2) in terms of given values of $\ddot{\theta}$ and $\dot{\theta}$.

33. An ellipse with major and minor semidiameters a and b is described by

$$r = \frac{b^2}{a(1 - e \cos \phi)}$$

where r is the radius measured from a focus F_1, ϕ is the polar angle about the focus, and $e = [1 - (b/a)^2]^{1/2}$. Find an expression for the angle v between the radius vector and the outward normal N. Refer to Fig. 14.

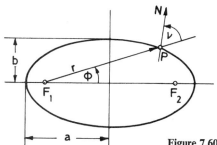

Figure 7.60-14 (Prob. 33).

34. An elliptic cam is pivoted at a focus A and acts on a flat-faced oscillating follower as in Fig. 15. Find the velocity ratio $\dot{\beta}/\dot{\theta}$ for an arbitrary position of the cam. Note that the equation of the cam profile is given in Problem 33.

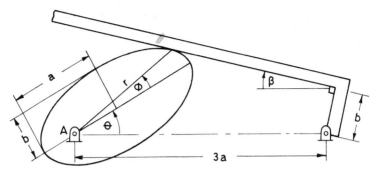

Figure 7.60-15 (Prob. 34).

8

FREEDOM AND CONSTRAINT IN KINEMATICS

8.10 LAGRANGIAN AND GENERALIZED COORDINATES

The motion of one or more *unconstrained* particles is conveniently described by the Cartesian or polar coordinates of each particle. However, a mechanism is a system of *interconnected* particles and rigid bodies whose motions are highly interdependent. For such *constrained* systems it would be inordinately cumbersome to restrict the choice of coordinates to the ordinary Cartesian or polar systems. Because a suitable choice of coordinates is so crucial to the analysis of constrained systems, we have taken some pains, in the following subsections, to carefully define the basis for such a choice. It will be seen that constrained systems are best described in terms of so-called *Lagrangian coordinates*, which include the well-known *generalized coordinates* as a subcategory.

8.11 Systems of Discrete Particles

We shall consider the motion of one or several particles with respect to a reference body in which we have embedded a Cartesian coordinate system, referred to henceforth as the **base coordinate system**. The **base coordinates** of a particle P_i will be designated as x_i, y_i, z_i. Since three coordinates uniquely define the position of P_i, we say that a particle has 3 degrees of freedom (DOF) in space. Henceforth we shall focus our attention on *planar motion* in which z_i is constant for each particle (the basic ideas developed are readily extended to spatial motion).

For planar motions, a particle has 2 DOF. If we are given a relation of the type

$$F(x_i, y_i, t) = 0 \tag{1}$$

we may use it to calculate y_i, at a given time t, from a knowledge of x_i.[1] Such a *constrained particle* will have only 1 DOF.

Equations such as Eq. (1), which reduce the freedom of a system, are called **equations of constraint**. At any given time, Eq. (1) represents a curve in the x, y plane. If t does not enter explicitly into the equation of constraint, the curve is fixed (a **stationary constraint**) and can be represented as a curved wire upon which the particle is constrained to slide.[2] If t does appear explicitly in Eq. (1), the "wire" is changing its form with time, and the system is said to have a **moving constraint**. Note that if, at a given time, we are given the arc distance s of the particle (measured from a fixed point on the curve), we can express Eq. (1) in its parametric form,

$$x_i = x_i(s, t)$$
$$y_i = y_i(s, t) \tag{2}$$

and thereby calculate the base coordinates from a single, more general, type of coordinate. Equations such as Eq. (2), which give the base coordinates of a particle in terms of other time-dependent variables, are called **coordinate transformation** equations. Coordinates (e.g., s) other than base coordinates are called **Lagrangian coordinates**.[3]

Now imagine that we have a system of two free particles P_1 and P_2 defined by the base coordinates (x_1, y_1) and (x_2, y_2); obviously this system has 4 DOF. However, the imposition of an equation of constraint of the type

$$F(x_1, y_1, x_2, y_2, t) = 0 \tag{3}$$

will reduce the DOF to 3, since Eq. (3) can be used to calculate one of the four base coordinates from a knowledge of the other three.

For example, suppose that the two particles are connected by a rigid rod of length a as in Fig. 1. Then Eq. (3) can be expressed in the form

$$F(x_i, y_i) \equiv (x_2 - x_1)^2 + (y_2 - y_1)^2 - a^2 = 0 \tag{4}$$

Any three of the four base coordinates (x_i, y_i) may be used to specify the configuration of the system. However, innumerable other sets of coordinates may be used. For example, we may use the polar coordinates (r, ϕ) of point P_1, and the angle θ shown in the figure, as a suitable set of three *Lagrangian Coordinates*; i.e., $\{\psi_1, \psi_2, \psi_3\} \equiv \{r, \phi, \theta\}$. With this choice, the *transformation equations* become

$$
\begin{aligned}
x_1 &= r \cos \phi = \psi_1 \cos \psi_2 \\
y_1 &= r \sin \phi = \psi_1 \sin \psi_2 \\
x_2 &= r \cos \phi + a \cos \theta = \psi_1 \cos \psi_2 + a \cos \psi_3 \\
y_2 &= r \sin \phi + a \sin \theta = \psi_1 \sin \psi_2 + a \sin \psi_3
\end{aligned} \tag{5}
$$

---

[1]Unless noted otherwise, we assume that equations of constraint can be solved for whatever variables are desired. For certain ranges of motion, this assumption may not be justified; such singular behavior is discussed further in Sec. **8.23**.

[2]When time is absent from an equation of constraint, the constraint is said to be **scleronomic**. If the time appears explicitly in the equation, the constraint is said to be **rheonomic**. The terms were coined by L. Boltzmann from the Greek roots *scleros* = "hard" and *rheos* = "flowing."

[3]The name *Lagrangian coordinates* has been suggested by Pars [1968]. Related terminology and other usage is discussed in Sec. **8.13**.

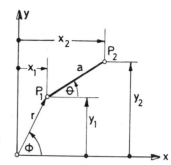

Figure 8.11-1. Particles connected by a rigid rod.

In general, we shall represent a set of M Lagrangian coordinates by the symbols $(\psi_1, \psi_2, \ldots, \psi_M)$, and the base coordinates for a set of N particles will be given by *transformation equations* of the form

$$x_i = f_i(\psi_1, \psi_2, \ldots, \psi_M, t) \tag{6a}$$

$$y_i = g_i(\psi_1, \psi_2, \ldots, \psi_M, t) \tag{6b}$$

where $i = 1, 2, 3, \ldots, N$.

Although the time t does not appear explicitly in Eqs. (5), it would indeed appear if, for example, the *moving constraint*

$$r - r_0 \sin pt = 0 \tag{7}$$

were imposed.

Note that Eq. (7) is a constraint involving Lagrangian variables. It is a special case of the more general form

$$F(\psi_1, \psi_2, \ldots, \psi_M, t) = 0 \tag{8}$$

Equations of contraint, such as Eq. (8), which contain *Lagrangian* coordinates (and possibly time) are called **explicit equations of constraint**. Equations of constraint which contain *base* coordinates [e.g., Eq. (4)] are implied by the transformation equations and will therefore be referred to as **implicit equations of constraint**.[4] Upon choosing a suitable set of Lagrangian coordinates for a given problem, one seldom has any reason to thereafter utilize the *implicit* equations of constraint.

Note that all equations of constraint considered so far (both implicit and explicit) have been expressed as *equations* relating *displacements* and perhaps the time. Such constraints are said to be **holonomic**. In Sec. **8.14** we shall consider constraints of a more general form (e.g., *inequalities* and relations between *velocity* components), but we restrict our discussion to *holonomic* constraints until further notice.

Any rigid body may be modeled, for kinematic purposes, by a finite number of constrained particles. For example, consider the link of a mechanism shown in Fig. 2(a). The kinematic function of the link is identical with that of the system of four particles P_1, P_2, P_3, P_4 shown in Fig. 2(b), provided that the interparticle distances remain fixed. This is achieved by enforcing the five (implicit) equations of constraint:

$$(x_i - x_j)^2 + (y_i - y_j)^2 - a_{ij}^2 = 0,$$
$$(i, j) = (1, 2), (2, 3), (3, 4), (4, 1), (2, 4) \tag{9}$$

---

[4]For example, Eq. (4) is readily derived from Eqs. (5). What we have termed *explicit* and *implicit* constraints have been called *residual* and *built-in* constraints by Kilmister and Reeve [1966, pp. 196, 197].

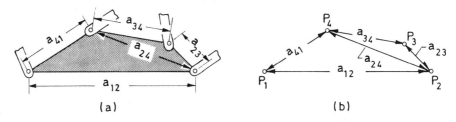

Figure 8.11-2. (a) Rigid lamina; (b) System of four particles.

We call the triangulated system of particles in Fig. 2(b) a **discrete particle model** of the original system.

Four *free* particles would have 8 DOF, but the imposition of the five constraint equations reduces the system DOF to 3, as it should be for a rigid link. In general, the degree of freedom F of a planar system of N particles related by C_i implicit equations of constraint is

$$F = 2N - C_i \tag{10}$$

Equation (10) is the basic equation for defining the degree of freedom of a planar system of discrete particles.

8.12 Systems of Rigid Bodies

It is unnecessary to replace all rigid bodies by their discrete particle models. It is usually preferable to view a rigid body as a continuous system. Despite the fact that a continuous lamina contains an infinite number of particles, its configuration is uniquely specified by only three Lagrangian coordinates. For example, in Fig. 1, the coordinates (x_Q, y_Q) of a reference point Q and the angle ϕ of an embedded reference line QS define the configuration of the lamina shown. A specific point P within the lamina is identified by *parameters* (ξ, η) referred to a Cartesian reference frame embedded in the lamina. Quantities such as ξ and η which do not vary with time will be called **parameters**, in distinction to *coordinates*, which do vary with time.

The base coordinates for particle P are given by the transformation equations

$$x_P = \psi_1 + \xi_P \cos \psi_3 - \eta_P \sin \psi_3$$
$$y_P = \psi_2 + \xi_P \sin \psi_3 + \eta_P \cos \psi_3 \tag{1}$$

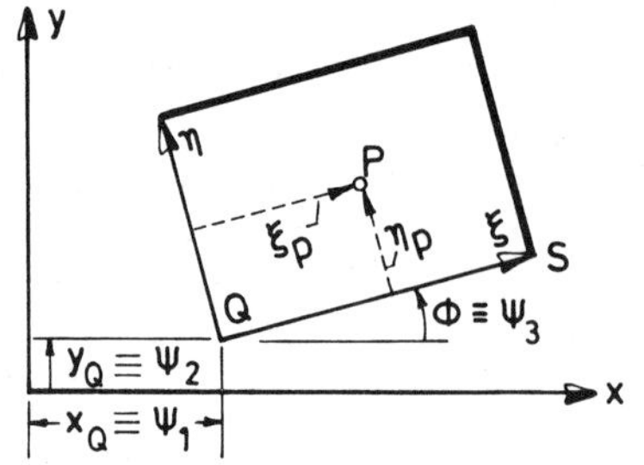

Figure 8.12-1. A lamina in plane motion.

Now consider a system of rigid bodies, such as in the triple pendulum of Fig. 2, where the base coordinates of arbitrary points, such as E and F, may be expressed in terms of three independent variables (ψ_1, ψ_2, ψ_3); e.g.,

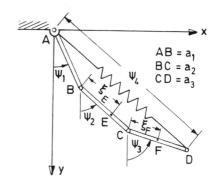

Figure 8.12-2. Triple pendulum.

$$x_E = a_1 \sin \psi_1 + \xi_E \sin \psi_2 \tag{2a}$$

$$y_F = a_1 \cos \psi_1 + a_2 \cos \psi_2 + \xi_F \cos \psi_3 \tag{2b}$$

In general, we may write transformation equations for any point P in a system of constrained rigid bodies in the form

$$\begin{aligned} x_P &= f_P(\psi_1, \psi_2, \ldots, \psi_M, t) \\ y_P &= g_P(\psi_1, \psi_2, \ldots, \psi_M, t) \end{aligned} \tag{3}$$

Frequently (particularly for closed-loop mechanisms) it will be found that the chosen Lagrangian coordinates are not independent but are related by *explicit equations of constraint*. For example, in the system of Fig. 3, the following two such equations ensure that loop $ABCDA$ remains closed:

$$\begin{aligned} a_1 \cos \psi_1 + a_2 \cos \psi_2 + a_3 \cos \psi_3 - f &= 0 \\ a_1 \sin \psi_1 + a_2 \sin \psi_2 + a_3 \sin \psi_3 &= 0 \end{aligned} \tag{4}$$

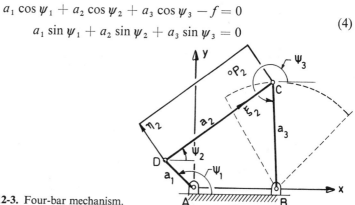

Figure 8.12-3. Four-bar mechanism.

In general, there will be N explicit equations of constraint associated with a given system; they may be expressed in the form

$$F_i(\psi_1, \psi_2, \ldots, \psi_M, t) = 0 \qquad (i = 1, 2, \ldots, N) \tag{5}$$

Lagrangian coordinates may be chosen in a wide variety of ways. However, for holonomic systems the minimum number of such variables needed to define the position of every particle in the system is, *by definition*, the *degree of freedom F*. Recall that F is a system characteristic which may be calculated from the equivalent *discrete particle model* via Eq. (**8.11**-10). Any set of M Lagrangian coordinates ψ_i will be said to be **complete** if the position of each point in the system is given in terms of the chosen ψ_i by *transformation equations* of the form (3).

Occasionally it is convenient to work with more than the minimum number of Lagrangian coordinates required for a complete set. For example, if a spring is stretched between points A and D in the 3-DOF system of Fig. 2, it may be desired to introduce a fourth position variable ψ_4 to simply describe the instantaneous length of the spring.

When we use more Lagrangian variables than the minimum required for a complete set, we say that we are using **redundant coordinates**. For each redundant coordinate introduced, there exists one explicit equation of constraint. For example, the constraint equation associated with the use of ψ_4 in Fig. 2 is

$$\left(\sum_{i=1}^{3} a_i \cos \psi_i\right)^2 + \left(\sum_{i=1}^{3} a_i \sin \psi_i\right)^2 - (\psi_4)^2 = 0 \tag{6}$$

If M Lagrangian variables are chosen for a system with a maximum of C independent equations of constraint among them, it is possible, in principle, to compute C of the coordinates from a knowledge of the other $(M - C)$ coordinates. Accordingly, the system has

$$F = M - C \tag{7}$$

degrees of freedom.

If there is any doubt about the number of independent[5] explicit constraint equations associated with a given choice of M, one can always calculate F directly from Eq. (**8.11**-10) (for the associated discrete particle model); then C may be found from

$$C = M - F \tag{8}$$

When $C = 0$, the M chosen coordinates constitute a set of **independent Lagrangian coordinates**, equal in number to the system's DOF.

8.13   *Generalized Coordinates*

Having chosen a complete set of M Lagrangian coordinates for a system with F degrees of freedom, we may specify F of the Lagrangian coordinates as **primary coordinates** and denote them by $(q_1, q_2, \ldots, q_F)$. Then the remaining position variables (all of which may be computed from the q_i) are called **secondary coordinates** and are denoted by $(\phi_1, \phi_2, \ldots, \phi_N)$, where

$$N = M - F$$

In classical dynamics, the *primary* coordinates are often called **generalized coordi-**

---

[5]A more formal determination of the number of independent equations of constraint is given by the rank of the Jacobian matrix formed from Eqs. (5). This interpretation follows from the discussion of Sec. **8.23**.

nates. In other words, *generalized coordinates are independent Lagrangian coordinates.*[6] Note that, by definition, the number of generalized coordinates equals the degree of freedom of the system. The rates of change with respect to time ($\dot{\psi}_1$ or $\dot{q}_i$) are called **Lagrangian** or **generalized velocities**.

For open-loop systems such as the triple pendulum of Fig. **8.12**-2, the three independent angles (ψ_1, ψ_2, ψ_3) represent a suitable set of generalized coordinates (q_1, q_2, q_3). As we have seen [Eqs. (**8.12**-2)], the base coordinates of any point in such a system are easily expressed explicitly in terms of the generalized coordinates. Therefore, explicit expressions for the velocity and acceleration components of all system points are readily found in terms of the generalized coordinates, velocities, and accelerations ($q_i, \dot{q}_i, \ddot{q}_i$). For example, differentiation of Eq. (**8.12**-2a) gives

$$\dot{x}_E = a_1 \dot{q}_1 \cos q_1 + \xi_E \dot{q}_2 \cos q_2 \tag{1}$$

In general, we shall have no difficulty in expressing the transformation equations for *open-loop* systems in the explicit form

$$\begin{aligned} x_i &= x_i(q_1, q_2, \ldots, q_F, t) \\ y_i &= y_i(q_1, q_2, \ldots, q_F, t) \end{aligned} \tag{2}$$

and it will seldom be necessary to introduce any secondary Lagrangian coordinates for such systems.

For a *closed-loop* system, with F degrees of freedom, we shall select F Lagrangian coordinates as generalized coordinates. For example, in the four-bar mechanism of Fig. **8.12**-3, we may select any one ($F = 1$) of the three angles ψ_1, ψ_2, ψ_3 as the single required generalized coordinate q. For the proportions shown (those of a rocker-crank), the crank angle ψ_1 is a particularly good choice for q since its range is unrestricted. If we choose rocker angle ψ_3 as the generalized coordinate, we must be careful to restrict q to the range determined by the extreme positions of link BC (shown dashed); otherwise the constraint equations (**8.12**-4) have no real solutions.

Having selected a subset of the Lagrangian coordinates ψ_i to serve as generalized coordinates q_i, we shall frequently find that the associated constraint equations cannot be solved *explicitly* for the remaining (secondary) coordinates. Even when they can be found (see Sec. **1.35** for the case of the four-bar mechanism), the expressions for the secondary variables in terms of the q_i will be very cumbersome for most closed-loop systems, and it will be difficult, if not impossible, to write the transformation equations in the form of Eqs. (2). Therefore it seldom pays to work *exclusively* with generalized coordinates when dealing with closed-loop systems. We have already seen in Chapter 7—and will see further examples in later chapters—that the retention of secondary variables can be very advantageous.

---

[6]NOTE ON TERMINOLOGY: What we (following Pars [1968]) refer to as *Lagrangian coordinates* were introduced by the symbols ($\xi, \psi, \theta, \ldots$) in the 1811 edition of Lagrange's *Mecanique Analytique*. In the older literature, e.g., Routh [1868, p. 321], they are referred to simply as *coordinates*. The earliest use of the term *generalized coordinates* that I've seen is by Kelvin and Tait [1879, Secs. 204, 313], whose definition we adopt. Many contemporary authors, e.g., Synge [1960, pp. 38, 59] and Greenwood [1965, p. 231], use the term *generalized coordinates* for any set of Lagrangian coordinates (independent or not). We are following the practice of those, e.g., Webster [1912, p. 108], Timoshenko and Young [1948, p. 194], and Lanczos [1966, p. 10], who use the term *generalized coordinates* to describe *independent* coordinates.

The examples just given show that the number of particles involved is irrelevant and, in fact, may be finite or infinite. What is important is that the position of every particle in the system is predictable in terms of the generalized coordinates $q_1, q_2, \ldots, q_F$ at all times.

Systems comprised of a finite number of particles or rigid bodies will be described by a finite number of generalized coordinates. However, for deformable bodies such as the beam in Fig. 1, the generalized coordinates are infinite in number. They may be

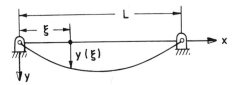

Figure 8.13-1. Deflected beam $(F = \infty)$.

chosen, for example, as the Fourier coefficients in a series of the type

$$y(\xi) = q_1 \sin \frac{\pi \xi}{L} + q_2 \sin \frac{2\pi \xi}{L} + \cdots \tag{3}$$

where, as usual, the q_i are functions of time and the parameter ξ serves to label the individual particles of the system.

8.14 Nonholonomic Constraints

In Sec. 8.11 we defined *holonomic constraints*[7] as those which can be expressed as *equations* relating *coordinates* and possibly the time. Any other type of constraint is said to be **nonholonomic**.

A common type of nonholonomic constraint is the so-called **unilateral** or **one-sided constraint**, where, for example, a piston can "slap" against its guides or a pin can rattle in a worn hole. Such constraints may be expressed in the form of inequalities; e.g., $x_{\min} \leq x \leq x_{\max}$. They may also be considered as *piecewise-holonomic*, with different holonomic constraints governing in different time intervals. Constraints expressed as equations rather than inequalities are sometimes called **bilateral constraints**.

Unilateral constraints are actually quite common in machinery. For example, disk cams are usually kept in contact by spring forces, and journal bearings for rail car wheels may consist of half a cylinder, concave upward, with the wheel axle held in place by the weight of the vehicle. In fact, Reuleaux [1876, pp. 226–246] viewed the tendency to replace unilateral constraints by bilateral constraints as a major evolutionary trend in the historical development of machinery. He referred to kinematic pairs with bilateral constraint as **closed pairs** and to those with unilateral constraint as **incomplete pairs** (also known today as **open pairs** or **unclosed pairs**). Where spring force, gravity, or other forces maintain closure of pairs, Reuleaux spoke of **force-closed pairs**.

Apart from unilateral constraints, nonholonomic constraints rarely occur in planar problems. However, an example of such a "rare occurrence" is shown in Fig. 1, where a rod moves over a plane in such a way that a point A of the rod always moves along the instantaneous direction AB of the rod. This motion may be enforced mechanically by a thin wheel hinged to point A, as in a "pizza slicer." A thin curved blade may replace the wheel, as in ice skates.

---

[7]The term was introduced by Hertz [1900] from the Greek *holos* = "whole (integral)" and *nomos* = "laws."

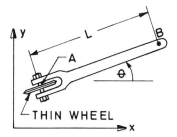

Figure 8.14-1. A nonholonomic system.

Introducing the Lagrangian coordinates

$$\{\psi_1, \psi_2, \psi_3\} = \{x_A, y_A, \theta\}$$

we see, from Fig. 1, that the velocity components of point A must satisfy the relation

$$\frac{\dot{y}_A}{\dot{x}_A} = \tan\theta \tag{1a}$$

or

$$dy_A - \tan\theta\, dx_A = 0 \tag{1b}$$

Equation (1b) shows that a small increment in y_A is *not* independent of small increments in x_A. If the system is forced to undergo a small increment dx_A, the value of dy_A will be given by Eq. (1b). Thus small motions of the system are uniquely specified by increments in two of the three Lagrangian coordinates $(dx_A, d\theta)$, and the system is seen to have only two **degrees of freedom in the small** (i.e., for *infinitesimal* displacements).

On the other hand, the system can undergo large-scale motions corresponding to any values of x_A, y_A, θ desired. To see this, perform the following mental (or physical) experiment. Start with $(x_A, y_A, \theta) = (0, 0, 0)$. Roll the wheel along the x axis to the configuration $(x_A, y_A, \theta) = (a, 0, 0)$. With point A fixed, turn the rod $90°$ and roll through distance b to the configuration $(a, b, 90°)$. Now keep point A fixed and rotate the rod through the angle $c - 90°$ to achieve the final configuration $(x_A, y_A, \theta) = (a, b, c)$, where a, b, and c are completely arbitrary. In other words, the system has *three* **degrees of freedom in the large** (i.e., for *finite* displacements).

It is characteristic of nonholonomic systems that they possess fewer **DOF in the small** than **in the large**. For analytical methods of distinguishing between holonomic and nonholonomic constraints, see Kilmister and Reeve [1966, p. 202] and Ince [1956, p. 54].

8.15 Problems

1. A system of N particles in a plane has C_i implicit equations of constraint. If the system is fully described by a minumum of M Lagrangian coordinates, how many independent explicit equations of constraint exist? Explain your reasoning.

2. A system of N rigid bodies in planar motion has C_i implicit equations of constraint. If the system has C_e explicit equations of constraint, how many Lagrangian coordinates have been chosen? Explain your reasoning.

3. Using the Lagrangian variables indicated in Fig. 1, find
   a. A suitable set of generalized coordinates q_i.
   b. The explicit equations of constraint.

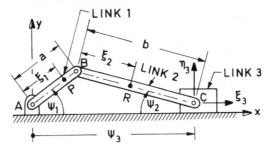

Figure 8.15-1. Slider-crank mechanism (Prob. 3).

  c. The equations of transformation from ψ_i to base coordinates (define any parameters you introduce).
  d. The implicit equations of constraint.

4. Repeat Problem 3 using Fig. 2.

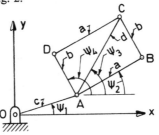

Figure 8.15-2. Four coplanar particles interconnected by rigid rods (Prob. 4).

5. Show that Eq. (**8.11**-4) may be deduced from Eqs. (**8.11**-5).

6. A string of total length L passes through a small hole in the floor (Fig. 3). The lower end carries a mass point P, and the upper end passes over a pulley to a handle H.
  a. Using S, R, and θ as Lagrangian coordinates, formulate a complete set of explicit constraint equations.
  b. Is the system holonomic? Explain.
  c. Is the system scleronomic or rheonomic? Explain.
  d. How many DOF does the system have? Explain.
  e. Given $S = S_0 \cos pt$, formulate a complete set of Lagrangian variables and the

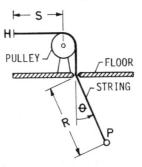

Figure 8.15-3 (Prob. 6).

corresponding explicit equations of constraint. Answer questions b, c, and d above for this system.

7. Assume that a point A on a boat can only move parallel to the boat's keel AB. To move the boat from position AB to the neighboring position $A'B'$ shown in Fig. 4, it is necessary to sail the boat on a path (shown dashed) which is far longer than the distance AA'. Explain this fact in terms of kinematic constraints on the boat and the DOF of holonomic versus nonholonomic systems.

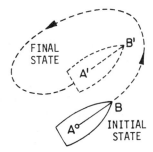

Figure 8.15-4 (Prob. 7).

8. Define the following terms:
   a. base coordinates
   b. equations of constraint
   c. stationary constraint
   d. moving constraint
   e. coordinate transformation equations
   f. Lagrangian coordinates
   g. explicit equations of constraint
   h. implicit equations of constraint
   i. holonomic and nonholonomic
   j. discrete particle model
   k. complete set of Lagrangian coordinates
   l. redundant coordinates
   m. generalized coordinates
   n. primary and secondary coordinates
   o. parameters (as opposed to coordinates)
   p. unilateral constraint
   q. bilateral constraint
   r. DOF in the small
   s. DOF in the large

8.20 MOBILITY ANALYSIS

One of the first questions that must be faced in the design or analysis of a mechanism is whether a given arrangement of links and joints will have the desired mobility or degree of freedom. Most mechanisms are designed to have a single degree of freedom (at least at low speeds, where elastic deformations are negligible), and a large body of literature exists on *criteria of constraint*[8] for determining whether or not a mechanism has *one* DOF. Typical of these rules are the criteria of Gruebler (Sec. **8.21**), which are based on a count of links and pairs of various types. These are useful rules, as we shall see, but their reliability is somewhat limited because of numerous *exceptional* cases which arise.

It is therefore desirable to develop an alternative to Gruebler's criteria, such as the

---

[8]Recall (Sec. **1.22**) that only mechanisms with 1 DOF are said to be *constrained* in the traditional literature on kinematics of machinery, but we use the word in a more general sense consistent with current and past usage in classical mechanics.

independent loop mobility criteria. This form of mobility analysis (Sec. **8.22**) is based on some simple ideas from the topology of networks.[9] By considering the independent loops in a network we are also led (Sec. **8.22**) to a better understanding of those *singular* cases for which Gruebler criteria (and some other forms of mobility analysis) break down. In subsequent chapters we shall see how the independent loops of a kinematic network may be further exploited to formulate the equations needed for a complete kinematic analysis.

8.21 Gruebler-Type Formulas

In Sec. **1.23** we saw that only *four* types of kinematic pairs occur in planar mechanisms, namely, *turning*, *sliding*, *rolling*, and *cam* pairs. Associated with each of these idealizations are certain *equations of constraint*, which may be formulated as follows:

1. For a *turning pair* [Fig. 1(a)], the displacements (x_{P1}, y_{P1}) of point P in body 1 must match the displacements (x_{P2}, y_{P2}) of point P in body 2; i.e.,

$$x_{P1} = x_{P2} \tag{1a}$$

$$y_{P1} = y_{P2}$$

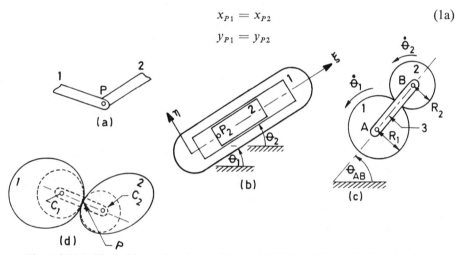

Figure 8.21-1. Planar kinematic pairs: (a) Hinge; (b) Slider; (c) Gear; (d) Cam (equivalent pairs and link shown by broken lines).

2. For *sliding pairs*, the slider and track must have the same angular orientation, and a point P_2 on the slider must remain on a line fixed in the track. In the nomenclature of Fig. 1(b),

$$\theta_1 = \theta_2 \tag{2a}$$

$$\eta_{P2} = 0 \tag{2b}$$

3. For *gear pairs* [Fig. 1(c)], we shall always assume that a link AB is present to maintain a fixed center distance.[10] Then the required equation of constraint is

---

[9]Appendix **D** contains all the concepts of network topology needed anywhere in this book.

[10]It is tempting to assume that the fixed center distance is enforced by a second equation of constraint rather than a physical link. This constraint, plus the no-slip constraint, would be equivalent to *hinging* the two gears together at the point of contact. In a sense this does happen, but because the hinge is in line with the centers A and B, the system forms a so-called *critical form* [see Fig. **8.23**-3(c)] which essentially removes one of the two constraints. This point is explained fully in Sec. **8.23**.

the epicyclic gear relation

$$(\dot{\theta}_2 - \dot{\theta}_{AB}) = -\frac{R_1}{R_2}(\dot{\theta}_1 - \dot{\theta}_{AB}) \qquad (3)$$

This equation is integrable; i.e., it is readily expressed in terms of the angular displacements rather than velocities [as in Eq. (7.30-19)].

4. In *cam pairs* the velocity component along the common normal at the contact point P must be the same in both bodies; i.e.,

$$V^n_{P1} = V^n_{P2} \qquad (4)$$

Although this single equation of constraint is expressed in terms of velocities, it is an integrable equation, since the constraint condition may also be expressed in terms of displacements, as in Sec. **7.43**.

Since the constraints for all four types of planar pairs may be formulated as *equations* in *position* variables, all planar pairs represent *holonomic* constraints.

We shall usually choose Lagrangian coordinates such that the equations of constraint for the two *lower* pairs (hinges and sliders) are *implicit*; i.e., they are implied by the choice of Lagrangian coordinates and need never be referred to explicitly. On the other hand, as we've seen in Secs. **7.30** and **7.40**, we shall always use the constraint equations for gears and noncircular cams *explicitly* as *auxiliary equations* in the Lagrangian variables.

For purposes of mobility analysis (i.e., for determining the DOF of a mechanism), it is sufficient to note that the introduction of a *lower* pair into an assemblage of links is equivalent to imposing *two* equations of constraint. That is, we reduce the DOF of a linkage by 2 each time we connect two members by a *simple*[11] lower pair. This idea leads to a simple **mobility criterion** for determining the DOF of a *linkage* (a mechanism with lower pairs only).

The DOF of N free links (in planar motion) is $3N$. If we form a kinematic chain by introducing P_s simple lower pairs, we reduce the DOF by $2P_s$. The kinematic *chain* so formed has DOF $= 3N - 2P_s$. If the chain is made into a mechanism by fixing one link, the DOF is reduced by 3, giving a DOF for the mechansim of

$$F = 3(N - 1) - 2P_s = 3M - 2P_s \qquad (5)$$

where $M = N - 1$ is the number of moving members.

The condition required for a linkage to have a single DOF is found by setting $F = 1$ in Eq. (5); i.e.,

$$3N - 2P_s - 4 = 0 \qquad (6)$$

This equation is usually called **Gruebler's criterion**, although Gruebler himself recognized the priority of both Chebyshev and Sylvester in obtaining the result (see Hartenberg and Denavit [1964] for historical references).

---

[11]Recall (Sec. **1.22**) that a *simple* pair connects two links. A joint of *multiplicity* k connects $k + 1$ links and may be considered as k simple pairs superposed.

In a linkage with P_1 simple pairs, P_2 double pairs, etc., the number of simple pairs is

$$P_s = P_1 + 2P_2 + \cdots + nP_n = \sum iP_i \qquad (7)$$

where n represents the maximum multiplicity of any pair in the linkage.

To illustrate the use of Eq. (5), consider the following example.

EXAMPLE 1

In Fig. 2, there are nine moving members ($N - 1 = 9$), and the pairs are classified as follows:

simple pairs at $B, C, D, E, F, G, H, I, K$: $P_1 = 9$

double pairs at A, J: $P_2 = 2$

total number of simple pairs: $P_s = P_1 + 2P_2 = 13$

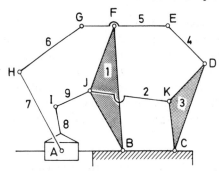

Figure 8.21-2. Buckeye valve gear mechanism.

Thus the DOF predicted by Eq. (5) is

$$F = 3(N - 1) - 2P_s = 3(9) - 2(13) = 1$$

Let us now generalize Eq. (5) to include systems with P_g *gear pairs* and P_c *cam pairs*, or a total of P_h *higher pairs*, where

$$P_h = P_g + P_c \qquad (8)$$

Because each higher pair introduces one additional equation of constraint [e.g., Eqs. (3) and (4)], 1 DOF is removed for each higher pair added to a planar mechanism. Thus Eq. (5) may be generalized to the form

$$F = 3(N - 1) - 2P_s - P_h \qquad (9)$$

For a system with 1 DOF, Eq. (9) assumes the form

$$3(N - 1) - 2P_s - P_h - 1 = 0 \qquad (10)$$

For brevity we shall henceforth refer to any equations based on Eq. (9) as **Gruebler-type formulas**.

The following examples illustrate the use of Eq. (9).

EXAMPLE 2

In Fig. **7.30**-2 there are four moving links *including the carrier arm CD;* i.e., $N - 1 = 4$. The kinematic pairs occur as follows:

> simple pairs A, B, C: $P_1 = 3$
>
> double pair at D: $P_2 = 1$
>
> gear pair at E: $P_g = 1$
>
> total simple pairs: $P_s = P_1 + 2P_2 = 5$
>
> total higher pairs: $P_h = P_g + P_c = 1$
>
> DOF: $F = 3(N - 1) - 2P_s - P_h = 3(4) - 2(5) - 1 = 1$

EXAMPLE 3

Figure 3(a) shows a rod (1) sliding over the ground at A and rolling at B on the disk (2), which in turn rolls over the ground at D. Before applying a Gruebler-type formula we add imaginary carrier arms CE and CF to keep the two rolling pairs *closed*, as shown in Fig. 3(b), where we think of the rolling pairs as gear pairs.

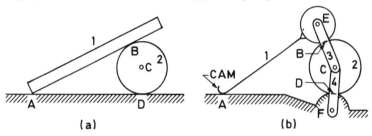

Figure 8.21-3. Example 3: (a) Sliding at A, rolling at B and D; (b) Equivalent mechanism.

In Fig. 3(b), we have distorted the "racks" AB and AD into finite radius gears to keep the length of the carrier arms finite, and we have represented the sharp corner of the rod at A by a cam surface. The equivalent system has four moving bodies ($N - 1 = 4$). The pairs in Fig. 3(b) are as follows:

> simple at E, F: $P_1 = 2$
> double at C: $P_2 = 1$ $\Big\}$ $P_s = P_1 + 2P_2 = 4$
>
> cam at A: $P_c = 1$
> gear at B, D: $P_g = 2$ $\Big\}$ $P_h = P_c + P_g = 3$
>
> $\therefore F = 3(N - 1) - 2P_s - P_h = 3(4) - 2(4) - 3 = 1$

This example illustrates that whenever cam pairs occur we have the option of replacing them by an *equivalent* set of lower pair systems or of counting them directly into the term P_h as we just did. In the author's experience, fewer blunders are made if cam pairs are replaced by their equivalent lower-pair systems before making a mobility analysis. For each gear pair, we *must* introduce one carrier link (e.g., links 3 and 4 in Fig. 3) otherwise the Gruebler formulas will be invalid.

It may have occurred to the reader that the term $N - 1$ in Gruebler's formulas could be replaced by a single term M representing the number of moving links.

Indeed, it can, but such a replacement can encourage the introduction of subtle errors. For example, if point A is fixed in Fig. 1(c), there are $M = N - 1 = 3$ moving members, two simple pairs, and one higher pair, so that Eq. (9) appears to give the result

$$F = 3M - 2P_s - P_h = 3(3) - 2(2) - 1 = 4$$

But we know that an epicycle gear train has 2 DOF. The error is caused by failure to recognize that the fixed frame on which A is mounted is part of the system. Therefore the joint at A connects three links of the system (1, 3, and frame) and constitutes a double pair. Hence $P_s = 3$, not 2, and

$$F = 3(N - 1) - 2P_s - P_h = 3(4 - 1) - 2(3) - 1 = 2$$

Exceptions to Gruebler-Type Formulas

We shall now discuss a recognizable class of *exceptions* to Gruebler's criterion. For example, the mechanism of Fig. 4 has $N - 1 = 2$, $P_s = 3$, $P_h = 0$. Therefore Eq. (9) predicts

$$F = 3(N - 1) - 2P_s - P_h = 3(2) - 2(3) = 0$$

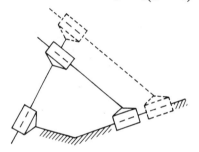

Figure 8.21-4. Prismatic loop.

This result suggests that the system is a structure. However, it is evident that the system may move, e.g., into the configuration indicated by broken lines. The linkage actually has 1 DOF.

Loops, such as in Fig. 4, in which all joints are prismatic pairs are called **prismatic loops**. The Gruebler-type formulas, given above, are invalid when prismatic loops are present. To find a suitable mobility criterion for systems with prismatic pairs only, we note that the links in such a system cannot experience any angular motion but can only *translate* relative to the ground link; hence the DOF of each unassembled translating body is 2. For N links, with one fixed, there are $2(N - 1)$ DOF. As each slider is assembled to its guide there will be introduced one equation of constraint similar to Eq. (2b). Let there be P_{sp} **simple prismatic pairs**, i.e., joints where one slider connects to one guide link. Subtracting the constraints from the DOF of the **unassembled** system, we find that the DOF for the assembled prismatic loop is

$$F_p = 2(N - 1) - P_{sp} \tag{11}$$

For the case of Fig. 4, Eq. (11) correctly predicts that

$$F_p = 2(N - 1) - P_{sp} = 2(2) - 3 = 1$$

EXAMPLE 4

For the coupled prismatic chains of Fig. 5(a), $N - 1 = 4$; simple pairs occur at A and D and double pairs at B and C, giving a total of $P_{sp} = 6$ simple prismatic pairs. Equation (11) predicts

$$F_p = 2(4) - 6 = 2$$

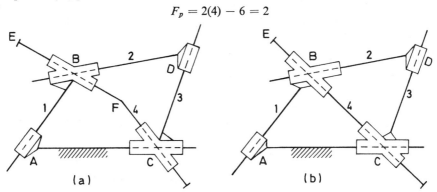

Figure 8.21-5. Coupled prismatic chains.

This is correct, since the distances AB and AC may be set at will; then all other key points, such as D and F, are uniquely located.

Suppose that the bent link 4 of Fig. 5(a) is straightened out as shown in Fig. 5(b). Then a third degree of freedom might appear to exist due to the possibility of link 4 sliding freely through the collinear sliders at B and C. However Eq. (11) still predicts that the DOF is 2. This prediction is correct because the fixing of point B uniquely sets the location of points C and D; however, link 4 may translate freely parallel to line BC.

A kinematic chain (or mechanism) will be called **regular** if it contains no prismatic loops; otherwise it is **nonregular**. The Gruebler-type formula (9) may be applied directly to find the DOF of a *regular* mechanism. To find the DOF for a *nonregular* mechanism, imagine that all prismatic loops (if any) are completely rigid and find the DOF of the resulting regular mechanism by Eq. (9). Then compute the additional degrees of freedom within each of the prismatic loops by means of Eq. (11) and add these to the previously calculated result. The procedure will be illustrated with the aid of Fig. 6.

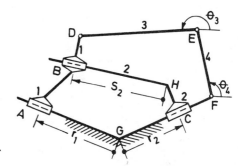

Figure 8.21-6. Nonregular mechanism.

EXAMPLE 5

If we "freeze" loop $ABHCGA$ in Fig. 6, we find that loop $DEFCHBD$ has $N - 1 = 2$ and $P_s = 3$ (pairs D, E, F), giving $F = 0$ according to Eq. (9). Now unfreeze the prismatic loop $ABHCGA$ and apply Eq. (11) to it, giving $F_p = 2(2) - 3 = 1$. Adding the two freedoms, we find that the DOF is $0 + 1 = 1$.

It is worth noting that the reasoning which led to the Gruebler-type formulas can be extended to the case of three-dimensional mechanisms (Bottema [1953]). In three dimensions, each free body has 6 DOF; thus there is a total of $6(N - 1)$ degrees of freedom for $N - 1$ moving free bodies. If there are p_k kinematic pairs at each of which C_k constraints are imposed (i.e., C_k DOF are removed), the mechanism as a whole will have

$$F = 6(N - 1) - \sum_k C_k p_k \tag{12}$$

degrees of freedom, barring singular cases. Equation (12) has been attributed by Hunt [1959] to Kutzbach and by Bagci [1971] to Maltysheff.

All of the Gruebler-type equations can be put into the compact form (see Problem **8.24**-15)

$$F = D(N - J - 1) + \sum F_i \tag{13}$$

where D, J, and F_i are interpreted appropriately (Freudenstein and Sandor [1964]).

When using any Gruebler-type formulas—even for *regular* mechanisms—one must be on guard for the possibility of the notorious "exceptional cases." We shall see in Sec **8.23**, these cases correspond to *singular states* where a certain important matrix has a vanishing determinant.

Gruebler-type formulas are useful in the classification of mechanisms according to the numbers of links and joints. The choice of a given number of links and joints to achieve a given design purpose is called **number synthesis** of mechanisms. To illustrate the use of Gruebler's formulas in number synthesis, consider plane mechanisms with 1 DOF, lower pairs, and no prismatic loops. Such mechanisms satisfy Eq. (6), which can be expressed as

$$P_s = \frac{3N}{2} - 2 \tag{14}$$

Equation (14) shows that the total number of links N is an even number. It then follows that the minimum number of links (including the fixed link) in a *regular closed linkage* is 4.

It may also be shown[12] that in regular closed linkages with a single DOF

1. The only linkage with four links has four *binary* links (e.g., the four-bar and slider-crank linkages).
2. Any six-link chain, with *simple* pairs only, must have two *ternary* links (i.e., members with three joints). In fact there are only two such chains; they are illustrated in Fig. 7 and in Fig. **8.24**-2.

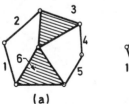

(a)   (b)

Figure 8.21-7. Six link chains: (a) Watt's linkage; (b) Stephenson's linkage. (Note: Any of the pairs shown may be revolute or prismatic pairs).

[12]Consult Rosenauer and Willis [1953, Chap. XI].

The number of possible arrangements increases rapidly when eight or more links are present. Catalogues of the various possibilities are given by Rosenauer and Willis [1953], Franke [1958], Hinkle [1960], Davies and Crossley [1966], Hain [1967], and Davies [1968]. Crossley [1964] and Pelecudi [1967] discuss systems with more than a single DOF; Soni [1971] discusses space configurations, and Artobolevskii [1975] shows schematic diagrams of 2000 mechanisms.

8.22 Loop Mobility Criteria

The most commonly used machines have one or more closed loops in their kinematic skeletons. We shall now show how the loop structure of such mechansims may be exploited to find their degree of freedom.

For the moment, we shall focus our attention on *linkages* (i.e., mechanisms with lower pairs only). Consider the shaper-crank mechanism of Fig. 1(a)—a typical multi-

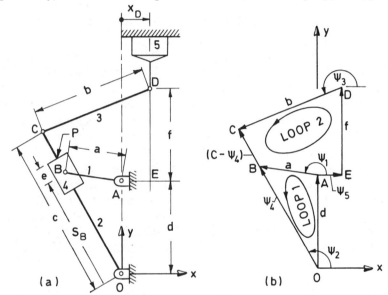

Figure 8.22-1. Shaper-crank mechanism: (a) Kinematic skeleton; (b) Vector polygons or oriented network.

loop linkage. In Fig. 1(b) we have drawn vector polygons describing two simple loops and have introduced the Lagrangian variables ψ_1, ψ_2, ψ_3 for the angles of links 1, 2, and 3 and ψ_4, ψ_5 for the distances s_B and x_D of the two sliders. We have *oriented* each side of the vector polygons, with arrowheads indicating the positive direction. Although the orientation adopted is arbitrary, it is convenient to locate the tail of each vector at the point where we mark its angle (e.g., point D for vector $\overrightarrow{DC}$). All angles are measured counterclockwise from the direction of the x axis.

The two closed loops indicated in Fig. 1 imply the two vector equations

$$\overrightarrow{AB} - \overrightarrow{OB} + \overrightarrow{OA} = 0 \tag{1a}$$

$$\overrightarrow{AE} + \overrightarrow{ED} + \overrightarrow{DC} - \overrightarrow{BC} - \overrightarrow{AB} = 0 \tag{1b}$$

Note that in these equations the sign in front of each vector is positive (negative) if the vector is traversed positively (negatively) when we traverse the loop in a definite sense (say counterclockwise). Each of these vector equations is equivalent to *two* scalar equations in the Lagrangian variables [as typified by Eqs. (7.10-2)]. Therefore, the two vector equations (1a) and (1b) are equivalent to four explicit *scalar equations of constraint* of the form

$$f_i(\psi_1, \psi_2, \ldots, \psi_M) = 0, \qquad (i = 1, 2, \ldots, N) \tag{2}$$

where $N = 4$.

We now raise the question, Is it possible, or necessary, to write additional constraint equations? It is certainly possible. For example, by tracing out loop *OAEDCBO*, we can formulate the equation

$$\overrightarrow{OA} + \overrightarrow{AE} + \overrightarrow{ED} + \overrightarrow{DC} - \overrightarrow{BC} - \overrightarrow{OB} = \mathbf{0} \tag{3}$$

To answer the second part of our question, we first view Fig. 1(b) as a network of *nodes* (O, A, B, C, D, E) and *directed edges* ($\overrightarrow{OA}, \overrightarrow{AB}, \ldots$, etc.).

From the discussion of network topology in Appendix **D.2**, we know that the vector loop equations corresponding to all the *simple circuits*[13] of a network form a *complete* set of *independent* loop equations for the network.

In other words, Eqs. (1a) and (1b) provide the *necessary and sufficient explicit equations of constraint associated with the set of Lagrangian coordinates chosen.*

That Eq. (3) is superfluous may be confirmed by noting that it results from the addition of Eqs. (1a) and (1b).

We shall return, in Sec. **9.22**, to a quantitative analysis of equations such as Eqs. (1) and (2). For the present, we merely remark that if the kinematic network of any linkage contains L *simple loops*, there exists a *complete* set of $2L$ explicit equations of constraint in the M Lagrangian coordinates which are necessary and sufficient to describe the loops. Accordingly, we may use these equations to find $2L$ of the coordinates in terms of the remaining $M - 2L$; hence the DOF of the mechanism is

$$F = M - 2L \tag{4}$$

Equation (4) is a simple, general, **loop mobility criterion**[14] which governs all non-singular planar *linkages*, including those with prismatic loops.

The following examples illustrate the use of Eq. (4).

EXAMPLE 1
In Fig. 1, $M = 5$, $L = 2$; hence

$$F = 5 - 2(2) = 1$$

EXAMPLE 2
In Fig. **8.21**-2, the simple circuits are

$$ABJIA, \quad BCKJB, \quad JKDEFJ, \quad JFGHAIJ \qquad (L = 4)$$

---

[13]This would be a good time to review the terminology of Appendix D, where a *simple circuit* has been defined as a closed loop which encloses exactly one polygon with nonintersecting sides.
[14]Introduced by Paul [1960].

Note that we may ignore edges *BF* and *CD*, which wouldn't even appear in the figure if ternary links 1 and 3 were represented only by bent lines (bell cranks) *BJF* and *CKD*. There are nine moving members; hence

$$F = M - 2L = 9 - 2(4) = 1$$

Frequently, the given kinematic skeleton will show that certain joints (e.g. A, B, C, D, in Fig. 2-a) are attached to the frame, but no information will be given on how

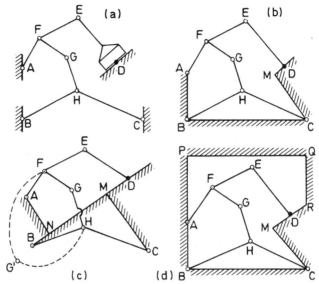

Figure 8.22-2. (a) Kinematic skeleton of a mechanism; (b) and (c) Alternative networks with three independent vector loops each. Note that frame edges form a topological tree. (d) Network with a spurious fourth loop, associated with a closed-loop frame.

these joints are physically interconnected. In order to complete the kinematic network, it is necessary to draw edges (e.g., AB, BC, CM, MD in Fig. 2-b) which complete the frame. There is no unique way to draw the required frame edges (as illustrated by the alternative frame construction of Fig. 2-c). The only rule to be observed is that the edges representing the frame should contain no closed loops. In other words, the collection of edges which form the frame should be represented as a *tree*[15] in the topological sense. This must be so because any loop consisting exclusively of fixed frame edges would contribute velocity loop equations of the useless form $0 + 0 + \cdots \ 0 = 0$. That is, a closed loop in the frame would not produce *independent* loop equations, and must not be counted as an independent loop.

To illustrate this rule, note that the network of Fig. 2-b has three simple circuits, and hence three independent vector loops. The closed loop representation of the frame shown in Fig. 2-d gives rise to an extraneous fourth loop, and should not be used to represent the network.

Note that the link *GH* crosses link *BD* in Fig. 2(c); however, this crossing can be

---

[15]"Tree" is defined in Appendix **D**.

eliminated by distorting the links *FG* and *GH* such that they occupy positions *FG'* and *G'H* (dashed lines). This is permissible for mobility analysis because topological properties of a network are invariant under distortions.

Treatment of Crossed Mechanisms

When it is possible to remove link crossings by metric distortions, we say that the mechanism is **mappable on a plane**[16]; however, not all mechanisms are mappable on a plane. For example, Fig. 3 shows a mechanism in which the crossings cannot[17] be

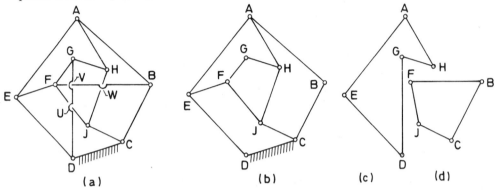

Figure 8.22-3. Crossed mechanism: (a) Original. (b) Base tier. (c) Second tier. (d) Third tier.

eliminated by a distortion of the edges; such systems will be called **crossed mechanisms**.

A crossed mechanism (e.g., Fig. 3-a) may be viewed as an uncrossed base tier (e.g., Fig. 3-b), plus additional uncrossed tiers (e.g., Figs. 3-c and 3-d) built up in succession upon previously created tiers. In drawing a tier we should consider all joints which are located on a previously created tier as fixed points or part of the frame upon which a later tier is built. The number of independent loops in a crossed mechanism equals the sum of the independent loops in each tier.

For example, the base tier of Fig. 3-b contributes four independent loops; the second tier (Fig. 3-c) and the third tier (Fig. 3-d) each contribute an additional loop—resulting in six independent loops for the linkage of Fig. 3-a.

Since the linkage has 13 movable members, its degree of freedom is $F = M - 2L = 13 - 2(6) = 1$.

Treatment of Prismatic Loops

The procedures just described are capable of handling *nonregular* mechanisms (i.e., containing prismatic loops), provided that we interpret M as the number of Lagrangian coordinates associated with the network[18]. The following examples illustrate the procedure.

---

[16]A mechanism is *mappable on a plane* when its kinematic network is a *planar graph* in the terminology of Appendix **D**.

[17]This is proved in Appendix **D** (see Fig. D-7) by a formal application of Kuratowski's theorem. Usually we can determine whether a graph is planar by less formal methods. If there is ever any doubt, assume that it is not planar and proceed with the method described in this section for nonplanar systems.

[18]Recall that for *regular linkages* M equals the number of moving links, but M will exceed that number if cam pairs or prismatic loops are present.

EXAMPLE 3

Three variables (s_1, s_2, s_3) are associated with the simple prismatic loop shown in Fig. 4. Accordingly, $M = 3, L = 1$, and $F = M - 2L = 1$.

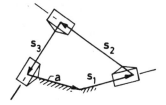

Figure 8.22-4. Simple prismatic loop.

EXAMPLE 4

The mechanism of Fig. **8.21**-5(b) is shown in Fig. 5 with its network diagram superposed. The six Lagrangian coordinates $(s_1, s_2, \ldots, s_6)$ are associated with this two-loop configuration. Thus $M = 6, L = 2$, and $F = M - 2L = 2$.

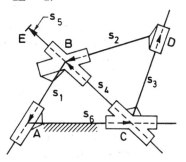

Figure 8.22-5. Coupled prismatic loop.

Treatment of Higher Pairs

So far we have only considered *linkages* (lower pairs only). The loop mobility criterion is readily generalized to cover mechanisms with higher pairs. It is only necessary to recall[19] that for each higher pair we must supplement the equations of constraint by one auxiliary equation. That is, each higher pair reduces the DOF by 1, so that the DOF for a system with P_h higher pairs is

$$F = M - 2L - P_h \tag{9}$$

EXAMPLE 5. Gear Mechanism

When dealing with gear mechanisms, it is essential to always count carrier arms as links, even when they are not explicitly shown (e.g., for gravity-closed gear pairs). Thus in Fig. 6, we must assume the presence of a link (shown dashed) between the centers C

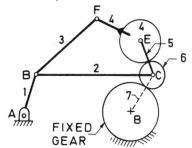

Figure 8.22-6. Gear mechanism.

---

[19]See the discussion related to Eqs. **(8.21**-3) and **(8.21**-4).

and B. Hence $M = 7$, $L = 2$, and $P_h = 2$, and Eq. (9) predicts that $F = 7 - 2(2) - 2 = 1$.

EXAMPLE 6. Cam Mechanism

In Fig. **7.43**-1, assume that points E and H are fixed, so that $L = 1$. Because this is not a *linkage*, M is not equal to the number of moving links. However, the six coordinates $(\theta_1, \phi_1, \theta_2, \phi_2, v_1, v_2)$ constitute a complete set of Lagrangian coordinates; i.e., $M = 6$. Since there is one higher pair, Eq. (9) predicts that

$$F = M - 2L - P_h = 6 - 2(1) - 1 = 3$$

Note that we did not choose to include r_1 and r_2 as Lagrangian coordinates, because they are considered to be known functions

$$r_1 = r_1(\phi_1) \tag{10a}$$
$$r_2 = r_2(\phi_2) \tag{10b}$$

of coordinates ϕ_1 and ϕ_2. Had we chosen to include r_1 and r_2 as Lagrangian coordinates we would have had $M = 8$. However, in that case, the two equations (10a) and (10b) would supplement the two loop equations and the one cam auxiliary equation [Eq. (**7.43**-6)], giving a total of $C = 5$ constraint equations; therefore

$$F = M - C = 8 - 5 = 3$$

The last example shows the consistency of Eq. (9) for cam mechanisms, but it also demonstrates an unnecessary degree of complexity in the mobility analysis. It is strongly recommended that cam pairs always be replaced by their lower-pair kinematic equivalents (see Fig. **1.54**-3) *before* performing a mobility analysis. If this is done for the case of Example 6, the system reduces to an ordinary six-bar linkage with $M = 5$, $L = 1$, and $F = 3$.

For future reference, we note that the complete set of N explicit constraint equations consists of the $2L$ loop equations plus the P_h auxiliary equations associated with gears and cams. In other terms the required constraint equations will be of the form

$$f_i(\psi_1, \psi_2, \ldots, \psi_M) = 0 \qquad (i = 1, 2, \ldots, N) \tag{11}$$
$$N = 2L + P_h \tag{12}$$

General Results

By considering the loops in a mechanism, it is possible to find the maximum number of joints J_m which may occur on *any single link* of a closed mechanism. Let us imagine that the link with J_m joints is the fixed link as indicated, by a polygon of J_m vertices, in Fig. 7 where only that part of the mechanism in the immediate neighborhood of the fixed link is shown.

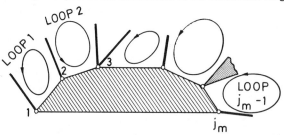

Figure 8.22-7. Part of a closed kinematic chain showing the link with the greatest number of joints.

From Fig. 7 we see that *at least* $J_m - 1$ independent loops must exist; i.e.,

$$L \geq J_m - 1 \tag{13}$$

For a regular linkage with N links (including the fixed link) $F = (N-1) - 2L$; hence

$$L = \tfrac{1}{2}(N - 1 - F) \geq J_m - 1 \tag{14}$$

For a single degree of freedom system, $F = 1$, and condition (14) becomes

$$J_m \leq \frac{N}{2} \tag{15}$$

Some interesting consequences of inequality (15) are described in Problems **8.24**-8 and **8.24**-9.

It is left as an exercise (Problem **8.24**-10) to show that the maximum multiplicity k_{max} of any joint in a regular linkage with N links and 1 DOF is given by

$$k_{max} \leq \frac{N}{2} \tag{16}$$

By utilizing Euler's theorem (Appendix **D**), it may be shown (Problem **8.24**-14) that Gruebler's formulas follow directly from the loop mobility criterion.

8.23 *Singular Configurations*

The explicit equations of constraint for any holonomic system can be expressed in the form of Eqs. (**8.22**-11); i.e.,

$$f_i(\psi_1, \psi_2, \ldots, \psi_M) = 0 \qquad (i = 1, 2, \ldots, N) \tag{1}$$

where $N = 2L + P_h$ for the general mechanism, and the DOF is given by

$$F = M - N \tag{2}$$

If Eqs. (1) are differentiated with respect to time, we find the **rate form** of the constraint equations:

$$\sum_{j=1}^{M} \frac{\partial f_i}{\partial \psi_j} \dot{\psi}_j = 0, \qquad (i = 1, \ldots, N) \tag{3}$$

As suggested in Sec. **8.13**, we may choose F of the Lagrangian coordinates as *primary* (or *generalized*) coordinates and designate them by $(q_1, q_2, \ldots, q_F)$; then the remaining (*secondary*) coordinates are designated by $(\phi_1, \phi_2, \ldots, \phi_N)$, and Eqs. (3) can be expressed in the form

$$\sum_{j=1}^{N} \frac{\partial f_i}{\partial \phi_j} \dot{\phi}_j + \sum_{k=1}^{F} \frac{\partial f_i}{\partial q_k} \dot{q}_k = 0 \qquad (i = 1, \ldots, N) \tag{4}$$

Since Eqs. (4) are linear in the *velocities* ($\dot{\phi}$ and $\dot{q}$), we may express them in the convenient matrix notation

$$\mathbf{A}\dot{\boldsymbol{\phi}} = \mathbf{B}\dot{\mathbf{q}} \tag{5}$$

where

$$\mathbf{A} \equiv [a_{ij}] = \left[\frac{\partial f_i}{\partial \phi_j}\right] \tag{6}$$

$$\mathbf{B} \equiv [b_{ij}] = -\left[\frac{\partial f_i}{\partial q_j}\right] \tag{7}$$

The square ($N \times N$) matrix $\mathbf{A}$ will henceforth be referred to as the **velocity matrix**. It is well known that the linear equations (5) can be solved for the unknown velocities $\dot{\phi}$ if, and only if, the determinant det $\mathbf{A}$ does not vanish.[20] If

$$\det \mathbf{A} = 0 \tag{8}$$

the matrix $\mathbf{A}$ is said to be **singular**, and a unique solution of Eq. (4) does not exist.

A mechanism for which det $\mathbf{A}$ vanishes is said to be in a **singular configuration** (or **singular state**). Some examples of such states will now be considered.

EXAMPLE 1. Redundant Structures Combined with Mobile Mechanisms

In Fig. 1 there are 5 simple loops ($L = 5$) and 11 movable members ($M = 11$).

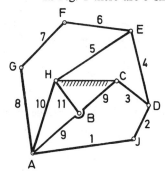

Figure 8.23-1. Mechanism including a redundant structure.

Hence it would appear that the DOF is

$$F = M - 2L = 11 - 2(5) = 1$$

However, it is readily seen that link 11 serves no kinematic function and merely makes a redundant structure out of the already rigid truss ACH. If we remove the redundant link (11), we see that the system has

$$F = 10 - 2(4) = 2$$

DOF rather than the single DOF predicted for the original structure. Note from Fig. 1 that only one driving link may be chosen from the links 1, 2, 3, 4, 5, since (1, 2, 3) and (3, 4, 5) form two four-bar mechanisms in tandem. At least one driving link must come from the group (6, 7, 8). With properly chosen driving links, a solution exists despite the fact that det$[A] = 0$. The mathematical basis for this behavior is discussed in Appendix **E.6**.

EXAMPLE 2. Critical Forms (Overclosed Kinematic Chains)

If no redundant structures are present and it is indicated by the loop mobility or Gruebler's criterion that zero degrees of freedom should exist (that is, the kinematic

---

[20]See Appendix E.

chain is just locked, forming a structure with statically determinate reactions), it is still possible that for certain configurations, known as **critical forms,** additional degrees of freedom can exist.

For example, Fig. 2 shows two four-bar mechanisms in tandem. It would appear that the two independent loops can provide four equations for the four moving members so that the DOF is

$$F = M - 2L = 4 - 2(2) = 0$$

Indeed, Fig. 2(a) does depict an immobile system. This lack of mobility can be verified by the observation that link 2 must rotate about its instant center I_1 for the four-bar mechanism $ABCD$, but it must also rotate about the instant center I_2 of the mechanism $CDEF$. Since a link cannot rotate simultaneously about two distinct centers, link 2 must be incapable of moving, and the system must be a rigid structure.

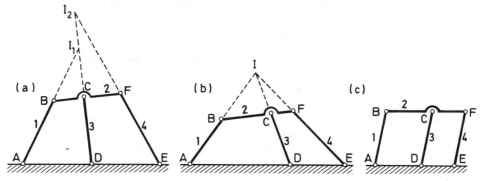

Figure 8.23-2. Topologically equivalent mechanisms with different degrees of freedom: (a) Structure; (b) Instantaneous critical form; (c) Permanent critical form.

On the other hand, Fig. 2(b) shows a configuration where the instant centers I_1 and I_2 momentarily coincide at point I, thus enabling link 2 to rotate through an *infinitesimal* angle about point I. Any *finite* motion would, of course, destroy the coincidence of points I_1 and I_2 and is therefore impossible.

If point I is always infinitely far from the fixed link, as in Fig. 2(c), link 2 is free to move (i.e., to translate) through finite displacements.

Critical forms of the type shown in Fig. 2(b) will be referred to as **instantaneous critical forms**, and those such as in Fig. 2(c) will be called **permanent critical forms.**

Although instantaneous critical forms cannot, *in principle*, undergo finite motions, it is easily seen that they can, *in practice*, undergo small finite motions due to clearance in the joints or due to the elasticity of the links.

Permanent critical forms are sometimes called **overclosed kinematic chains**; further examples are given in the problems of Sec. **8.24.**

It is worth noting that a gear pair (or rolling pair) may be considered as a permanent critical form. To see this we first note that the three-hinge truss shown in Fig. 3(a) collapses into an instantaneous critical form when its links are collinear, as shown in Fig. 3(b), because point C can undergo infinitesimal vertical motions without stretching the links.

If we regard a gear as a cluster of spokes, we see, from Fig. 3(c), that mating spokes

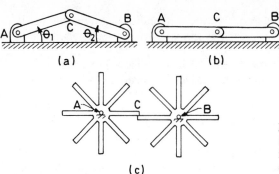

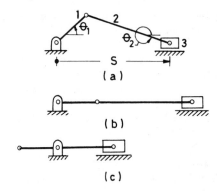

Figure 8.23-3. Systems with three links (including frame): (a) Truss; (b) Critical form of truss; (c) Spoke clusters—analog of gear set.

constitute a critical form *ACB* similar to that of Fig. 3(b). Passing to the limit of an infinite number of spokes, the system reduces to a permanent critical form. This explains why we emphasized in Sec. **8.21** that a gear pair contributes only one effective constraint equation, despite the fact that two constraints are implied by the condition of no slip at the pitch point.

EXAMPLE 3. Dead Center, Locked Configurations, Change-Points

If the piston is the driving link in Fig. 4, we see that the direction of motion of the crank (link 1) becomes indeterminate at a dead center. Hence the velocity matrix **A** must be singular for this case. Similarly, the motion of a parallelogram four-bar mechanism becomes indeterminate when all links are aligned (see Sec. **1.33**). Indeterminate velocities are possible only for singular velocity matrices.

With link 1 driving, the mechanism of Fig. 5 will lock in the position shown. Loss of a degree of freedom implies that the velocity matrix **A** is singular.

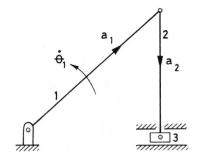

Figure 8.23-4. Slider crank mechanism: (a) Nonsingular configuration; (b) Right dead center (singular configuration); (c) Left dead center (singular configuration).

Figure 8.23-5. Singular configuration (locked mechanism) when $a_1 > a_2$.

Summary

In all cases, the *rate form* of the constraint equations [Eq. (3)] is given by

$$\sum_{j=1}^{M} C_{ij}\dot{\psi}_j = 0 \qquad (i = 1, 2, \ldots, N) \tag{9}$$

where $C_{ij} = \partial f_i / \partial \psi_j$. In matrix terms,

$$\mathbf{C}\dot{\psi} = \mathbf{0} \tag{10}$$

where $\mathbf{C} \equiv [\partial f_i/\partial \psi_j]$ is a rectangular matrix, of order $N \times M$, called the **Jacobian matrix** of the constraint equations (1).

If the rank of $\mathbf{C}$ is denoted by r, we know (from Theorem 2 of Appendix **E.6**) that $M - r$ of the variables ψ_i may be specified at pleasure, and the others will be uniquely determined. In kinematic terms, the DOF of the system[21] is

$$F = M - r \tag{11}$$

It is worth emphasizing that Eq. (11) is valid for all cases (singular, nonsingular, prismatic loops, etc.).

For nonsingular systems where the rank of $\mathbf{C}$ equals the total number of equations

$$r = 2L + P_h \tag{12}$$

and

$$F = M - 2L - P_h \tag{13}$$

in agreement with Eq. (**8.22**-9).

8.24 Problems

1. Classify each of the pairs shown in Fig. 1 into one of the four basic types of planar pairs (shown in Fig. **8.21**-1). What will be the effect of replacing the circular arcs in Fig. 1(a) by curved noncircular arcs? Explain and give an example.

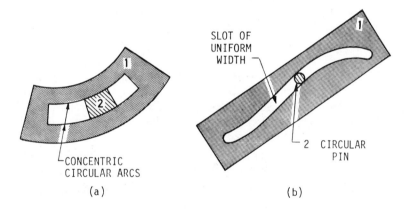

Figure 8.24-1. (Prob. 1).

2. Using θ_1 and θ_2 as Lagrangian coordinates for Fig. **8.23**-3(a),
   a. Find a complete set of constraint equations.
   b. Find the matrix $\mathbf{C}$ of Eq. (**8.23**-10).
   c. Show that the matrix $\mathbf{C}$ is singular for the case of Fig. **8.23**-3(b).

3. Derive the velocity matrix $\mathbf{A}$ [see Eq. (**8.23**-5)] for the slider-crank mechanism of Fig. **8.23**-4(a) when (a) the piston is driving and (b) the crank is driving. If the crank is shorter than the connecting rod, show that the only singularities in $\mathbf{A}$ occur for the dead center configurations when the piston is driving.

---

[21]The first explicit statement of this result appears to be that of Freudenstein [1962].

4. Derive the velocity matrix and show that it becomes singular for the configuration shown in Fig. **8.23**-5 when link 1 is the driver. Is the same configuration singular when the piston is the driving member? Discuss the nature of the transition through the singular configuration for both cases.

5. Number the links on the mechanisms shown in Fig. 2 so that the links correspond to either one of the mechanisms in Fig. **8.21**-7. Indicate for each mechanism whether it is a Watt or Stephenson chain or neither.

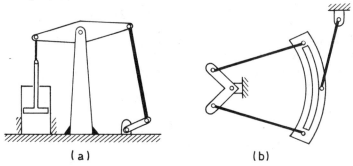

(a)   (b)

Figure 8.24-2. (Prob. 5).

6. Show that in a chain with only binary links the total number of links is $E = \frac{1}{2}(2P_1 + 3P_2 + 4P_3 + \cdots)$, where P_i is the number of joints where $i + 1$ links are connected. Use this result to prove that such a linkage has an even number of joints where an odd number of links are connected. This corresponds to the following theorem from topology: *A graph has an even number of odd vertices.*

7. Show that all regular mechanisms with lower pairs, four links, and 1 DOF must have
   a. Four simple lower pairs.
   b. All binary links.

8. a. Using the results of Problem 7, show that there are exactly five types of regular mechanisms with four links and four lower pairs. (Examples are slider-crank mechanism, Scotch yoke, etc.) Sketch an example of each chain.
   b. Show that one of the five types is kinematically equivalent to a certain kinematic chain with only three links.

9. Show that a regular mechanism with 1 DOF, six links, and only simple joints must consist of two ternary and four binary links. Sketch an example.

10. Show that a maximum of $N/2$ members can be attached to any joint in a regular mechanism with N links and 1 DOF.

11. Suppose that a system with N links, P_s simple pairs, and P_c cam pairs has F DOF. Imagine that each cam pair is replaced by a kinematically equivalent set of links and hinges. Use Eq. (**8.21**-5) to show that the DOF of the equivalent mechanism is $F = 3(N - 1) - 2P_s - P_c$.

12. Find the apparent DOF for the linkage in Fig. **8.23**-1 using a Gruebler formula.

13. Find the DOF of the mechanism in Fig. **8.21**-3(a). Apply the loop mobility criterion in the following ways,
   a. Replace the cam pair at A in Fig. **8.21**-3(b) by its equivalent lower-pair system.
   b. Do *not* replace the cam pair by its lower-pair equivalent.
   In both cases, clearly describe your choice of Lagrangian coordinates and the corresponding values of L, M, P_h.

14. For a regular linkage with N members (including the fixed link), L independent loops, E edges, J joints, and a total of P_s simple pairs, derive Gruebler's formula $F = 3(N - 1) - 2P_s$ from the loop mobility criterion $F = (N - 1) - 2L$. [Hints: (a) Use Euler's equation (D-4) to express F in terms of J and E. (b) If there are J_i joints where i members meet, show that $P_s = \sum_i (i - 1)J_i$ and that the number of pair elements is $e = \sum iJ_i$. (c) If there are N_k links with k joints attached, show that $E = \sum kN_k - N$ and $e = \sum kN_k$.]

15. A mechanism is to be viewed as a collection of N members joined together in pairs at J joints where exactly two members are connected. Let F_i represent the DOF of the relative motion between the two members connected at joint i. Show that the DOF of the mechanism is given by

$$F = D(N - J - 1) + \sum_{i=1}^{J} F_i$$

where
a. $D = 3$ for plane mechanisms with turning or sliding pairs, excluding case b.
b. $D = 2$ for plane mechanisms with prismatic pairs only.
c. $D = 6$ for general spatial mechanisms.
d. $D = 3$ for spatial mechanisms with prismatic pairs only.
e. $D = 3$ for spatial mechanisms with turning pairs only.
f. $D = 2$ for spatial mechanisms whose members all are constrained to slide over the surface of a sphere.

16–45. You are to number all moving links and find the degree of freedom F for each of the mechanisms shown in Figs. 3 through 32—using either or both of the following methods (in both cases, be sure to mention the nature of any singular cases):

Method a. Use *Gruebler-type* formulas. It is recommended that you present your results in the format of the examples in Sec. **8.21**. Be sure to state (1) the number of simple pairs at each lower joint, (2) which are the higher pairs, and (3) which form of Gruebler formula you are using.

Method b. Use loop mobility criteria, and give the following information: (1) a complete set of independent loops indicated by the sequence of joints in each loop (e.g., *ABCDEA*, *BOAB*, etc.); (2) the number M of independent Lagrangian coordinates; if M is different from the number of moving links, specifically show your choice of coordinates; and (3) the form of loop mobility criterion used, e.g., Eq. (**8.22**-4) or (**8.22**-9).

NOTE: Label as many reference points in the frame [e.g., points M and N in Fig. **8.22**-2(c)] as you find necessary.

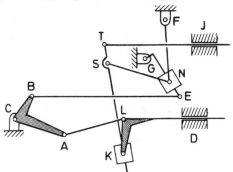

Figure 8.24-3. (Prob. 16) (HEUSINGER*).

*Valve gear diagrams, after Wittenbauer [1923]. Some curved links are shown as straight.

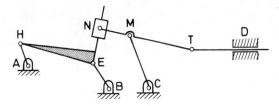

Figure 8.24-4. (Prob. 17) (FINK*).

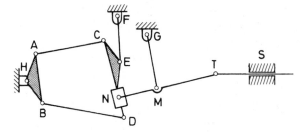

Figure 8.24-5. (Prob. 18) (GOOCH*).

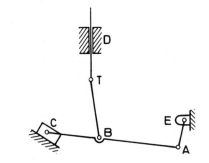

Figure 8.24-6. (Prob. 19) (HACKWORTH*).

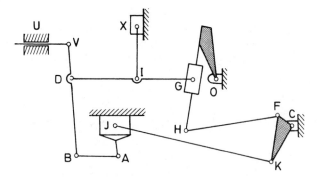

Figure 8.24-7. (Prob. 20) (WALSCHAERT).

*Valve gear diagrams, after Wittenbauer [1923]. Some curved links are shown as straight.

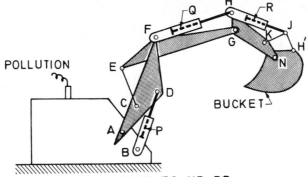

HYDRAULIC CYLINDERS: FQ, HR, BP

Figure 8.24-8. (Prob. 21) Excavator with hydraulic actuators.

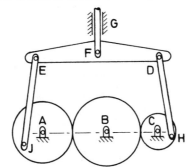

Figure 8.24-9. (Prob. 22).

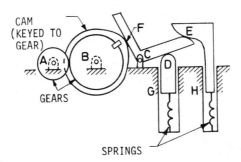

Figure 8.24-10. (Prob. 23).

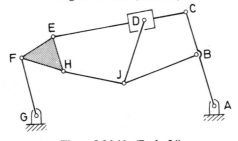

Figure 8.24-11. (Prob. 24).

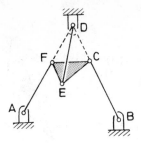

Figure 8.24-12. (Prob. 25).

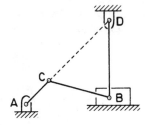

Figure 8.24-13. (Prob. 26).

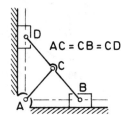

$AC = CB = CD$

Figure 8.24-14. (Prob. 27).

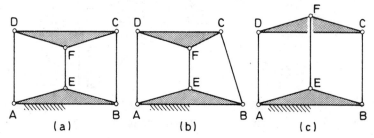

(a)   (b)   (c)

Figure 8.24-15. (Prob. 28 a, b, c).

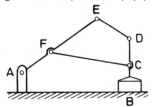

Figure 8.24-16. (Prob. 29).

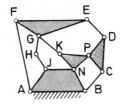

Figure 8.24-17. (Prob. 30).

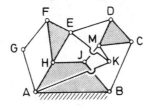

Figure 8.24-18. (Prob. 31).

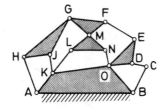

Figure 8.24-19. (Prob. 32).

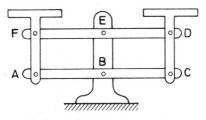

Figure 8.24-20. (Prob. 33) (Scale of Roberval).

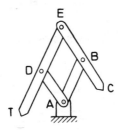

Figure 8.24-21. (Prob. 34) (Inverting Pantograph).

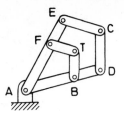

Figure 8.24-22. (Prob. 35) (Pantograph).

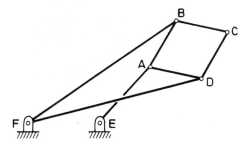

Figure 8.24-23. (Prob. 36) (Peaucellier's Linkage).

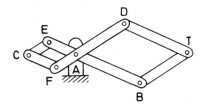

Figure 8.24-24. (Prob. 37) (Pantograph).

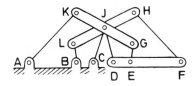

Figure 8.24-25. (Prob. 38) (Kempe's Linkage).

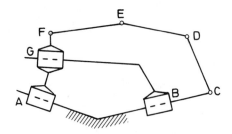

Figure 8.24-26. (Prob. 39).

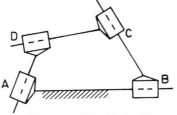

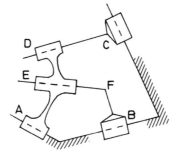

Figure 8.24-27. (Prob. 40).

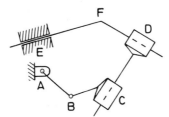

Figure 8.24-28. (Prob. 41).

Figure 8.24-29. (Prob. 42).

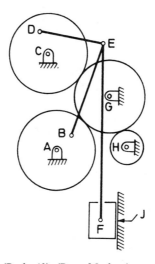

Figure 8.24-30. (Prob. 43). (Press Mechanism; after Hall [1966]).

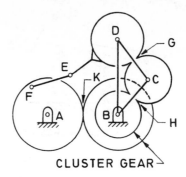

CLUSTER GEAR

Figure 8.24-31. (Prob. 44). After Oleksa and Tesar [1971].

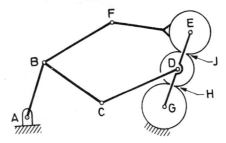

Figure 8.24-32. (Prob. 45).

POSITION ANALYSIS FOR CLOSED MECHANISMS

9.10   GENERAL CONSIDERATIONS

We have seen in Secs. **1.35** and **1.40** that we can find *explicit* algebraic expressions for the position of the driven links in four-bar and slider-crank mechanisms. Unfortunately this is not the general rule for mechanisms with one or more closed loops. We shall see below that for many such mechanisms, the displacements are governed by systems of tightly coupled nonlinear equations, and *explicit* formulas for the unknown displacements cannot be found. An example of such a situation is shown in Fig. **9.22**-1 where link AD is the driving link.

The most obvious solution to this problem is to make a scale drawing and to find the unknown displacements graphically. Frequently the graphical method itself will require a trial and error procedure, as in the case of Fig. **9.22**-1.

When greater accuracy is required, it will be necessary to introduce numerical methods. Once mastered, the numerical procedure has the additional advantages of generality, uniformity, and economy. The method we shall use is based on the position loop equations and auxiliary equations which, as we have seen [cf. Eq. (**8.22**-12)] are of the form

$$f_i(\psi_1, \psi_2, \ldots, \psi_M) = 0 \qquad (i = 1, 2, \ldots, N) \tag{1}$$

where the ψ_i represent a complete set of Lagrangian coordinates. For a system with F degrees of freedom, we may select F of the M variables ψ_i as inputs, or generalized coordinates, designated by $q_1, q_2, \ldots, q_F$. The remaining secondary variables are designated by $(\phi_1, \phi_2, \ldots, \phi_N)$, so that Eqs. (1) can be expressed in the form

$$f_i(\phi_1, \ldots, \phi_N; q_1, \ldots, q_F) = 0 \qquad (i = 1, 2, \ldots, N) \qquad (2)$$

In problems of kinematics, the driving variables q_i will be specified, and the problem of position analysis boils down to solving the N algebraic equations (2) for the N unknown position variables $\phi_1, \ldots, \phi_N$.

In the next section, we shall review a recommended numerical procedure (Newton-Raphson) for solving Eqs. (2) and illustrate its use with examples. Other methods that could be used are mentioned in Sec. **9.30**.

9.20 POSITION ANALYSIS BY THE NEWTON-RAPHSON METHOD

The numerical method recommended for position analysis of kinematics problems is the Newton-Raphson method.[1] This iterative technique for solving sets of non-linear algebraic equations is fully described in Appendix **F**. The essential features of the method will be reviewed here for application to the set of constraint equations (**9.10**-2), expressed in the form

$$f_i(\phi_1, \phi_2, \ldots, \phi_N) = 0 \qquad (i = 1, 2, \ldots, N) \qquad (1)$$

where explicit reference to the *known* driving parameters $q_1, q_2, \ldots, q_F$ has been suppressed for the sake of brevity.

We may solve Eqs. (1) for the unknown solution vector $\phi = \{\phi_1, \phi_2, \ldots, \phi_N\}$ by the following procedure:

1. Make an initial estimate of the desired solution vector (e.g., based on a rough sketch) and denote it by

$$\phi^1 \equiv \{\phi_1, \phi_2, \ldots\}^1$$

where the superscript represents an *iteration counter*.[2]

2. Evaluate the *Jacobian matrix* $\mathbf{A} \equiv [\partial f_i/\partial \phi_j]$ for the values ϕ_i^1 and call this matrix $\mathbf{A}^1$; i.e.,

$$\mathbf{A}^1 \equiv \left[\frac{\partial f_i}{\partial \phi_j}\right]_{\phi=\phi^1} \qquad (2)$$

3. Use Eqs. (1) to evaluate the **residual vector**

$$\mathbf{f}^1 \equiv \{f_1, f_2, \ldots, f_N\}_{\phi=\phi^1} \qquad (3)$$

at the point $\phi = \phi^1$.

---

[1] This numerical method was specifically suggested for kinematic analysis by Molian [1968a]. The method used by Uicker et al. [1964] could be viewed as a variant of the Newton-Raphson method. James et al. [1968, p. 145] solved a four-bar linkage problem by utilizing the Newton-Raphson method for a single equation with one unknown.

[2] In problems where iteration counters might become confused with exponents or other types of superscripts, they should be enclosed in parentheses.

4. Solve for the **correction vector** $\Delta\phi = \{\Delta\phi_1, \Delta\phi_2, \ldots\}$ from the linear equation

$$\mathbf{A}^1 \Delta\phi = -\mathbf{f}^1 \tag{4}$$

5. Make an improved estimate ϕ^2 of the solution vector of the form

$$\phi^2 \equiv \phi^1 + \Delta\phi = \{\phi_i^1 + \Delta\phi_i\} \tag{5}$$

6. Test whether the improved estimates of ϕ_i satisfy Eqs. (1) within a specified tolerance f_{tol}; i.e., test whether

$$|f_i| \leq f_{\text{tol}} \qquad (i = 1, 2, \ldots, N) \tag{6}$$

If inequalities (6) are satisfied, the problem is solved. If not, repeat steps 2 through 6, with ϕ^2 playing the role formerly played by ϕ^1.

The general algorithm is

$$\mathbf{A}^i \Delta\phi = -\mathbf{f}^i \tag{7}$$
$$\phi^{i+1} = \phi^i + \Delta\phi \tag{8}$$

where i is the iteration counter. The procedure is stopped as soon as inequalities (6) are satisfied. Alternatively, the procedure may be stopped whenever all $|\Delta\phi_i|$ are sufficiently small.

Since the Newton-Raphson method will not always converge, it is essential to terminate the procedure after a finite number of iterations (e.g., ten). In problems of kinematics, failure to converge can be due to any of the following causes:

1. The initial guesses for ϕ_i are too far away from the true solution. This situation should not occur if the initial guess is made with moderate care. In fact, the method will usually work if the initial ϕ_i are within 10% of their total range of motion and will often work for considerably greater deviations.
2. A numerical blunder has been made in specifying the constants of the system, or an algebraic blunder has been made in establishing the equations for $\mathbf{f}$ and $\mathbf{A}$.
3. The system configuration is very close to a singular configuration.

Note that the Jacobian matrix $\mathbf{A}$ is precisely the *velocity matrix* defined by Eq. (8.23-5). Therefore, its determinant vanishes for any of the kinematically singular configurations discussed in Sec. **8.23**, and it will not be possible to solve Eq. (4) at (or near) such singular states.

One way to identify singular configurations is merely to calculate the determinant J of matrix $\mathbf{A}$. If $|J|$ is within a specified tolerance of zero, the mechanism is very close to a singular configuration, and a loss of accuracy in the solution of Eqs. (1) is likely to occur. As a matter of fact, most well-designed subroutines for the solution of linear equations will automatically calculate J as a by-product of the solution procedure and will print out a warning when $|J|$ sinks below a specified tolerance J_{tol}.[3]

---

[3]See, for example, Sec. E.8 of Appendix.

9.21 Application to Four-Bar Mechanism

In Sec. **1.35**, we derived explicit expressions for the position of follower links when a side link is the driver. Thus there is no compelling need to use an iteration method for this linkage; however, it serves well to explain the details of the Newton-Raphson algorithm.

The loop displacement equations for the four-bar mechanism of Fig. 1 are

$$f_1 = a_1 \cos \psi_1 + a_2 \cos \psi_2 + a_3 \cos \psi_3 - a_4 = 0$$
$$f_2 = a_1 \sin \psi_1 + a_2 \sin \psi_2 + a_3 \sin \psi_3 = 0 \tag{1}$$

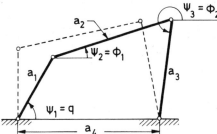

Figure 9.21-1. Four-bar mechanism. Solid line, $q = 60°$; broken line, $q = 90°$.

If we let ψ_1 be the driving angle, designated by q, and denote the secondary angles ψ_2, ψ_3 by ϕ_1, ϕ_2, the residual functions f_1 and f_2 may be expressed in the form

$$f_1 = a_1 \cos q + a_2 \cos \phi_1 + a_3 \cos \phi_2 - a_4$$
$$f_2 = a_1 \sin q + a_2 \sin \phi_1 + a_3 \sin \phi_2 \tag{2}$$

The associated Jacobian matrix is therefore given by

$$\mathbf{A} = \begin{bmatrix} \partial f_1/\partial \phi_1 & \partial f_1/\partial \phi_2 \\ \partial f_2/\partial \phi_1 & \partial f_2/\partial \phi_2 \end{bmatrix} = \begin{bmatrix} -a_2 \sin \phi_1 & -a_3 \sin \phi_2 \\ a_2 \cos \phi_1 & a_3 \cos \phi_2 \end{bmatrix} \tag{3}$$

and the general algorithm, Eqs. (**9.20**-7) and (**9.20**-8), assumes the form

$$\mathbf{A}^i \Delta\phi = \begin{bmatrix} -a_2 \sin \phi_1 & -a_3 \sin \phi_2 \\ a_2 \cos \phi_1 & a_3 \cos \phi_2 \end{bmatrix}^i \begin{bmatrix} \Delta\phi_1 \\ \Delta\phi_2 \end{bmatrix} = \begin{bmatrix} -f_1 \\ -f_2 \end{bmatrix}^i \tag{4}$$

$$\phi_1^{i+1} = \phi_1^i + \Delta\phi_1 \tag{5}$$

$$\phi_2^{i+1} = \phi_2^i + \Delta\phi_2 \tag{6}$$

For most mechanisms, Eqs. (4) are best solved by a numerical routine, such as the Gauss elimination method. However, when there are only two equations involved, it is simpler to solve them explicitly, by Cramer's rule, which in this case gives

$$\Delta\phi_1 = \frac{f_1 \cos \phi_2 + f_2 \sin \phi_2}{a_2[\sin (\phi_1 - \phi_2)]} \tag{7}$$

$$\Delta\phi_2 = \frac{f_1 \cos \phi_1 + f_2 \sin \phi_1}{a_3[\sin (\phi_2 - \phi_1)]} \tag{8}$$

where it is understood that the current values of ϕ_1 and ϕ_2 are to be used in evaluating all functions.

To illustrate the numerical procedure, let us assume the following data: $a_1 = 5$, $a_2 = 9$, $a_3 = 7$, $a_4 = 10$ (units arbitrary), and find ϕ_1 and ϕ_2 for $q = 60°$. From Fig. 1, rough starting values of $\phi_1 = 19°$ and $\phi_2 = 263°$ can be measured with a protractor. All details of the subsequent calculations are shown in Fig. 2.

q	Iter'n.	ϕ_1	ϕ_2	f_1	f_2	$\Delta\phi_1$	$\Delta\phi_2$
$(60°)$ 1.0472	1	$(19°)$ .3316	$(263°)$ 4.5902	.1566	.3124	-.0407	-.0381
	2	.2909	4.5521	.0048	.0013	-.0002	-.0008
	3	.2907 $(16.65°)$	4.5513 $(260.77°)$	4E-8	2E-6	-2E-7	-1E-7
$(90°)$ 1.5708	1	$(16.65°)$ .2907	$(260.77°)$ 4.5513	-2.5000	.6699	-.0321	.3566
	2	.2585	4.9079	.0609	.4342	-.0461	-.0216
	3	.2124	4.8863	.0089	.0030	-.0002	-.0013
	4	.2123 $(12.16°)$	4.8850 $(279.89°)$	-1E-6	6E-6	-7E-7	-4E-8

Figure 9.21-2. Detailed results for example No. 1; angles in radians. Angles in degrees shown in parenthesis.

The following points are worth noting from Fig. 2:

1. All calculation should be done with angles expressed in radians, since this is a calculus-based method. Angles, in degrees, are shown in parentheses.
2. The notation 4E-8 signifies 4×10^{-8}. We shall assume that convergence has occurred when f_1 and f_2 are less than 1E-4 in magnitude.
3. For $q = 60°$, the solution has converged by the third iteration. Perhaps this rapid convergence is not surprising since we started from guesses within 3° of the correct values.
4. To show that considerably coarser initial estimates are feasible, we have used the final results for $q = 60°$, to estimate ϕ_1 and ϕ_2 for the case of $q = 90°$. Now convergence to an acceptable level occurs within four iterations. This technique —of using the last calculated output as input for a subsequent step—is particularly well suited to programmed digital calculation. That we can work with huge step sizes, of the order of 30°, is fortunately typical of problems in kinematic analysis.
5. Far more detail is shown in Fig. 2 than would normally be recorded in the course of calculation. The more compact arrangement shown in Fig. 3 is all that need be recorded for a complete cycle of motion. In fact, the results shown in Fig. 3 were calculated with a programmable electronic calculator (Texas Instrument SR-52). The program used is valid for any four-bar linkage and required a total elapsed time of 75 sec for each value of q used; this includes the time needed to record by hand the information shown in the table. The same program will compute a pair of coupler point coordinates via Eqs. (7.11-2) within a few additional seconds.

q	30	60	90	120	150	180
ϕ_1	27.07	16.65	12.16	10.51	11.61	17.85
ϕ_2	250.43	260.77	279.89	301.44	321.98	336.79

q	210	240	270	300	330	360
ϕ_1	31.41	48.73	65.29	76.65	74.66	50.70
ϕ_2	341.77	339.65	330.02	320.77	298.02	264.26

Figure 9.21-3. Complete results for example No. 1. All angles shown in degrees.

Although Eqs. (4)–(8) illustrate a general procedure applicable to all mechanisms, other specialized techniques may be devised for particular mechanisms. For example, by eliminating the coupler angle ψ_2 from Eqs. (1), (Problem **9.40**-3), one arrives at **Freudenstein's equation**[4]:

$$R_1 \cos \psi_1 + R_2 \cos \psi_3 - R_3 - \cos(\psi_1 - \psi_3) = 0 \qquad (9)$$

$$R_1 = \frac{a_4}{a_3}, \qquad R_2 = \frac{a_4}{a_1}, \qquad R_3 = \frac{a_1^2 - a_2^2 + a_3^2 + a_4^2}{2a_1 a_3} \qquad (10)$$

If either ψ_1 or ψ_3 is given, the other may be found[5] from Eq. (9) by means of the Newton-Raphson algorithm for a single equation [see Eq. (**F.1**-11)].

9.22 Application to Quick-Return Mechanism

In the case of single-loop mechanisms, such as the four-bar linkage, we had the option of using direct or iterative methods of solution. However, for *strongly coupled*[6] multiloop systems, the constraint equations will usually be so highly nonlinear that there is no hope for an explicit algebraic solution. Therefore we must be content either to find the ϕ_j from a large scale drawing or else to utilize appropriate numerical methods. For example, consider the quick-return mechanism of Fig. 1.

From loop *CBEC* we find the two displacement equations

$$\begin{aligned} a_4 \cos \theta_4 - r_5 &= 0 \\ a_4 \sin \theta_4 - L - r_1 &= 0 \end{aligned} \qquad (1)$$

From loop *CADC* we find the equations

$$\begin{aligned} r_3 \cos \theta_4 - a_2 \cos \theta_2 &= 0 \\ r_3 \sin \theta_4 - a_2 \sin \theta_2 - r_1 &= 0 \end{aligned} \qquad (2)$$

If we assume that the driving variable is $\theta_2 \equiv q$ and introduce the notation

$$\{r_1, r_3, \theta_4, r_5\} \equiv \{\phi_1, \phi_2, \phi_3, \phi_4\} \qquad (3)$$

---

[4]Freudenstein [1955].
[5]See Problem **9.40**-4 or James et al. [1968, pp. 145–151].
[6]The distinction between *strong* and *weak* coupling is discussed in Sec. **10.30**.

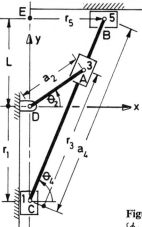

Figure 9.22-1. Quick-return mechanism.
$\{\phi_1, \phi_2, \phi_3, \phi_4, q\} \equiv \{r_1, r_3, \theta_4, r_5, \theta_2\}$.

the four loop equations (1) and (2) provide the *residual functions*

$$
\begin{aligned}
f_1 &= a_4 \cos \phi_3 - \phi_4 = 0 \\
f_2 &= a_4 \sin \phi_3 - L - \phi_1 = 0 \\
f_3 &= \phi_2 \cos \phi_3 - a_2 \cos q = 0 \\
f_4 &= \phi_2 \sin \phi_3 - a_2 \sin q - \phi_1 = 0
\end{aligned}
\tag{4}
$$

The Jacobian matrix for Eq. (4) is given by

$$
\mathbf{A} \equiv \left[\frac{\partial f_i}{\partial \phi_j} \right] =
\begin{bmatrix}
0 & 0 & -a_4 \sin \phi_3 & -1 \\
-1 & 0 & a_4 \cos \phi_3 & 0 \\
0 & \cos \phi_3 & -\phi_2 \sin \phi_3 & 0 \\
-1 & \sin \phi_3 & \phi_2 \cos \phi_3 & 0
\end{bmatrix}
\tag{5}
$$

At this point, one may apply the Newton-Raphson algorithm as outlined in Sec. **9.20**. However, it will be necessary to solve the 4×4 system of algebraic equations $\mathbf{A}\Delta\phi = -\mathbf{f}$ at each iteration. For this reason, the method is practical only if a digital computer or calculator is available which is programmed to solve sets of four linear equations. Therefore, further discussion of this problem will be within the context of computer implementation.

9.23 Computer Implementation

In Appendix **F**, a general-purpose Newton-Raphson subroutine called NEWTR is described. This subroutine may be embedded in a main program designed to be as specific or general as the user desires. For example, Paul and Hud [1975] have utilized NEWTR in a general-purpose FORTRAN program for kinematics analysis called DKINAL (Double Precision KINematics ANALysis Program). This program[7] will calculate position, velocity, and acceleration for all secondary variables ϕ_i as well as for a number of user-defined points of interest (e.g., coupler points). Whether one

---

[7]A full description of DKINAL, together with a program listing, will be found in Sec. **10.43**.

uses such a program or writes a personalized main program to utilize NEWTR, it is necessary to supply FORTRAN expressions for calculating the residual functions f_i and the Jacobian matrix $[A_{ij}]$. These expressions are most conveniently generated in subroutines named RESFNC and MATRIX, respectively, in which we shall use the following FORTRAN names:

$$Q = q, \qquad\qquad\qquad \text{NEQS} = \text{no. of equations}$$

$$\text{PHI(I)} = \phi_i, \qquad\qquad\qquad \text{FTOL} = f_{\text{tol}}$$

$$A(I, J) = \frac{\partial f_i}{\partial \phi_j}, \qquad\qquad\qquad \text{ITER} = \text{iteration counter}$$

$$\text{CN(I)} = \text{input constants, as needed}$$

$$\text{F(I)} = f_i$$

For the problem of Fig. **9.22**-1 the residual functions are defined by Eqs. (**9.22**-4), and the required subroutine RESFNC may be written as follows:

```
C
      SUBROUTINE RESFNC
C.......... THIS ROUTINE CALCULATES RESIDUALS, F(I)......
      IMPLICIT REAL*8(A-H,O-Z)
      COMMON//A(20,20),PHI(30),DPHI(30),DDPHI(30),F(20),CN(20)
     1,Q,DQ,DDQ,NEQS,NVI,NPHI,IND
      COMMON/PROB/SC,CQ,S3,C3
C
      SQ=DSIN(Q)
      CQ=DCOS(Q)
      S3=DSIN(PHI(3))
      C3=DCOS(PHI(3))
      F(1)=CN(6)*C3-PHI(4)
      F(2)=CN(6)*S3-PHI(1)-CN(7)
      F(3)=PHI(2)*C3-CN(5)*CQ
      F(4)=PHI(2)*S3-CN(5)*SQ-PHI(1)
      RETURN
      END
C
```

The IMPLICIT statement, which automatically makes all quantities (other than integers) double precision, is recommended only for computers (e.g., IBM 370) with small word length. Note that the COMMON block includes some variables (DPHI, DDPHI, DQ, DDQ, NVI, NPHI, IND) which are not needed here and may be ignored for now. They will be used later (in Sec. **10.43**) when we wish to compute velocities and accelerations. The variables SQ, CQ, S3, and C3 are convenient for this problem only and will automatically be available to any other subprogram which contains the COMMON/PROB/... statement.

Prior to using this subroutine it is necessary to read in (e.g., in the main program) the constants:

$$\text{CN(5)} = a_2, \qquad \text{CN(6)} = a_4, \qquad \text{CN(7)} = L$$

The FORTRAN expressions for F(I) correspond exactly to Eqs. (**9.22**-4).

Using the same notation, the subroutine MATRIX corresponding to Eq. (**9.22**-5) can be expressed in the form

```
C
        SUBROUTINE MATRIX
C......THIS ROUTINE CALCULATES JACOBIAN MATRIX A(I,J).....
C
        IMPLICIT REAL*8(A-H,O-Z)
        COMMON//A(20,20),PHI(30),DPHI(30),DDPHI(30),F(20),CN(20)
       1,Q,DQ,DDQ,NEQS,NVI,NPHI,IND
        COMMON/PROB/SC,CQ,S3,C3
        DO 20 I=1,NEQS
        DO 10 J=1,NEQS
     10 A(I,J)=0.0
     20 CONTINUE
        A(1,3)=-CN(6)*S3
        A(1,4)=-1.0D0
        A(2,3)=CN(6)*C3
        A(2,1)=-1.0D0
        A(3,2)=C3
        A(3,3)=-PHI(2)*S3
        A(4,1)=-1.0D0
        A(4,2)=S3
        A(4,3)=PHI(2)*C3
        RETURN
        END
C
```

The main program for this problem would normally contain instructions to read in the numerical data (a_2, a_4, L), the starting value of the driving variable Q, initial guesses of the secondary variables PHI(1), ..., PHI(4), and the maximum permitted magnitude of the residuals FTOL (typically 0.0001 in.). If several values of Q are to be used, one should read in the desired increment DELQ (Δq) and the maximum value QMAX of Q. Suitable instructions should be given for printing the input data. The instructions for printing, or otherwise processing (e.g., plotting), the calculated results may be assigned to a subroutine called OUTPUT. The essential parts of such a main program are given in the following listing:

```
        MAIN PROGRAM LISTING FOR DISPLACEMENT ANALYSIS

C
        IMPLICIT REAL*8(A-H,O-Z)
        DIMENSION AV(400),AA(400),TITLE(20)
        COMMON//A(20,20),PHI(30),DPHI(30),DDPHI(30),F(20),CN(20)
       1,Q,DQ,DDQ,NEQS,NVI,NPHI,IND
C
C       ...EPS IS RELATIVE TOLERANCE REQUIRED FOR SUBROUTINE DGELG
C          USER MAY ALTER IF DESIRED
C
        EPS=1.0D-10
C
C       ...READ INPUT DATA...
C
      1 READ(5,2,END=150)TITLE
      2 FORMAT(20A4)
     70 CONTINUE
C       ...SOLVE FOR DISPLACEMENT...
C
        CALL NEWTR(PHI,A,NEQS,FTOL,ITER,IER,F)
        IF(IER.NE.0)GO TO 130
        IF(ITER.LT.0)GO TO 110
    105 CALL OUTPUT
C
C       ....UPDATE THE INDEPENDENT VARIABLE...
```

```
C
      IF(Q.GE.QMAX)GO TC 1
      Q=Q+DELQ
      GO TO 70
110 WRITE(6,115)
115 FORMAT(/5X,'...NO CONVERGENCE WITH PRESCRIBED ITERATIONS..F(I)=')
      WRITE(6,120)(F(I),I=1,NEQS)
120 FORMAT(10X,D14.6)
      GO TO 1
130 WRITE(6,135)
135 FORMAT(/10X,'***MATRIX APPEARS TO BE SINGULAR, CHECK FOR DEAD ',
     1'POINTS, CHANGE POINTS, ETC ***')
140 GO TO 1
150 WRITE(6,155)
155 FORMAT(//'......LAST CASE FINISHED....')
      STOP
      END
```

Note that the real calculation work begins with the CALL to subroutine NEWTR.[8]

If NEWTR returns a nonzero value of an index IER, the matrix is (numerically) singular, and an appropriate message is printed. In that case, the user should inspect for critical forms, change points, or other anomalies.

If NEWTR returns a negative value of ITER, convergence has not occurred within ten iterations, and the user should examine for one of the difficulties described at the end of Sec. **9.20**.

If neither of these two situations arises, a good solution has been found and will be printed, stored, or otherwise processed by the user's OUTPUT subroutine.

If, at this point, q has not exceeded its upper limit Q_{max}, it will be incremented by Δq, and the process will be repeated with the last calculated values of ϕ_i serving as initial guesses for the next iteration.

The technique has been described so that readers who desire to develop their own programs can do so with a moderate amount of programming effort.

The main program just given is an excerpt from the more general program DKINAL, given in Sec. **10.43**, which finds velocities and accelerations as well as displacements. Some numerical output for the mechanism of Fig. **9.22**-1 will be found in Sec. **10.43**.

9.30 OTHER METHODS FOR POSITION ANALYSIS

If the reader cares to experiment with some other techniques for solving sets of nonlinear algebraic equations, he may consult the substantial body of literature on that topic.

---

[8]Full details on NEWTR, including a FORTRAN listing, are given in Appendix **F**. For now it is sufficient to know that the Newton-Raphson algorithm will be activated by a call to NEWTR.

For surveys of pertinent numerical techniques, see Rabinowitz [1970], Broyden [1970], and Ortega and Rheinboldt [1970]. Other pertinent references include Freudenstein and Roth [1963], Uicker et al. [1964], Molian [1968a], and Kane [1968].

Extensive discussions of the so-called *descent* or *minimization methods* will be found in the literature on optimization methods, e.g., Fox [1971], Hamming [1971, Chap. 9], and Wilde and Beightler [1967].

Gupta et al. [1972] have formulated kinematics problems in terms of the *implicit*[9] *equations of constraint*. In their procedure, they recommend certain modifications to the Newton-Raphson procedure to eliminate convergence difficulties that arise. It should be noted that no such difficulties plague the Newton-Raphson method when the (*explicit*) loop constraint equations are used, as illustrated in this chapter.

Finally, it should be mentioned that problems of kinematics can be solved by the so-called *method of differential equations*, described in Sec. **10.42**. However, that method is computationally less efficient than the Newton-Raphson method in strictly *kinematics* (as opposed to *dynamics*) problems and will therefore not be considered further in this chapter.

9.40 PROBLEMS

1. Denote the three angles of the four-bar mechanisms of Fig. **9.21**-1 by ψ_l, ψ_m, ψ_n, where the indices (l, m, n) may assume any permutation of the integers (1, 2, 3). Show that with ψ_n ($n = 1, 2,$ or 3) given, the corrections to be used in the Newton-Raphson algorithm [analogous to Eqs. (**9.21**-7) and (**9.21**-8)] are given by

$$\Delta\psi_l = \frac{f_1 \cos \psi_m + f_2 \sin \psi_m}{a_l \sin (\psi_l - \psi_m)}$$

where

$$f_1 = \sum_{i=1}^{3} a_i \cos \psi_i - a_4$$

$$f_2 = \sum_{i=1}^{3} a_i \sin \psi_i$$

and l and m are any two distinct indices other than n.

2. If a scalar variable x satisfies the single equation $f(x) = 0$, show that the Newton-Raphson algorithm reduces to

$$x^{r+1} = \left[x - \frac{f(x)}{df/dx} \right]^r$$

Interpret this result in terms of the tangent to the curve $y = f(x)$ at the point (x^r, y^r).

3. Derive Freudenstein's equation (**9.21**-9).

4. Use Freudenstein's equation (**9.21**-9) and the Newton-Raphson method for one equa-

---

[9]Recall the definition in Sec. **8.11**.

tion [see Eq. (F-11)] to find ψ_3 for $\psi_1 = 0°$ and $30°$. Assume $a_1 = 6$, $a_2 = 10$, $a_3 = 8$, $a_4 = 11$ (units arbitrary).

5. If the four-bar linkage of Fig. **9.21**-1 has $a_1 = a_3$ and $a_2 = a_4$,
   a. Show that if the coupler link (a_2) is chosen as the driving link, the Jacobian matrix **A** will be singular at all times when the linkage is in the parallelogram mode. But **A** will only be singular for two discrete configurations when the linkage is in the bow-tie mode.
   b. Show that if a side link $(a_1$ or $a_3)$ is selected as the driving link, the matrix **A** will be nonsingular except for two discrete configurations.
   c. What name has been given to the singular states referred to in a and b?

6. For the quick-return mechanism of Fig. **9.22**-1, use a graphical method to find the position variables $(r_1, r_3, r_5, \theta_4)$ for $\theta_2 = 0°$, $\pm 30°$, $\pm 60°$, $\pm 90°$. Use the following numerical data: $L = 6$, $a_2 = 4$, $a_4 = 18$ (units arbitrary). Can you solve this problem without a trial and error positioning of link CB? Reminder: You are not limited to Euclid's tool kit of compass and straightedge.

7. The linkage shown in Fig. 1 is used to advance a film strip intermittently by the motion of point P which periodically engages and disengages sprocket holes in the film strip as crank AD rotates counterclockwise at 1000 rpm. Dimensions shown are in centimeters.
   a. Write the equations which express the coordinates (x_P, y_P) of point P in terms of $\theta_1, \theta_2, \theta_3$, and known dimensions.
   b. Calculate $\theta_2, \theta_3, x_P, y_P$ for the given values $\theta_1 = 0°$, $20°$, $40°, \ldots, 360°$.
   c. Plot the path of coupler point P.
   d. How far apart should the sprocket holes be placed in the film?
   e. What is the mean speed of the film strip?

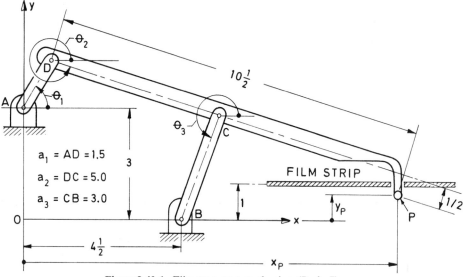

Figure 9.40-1. Film transport mechanism (Prob. 7).

8. Links 4 and 5 have been added to the internal combustion engine shown in Fig. 2 to provide support for the strain gage wires shown by the broken curve.
   a. Write displacement loop equations for each independent loop in the system.

b. Devise a procedure for finding angles θ_4 and θ_5 for specified values of the crank angle θ_1. Explain your method, and give all equations that will be needed.

c. Find the angles θ_4, θ_5 for 30° increments of the crank angle ($\theta_1 = 0, 30°, \ldots, 360°$). Use the following dimensions (inches) for link lengths: $a_1 = 5$, $a_2 = 10$, $b_2 = 5$, $a_4 = 8$, $a_5 = 8$, $g = 10$, $h = 12$.

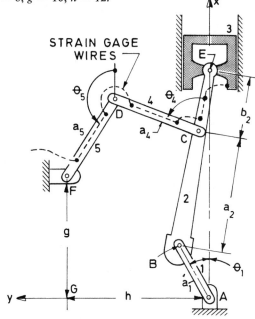

Figure 9.40-2. (Prob. 8).

9. Figure 3 shows a web cutter[10] in which blade tips attached to the coupler link and follower link of a rocker-crank mechanism move in a scissor-like fashion to periodically cut a moving strip or web. The arrangement gives a horizontal (x) velocity component to the blades which closely matches the web velocity at the time of cutting. The following

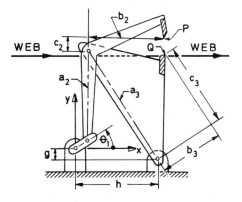

Figure 9.40-3. Web cutting mechanism (Prob. 9).

[10]Hall [1966, p. 6] and Chironis [1966, p. 7].

dimensions are given in inches: $a_1 = 4$, $a_2 = 28$, $a_3 = 40$, $g = 4$, $h = 26$, $b_2 = 24$, $c_2 = 4$, $b_3 = 20$, $c_3 = 26$.

   a. Find the x, y coordinates of the tips P and Q of the two blades for $\theta_1 = 0°$, $20°$, $640°$, $..$, $360°$.

   b. Plot $y_P - y_Q$ versus θ_1. At what value of θ_1 will cutting occur?

   c. Plot x_P and x_Q versus θ_1. Can you conclude from your plots that the system meets its design objective? Explain.

10. The tip C of bar BC in Fig. 4 moves along the curve $y = f(x)$, where $f(x)$ is a given function of x.

   a. Find explicit algebraic expressions for the angles θ and ϕ in terms of s and $f(s)$. [Hint: Express θ and ϕ in terms of the angle $\lambda = \angle CAD$, and find λ in terms of s and $h = f(s)$].

   b. If $a = 4$ in., $b = 12$ in., and $f(x) = \sin x$ (x in radians), evaluate θ and ϕ for $s = 11$ in.

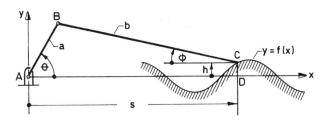

Figure 9.40-4. (Probs. 10–13).

11. If θ is the input variable for the system of Fig. 4 and $h = f(s)$ is a given smooth function of s,

   a. Find explicit expressions for the corrections $\Delta\phi$ and Δs given by the Newton-Raphson method for refining initial estimates of ϕ and s.

   b. Find ϕ and s for $\theta = 60°$ and $90°$ using $a = 4$ in., $b = 12$ in., if $f(x) = \sin s$ (s in radians).

12. If θ is the input variable for the system of Fig. 4 and $y = f(x)$ is a given smooth function of x,

   a. Find an expression of the form $F(\phi, \theta) = 0$, and give an explicit expression for finding the correction $\Delta\phi$ associated with the *Newton-Raphson method for one equation in one unknown*.

   b. Use the results of part a to find ϕ and s for $\theta = 60°$ and $90°$ if $f(x) = \sin x$, $a = 4$ in., and $b = 12$ in. (x in radians).

13. If $f(x)$ is given and θ is specified as input angle in Fig. 4,

   a. Find an expression of the form $F(s, \theta) = 0$, and give the correction Δs predicted by the *Newton-Raphson method for one equation in one unknown*.

   b. Use the results of part a to find s and ϕ for $\theta = 60°$ and $90°$, with $f(x) = \sin x$, $a = 4$ in., and $b = 12$ in. (x in radians).

14. In the Whitworth quick-return mechanism of Fig. **10.20-1**, angle $\theta_1 \equiv q$ is given. Find explicit expressions for calculating the unknown position variables r_2, θ_3, θ_4, r_5.

15. Find the angles β, γ, and ϕ in Fig. 5 for (a) $\theta = 30°$ and (b) $\theta = 60°$. $AE = 3$.

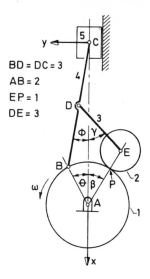

Figure 9.40-5. (Prob. 15).

16. Given $\theta_3 = 100°$ in Fig. 6, find θ_1, θ_2, r_1, and r_4.

17. In the punch press of Fig. 6, find the actuator position (r_1) when the ram position is given by $r_4 = 3.857$ in.

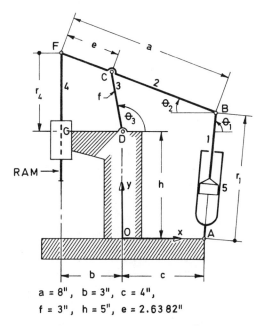

$a = 8''$, $b = 3''$, $c = 4''$,
$f = 3''$, $h = 5''$, $e = 2.6382''$

Figure 9.40-6. (Probs. 16, 17, 18).

18. Given the position variable r_1 of the pneumatic actuator in Fig. 6, establish the displacement loop equations and associated Jacobian matrix needed to find the ram position r_4.

19. In the quick-return mechanism of Fig. **10.60**-3, $AB = 2.5$ in., $b = 5.0$ in., and $c = 4.5$ in. Find the position variables ϕ, r, and x_P (the displacement of point P) when $\theta = 60°$.

20. In the shaper mechanism of Fig. **10.60**-4, $a_1 = 2.5$ in., $a_2 = 9$ in., $a_3 = 6$ in., $b = 5$ in., $c = 10$ in., and $q = 120°$. Find ϕ_1, ϕ_2, ϕ_3, ϕ_4.

VELOCITY AND ACCELERATION ANALYSIS

10.10 GENERAL CONSIDERATIONS

For an *open* mechanism, such as the multiple pendulum of Fig. 1, it is easy to express the base coordinates of any point in the system explicitly in terms of generalized coordinates. For example, the base coordinates (x_3, y_3) of point P_3 are given by the expressions

$$\begin{Bmatrix} x_3 \\ y_3 \end{Bmatrix} = a_1 \begin{Bmatrix} \cos q_1 \\ \sin q_1 \end{Bmatrix} + a_2 \begin{Bmatrix} \cos q_2 \\ \sin q_2 \end{Bmatrix} + \xi_3 \begin{Bmatrix} \cos q_3 \\ \sin q_3 \end{Bmatrix} \tag{1}$$

From these equations, the velocity components of point P_3 are readily found, by differentiation, to be

$$\begin{Bmatrix} \dot{x}_3 \\ \dot{y}_3 \end{Bmatrix} = a_1 \dot{q}_1 \begin{Bmatrix} -\sin q_1 \\ \cos q_1 \end{Bmatrix} + a_2 \dot{q}_2 \begin{Bmatrix} -\sin q_2 \\ \cos q_2 \end{Bmatrix} + \xi_3 \dot{q}_3 \begin{Bmatrix} -\sin q_3 \\ \cos q_3 \end{Bmatrix} \tag{2}$$

and a second differentiation gives the acceleration components

$$\begin{Bmatrix} \ddot{x}_3 \\ \ddot{y}_3 \end{Bmatrix} = a_1 \ddot{q}_1 \begin{Bmatrix} -\sin q_1 \\ \cos q_1 \end{Bmatrix} + a_2 \ddot{q}_2 \begin{Bmatrix} -\sin q_2 \\ \cos q_2 \end{Bmatrix} + \xi_3 \ddot{q}_3 \begin{Bmatrix} -\sin q_3 \\ \cos q_3 \end{Bmatrix}$$
$$- a_1 \dot{q}_1^2 \begin{Bmatrix} \cos q_1 \\ \sin q_1 \end{Bmatrix} - a_2 \dot{q}_2^2 \begin{Bmatrix} \cos q_2 \\ \sin q_2 \end{Bmatrix} - \xi_3 \dot{q}_3^2 \begin{Bmatrix} \cos q_3 \\ \sin q_3 \end{Bmatrix} \tag{3}$$

In general, we shall always be able to express the base coordinates (x_P, y_P) of any

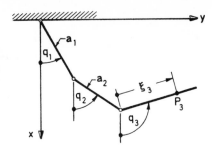

Figure 10.10-1. Triple pendulum.

point P in an *open mechanism* by explicit transformation equations of the type

$$x_P = x_P(q_1, q_2, \ldots, q_F, t)$$
$$x_P = y_P(q_1, q_2, \ldots, q_F, t)$$

(4)

where F is the DOF of the system and q_i are generalized coordinates. Then, by a straightforward process of differentiation, we can find expressions for velocity and acceleration analogous to Eqs. (2) and (3). Since the q_i are independent quantities, they may all be utilized as driving variables. Thus equations such as Eqs. (1), (2), and (3) give the complete kinematic response to a system with F degrees of freedom.

On the other hand, we have seen that for *closed* mechanisms we can seldom write explicit equations of the type (4) and must usually introduce a redundant set of Lagrangian coordinates $(\psi_1, \psi_2, \ldots, \psi_M)$, where $M > F$. Then we shall find, as in Eq. **(7.11-2)**, that the base coordinates of an arbitrary point can be expressed explicitly in the form

$$x_P = x_P(\psi_1, \psi_2, \ldots, \psi_M)$$
$$y_P = y_P(\psi_1, \psi_2, \ldots, \psi_M)$$

(5)

These equations may be differentiated to give velocities and accelerations in the form[1]

$$\begin{Bmatrix} \dot{x}_P \\ \dot{y}_P \end{Bmatrix} = \sum_{i=1}^{M} \dot{\psi}_i \frac{\partial}{\partial \psi_i} \begin{Bmatrix} x_P \\ y_P \end{Bmatrix}$$

(6)

$$\begin{Bmatrix} \ddot{x}_P \\ \ddot{y}_P \end{Bmatrix} = \sum_{i=1}^{M} \ddot{\psi}_i \frac{\partial}{\partial \psi_i} \begin{Bmatrix} x_P \\ y_P \end{Bmatrix} + \sum_{j=1}^{M} \sum_{i=1}^{M} \dot{\psi}_j \dot{\psi}_i \frac{\partial}{\partial \psi_j} \frac{\partial}{\partial \psi_i} \begin{Bmatrix} x_P \\ y_P \end{Bmatrix}$$

(7)

Thus, the velocities (and accelerations) of points in a mechanism are readily found in terms of the Lagrangian velocities (and accelerations).

We have seen in Chapters 8 and 9 that for closed mechanisms the coordinates ψ_j are *not* independent but are related by N equations of constraint of the form

$$f_i(\psi_j) = 0 \qquad (i = 1, 2, \ldots, N)$$

(8)

---

[1]Notation:

$$\dot{\psi} \frac{\partial}{\partial \psi} \begin{Bmatrix} x \\ y \end{Bmatrix} \equiv \begin{Bmatrix} \dot{\psi} \partial x / \partial \psi \\ \dot{\psi} \partial y / \partial \psi \end{Bmatrix}$$

These equations express the constraint of loop closure (two equations per loop) and any auxiliary constraints due to the presence of gear or cam pairs (one equation per higher pair). When Eqs. (8) are differentiated with respect to time we find the velocity loop equations

$$\sum_{j=1}^{M} \frac{\partial f_i}{\partial \psi_j} \dot{\psi}_j = 0 \qquad (i = 1, 2, \ldots, N) \tag{9}$$

If, as in Sec. **8.23**, we designate F of the Lagrangian coordinates as *primary* (or *generalized*) coordinates $(q_1, q_2, \ldots, q_F)$ and denote the remaining N *secondary* coordinates by $(\phi_1, \phi_2, \ldots, \phi_N)$, Eq. (9) takes the form

$$\sum_{j=1}^{N} \frac{\partial f_i}{\partial \phi_j} \dot{\phi}_j = -\sum_{k=1}^{F} \frac{\partial f_i}{\partial q_k} \dot{q}_k \equiv R_i \tag{10a}$$

or, in matrix notation,

$$\mathbf{A}\dot{\boldsymbol{\phi}} = \mathbf{B}\dot{\mathbf{q}} \equiv \mathbf{R} \tag{10b}$$

where

$$\mathbf{A} \equiv \left[\frac{\partial f_i}{\partial \phi_j} \right], \qquad \text{an } N \times N \text{ velocity matrix} \tag{11}$$

$$\mathbf{B} \equiv -\left[\frac{\partial f_i}{\partial q_j} \right], \qquad \text{an } N \times F \text{ matrix} \tag{12}$$

$$\mathbf{R} \equiv \mathbf{B}\dot{\mathbf{q}} = -\left\{ \sum_{j=1}^{F} \frac{\partial f_i}{\partial q_j} \dot{q}_j \right\} \equiv \{R_1, R_2, \ldots, R_N\} \qquad (N \times 1) \text{ matrix} \tag{13}$$

The elements of matrices $\mathbf{A}$ and $\mathbf{B}$ are functions of the position variables ψ_i, which we shall consider known. That is, it will be necessary to find the ψ_i (e.g., by the methods of Chapter 9 or by a graphical method) prior to calculating velocities or accelerations.

If the mechanism is not in a singular configuration (see Sec. **8.23**), the velocity matrix $\mathbf{A}$ will be nonsingular, and Eqs. (10) can be solved for all the unknown secondary velocities $\dot{\phi}_i$, associated with any given set of **driving velocities** $(\dot{q}_1, \ldots, \dot{q}_F)$.

To calculate the secondary accelerations $\ddot{\phi}_i$, it is only necessary to differentiate the velocity equations (10) to obtain the equations

$$\sum_{j=1}^{N} \left[\frac{\partial f_i}{\partial \phi_j} \ddot{\phi}_j + \dot{\phi}_j \frac{d}{dt} \frac{\partial f_i}{\partial \phi_j} \right] = -\sum_{k=1}^{F} \left[\frac{\partial f_i}{\partial q_k} \ddot{q}_k + \dot{q}_k \frac{d}{dt} \frac{\partial f_i}{\partial q_k} \right] \equiv \dot{R}_i \tag{14a}$$

or

$$\sum_{j=1}^{N} \frac{\partial f_i}{\partial \phi_j} \ddot{\phi}_j \equiv \sum_{j=1}^{N} A_{ij} \ddot{\phi}_j = S_i \tag{14b}$$

where

$$S_i \equiv \dot{R}_i - \sum_{j=1}^{N} \dot{\phi}_j \dot{A}_{ij}$$

$$= -\sum_{k=1}^{F} \left[\frac{\partial f_i}{\partial q_k} \ddot{q}_k + \dot{q}_k \frac{d}{dt} \frac{\partial f_i}{\partial q_k} \right] - \sum_{j=1}^{N} \dot{\phi}_j \frac{d}{dt} \frac{\partial f_i}{\partial \phi_j} \tag{15a}$$

In matrix notation,

$$\mathbf{A}\ddot{\boldsymbol{\phi}} = \mathbf{S} \tag{14c}$$

where

$$\mathbf{S} \equiv \mathbf{B}\ddot{\mathbf{q}} + \dot{\mathbf{B}}\dot{\mathbf{q}} - \dot{\mathbf{A}}\dot{\boldsymbol{\phi}} = \dot{\mathbf{R}} - \dot{\mathbf{A}}\dot{\boldsymbol{\phi}} \tag{15-b}$$

$$\dot{\mathbf{A}} \equiv [\dot{A}_{ij}] \equiv \left[\frac{d}{dt}\frac{\partial f_i}{\partial \phi_j}\right], \qquad \dot{\mathbf{B}} = [\dot{B}_{ij}] = -\left[\frac{d}{dt}\frac{\partial f_i}{\partial q_j}\right] \tag{16}$$

Note that the right-hand side of Eqs. (14) depends on the **driving accelerations** $\ddot{q}_i$ as well as on the driving velocities $\dot{q}_i$ and on previously calculated values of secondary velocities $\dot{\phi}_i$.

In summary, we see that velocities and accelerations in *open* mechanisms are easily found from equations analogous to Eqs. (2) and (3), whereas for *closed* mechanisms, the general procedure of analysis is as follows:

1. Find all secondary displacements $\boldsymbol{\phi}$ from an accurate drawing or by the numerical methods of Chapter 9.
2. Find all secondary velocities $\dot{\boldsymbol{\phi}}$ from Eq. (10b).
3. Find all secondary accelerations $\ddot{\boldsymbol{\phi}}$ from Eq. (14c).
4. Find displacements, velocities, and accelerations of points of interest from Eqs. (5), (6), and (7).

It is worth noting that steps 1, 2, and 3 all require the solution of linear equations of the general form $\mathbf{A}\mathbf{x} = \mathbf{y}$, where matrix $\mathbf{A}$ is the same for all three steps. Thus, each of the three sets of equations will have a unique solution when the velocity matrix $\mathbf{A}$ is nonsingular. If desired, $\dddot{\phi}_i$ and higher derivatives may be found by a simple extension of the procedure outlined.

The general procedure just described will be illustrated by specific examples in the following sections.

10.20 WEAKLY COUPLED LOOPS

For a mechanism with L independent loops, it will be necessary to solve a set of at least $2L$ linear equations for the secondary velocities and accelerations. However, these equations will not always be tightly coupled, and we may be able to reduce the amount of numerical work required. This possibility will be illustrated by an example and then considered more generally.

EXAMPLE 1. Whitworth Quick-Return Mechanism

In the mechanism of Fig. 1, the driving link 1 revolves at a uniform speed $\dot{\theta}_1$. The mechanism has two obvious independent loops, namely $ABCA$ and $CDEC$. The first loop is a swinging block mechanism, and the second loop is an ordinary slider-crank mechanism, driven by member 3, which is common to both loops. With five moving links and two independent loops, the DOF is $F = M - 2L = 1$. The single driving variable is $\theta_1 = q$, with $\dot{\theta}_1 = \dot{q} = $ constant and $\ddot{\theta}_1 = \ddot{q} = 0$. The secondary variables are

$$\{\phi_1, \phi_2, \phi_3, \phi_4\} \equiv \{r_2, \theta_3, \theta_4, r_5\} \tag{1}$$

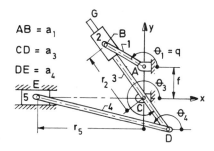

$$AB = a_1$$
$$CD = a_3$$
$$DE = a_4$$

Figure 10.20-1. Whitworth mechanism.

For loop $ABCA$ the displacement loop equations are

$$a_1 \cos q + r_2 \cos \theta_3 = 0 \tag{2}$$

$$a_1 \sin q + r_2 \sin \theta_3 + f = 0 \tag{3}$$

These equations may be readily solved for r_2 and θ_3 (Problem **9.40**-14). The displacement equations for loop $CDEC$ are

$$a_3 \sin \theta_3 + a_4 \sin \theta_4 = 0 \tag{4}$$

$$a_3 \cos \theta_3 + a_4 \cos \theta_4 + r_5 = 0 \tag{5}$$

from which the remaining unknowns θ_4 and r_5 follow in succession. The velocity equations corresponding to Eqs. (2) and (3) are

$$\dot{r}_2 \cos \theta_3 - r_2 \dot{\theta}_3 \sin \theta_3 = a_1 \dot{q} \sin q \tag{6}$$

$$\dot{r}_2 \sin \theta_3 + r_2 \dot{\theta}_3 \cos \theta_3 = -a_1 \dot{q} \cos q \tag{7}$$

which can easily be solved by Cramer's rule to give

$$\dot{r}_2 = a_1 \dot{q} \sin (\theta_1 - \theta_3) \tag{8}$$

$$\dot{\theta}_3 = -\left(\frac{a_1}{r_2}\right) \dot{q} \cos (\theta_1 - \theta_3) \tag{9}$$

Similarly, the velocity equations corresponding to Eqs. (4) and (5) are

$$a_4 \dot{\theta}_4 \cos \theta_4 = -a_3 \dot{\theta}_3 \cos \theta_3 \tag{10}$$

$$\dot{r}_5 = a_3 \dot{\theta}_3 \sin \theta_3 + a_4 \dot{\theta}_4 \sin \theta_4 \tag{11}$$

which gives the remaining secondary velocities, $\dot{\theta}_4$ and $\dot{r}_5$, in terms of previously calculated quantities.

The main point of interest in this example is that we were able to solve for the unknowns which occur in the first loop without even considering the second loop. Then we found the remaining unknowns by considering the second loop alone. In other words, we were able to reduce the problem from a 4×4 system of linear equations to two 2×2 systems of linear equations—a much simpler computational task. This reduction in effort will be possible whenever the constraint equations can be formulated in pairs such that each successive pair of equations contains no more than two secondary variables which have not been previously solved for. For linkages, this requires that each successive loop introduce at most two new unknowns.

Note that the complete set of velocity equations (6), (7), (10), and (11) can be written in the matrix form

$$
\begin{bmatrix}
C_3 & -r_2 S_3 & 0 & 0 \\
S_3 & r_2 C_3 & 0 & 0 \\
\hline
0 & a_3 C_3 & a_4 C_4 & 0 \\
0 & a_3 S_3 & a_4 S_4 & -1
\end{bmatrix}
\begin{bmatrix}
\dot{r}_2 \\
\dot{\theta}_3 \\
\dot{\theta}_4 \\
\dot{r}_5
\end{bmatrix}
=
\begin{bmatrix}
a_1 S_1 \\
-a_1 C_1 \\
\hline
0 \\
0
\end{bmatrix}
\dot{q}
\tag{12}
$$

where $C_i \equiv \cos \theta_i$ and $S_i \equiv \sin \theta_i$.

The broken lines in Eq. (12) indicate that the matrices may be *partitioned* into the form

$$
\begin{bmatrix}
\mathbf{A}_{11} & \mathbf{0} \\
\mathbf{A}_{21} & \mathbf{A}_{22}
\end{bmatrix}
\begin{bmatrix}
\dot{\boldsymbol{\phi}}_1 \\
\dot{\boldsymbol{\phi}}_2
\end{bmatrix}
=
\begin{bmatrix}
\mathbf{B}_1 \\
\mathbf{B}_2
\end{bmatrix}
\dot{q}
\tag{13}
$$

where

$$
\dot{\boldsymbol{\phi}}_1 \equiv \{\dot{\phi}_1, \dot{\phi}_2\} \equiv \{\dot{r}_2, \dot{\theta}_3\}
$$
$$
\dot{\boldsymbol{\phi}}_2 \equiv \{\dot{\phi}_3, \dot{\phi}_4\} \equiv \{\dot{\theta}_4, \dot{r}_5\}
$$

and similarly $\mathbf{A}_{ij}$ and $\mathbf{B}_i$ are defined by the submatrices in Eq. (12) with matching positions in Eq. (13).

In general, we shall say that a system is **weakly coupled** whenever the velocity equations may be written in the partitioned form

$$
\begin{bmatrix}
\mathbf{A}_{11} & \mathbf{0} & \mathbf{0} & \cdots & \mathbf{0} \\
\mathbf{A}_{21} & \mathbf{A}_{22} & \mathbf{0} & \cdots & \mathbf{0} \\
\cdots & & & & \\
\mathbf{A}_{n1} & \mathbf{A}_{n2} & \mathbf{A}_{n3} & \cdots & \mathbf{A}_{nn}
\end{bmatrix}
\begin{bmatrix}
\dot{\boldsymbol{\phi}}_1 \\
\dot{\boldsymbol{\phi}}_2 \\
\cdots \\
\dot{\boldsymbol{\phi}}_n
\end{bmatrix}
=
\begin{bmatrix}
\mathbf{B}_1 \\
\mathbf{B}_2 \\
\cdots \\
\mathbf{B}_n
\end{bmatrix}
\dot{\mathbf{q}}
\tag{14}
$$

When the velocity matrix $\mathbf{A}$ can be expressed as such a *triangular* supermatrix, it will be possible to first solve the linear subsystem

$$
\mathbf{A}_{11} \dot{\boldsymbol{\phi}}_1 = \mathbf{B}_1 \dot{\mathbf{q}}
\tag{15}
$$

for the group of unknown secondary velocities $\dot{\boldsymbol{\phi}}_1 \equiv \{\dot{\phi}_1, \dot{\phi}_2, \ldots, \dot{\phi}_k\}$, where k is the order of square matrix $\mathbf{A}_{11}$. After $\dot{\boldsymbol{\phi}}_1$ is found, we may then solve the system

$$
\mathbf{A}_{22} \dot{\boldsymbol{\phi}}_2 = \mathbf{B}_2 \dot{\mathbf{q}} - \mathbf{A}_{21} \dot{\boldsymbol{\phi}}_1
\tag{16}
$$

for the vector $\dot{\boldsymbol{\phi}}_2$. The process is then repeated for $\dot{\boldsymbol{\phi}}_3, \ldots, \dot{\boldsymbol{\phi}}_n$.

For a system with continuously prescribed acceleration inputs, it follows from Eq. (**10.10**-14c) that if the secondary velocities $\dot{\boldsymbol{\phi}}$ are weakly coupled, so too are the accelerations $\ddot{\boldsymbol{\phi}}$. Other aspects of weakly coupled systems are illustrated by the following example.

EXAMPLE 2

Figure 2 shows a schematic diagram of a Stirling cycle external combustion engine. It may be verified that the system has nine configuration variables and four inde-

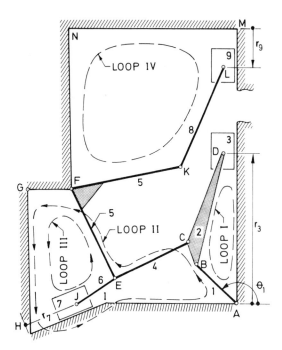

Figure 10.20-2. Kinematic skeleton of Stirling cycle engine.

pendent loops (e.g., those shown by broken lines). The single driving variable is conveniently taken as the crank angle $\theta_1 \equiv q$. It is apparent that the slider-crank mechanism of loop I may be solved independently of the remainder of the system, giving $\dot{\theta}_2$ and $\dot{r}_3$ (where $\dot{\theta}_i$ and $\dot{r}_i$ will be used to signify angular and linear velocities, respectively, for link i). With $\dot{\theta}_2$ known, loop II provides two equations for determining $\dot{\theta}_4$ and $\dot{\theta}_5$. Loop III then provides the two equations needed for determination of $\dot{\theta}_6$ and $\dot{r}_7$. Finally, $\dot{\theta}_8$ and $\dot{r}_9$ are found from loop IV.

Note that it is never necessary to solve a set of more than two simultaneous equations for the system. The "weakness" of the coupling is a strong function of

1. Which members are selected as drivers
2. How the independent loops are formed

For example, it should be verified that if link 9 is specified as the driving link and the same loops are used as in Fig. 2, the equations of loops I and II will be fully coupled, and four simultaneous equations will be needed to solve for $\dot{\theta}_4, \dot{\theta}_2, \dot{\theta}_1, \dot{r}_3$.

It should also be observed that had we used the "outer" loop *ABCEJHIA* instead of loop II, it would have been necessary to solve for $\dot{\theta}_4, \dot{\theta}_5, \dot{\theta}_6, \dot{r}_7$ simultaneously. The latter choice of loops was in fact made in the analysis of this engine by Paul and Krajcinovic [1970a], who set the problem up for solution on a digital computer.

It is a great convenience to explicitly utilize weak coupling whenever possible if one is solving a problem algebraically, graphically, or by means of "hand computations." However, the advantages of utilizing weak coupling are rather minor for all but the most complicated systems when using digital computers for the solution of linear equations.

10.30 STRONGLY COUPLED LOOPS

A linkage will be said to be **strongly coupled** if, for the given input variables, it is not possible to select a system of loops which makes it weakly coupled in the sense of Sec. **10.20**. The so-called *complex mechanisms* which require special considerations in their graphical analysis (see Sec. **6.11**) fall into this category. The solution of such problems by the current analytical method requires no special considerations and follows the pattern of Sec. **10.10** without modification. As shown by the last example, one can verify, by inspection of the kinematic network diagram, whether or not a system is strongly coupled; it is not necessary to explicitly formulate the loop equations for this purpose.

For strongly coupled systems with a single DOF, it may be possible to effectively decouple the velocity equations by the following procedure. The velocity equations are expressed in the form

$$\sum_{j=1}^{N} A_{ij}\dot{\phi}_j = \dot{q}b_i \quad \text{or} \quad \mathbf{A}\dot{\boldsymbol{\phi}} = \dot{q}\mathbf{b} \tag{1}$$

where $\dot{q}$ is the given driving velocity. If we define the **velocity ratios** by

$$\frac{\dot{\phi}_j}{\dot{q}} = k_j \tag{2}$$

Eqs. (1) assume the form

$$\sum_{j=1}^{N} A_{ij}k_j = b_i \quad \text{or} \quad \mathbf{A}\mathbf{k} = \mathbf{b} \tag{1a}$$

from which we can find the velocity ratios k_j independently of any specific value assigned to the driving velocity $\dot{q}$. Then Eqs. (2) give the unknown velocities $\dot{\phi}_i$ for any desired choice of $\dot{q}$. The procedure is illustrated by the following example.

EXAMPLE 1. Use of Velocity Ratios for Decoupling Velocity Equations

Figure 1 shows the quick-return mechanism (discussed in Sec. **9.22**) where crank 2 moves with known velocity $\dot{\theta}_2 = \dot{q}$. There is no obvious loop which involves θ_2 and at

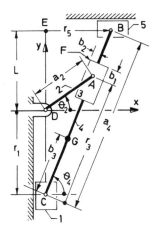

Figure 10.30-1. Quick-return mechanism $\{\phi_1, \phi_2, \phi_3, \phi_4, q\} \equiv \{r_1, r_3, \theta_4, r_5, \theta_2\}$.

CHAPTER 10 VELOCITY AND ACCELERATION ANALYSIS

322

most two other variables. Hence the problem is strongly coupled if link 2 is the driver. Note, however, that loop $CBEC$ involves only the variables r_5, r_1, and θ_4, any two of which may be found in terms of the third. Therefore the system is weakly coupled if any one of the links 5, 1, or 4 is chosen as drivers. If, for example, we choose $\dot{r}_5$ as the driving velocity, we can find the velocity ratios $\dot{r}_1/\dot{r}_5$ and $\dot{\theta}_4/\dot{r}_5$ from loop $CBEC$. Then we can find $\dot{\theta}_2/\dot{r}_5$ and $\dot{r}_3/\dot{r}_5$ from loop $ADCA$. At this point one can express all velocities in terms of $\dot{\theta}_2$ by means of identities such as

$$\dot{r}_3 = \frac{\dot{r}_3/\dot{r}_5}{\dot{\theta}_2/\dot{r}_5}\dot{\theta}_2, \qquad \dot{\theta}_4 = \frac{\dot{\theta}_4/\dot{r}_5}{\dot{\theta}_2/\dot{r}_5}\dot{\theta}_2$$

This example shows that although the velocity equations for a single DOF mechanism may be strongly coupled for one choice of driving velocity, it may still be possible to partially uncouple the equations by temporarily assigning a different driving link. More generally, let us consider a system with five configuration variables (the discussion is easily extended to M variables) where the velocity equations can be expressed in the partitioned matrix form

$$\begin{bmatrix} a_{11} & a_{12} & 0 & 0 \\ a_{21} & a_{22} & 0 & 0 \\ \hline a_{31} & a_{32} & a_{33} & a_{34} \\ a_{41} & a_{42} & a_{43} & a_{44} \end{bmatrix} \begin{bmatrix} \dot{\psi}_1 \\ \dot{\psi}_2 \\ \hline \dot{\psi}_4 \\ \dot{\psi}_5 \end{bmatrix} = \dot{\psi}_3 \begin{bmatrix} b_1 \\ b_2 \\ \hline b_3 \\ b_4 \end{bmatrix} \tag{3}$$

With $\dot{\psi}_3$ chosen as driving velocity, these equations are weakly coupled, and we may solve the first two of them for $\dot{\psi}_1$ and $\dot{\psi}_2$. Then the remaining two equations are readily solved for $\dot{\psi}_4$ and $\dot{\psi}_5$.

If, however, we are given $\dot{\psi}_5$ instead of $\dot{\psi}_3$, the equations do not appear to uncouple. But if both sides of Eq. (3) are divided by $\dot{\psi}_3$, the equation may be solved directly for the **velocity ratios** (or **influence coefficients**)

$$k_{i3} \equiv \frac{\dot{\psi}_i}{\dot{\psi}_3} \qquad (i = 1, 2, 4, 5) \tag{4}$$

Then with $\dot{\psi}_5$ given, we can find

$$\dot{\psi}_1 = \frac{\dot{\psi}_1/\dot{\psi}_3}{\dot{\psi}_5/\dot{\psi}_3}\dot{\psi}_5 = \frac{k_{13}}{k_{53}}\dot{\psi}_5 \tag{5}$$

Similarly,

$$\dot{\psi}_2 = \frac{k_{23}}{k_{53}}\dot{\psi}_5, \qquad \dot{\psi}_4 = \frac{k_{43}}{k_{53}}\dot{\psi}_5, \qquad \dot{\psi}_3 = \frac{\dot{\psi}_5}{\dot{\psi}_5/\dot{\psi}_3} = \dot{\psi}_5/k_{53} \tag{6}$$

So far, we have shown that a system which is weakly coupled for any choice of driving velocity is *effectively* weakly coupled for any other choice of driving velocity that does not lead to a singular state.

However, the *accelerations* will not, in general, be decoupled in this fashion. To see this, we note that the first two acceleration equations corresponding to Eq. (3) are of the form

$$A_1 \begin{Bmatrix} \ddot{\psi}_1 \\ \ddot{\psi}_2 \end{Bmatrix} = B_1 \ddot{\psi}_3 + \dot{B}_1 \dot{\psi}_3 - \dot{A}_1 \begin{Bmatrix} \dot{\psi}_1 \\ \dot{\psi}_2 \end{Bmatrix} \tag{7}$$

If $\ddot{\psi}_5$ is given rather than $\ddot{\psi}_3$, we cannot solve Eq. (7) directly. Moreover, we may not be able to find the acceleration ratios from the equation

$$A_1 \begin{Bmatrix} \ddot{\psi}_1/\ddot{\psi}_3 \\ \ddot{\psi}_2/\ddot{\psi}_3 \end{Bmatrix} = B_1 + \dot{B}_1 \dot{\psi}_3/\ddot{\psi}_3 - \dot{A}_1 \begin{Bmatrix} \dot{\psi}_1 \\ \dot{\psi}_2 \end{Bmatrix}/\ddot{\psi}_3 \tag{8}$$

because the unknown quantity $\ddot{\psi}_3$ will still appear on the right-hand side of Eq. (8) (unless $\dot{B}_1$ and $\dot{A}_1$ both vanish). In short, the attempt to partially uncouple the acceleration equations by a conceptual change of input link will not be successful in general.

This reasoning explains why it is necessary to use a trial and error procedure or some technique involving *auxiliary points* in graphical solutions where the acceleration is specified for a point in a strongly coupled loop. If, for example, angular acceleration is specified for link 4 in the mechanism of Fig. 2, the simple graphical methods of acceleration analysis given in Chapter 6 are not applicable.

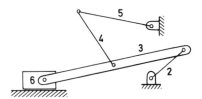

Figure 10.30-2. Strong coupling exists if $\ddot{\theta}_4$ is specified.

10.40 NUMERICAL METHODS OF KINEMATIC ANALYSIS

Although the methods discussed below may be carried out by manual calculations, we shall assume that a digital computer or programmable calculator is available. We have already seen, in Chapter 9, how to calculate the position variables for arbitrary plane mechanisms. In Sec. **10.41** we shall show how the position analysis may be extended to a velocity and acceleration analysis based on certain sets of linear *algebraic* equations. In Sec. **10.42** an alternative method will be given for finding positions by integration of certain *differential* equations. A computer program, DKINAL, which performs a complete kinematic analysis of planar linkages is described and listed in Sec. **10.43**.

10.41 *Method of Algebraic Equations*

For a mechanism with F input (primary) coordinates q_i and N output (secondary) coordinates ϕ_i, we write the constraint equations (two per loop, one per higher pair) in the form

$$f_i(\phi_j; q_k) = 0 \qquad (i = 1, 2, \ldots, N) \tag{1}$$

and form the Jacobian matrix

$$\mathbf{A} = \left[\frac{\partial f_i}{\partial \phi_j} \right] \qquad (2)$$

If velocities are required, we shall need the velocity equations [see Eq. (**10.10**-10)]

$$\mathbf{A}\dot{\boldsymbol{\phi}} = \mathbf{R} = \left[-\frac{\partial f_i}{\partial q_j} \right] \dot{\mathbf{q}} \qquad (3)$$

If accelerations are required, we form the equations

$$\mathbf{A}\ddot{\boldsymbol{\phi}} = \mathbf{S} \qquad (4)$$

where $\mathbf{S}$ is defined by Eq. (**10.10**-15).

For given values of $\mathbf{q}$, $\dot{\mathbf{q}}$, and $\ddot{\mathbf{q}}$, the unknown position variables are initially estimated and then refined by applying the Newton-Raphson method to Eqs. (1), as described in Sec. **9.23**. There it was seen that it is desirable to organize the work by using the following subroutines:

RESFNC: calculates residual functions f_i from Eqs. (1)

MATRIX: calculates elements of Jacobian matrix $\mathbf{A}$ from Eq. (2)

DGELG: solves sets of linear algebraic equations (Appendix **E**)

NEWTR: calls upon RESFNC, MATRIX, and DGELG to solve Eqs. (1) for the unknowns ϕ_i (Appendix **F**)

After the ϕ_i have been calculated, the right-hand side $\mathbf{R}$ of Eq. (3) may be found, and the velocities $\dot{\phi}_i$ may be found by calling MATRIX and DGELG to solve Eqs. (3). Then the right-hand side $\mathbf{S}$ of Eq. (4) may be calculated, and the accelerations $\ddot{\phi}_i$ may be found by calling upon MATRIX and DGELG to solve Eqs. (4).

With all Lagrangian coordinates, velocities, and accelerations known it is a straightforward matter to find the kinematic state $(x, y, \dot{x}, \dot{y}, \ddot{x}, \ddot{y})$ of any desired points of interest from Eqs. (**10.10**-5)–(**10.10**-7).

It proves convenient to organize the remaining calculations into the following subroutines:

RHSVEL: calculates the right-hand side vector $\mathbf{R}$ of the velocity equations (3)

RHSACL: calculates the right-hand side vector $\mathbf{S}$ of the acceleration equations (4)

VARINT: calculates the displacements (x, y), velocities $(\dot{x}, \dot{y})$, and accelerations $(\ddot{x}, \ddot{y})$ of points of interest, as defined by Eqs. (**10.10**-5)–(**10.10**-7)

Examples of such subroutines will now be given for the mechanism of Fig. **10.30**-1, with $\theta_2 = q$ specified as the driving variable. It will be recalled that a position analysis for this mechanism was done in Secs. **9.22** and **9.23**, where the subroutines RESFNC and MATRIX were given. To find velocities we must differentiate the residual functions f_i, given in Eqs. (**9.22**-4), to give the vector

$$\mathbf{R} = \left\{ -\frac{\partial f_i}{\partial q} \right\} \dot{q} = \{0, 0, -a_2\dot{q}\sin q, a_2\dot{q}\cos q\} \qquad (5)$$

Similarly, the vector **S**, defined by Eq. (**10.10**-15), follows from Eqs. (5) and (**9.22**-5) in the form

$$S_1 = a_4\dot{\phi}_3^2 c_3$$
$$S_2 = a_4\dot{\phi}_3^2 s_3$$
$$S_3 = -a_2(\ddot{q}s_q + \dot{q}^2 c_q) + 2\dot{\phi}_2\dot{\phi}_3 s_3 + \phi_2\dot{\phi}_3^2 c_3 \qquad (6)$$
$$S_4 = a_2(\ddot{q}c_q - \dot{q}^2_q) - 2\dot{\phi}_2\dot{\phi}_3 c_3 + \phi_2\dot{\phi}_3^2 s_3$$

where

$$\{\phi_1, \phi_2, \phi_3, \phi_4\} \equiv \{r_1, r_3, \theta_4, r_5\} \qquad (9.22\text{-}3)$$

$$s_3 = \sin\phi_3, \qquad c_3 = \cos\phi_3, \qquad s_q = \sin q, \qquad c_q = \cos q \qquad (7)$$

The base coordinates for points of interest F and G (on links 3 and 4) are found from inspection of Fig. **10.31** to be

$$x_F = a_2 c_q + b_1 c_3 - b_2 s_3 \equiv \phi_5$$
$$y_F = a_2 s_q + b_1 s_3 + b_2 c_3 \equiv \phi_6$$
$$x_G = a_2 c_q - (\phi_2 - b_3)c_3 \equiv \phi_7 \qquad (8)$$
$$y_G = a_2 s_q - (\phi_2 - b_3)s_3 \equiv \phi_8$$

Note that we have *assigned* the names $\phi_5, \ldots, \phi_7$ to the *displacements of interest*. This will enable us to have the output for points of interest printed in a uniform format along with the secondary variables $\phi_1, \ldots, \phi_4$. The velocities and accelerations for the point of interest F are readily seen to be

$$\dot{x}_F = -a_2\dot{q}s_q - \dot{\phi}_3(b_1 s_3 + b_2 c_3) \equiv \dot{\phi}_5$$
$$\dot{y}_F = a_2\dot{q}c_q + \dot{\phi}_3(b_1 c_3 - b_2 s_3) \equiv \dot{\phi}_6 \qquad (9)$$

$$\ddot{x}_F = -a_2\ddot{q}s_q - \ddot{\phi}_3(b_1 s_3 + b_2 c_3) - a_2\dot{q}^2 c_q - \dot{\phi}_3^2(b_1 c_3 - b_2 s_3) = \ddot{\phi}_5$$
$$\ddot{y}_F = a_2\ddot{q}c_q + \ddot{\phi}_3(b_1 c_3 - b_2 s_3) - a_2\dot{q}^2 s_q - \dot{\phi}_3^2(b_1 s_3 + b_2 c_3) = \ddot{\phi}_6 \qquad (10)$$

Equations (5), (6), and (8)–(10) have been programmed in the listings given below for subroutines RHSVEL, RHSACL, and VARINT. The FORTRAN notation is

$$\text{PHI(I)} = \phi_i \qquad \text{DPHI(I)} = \dot{\phi}_i \qquad \text{DDPHI(I)} = \ddot{\phi}_i$$
$$\text{Q} = q \qquad \text{DQ} = \dot{q} \qquad \text{DDQ} = \ddot{q}$$
$$\text{S3} = s_3 \qquad \text{C3} = c_3 \qquad \text{SQ} = s_q$$
$$\text{CQ} = c_q \qquad \text{B1} = b_1 \qquad \text{B2} = b_2$$
$$\text{B3} = b_3 \qquad \text{CN(5)} = a_2 \qquad \text{CN(6)} = a_4$$

$$\text{F(I)} = R_i \text{ (in subroutine RHSVEL)}; \qquad \text{F(I)} = S_i \text{ (in subroutine RHSACL)}$$

Note that the programs are dimensioned to allow for up to 20 secondary coordinates ϕ_i plus 10 additional values of ϕ_i representing coordinates for points of interest; i.e., PHI, DPHI, and DDPHI are given the dimension of 30.

The way in which these subroutines are utilized by a main program is described in Sec. **10.43**. The procedure illustrated here for the mechanism of Fig. **10.30**-1 is readily adapted to any other mechanism.

```
      SUBROUTINE RHSVEL
C.......... THIS ROUTINE CALCULATES RIGHT-HAND SIDE OF VELOCITY EQS.
      IMPLICIT REAL*8 (A-H,O-Z)
      COMMON//A(20,20),PHI(30),DPHI(30),DDPHI(30),F(20),CN(20)
     1,Q,DQ,DDQ,NEQS,NVI,NPHI,IND
      COMMON/PROB/SC,CQ,S3,C3
      F(1)=0.0D0
      F(2)=0.0D0
      F(3)=-CN(5)*SC*DQ
      F(4)=CN(5)*CQ*DQ
      RETURN
      END


      SUBROUTINE RHSACL
C......THIS ROUTINE CALCULATES RIGHT-HAND SIDE OF ACCELERATION EQUATIONS
C
      IMPLICIT REAL*8 (A-H,O-Z)
      COMMON//A(20,20),PHI(30),DPHI(30),DDPHI(30),F(20),CN(20)
     1,Q,DQ,DDQ,NEQS,NVI,NPHI,IND
      COMMON/PROB/SC,CQ,S3,C3
      F(1)=CN(6)*DPHI(3)*DPHI(3)*C3
      F(2)=CN(6)*DPHI(3)*DPHI(3)*S3
      F(3)=2.0D0*DPHI(3)*DPHI(2)*S3+PHI(2)*DPHI(3)*DPHI(3)*C3
     1-CN(5)*DDQ*SQ-CN(5)*DQ*DQ*CQ
      F(4)=-2.0D0*DPHI(2)*DPHI(3)*C3+PHI(2)*DPHI(3)*DPHI(3)*S3
     1+CN(5)*DDQ*CQ-CN(5)*DQ*DQ*SQ
      RETURN
      END


      SUBROUTINE VARINT
C     THIS SUBROUTINE COMPUTES DISPLACEMENT, VELOCITY, AND
C     ACCELERATION OF VARIABLES OF INTEREST.
      IMPLICIT REAL*8 (A-H,O-Z)
      COMMON//A(20,20),PHI(30),DPHI(30),DDPHI(30),F(20),CN(20)
     1,Q,DQ,DDQ,NEQS,NVI,NPHI,IND
      COMMON/PROB/SC,CQ,S3,C3
      B1=0.3D0
      B2=0.1D0
      B3=1.0D0
      PHI(5)=CN(5)*CQ+B1*C3-B2*S3
      PHI(6)=CN(5)*SQ+B1*S3+B2*C3
      DPHI(5)=-CN(5)*SQ*DQ-B1*S3*DPHI(3)-B2*C3*DPHI(3)
      DPHI(6)=CN(5)*CQ*DQ+B1*C3*DPHI(3)-B2*S3*DPHI(3)
      DDPHI(5)=-CN(5)*CQ*DQ*DQ-CN(5)*SQ*DDQ-B1*C3*DPHI(3)*DPHI(3)
     1-B1*S3*DDPHI(3)+B2*S3*DPHI(3)*DPHI(3)-B2*C3*DDPHI(3)
      DDPHI(6)=-CN(5)*SQ*DQ*DQ+CN(5)*CQ*DDQ-B1*S3*DPHI(3)*DPHI(3)
     1+B1*C3*DDPHI(3)+B2*C3*DPHI(3)*DPHI(3)-B2*S3*DDPHI(3)
      PHI(7)=CN(5)*CQ-(PHI(2)-B3)*C3
      PHI(8)=CN(5)*SQ-(PHI(2)-B3)*S3
      RETURN
      END
```

10.42 The Method of Differential Equations

By solving the nonlinear *algebraic* displacement equations

$$f_i(\phi_j; q_k) = 0 \tag{1}$$

for ϕ_j and then using the results to solve the linear algebraic velocity equation

$$\mathbf{A}\dot{\boldsymbol{\phi}} = \mathbf{R} \tag{2}$$

for $\dot{\phi}_j$, we are paralleling the sequence of operations followed in the graphical procedures of the precomputer era.

An alternative approach was proposed by Paul and Krajcinovic [1970a], who viewed Eqs. (2) as a set of coupled nonlinear *differential* equations for the displacements ϕ_i. Such equations may readily be solved by numerical integration using any of the readily available library routines for systems of nonlinear ordinary differential equations with specified initial conditions. For the initial state we may take any convenient configuration of the system and then work forward or backward in time about the initial positions. Having found the displacements by integration of Eqs. (2), we may utilize Eqs. (1) for a check on the roundoff error.

In using this method it is necessary to realize that a numerical subroutine for solving sets of ordinary differential equations may be viewed[2] as a black box into which one feeds known values of ϕ_i and $\dot{\phi}_i$ at a given time t, whereupon the black box prints out values of the displacements ϕ_i evaluated at the time $t + h$, where h is called the *time step*. We may find the numerical values of the velocities $\dot{\phi}_i$, at time t, by viewing Eq. (2) as a linear system of *algebraic* equations in $\dot{\phi}_i$ which may readily be solved for by means of a Gauss elimination procedure.

Thus we see that the successful integration of Eq. (2) requires us to have available two standard types of subroutines—one for solving sets of linear *algebraic* equations and another for solving sets of ordinary *differential* equations with specified initial conditions.[3]

Suitable logic for the complete kinematic analysis of an arbitrary plane mechanism is given in Fig. 1.

It is important in using this method that the *initial conditions* for secondary variables $\phi_i(0)$ (which may have come from a scale drawing) be accurate. Errors in these initial values represent a violation of the loop closure equations (1) and can cause the numerical solution to gradually drift away from the desired results. One way to ensure initial satisfaction of the displacement equations is to calculate the *initial* values of ϕ_i by means of a Newton-Raphson procedure, as in Sec. **9.23**. Note that the Newton-Raphson algorithm is needed only to get the method of differential equations started and is not called upon for $t > 0$.

Comparison of Methods

In numerical experiments reported by Hud [1976], it was found that the method of *differential* equations will require about 50 % more computer time than the method of *algebraic* equations for step sizes corresponding to $\Delta q = 10°$ and 100 % more time for $\Delta q = 20°$ in typical mechanisms. We can therefore conclude that, for purposes of *kinematics* analysis, the (iterative) algebraic method of Sec. **10.41** is more efficient than the method of differential equations described here.

---

[2]See Appendix **G**.
[3]Typical subroutines are DGELG (Appendix **E**) for linear algebraic equations and DRKGS (Appendix **G**) for differential equations.

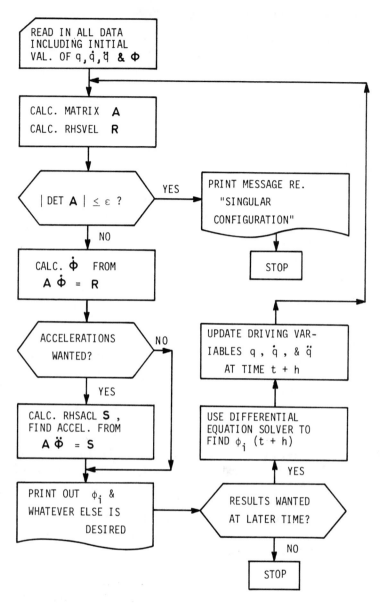

Figure 10.42-1. Flow chart for method of differential equations ϵ is tolerance on determinant of **A**.

However, when the kinematics analysis is merely the first step in a complete dynamic analysis of a mechanism (where numerical integration of the differential equations of motion is an inherent part of the total problem), the method of differential equations is more efficient for the total *dynamics* problem than the iterative algebraic method. We shall return to this point in our discussion of numerical methods for dynamics of machinery in Sec. **14.20**.

SECTION 10.42 THE METHOD OF DIFFERENTIAL EQUATIONS

In Sec. **9.23** we saw how to prepare the subroutines RESFNC and MATRIX, which provide the functions $f_i(\phi_j)$ and the Jacobian matrix needed for the displacement analysis of a mechanism. Then, in Sec. **10.41**, we saw how to prepare the subroutines RHSVEL, RHSACL, and VARINT, which enable us to find the secondary velocities, and accelerations, and the kinematic state for any points of interest. In Fig. 1, we give the FORTRAN listing for a general-purpose program, DKINAL (Double-Precision KINematic anALysis), which ties all these subroutines together and performs a complete kinematic analysis for planar linkages with 1 DOF (Paul and Hud [1975]).

It should be noted that this program executes with great efficiency, but it does require the user to invest some time in preparation of the subroutines referred to above. Other programs exist in which the computer generates the required vectors and matrices, thereby reducing or eliminating the burden of coding by the user. One such program is KINMAC[4] (KINematics of MAChinery), in which the user merely supplies numerical and logical information sufficient to describe the geometry and topology (loop structure) of the system. KINMAC is related to a more general program DYMAC[5] (DYnamic Analysis of MAChinery), which is described in Sec. **14.40**.

The listing for DKINAL is largely self-documented, and users are urged to study the introductory comments given in the listing. The following supplementary comments may also be useful.

The program version (version C) is given in double precision primarily because it was used on IBM and UNIVAC computers which have a relatively small word length; for most other machines, a single-precision version is adequate. The version given utilizes the standard subroutines DGELG (see Appendix E), NEWTR (see Appendix F), and OUTPUT (included in listing). Examples of how to prepare the user-supplied subroutines (RESFNC, MATRIX, RHSVEL, RHSACL, VARINT) have already been given (in Secs. **9.23** and **10.41**). The number of unknown position variables ϕ_i [PHI(I)] exceeds the number of constraint equations (NEQS) because we utilize ϕ_i ($i >$ NEQS) to represent the displacements of points of interest. For example, in Sec. **10.41**, NEQS = 4 and $\phi_5 = x_F$, $\phi_6 = y_F$, $\phi_7 = x_G$, $\phi_8 = y_G$. The number NVI of such variables of interest is 4 in this example but may range up to 10.[6] The user may specify up to 20 constants stored in the array CN(I); of these, the first NEQS values represent the *initial* estimates of $\phi_1, \phi_2, \ldots, \phi_{NEQS}$, which will be refined by the Newton-Raphson process. The total number of constants CN(I) to be read in is denoted by INCNS. The user must set the *initial* values of the driving variables q, $\dot{q}$, $\ddot{q}$ (Q, DQ, DDQ) and the desired increment Δq (DELQ) to be used for successive states until q reaches the value QMAX. If $\dot{q}$ and $\ddot{q}$ are not constants, their updated values should be specified in subroutines RHSVEL and RHSACCL where needed.

The input data must be supplied in the format indicated by the comment statements following INPUT DATA ARRANGEMENT.

The output is arranged as shown in Fig. 2, where the numerical data are given for

---

[4]Paul and Amin [1978a]. Program KINMAC is available from the author.
[5]Paul [1977] and Paul and Hud [1976].
[6]For this reason, in the DIMENSION statement of subroutine NEWTR (Appendix F), change X(20) to X(30) for compatibility with DKINAL-C.

```
C       PROGRAM  DKINAL-C
C       BY  B. PAUL AND A. AMIN,  MARCH 1976
C
C       PURPOSE...
C          KINEMATIC ANALYSIS OF PLANAR MACHINES
C
C       PROGRAM DESCRIPTION...
C          THIS DOUBLE PRECISION PROGRAM SOLVES FOR POSITION, VELOCITY,
C       AND ACCELERATION VARIABLES IN  PLANAR MECHANISM. IT IS DIMENSIONED
C       FOR UPTO 20 EQUATIONS OF CONSTRAINT (LOOP EQS.) AND 10 VARIABLES
C       OF INTEREST (E.G. POSITION, VELOCITY, OR ACCELERATION OF POINTS OF
C       INTEREST). NOTE THAT ANGLES MUST BE EXPRESSED IN RADIANS.
C          THE PROGRAM CONSISTS OF A MAIN PROGRAM , THREE SUBROUTINES
C       (NEWTR, DGELG, OUTPUT) AND FIVE USER-SUPPLIED SUBROUTINES (RESFNC,
C       MATRIX, RHSVEL, RHSACL, VARINT).
C
C       USE OF PROGRAM AND SUBROUTINES...
C
C       STANDARD ROUTINES...
C       MAIN     READS AND WRITES INPUT DATA & MESSAGES, COORDINATES ROUTINES
C       OUTPUT   PRINTS RESULTS AT EACH INTERVAL
C       NEWTR    SOLVES NONLINEAR EQS. BY NEWTON-RAPHSON METHOD
C       DGELG    SOLVES LINEAR SIMULTANEOUS EQS.BY GAUSS ELIMINATION METHOD
C                REFERENCE: IBM SSP MANUAL H20-0205
C
C       USER SUPPLIED SUBROUTINES ...
C       RESFNC   EVALUATES RESIDUAL FUNCTIONS F(I) FROM DISPLACEMENT EQS.
C       MATRIX   EVALUATES ANALYTICAL FIRST DERIVATIVES OF DISP. EQS.
C       RHSVEL   COMPUTES RHS OF VELOCITY EQS.
C       RHSACL   COMPUTES RHS OF ACL. EQS.
C       VARINT   COMPUTES DISP., VELOCITY, & ACCEL. OF VARIABLES OF INTEREST
C
C       DESCRIPTION OF INPUT VARIABLES...
C
C       TITLE    ANY 80 CHARACTER TITLE TO IDENTIFY PROBLEM
C       NEQS     NO. OF CONSTRAINT EQUATIONS (LOOP EQS.)
C       NVI      NO. OF VARIABLES OF INTEREST
C                SEE   COMMENTS IN SUBROUTINE VARINT
C       INCNS    NO. OF INPUT CONSTANTS
C       LEVEL    ANALYSIS RESTRICTION LEVEL INDEX
C                =0 FOR DISPLACEMENT ANALYSIS
C                =1 FOR DISPLACEMENT & VELOCITY ANALYSIS
C                =2 FOR DISPLACEMENT, VELOCITY, & ACCELERATION ANALYSIS
C       FTOL     TOLERANCE FOR RESIDUALS
C       Q        INITIAL VALUE OF INPUT (DRIVING) VARIABLE
C       DELQ     INCREMENT IN Q
C       QMAX     MAXIMUM VALUE OF Q
C       DQ       DRIVING VELOCITY, DQ/DT
C       DDQ      DRIVING ACCELERATION, D(DQ/DT)/DT
C       CN(I)    A VECTOR OF CONSTANTS. THE INITIAL ESTIMATES OF
C                UNKNOWN POSITION VARIABLES PHI(I) ARE READ IN AS
C                CN(1), CN(2),...CN(NEQS). THE REMAINING ENTRIES
C                IN CN(I) ARE OTHER PARAMETERS DESIRED BY USER.
C
C       DESCRIPTION OF FORTRAN VARIABLES USED INTERNALLY...
C
C       PHI(I)   UNKNOWN POSITION VARIABLES
C       DPHI(I)  RATE OF CHANGE OF PHI (VELOCITY)
C       DDPHI(I)   ''        ''    DPHI (ACCELERATION)
C       F(I)     RESIDUAL FUNCTIONS. ALSO USED TO DENOTE RHS OF EQS. SOLVED
C                BY DGELG
C       A(I,J)   JACOBIAN MATRIX OF FUNCTION SET F(I)
C       NPHI     NEQS+NVI  TOTAL NO. OF POSITION VARIABLES TO BE SOLVED FOR
C       IND      COUNTER WHICH INDICATES STATE OF SYSTEM
```

Figure 10.43-1. Computer program DKINAL.

```
C      ITER     SEE SUBROUTINE NEWTR
C      EPS      SEE SUBROUTINE DGELG
C      IER      SEE SUBROUTINE DGELG
C      DET      SEE SUBROUTINE DGELG
C
C      INPUT DATA ARRANGEMENT...
C
C      CARD TYPE   NO. OF CARDS    FORMAT       VARIABLES
C         1            1           (20A4)       TITLE
C         2            1           (4I2,7F10.0) NEQS,NVI,LEVEL,INCNS,FTOL
C                                               Q,DELQ,QMAX,DQ,DDQ
C         3         INCNS/8        (8F10.0)     CN(I),I=1,INCNS
C
C      ...........................................................
C
       IMPLICIT REAL*8(A-H,O-Z)
       DIMENSION AV(400),AA(400),TITLE(20)
       COMMON//A(20,20),PHI(30),DPHI(30),DDPHI(30),F(20),CN(20)
      1,Q,DQ,DDQ,NEQS,NVI,NPHI,IND
       EQUIVALENCE (A,AV),(F,TITLE)
C
C      ...EPS IS RELATIVE TOLERANCE REQUIRED FOR SUBROUTINE DGELG
C         USER MAY ALTER IF DESIRED
C
       EPS=1.0D-10
C
C      ...READ INPUT DATA...
C
     1 READ(5,2,END=150)TITLE
     2 FORMAT(20A4)
       READ(5,5)NEQS,NVI,LEVEL,INCNS,FTOL,Q,DELQ,QMAX,DQ,DDQ
     5 FORMAT(4I2,7F10.0)
       READ(5,10)(CN(I),I=1,INCNS)
    10 FORMAT(8F10.0)
       NPHI=NEQS+NVI
       DO 15 I=1,NPHI
       PHI(I)=0.0
       DPHI(I)=0.0
       DDPHI(I)=0.0
    15 CONTINUE
C
C      ...REPRINT INPUT DATA...
C
       WRITE(6,20)(TITLE(I),I=1,20)
    20 FORMAT('1', 9X,'PROGRAM DKINAL-C'/10X,20A4)
       WRITE(6,30)NEQS,NVI,INCNS,LEVEL,FTOL,Q,DELQ,QMAX,DQ,DDQ
    30 FORMAT(//10X, 'INPUT DATA...'/
      1/10X,'NO. OF EQUATIONS(NEQS)=',I2,
      2/10X,'NO. OF VARIABLES OF INTEREST(NVI)=',I2,
      3/10X,'NO. OF CONSTANTS(INCNS)=',I2,
      4/10X,'ANALYSIS RESTRICTION LEVEL(LEVEL)=',I2,
      5/10X,'TOLERANCE FOR  RESIDUALS(FTOL)=',D14.6,
      6/10X,'INITIAL VALUE OF INPUT(DRIVING) VARIABLE(Q)=',D14.6,
      7/10X,'INCREMENTAL VALUE FOR INPUT VARIABLE(DELQ)=',D14.6,
      8/10X,'MAXIMUM VALUE OF  INPUT VARIABLE(QMAX)=',D14.6
      A/10X,'INITIAL VALUE OF INPUT VARIABLE VELOCITY(DQ)=',D14.6,
      9/10X,'INITIAL VALUE OF INPUT VARIABLE ACCELERATION(DDQ)=',D14.6,
      B///10X,'ESTIMATES OF SECONDARY DISPLACEMENT VARIABLES' /)
       WRITE(6,40)(I,CN(I),I=1,NEQS)
    40 FORMAT( 10X,'PHI(',I2,')=',D14.6)
       WRITE(6,50)
    50 FORMAT( /10X,'OTHER CONSTANTS FOR GEOMETRY OF MACHINE'/)
       I=NEQS+1
       WRITE(6,60)(J,CN(J),J=I,INCNS)
```

Figure 10.43-1. (Cont.)

```
       60 FORMAT( 10X,'CN(',I2,')=',D14.6)
          DO 65I=1,NEQS
       65 PHI(I)=CN(I)
          IND=0
       70 CONTINUE
C
C          ...SOLVE FOR DISPLACEMENT...
C
          CALL NEWTR(PHI,A,NEQS,FTOL,ITER,IER,F)
          IF(IER.NE.0)GO TO 130
          IF(ITER.LT.0)GO TO 110
          IF(LEVEL.EQ.0)GO TO 100
C
C          ...SOLVE FOR VELOCITIES...
C
          CALL MATRIX
          CALL RHSVEL
          K=1
          DO 75 I=1,NEQS
          DO 75 J=1,NEQS
          AA(K)=A(J,I)
          AV(K)=A(J,I)
       75 K=K+1
       80 CALL DGELG(F,AV,NEQS,1,EPS,IER,DET)
          IF(IER.NE.0)GO TO 130
          DO 85 K=1,NEQS
       85 DPHI(K)=F(K)
          IF(LEVEL.EQ.1)GO TO 100
C
C          ....SOLVE FOR ACCELERATIONS...
C
          CALL RHSACL
          CALL DGELG(F,AA,NEQS,1,EPS,IER,DET)
          DO 90K=1,NEQS
       90 DDPHI(K)=F(K)
      100 IF(NVI.EQ.0)GO TO 105
C
C          ....SOLVE FOR VARIABLES OF INTEREST...
C
          CALL VARINT
      105 CALL OUTPUT
C
C          ....UPDATE THE INDEPENDENT VARIABLE...
C
          IF(Q.GE.QMAX)GO TO 1
          Q=Q+DELQ
          GO TO 70
      110 WRITE(6,115)
      115 FORMAT(/5X,'...NO CONVERGENCE WITH PRESCRIBED ITERATIONS..F(I)=')
          WRITE(6,120)(F(I),I=1,NEQS)
      120 FORMAT(10X,D14.6)
          GO TO 1
      130 WRITE(6,135)
      135 FORMAT(/10X,'***MATRIX APPEARS TO BE SINGULAR, CHECK FOR DEAD ',
         1'POINTS, CHANGE POINTS, ETC ***')
      140 GO TO 1
      150 WRITE(6,155)
      155 FORMAT(//'.....LAST CASE FINISHED....')
          STOP
          END
C          ...............................................................
C
          SUBROUTINE OUTPUT
C
```

Figure 10.43-1. (Cont.)

```
C      THIS SUBROUTINE DETERMINES DISPLAY OF OUTPUT
C
       IMPLICIT REAL*8(A-H,O-Z)
       COMMON//A(20,20),PHI(30),DPHI(30),DDPHI(30),F(20),CN(20)
      1,Q,DQ,DDQ,NEQS,NVI,NPHI,IND
       IND=IND+1
       IF(IND.NE.1) GO TO 10
       WRITE(6,5)
     5 FORMAT('1',10X,'STATE',4X,'INPUT,Q',6X,'VARI-',4X,'POSITION'
      1,8X,'VELOCITY',6X,'ACCELERATION'/33X,'ABLE'/)
    10 WRITE(6,15)IND,Q
    15 FORMAT(12X,I2,2X,D14.6)
       WRITE(6,20)(I,PHI(I),DPHI(I),DDPHI(I),I=1,NPHI)
    20 FORMAT('+',T34,I2,3X,D14.6,2X,D14.6,2X,D14.6/)
       WRITE(6,25)
    25 FORMAT('+',9X,'-------------------------------------------------'
      1,'----------------------' )
       RETURN
       END
C
```

Figure 10.43-1. (Cont.).

```
PROGRAM DKINAL-C
DOUBLE SLIDER QUICK-RETURN MECHANISM


INPUT DATA...

NO. OF EQUATIONS(NEQS)= 4
NO. OF VARIABLES OF INTEREST(NVI)= 4
NO. OF CONSTANTS(INCNS)= 7
ANALYSIS RESTRICTION LEVEL(LEVEL)= 2
TOLERANCE FOR  RESIDUALS(FTOL)=  0.100000D-08
INITIAL VALUE OF INPUT(DRIVING) VARIABLE(Q)=  0.000000D 00
INCREMENTAL VALUE FOR INPUT VARIABLE(DELQ)=  0.349066D 00
MAXIMUM VALUE OF  INPUT VARIABLE(QMAX)=  0.628319D 01
INITIAL VALUE OF  INPUT VARIABLE VELOCITY(DQ)=  0.100000D 01
INITIAL VALUE OF INPUT VARIABLE ACCELERATION(DDQ)=  0.000000D 00


ESTIMATES OF SECONDARY DISPLACEMENT VARIABLES

PHI( 1)=  0.110000D 01
PHI( 2)=  0.140000D 01
PHI( 3)=  0.120000D 01
PHI( 4)=  0.850000D 00

OTHER CONSTANTS FOR GEOMETRY OF MACHINE

CN( 5)=  0.500000D 00
CN( 6)=  0.300000D 01
CN( 7)=  0.120000D 01
```

Figure 10.43-2. Computer output for example of Fig. 10.30-1 with crank speed $\dot{q} = 1$ rad/sec.

STATE	INPUT,Q	VARI- ABLE	POSITION	VELOCITY	ACCELERATION
1	0.000000D 00	1	0.167460D 01	0.817379D-01	0.182759D 00
		2	0.174765D 01	0.557421D 00	0.479207D-01
		3	0.128064D 01	0.952330D-01	0.243308D 00
		4	0.858294D 00	-0.273757D-01	-0.707197D 00
		5	0.490009D 00	-0.301003D-01	-0.576812D 00
		6	0.316070D 00	0.499049D 00	-0.529733D-02
		7	0.286098D 00	0.000000D 00	0.000000D 00
		8	-0.716401D 00	0.000000D 00	0.000000D 00
2	0.349066D 00	1	0.171063D 01	0.115299D 00	0.214170D-01
		2	0.193942D 01	0.526285D 00	-0.210152D 00
		3	0.132610D 01	0.158643D 00	0.130259D 00
		4	0.726785D 00	_(obscured)_	_(obscured)_
		5	0.6455_(obscured)_	_(obscured)_	_(obscured)_
				0.000000D 00	0.000000D 00
		8	-0.707431D 00	0.000000D 00	0.000000D 00
19	0.628319D 01	1	0.167460D 01	0.817384D-01	0.182758D 00
		2	0.174765D 01	0.557422D 00	0.479183D-01
		3	0.128064D 01	0.952336D-01	0.243307D 00
		4	0.858294D 00	-0.273759D 00	-0.707194D 00
		5	0.490009D 00	-0.301018D-01	-0.576811D 00
		6	0.316071D 00	0.499049D 00	-0.529872D-02
		7	0.286098D 00	0.000000D 00	0.000000D 00
		8	-0.716401D 00	0.000000D 00	0.000000D 00

Figure 10.43-2. (Cont.).

the mechanism of Fig. **10.30**-1, and excerpts of the computed results are shown. The initial guesses for $\phi_1, \phi_2, \phi_3, \phi_4$ were given the values of CN(1), ..., CN(4) shown in Fig. 2, and the numerical values of a_2, a_4, and L were entered as the constants CN(5), CN(6), and CN(7), respectively.

10.50 VELOCITY AND ACCELERATION COEFFICIENTS

The concept of the *velocity ratio* or *influence coefficient* was utilized in Sec. **10.30** as an aid in the analysis of weakly coupled systems. Such coefficients, and similar ones related to accelerations, are the basis of semigraphical methods of kinematic analysis described by Goodman [1958] and Modrey [1959]. From our point of view, the influence coefficients will be of greatest value when we discuss *dynamics* of machinery. Therefore we shall show how to calculate such kinematic influence coefficients and establish some of their properties at this point, but their real value will become apparent when they are used in later chapters.

In Sec. **10.51** we shall show that for systems with 1 DOF the secondary velocities and accelerations are related to the primary velocity and primary acceleration by relations of the form

$$\dot{\phi}_i = k_i \dot{q}$$
$$\ddot{\phi}_i = k_i \ddot{q} + k_i' \dot{q}^2 \tag{1}$$

Similarly, the angular velocity and acceleration for link i are given by

$$\dot{\theta}_i = \omega_i \dot{q}, \qquad \ddot{\theta}_i = \omega_i \ddot{q} + \omega_i' \dot{q}^2 \tag{2}$$

and the velocity and acceleration components of any point of interest (with identification number i) in the system are given by

$$\dot{x}_i = u_i \dot{q}, \qquad \ddot{x}_i = u_i \ddot{q} + u_i' \dot{q}^2 \tag{3}$$

$$\dot{y}_i = v_i \dot{q}, \qquad \ddot{y}_i = v_i \ddot{q} + v_i' \dot{q}^2 \tag{4}$$

The **velocity coefficients** $(k_i, \omega_i, u_i, v_i)$ and the **centripetal coefficients** $(k_i', \omega_i', u_i', v_i')$ are functions of the instantaneous *configuration* of the system but not of the velocities of the moving parts.

A generalization of these results to systems with several degrees of freedom is given in Sec. **10.52**.

The detailed procedures for calculating such influence coefficients, given in Secs. **10.51** and **10.52**, may be bypassed at this point and returned to when needed in later chapters for dynamics analyses.

10.51 Systems with One Degree of Freedom

Velocity Coefficients

We have seen in Sec. **10.10** that the secondary velocities $\dot{\phi}_i$ in a constrained system with 1 DOF are related by N linear constraint equations of the form

$$\sum_{j=1}^{N} A_{ij} \dot{\phi}_j = b_i \dot{q} \tag{1}$$

where the terms A_{ij} and b_i are known functions of the secondary coordinates ϕ_i and the primary variable q. Since Eq (1) only defines the ratios $\dot{\phi}_j/\dot{q}$, it is convenient to introduce the *velocity ratios*

$$k_j \equiv \frac{\dot{\phi}_j}{\dot{q}} = \frac{d\phi_j/dt}{dq/dt} = \frac{d\phi_j}{dq} \tag{2}$$

and express Eq. (1) in the form

$$\sum_{j=1}^{N} A_{ij} k_j = b_i \quad \text{or} \quad \mathbf{Ak} = \mathbf{b} \tag{3}$$

After solving the $N \times N$ system for the influence coefficients $k_i \equiv \dot{\phi}_i/\dot{q}$, one can readily find the angular velocity ratios $\omega_i \equiv \dot{\theta}_i/\dot{q}$. For example, in the case of the four-bar linkage of Fig. **7.10**-1 (a), with the variables selected as

$$\{\dot{\theta}_i, \dot{\theta}_j, \dot{\theta}_k\} = \{\dot{\phi}_1, \dot{\phi}_2, \dot{q}\} \tag{4a}$$

where (i, j, k) represent $(1, 2, 3)$, $(2, 3, 1)$, or $(3, 1, 2)$, it follows that

$$\{\omega_i, \omega_j, \omega_k\} = \frac{1}{\dot{q}}\{\dot{\phi}_1, \dot{\phi}_2, \dot{q}\} = \{k_1, k_2, 1\} \tag{4b}$$

Thus we see that, with θ_k driving,

$$\omega_k = 1 \tag{5a}$$

and the remaining velocities may be expressed, from Eq. (**7.10**-8), in the compact form

$$\omega_i \equiv \frac{\dot{\theta}_i}{\dot{\theta}_k} = \frac{a_k \sin(\theta_j - \theta_k)}{a_i \sin(\theta_i - \theta_j)} \qquad (i \neq k) \tag{5b}$$

To find ω_j, it is only necessary to interchange indices i and j in Eq. (5b)

Once the velocity ratios k_i and ω_i are known, all velocities in the system are easily calculated for any desired value of $\dot{q}$. To see this, consider a point of interest whose coordinates are given by known expressions of the form

$$\begin{aligned} x_i &= x_i(\phi_j, q) \\ y_i &= y_i(\phi_j, q) \end{aligned} \tag{6}$$

Whenever we seek to differentiate a function of the form $F(\phi_j, q)$, we should first observe that

$$dF = \frac{\partial F}{\partial q} dq + \sum_{j=1}^{N} \frac{\partial F}{\partial \phi_j} d\phi_j \tag{7}$$

Dividing all terms in this equation by dq, we find the total derivative with respect to q in the form

$$F' \equiv \frac{dF}{dq} = \frac{\partial F}{\partial q} + \sum_{j=1}^{N} \frac{\partial F}{\partial \phi_j} \frac{d\phi_j}{dq} = \frac{\partial F}{\partial q} + \sum_{j} k_j \frac{\partial F}{\partial \phi_j} \tag{8}$$

where use has been made of Eq. (2). Similarly, by dividing Eq. (7) throughout by dt, we find the time rate of F in the form

$$\dot{F} = \frac{\partial F}{\partial q} \dot{q} + \sum_{j=1}^{N} \frac{\partial F}{\partial \phi_j} \dot{\phi}_j = \frac{dF}{dq} \frac{dq}{dt} = F'\dot{q} \tag{9}$$

By applying this general result to Eqs. (6), the velocities of interest are seen to be

$$\begin{aligned} \dot{x}_i &= x_i'\dot{q} = u_i\dot{q} \\ \dot{y}_i &= y_i'\dot{q} = v_i\dot{q} \end{aligned} \tag{10}$$

where we have introduced the symbols u_i, v_i for the velocity coefficients:

$$\begin{Bmatrix} u_i \\ v_i \end{Bmatrix} \equiv \begin{Bmatrix} x_i' \\ y_i' \end{Bmatrix} = \frac{\partial}{\partial q} \begin{Bmatrix} x_i \\ y_i \end{Bmatrix} + \sum_{j=1}^{N} k_j \frac{\partial}{\partial \phi_j} \begin{Bmatrix} x_i \\ y_i \end{Bmatrix} \tag{11}$$

Although the velocity coefficients may always be calculated by a direct application of Eq. (11), it is desirable to develop an alternative method, based on the geometry of Fig. 1, which shows one link with identification number i.

An arbitrary point P_i in link i is defined by rectangular coordinates (ξ_i, η_i) with respect to **local axes** (ξ, η) which emanate from a reference point R fixed in the link. The coordinates of P_i with respect to fixed **global coordinates** (x, y) are denoted by (x_i, y_i), and the angle between the ξ and x axes is denoted by θ_i.

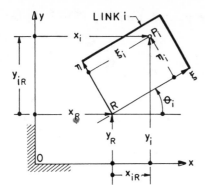

Figure 10.51-1. Typical link.

We shall denote the components of vector $\overrightarrow{RP_i}$ by

$$\{x_{P_iR}, y_{P_iR}\} \equiv \{x_i - x_R, y_i - y_R\}$$

When it is understood that P_i represents a general point on link i we can simplify the subscript P_i to i and write

$$
\begin{aligned}
x_{iR} &\equiv x_i - x_R = \xi_i \cos \theta_i - \eta_i \sin \theta_i \\
y_{iR} &\equiv y_i - y_R = \xi_i \sin \theta_i + \eta_i \cos \theta_i
\end{aligned}
\tag{12}
$$

It then follows from Eq. (**5.31**-6) that the velocity of point P_i is given by

$$
\begin{aligned}
\dot{x}_i &= \dot{x}_R - y_{iR}\dot{\theta}_i \\
\dot{y}_i &= \dot{y}_R + x_{iR}\dot{\theta}_i
\end{aligned}
\tag{13}
$$

Utilizing the definitions

$$
\dot{\theta}_i = \omega_i \dot{q}, \qquad
\begin{Bmatrix} \dot{x}_R \\ \dot{y}_R \end{Bmatrix} \equiv \begin{Bmatrix} u_R \\ v_R \end{Bmatrix} \dot{q}
\tag{14}
$$

in Eqs. (13), we find that

$$
\begin{Bmatrix} \dot{x}_i \\ \dot{y}_i \end{Bmatrix} = \begin{Bmatrix} u_R - y_{iR}\omega_i \\ v_R + x_{iR}\omega_i \end{Bmatrix} \dot{q}
\tag{15}
$$

from which it follows that

$$
\begin{Bmatrix} u_i \\ v_i \end{Bmatrix} \equiv \frac{1}{\dot{q}} \begin{Bmatrix} \dot{x}_i \\ \dot{y}_i \end{Bmatrix} = \begin{Bmatrix} u_R \\ v_R \end{Bmatrix} + \begin{Bmatrix} -y_{iR} \\ x_{iR} \end{Bmatrix} \omega_i
\tag{16}
$$

Equation (16) is useful in finding the velocity coefficients for points of interest in chains of links. For example, in the four-bar linkage of Fig. 2, we see that $\dot{x}_A = \dot{y}_A = 0$; hence

$$
\{u_A, v_A\} = \frac{\{\dot{x}_A, \dot{y}_A\}}{\dot{q}} = \{0, 0\}
\tag{17}
$$

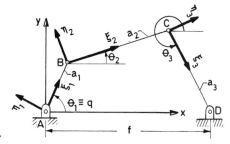

Figure 10.51-2. Four-bar linkage.

It then follows that for link 1, where R represents point A, $\{u_R, v_R\} = \{u_A, v_A\} = \{0, 0\}$ and Eq. (16) predicts

$$\begin{Bmatrix} u_1 \\ v_1 \end{Bmatrix} = \begin{Bmatrix} -y_{1A} \\ x_{1A} \end{Bmatrix} \omega_1 \tag{18}$$

Since point B is on link 1, it follows that

$$\begin{Bmatrix} u_B \\ v_B \end{Bmatrix} = \begin{Bmatrix} -(y_B - y_A) \\ x_B - x_A \end{Bmatrix} \omega_1 \equiv \begin{Bmatrix} -y_{BA} \\ x_{BA} \end{Bmatrix} \omega_1 \tag{18a}$$

For link 2, R represents point B, and Eqs. (16) and (18a) lead to the result

$$\begin{Bmatrix} u_2 \\ v_2 \end{Bmatrix} = \begin{Bmatrix} -y_{BA} \\ x_{BA} \end{Bmatrix} \omega_1 + \begin{Bmatrix} -y_{2B} \\ x_{2B} \end{Bmatrix} \omega_2 \tag{19}$$

Finally, for link 3, we may let R represent point D and observe that the vanishing of $\dot{x}_D$ and $\dot{y}_D$ implies $u_D = v_D = 0$; then Eq. (16) takes the form

$$\begin{Bmatrix} u_3 \\ v_3 \end{Bmatrix} = \begin{Bmatrix} -y_{3D} \\ x_{3D} \end{Bmatrix} \omega_3 = \begin{Bmatrix} -y_{3C} + y_{DC} \\ x_{3C} - x_{DC} \end{Bmatrix} \omega_3 \tag{20}$$

where the last result follows from

$$\begin{Bmatrix} -y_{3D} \\ x_{3D} \end{Bmatrix} = \begin{Bmatrix} -(y_3 - y_D) \\ x_3 - x_D \end{Bmatrix} \equiv \begin{Bmatrix} -(y_3 - y_C) + (y_D - y_C) \\ (x_3 - x_C) - (x_D - x_C) \end{Bmatrix} \tag{21}$$

Equations (18), (19), and (20) are compact expressions for the velocity coefficients in any four-bar mechanism. To show these coefficients *in extenso* for one of the six possible permutations of the indices (i, j, k), consider the case where link 1 is the driving link and

$$\{\phi_1, \phi_2, q\} \equiv \{\theta_2, \theta_3, \theta_1\} \tag{22}$$

Then Eqs. (5) reduce to

$$\frac{\dot{\theta}_1}{\dot{\theta}_1} \equiv \omega_1 = 1 \tag{23}$$

$$k_1 = \frac{\dot{\phi}_1}{\dot{\theta}_1} = \frac{\dot{\theta}_2}{\dot{\theta}_1} \equiv \omega_2 = \frac{a_1 \sin (\theta_3 - \theta_1)}{a_2 \sin (\theta_2 - \theta_3)} \tag{24}$$

$$k_2 = \frac{\dot{\phi}_2}{\dot{\theta}_1} = \frac{\dot{\theta}_3}{\dot{\theta}_1} \equiv \omega_3 = \frac{a_1 \sin (\theta_2 - \theta_1)}{a_3 \sin (\theta_3 - \theta_2)} \tag{25}$$

and Eqs. (18), (19), and (20) give

$$\begin{Bmatrix} u_1 \\ v_1 \end{Bmatrix} = \begin{Bmatrix} -\xi_1 S_1 - \eta_1 C_1 \\ \xi_1 C_1 - \eta_1 S_1 \end{Bmatrix} \tag{26}$$

$$\begin{Bmatrix} u_2 \\ v_2 \end{Bmatrix} = \begin{Bmatrix} -a_1 S_1 \\ a_1 C_1 \end{Bmatrix} + \begin{Bmatrix} -\xi_2 S_2 - \eta_2 C_2 \\ \xi_2 C_2 - \eta_2 S_2 \end{Bmatrix} \omega_2 \tag{27}$$

$$\begin{Bmatrix} u_3 \\ v_3 \end{Bmatrix} = \begin{Bmatrix} -\xi_3 S_3 - \eta_3 C_3 + a_3 S_3 \\ \xi_3 C_3 - \eta_3 S_3 - a_3 C_3 \end{Bmatrix} \omega_3 \tag{28}$$

where $C_i \equiv \cos \theta_i$, $S_i \equiv \sin \theta_i$.

Acceleration Coefficients

We have defined the velocity coefficients k_i and ω_i by the expressions

$$\dot{\phi}_i = k_i \dot{q}, \qquad \dot{\theta}_i = \omega_i \dot{q} \tag{29}$$

which may be differentiated to give

$$\ddot{\phi}_i = k_i \ddot{q} + \frac{dk_i}{dq} \frac{dq}{dt} \dot{q} = k_i \ddot{q} + k_i' \dot{q}^2 \tag{30}$$

$$\ddot{\theta}_i = \omega_i \ddot{q} + \frac{d\omega_i}{dq} \frac{dq}{dt} \dot{q} = \omega_i \ddot{q} + \omega_i' \dot{q}^2 \tag{31}$$

This shows that any accelerations may be expressed as a *linear combination* of $\ddot{q}$ and $\dot{q}^2$. The coefficient (k_i' or ω_i') of $\dot{q}^2$ in such an expression is called a **centripetal coefficient**. For systems with at most two or three secondary variables we can find expressions for the centripetal coefficients by direct differentiation. For example, in the case of the four-bar linkage, with link k driving, Eq. (5a) gives

$$\omega_k' = 0 \tag{32}$$

and Eq. (5b) can be expressed in the form

$$\ln \omega_i = \ln \left(\frac{a_k}{a_i} \right) + \ln \sin (\theta_j - \theta_k) - \ln \sin (\theta_i - \theta_j)$$

Differentiating this expression gives

$$\frac{\omega_i'}{\omega_i} = \frac{(\theta_j' - \theta_k') \cos (\theta_j - \theta_k)}{\sin (\theta_j - \theta_k)} - \frac{(\theta_i' - \theta_j') \cos (\theta_i - \theta_j)}{\sin (\theta_i - \theta_j)}$$

We now use the definition

$$\theta'_i = \frac{d\theta_i}{dq} = \frac{d\theta_i/dt}{dq/dt} \equiv \omega_i$$

to find the desired result:

$$\omega'_i = \omega_i[(\omega_j - 1)\cot(\theta_j - \theta_k) - (\omega_i - \omega_j)\cot(\theta_i - \theta_j)] \qquad (i, j \neq k) \qquad (33)$$

From Eq. (4b) we also see that

$$k'_1 = \omega'_i, \qquad k'_2 = \omega'_j \qquad (34)$$

When gear pairs are present, as in Fig. **7.30**-2, they give rise to velocity relations such as Eq. (**7.30**-13), which may be expressed in the form

$$k_4 \equiv \frac{\dot\theta_4}{\dot\theta_3} = \left(1 + \frac{R_2}{R_4}\right) - \frac{R_2}{R_4}\frac{\dot\theta_2}{\dot\theta_3} = \left(1 + \frac{R_2}{R_4}\right) - \frac{R_2}{R_4}k_2$$

where $\dot\theta_3$ is considered the primary velocity $\dot q$. In this example

$$k'_4 = -\frac{R_2}{R_4}k'_2$$

where k'_2 is given by the second of Eqs. (34).

When there are more than two or three secondary velocities to be found, the linear equations (3) are best solved for numerical values of k_i, and algebraic expressions for k_i will not be available. Therefore we shall not have algebraic expressions for k'_i. To find numerical values of the k'_i, we start by differentiating Eqs. (3) with respect to q, obtaining

$$\sum_{j=1}^{N} (A_{ij}k'_j + A'_{ij}k_j) = b'_i \qquad (35)$$

Thus, numerical values of k'_j may be found by solving the system of linear equations

$$\mathbf{A}k' = \mathbf{\beta}' = \{\beta'_i\} \qquad (36)$$

where

$$\beta'_i \equiv b'_i - \sum_{j=1}^{N} A'_{ij}k_j \qquad (37)$$

The terms b'_i and A'_{ij} may be evaluated, via Eqs. (8), from the explicit expressions for A_{ij} and b_i that are always available.

For example, in the case of the four-bar linkage with

$$\{\dot\phi_1, \dot\phi_2, \dot q\} = \{\dot\theta_i, \dot\theta_j, \dot\theta_k\} \qquad (38a)$$

or

$$\{k_1, k_2, 1\} = \{\omega_i, \omega_j, \omega_k\} \qquad (38b)$$

the velocity loop equations (7.10−3) take the form

$$\mathbf{Ak} = \begin{bmatrix} a_i s_i & a_j s_j \\ a_i c_i & a_j c_j \end{bmatrix} \begin{Bmatrix} k_1 \\ k_2 \end{Bmatrix} = \begin{bmatrix} -a_k s_k \\ -a_k c_k \end{bmatrix} \tag{39}$$

where $s_i = \sin \theta_i$, $c_i = \cos \theta_i$. Differentiating across Eq. (39) with respect to $q \equiv \theta_k$, we find

$$\begin{bmatrix} a_i s_i & a_j s_j \\ a_i c_i & a_j c_j \end{bmatrix} \begin{Bmatrix} k_1' \\ k_2' \end{Bmatrix} = \begin{bmatrix} -a_k c_k \\ a_k s_k \end{bmatrix} - \begin{bmatrix} a_i c_i k_1 & a_j c_j k_2 \\ -a_i s_i k_1 & -a_j s_j k_2 \end{bmatrix} \begin{bmatrix} k_1 \\ k_2 \end{bmatrix} = \begin{bmatrix} \beta_1 \\ \beta_2 \end{bmatrix} \tag{40}$$

Solving by Cramer's rule, we find

$$k_1' = \frac{\beta_1 c_j - \beta_2 s_j}{a_i \sin (\theta_i - \theta_j)}$$

$$k_2' = \frac{\beta_1 c_i - \beta_2 s_i}{a_j \sin (\theta_j - \theta_i)} \tag{41}$$

To find the accelerations of points of interest, we differentiate Eqs. (10) to get

$$\ddot{x}_i = u_i \ddot{q} + \dot{q} \frac{du_i}{dq} \dot{q} = u_i \ddot{q} + u_i' \dot{q}^2$$

$$\ddot{y}_i = v_i \ddot{q} + \dot{q} \frac{dv_i}{dq} \dot{q} = v_i \ddot{q} + v_i' \dot{q}^2 \tag{42}$$

The expressions for u_i' and v_i' may be found by differentiating expressions for u_i and v_i. For example, the required functions u_1', v_1' for a four-bar mechanism with $\theta_1 \equiv q$ are given by differentiating across Eqs. (26)–(28). It may be verified (problem 10.60-20) that the resulting centripetal coefficients are

$$\begin{Bmatrix} u_1' \\ v_1' \end{Bmatrix} = \begin{Bmatrix} -x_{iA} \\ -y_{iA} \end{Bmatrix} \tag{43}$$

$$\begin{Bmatrix} u_2' \\ v_2' \end{Bmatrix} = \begin{Bmatrix} -a_1 c_1 \\ -a_1 s_1 \end{Bmatrix} + \begin{Bmatrix} -x_{2B} \\ -y_{2B} \end{Bmatrix} \omega_2^2 + \begin{Bmatrix} -y_{2B} \\ x_{2B} \end{Bmatrix} \omega_2' \tag{44}$$

$$\begin{Bmatrix} u_3' \\ v_3' \end{Bmatrix} = \begin{Bmatrix} a_3 c_3 - x_{3c} \\ a_3 s_3 - y_{3c} \end{Bmatrix} \omega_3^2 + \begin{Bmatrix} -y_{3c} + a_3 s_3 \\ x_{3c} - a_3 c_3 \end{Bmatrix} \omega_3' \tag{45}$$

where use has been made of the compact notation of Eq. (12), and ω_i and ω_i' are given by Eqs. (24), (25), and (33).

An alternative way to find coefficients u_i', v_i' follows from the expressions (5.32-15) for acceleration of a generic point on the moving link i of Fig. 1; i.e.,

$$\ddot{x}_i = \ddot{x}_R - y_{iR} \ddot{\theta}_i - x_{iR} \dot{\theta}_i^2$$

$$\ddot{y}_i = \ddot{y}_R + x_{iR} \ddot{\theta}_i - y_{iR} \dot{\theta}_i^2 \tag{46}$$

Letting $i = R$ in Eq. (42) and using Eq. (10.50-2), we may express Eq. (46) in the form

$$\left\{ \begin{matrix} \ddot{x}_i \\ \ddot{y}_i \end{matrix} \right\} = \left\{ \begin{matrix} u_R \\ v_R \end{matrix} \right\} \ddot{q} + \left\{ \begin{matrix} u'_R \\ v'_R \end{matrix} \right\} \dot{q}^2 + \left\{ \begin{matrix} -y_{iR} \\ x_{iR} \end{matrix} \right\} (\omega_i \ddot{q} + \omega'_i \dot{q}^2) - \left\{ \begin{matrix} x_{iR} \\ y_{iR} \end{matrix} \right\} (\omega_i \dot{q})^2$$

Upon comparing this expression with the desired form

$$\left\{ \begin{matrix} \ddot{x}_i \\ \ddot{y}_i \end{matrix} \right\} = \left\{ \begin{matrix} u_i \\ v_i \end{matrix} \right\} \ddot{q} + \left\{ \begin{matrix} u'_i \\ v_i \end{matrix} \right\} \dot{q}^2$$

we see that

$$\left\{ \begin{matrix} u_i \\ v_i \end{matrix} \right\} = \left\{ \begin{matrix} u_R \\ v_R \end{matrix} \right\} + \left\{ \begin{matrix} -y_{iR} \\ x_{iR} \end{matrix} \right\} \omega_i \tag{47}$$

$$\left\{ \begin{matrix} u'_i \\ v'_i \end{matrix} \right\} = \left\{ \begin{matrix} u'_R \\ v'_R \end{matrix} \right\} + \left\{ \begin{matrix} -y_{iR} \\ x_{iR} \end{matrix} \right\} \omega'_i - \left\{ \begin{matrix} x_{iR} \\ y_{iR} \end{matrix} \right\} \omega_i^2 \tag{48}$$

The first result confirms Eq. (16), but the second gives us a new tool which is useful for chains of links.

When applied to the links of the four-bar mechanism of Fig. 2, Eq. (48) gives the following results:

Link 1 $(i = 1, R = A, u'_A = v'_A = 0)$:

$$\left\{ \begin{matrix} u'_1 \\ v'_1 \end{matrix} \right\} = \left\{ \begin{matrix} -y_{1A} \\ x_{1A} \end{matrix} \right\} \omega'_1 - \left\{ \begin{matrix} x_{1A} \\ y_{1A} \end{matrix} \right\} \omega_1^2 \tag{49}$$

Link 2 $(i = 2, R = B)$: From Eq. (49) we first find $\{u'_1, v'_1\}$ for point B; i.e.:

$$\left\{ \begin{matrix} u'_B \\ v'_B \end{matrix} \right\} = \left\{ \begin{matrix} -y_{BA} \\ x_{BA} \end{matrix} \right\} \omega'_1 - \left\{ \begin{matrix} x_{BA} \\ y_{BA} \end{matrix} \right\} \omega_1^2$$

Then Eq. (48) gives

$$\left\{ \begin{matrix} u'_2 \\ v'_2 \end{matrix} \right\} = \left\{ \begin{matrix} -y_{BA} \\ x_{BA} \end{matrix} \right\} \omega'_1 - \left\{ \begin{matrix} x_{BA} \\ y_{BA} \end{matrix} \right\} \omega_1^2$$
$$+ \left\{ \begin{matrix} -y_{2B} \\ x_{2B} \end{matrix} \right\} \omega'_2 - \left\{ \begin{matrix} x_{2B} \\ y_{2B} \end{matrix} \right\} \omega_2^2 \tag{50}$$

Link 3 $(i = 3, R = D, u'_D = v'_D = 0)$:

$$\left\{ \begin{matrix} u'_3 \\ v'_3 \end{matrix} \right\} = \left\{ \begin{matrix} -y_{3D} \\ x_{3D} \end{matrix} \right\} \omega'_3 - \left\{ \begin{matrix} x_{3D} \\ y_{3D} \end{matrix} \right\} \omega_3^2 \tag{51}$$

Recall that Eqs. (5), (32), and (33) give the required coefficients ω_i and ω'_i.

For the special case where $\theta_1 \equiv q$, it follows that $\omega_1 = 1$, $\omega'_1 = 0$, and Eqs. (49)–(51) may readily be seen to coincide with Eqs. (43)–(45).

10.52 Systems with Several Degrees of Freedom

We now wish to extend the results of the previous section to the case of several generalized coordinates $q_1, q_2, \ldots, q_F$. In particular, we shall show how to develop sets of influence coefficients which relate the secondary quantities $\dot{\phi}_i$, $\ddot{\phi}_i$ to the primary quantities $\dot{q}_i$ and $\ddot{q}_i$. The relationships we seek have the following form:

$$\dot{\phi}_i = \sum_{j=1}^{F} K_{ij} \dot{q}_j \tag{1}$$

$$\ddot{\phi}_i = \sum_{j=1}^{F} K_{ij} \ddot{q}_j + K_i \tag{2}$$

where the *velocity coefficients* K_{ij} are functions of *position* only and the term K_i depends on positions and *velocities*. Note that we are using capital letters for the doubly subscripted velocity coefficients to distinguish them from the singly subscripted coefficients k_i used for single-freedom systems in the previous section. The velocity-dependent term K_i is the generalization of the centripetal acceleration term $k_i' q^2$ used for single-freedom systems.

Furthermore, the velocities and accelerations of points of interest will be shown to be of the form

$$\begin{Bmatrix} \dot{x}_i \\ \dot{y}_i \end{Bmatrix} = \sum_{j=1}^{F} \begin{Bmatrix} U_{ij} \\ V_{ij} \end{Bmatrix} \dot{q}_j \tag{3}$$

$$\begin{Bmatrix} \ddot{x}_i \\ \ddot{y}_i \end{Bmatrix} = \sum_{j=1}^{F} \begin{Bmatrix} U_{ij} \\ V_{ij} \end{Bmatrix} \ddot{q}_j + \begin{Bmatrix} U_i \\ V_i \end{Bmatrix} \tag{4}$$

where i is an identification number for the point of interest, the velocity coefficients U_{ij}, V_{ij} depend on position only, and the centripetal terms U_i, V_i depend on both position and velocity.

In addition, we shall show that the angular velocity $\dot{\theta}_i$ of link i may be expressed in the form

$$\dot{\theta}_i = \sum_{j=1}^{F} \Omega_{ij} \dot{q}_j \tag{5}$$

and the angular acceleration $\ddot{\theta}_i$ in the form

$$\ddot{\theta}_i = \sum_{j=1}^{F} \Omega_{ij} \ddot{q}_j + \Omega_i \tag{6}$$

where the double-subscript coefficients Ω_{ij} depend only on position and the single-subscript coefficients Ω_i depend on both position and velocity. These are the principal kinematic results that will be needed to perform a complete dynamic analysis for multifreedom systems.

Velocity Coefficients

For a system with F degrees of freedom and N secondary variables, the velocity equations (**10.10**-10) are of the form

$$\mathbf{A}\dot{\boldsymbol{\phi}} = \mathbf{B}\dot{\mathbf{q}} \tag{7}$$

or

$$\sum_{j=1}^{N} A_{ij}\dot{\phi}_j = \sum_{j=1}^{F} B_{ij}\dot{q}_j, \qquad (i = 1, 2, \dots, N) \tag{8}$$

Hence

$$\dot{\phi} = \mathbf{A}^{-1}\mathbf{B}\dot{\mathbf{q}} = \mathbf{K}\dot{\mathbf{q}} \tag{9}$$

or

$$\dot{\phi}_i = \sum_{j=1}^{F} K_{ij}\dot{q}_j \qquad (i = 1, 2, \dots, N) \tag{10}$$

where $\mathbf{K}$ is a matrix of order $N \times F$ defined by

$$\mathbf{K} = \mathbf{A}^{-1}\mathbf{B} \equiv [K_{ij}] \tag{11}$$

The terms K_{ij} are the **velocity coefficients** or **influence coefficients.** If we only wish to find numerical values of the velocities $\dot{\phi}_j$, it is computationally efficient to solve for $\dot{\phi}_j$ by using Gaussian elimination on Eqs. (8). On the other hand, it pays to find the inverse matrix $\mathbf{A}^{-1}$ and use Eq. (11) for $\mathbf{K}$ if accelerations are also needed.[7]

The angular velocity $\dot{\theta}_i$ of link i will be a member of the set $\dot{q}_i$ or of the set $\dot{\phi}_i$. Since the $\dot{\phi}_i$ are linear combinations of the $\dot{q}_i$, we can express all link angular velocities in the form

$$\dot{\theta}_i = \sum_{j=1}^{F} \Omega_{ij}\dot{q}_j \tag{12}$$

where the **angular velocity coefficients** Ω_{ij} are either 0, ± 1, or a linear combination of the coefficients K_{ij}.

To find the velocity components $(\dot{x}, \dot{y})$ of point of interest i, we must differentiate the transformation equations

$$\left. \begin{array}{l} x_i = x_i(\phi_s, q_j) \\ y_i = y_i(\phi_s, q_j) \end{array} \right\} \qquad (s = 1, \dots, N; j = 1, \dots, F) \tag{13}$$

First, however, we note that a function F of the variables ϕ_j and q_r expressed in the explicit form

$$F(\phi, q) \tag{14}$$

has a total derivative with respect to time given by

$$\dot{F} = \sum_{j=1}^{F} \frac{\partial F}{\partial q_j}\dot{q}_j + \sum_{s=1}^{N} \frac{\partial F}{\partial \phi_s}\dot{\phi}_s \tag{15}$$

But since $\dot{\phi}_s = \sum_{j=1}^{F} K_{sj}\dot{q}_j$ by Eq. (10),

$$\dot{F} = \sum_{j=1}^{F} \left(\frac{\partial F}{\partial q_j} + \sum_{s=1}^{N} K_{sj}\frac{\partial F}{\partial \phi_s} \right)\dot{q}_j \equiv \sum_{j=1}^{F} \frac{DF}{Dq_j}\dot{q}_j \tag{16}$$

---

[7]See Note on "efficiency of inversion" in Appendix E.5.

where

$$\frac{D}{Dq_j} = \frac{\partial}{\partial q_j} + \sum_{s=1}^{N} K_{sj} \frac{\partial}{\partial \phi_s} \tag{17}$$

is a generalization of the operator denoted d/dq in Eq. (10.51-8). Applying the operation of Eq. (16) to Eqs. (13), we find

$$\begin{Bmatrix} \dot{x}_i \\ \dot{y}_i \end{Bmatrix} = \sum_{j=1}^{F} \begin{Bmatrix} U_{ij} \\ V_{ij} \end{Bmatrix} \dot{q}_j \tag{18}$$

where

$$\begin{Bmatrix} U_{ij} \\ V_{ij} \end{Bmatrix} \equiv \frac{D}{Dq_j} \begin{Bmatrix} x_i(\phi, q) \\ y_i(\phi, q) \end{Bmatrix} = \frac{\partial}{\partial q_j} \begin{Bmatrix} x_i \\ y_i \end{Bmatrix} + \sum_{j=1}^{N} K_{sj} \frac{\partial}{\partial \phi_s} \begin{Bmatrix} x_i \\ y_i \end{Bmatrix} \tag{19}$$

All derivatives needed in Eq. (19) may be evaluated from the explicit transformation equations (13).

Acceleration Coefficients

Using the notation already defined, we may differentiate Eq. (10) to find

$$\ddot{\phi}_i = \sum_{j=1}^{F} K_{ij} \ddot{q}_j + K_i \tag{20}$$

where

$$K_i \equiv \sum_{j=1}^{F} \dot{K}_{ij} \dot{q}_j = \sum_{j=1}^{F} \dot{q}_j \sum_{r=1}^{F} K_r^{ij} \dot{q}_r \tag{21}$$

$$K_r^{ij} \equiv \frac{DK_{ij}}{Dq_r} = \frac{\partial K_{ij}}{\partial q_r} + \sum_{s=1}^{N} K_{sr} \frac{\partial K_{ij}}{\partial \phi_s} \tag{22}$$

Equation (21) shows that K_i is a quadratic form, in the generalized velocities, which represents a generalized centripetal acceleration. In practice, we shall not usually be able to calculate the **centripetal coefficients**[8] K_r^{ij} from Eq. (22) because we shall not have explicit algebraic expressions $K_{ij}(\phi, q)$ available to differentiate. However, numerical values of the triple-index terms K_s^{ij} may be found as follows. Equation (11) is written in the form $\mathbf{AK} = \mathbf{B}$, or

$$\sum_{p=1}^{N} A_{ip} K_{pj} = B_{ij} \tag{23}$$

Applying the operator D/Dq_r to each term in Eq. (23) results in

$$\sum_{p=1}^{N} \left(A_{ip} \frac{DK_{pj}}{Dq_r} + \frac{DA_{ip}}{Dq_r} K_{pj} \right) = \frac{DB_{ij}}{Dq_r} \tag{24}$$

Hence

$$\sum_{p=1}^{N} A_{ip} K_r^{pj} = B_r^{ij} \tag{25}$$

---

[8]Note the triple indices which distinguish the centripetal coefficients K_r^{ij} from the doubly indexed velocity coefficients K_{ij}.

where

$$B_r^{ij} = \frac{DB_{ij}}{Dq_r} - \sum_{p=1}^{N} K_{pj} \frac{DA_{ip}}{Dq_r} \tag{26}$$

Since explicit expressions will always be available for the coefficients $B_{ij}(\phi, q)$ and $A_{ip}(\phi, q)$, we can evaluate the terms B_r^{ij} from Eq. (26) and then solve the linear equations (25) for the desired unknowns K_r^{pj}.

For a fixed value of r we may express B_r^{ij} and K_r^{ij} as matrices

$$\mathbf{B}_r \equiv [B_r^{ij}], \qquad \mathbf{K}_r \equiv [K_r^{ij}] \tag{27}$$

Then Eq. (25) may be expressed in the form

$$\mathbf{A}\mathbf{K}_r = \mathbf{B}_r \tag{28}$$

Hence

$$\mathbf{K}_r = \mathbf{A}^{-1}\mathbf{B}_r \tag{29}$$

$$K_r^{ij} = \sum_{s=1}^{N} \alpha_{is} B_r^{sj} \tag{30}$$

where α_{is} represents a typical element of $\mathbf{A}^{-1}$.

Having found numerical values for the coefficients K_{ij} and K_r^{ij}, it is easy to find expressions for the acceleration components (x_i, y_i) of points of interest identified by index i. By differentiating Eqs. (18) we find

$$\begin{Bmatrix} \ddot{x}_i \\ \ddot{y}_i \end{Bmatrix} = \sum_{j=1}^{F} \begin{Bmatrix} U_{ij} \\ V_{ij} \end{Bmatrix} \ddot{q}_j + \sum_{j=1}^{F} \dot{q}_j \sum_{r=1}^{F} \dot{q}_r \frac{D}{Dq_r} \begin{Bmatrix} U_{ij} \\ V_{ij} \end{Bmatrix} \tag{31}$$

These equations are of the desired form (4) where the quadratic forms in $\dot{q}_j \dot{q}_r$ are given by

$$\begin{Bmatrix} U_i \\ V_i \end{Bmatrix} = \sum_{j=1}^{F} \sum_{r=1}^{F} \dot{q}_j \dot{q}_r \begin{Bmatrix} U_r^{ij} \\ V_r^{ij} \end{Bmatrix} \tag{32}$$

and the triple-index symbols are defined, via Eq. (19), by

$$\begin{Bmatrix} U_r^{ij} \\ V_r^{ij} \end{Bmatrix} = \frac{D}{Dq_r} \frac{D}{Dq_j} \begin{Bmatrix} x_i(\phi, q) \\ y_i(\phi, q) \end{Bmatrix} \tag{33}$$

Fortunately, the tedium of evaluating the necessary derivatives in Eq. (33) is relieved because many of the matrices involved are **sparse**; i.e, many of the matrix elements are zero.

The angular acceleration of link i is found from Eq. (12) to be

$$\ddot{\theta}_i = \sum_{j=1}^{F} \Omega_{ij} \ddot{q}_j + \Omega_i \tag{34}$$

where

$$\Omega_i \equiv \sum_{s=1}^{F} \sum_{j=1}^{F} \Omega_s^{ij} \dot{q}_s \dot{q}_j \tag{35}$$

and

$$\Omega_s^{ij} \equiv \frac{D\Omega_{ij}}{Dq_s} \qquad (36)$$

are formed from the appropriate linear combination of K_s^{ij}.

Given the influence coefficients for a point R fixed on link i, we may find those for any point P_i on the same link by starting with the general acceleration equations (10.51-46),

$$\begin{Bmatrix} \ddot{x}_i \\ \ddot{y}_i \end{Bmatrix} = \begin{Bmatrix} \ddot{x}_R \\ \ddot{y}_R \end{Bmatrix} + \begin{Bmatrix} -y_{iR} \\ x_{iR} \end{Bmatrix} \ddot{\theta}_i - \begin{Bmatrix} x_{iR} \\ y_{iR} \end{Bmatrix} \dot{\theta}_i^2$$

and using Eqs. (5), (6), and (4) to eliminate $\dot{\theta}_i$, $\ddot{\theta}_i$, $\ddot{x}_R$, and $\ddot{y}_R$. The final results are

$$\begin{Bmatrix} U_{ij} \\ V_{ij} \end{Bmatrix} = \begin{Bmatrix} U_{Rj} \\ V_{Rj} \end{Bmatrix} + \begin{Bmatrix} -y_{iR} \\ x_{iR} \end{Bmatrix} \Omega_{ij} \qquad (37)$$

$$\begin{Bmatrix} U_i \\ V_i \end{Bmatrix} = \begin{Bmatrix} U_R \\ V_R \end{Bmatrix} + \begin{Bmatrix} -y_{iR} \\ x_{iR} \end{Bmatrix} \Omega_i - \begin{Bmatrix} x_{iR} \\ y_{iR} \end{Bmatrix} \dot{\theta}_i^2 \qquad (38)$$

10.60 PROBLEMS[9]

1. Define and sketch a suitable set of configuration variables θ_i and r_i for the Stirling cycle engine of Fig. **10.20**-2.
   a. Write out the displacement loop equations utilizing the abbreviations $c_i = \cos\theta_i$, $s_i = \sin\theta_i$.
   b. Write out the velocity loop equations.
   c. Solve for all secondary velocities, assuming that $\dot{\theta}_1 = \dot{q}$ is known.
   d. Write out the acceleration loop equations, assuming that $\ddot{\theta}_1 = \ddot{q}$ is known.

2. Find expressions for the unknown accelerations in the Whitworth mechanism of Fig. **10.20**-1, assuming that the velocity $\dot{\theta}_1$ and the acceleration $\ddot{\theta}_1$ are given.

3. Determine whether the system of Fig. 1 is *weakly* or *strongly coupled*, when
   a. The slider is the driving link.
   b. *AB* is the driving link.
   Explain your reasoning.

4. In Fig. 1, point D moves with velocity $\dot{y}_D = 2$ ft/sec. Find
   a. The angular velocity of link *CF*.
   b. The velocity components $(\dot{x}_F, \dot{y}_F)$ of point F.

5. With link *AB* of Fig. 1 revolving counterclockwise at a constant speed of 0.8404 rad/sec, find
   a. The acceleration components $(\ddot{x}_C, \ddot{y}_C)$ of point C.
   b. The angular acceleration of link *CB*.

---

[9]Many of the problems in this section have multiple parts. It is possible (and desirable) to assign only part of a problem, but bear in mind that subsequent parts will often depend on the results of previous parts.

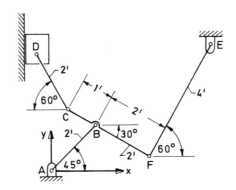

Figure 10.60-1. (Probs. 3, 4, 5, 6).

6. If link AB in Fig. 1 is rotating at an unknown constant speed and the velocity of the slider is measured at $\dot{y}_D = 2$ ft/sec, find the instantaneous values of
   a. The acceleration component $(\ddot{y}_D)$ of point D.
   b. The acceleration components $(\ddot{x}_F, \ddot{y}_F)$ of point F.

7. In Fig. **9.40**-5, lengths are given in inches, and $\theta = 30°$, $\phi = 9.59°$, $\beta = 32.09°$, and $\gamma = 44.26°$, and gear 2 meshes with gear 1, which rotates counterclockwise at the uniform speed $\omega = 10$ rad/sec. Find
   a. The velocity components $(\dot{x}_E, \dot{y}_E)$ of point E.
   b. The acceleration components $(\ddot{x}_E, \ddot{y}_E)$ of point E.

8. In Fig. 2, $a = 1$ in., $b = 0.7$ in., and point P has velocity $\dot{y}_P = 10$ in./sec and acceleration $\ddot{y}_P = 0$. Find
   a. Velocity components $(\dot{x}_D, \dot{y}_D)$ of point D.
   b. Acceleration components $(\ddot{x}_D, \ddot{y}_D)$ of point D.

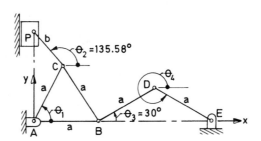

Figure 10.60-2. (Prob. 8).

9. In the quick-return mechanism of Fig. 3, link 1 rotates at constant velocity $\dot{\theta} = \omega$, and the yoke on member 2 drives the pin P, which is fixed to the reciprocating ram.
   a. Find expressions for the velocity $(\dot{x}_P)$ of the ram and the velocity $(\dot{r})$ of point A relative to link 2.
   b. Find expressions for the accelerations $\ddot{x}_P$ and $\ddot{r}$.
   c. If $AB = 2.5$ in., $b = 5.0$ in., $c = 4.5$ in., and $\theta = 60°$, verify that $r = 6.614$ in., $\phi = 19.11°$, and $x_P = -3.291$ in.
   d. Using parts a, b, and c, find $\dot{x}_P$ and $\dot{r}$ if $\omega = 30$ rpm.
   e. Using parts a–d, find $\ddot{x}_P$ and $\ddot{r}$.

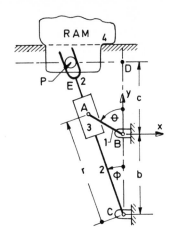

Figure 10.60-3. (Prob. 9).

10. In the shaper mechanism of Fig. 4, link CB rotates with constant velocity $\dot{q}$. Use the notation indicated in the figure.
   a. Write a complete set of displacement loop equations.
   b. Formulate the velocity loop equations and find explicit expressions for all $\dot{\phi}_i$.
   c. Formulate the equations needed to find all $\ddot{\phi}_i$.
   d. If $a_1 = 2.5$ in., $a_2 = 9$ in., $a_3 = 6$ in., $b = 5$ in., $c = 10$ in., and $q = 120°$, verify that $\phi_1 = 30°$, $\phi_2 = 4.330$ in., $\phi_3 = 68.430°$, and $\phi_4 = 10.080$ in.
   e. Using parts b and d of this problem, find $\dot{\phi}_1, \dot{\phi}_2, \dot{\phi}_3, \dot{\phi}_4$ if $\dot{q} = 5$ rad/sec.
   f. Using parts c–e of this problem, find $\ddot{\phi}_1, \ddot{\phi}_2, \ddot{\phi}_3, \ddot{\phi}_4$.

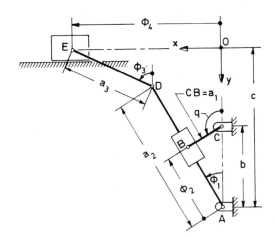

Figure 10.60-4. (Prob. 10).

11. For the machine tool drive shown in Fig. 5, link 2 rotates with constant angular velocity $\dot{\theta}_2 = \dot{q} = 10$ rad/sec.
   a. Formulate a complete set of velocity loop equations.
   b. Find *explicit* expressions for $\dot{r}_1$, $\dot{r}_3$, and $\dot{\theta}_5$ in terms of $\dot{q}$. (Hint: Try to partially decouple the unknown velocities.)

c. For the dimensions shown, verify that when $\theta_2 = 60°$, $\theta_4 = 20.84°$, $\theta_5 = 113.92°$, $r_1 = 7.13$ in., and $r_3 = 6.59$ in.

d. Evaluate $\dot{r}_1$ when $\theta_2 = 60°$.

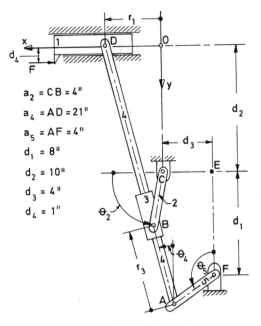

$a_2 = CB = 4''$

$a_4 = AD = 21''$

$a_5 = AF = 4''$

$d_1 = 8''$

$d_2 = 10''$

$d_3 = 4''$

$d_4 = 1''$

Figure 10.60-5. (Prob. 11).

12. In Fig. **10.30-1**, $a_2 = 0.5$ in., $a_4 = 3$ in., and $L = 1.2$ in. If $q = \theta_2 = 20°$ and $\dot{q} = \dot{\theta}_2 = 30$ rpm,

a. Verify that $\theta_4 = 75.98°$, $r_1 = 1.711$ in., $r_3 = 1.939$ in., and $r_5 = 0.7268$ in.

b. Find the velocities $\dot{\theta}_4$, $\dot{r}_1$, $\dot{r}_3$, $\dot{r}_5$.

13. Compressed air drives the piston (on link 1) within the cylinder 5 of the pneumatic riveter (or punch press) shown in Fig. **9.40-6**. With the actuator position given as $r_1 = 6.142$ in.,

a. Verify that $r_4 = 3.857$ in., $\theta_1 = 85.17°$, $\theta_2 = 20.00°$, and $\theta_3 = 100°$.

b. Find the ratio $-(\dot{r}_1/\dot{r}_4)$ for the given position (i.e., the *mechanical advantage* of the linkage).

14. The *floating link* or *single flyer ABC* of Fig. 6 has three links pinned to it whose end points D, E, F have known velocities $V_D = (\dot{x}_D, \dot{y}_D)$, $V_E = (\dot{x}_E, \dot{y}_E)$, $V_F = (\dot{x}_F, \dot{y}_F)$. Show that the velocities $\dot{\theta}_1$, $\dot{\theta}_2$, $\dot{\theta}_3$, $\dot{\theta}_4$ are given by the solution of the equations

$$\begin{bmatrix} a_1 C_1 & a_2 C_2 & -a_3 C_3 & 0 \\ a_1 S_1 & a_2 S_2 & -a_3 S_3 & 0 \\ a_1' \cos(\theta_1 + \alpha) & a_2 C_2 & 0 & a_4 C_4 \\ a_1' \sin(\theta_1 + \alpha) & a_2 S_2 & 0 & a_4 S_4 \end{bmatrix} \begin{bmatrix} \dot{\theta}_1 \\ \dot{\theta}_2 \\ \dot{\theta}_3 \\ \dot{\theta}_4 \end{bmatrix} = \begin{bmatrix} \dot{y}_F - \dot{y}_D \\ \dot{x}_D - \dot{x}_F \\ \dot{y}_E - \dot{y}_D \\ \dot{x}_D - \dot{x}_E \end{bmatrix}$$

where $C_k \equiv \cos\theta_k$, $S_k \equiv \sin\theta_k$.

(Hint: Write the loop equation for loops $DABFD$ and $DACED$.)

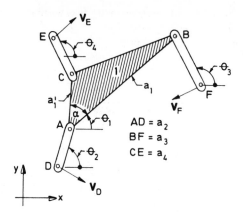

Figure 10.60-6 (Prob. 14).

15. Find the influence coefficients k_i and k'_i for the mechanism of Fig. **7.41-1**.

16. Find the influence coefficients k_i and k'_i for the mechanism of Fig. **7.30-1**.

17. Find the influence coefficients k_i and k'_i for the mechanism of Fig. **7.30-2**.

18. Find the centripetal coefficients for the slider-crank mechanism shown in Fig. 7 with
    a. The crank, 1, driving.
    b. The piston, 3, driving.

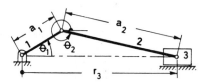

Figure 10.60-7 (Prob. 18).

19. Starting from Eq. (**10.10**-15), develop Eqs. (**10.41**-6).

20. Verify Eqs. (**10.51**-43)–(**10.51**-45).

21. For the five-bar mechanism of Fig. **7.10**-11(c), let $\{q_1, q_2, \phi_1, \phi_2\} = \{\theta_1, \theta_4, \theta_2, \theta_3\}$, and find the matrices **A** and **B** explicitly in terms of θ_i (where $\mathbf{A}\dot{\phi} = \mathbf{B}\dot{q}$).

22. For the five-bar mechanism of Fig. **7.10**-1(c), let $\{q_1, q_2, \phi_1, \phi_2\} = \{\theta_1, \theta_4, \theta_2, \theta_3\}$, and find the matrix **K** of velocity coefficients, explicitly in terms of the angles θ_i, by
    a. Solving Eq. (**10.52**-7) by Cramer's rule and comparing your solution with Eq. (**10.52**-10), or
    b. Inverting matrix **A**, and using Eq. (**10.52**-11).

23. Show that the operator defined in Eq. (**10.52**-33) can be expanded into the form

$$\frac{D}{Dq_s}\frac{D}{Dq_j} \equiv \frac{\partial^2}{\partial q_s \, \partial q_j} + \sum_{r=1}^{N} \left(k_{rs}\frac{\partial^2}{\partial \phi_r \, \partial q_j} + k_{rj}\frac{\partial^2}{\partial \phi_r \, \partial q_s} + k_s^{rj}\frac{\partial}{\partial \phi_r} \right.$$
$$\left. + \sum_{p=1}^{N} k_{rj}k_{ps}\frac{\partial^2}{\partial \phi_r \, \partial \phi_p} \right)$$

24–27. Indicate whether the designated mechanisms are weakly or strongly coupled when the driving link is chosen as specified; explain your reasoning.

Problem No.	Fig. No.	Driving Link
24	**9.40**-2	5
	9.40-2	3
25	**9.40**-5	4
	9.40-5	*AE*
26	**9.40**-6	4
	9.40-6	1
27	**10.60**-5	2
	10.60-5	5

28. For the system (and dimensions) given in Problem **9.40**-8, find the maximum acceleration a_D of point D when $\dot{\theta}_1 = 1000$ rpm and $\ddot{\theta}_1 = 0$. [Suggestion: Use a computer to evaluate $a_D = (\ddot{x}_D^2 + \ddot{y}^2)^{1/2}$ for various crank angles, and pick max a_D from a plot of the results.]

29. For the strain measurement system of Fig. **9.40**-2, find *explicit* expressions for the angular velocities of links 4 and 5, for a given crank speed $\dot{\theta}_1 \equiv \dot{q} = $ constant.

30. For the system of Prob. 29, find explicit expressions for the acceleration components $(\ddot{x}_D, \ddot{y}_D)$ of point D.

PART 3

ANALYTICAL DYNAMICS

"The mechanics of Newton are purely *geometrical*. The term 'analytical mechanics,' which is contrasted with the synthetical, or geometrical, mechanics of Newton, is the exact equivalent of the phrase 'algebraical mechanics.' . . . Analytical mechanics, however, was brought to its highest degree of perfection by LAGRANGE. . . . Every case that presents itself can now be dealt with by a very simple, highly symmetrical and perspicacious schema. . . . The mechanics of Lagrange is a stupendous contribution to the economy of thought."

—**Ernst Mach [1960, pp. 560–562]**

Vector dynamics (the approach to dynamics based directly on Newton's laws) turns out to be less attractive in many ways than **analytical dynamics** (an approach based on the principle of virtual work). The latter treatment of the subject first appeared in Lagrange's *Mécanique Analytique* of 1788—an "epoch-making" book, which is ". . . rightly regarded as one of the outstanding intellectual achievements of mankind."* Because of the beauty, depth, and elegance of Lagrange's methods, he has been called the "Shakespeare of mathematics" by Sir William Hamilton, who referred to his work as "a kind of scientific poem."**

Critics of *analytical mechanics* have sometimes charged that its procedures are unduly abstract. Ironically, this reputation is due in large measure to Lagrange's unbending refusal to draw even a single picture in the *Mécanique*. Removal of this self-imposed handicap tends to remove the aura of abstractness and permits one to better appreciate the great generality and the uniformity of approach embodied in analytical mechanics. Indeed, we shall find that this approach is ideally suited to the highly constrained systems typical of machinery and perfectly matched to the digital methods of calculation called for by modern computing equipment.

Following Lagrange, we shall begin with a discussion of *statics* in Chapter 11, where we prove and make extensive use of the principle of virtual work. This principle is the fundamental basis of analytical statics and when extended by the use of d'Alembert's principle, becomes the basis for our further discussions of *kinetics*. However, before undertaking a discussion of the most general problems of dynamics, we treat the special case (an extremely important one for machinery) of systems with a single degree of freedom. In Chapter 12, we show how to find the motion of any given single-freedom system (e.g., a multicylinder reciprocating engine) when the applied forces (e.g., gas pressures) are specified. Then, in Chapter 13, we show how one may design or alter such machines so as to minimize, or eliminate, the *shaking forces* which they transmit to the ground, i.e., how to *balance* machinery.

Finally, in Chapter 14, methods are described for the dynamic analysis of machines with any number of degrees of freedom; the basis is established for the development of powerful general-purpose computer programs for the dynamic analysis of realistic machines, and a brief survey of existing programs is given.

---

*Quoted from Pars [1968, p. 76].
**Lanczos [1966, p. 347]; also see Bell [1965, pp. 153–171], which is a minibiography of Lagrange. For a microbiography, see Timoshenko [1953, pp. 37–40].

11

STATICS

11.10 VECTOR STATICS

If the resultant force vector and resultant moment vector (about any point) applied to a rigid body both vanish, we shall say that the body is in a state of *generalized*[1] *equilibrium*. We know, from elementary mechanics, that the mass center of such a body moves in uniform rectilinear motion (relative to an inertial coordinate frame), and the body undergoes at most a uniform angular velocity. If such a body was initially at rest, it will remain at rest under the action of such a null force system and is said to be in a state of *equilibrium*.

The conditions of vanishing force and moment vectors, together with the principle of action and reaction at points of contact between rigid bodies, are the fundamental principles for **vector statics**, as opposed to **analytical statics**, which is based on the principle of virtual work.

To solve problems using vector statics it is necessary to consider each rigid body of the system separately (i.e., as a *free body*) and to write equations of equilibrium, for each body, which include all of the unknown reactions and contact forces acting on the bodies. At best, this method is cumbersome for systems of several bodies. To see this, consider the following *example*.

A four-bar linkage is shown in Fig. 1(a), where horizontal and vertical forces H and V act, respectively, at the midpoints of links AB and BC. We seek to find the

---

[1]The word *generalized* is explained in depth in Sec. **11.71**. It extends the definition of *equilibrium* to cover rigid bodies which are rotating with constant speed about a fixed axis. For example, if a rigid hockey puck moves along the ice with constant velocity, it will be in generalized equilibrium even if it is spinning about its vertical axis with constant angular speed.

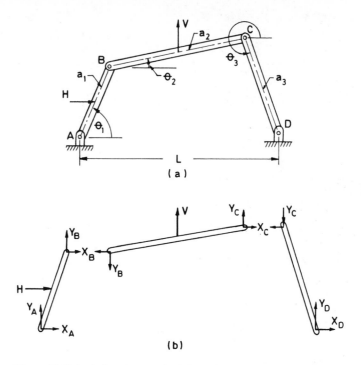

Figure 11.10-1. A four-bar mechanism in equilibrium under external forces: (a) Linkage diagram; (b) Free-body diagrams.

configuration assumed by the linkage under the influence of the given forces. In other words, we seek to find an algebraic relationship of the type

$$F(H, V, \theta_1, \theta_2, \theta_3) = 0 \tag{1}$$

which, together with the two kinematic loop equations

$$a_1 \cos \theta_1 + a_2 \cos \theta_2 + a_3 \cos \theta_3 = L \tag{2}$$
$$a_1 \sin \theta_1 + a_2 \sin \theta_2 + a_3 \sin \theta_3 = 0 \tag{3}$$

will enable us to solve for all three angles corresponding to given values of H and V.

Using vector methods, we must write 3 equations of equilibrium for each of the three free bodies shown in Fig. 1(b). That is, we must write out 9 equations, of which the first 3 are shown explicitly below:

$$\sum X = X_A + H + X_B = 0$$
$$\sum Y = Y_A + Y_B = 0$$
$$\sum M_A = H \frac{a_1}{2} \sin \theta_1 + X_B a_1 \sin \theta_1 - Y_B a_1 \cos \theta_1 = 0$$

. . . , etc.

These 9 equilibrium equations, together with Eqs. (2) and (3), comprise a set of 11 equations for the 11 unknowns $(X_A, Y_A, X_B, Y_B, X_C, Y_C, X_D, Y_D, \theta_1, \theta_2, \theta_3)$, which are thereby determined *for given values of H and V.* If we are not specifically interested in knowing the reactions $(X_A, \ldots, Y_D)$, we may eliminate the 8 unknown reactions from the 9 equilibrium equations, which will, in the process, reduce to the single desired equation (1).

This seems like an unreasonably awkward way to find the single relationship dictated by equilibrium considerations, and a more straightforward and efficient approach would be most welcome. Indeed we shall see in subsequent sections that the methods of analytical statics, based on the principle of virtual work, will enable us to quickly reach the desired goal.

One important result from vector statics that we shall draw upon frequently is that any system of forces acting on a rigid body can be reduced to a resultant force $\mathbf{F}_A$ acting through *any* chosen point A on the body, plus a couple M_A, as shown in Fig. 2(a). To show that point A is arbitrary, consider the two equal and opposite forces

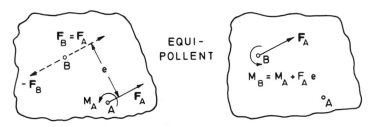

Figure 11.10-2. Statically equipollent force systems.

$\mathbf{F}_B = \mathbf{F}_A$ and $-\mathbf{F}_B$ acting at any point B, as shown by the dashed vectors in Fig. 2(a). Addition of this *null system* $(\mathbf{F}_B + -\mathbf{F}_B = \mathbf{0})$ to the original forces cannot influence the equilibrium of the body. The forces $\mathbf{F}_A$ and $-\mathbf{F}_B$ have zero resultant and may be replaced by a pure couple of moment $|\mathbf{F}_A|e$, where e is the distance between the lines of action of $\mathbf{F}_A$ and $\mathbf{F}_B$. The total system of forces now consists of the force $\mathbf{F}_A$ acting at B plus the resultant couple $M_B \equiv M_A + |\mathbf{F}_A|e$. This system of forces [Fig. 2(b)] is fully equivalent to the original system, *insofar as rigid body statics is concerned,*[2] and we say that the two force systems are **statically equipollent.**

11.20 PRELIMINARY CONCEPTS FOR THE PRINCIPLE OF VIRTUAL WORK

Analytical mechanics rests upon the principle of virtual work. Like most great ideas, this principle did not spring, fully developed, from the mind of a single genius. Some remarks on its evolution will be found in Sec. **11.30.** The principle can be stated in a few words, provided that certain key terms, such as *virtual displacements, generalized forces, virtual work, ideal systems,* etc., are fully understood. It is the purpose of the following subsections to explain these necessary preliminaries.

---

[2]The two systems are not at all equivalent if we wish to study deformations or internal stresses in the body.

11.21   *Virtual Displacements*

If a mechanical system with Lagrangian coordinates ψ_i is subjected to a set of kinematic constraints of the form[3]

$$f_i \ (\psi_1, \psi_2, \ldots, \psi_M) = 0 \qquad (i = 1, 2, \ldots, c) \tag{1}$$

the Lagrangian velocities $\dot{\psi}_i$ must satisfy the relations

$$\dot{f}_i = \sum_{j=1}^{M} \frac{\partial f_i}{\partial \psi_j} \dot{\psi}_j = 0 \tag{2}$$

Any set of $\dot{\psi}_j$ which satisfy Eq. (2) is called a set of **virtual velocities**. The infinitesimal displacements

$$\delta \psi_j = \dot{\psi}_j \, \delta t \tag{3}$$

produced by the virtual velocities in an infinitesimal time increment δt are called **virtual displacements**; they will of course satisfy the equations

$$\delta f_i = \sum_{j=1}^{M} \frac{\partial f_i}{\partial \psi_j} \delta \psi_j = 0 \tag{4}$$

In words: *Any set of infinitesimal displacements consistent with the kinematic constraints on a system are called virtual displacements.*

In a system with F DOF we have seen that it is advantageous to designate F of the Lagrangian   variables as *generalized* coordinates $(q_1, \ldots, q_F)$ and the remaining coordinates $(\phi_1, \phi_2, \ldots, \phi_N)$ as *secondary* coordinates. Then Eqs. (1), (2), and (4) take the forms

$$f_i(\phi, q) = 0 \tag{5}$$

$$\sum_{j=1}^{N} \frac{\partial f_i}{\partial \phi_j} \dot{\phi}_j = -\sum_{j=1}^{F} \frac{\partial f_i}{\partial q_j} \dot{q}_j \tag{6}$$

$$\sum_{j=1}^{N} \frac{\partial f_i}{\partial \phi_j} \delta \phi_j = -\sum_{j=1}^{F} \frac{\partial f_i}{\partial q_j} \delta q_j \tag{7}$$

The last equation shows that the secondary virtual displacements $\delta \phi_j$ may be calculated for any given set of the primary (or generalized) virtual displacements δq_j.

Virtual displacements (and virtual velocities) do not necessarily coincide with the actual displacements (and velocities) which the system experiences. They are to be viewed as *test* quantities which are used to probe the nature of a mechanical system rather than the actual response of such a system to forces acting on it. The word *virtual* connotes that which exists in effect, although not in fact.

In Secs. **10.51** and **10.52** it was shown that the velocity components $(\dot{x}_i, \dot{y}_i)$ of a typical particle P_i can be expressed as a linear combination of the generalized velocities in the form

---

[3] Note that Eqs. (1) represent *stationary holonomic* constraints, as defined in Sec. **8.11**. This is not the only possible type of constraint, but it is the type that usually occurs in problems of machinery, and we shall assume such constraints unless otherwise noted.

$$\begin{Bmatrix} \dot{x}_i \\ \dot{y}_i \end{Bmatrix} = \sum_{j=1}^{F} \begin{Bmatrix} U_{ij} \\ V_{ij} \end{Bmatrix} \dot{q}_j \tag{8}$$

where the velocity coefficients (U_{ij}, V_{ij}) depend on the instantaneous configuration. Since the $\dot{q}_j$ are independent, any set of velocities $(\dot{x}_i, \dot{y}_i)$ satisfying Eqs. (8) satisfy the definition of virtual velocities, and the corresponding virtual displacements $(\delta x_i, \delta y_i)$ satisfy the equations

$$\begin{Bmatrix} \delta x_i \\ \delta y_i \end{Bmatrix} = \sum_{j=1}^{F} \begin{Bmatrix} U_{ij} \\ V_{ij} \end{Bmatrix} \delta q_j \tag{9}$$

for arbitrary δq_j.

Some examples illustrating the concept of virtual displacements follow.

EXAMPLE 1. Open-Loop System

Given the system of Fig. 1, with generalized coordinates q_1, q_2 as shown we seek an acceptable set of virtual displacements $(\delta x_2, \delta y_2)$ for an arbitrary point P with coordinates (ξ_2, η_2) on link 2.

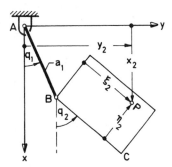

Figure 11.21-1. Double pendulum.

By inspection, the displacements of point P are:

$$\begin{aligned} x_2 &= a_1 \cos q_1 + \xi_2 \cos q_2 - \eta_2 \sin q_2 \\ y_2 &= a_1 \sin q_1 + \xi_2 \sin q_2 + \eta_2 \cos q_2 \end{aligned} \tag{10}$$

By forming differentials, we find the desired virtual displacements:

$$\begin{aligned} \delta x_2 &= -a_1 \sin q_1 \, \delta q_1 + (-\xi_2 \sin q_2 - \eta_2 \cos q_2) \, \delta q_2 \\ \delta y_2 &= a_1 \cos q_1 \, \delta q_1 + (\xi_2 \cos q_2 - \eta_2 \sin q_2) \, \delta q_2 \end{aligned} \tag{11}$$

Expressing these results in the form of Eq. (9), we find

$$\begin{Bmatrix} \delta x_2 \\ \delta y_2 \end{Bmatrix} = \begin{bmatrix} U_{21} & U_{22} \\ V_{21} & V_{22} \end{bmatrix} \begin{Bmatrix} \delta q_1 \\ \delta q_2 \end{Bmatrix} = \begin{bmatrix} -a_1 S_1 & -(\xi_2 S_2 + \eta_2 C_2) \\ a_1 C_1 & \xi_2 C_2 - \eta_2 S_2 \end{bmatrix} \begin{Bmatrix} \delta q_1 \\ \delta q_2 \end{Bmatrix} \tag{12}$$

where $(S_i, C_i) \equiv (\sin q_i, \cos q_i)$.

As is typical with open-loop systems, the virtual displacements of any point are readily expressed directly in terms of generalized coordinates, and there is no need to introduce any secondary coordinates.

As an alternative, one may find the velocity coefficients (U_{ij}, V_{ij}) directly by recognizing that

$$\dot{x}_2 = \dot{x}_B + (\dot{x}_P - \dot{x}_B) \equiv \dot{x}_{BA} + \dot{x}_{PB}$$
$$\dot{y}_2 = \dot{y}_B + (\dot{y}_P - \dot{y}_B) \equiv \dot{y}_{BA} + \dot{y}_{PB}$$

(13a)

where $(x_P, y_P) \equiv (x_2, y_2)$, $x_{PB} \equiv x_P - x_B$, etc.

With the help of Eqs. (5.31-6) and (5.31-7), Eqs. (13a) can be expressed as

$$\dot{x}_2 = -y_{BA}\dot{q}_1 - y_{PB}\dot{q}_2$$
$$\dot{y}_2 = x_{BA}\dot{q}_1 + x_{PB}\dot{q}$$

(13b)

Comparing this result with Eq. (8), we see that

$$\begin{bmatrix} U_{21} & U_{22} \\ V_{21} & V_{22} \end{bmatrix} = \begin{bmatrix} -y_{BA} & -y_{PB} \\ x_{BA} & x_{PB} \end{bmatrix}$$

(14)

It is readily verified that these velocity coefficients are identical to those in Eq. (12).

EXAMPLE 2. A Rigid Lamina

One corner A of a rectangular plate slides along the y axis, and a second corner B slides along the x axis of a rectangular Cartesian coordinate frame. The segment AB has length L and makes an angle θ with the x axis (measured as shown in Fig. 2). For

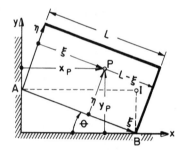

Figure 11.21-2. Rigid lamina.

an arbitrary point P (located by rectangular Cartesian coordinates ξ, η in the moving plane), find increments δx_P and δy_P, which constitute valid virtual displacements.

From the geometry of Fig. 2, it is readily verified that

$$x_P = \eta \sin \theta + \xi \cos \theta$$
$$y_P = (L - \xi) \sin \theta + \eta \cos \theta.$$

(15)

Equations (15) indicate that x_P and y_P cannot be varied at will if the specified constraints are to be maintained. On the other hand, the angle θ constitutes a suitable generalized coordinate (for this single degree of freedom system) which may be varied at will. When θ changes by an increment $\delta\theta$ it follows by differentiation of Eqs. (15) that x_P and y_P undergo virtual displacements

$$\delta x_P = (\eta \cos \theta - \xi \sin \theta)\, \delta\theta = (y_P - L \sin \theta)\, \delta\theta$$
$$\delta y_P = [(L - \xi) \cos \theta - \eta \sin \theta]\, \delta\theta = (L \cos \theta - x_P)\, \delta\theta$$

(16)

It may be verified that these equations constitute instructions to rotate the movable plane about the instant center I through an angle $\delta\theta$.

EXAMPLE 3. Four-Bar Linkage

For the single DOF system shown in Fig. 3, any one of the three angles $(\theta_1, \theta_2, \theta_3)$ can serve as a generalized coordinate. With θ_k chosen as generalized coordinate, virtual displacements for the other angles are given by

$$\delta\theta_i = \omega_i \, \delta\theta_k \tag{17}$$

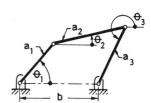

Figure 11.21-3. Four-bar linkage.

where ω_i is given by Eq. (**10.51**-5b) as

$$\omega_i = \frac{a_k}{a_i} \frac{\sin(\theta_j - \theta_k)}{\sin(\theta_i - \theta_j)} \qquad (i \neq k) \tag{18}$$

and (i, j, k) is any cyclic permutation of 1, 2, 3.

11.22 Virtual Work and Generalized Forces

If a particle undergoes a vector displacement $\Delta\mathbf{r}$ while a force acts upon it, the work done is

$$\Delta W = \mathbf{F} \cdot \Delta\mathbf{r}$$

where $\mathbf{F}$ is the mean force on the particle during the displacement. If the displacement takes place in time Δt, the mean *rate* of work or mean *power* is $\Delta W/\Delta t$. The limit, as $\Delta t \rightarrow 0$, of this ratio is the instantaneous power of the force, given by

$$P = \lim_{\Delta t \to 0} \mathbf{F} \cdot \frac{\Delta\mathbf{r}}{\Delta t} = \mathbf{F} \cdot \dot{\mathbf{r}} \tag{1}$$

where $\mathbf{F}$ is the force acting on the particle in position $\mathbf{r}$.

If, in a system of n particles, the jth particle is located in a base coordinate system (x, y) by the vector $\mathbf{r}_j = (x_j, y_j)$ and is acted on by a force $\mathbf{F}_j = (X_j, Y_j)$, the power of the forces for the system is

$$P = \sum_{j=1}^{n} \mathbf{F}_j \cdot \dot{\mathbf{r}}_j = \sum_{j=1}^{n} (X_j \dot{x}_j + Y_j \dot{y}_j) \tag{2}$$

Equation (2) is useful for a system with a discrete number of particles and forces. For the continuous system of particles comprising a rigid body, Eq. (2) can be transformed into a more convenient form. For example, consider the rigid lamina of Fig. 1, where coplanar forces X_j and Y_j act on a typical particle P_j situated at point (x_j, y_j). The 3 DOF of the lamina are described by coordinates (x_A, y_A, θ), where x_A, y_A are Cartesian coordinates of a point A, fixed in the lamina, and θ is the angle of rotation of the lamina. During an arbitrary motion of the lamina, the forces X_j, Y_j do

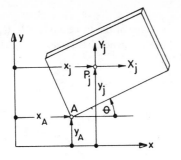

Figure 11.22-1. Rigid lamina.

work at a rate given by

$$P = \sum_{j=1}^{\infty} (X_j \dot{x}_j + Y_j \dot{y}_j) \tag{3}$$

From kinematics, we know that

$$\begin{aligned} \dot{x}_j &= \dot{x}_A - (y_j - y_A)\dot{\theta} \\ \dot{y}_j &= \dot{y}_A + (x_j - x_A)\dot{\theta} \end{aligned} \tag{10.51-13}$$

Upon substitution of these values into Eq. (3) we find

$$P = X\dot{x}_A + Y\dot{y}_A + M_A\dot{\theta} \tag{4a}$$

where

$$X \equiv \sum_{j=1}^{\infty} X_j \tag{4b}$$

$$Y \equiv \sum_{j=1}^{\infty} Y_j \tag{4c}$$

$$M_A \equiv \sum_{j=1}^{\infty} [Y_j(x_j - x_A) - X_j(y_j - y_A)] \tag{4d}$$

are the resultant force components applied to the system and the resultant moment of the applied forces about point A.

If we have a system of M such laminas (e.g., a mechanism), we may select a reference point P_i fixed in lamina i and represent the most general distribution of loading on the lamina by a set of forces (X_i, Y_i) acting through P_i together with a couple M_i; for brevity we refer to (X_i, Y_i, M_i) as the **force set** at P_i.

The power of all the forces is given by Eq. (4a) in the form

$$P = \sum_{i=1}^{M} (X_i \dot{x}_i + Y_i \dot{y}_i + M_i \dot{\theta}_i) \tag{5}$$

For a constrained system with F degrees of freedom, we may express any set of velocities which are consistent with the constraints (i.e., a set of virtual velocities) in terms of the generalized velocities via Eqs. (10.52-3) and (10.52-5); i.e.,

$$\begin{Bmatrix} \dot{x}_i \\ \dot{y}_i \\ \dot{\theta} \end{Bmatrix} = \sum_{j=1}^{F} \begin{Bmatrix} U_{ij} \\ V_{ij} \\ \Omega_{ij} \end{Bmatrix} \dot{q}_j \tag{6a}$$

Upon substituting these virtual velocities into Eq. (5), we find the **virtual power**:

$$P = \sum_{i=1}^{M} \sum_{j=1}^{F} (X_i U_{ij} + Y_i V_{ij} + M_i \Omega_{ij}) \dot{q}_j$$

Interchanging the order of summation and collecting all terms which multiply $\dot{q}_j$, we find

$$P = \sum_{j=1}^{F} Q_j \dot{q}_j \tag{7}$$

where the terms

$$Q_j \equiv \sum_{i=1}^{M} (X_i U_{ij} + Y_i V_{ij} + M_i \Omega_{ij}) \tag{8a}$$

are called the components of **generalized force** associated with (or **conjugate** to) the generalized coordinates q_j.

For a single-freedom system, there is only one generalized coordinate q, and the double-subscript velocity coefficients in Eq. (6a) may be replaced by single-subscript coefficients as in Eqs. (**10.51**-2)–(**10.51**-4), i.e.,

$$\begin{Bmatrix} \dot{x}_i \\ \dot{y}_i \\ \dot{\theta}_i \end{Bmatrix} = \begin{Bmatrix} u_i \\ v_i \\ \omega_i \end{Bmatrix} \dot{q} \tag{6b}$$

and the single generalized force corresponding to Eq. (8a) becomes

$$Q = \sum_{i=1}^{M} (X_i u_i + Y_i v_i + M_i \omega_i) \tag{8b}$$

If both sides of Eq. (7) are multiplied by the infinitesimal time increment δt, we find an expression for the **virtual work** of the force system (X_i, Y_i, M_i); viz.,

$$\delta W = P \delta t = \sum_{j=1}^{F} Q_j \, \delta q_j \tag{9}$$

From Eq. (9) we see that the virtual work of the forces (X_i, Y_i, M_i) can always be expressed as the *scalar product* of the *vector*

$$\mathbf{Q} \equiv (Q_1, Q_2, \ldots, Q_F)$$

by the incremental *vector* of generalized coordinates

$$\delta \mathbf{q} \equiv (\delta q_1, \delta q_2, \ldots, \delta q_F)$$

Note, from Eqs. (8), that the generalized force consists of a linear combination of the given Cartesian force components (X_i, Y_i, M_i). However, the generalized forces do not necessarily have the dimension of force but may have dimensions of pressure, torque, line load, etc.; in any case, the product of a given Q_i and δq_i (or $\dot{q}_i$) must always have the dimensions of work (or power).

In practice, it is not necessary to utilize the formal expressions (8) in order to find the generalized forces. Instead, one may merely express the work done on a set of increments in q_i, and then identify Q_i as the multiplier of δq_i in an expression of type (9). This will be illustrated by some examples. In these examples, the forces (X_i, Y_i, M_i) are given (active) forces.

EXAMPLE 1

The pulley system of Fig. 2 consists of three fixed pulleys and three movable pulleys. The latter are connected to the former by means of six vertical strands of equal length z (measured between the axles of the two pulley sets). The movable pulleys carry a weight G, and a vertical force P is applied to the hanging strand of length s. We are to find the generalized force Q corresponding to the coordinate s.

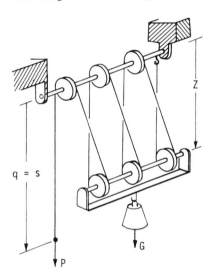

Figure 11.22-2. Pulley system.

When the force P moves through a distance δs (positive downward) the weight G will move through a distance δz (positive downward), and the net work done by P and G will be

$$\delta W = P\,\delta s + G\,\delta z$$

Since the total length of cable is constant, we see from the figure that

$$s + 6z = \text{constant}$$

Hence

$$\delta s + 6\delta z = 0 \quad \text{or} \quad \delta z = -\frac{\delta s}{6}$$

Thus, the expression for the work becomes

$$\delta W = \left(P - \frac{G}{6}\right)\delta s \equiv Q\,\delta q$$

If we interpret s as the single generalized coordinate q for this single degree of freedom system, we see that the corresponding generalized force is

$$Q \equiv P - \frac{G}{6} \tag{10}$$

EXAMPLE 2

A four-bar linkage has given forces H and V acting as shown in Fig. 3.

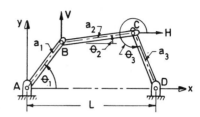

Figure 11.22-3. Four-bar mechanism.

We seek the generalized force Q corresponding to the generalized coordinate $q \equiv \theta_1$. The virtual work is given by

$$\delta W = V \, \delta y_B + H \, \delta x_C$$

where

$$y_B = a_1 \sin \theta_1, \qquad \therefore \; \delta y_B = a_1 \, \delta\theta_1 \cos \theta_1$$
$$x_C = L - a_3 \cos \theta_3, \qquad \therefore \; \delta x_C = a_3 \, \delta\theta_3 \sin \theta_3$$

Hence

$$\delta W = \left(V a_1 \cos \theta_1 + H a_3 \sin \theta_3 \frac{\delta\theta_3}{\delta\theta_1} \right) \delta\theta_1 \equiv Q \, \delta\theta_1 = Q \, \delta q$$

It follows that the parenthesized term in the last equation represents the required generalized force Q. It is seen that Q depends on the velocity ratio

$$\frac{\delta\theta_3}{\delta\theta_1} = \frac{\dot{\theta}_3}{\dot{\theta}_1}$$

The velocity ratio is readily found from the complete loop equations for velocity and may be verified to be [see Eq. (**11.21**-18)]

$$\omega_3 \equiv \frac{\delta\theta_3}{\delta\theta_1} = \frac{a_1 \sin (\theta_1 - \theta_2)}{a_3 \sin (\theta_2 - \theta_3)}$$

Hence

$$Q \equiv a_1 \left[V \cos \theta_1 + H \sin \theta_3 \frac{\sin (\theta_1 - \theta_2)}{\sin (\theta_2 - \theta_3)} \right] \tag{11}$$

The same result is given by the formal application of Eq. (8b).

EXAMPLE 3

For the reciprocating engine shown in Fig. 4, the gas force P does work on the piston at the rate

$$-P\dot{s} = -P \frac{\dot{s}}{\dot{\theta}} \dot{\theta} \equiv Q\dot{\theta}$$

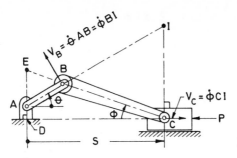

Figure 11.22-4. Reciprocating engine.

where

$$Q \equiv -P\frac{\dot{s}}{\dot{\theta}} \tag{12}$$

is the generalized force corresponding to the crank angle θ. A simple expression for the velocity ratio $\dot{s}/\dot{\theta}$ follows after noting that I is the instantaneous center of the connecting rod BC [see Fig. **5.13**-4(c)] and hence

$$\dot{s} = V_C = -\dot{\phi}CI \quad \text{and} \quad V_B = \dot{\phi}BI$$

Therefore

$$\frac{\dot{s}}{V_B} = -\frac{CI}{BI}$$

but since $V_B = \dot{\theta}AB$,

$$\frac{\dot{s}}{\dot{\theta}} = -\frac{CI}{BI}AB$$

Finally, we note from the similar triangles BAE and BIC that $CI/BI = AE/AB$, so that the desired velocity ratio becomes

$$\frac{\dot{s}}{\dot{\theta}} = -AE \tag{13}$$

When Eq. (13) is compared to Eq. (12) we see that the generalized force is

$$Q = P\,AE \tag{14}$$

We see that the variable distance AE plays the role of an effective moment arm for the gas force P in its effort to turn the crank. Engine designers call Q the *turning moment* of the gas force.

11.23 Ideal (Workless) Constraints

The members (links) of a machine are constrained by kinematic pairs such as sliders, rollers, pivots, knife edges, ball joints, screw pairs, etc. Those pairs with surface contact will be called **ideal pairs** if no friction exists and the mutual contact forces act normal to the contacting surface elements.

Let us denote the mutual contact forces on two contacting pair elements[4] by $\mathbf{F}_1$,

---

[4]Curved contact surfaces may be divided into an arbitrarily large number of pairs of surface elements with infinitesimal plane areas.

F_2, and the velocities of the two elements by v_1, v_2. For plane systems, the mutual contact moments are given by M_1, M_2, and the corresponding angular velocities by $\dot\theta_1, \dot\theta_2$. The rate at which the contact forces do work at the joint in question is

$$P = \mathbf{F}_1 \cdot \mathbf{v}_1 + \mathbf{F}_2 \cdot \mathbf{v}_2 + M_1\dot\theta_1 + M_2\dot\theta_2 \tag{1}$$

where, by Newton's third law,

$$\mathbf{F}_2 = -\mathbf{F}_1, \qquad M_2 = -M_1 \tag{2}$$

Therefore, the power of the constraint forces is

$$P = \mathbf{F}_2 \cdot (\mathbf{v}_2 - \mathbf{v}_1) + M_2(\dot\theta_2 - \dot\theta_1) \tag{3}$$

When the velocities are consistent with the constraints, Eq. (3) represents the *virtual power* of the pair forces.

For members with *surface* contact (e.g., sliders), $\dot\theta_2 = \dot\theta_1$, and $P = \mathbf{F}_2 \cdot (\mathbf{v}_2 - \mathbf{v}_1)$. If the pair is *ideal* (frictionless), $\mathbf{F}_2$ is *normal* to the relative velocity $(\mathbf{v}_2 - \mathbf{v}_1)$, and $P = 0$.

In the case of *point* contact (e.g., hinges and pivots) the moments M_1 and M_2 vanish, and $\mathbf{v}_2 = \mathbf{v}_1$; thus the virtual power given by Eq. (3) also vanishes in this case.

So far, we have considered specific examples of constraints which produce no work. We shall now generalize the definition of an **ideal constraint** to include *any arrangement where the forces of constraint do no net work during a virtual displacement of the system.* Later on, we shall discuss how to take into account friction, elastic springs, and other devices which may dissipate or store energy.

11.24 Problems

1. A moment M is applied to the crank of a slider-crank mechanism, and a force P to the slider (Fig. 1). The generalized coordinate is the crank angle ϕ. Find an expression for the associated generalized force.

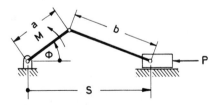

Figure 11.24-1 (Probs. 1, 2).

2. Repeat Problem 1 but treat the slider distance s as the generalized coordinate.

3. The flyball governor mechanism has forces F (centrifugal force) and S (spring force) acting as shown in Fig. 2. Choose the angle ϕ as the generalized coordinate, and find an expression for the associated generalized force.

4. Repeat Problem 3 but consider the distance AC as the generalized coordinate.

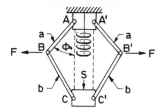

Figure 11.24-2 (Probs. 3, 4).

5. The three-bar open chain (multiple pendulum) has links of equal length a and is acted upon by equal vertical forces V at each hinge as well as by a single horizontal force H at the lowest point (Fig. 3). Choose the angles ϕ_1, ϕ_2, ϕ_3 as generalized coordinates, and find expressions for the associated generalized forces.

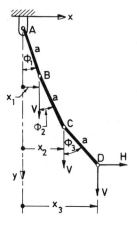

Figure 11.24-3 (Probs. 5, 6).

6. For the system treated in Problem 5, choose the horizontal distances x_1, x_2, x_3 as generalized coordinates, and find Q_1, Q_2, Q_3. (Assume $|\phi_i| \leq \pi/2, i = 1, 2, 3$)

7. The hinged rhombus $ABCD$, of side length a, is suspended from the fixed hinge at A (Fig. 4). Vertical forces V act at hinges B and D, and a vertical force P acts at C. Find the generalized forces corresponding to coordinates ϕ and α.

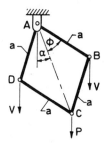

Figure 11.24-4 (Prob. 7).

8. An ellipse of revolution (football bladder) has volume V and is subjected to an internal pressure p (Fig. 5). Choose the volume V as generalized coordinate, and find the associated generalized force.

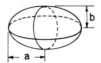

Figure 11.24-5 (Prob. 8).

9. A rectangular block has length a, width b, and thickness c normal to the plane of paper. Uniform tensile stresses σ_x and σ_y (pounds per square inch) act as shown in Fig. 6. Consider the strains $\epsilon_x = \Delta a/a$ and $\epsilon_y = \Delta b/b$ as generalized coordinates, and find corresponding generalized forces.

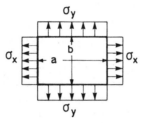

Figure 11.24-6 (Prob. 9).

10–14. (Block and tackle problems) Refer to Fig. 7. If $\dot{q}$ represents the downward velocity of the load P and F represents the force applied to the free end of the cable, find

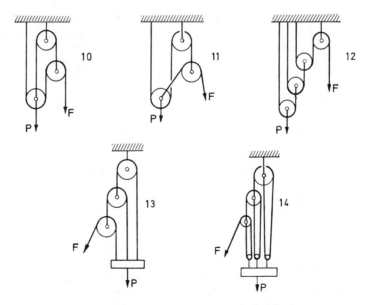

Figure 11.24-7 (Probs. 10, 11, 12, 13, 14).

the generalized force Q corresponding to $\dot{q}$. Consider all strands to be vertical if not otherwise indicated and all pulleys as frictionless.

15–16. Refer to Fig. 8. If $\dot{q}$ represents the common angular velocity of the winding and unwinding drums of the Chinese windlass shown, find the generalized force Q corresponding to $\dot{q}$ when the torque M acts on the handwheel and the weight P is being raised.

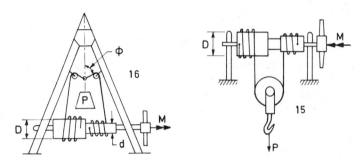

Figure 11.24-8 (Probs. 15, 16).

17. For the four-bar mechanism shown in Fig. **10.51**-2, let the resultant force (X_i, Y_i) act through the point P_i on link i of the mechanism, and let moment M_i act about P_i. The coordinates of P_i are ξ_i and η_i relative to the link i. Find an expression for the generalized force corresponding to the generalized coordinate q, where

$$\text{a. } q = \theta_1, \quad \text{b. } q = \theta_2, \quad \text{and} \quad \text{c. } q = \theta_3.$$

Note: You may express your answer in terms of the velocity coefficients used in Sec. **10.51**.

18. A rigid lamina has angular velocity $\dot{\phi}$. Show that the generalized force corresponding to angle ϕ is the moment of the applied force system about the instantaneous center of the lamina.

19. In a system of particles, $P_1, P_2, \ldots, P_N$, where the force exerted on P_i by P_j is $\mathbf{F}_{ij} = -\mathbf{F}_{ji}$, let $\mathbf{r}_i$ represent the position vector of P_i.

a. Show that the virtual work of the given force system is

$$\delta W = \tfrac{1}{2} \sum_{i,j=1}^{N} \mathbf{F}_{ij} \cdot \delta(\mathbf{r}_i - \mathbf{r}_j) \qquad (i \neq j)$$

b. Assume that $\mathbf{F}_{ij}$ is directed along the line from P_i to P_j (central forces); i.e., $\mathbf{F}_{ij} = A_{ij}(\mathbf{r}_i - \mathbf{r}_j)$, where $A_{ij} = A_{ji}$ is a scalar function of position. Then show that in a rigid body $\delta W = 0$.

c. What does this result say about the work done by internal forces (stresses) in a rigid body?

20. In a mechanism with M members, let the applied forces on link i be equipollent to a force of magnitude F_i acting along a directed line L_i plus a couple M_i. Also let h_i be the distance from the instant center of the link to the line of action of the force.

a. Show that in a single-freedom mechanism, the generalized force Q corresponding to generalized coordinate q is

$$Q = \sum_{i=1}^{M} \omega_i(F_i h_i + M_i) \qquad \text{(i)}$$

where the angular velocity of link i is $\dot{\theta}_i = \omega_i \dot{q}$.

b. Show that $\omega_i h_i = \sigma_i$, where $\sigma_i \dot{q}$ is the velocity component, of any point on the line L_i, in the positive sense of the directed line L_i, and that

$$Q = \sum_{i=1}^{M} (F_i \sigma_i + \omega_i M_i) \qquad \text{(ii)}$$

Why is Eq. (ii) preferable to Eq. (i) in a mechanism where one or more links undergo pure translation?

21. Along the edges AB, BC, CD, ... of a rigid convex plane polygon there are forces represented by the vectors $\overrightarrow{AB}$, $\overrightarrow{BC}$, $\overrightarrow{CD}$, ... Show that the virtual work of these forces during *any* virtual displacement of the polygon (in its plane) is $\delta W = 2A\,\delta\theta$, where A is the area enclosed by the polygon and $\delta\theta$ is the rotation of the plane. Can this result be generalized to express the work as $2A\theta$ for any *finite* displacement θ in the plane of the polygon? Explain.

22. a. Show that the result proved in Problem 21 is valid for any plane polygon (convex or not) whose sides do not intersect.

   b. Show that the result of Problem 21 is valid for any plane polygon (including those with self-intersecting sides) provided that the net area A is defined with due regard to the sign of certain subareas within the given polygon. Give the rule for interpreting signs.

11.30 THE PRINCIPLE OF VIRTUAL WORK (STATEMENT)

The *method of virtual work* has the great advantage of allowing us to formulate the required equations of static equilibrium *without having to consider the mutual reactions exerted upon one another by the bodies of the system*. The great value of such a procedure was recognized early in the history of mechanics.[5] Faint shadows of the method of virtual work appeared in the writings of Aristotle (384–322 B.C.), Galileo (1594), Stevinus (1608), and Jean Bernoulli (1717). However, it was J. L. Lagrange who developed it into an all-powerful instrument which was made the foundation of both statics and dynamics in his profound work *Mecanique Analytique* (1788).

The principle of virtual work may be viewed as a fundamental postulate from which it is possible to deduce all the laws of *vector statics*. Alternatively one may prove the principle of virtual work starting from vector statics. To avoid obfuscation of the simple manner in which the principle applies, let us accept its validity at this point and use it to solve several problems. In Sec. **11.70**, we shall return to deduce the principle from Newton's laws.

---

[5]The tortuous development of the principle of virtual work can be followed in the historical survey of Lanczos [1966, pp. 341–351] and in the review article by Oravas and McLean [1966]. It is also discussed in the well-known interpretive history of Mach [1960] and in the standard history of mechanics by Dugas [1955].

The principle is simply stated, but each word in the statement must be understood in light of previously defined concepts, which are summarized below:

1. **Virtual displacements** = a set of infinitesimal displacements which are consistent with the constraints
2. **Virtual work** = work done by specified forces on virtual displacements
3. **Active forces**[6] = all forces which do nonzero virtual work
4. **Ideal mechanical system** = system where constraints do no work
5. **Generalized forces** = terms which multiply the virtual generalized displacements in the expression for virtual work of active forces
6. **Equilibrium** = a state where the resultant force on each particle of the system vanishes
7. **Generalized equilibrium** = a state where all the generalized forces of a system vanish; equivalent to *equilibrium* for systems "at rest" (see Sec. **11.70**).

With the above definitions in mind we may formulate the principle of virtual work in the following two-part statement:

1. **If an ideal mechanical system is in equilibrium, the net virtual work of all the active forces vanishes for every set of virtual displacements.**
2. **If the net virtual work of all the active forces vanishes, for every set of virtual displacements, an ideal mechanical system is in a state of generalized equilibrium.**

Many writers combine both statements into a single sentence by replacing the word *if* in statement 1 with the words *if and only if*. For reasons discussed in the proofs, we prefer the form given.

Note that the statement of the principle remains valid if the terms *virtual work* and *virtual displacements* are replaced by the terms *virtual power* and *virtual velocities*, respectively. We may therefore use *virtual displacements* and *virtual velocities* interchangeably for most practical purposes.

From Eq. (**11.22**-9), we see that the virtual work for a system with F degrees of freedom may be expressed in generalized coordinates as

$$\delta W = Q_1 \, \delta q_1 + Q_2 \, \delta q_2 + \cdots + Q_F \, \delta q_F$$

Hence, the principle of virtual work tells us that a necessary and sufficient condition for generalized equilibrium is that each of the generalized forces Q_i must vanish; i.e.,

$$Q_1 = 0, \quad Q_2 = 0, \quad \ldots, \quad Q_F = 0 \tag{1}$$

When a *rigid* lamina is subject to forces X, Y and a moment M about an arbitrary point, Eqs. (1) and Eqs. (**11.22**-4) show that the conditions

$$X = 0, \qquad Y = 0, \qquad M = 0 \tag{2}$$

are not only *necessary* for equilibrium, but they are also *sufficient* for generalized

---

[6]Some writers use the terms *applied*, *external*, or *given* for what is here designated by *active*.

equilibrium of a *rigid* lamina. This result is easily generalized to the three-dimensional case for any rigid body.

It follows from Eq. (1) and Eq. (**11.22**-8a) and (**11.22**-8b) that *any mechanism consisting of M rigid members and having F degrees of freedom is in generalized equilibrium if, and only if, the applied loads satisfy the F linear relations*

$$Q_j = \sum_{i=1}^{M} (X_i U_{ij} + Y_i V_{ij} + M_i \Omega_{ij}) = 0 \qquad (j = 1, 2, \ldots, F) \qquad (3a)$$

which, for the case of $F = 1$, reduce to

$$Q = \sum_{i=1}^{M} (X_i u_i + Y_i v_i + M_i \omega_i) = 0 \qquad (3b)$$

These equations will be called the **system equilibrium equations** to distinguish them from the **member equilibrium equations** which govern the individual links of the mechanism.

Some examples will now be given to illustrate the application of the principle of virtual work to problems of statics.

EXAMPLE 1

The pulley system of Fig. **11.22**-2 will be in equilibrium when $Q_1 = 0$, or, according to Eq. (**11.22**-10), when $P = G/6$.

EXAMPLE 2

The four-bar mechanism of Fig. **11.22**-3 is in equilibrium when $Q_1 = 0$, or, according to Eq. (**11.22**-11), when

$$V = H \frac{\sin \theta_3 \sin (\theta_2 - \theta_1)}{\cos \theta_1 \sin (\theta_2 - \theta_3)}$$

EXAMPLE 3

Figure 1 shows an open kinematic chain (or multiple pendulum) with three rigid links. A typical link has length a_i and weight Z_i, and the center of gravity of each link

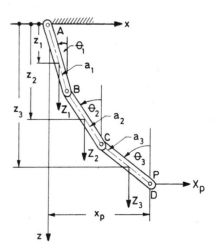

Figure 11.30-1. An open kinematic chain.

is located midway between the hinges on the link. A horizontal force X_P acts at P—the outermost point of the chain. The angles θ_i, between the center lines of the links and the vertical z direction, serve as suitable generalized coordinates, and the vertical distances from the topmost hinge to the centers of gravity of the links are denoted by z_1, z_2, z_3.

When the generalized coordinates θ_i undergo virtual displacements $\delta\theta_i$, the points of application of the gravity forces move vertically through increments δz_i, and the point P moves through the horizontal distance δx_P; hence the virtual work of the external forces is given by

$$\delta W = Z_1\,\delta z_1 + Z_2\,\delta z_2 + Z_3\,\delta z_3 + X_P\,\delta x_P = 0 \tag{4}$$

Because δz_1, δz_2, δz_3, and δx_P are not independent of one another, we cannot equate their coefficients, in Eq. (3), to zero. Therefore we seek to express δz_i and δx_P in terms of the generalized coordinates $\delta\theta_i$. This is conveniently done by means of the open-loop displacement equations:

$$z_1 = \frac{a_1}{2}\cos\theta_1$$

$$z_2 = a_1\cos\theta_1 + \frac{a_2}{2}\cos\theta_2$$

$$z_3 = a_1\cos\theta_1 + a_2\cos\theta_2 + \frac{a_3}{2}\cos\theta_3$$

$$x_P = a_1\sin\theta_1 + a_2\sin\theta_2 + a_3\sin\theta_3$$

By taking differentials across these equations, we find

$$\delta z_1 = -\frac{a_1}{2}S_1\,\delta\theta_1$$

$$\delta z_2 = -a_1 S_1\,\delta\theta_1 - \frac{a_2}{2}S_2\,\delta\theta_2$$

$$\delta z_3 = -a_1 S_1\,\delta\theta_1 - a_2 S_2\,\delta\theta_2 - \frac{a_3}{2}S_3\,\delta\theta_3 \tag{5}$$

$$\delta x_P = a_1 C_1\,\delta\theta_1 + a_2 C_2\,\delta\theta_2 + a_3 C_3\,\delta\theta_3$$

where $S_i \equiv \sin\theta_i$ and $C_i \equiv \cos\theta_i$. Upon substitution of Eqs. (5) into Eq. (4) we find the virtual work in the form

$$\delta W = Q_1\,\delta\theta_1 + Q_2\,\delta\theta_2 + Q_3\,\delta\theta_3 \tag{6}$$

where

$$Q_1 \equiv X_P a_1 C_1 - \tfrac{1}{2}Z_1 a_1 S_1 - Z_2 a_1 S_1 - Z_3 a_1 S_1$$
$$Q_2 \equiv X_P a_2 C_2 - \tfrac{1}{2}Z_2 a_2 S_2 - Z_3 a_2 S_2 \tag{7}$$
$$Q_3 \equiv X_P a_3 C_3 - \tfrac{1}{2}Z_3 a_3 S_3$$

According to Eq. (1), the system will be in equilibrium if

$$Q_1 = 0, \qquad Q_2 = 0, \qquad Q_3 = 0 \tag{8}$$

Equations (8) and definitions (7) constitute the required set of equilibrium equations for the problem. It so happens that this particular problem may be solved *explicitly* for the configuration variables $\theta_1, \theta_2, \theta_3$ in terms of the applied forces; i.e., Eqs. (7) and (8) lead to the expressions

$$\tan \theta_3 = \frac{2X_P}{Z_3}$$

$$\tan \theta_2 = \frac{2X_P}{Z_2 + 2Z_3} \qquad\qquad (9)$$

$$\tan \theta_1 = \frac{2X_P}{Z_1 + 2Z_2 + 2Z_3}$$

If all links have the same weight, $Z_1 = Z_2 = \cdots \equiv Z$; then $\tan \theta_i = 2X_P/[(2i-1)Z]$.

It may easily be shown (Problem **11.31**-19) that explicit expressions can be obtained for the angles of any link in any *open* network (i.e., a topological *tree*) of hinged links subjected to arbitrary loading.

For *closed* kinematic networks, the equilibrium equations will be transcendental in the displacement variables and will have to be solved by a numerical or graphical procedure. The formulation of the necessary equations for such systems is illustrated in the following example.

EXAMPLE 4

Four weightless rods of equal length a are hinged together to form a hanging chain as shown in Fig. 2. It is required to find the configuration of this closed kinematic chain for given vertical loads P_A, P_B, P_C applied at the hinges. In other words, we wish to find the four angles $\theta_1, \theta_2, \theta_3, \theta_4$.

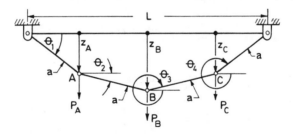

Figure 11.30-2. A closed kinematic chain.

The virtual work of the applied loads is given by

$$\delta W = P_A \delta z_A + P_B \delta z_B + P_C \delta z_C = 0 \qquad\qquad (10)$$

where z_A, z_B, z_C represent the vertical coordinates of the load points and are given by

$$z_A = a \sin \theta_1$$
$$z_B = a(\sin \theta_1 + \sin \theta_2) \qquad\qquad (11)$$
$$z_C = a(\sin \theta_1 + \sin \theta_2 + \sin \theta_3)$$

By taking differentials across Eqs. (11) we find that

$$\delta z_A = a C_1 \, \delta\theta_1$$
$$\delta z_B = a(C_1 \, \delta\theta_1 + C_2 \, \delta\theta_2) \tag{12}$$
$$\delta z_C = a(C_1 \, \delta\theta_1 + C_2 \, \delta\theta_2 + C_3 \, \delta\theta_3)$$

where $C_i = \cos\theta_i$. Upon substitution of Eqs. (12) into Eq. (10) we find

$$\delta W = a C_1 (P_A + P_B + P_C) \, \delta\theta_1$$
$$+ a C_2 (P_B + P_C) \, \delta\theta_2 + a C_3 P_C \, \delta\theta_3 = 0 \tag{13}$$

The system has 2 DOF; hence θ_1 and θ_2 will serve as suitable generalized coordinates. We therefore seek to express $\delta\theta_3$ in terms of $\delta\theta_1$ and $\delta\theta_2$. This is a strictly kinematic problem which is easily solved by writing the complete loop displacement equations:

$$a(C_1 + C_2 + C_3 + C_4) = L \tag{14}$$
$$a(S_1 + S_2 + S_3 + S_4) = 0 \tag{15}$$

Taking differentials across Eqs. (14) and (15) and placing terms with $\delta\theta_1$ and $\delta\theta_2$ on the right-hand side, we find

$$-S_3 \, \delta\theta_3 - S_4 \, \delta\theta_4 = S_1 \, \delta\theta_1 + S_2 \, \delta\theta_2$$
$$C_3 \, \delta\theta_3 + C_4 \, \delta\theta_4 = -C_1 \, \delta\theta_1 - C_2 \, \delta\theta_2 \tag{16}$$

Equations (16) are readily solved for $\delta\theta_3$ in the form

$$\delta\theta_3 = \frac{(S_1 C_4 - C_1 S_4)\,\delta\theta_1 + (S_2 C_4 - C_2 S_4)\,\delta\theta_2}{S_4 C_3 - C_4 S_3} \equiv k_{31}\,\delta\theta_1 + k_{32}\,\delta\theta_2 \tag{17}$$

where we have defined the coefficients of $\delta\theta_1$ and $\delta\theta_2$ by the symbols

$$k_{31} \equiv \frac{\sin(\theta_1 - \theta_4)}{\sin(\theta_4 - \theta_3)}, \qquad k_{32} \equiv \frac{\sin(\theta_2 - \theta_4)}{\sin(\theta_4 - \theta_3)} \tag{18}$$

Upon substitution of Eq. (17) into Eq. (13) we find that

$$\delta W = [a C_1 (P_A + P_B + P_C) + a C_3 P_C k_{31}]\,\delta\theta_1$$
$$+ [a C_2 (P_B + P_C) + a C_3 P_C k_{32}]\,\delta\theta_2 \equiv Q_1 \, \delta\theta_1 + Q_2 \, \delta\theta_2 \tag{19}$$

Hence the equations of equilibrium are

$$Q_1 = (P_A + P_B + P_C)\cos\theta_1 + k_{31} P_C \cos\theta_3 = 0 \tag{20}$$
$$Q_2 = (P_B + P_C)\cos\theta_2 + k_{32} P_C \cos\theta_3 = 0 \tag{21}$$

Equations (20) and (21) together with Eqs. (14) and (15) comprise four independent equations which can be used to find the four angles $\theta_1, \theta_2, \theta_3, \theta_4$. Unlike the case of the open kinematic chain discussed in Example 3, we cannot write explicit expressions for the configuration variables, but we can solve the governing set of algebraic equations by graphical or numerical methods (see Problem **11.31**-18).

Of course, if the configuration is given, Eqs. (20) and (21) may be solved directly to find the ratios of the load components needed to maintain equilibrium in the given configuration.

11.31   Problems

1. For the slider-crank mechanism of Fig. **11.24**-1, find an expression for the crank moment M that will hold piston force P in equilibrium.

2. For the flyball governor of Fig. **11.24**-2, what spring force S is necessary to equilibrate the forces $F = 20$ lb if $\phi = 45°$, $a = 1$ in., and $b = 1.5$ in.?

3. Find the angles ϕ_1, ϕ_2, ϕ_3 for the triple pendulum of Fig. **11.24**-3 if $H = 2V$.

4. In Fig. **11.30**-2, find the forces P_A and P_C needed to equilibrate a force of $P_B = 100$ lb if $\theta_1 = 60°$, $\theta_4 = 315°$, and $L = 3a$.

5. For the four-bar mechanism of Fig. **11.10**-1, find the horizontal force H, applied at the midpoint of bar AB, which will equilibrate a vertical force $V = 50$ lb applied at the midpoint of bar BC. Use the following data: $a_1 = 4$ in., $a_2 = 6$ in., $a_3 = 7$ in., $L = 11.723$ in., $\theta_1 = 70°$, $\theta_2 = 15°$, $\theta_3 = 310.64°$.

6. Find the moment M_A applied to the crank AB of the engine in Fig. **11.22**-4 that will equilibrate a piston force $P = 1000$ lb when the piston travel $s = 10$ in. and angle $\phi = 30°$. Assume an offset $DA = 1$ in.

7. What moment M_A must be applied to the crank AB of the Whitworth quick-return mechanism of Fig. **10.20**-1 in order to equilibrate a piston force P, acting in the x direction, on the slider link 5. Express your results in terms of the configuration variables $(\theta_1, r_2, \theta_3, \theta_4, r_5)$ and parameters (a_1, a_3, a_4, f) shown in the figure. (Hint: Use the information developed in Example 1 of Sec. **10.20**.)

8. What vertical force Y_D must be applied to the slider D of Fig. **10.60**-1 in order to balance a horizontal force $X_F = 100$ lb applied to point F.

9. Find the moment M_C that must be applied to link BC of Fig. **10.60**-5 in order to drive the carriage 1 against a horizontal force $F = 200$ lb when $\theta_2 = 90°$, $\theta_4 = 9.118°$, $\theta_5 = 133.1°$, $r_1 = 2.247$ in., and $r_3 = 6.821$ in. Use the dimensions shown in the figure.

10–16. Find the mechanical advantage (i.e., the ratio P/F or P/M) for the lifting rigs associated with problems **11.24**-10–16. Use the method of virtual work.

17. For the hinged rhombus of Fig. 1,
    a. Find expressions for θ and ψ in terms of the vertical forces V and P and the horizontal force H.
    b. Find θ and ψ if $P = H$, $V = 2H$.

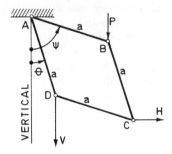

Figure 11.31-1 (Prob. 17).

18. In Fig. **11.30**-2, let $P_A = P_C$, so that symmetry prevails, and $\theta_3 = 2\pi - \theta_2$, $\theta_4 = 2\pi - \theta_1$.

   a. Show that two equations for finding θ_1 and θ_2 are of the form

$$\cos \theta_1 + \cos \theta_2 - \frac{L}{2a} = 0$$

$$\tan \theta_1 - \left(1 + 2\frac{P_A}{P_B}\right) \tan \theta_2 = 0$$

   b. Find θ_1 and θ_2 when $P_A = 2P_B$, $L = 3a$.

19. Figure 2 shows a typical bar AB in an *open* kinematic network of hinged links (a *tree*). The forces exerted by the adjacent link (or links) at hinge B are denoted by H_B, V_B, and the force set applied directly to link AB at a reference point C is denoted by (X_C, Y_C, M_C).

   a. Find an expression for angle θ in terms of the given quantities: X_C, Y_C, M_C, H_B, V_B, a, b, β.

   b. Show that (H_B, V_B) represent the components of the resultant of all active forces applied to the links on the branch of the tree attached to B.

   c. Use the results of parts a and b to show that explicit expressions for θ may be found for each link in an *open* network (no closed loops).

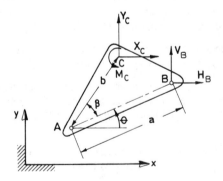

Figure 11.31-2 (Prob. 19).

20. Each of the three uniform bars shown in Fig. 3 have weight P and length a, and the slider at D is weightless. Find the horizontal forces F and R that will keep the bars in the configuration shown. Bar BC is parallel to the vertical y axis.

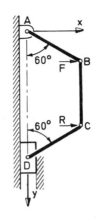

Figure 11.31-3 (Prob. 20).

11.40 WORKING CONSTRAINTS (CONSERVATIVE)

Up to now, we have assumed that the reaction forces due to constraints neither produce nor consume work. This restriction has prohibited us from treating systems with nonrigid links (such as springs or elastic bodies) or systems with energy-dissipating devices (such as viscous or Coulomb dampers). The effects of such *working constraints* are readily included if we consider that the net effect of such constraints is to produce certain pairs of internal forces which we may treat exactly as we would any other active forces. An example illustrating this treatment of spring forces is given in Sec. **11.41**. In Sec. **11.42** it is shown that spring forces belong to a wider class of so-called *conservative* force systems for which the idea of a *potential function* proves very fruitful. Frictional forces, and other velocity-dependent forces, are of greater interest in dynamics than in statics; hence we shall defer our discussion of them to later sections.

11.41 Spring Forces

Springs may always be removed from a system (conceptually) provided that the equal, opposite, and collinear forces which they exert upon their points of attachment are taken into consideration. This procedure will be illustrated by the following example.

EXAMPLE 1. A System with 2 DOF and Internal Spring Forces
Figure 1(a) shows two uniform rods, AB and CD, suspended from fixed pivots at A and D which are at the same height a distance L apart. A spring of length s connects points B and C. The length and weight of the two rods are a_1, P_1 and a_2, P_2, respectively. The system has 2 DOF which may be represented by the coordinates θ_1 and θ_2.

The spring BC may have a nonlinear relationship between tensile force S and extension $s - s_f$, as shown in Fig. 1(c), where s_f denotes the free (unstrained) length of the spring. For a linear spring with stiffness k (force per unit extension) the spring force is given by $S = k(s - s_f)$.

We seek to find the equilibrium configuration of the system. Toward this end we

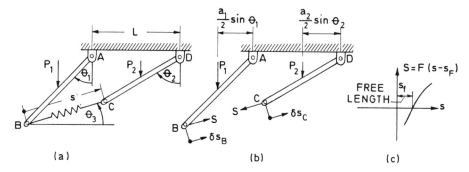

Figure 11.41-1. Spring coupled pendula: (a) Arbitrary configuration; (b) Spring forces S shown active; (c) Force vs. extension curve.

"remove" the spring and replace it by the two equal but opposite spring forces S acting as shown at points B and C in Fig. 1(b). Note that the forces have been given the directions corresponding to tension in the spring, so that a negative value of S will indicate a compressive force.

We now imagine that virtual displacements $\delta\theta_1$ and $\delta\theta_2$ have occurred and calculate the corresponding virtual work for the system shown in Fig. 1(b).

The moment of gravity force P_1 about point A is $\frac{1}{2}P_1a_1 \sin \theta_1$, positive counterclockwise. Hence the work of this moment on the clockwise angular increment $\delta\theta$ is given by

$$\delta W_1 = -\tfrac{1}{2}P_1a_1 \sin \theta_1 \, \delta\theta_1$$

Similarly, the work of P_2 is

$$\delta W_2 = -\tfrac{1}{2}P_2a_2 \sin \theta_2 \, \delta\theta_2$$

If we denote by δs_B the displacement component of point B in the direction BC, the work of the force S acting at point B is clearly

$$\delta W_B = S \, \delta s_B$$

Similarly, the work of the force S acting at point C is

$$\delta W_C = -S \, \delta s_C$$

where the negative sign is used because the direction of positive S at point C is opposite to that assumed for positive values of δs_C.

The net work due to the spring force is

$$\delta W_s \equiv \delta W_B + \delta W_C = S(\delta s_B - \delta s_C) = -S\delta s$$

where the net extension of the spring δs is defined by

$$\delta s \equiv \delta s_C - \delta s_B$$

The virtual work of all the active forces is therefore given by

$$\delta W = \delta W_1 + \delta W_2 + \delta W_s$$
$$= -\tfrac{1}{2}P_1a_1 \sin \theta_1 \, \delta\theta_1 - \tfrac{1}{2}P_2a_2 \sin \theta_2 \, \delta\theta_2 - S \, \delta s \qquad (1)$$

To relate δs to the generalized coordinates, we form the two displacement equations for loop $ABCD$:

$$-a_1 \sin \theta_1 + s \cos \theta_3 + a_2 \sin \theta_2 - L = 0 \tag{2}$$

$$a_1 \cos \theta_1 - s \sin \theta_3 - a_2 \cos \theta_2 = 0 \tag{3}$$

After taking differentials *across* Eqs. (2) and (3), we find

$$\begin{bmatrix} -C_3 & sS_3 \\ S_3 & sC_3 \end{bmatrix} \begin{Bmatrix} \delta s \\ \delta\theta_3 \end{Bmatrix} = \begin{Bmatrix} -a_1 C_1 \, \delta\theta_1 + a_2 C_2 \, \delta\theta_2 \\ -a_1 S_1 \, \delta\theta_1 + a_2 S_2 \, \delta\theta_2 \end{Bmatrix} \tag{4}$$

where $C_i = \cos \theta_i$, $S_i = \sin \theta_i$.

Equations (4) are readily solved for δs in the form

$$\delta s = \frac{(-a_1 C_1 \, \delta\theta_1 + a_2 C_2 \, \delta\theta_2)sC_3 - (-a_1 S_1 \, \delta\theta_1 + a_2 S_2 \, \delta\theta_2)sS_3}{-sC_3^2 - sS_3^2} \tag{5}$$

After regrouping terms and utilizing some trigonometric identities one may rewrite Eq. (5) in the simplified form

$$\delta s = k_{s1} \, \delta\theta_1 + k_{s2} \, \delta\theta_2 \tag{6}$$

where the coefficients are defined by

$$
\begin{aligned}
k_{s1} &\equiv a_1 \cos (\theta_1 + \theta_3) \\
k_{s2} &\equiv -a_2 \cos (\theta_2 + \theta_3)
\end{aligned}
\tag{7}
$$

These results could have been derived more directly, but it is our intent to illustrate a standard procedure which is valid in the most complicated geometrical situation.

Upon substitution of Eq. (6) into Eq. (1) we find that the virtual work is given by

$$\delta W = \underbrace{[-\tfrac{1}{2}P_1 a_1 \sin \theta_1 - Sk_{s1}]}_{\equiv \, Q_1} \delta\theta_1 + \underbrace{[-\tfrac{1}{2}P_2 a_2 \sin \theta_2 - Sk_{s2}]}_{\equiv \, Q_2} \delta\theta_2 \tag{8}$$

If we now make use of the given spring force characteristic

$$S = F(s - s_f) = k(s - s_f) \tag{9}$$

(where the latter form is to be used only for a linear spring), we may, in the usual fashion, set each of the generalized forces Q_1 and Q_2 equal to zero and obtain the equilibrium equations in the final form

$$Q_1 = -a_1 S \cos (\theta_1 + \theta_3) - \tfrac{1}{2}P_1 a_1 \sin \theta_1 = 0 \tag{10}$$

$$Q_2 = a_2 S \cos (\theta_2 + \theta_3) - \tfrac{1}{2}P_2 a_2 \sin \theta_2 = 0 \tag{11}$$

The two equilibrium equations (10) and (11) together with the two kinematic conditions (2) and (3) constitute a set of four equations for the four unknown quantities $\theta_1, \theta_2, \theta_3, s$. Once again we see that the governing equations are linear in the forces, but nonlinear in the displacements. If displacements are unknown, they will frequently have to be solved for by numerical or graphical methods. For numerical examples, see Problems **11.43-6**, 7 and 8.

11.42 Potential Energy

Spring forces belong to the wider class of so-called *conservative forces* for which the concept of *potential energy* proves to be very useful.

Let $\mathbf{F}_i = (X_i, Y_i)$ represent the set all active forces applied at points with *base* coordinates $\mathbf{r}_i = (x_i, y_i)$ as the system moves from an initial configuration $\mathbf{r}_i^A$ to a second configuration $\mathbf{r}_i^B$. If the *work* done by the forces $\mathbf{F}_i$ during this motion depends only on the coordinates $(x_i^A, y_i^A, x_i^B, y_i^B)$ of the end states, the *system* is said to be **conservative**.

Obviously the *forces* in a conservative system cannot depend arbitrarily[7] on time, velocities, or accelerations, because then the work would depend on whether the system moved from A to B slowly or quickly. The forces in a *conservative system* which *depend only on the displacements* are defined to be **conservative forces**.

Not all displacement-dependent forces are conservative. As an example, suppose that a particle in the x–y plane is subjected to forces $X = ky$, $Y = 0$ in moving from the origin to the point $P(x = 1, y = 1)$. If the particle follows the path OSP in Fig. 1,

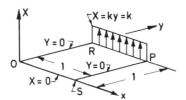

Figure 11.42-1. Nonconservative force system.

the work done is *zero*. But if the particle follows the path ORP, the work done is

$$\int_0^{y_R} Y\, dy + \int_{x_R}^{x_P} X\, dx = \int_0^1 0 \cdot dy + \int_0^1 k(1)\, dx = k \neq 0$$

Thus we see that, although forces in a conservative system must depend only on the configuration, not all configuration-dependent forces give rise to a conservative system (see Problem **11.43**-15).

In a conservative system, the work of the forces during the passage from a datum configuration $\mathbf{r}_i^A$ to any other state $\mathbf{r}_i$ may be represented by a **work function**,

$$W(\mathbf{r}_i, \mathbf{r}_i^A) = W(x_i, y_i, x_i^A, y_i^A) \tag{1}$$

Since the base coordinates may be expressed in terms of generalized coordinates by transformation expressions of the form

$$\begin{aligned} x_i &= x_i(q_1, \ldots, q_F, t) \\ y_i &= y_i(q_1, \ldots, q_F, t) \end{aligned} \tag{2}$$

---

[7]There exist certain systems of velocity-dependent forces which produce no work. They have been called **gyroscopic** forces (Ziegler [1968, p. 29]), and they play an important role in questions of dynamic stability. An example of such gyroscopic forces is the whirl-producing force described in Example 2 of Sec. **2.20**; also see Problems **11.43**-18 and 19.

it follows that Eq. (1) can be expressed in the form

$$W(\mathbf{r}_i, \mathbf{r}_1) = W(q_1, q_2, \ldots, q_F; q_1^A \cdots q_F^A, t) \tag{3}$$

Thus we see that although the work function is, by definition, an explicit function of the *base coordinates* (x_i, y_i), it can also be a function of the time and generalized coordinates when the transformation equations involve the time explicitly (i.e., are rheonomic).

Usually we shall deal with a scleronomic system where t does not appear explicitly in Eqs. (2) or (3). In any case, the virtual work corresponding to a set of virtual displacement δq_i is defined by

$$\delta W = \frac{\partial W}{\partial q_1} \delta q_1 + \cdots + \frac{\partial W}{\partial q_n} \delta q_n \tag{4}$$

However, by definition, the change in work is also given by

$$\delta W = Q_1 \, \delta q_1 + \cdots + Q_n \, \delta q_n \tag{5}$$

Subtracting Eq. (4) from Eq. (5), we see that

$$\left(Q_1 - \frac{\partial W}{\partial q_1}\right)\delta q_1 + \cdots + \left(Q_n - \frac{\partial W}{\partial q_n}\right)\delta q_n = 0 \tag{6}$$

Since the δq_i are arbitrary, it follows that

$$Q_i = \frac{\partial W}{\partial q_i} \qquad (i = 1, 2, \ldots, n) \tag{7}$$

In other terms, *the forces in a conservative system are given by the derivatives of a work function*.

It might appear from Eq. (3) that the generalized force Q_i depends on the reference configuration A from which the work is measured. However, it is readily seen that the work function $W(q_i, q_i^B)$ defined with respect to a different datum configuration q_i^B differs from $W(q_i, q_i^A)$ by an additive constant; i.e.,

$$W(q_i, q_i^A) = W_{AB} + W(q_i, q_i^B)$$

where W_{AB} (the work done in going from q_i^A to q_i^B) is a function of the datum points A and B but is independent of q_i. Therefore

$$\frac{\partial W(q_i, q_i^A)}{\partial q_i} = \frac{\partial W(q_i, q_i^B)}{\partial q_i}$$

and the generalized force is given by Eq. (7) for any datum configuration.

For historical reasons, the *negative* of the work function is called the **potential** or **potential energy** of the system, and it is usually denoted by V; i.e.,

$$V \equiv -W \tag{8}$$

With this notation, the principle of virtual work assumes the form

$$\delta V = \frac{\partial V}{\partial q_1} \delta q_1 + \cdots + \frac{\partial V}{\partial q_n} \delta q_n = -\delta W = 0 \qquad (9)$$

Equation (9) may be interpreted to read "*A necessary and sufficient condition for (generalized) equilibrium of a conservative system is that the potential energy be stationary.*"[8] Therefore a single-freedom conservative system will be in equilibrium if

$$\frac{dV}{dq} = 0 \qquad (10)$$

Later we will see that such a system will be in a *stable* equilibrium position where V is a relative minimum; i.e., if[9]

$$\frac{d^2 V}{dq^2} > 0 \qquad (11)$$

For systems with several degrees of freedom, generalized equilibrium occurs where

$$\frac{\partial V(q_1, \ldots, q_F)}{\partial q_i} = 0 \qquad (i = 1, 2, \ldots, F) \qquad (12)$$

The equilibrium will be stable if

$$\sum_{i, j = 1}^{F} \frac{\partial^2 V}{\partial q_i \, \partial q_j} \delta q_i \, \delta q_j > 0 \qquad (13)$$

for all possible choices of δq_i and δq_j.

From Eqs. (7)–(9) we see that the generalized forces satisfy

$$Q_i = -\frac{\partial V(q_1, \ldots, q_n)}{\partial q_i} \qquad (14)$$

In other terms, we may say that the "force" **Q** is the *negative gradient* of the potential V.

Gravity Force

A familiar example of a conservative system is provided by a field of parallel force vectors. For example, the force of gravity on a particle may be assumed to be constant and to act in a fixed direction (the local vertical) in the neighborhood of a point on the surface of the earth. Let y_i represent the vertical elevation (measured from an arbitrary horizontal datum plane) of a particle of weight P_i. The work done by gravity on a finite upward displacement y_i is therefore

$$W_i = -P_i y_i \equiv -V_i$$

---

[8] Recall that a function of several variables is said to be *stationary* when all of its first-order partial derivatives vanish.

[9] Stability of equilibrium is discussed in Sec. **12.52**. For those cases where $d^2 V/dq^2 = 0$, Inequality (11) should be supplanted by $d^n V/dq^n > 0$, where all derivatives up to order n vanish.

where V_i, the negative of the work, is the potential of the particle. For a collection of particles, the total potential is

$$V = \sum_i V_i = \sum_i P_i y_i \qquad (15)$$

Recall that the elevation of the *center of gravity* is defined by

$$\bar{y} = \frac{1}{P} \sum P_i y_i \qquad (16)$$

where $P = \sum P_i$ is the weight of all particles. Accordingly, the potential energy for the system is equal to

$$V = P\bar{y} \qquad (17)$$

Since V must be a minimum in a position of stable equilibrium, we may conclude that *in a uniform gravity field a system of particles will be in stable equilibrium when its center of gravity is at the lowest elevation consistent with whatever kinematic constraints are imposed.*

This important result can be used to prove some interesting theorems in pure geometry, as in the following example.[10]

EXAMPLE 1

Show that the tangent to an ellipse makes equal angles with the focal radii at the point of tangency.

Proof:

If A and B are foci, an arbitrary point P of the ellipse (shown in Fig. 2) satisfies the condition $AP + PB = L$, where L is a constant. The distances AP and BP are called focal radii. Let a weightless string of length L be threaded through a bead P and have its ends supported at two fixed points A and B such that AB makes an arbitrary angle with the horizontal direction as shown. According to the principle of minimum potential energy, point P (the center of gravity of the single-particle system) will descend to its lowest possible position, pulling the string into a taut triangle APB lying in the vertical plane, as shown.

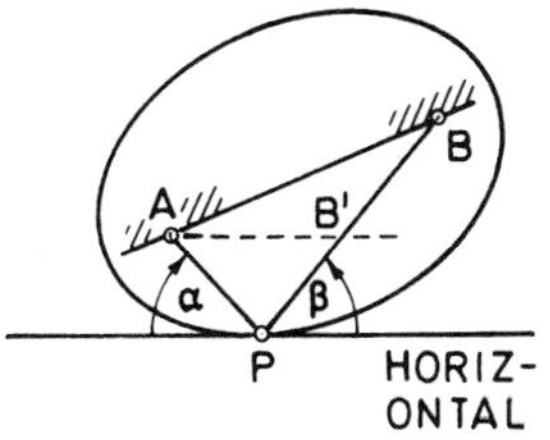

Figure 11.42-2. Ellipse.

If the bead is moved in the vertical plane APB so as to keep the string taut, it will trace out an ellipse (with A and B as foci) whose tangent is horizontal at the lowest point P (because P is at a relative minimum). If the string is supported at a point B' on PB which is at the same elevation as A, symmetry requires that APB' is an isosceles triangle with equal angles at A and B'. Therefore $\alpha = \angle PAB' = \angle PB'A = \beta$, and the theorem is proved.

---

[10]For other examples, see Uspenskii [1961].

Spring Forces

When the system contains elastic members, the work U stored in the "springs" due to the motion of the system is called the **strain energy** of the elastic bodies. The strain energy U of a one-dimensional spring may be readily found from the force-extension curve of the spring as shown in Fig. 3.

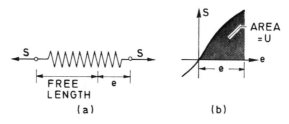

Figure 11.42-3. (a) Stretched spring; (b) Force-extension curve.

From Fig. 3(b) we see that the work done *on* the spring is

$$U = \int_0^e S\, de \tag{18}$$

where e is the extension of the spring measured from the unstrained state and S is the tensile force in the spring.

In passing, it is worth observing that Eq. (18) is equivalent to the statement that the spring force may be derived from the strain energy in the form

$$S = \frac{dU}{de} \tag{19}$$

For a **linear spring** of **stiffness** k, $S = ke$, and

$$U = \int_0^e S\, de = \int_0^e ke\, de = \tfrac{1}{2}ke^2 \tag{20}$$

Note that the work done *by* the spring on its terminal points is the negative of that done *on* the spring and can be expressed as

$$W_s \equiv -U \tag{21}$$

Hence the potential energy contributed to the system "surrounding" the spring is

$$V_s \equiv -W_s = U \tag{22}$$

In other words, the potential energy added to the surrounding system due to the presence of the spring is equal to the strain energy in the spring. The above statement is true if *elastic body* is substituted for *spring*; the term **elastic body** means a body whose deformations are unique functions of the applied stresses. To illustrate the present point of view, let us reconsider Example 1 of Sec. **11.41**.

EXAMPLE 2

We seek the equilibrium configuration of the system shown in Fig. 4. The work of P_1 and P_2 (measured from $\theta_i = 90°$) is

$$W_1 + W_2 = P_1 \frac{a_1}{2} \cos \theta_1 + P_2 \frac{a_2}{2} \cos \theta_2$$

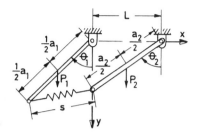

Figure 11.42-4. Example 2.

The potential of the spring is U, and the total potential is $V = -W_1 - W_2 + U$. The equation of equilibrium is $\delta V = -\delta W_1 - \delta W_2 + \delta U = 0$. Therefore,

$$P_1 \frac{a_1}{2} \sin \theta_1 \, \delta\theta_1 + P_2 \frac{a_2}{2} \sin \theta_2 \, \delta\theta_2 + \frac{\partial U}{\partial s} \delta s = 0 \qquad (23)$$

Upon utilizing Eq. (19), we see that the spring force S is given by $\partial U/\partial s$; hence Eq. (23) is equivalent to Eq. (**11.41**-1). The remainder of the solution is the same as in Sec. **11.41**.

EXAMPLE 3. Illustrating Elastic and Gravitational Potential

A uniform platform AB of length b and weight P is pivoted at A and supported at B by a linear elastic spring BD of stiffness k which is tied to a fixed point D situated at a distance $\frac{3}{2}b$ vertically above A (Fig. 5). The angle between AB and the horizontal x

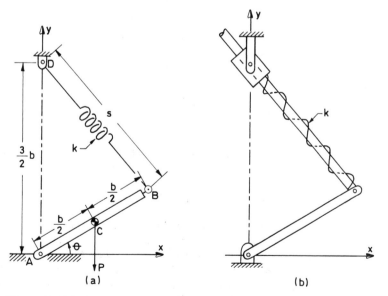

(a)   (b)

Figure 11.42-5. Spring suspended platform: (a) Given form; (b) Equivalent mechanism.

axis is denoted by θ, and the length of the spring DB is denoted by s. When the spring is unstrained, its length is s_o, and the corresponding value of θ is θ_o. We seek to find the potential energy V of the system as a function of angle θ and to locate the equilibrium positions of the system.

The given problem is kinematically equivalent to the slider-crank inversion (with added coil spring) shown in Fig. 5(b). Thus we recognize that the system has a single degree of freedom, and we choose θ as the generalized coordinate. The only active forces applied to the ideal mechanism of Fig. 5(b) are the gravity force and the spring force, both of which are conservative.

In calculating the potential energy of the system, let us use the unstrained position of the spring as a datum; i.e., $V(\theta)$ is measured from $\theta = \theta_o$.

Therefore the potential energy V_g of the gravity force is the negative of the work done by P in moving through the vertical distance $(b/2)(\sin\theta - \sin\theta_o)$; i.e.,

$$V_g = \frac{Pb}{2}(\sin\theta - \sin\theta_o)$$

The extension of the spring, measured from its datum position, is $e = s - s_o$; hence Eq. (20) shows that the spring's potential V_s is

$$V_s = \tfrac{1}{2}ke^2 = \tfrac{1}{2}k(s - s_o)^2$$

To express V_s in terms of θ, we note from the law of cosines applied to triangle ABD that

$$s^2 = \left(\frac{3b}{2}\right)^2 + b^2 - 2\left(\frac{3b}{2}\right)b\cos(90° - \theta)$$

or

$$s = \frac{b}{2}(13 - 12\sin\theta)^{1/2}$$

This equation gives the free length of the spring as

$$s_o = \frac{b}{2}(13 - 12\sin\theta_o)^{1/2} \tag{24}$$

The total potential energy is given by

$$V = V_s + V_g \tag{25}$$

Now let us assign the following numerical data: $b = 6$ ft, $P = 50$ lb, $k = 10$ lb/ft, $\theta_o = 30°$. Therefore Eq. (24) gives $s_o = \sqrt{63}$, and Eq. (25) gives

$$V = 5[(117 - 108\sin\theta)^{1/2} - \sqrt{63}]^2 + 150\sin\theta - 75 \tag{26}$$

This expression for $V(\theta)$ is plotted in Fig. 6, which shows four places (marked M_1, M_2, M_3, M_4) where $dV/d\theta$ vanishes and hence equilibrium exists. A configuration corresponding to the *minimum* of V occurs at M_2, where $\theta = -2°$; this is a *stable* equilibrium position. Point M_1 at $\theta = -90°$ corresponds to the platform hanging vertically downward like a pendulum at rest. However, V is a *maximum* at M_1, and the pendulum will "snap over" to the right or left if it is disturbed ever so slightly from the downward vertical position. Similarly, point M_3 at $\theta = 90°$ corresponds to an unstable equilibrium

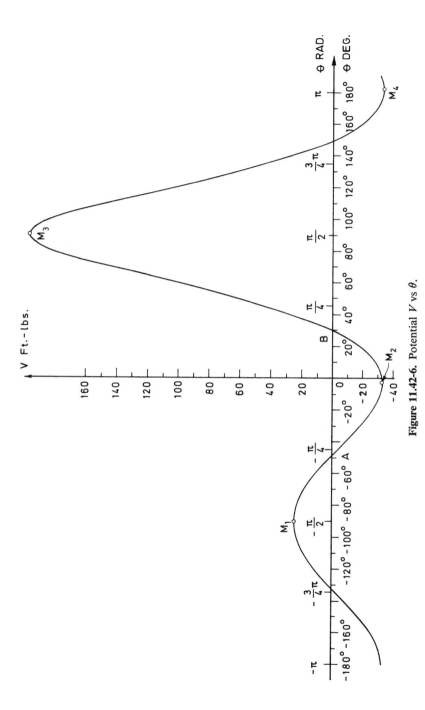

Figure 11.42-6. Potential V vs θ.

configuration with the platform balanced vertically upward (an inverted pendulum). Point M_4 at $\theta = 182°$ is a stable equilibrium position which is the mirror image, about the vertical, of the configuration M_2.

In this problem, we could have found explicit expressions for all of the equilibrium positions (see Problem **11.43**-8). In general, however, the solutions of the equilibrium equation $dV/dq = 0$ will have to be found by graphing $V(q)$ or by some equivalent numerical method.

EXAMPLE 4. Potential of Attracting Masses

According to Newton's universal law of gravitation, two particles of mass m and M, respectively, which are separated by a distance r experience a mutual attractive force F given by

$$F = \frac{\mu m M}{r^2} \tag{27}$$

where μ is a constant. The work of the *attraction F* on an incremental relative *separation dr* is

$$dW = -F\,dr = -\frac{\mu m M\,dr}{r^2}$$

The work of F in going from an initial separation r_0 to a final value r is therefore

$$W = -\mu m M \int_{r_0}^{r} \frac{dr}{r^2} = \mu m M\left(\frac{1}{r} - \frac{1}{r_0}\right)$$

It is convenient to take the reference separation r_0 as ∞, so that the potential energy is given by

$$V = -W = -\frac{\mu m M}{r} \tag{28}$$

If M represents the mass of the earth, the potential for a particle of mass m in the earth's gravitational field is therefore

$$V = -\frac{W_e r_e^2}{r} \tag{29}$$

where r_e is the radius of the earth, and

$$W_e \equiv \frac{\mu m M}{r_e^2} \tag{30}$$

is the weight of mass m on the earth's surface.

From Eq. (28) it may be shown that an elongated body in orbit about the earth experiences a *gravity-gradient torque* which tends to align the long axis of the body with the local vertical direction, i.e., along the line joining the mass center of the body to that of the earth. It is this earth-pointing tendency which accounts for the fact that the moon (which is slightly elongated due to the pull of the earth) always presents the same face to the earth. The gravitational torque can also be used to design a communication satellite with accurately aimed, earth-pointing antennas. A detailed description of such a design is given by Paul et al. [1963].

To calculate the gravity-induced torque, let us model the elongated satellite by a *dumbbell*, consisting of two tip masses m_1 and m_2 separated by a long rod AB of length s, as shown in Fig. 7. The distance of mass m_i from E (the center of the earth) is r_i, and the distance between C (the mass center of the satellite) and E is denoted by r. The gravitational potential of each mass is given by Eq. (28), and the total potential is therefore

$$V = V_1 + V_2 = -\mu M \left(\frac{m_1}{r_1} + \frac{m_2}{r_2} \right) \tag{31}$$

where M represents the mass of the earth.

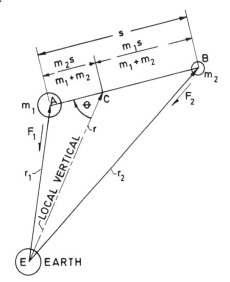

Figure 11.42-7. Model of a dumbbell satellite.

From the law of cosines applied to triangle EAC we find

$$r_1^2 = r^2 + \left(\frac{m_2 s}{m} \right)^2 - 2r \left(\frac{m_2 s}{m} \right) \cos \theta \tag{32}$$

where $m \equiv m_1 + m_2$ is the total satellite mass and θ is angle ECA. Similarly,

$$r_2^2 = r^2 + \left(\frac{m_1 s}{m} \right)^2 + 2r \left(\frac{m_1 s}{m} \right) \cos \theta \tag{33}$$

From Eq. (32) we may write

$$\frac{1}{r_1} = \frac{1}{r} \frac{r}{r_1} = \frac{1}{r} \left(\frac{r_1}{r} \right)^{-1} = \frac{1}{r} \left[1 + \left(\frac{m_2}{m} \frac{s}{r} \right)^2 - \frac{2m_2}{m} \frac{s}{r} \cos \theta \right]^{-1/2} \tag{34}$$

with a similar expression for $1/r_2$.

The dumbbell length s is typically on the order of 100 ft, whereas the geocentric altitude r is typically about 10,000 miles for a medium-altitude communication satellite. Therefore $s/r \ll 1$, and we may utilize the binomial theorem to expand the expressions for $1/r_1$ and $1/r_2$ in a power series in s/r; then the resulting expression may be substituted into Eq. (31) to find the potential V. The detailed algebra is left as an exercise (or see Paul [1963]); the final result is

$$V = -\mu m \frac{M}{r} - \frac{\mu m \bar{m}}{r} \frac{1}{2} \left(\frac{s}{r}\right)^2 \left(1 + \frac{3}{2}\cos 2\theta\right) + \cdots \qquad (35)$$

where

$$\bar{m} \equiv \frac{m_1 m_2}{m_1 + m_2} \qquad (36)$$

and terms of order higher than $(s/r)^2$ have been neglected. The potential is plotted schematically in Fig. 8.

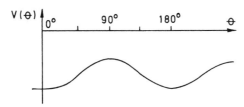

Figure 11.42-8. Potential for dumbbell satellite.

Note that the first term of the potential is precisely that of a point mass equal to the total mass of the satellite. The second term, however, involves the angular misalignment θ, which is absent for the case of a point mass (or *spherical* body) satellite.

If we assume that the mass center C moves with uniform speed in a circular orbit of radius r, we may treat θ as a generalized coordinate. The corresponding generalized force Q_θ is the *torque* produced by the gravity forces F_1 and F_2 about the mass center C. The fact that gravity can produce a torque may seem surprising at first, but it results from the *very small* differences in the gravity forces F_1 and F_2 due to the fact that m_1 and m_2 are at different distances from the center of the earth. The value of this so-called *gravity-gradient* torque is given by

$$Q_\theta = -\frac{\partial V}{\partial \theta} = -\frac{3}{2}\frac{\mu m \bar{m} s^2}{r^3} \sin 2\theta \qquad (37)$$

For a typical medium-altitude communication satellite, the gravity-gradient torques are typically on the order of only 0.01 ft-lb (for $\theta = 15°$). Although exceedingly small, by our usual earthbound standards, such tiny torques are adequate, in the "weightless" environment of outer space, to provide a restoring action if the satellite axis AB drifts away from the local vertical direction by an angle θ which is less than 90° in magnitude.

Figure 8 indicates that Q_θ vanishes for $\theta = 0°$ and for $\theta = 90°$. The former position is the desired *earth-pointing* equilibrium configuration. Fortunately, the equilibrium position with $\theta = 0°$ occurs at a relative minimum of the potential energy, whereas that at $\theta = 90°$ corresponds to a maximum of $V(\theta)$. Thus the earth-pointing configuration is stable, and the unwanted horizontal equilibrium configuration is unstable. See problem **11.43**-20 for further discussion.

11.43 Problems

1. In Fig. 1, the spring with stiffness k lb/in. is unstrained when $\theta = 90°$. Find the angle θ when a weight P acts as shown. Neglect other weights in the system. Express your answer in terms of P, k, and a. Evaluate θ for $P = 1000$ lb, $k = 10$ lb/in., and $a = 50$ in.

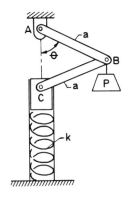

Figure 11.43-1. (Prob. 1).

2. In Fig. 2, the inclined door of weight P is held open by a spring-restrained cable. The stiffness of the spring is k lb/in., and the spring force is S_0 when $\theta = 90°$. Compute θ for $k = 5$ lb/ft, $P = 100$ lb, $S_0 = 5$ lb, and $L = 4$ ft. Assume that the pulley diameter is negligible.

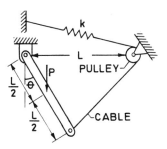

Figure 11.43-2. (Prob. 2).

3. In Fig. 3, the weight P hangs from the top B of an *A-frame* with equal legs of length L. A linear spring of stiffness k lb/in. connects the center of the roller at A to point C of the leg BC. Find the angle θ if the tension in the spring is 20 lb when $\theta = 30°$, $P = 150$ lb, $L = 6$ ft, and $k = 10$ lb/in.

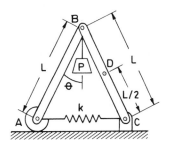

Figure 11.43-3. (Prob. 3).

4. If the spring in Problem 3 is unhooked at C and reattached to the midpoint D of leg BC, find the angle θ of the new equilibrium position.

5. Find the ratio P_1/P_2 in Fig. **11.41**-1 if $a_1 = a_2 = L = 12$ in., $\theta_1 = 45°$, and $\theta_2 = 30°$.

6. Find the configuration variables $(\theta_1, \theta_2, \theta_3, s)$ in Fig. **11.41**-1 if the spring force is given by $S = k(s - s_f)$, and $k = 1.5$ lb/in., $s_f = 6$ in., $a_1 = a_2 = L = 12$ in., $P_1 = 10$ lb, and $P_2 = 15$ lb. [Hint: A numerical procedure such as Newton-Raphson's will be needed.]

7. In Fig. 4, the spring is linear with free length a and stiffness k. The applied force is $P = ka/2$, acting normal to BC.
    a. Find the four equations needed to calculate the configuration variables $(\theta_1, \theta_2, \theta_3, s)$.
    b. Use a numerical procedure (e.g., Newton-Raphson) to calculate the configuration variables.

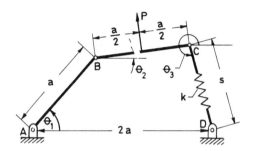

Figure 11.43-4. (Prob. 7).

8. a. For the example illustrated by Fig. **11.42**-5, show that $dV/d\theta = \frac{1}{2}b \cos \theta\{P - 3bk[(s - s_0)/s]\}$ and hence that the three equilibrium positions are

$$\theta_1 = 90°, \qquad \theta_2 = -90°, \qquad \theta_3 = \arcsin\left[\frac{13 - (2s_3/b)^2}{12}\right]$$

where $s_3 = -s_0/[(P/3bk) - 1]$ and $s_0 = (b/2)(13 - 12 \sin \theta_0)^{1/2}$.
    b. If $b = 6$ ft, $P = 50$ lb, $k = 10$ lb/ft, and $\theta_o = 30°$, verify that $\theta_3 = -2.006°$.
    c. Find an expression for $d^2V/d\theta^2$, and verify that the equilibrium is stable at θ_3 but unstable at θ_1 and θ_2 for the numerical values given.

9. Find the *maximum* value of the spring constant k for the system of Fig. **11.42**-5 such that the equilibrium position $\theta = -90°$ is stable.

10. If a conservative single degree of freedom system has stable equilibrium configurations at two positions denoted by q_1 and q_2, show that an unstable equilibrium position must occur at a third position q_3, where $q_1 < q_3 < q_2$.

11. A spherical communication satellite weighs P_e lb on the surface of the earth. Show that its potential energy due to gravity at a height h above the surface of the earth (which may be considered a uniform sphere of radius r_e) is $V = -P_e r_e/(1 + h/r_e)$.

12. Verify that the potential for the dumbbell satellite shown in Fig. **11.42**-7 is given by Eq. (**11.42**-35).

13. A particle moves in the x, y plane subject to forces $X = ky$ and $Y = cx$, where k and c are constants. If the particle moves clockwise along a simple closed curve which encloses area A, show that the work done in one complete circuit is given by $W = (k - c)A$. Is the system conservative?

14. Show that the necessary and sufficient condition for a planar force field with components $X(x, y)$, $Y(x, y)$ to be conservative is that $\partial X/\partial y = \partial Y/\partial x$. Use this result to show that the force field given by $X = a_{11}x + a_{12}y$, $Y = a_{21}x + a_{22}y$ is conservative if $a_{12} = a_{21}$, where all a_{ij} are constants.

15. Consider a system where the generalized forces Q_i are linear functions of the generalized coordinates q_i; i.e.,

$$Q_i = \sum_{j=1}^{F} a_{ij}q_j \qquad (i = 1, 2, 3)$$

where a_{ij} are all constants. Show that the system is conservative if, and only if, the matrix $[a_{ij}]$ is symmetric; i.e., $a_{ij} = a_{ji}$. And show that the potential energy measured from an arbitrary state is given by $V = -\frac{1}{2}\sum_{i,j=1}^{F} a_{ij}q_iq_j$.

16. A force $\mathbf{F}$ is applied to a rigid body in such a way that it always points along a line fixed in the moving body. Show that the force is not conservative.[11]

17. A force $\mathbf{F}$ always acts along the axis CB of the double pendulum ABC (Fig. 5). Show that the force $\mathbf{F}$ is not conservative.[11]

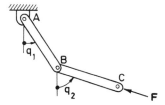

Figure 11.43-5. (Prob. 17).

18. In Example 2 of Sec. **2.20**, it was shown that the force $(X, Y) \equiv mk(-\dot{y}, \dot{x})$ produces a whirling motion (circular orbit). Show that the force system is workless; i.e., it belongs to the class of *gyroscopic conservative systems*, as defined in Problem 19.

19. Suppose that the generalized forces Q_i are linear functions of the generalized velocities; i.e.,

$$Q_i = \sum_{j=1}^{F} b_{ij}\dot{q}_j$$

where the b_{ij} are all constants. Show that a necessary and sufficient condition for the forces Q_i to be **gyroscopic** (i.e., to produce zero net work in any motion) is that the matrix $[b_{ij}]$ be skew symmetric; i.e., $b_{ij} = -b_{ji}$ for all i, j.

20. What is the significance of the stable equilibrium position at $\theta = 180°$ in Fig. **11.42**-8, and what implications does this have for launching a communication satellite which carries a narrow beam transmitting antenna attached to the mass m_1?

11.50 CALCULATING REACTIONS

In Sec. **11.10**, we noted that a static analysis by vector methods entails the formulation of three equilibrium equations for each rigid body in a mechanism, thereby providing a rather substantial number of equations to be solved for the many unknown

---

[11]This is an example of a **circulatory force system**, which is of interest in problems of structural and dynamic stability (Ziegler [1968, pp. 29–33]).

joint reactions. We then saw that the method of virtual work provides us with a much smaller set of equilibrium equations (one per DOF) in which the joint reactions do not occur. Therefore, the latter approach is far more efficient if our goal is to determine the configuration of the system under the action of known applied forces or to find a set of unknown applied forces. However, if we also seek the unknown joint reactions, the method of virtual work must be supplemented[12] by consideration of each member as a free body in equilibrium.

In Sec. **11.52** we shall see that this task is straightforward after the method of virtual work has been used to find the geometric configuration or to satisfy the overall *system* equations of equilibrium. Until further notice, we shall consider only ideal (frictionless) systems. The influence of friction will be discussed in Sec. **11.60**.

11.51 *Notational Conventions*

Some convenient notational practices will now be described with the aid of Fig. 1, which shows a typical free-body diagram for a member denoted as link i. Joints A and B are hinges which exert reactions (X_A, Y_A) and (X_B, Y_B) on member i. The direc-

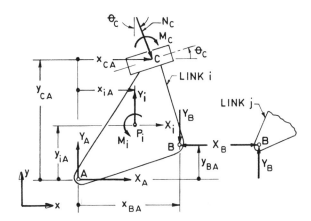

Figure 11.51-1 Free-body diagram for typical link, i.

tions assumed for these reactions are completely arbitrary. However, once they are assigned, we must show equal but opposite reactions on contacting members, as shown at B in Fig. 1 on the neighboring link j. When three or more members are hinged at a single point or when a hinge is loaded by external forces (H, V), as shown in Fig. 2(a), the simplest procedure is to use separate symbols for the reactions on each link, as shown in Fig. 2(b). For equilibrium of the pin which connects the three elements at B, the reactions must be related by the two equations

---

[12]The principle of virtual work may be used to find any individual joint reaction if we imagine that the joint is "broken" and the constraint is maintained by unknown active forces applied to a mechanism with two additional DOF (due to the opening of the joint). Closely related to this approach is Lagrange's *method of undetermined multipliers*. Both of these techniques are described by Paul [1975] but will not be discussed further in this book.

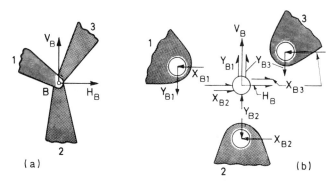

Figure 11.51-2 (a) Double-hinge pair at joint B; (b) Free-body diagram of hinge pin at B.

$$X_{B1} + X_{B2} + X_{B3} + H_B = 0$$
$$Y_{B1} + Y_{B2} + Y_{B3} + V_B = 0 \tag{1}$$

At a sliding pair element, such as C in Fig. 1, the reactions consist of a moment M_C plus a force N_C which is normal to the slider axis.

The active force on link i will be represented by the **force set** (X_i, Y_i, M_i) acting through a reference point P_i fixed in the link. The coordinates of point P_i relative to any other point (e.g., point A) are denoted in the following fashion:

$$x_{iA} \equiv x_i - x_A, \qquad y_{iA} \equiv y_i - y_A \tag{2a}$$

Similarly, the coordinates of point B relative to point A are denoted by

$$x_{BA} \equiv x_B - x_A, \qquad y_{BA} \equiv y_B - y_A \tag{2b}$$

It is important to recognize the signed nature of these quantities; e.g., $x_{BA} = -x_{AB}$.

We seek three independent equilibrium equations for each link which reflect the fact that the resultant force, and the resultant moment about any convenient point, must vanish. By using point A as the moment center for link i in Fig. 1, these equations may be expressed in the form

$$\sum X = X_A - X_B + N_C \sin \theta_C + X_i = 0 \tag{3}$$
$$\sum Y = Y_A - Y_B - N_C \cos \theta_C + Y_i = 0 \tag{4}$$
$$\sum M_A = X_B y_{BA} - Y_B x_{BA} - N_C \sin \theta_C \, y_{CA} - N_C \cos \theta_C \, x_{CA} + M_{iA} = 0 \tag{5}$$

In Eq. (5) we have introduced the notation M_{iA} for the moment of the active force set (X_i, Y_i, M_i) about point A; i.e.,

$$M_{iA} \equiv M_i + Y_i x_{iA} - X_i y_{iA} \tag{6}$$

Although Fig. 1 shows the general case, we should note the possibility for a simplification when a binary link (e.g., bar AB in Fig. 3) is not loaded by any active forces. For this special case, equilibrium requires that the end reactions on AB, denoted by F_{AB} in Fig. 3, must be equal, opposite, and directed along the line AB.

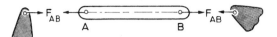

Figure 11.51-3. Link without active loading experiences purely axial reactions at ends.

By recognizing the presence of such *axially loaded* members we automatically satisfy the equilibrium equations for those members and reduce the number of unknown reactions to be solved for.

At the contact points of a (frictionless) *cam pair* reaction forces act along the common normal to the contacting surfaces, as illustrated by N_1 and N_2 in Fig. 4.

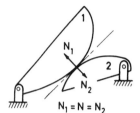

Figure 11.51-4. Reactions on cam surfaces.

For the case of a gear pair, the common normal to the contacting gear teeth is inclined to the common tangent of the pitch circles by the *pressure angle* ϕ, as shown in Fig. 5. For involute gears, ϕ is a constant—usually 20° in modern gears.[13] For frictionless contact, the tooth reactions N lie along the common normal, which accounts for its name, *line of action*. Therefore each gear in a spur gear set can be considered to be loaded by a contact force N inclined at the pressure angle ϕ, as shown in Fig. 5. Alternatively, for purposes of static analysis, we can replace involute gears by pulleys having diameters equal to the base circles of the gears, as suggested in Fig. 5.

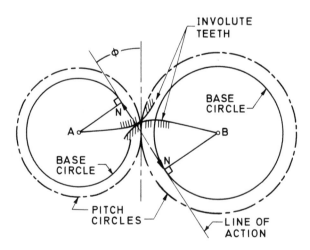

Figure 11.51-5. Line of action of contact force N inclined at pressure angle ϕ to common tangent of pitch circles. Equivalent pulleys formed by base circles.

---

[13]See Sec. **3.43**.

11.52 Reactions in Ideal Mechanisms

To find joint reactions, we may treat each member in a system as a free rigid body and write *three* independent **equations of member equilibrium** ($\sum X = 0$, $\sum Y = 0$, $\sum M = 0$) for each body. In a regular linkage with M members and P_s simple pairs, this procedure gives rise to $3M$ equations involving the $2P_s$ joint reactions. However, not all of the *member* equilibrium equations are independent, because the active forces which appear in them are related by the F *system equations of equilibrium*

$$Q_j = \sum_{i=1}^{M} (X_i U_{ij} + Y_i V_{ij} + M_i \Omega_{ij}) = 0 \qquad (j = 1, 2, \ldots, F) \qquad \textbf{(11.30-3a)}$$

where F is the DOF of the system.

Thus we have a total of $3M - F$ *independent* equilibrium equations involving $2P_s$ joint reactions. But according to Gruebler's criterion for regular linkages,

$$F = 3M - 2P_s \qquad \textbf{(8.21-5)}$$

Thus the *number of independent member equilibrium equations matches exactly the number of unknown joint reactions.*

If desired, we can altogether omit the use of the system equilibrium equations ($Q_1 = 0, \ldots, Q_F = 0$). Then the $3M$ member equilibrium equations suffice to solve for the $2P_s$ joint reactions plus F of the active force components (X_i, Y_i, M_i).

We shall see that all member equilibrium equations are linear in the joint reactions, so the mathematical problem boils down to the solution of sets of linear algebraic equations. Fortunately, these equations are usually so weakly coupled that we seldom have to solve more than two of them simultaneously, provided that we take them in a suitable order. The general procedure will be illustrated for specific mechanisms.

EXAMPLE 1. Slider-Crank Mechanism

A slider-crank mechanism with the most general possible loading is shown in Fig. 1, together with free-body diagrams of each link. We wish to find all of the eight reactions (X_B, Y_B, N_B, M_B, X_A, Y_A, X_O, Y_O) in terms of the applied loads (X_i, Y_i, M_i, $i = 1$, 2, 3).

From the free-body diagram for the slider [Fig. 1(b)] we see that

$$X_B = X_3 \qquad (1)$$

Equilibrium of moments about point A on the connecting rod requires [Fig. 1(c)] that

$$X_B y_{AB} + Y_B x_{BA} + M_{2A} = 0 \qquad (1a)$$

where

$$M_{2A} \equiv M_2 + Y_2 x_{2A} - X_2 y_{2A} \qquad (1b)$$

is the moment about A of the applied load set (X_2, Y_2, M_{2A}). Using Eq. (1), we can solve Eq. (1a) for Y_B in the form

$$Y_B = -\frac{M_{2A} + X_B y_{AB}}{x_{BA}} \qquad (2)$$

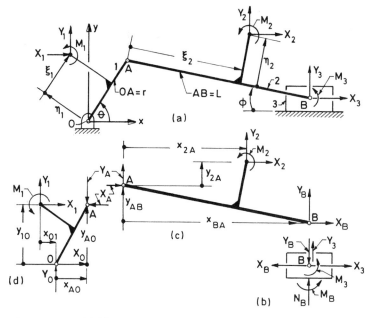

Figure 11.52-1. (a) Slider-crank mechanism; (b) Free body diagram (F.B.D.) of slider; (c) F.B.D. of connecting rod; (d) F.B.D. of crank.

The remaining equations of equilibrium for the slider give

$$N_B = Y_B - Y_3 \tag{3}$$

$$M_B = -M_3 \tag{4}$$

Force equilibrium for the connecting rod gives [Fig. 1(c)]

$$X_A = -(X_B + X_2) \tag{5}$$

$$Y_A = -(Y_B + Y_2) \tag{6}$$

and force equilibrium of the crank [Fig. 1(d)] requires that

$$X_O = X_A - X_1 \tag{7}$$

$$Y_O = Y_A - Y_1 \tag{8}$$

Thus we have *found all eight reactions* in terms of applied loads. Note that the order in which we used the equations made it unnecessary to solve for more than one unknown at a time. That is, the equations were fully decoupled.

It is worth observing that no use was made of the equation of moment equilibrium about point O on the crank. This equation gives the net moment about the crankshaft axis of the *active* loads on the crank in the form

$$M_{1O} \equiv M_1 - Y_1 x_{O1} - X_1 Y_{1O} + X_A y_{AO} - Y_A x_{AO} = 0 \tag{9}$$

Equation (9) may be used to find the crank couple M_1 which equilibrates the other applied loads. On the other hand, if the active loads are known to satisfy the *system* equation of equilibrium

$$Q = \sum_{i=1}^{3} (X_i u_i + Y_i v_i + M_i \omega_i) = 0 \tag{10}$$

then Eq. (9) is satisfied identically by the active loads.

For later reference purposes, we record here alternative forms for the various lengths needed:

$$(x_{BA}, y_{AB}) = L(\cos \phi, \sin \phi) \tag{11a}$$

$$(x_{AO}, y_{AO}) = r(\cos \theta, \sin \theta) \tag{11b}$$

$$x_{2A} = \xi_2 \cos \phi + \eta_2 \sin \phi \tag{11c}$$

$$y_{2A} = -\xi_2 \sin \phi + \eta_2 \cos \phi \tag{11d}$$

$$x_{O1} = -x_{1O} = -(\xi_1 \cos \theta - \eta_1 \sin \theta) \tag{11e}$$

$$y_{1O} = \xi_1 \sin \theta + \eta_1 \cos \theta \tag{11f}$$

EXAMPLE 2. Four-Bar Mechanism

Consider a four-bar mechanism subjected to a general set of loads, as shown in Fig. 2(a).

Equilibrium of moments about joint A on link 1 requires that

$$M_A = Y_B x_{BA} - X_B y_{BA} + M_{1A} = 0 \tag{12}$$

Similarly, equilibrium of moments about joint C on link 2 gives

$$M_C = Y_B x_{CB} - X_B y_{CB} + M_{2C} = 0 \tag{13}$$

These two equations involve two unknowns which are readily found, from Cramer's rule, to be

$$Y_B = \frac{M_{1A} y_{CB} - M_{2C} y_{BA}}{-x_{BA} y_{CB} + x_{CB} y_{BA}}$$

$$X_B = \frac{-x_{BA} M_{2C} + x_{CB} M_{1A}}{-x_{BA} y_{CB} + x_{CB} y_{BA}} \tag{14}$$

With (X_B, Y_B) known, we may easily obtain the remaining unknowns from the equations of force equilibrium for each of the three free bodies shown; i.e.,

$$X_A = X_B + X_1, \qquad Y_A = Y_B + Y_1 \tag{15}$$

$$X_C = X_B - X_2, \qquad Y_C = Y_B - Y_2 \tag{16}$$

$$X_D = X_C - X_3, \qquad Y_D = Y_C - Y_3 \tag{17}$$

Note that we have not explicitly used the equation of moment equilibrium (about point D) for link 3; i.e.,

$$M_{3D} \equiv M_3 + Y_3 x_{3D} - X_3 y_{3D} = -(X_C y_{CD} + Y_C x_{DC}) \tag{18}$$

If the given active forces satisfy the *system equation of equilibrium* (10), Eq. (18) will be satisfied identically. However, we may use Eq. (18) *in lieu of* Eq. (10) if equilibrium values of either M_3, X_3, or Y_3 are not given *a priori*.

Similarly, if one of the three quantities M_1, X_1, Y_1 is unknown, it may be found, together with all joint reactions, by taking moments about the centers D, B, A in

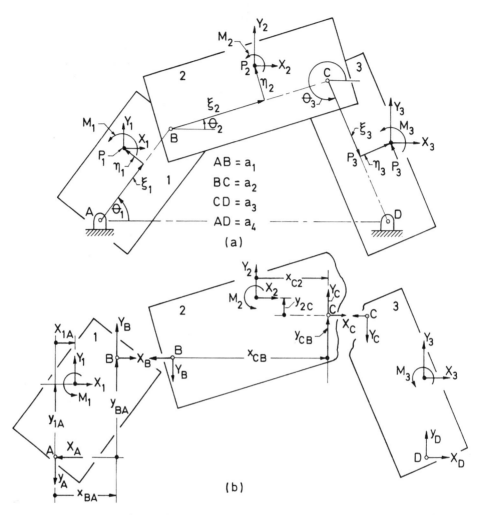

Figure 11.52-2. (a) Four bar mechanism under general loading; (b) Free-body diagrams of links.

sequence. Then (Problem **11.53**-12) the unknown forces can be calculated in the following order:

$$X_C = \frac{M_{2B}x_{DC} - M_{3D}x_{CB}}{y_{CB}x_{DC} + y_{CD}x_{CB}} \tag{19a}$$

$$Y_C = \frac{-y_{CB}M_{3D} - y_{CD}M_{2B}}{y_{CB}x_{DC} + y_{CD}x_{CB}} \tag{19b}$$

$$X_B = X_C + X_2, \qquad Y_B = Y_C + Y_2 \tag{20}$$

$$M_{1A} \equiv M_1 + Y_1x_{1A} - X_1y_{1A} = -(Y_Bx_{BA} - X_By_{BA}) \tag{21}$$

Equations (15) and (17) may then be used to find X_A, Y_A, X_D, Y_D.

For reference purposes, we note that the terms appearing in Eqs. (14) and (19) can be expressed in the explicit form

$$M_{iP} = M_i - X_iy_{iP} + Y_ix_{iP} \qquad (i = 1, 2, 3; P = A, B, C, D) \tag{22}$$

$$(x_{BA}, y_{BA}) = a_1(\cos\theta_1, \sin\theta_1) \tag{23a}$$

$$(x_{CB}, y_{CB}) = a_2(\cos\theta_2, \sin\theta_2) \tag{23b}$$

$$(x_{DC}, y_{DC}) = a_3(\cos\theta_3, \sin\theta_3) \tag{23c}$$

$$(x_{1A}, y_{1A}) = \xi_1(\cos\theta_1, \sin\theta_1) + \eta_1(-\sin\theta_1, \cos\theta_1) \tag{24a}$$

$$(x_{2B}, y_{2B}) = \xi_2(\cos\theta_2, \sin\theta_2) + \eta_2(-\sin\theta_2, \cos\theta_2) \tag{24b}$$

$$(x_{2C}, Y_{2C}) = (\xi_2 - a_2)(\cos\theta_2, \sin\theta_2) + \eta_2(-\sin\theta_2, \cos\theta_2) \tag{24c}$$

$$(x_{3D}, y_{3D}) = (a_3 - \xi_3)(-\cos\theta_3, -\sin\theta_3) + \eta_3(-\sin\theta_3, \cos\theta_3) \tag{24d}$$

The above equations provide a complete solution for general cases of loading. The following example illustrates a simplification which is possible when at least one link is free of active loads.

EXAMPLE 3. Simplifications Possible when Some Links Have No Active Loading
    In Fig. 3, horizontal and vertical loads (H and V) act at the midpoints of bars AB and BC, but there are no active loads on link CD. Therefore equilibrium of link CD requires that the reactions (F) at joints C and D must be equal and collinear with line CD, as shown in Fig. 3(b). By taking moments about point B on link BC, we find that

$$\Sigma M_B = -(F\cos\theta_C)x_{CB} - (F\sin\theta_C)y_{CB} + Vx_{EB} = 0$$

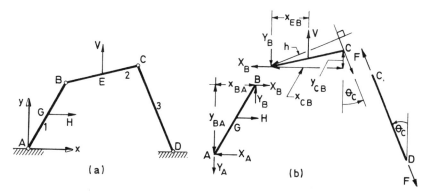

Figure 11.52-3. Mechanism with no active load on one link: (a) Mechanism; (b) Free-body diagrams of links.

Hence

$$F = \frac{Vx_{EB}}{x_{CB}\cos\theta_C + y_{CB}\sin\theta_C} \equiv \frac{Vx_{EB}}{h}$$

where h is the moment arm of F with respect to point B. It then follows immediately from the summation of forces on links 2 and 1 that

$$X_B = F\sin\theta_C$$

$$Y_B = V - F\cos\theta_C$$

$$X_A = H + X_B$$

$$Y_A = Y_B$$

This example illustrates that the coupling of equilibrium equations can be reduced or eliminated if we recognize the presence of axially loaded links.

Note that this solution makes no apparent use of the fact that H acts at the mid-point of bar AB. However, this information would have been used in a prior determination of the configuration. Alternatively, if the angle θ_C is given (thereby fixing all angles in this single-freedom mechanism), we can use the one remaining equilibrium equation for link 1 in the form

$$\sum M_A = -H\frac{y_{BA}}{2} - X_B y_{BA} + Y_B x_{BA} = 0$$

to find the value of H needed to equilibrate the applied force V; i.e.,

$$H = \frac{2(Y_B x_{BA} - X_B y_{BA})}{y_{BA}}$$

EXAMPLE 4. Multiloop Machine with Sliding Pairs

As a final example, which shows how to include the effects of multiple loops, consider the quick-return mechanism of Fig. 4(a), where the joint reactions are defined as shown in Fig. 4(b).

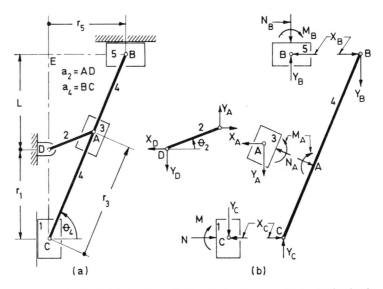

Figure 11.52-4. (a) Mechanism; (b) Free-body diagrams. Note: Active loads (X_i, Y_i, M_i), not shown, are applied to all links.

To reduce clutter on the free-body diagrams, the load sets (X_i, Y_i, M_i) are not shown, but they are assumed to act at conveniently chosen reference points on each of the five moving links.

Because of the large number of equilibrium equations that must be generated, it is helpful to organize the analysis in tabular form, where each line contains the three equations for a given link. In Table 1, the first column gives the link number, the next column specifies the moment center for the equation of moment equilibrium, and the third column contains that equation. The last two columns give the equations for force equilibrium in the x and y directions.

TABLE 11.52-1 *Equilibrium Equations Corresponding to Fig. 4*

Body	Mom. Ctr	$\sum M$	$\sum X$	$\sum Y$
1	C	$\overset{\checkmark}{M_C} = M_{1C}$	$X_C - N_C = X_1$	$\overset{\checkmark}{Y_C} = Y_1$
2	D	$Y_A a_2 C_2 - X_A a_2 S_2$ $= -M_{2D}$	$X_D - X_A = X_2$	$Y_D - Y_A = Y_2$
3	A	$\overset{\checkmark}{M_A} = -M_{3A}$	$X_A + N_A S_4 = X_3$	$Y_A - N_A C_4 = Y_3$
4	C	$N_A r_3 + Y_B r_5 =$ $M_{4C} - X_B(r_1 + L) - M_A$	$X_C + N_A S_4 = -X_4 - \overset{\checkmark}{X_B}$	$N_A C_4 + Y_B = Y_4 + \overset{\checkmark}{Y_C}$
5	B	$\overset{\checkmark}{M_B} = M_{5B}$	$\overset{\checkmark}{X_B} = X_5$	$N_B - Y_B = Y_5$

Note: $(C_i, S_i) \equiv (\cos \theta_i, \sin \theta_i)$.

By scanning the table we see (from lines 1, 3, and 5) that certain reactions $(M_C, Y_C, M_A, M_B, X_B)$ are given explicitly. A check is placed above these quantities, and they are considered as knowns wherever else they appear (as in line 4). The system of equations is coupled in the nine unchecked variables $(X_C, N_C, Y_A, N_A, X_A, N_B, Y_B, X_D, Y_D)$. Fortunately, however, we are not forced to solve nine simultaneous equations, because we see that the first and third equations for body 4 involve only the unknowns N_A and Y_B. These two equations may be expressed in the form

$$\begin{bmatrix} r_3 & r_5 \\ \cos \theta_4 & 1 \end{bmatrix} \begin{bmatrix} N_A \\ Y_B \end{bmatrix} = \begin{bmatrix} M_{4C} - X_B(r_1 + L) - M_A \\ Y_4 + Y_C \end{bmatrix} \equiv \begin{bmatrix} z_1 \\ z_2 \end{bmatrix}$$

and are easily solved to give

$$N_A = \frac{z_1 - z_2 r_5}{r_3 - r_5 \cos \theta_4}$$

$$Y_B = \frac{r_3 z_2 - z_1 \cos \theta_4}{r_3 - r_5 \cos \theta_4}$$

From this point, the remaining equations are completely uncoupled and may be solved in sequence as follows:

$$X_C = -(X_4 + X_B + N_A \sin \theta_4)$$
$$N_B = Y_5 + Y_B$$
$$Y_A = Y_3 + N_A \cos \theta_4$$
$$X_A = X_3 - N_A \sin \theta_4$$
$$X_D = X_2 + X_A$$
$$Y_D = Y_2 + Y_A$$
$$N_C = X_C - X_1$$

In solving for reactions in mechanisms, the level of complexity will seldom exceed that of this example. At worst, one will have to solve sets of three or more simultaneous linear equations—a relatively trivial task when digital computing aids are available.

11.53 Problems

General note: Ignore friction in all problems of this section. Use the general notational conventions of Sec. **11.51** plus any specific nomenclature given in the figures referred to. Assume that all active forces (X_i, Y_i, M_i) vanish except for those specifically mentioned in the problem statement.

Note for Problems 1–5: This group of problems deals with the slider-crank mechanism of Fig. **11.52**-1, where $OA = 2$ in., $AB = 6$ in., $\theta = 60°$, and $\phi = 16.78°$.

1. If the piston force $X_3 = -400$ lb, what resisting torque (M_1) must be applied to the crankshaft to maintain equilibrium? Find all internal reactions.

2. If a torque $M_1 = -200$ in.-lb acts on the crank, find the equilibrating piston force X_3 and all internal reactions.

3. The midpoint of connecting rod AB is acted upon by forces $X_2 = 100$ lb and $Y_2 = 150$ lb, and $X_3 = -400$ acts on the piston. What torque M_1 must act on the crank OA to maintain equilibrium, and what are the associated reactions at all joints?

4. The midpoint of connecting rod AB is loaded by forces $X_2 = -100$ lb and $Y_2 = -150$ lb, and a torque $M_1 = 300$ in.-lb acts on crank OA. Find the equilibrating piston force X_3 and the associated reactions at all joints.

5. The force set $(X_2, Y_2, M_2) = (100$ lb, 150 lb, 50 in.-lb) acts at the point $(\xi_2, \eta_2) = (2$ in., 1 in.) on link AB, and force $X_3 = -200$ lb acts on the piston. What equilibrating force must act normal to the crank AB at the point $(\xi_1, \eta_1) = (1.5$ in., $0)$ on the crank, and what are the corresponding joint reactions?

Note for Problems 6–11: This group of problems deals with the four-bar mechanism of Fig. **11.52**-2, where $a_1 = 1.9$ in., $a_2 = 2.7$ in., $a_3 = 2.45$ in., $a_4 = 4.72$ in., $\theta_1 = 52°$, $\theta_2 = 16.5°$, and $\theta_3 = 293°$.

6. What torque M_1 must be applied to link 1 in order to equilibrate the force set $(X_3 = 100$ lb, $Y_3 = 400$ lb) applied at the midpoint of bar DC? What are the magnitudes of the associated joint reactions? What is the resultant force and moment about A transmitted to the frame?

7. What torque M_1 must be applied to link 1 in order to equilibrate the force sets $(X_3, Y_3) = (100$ lb, 400 lb) and $(X_2, Y_2) = (-200$ lb, 200 lb) acting at the midpoints of bars DC and CB, respectively? Find the magnitude of all associated joint reactions, and find the resultant force and moment about point A transmitted to the frame.

8. A couple $M_3 = -1000$ in.-lb acts on link 3, and the force set $(X_2, Y_2, M_2) = (-200$ lb, -200 lb, 300 in.-lb) acts at the midpoint of bar BC. Find the equilibrating couple M_1 which must act on link 1 to maintain equilibrium. Also find the magnitudes of all joint reactions.

9. A couple $M_1 = 600$ in.-lb acts on link 1, and a couple $M_2 = -600$ in.-lb acts on link 2. Find the force P that must act normal to link 3, through its midpoint, in order to maintain equilibrium. Find all corresponding joint reactions.

10. What force Y_2 must act through the point $(\xi_2, \eta_2) = (1.5, 1.0)$ on link 2 in order to equilibrate couples $M_1 = 100$ in.-lb and $M_3 = 200$ in.-lb acting on links 1 and 2? Find the corresponding joint reactions.

11. Find the force X_3 acting at the midpoint of bar CD required to equilibrate the force $Y_2 = 400$ lb acting at the midpoint of bar BC. Calculate all the associated joint reactions.

12. Derive Eqs. (**11.52**-19a) and (**11.52**-19b).

13. For the mechanism shown in Fig. **10.60**-4, assume the data given in Problem **10.60**-10. Find the torque that must be applied to bar CB in order to equilibrate a force $X_4 = -100$ lb applied to the piston at E.

14. For the mechanism of Fig. **10.60**-5, with the data given in Problem **10.60**-11, find the torque that must be applied to crank CB in order to equilibrate the cutting force $F = 100$ lb acting as shown. Find all joint reactions.

15. For the mechanism of Fig. **10.30**-1, with the data of Problem **10.60**-12,
   a. Find the force X_5 that must be applied to point B of the piston 5 in order to equilibrate a torque of $M_2 = 200$ in.-lb applied to crank DA.
   b. Find all associated reactions.

11.60 STATIC EFFECTS OF FRICTION

Much new knowledge on friction has accumulated in recent years due to a burgeoning of activity in the field of **tribology**—i.e., the study of friction, wear, and lubrication. These developments are described in Moore [1975], Rabinowicz [1965], and Bowden and Tabor [1954, Part 2, 1956].

For the dynamic analysis of machinery, it usually suffices to rely on a few empirical rules which describe the friction between sliding bodies in terms of experimentally determined *coefficients of friction*. For discussions on the development of such coefficients the reader may consult the references just cited.

When two bodies are pressed together, they make contact at the tips (asperities) of microscopic surface irregularities. Unless the most painstaking precautions are taken, the spaces between the asperities will contain foreign matter such as oxide coatings, lubricant, or other forms of "contaminant." When the surfaces are made to undergo relative sliding, the *tangential* forces, or *frictional* forces, transmitted across the surfaces are very sensitive to the nature of the surface contaminants. The state of sliding surfaces may be categorized as follows:

1. **Full film** lubrication occurs when sufficient lubricant is present to completely separate the asperities of the two surfaces. Usually this term connotes the presence of a fluid film, but **solid lubricants** such as graphite, molybdenum disulfide, lead, etc., may be used. For *full film fluid* lubricants, the friction developed depends on the relative sliding speed and on bulk properties (especially viscosity) of the fluid. In general, the drag force in a fluid film bearing is a nonlinear function of sliding velocity. For methods of calculating such drag forces and torques, see Cameron [1966, 1976], O'Connor [1968], and Raimondi and Boyd [1958].

2. **Boundary lubrication** is said to exist when a significant amount of fluid lubricant is present, but opposing asperities penetrate the fluid film and come into contact. The bulk properties of the fluid play an insignificant role in this state. A detailed survey of boundary lubrication is given by Ling et al. [1969].

3. **Dry friction** or **Coulomb friction** is said to exist when no significant amount of lubricant covers the interface. This type of friction was studied by Leonardo da Vinci (1452–1519),[14] but like most of his work, it lay unused until his notebooks became known in the nineteenth century.

---

[14]For a brief review of the history and current status of the subject of dry friction, see Brown et al. [1969].

In 1699, Guillaume Amontons published some experimental data (and extrapolations thereof) on dry friction; he concluded that frictional force is always approximately one-third of the normal force and independent of the contact area. Although other distinguished scientists (e.g., de La Hire, Parent, Euler, and Leibnitz) took up the subject, the first comprehensive and conclusive experiments on the subject were done by Coulomb in 1779.[15] Based largely on these experiments, the *classical laws* of dry friction may be stated as follows:

1. For bodies at rest, the maximum friction force F occurs when motion impends and is given by

$$F = \mu_s N \tag{1}$$

where N is the normal force between surfaces and μ_s is called the **coefficient of static friction**.

2. For bodies in relative motion, the friction force is given by

$$F = \mu_k N \tag{2}$$

where μ_k is called the **coefficient of kinetic friction**.

3. Coefficients of friction are independent of apparent contact area and of the normal force.

4. μ_k is less than μ_s and independent of sliding speed.

For given surfaces—under given conditions of temperature, humidity, and surrounding medium (air, water, oil, vacuum, etc.)—the coefficients are approximately constant; however, scatter of more than 20% is not unusual. In particular, μ_k can be sensitive to surface speed and normal force (especially at low force levels). However, these "laws" are sufficiently accurate for most materials, in most situations of interest in machine design, and will be accepted as correct for purposes of this book. To a first approximation, these laws of dry friction are applicable to *boundary lubrication* and *solid film lubrication*.

Extensive tables of friction coefficients for a variety of materials are given by Bowden and Tabor [1954] and by Fuller [1958]. For steel surfaces coated with paraffin oil, we can assume (following Bowden and Tabor) that $\mu_k \doteq 0.15$, $\mu_s \doteq 0.2$; and $\mu_s \doteq \mu_k \doteq 0.1$ in the presence of fatty acids.

Usually, problems involving dry friction fall into one of the three following classes:

1. Contacting surfaces are known to be *undergoing relative motion*. Most problems of *kinetics* are of this type, and the coefficient of kinetic friction should be used for all sliding surfaces.

2. A mechanism is at rest but is known to be in a state of **impending motion** (i.e., motion will occur under specified loads if a coefficient of friction is decreased

---

[15]For a study of Coulomb's life and work, see Gillmor [1971], who traces the historical development of friction studies before, and after, Coulomb's contributions; also see Dowson [1978].

infinitesimally). We are asked to find one or more relationships among forces, friction coefficients, and geometrical parameters. The friction forces attain their maximum values, and coefficients of static friction should be used, on those surfaces where sliding impends.

3. We are asked to determine whether a system is in static equilibrium under known forces. The friction forces F are to be calculated (from equations of statics alone, if possible) and compared with their maximum attainable values ($\mu_s N$). If $F \le \mu_s N$, the system is in equilibrium.

In statics, we generally deal with cases 2 or 3, and we shall use the symbol μ to represent the coefficient of static friction unless otherwise noted.

In cases of (actual or impending) motion, the frictional force F acting on a surface element will have the direction opposite to the relative velocity of that element. Thus, the sliding block in Fig. 1 experiences frictional and normal forces (F, N) acting in the directions shown for the different directions of the relative velocity $\dot{x}$.

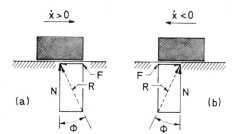

Figure 11.60-1. Direction of friction opposes relative velocity of slider.

If we regard F and N as signed quantities, we can write

$$F = \mu' |N| \tag{3}$$

where $\mu' = \mu \operatorname{sgn} \dot{x}$. In other terms, μ' is a *signed* coefficient of friction which establishes the direction of the frictional force associated with a given choice of relative motion.

The angle ϕ shown in Fig. 1 and defined by the expression

$$\phi = \tan^{-1} \mu \tag{4}$$

is called the **angle of friction**; it represents the inclination of the resultant contact force R to the direction of the normal.

Problems of dry friction for single machine elements (e.g., wedges, screws, belts, etc.) have been well covered in books[16] on statics or machine design and will not be discussed at any length here. We are primarily interested in the effects of friction in systems of links and elements and in finding general procedures for the static analysis of machines with friction at the joints. The influence of friction in sliding pairs is discussed in Sec. **11.61** and that in revolute pairs is covered in Sec. **11.62**.

---

[16]See any standard text on statics, e.g., Beer and Johnston [1972], or on machine design, e.g., Green [1956]. Goodman and Warner [1961] give a good brief discussion of the underlying physics of friction as well as tables of coefficients and numerous solved problems.

11.61 Slider Friction

Figure 1 shows all possible orientations of a slider within a guide having small clearance.

It is seen that there are five possible **modes of contact**. In modes 1 and 2, contact occurs along a side of the slider, and in modes 3 and 4 it occurs at diagonally opposite corners. The fifth mode, which represents an unconstrained body without contact forces, is of little interest in problems of statics and will not be considered further here. However, it will be essential for us to determine which of the first four modes will occur when the slider makes contact with its guide.

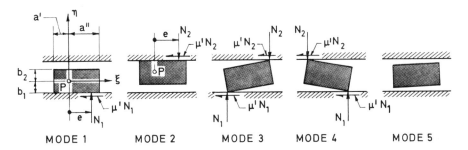

Figure 11.61-1. Possible orientations of a slider, and associated reactions; $N_1 \geq 0$; $N_2 \geq 0$; $\mu' \equiv \mu \text{ sgn } V_\xi$.

The normal contact forces (N_1, N_2) shown in Fig. 1 must act in the directions shown since the guide cannot exert tension on the slider. The corresponding friction forces (F_1, F_2) are treated as signed quantities given by

$$F_1 = \mu' N_1, \qquad F_2 = \mu' N_2 \qquad (1)$$

where

$$\mu' = \mu \text{ sgn } V_\xi \qquad (2)$$

and V_ξ is the slider velocity, positive in the direction of the ξ axis. When the slider velocity V_ξ is positive (negative) the friction forces will act in (opposite to) the direction shown.

Because the lateral contact pressures are necessarily compressive, the line of action of the forces N_1 and N_2 must fall within the length of the slider. That is, the eccentricity e of the reactions must satisfy the relations

$$-a' \leq e \leq a'' \qquad (3)$$

where a' and a'' are the distances of the slider's end planes from a fixed reference point (P) on the slider and e is a signed quantity measured positively from P in the direction of the ξ axis.

In general, we shall not know *a priori* which mode occurs in a given problem, and we shall have to assume a mode in order to begin the analysis. If the wrong mode has been chosen, the equations of equilibrium will predict a wrong direction for contact force N_i, or else inequality (3) will be violated. Then the problem must be solved again with a different choice of mode, and the process may be repeated until the correct mode is found. To avoid blind guessing it is desirable (especially for problems with several sliders) to formulate a technique which guides us in the provisional choice of mode.

To formulate such a guide, we tentatively assume that friction is absent and use the methods of Sec. **11.50** to calculate the reactive force N_o and moment M_o (about reference point P) exerted on the slider. N_o is considered positive in the direction of the η axis of Fig. 1, and M_o is positive counterclockwise. These quantities are statically equipollent to the force sets shown in Fig. 1 (with μ' set at zero) provided that

$$N_1 = N_o, \qquad\qquad e = \frac{M_o}{N_o} \qquad\qquad \text{(for mode 1)} \qquad (4a)$$

$$N_2 = -N_o, \qquad\qquad e = \frac{M_o}{N_o} \qquad\qquad \text{(for mode 2)} \qquad (4b)$$

$$N_1 = \frac{a''N_o - M_o}{a' + a''}, \qquad N_2 = \frac{-a'N_o - M_o}{a' + a''} \qquad \text{(for mode 3)} \qquad (4c)$$

$$N_1 = \frac{a'N_o + M_o}{a' + a''}, \qquad N_2 = \frac{-a''N_o + M_o}{a' + a''} \qquad \text{(for mode 4)} \qquad (4d)$$

To decide upon the mode that governs in the absence of friction, we form $e \equiv M_o/N_o$. When $-a'' \le e \le a''$, we know (see Fig. 1) that mode 1 prevails if $N_o > 0$ and that mode 2 prevails if $N_o < 0$ (i.e., the contact force, N_1 or N_2, must be positive). When $e > a''$ or $e < -a''$, we can conclude (see Fig. 1) that mode 3 prevails if $M_o < 0$ and that mode 4 prevails if $M_o > 0$.

The logic for determining the prevailing mode in the absence of friction is summarized in Fig. 2.

The mode so determined is strictly valid only in the limit of vanishing μ, but it will

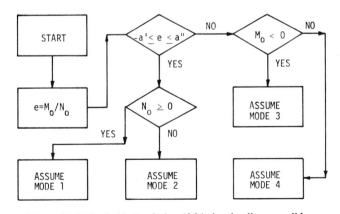

Figure 11.61-2. Guide to choice of friction loadings on sliders.

also be correct for sufficiently small values of μ and will usually be correct for the practical values of μ used in machinery. Having made a provisional choice of the prevailing mode, the problem may now be solved for the contact forces N_i in the presence of friction; in rare cases these recalculated values of N_i will be negative [or inequality (3) will be violated], and the problem will have to be solved again for a different mode of contact.

Special Case. Moment-Free Sliders

Frequently, all the external loads on a slider (excluding the reactions on the sliding surfaces) will pass through a fixed reference point P on the slider. For example, in a conventional engine, the gas load and connecting rod load pass through the wrist pin. For brevity, we refer, in such cases, to a **moment-free slider**.

Moment equilibrium requires that the reactive moment (M_P) about P must vanish for a moment-free slider whether or not friction is present. Thus when $\mu = 0$, the reactive moment M_o vanishes and the logic of Fig. 2 results in a provisional choice of mode 1 when $N_o \geq 0$ and of mode 2 when $N_o < 0$. If mode 1 prevails and friction is taken into account, the reactive moment about P is (see Fig. 1)

$$M_P = N_1 e - \mu' N_1 b_1$$

But M_P must vanish for a *moment-free* slider; hence

$$e = \mu' b_1 \qquad \text{(for mode 1 contact)} \tag{5a}$$

Similarly, it follows from Fig. 1 that for mode 2 contact, $M_P = -N_2 e + \mu' N_2 b_2 = 0$; hence

$$e = \mu' b_2 \qquad \text{(for mode 2 contact)} \tag{5b}$$

Since e must fall in the range $(-a', a'')$, it follows that for *moment-free sliders*

$$-a' \leq \mu' b_1 \leq a'' \qquad \text{(for mode 1)} \tag{5c}$$

$$-a' \leq \mu' b_2 \leq a'' \qquad \text{(for mode 2)} \tag{5d}$$

It follows that for *forward sliding* ($\mu' = \mu$)

$$\mu b_i \leq a'' \quad \text{(in mode } i) \tag{5e}$$

and for *backward sliding* ($\mu' = -\mu$)

$$\mu b_i \leq a' \quad \text{(in mode } i) \tag{5f}$$

In other words, a *moment-free slider will tilt if the axial extension of the slider is too short relative to its width*, and inequalities (5e) and (5f) give the required ratios of "length" to "width" needed to prevent tilting.

The following examples illustrate the use of these general ideas.

EXAMPLE 1. Slider-Crank Mechanism

Find all internal reactions in the mechanism of Fig. 3 if the slider has friction coefficient $\mu = \tan 20° = 0.364$ and is moving to the right. Ignore friction in the turning

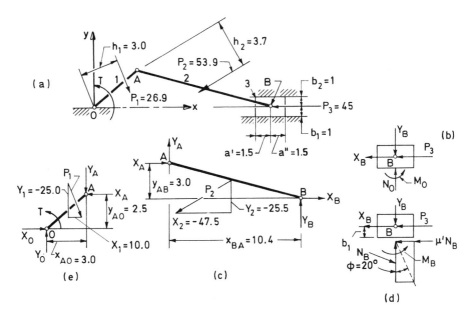

Figure 11.61-3. (a) Slider-crank mechanism; (b) Slider without friction; (c) Free-body of connecting rod; (d) Free-body of slider with $\mu = \tan \phi$; (e) free-body of crank.

pairs. Also find the torque T needed on link OA to equilibrate the applied loads. Dimensions and applied force levels are specified numerically in the figure in arbitrary units.

If $\mu = 0$, the free-body diagram for the slider would be as in Fig. 3(b), and the corresponding equilibrium equations would be

$$Y_B = N_o, \qquad X_B = -P_3, \qquad M_o = 0 \tag{6}$$

Taking moments about point A in the free-body diagram of the connecting rod [Fig. 3(c)], we see that

$$-P_2h_2 + X_By_{AB} + Y_Bx_{BA} = 0 \tag{7}$$

Then, using Eqs. (6) and (7), we find

$$N_o = Y_B = \frac{1}{x_{BA}}(P_2h_2 + P_3y_{AB}) = 32.16 \tag{8}$$

Referring to Fig. 2, we see that $-a' \le e = 0 \le a''$ and since $N_o > 0$, mode 1 is to be assumed. Thus the free-body diagram for the slider, with friction, is as in Fig. 3(d), and the associated equilibrium equations are

$$X_B = -(P_3 + \mu'N_B), \qquad Y_B = N_B, \qquad M_B = \mu'N_Bb_1 \tag{9}$$

where $\mu' = \mu$, because motion is to the right [cf. Eq.(2)].

These forces may be substituted into Eq. (7) (which is valid for any value of μ') to give

$$-P_2h_2 - (P_3 + \mu'N_B)y_{AB} + N_Bx_{BA} = 0$$

This equation can be solved to find

$$N_B = \frac{P_2 h_2 + P_3 y_{AB}}{x_{BA} - \mu' y_{AB}} = \frac{(53.9)(3.7) + (45)(3)}{10.4 - (0.364)(3)} = 35.93 \tag{9}$$

Since $N_B \geq 0$, our *provisional* choice of mode 1 appears to be correct. However, we must also check the eccentricity (e) of the slider reaction. Since we are dealing with a *moment-free* slider in mode 1 with forward sliding, inequality (5e) must be satisfied; i.e., mode 1 is valid only if $\mu \leq a''/b_1$. For the present case ($\mu = \mu' = 0.364$, $a'' = 1.5$, $b_1 = 1$), this condition is satisfied, and the assumption of mode 1 is justified.

From Eqs. (9) we now find

$$X_B = -58.08, \qquad Y_B = 35.93, \qquad M_B = 13.08$$

From the free-body diagram of Fig. 3(c), we see that

$$X_A = -X_B - X_2 = 58.08 + 47.5 = 105.58$$
$$Y_A = -Y_2 - Y_B = +25.5 - 35.93 = -10.43$$

By summing the forces in Fig. 3(e), we find

$$X_O = X_A - X_1 = 105.58 - 10 = 95.58$$
$$Y_O = Y_A - Y_1 = -10.43 + 25 = 14.57$$

Taking moments about point O in Fig. 3(e), we find the required equilibrating torque,

$$T = P_1 h_1 + Y_A x_{AO} - X_A y_{AO}$$
$$= (26.9)(3) + (-10.43)(3) - (105.6)(2.5) = -214.6$$

which completes the problem.

All problems involving slider friction may be approached in a similar fashion. However, situations may arise where the assumption of impending motion leads to contradictions. In such cases, we must conclude that the slider cannot move; i.e., the slider is **locked**, as illustrated in the following example. Other aspects of the slider locking phenomenon are discussed by Kleven [1976].

EXAMPLE 2. Friction Locking

We seek to find the load P, acting to the right, which will cause impending motion (to the right) of the slider shown in Fig. 4(a) when the coefficient of friction μ and the force W are given.

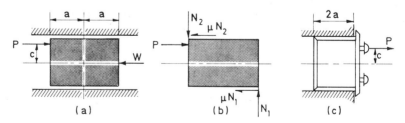

Figure 11.61-4. Example 2: (a) Slider; (b) Free-body diagram; (c) Drawer.

For low values of μ, the diagonal tipping mode of Fig. 4(b) seems plausible, and the corresponding equilibrium equations are

$$\Sigma \ Y: \quad N_1 = N_2 = N$$
$$\Sigma \ X: \quad P = W + 2\mu N \qquad\qquad (11)$$
$$\Sigma \ M: \quad Pc = 2Na$$

Upon eliminating N between the last two equations, we find

$$P = \frac{W}{1 - \mu(c/a)} \qquad\qquad (12)$$

It is easily seen that P, N_1, and N_2 are all positive if

$$\mu \leq \frac{a}{c} \qquad\qquad (13)$$

thus justifying the choice of mode and validating the solution. Note that P grows indefinitely large as μ approaches a/c from below. In other words, no force is large enough to move the slider when $\mu = a/c$. What happens when $\mu > a/c$? Then Eqs. (12) and (11) predict negative values for the reactions $N_1 = N_2$, and the assumed contact mode is untenable. It is easily seen that contact mode 3 (Fig. 1) cannot satisfy moment equilibrium and that modes 1 and 2 violate force equilibrium. We must conclude that motion *cannot* impend and the slider is jammed when $\mu \geq a/c$. This situation is sometimes described as **friction lock**.

When the opposing load W vanishes, the problem reduces to the familiar one, shown in Fig. 4(c), of pulling out a drawer by one of two symmetrically located handles. We see from Eq. (12) that as long as $a \geq \mu c$ the drawer will open with negligible effort, whereas we know it will lock if $a \leq \mu c$.

11.62 *Friction in Hinge Joints*

Figure 1 shows a journal (axle) rotating clockwise with angular speed ω in a bearing having a small clearance.

Due to a force (R) applied at the journal center, the center (J) of the journal will be displaced from the center of the bearing, and contact will be made at point C on

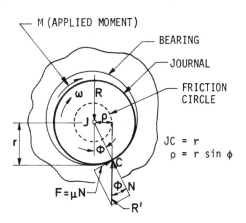

Figure 11.62-1. Dry journal bearing radius of friction circle $\rho = r \sin \phi$.

the line of centers. The resultant (R') of the normal force (N) and friction force (F) at the contact point must be equal and parallel to the applied force R at J for equilibrium of forces. For this to happen, the journal must "roll up" the bearing to a point such that JC makes an angle ϕ with the applied load axis. But $\tan \phi = F/N = \mu$. That is, the angle between the applied load axis and the line of centers JC is the friction angle

$$\phi = \tan^{-1} \mu \tag{1}$$

Note that the applied load and its reaction produce a net **frictional couple**

$$M_f = R\rho \tag{2}$$

which opposes the direction of journal rotation, where

$$\rho = JC \sin \phi = r \sin \phi \tag{3}$$

and r is the journal radius. To equilibrate this resisting couple, it is necessary for a moment of magnitude M_f to be applied to the shaft in the direction of rotation (or impending rotation).

Another way of describing the situation is to say that the resultant reaction on the journal always acts tangent to a circle of radius ρ and center J in such a direction as to oppose the assumed rotation of the journal relative to the bearing. This circle of radius ρ is called the **friction circle**, and ρ will be called the **friction radius**.

For *small* coefficients of friction ($\phi \ll 1$) we can use the approximation

$$\sin \phi \doteq \tan \phi = \mu \qquad (\phi \ll 1)$$

and Eq. (3) can be expressed in the approximate form

$$\rho \doteq r\mu \qquad (\text{if } \mu \ll 1) \tag{4}$$

The error of the approximation varies from 2 % to 12 % as μ varies from 0.2 to 0.5.

One conclusion we draw from the concept of the friction circle is that the directions of forces in machine elements will seldom be changed significantly due to the presence of hinge friction. To see this, consider the link AB of Fig. 2, which in the absence of friction would be loaded by a purely axial force.

Due to friction, the resultant force S becomes inclined to the center line AB by an angle α. If the friction circles at A and B have radii ρ_A and ρ_B, we see from Fig. 2 that

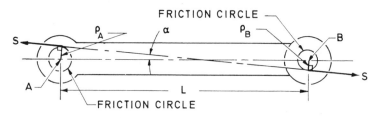

Figure 11.62-2. Link with hinge friction.

the angle α must satisfy the relation

$$\sin \alpha = \frac{\rho_A + \rho_B}{L} = \frac{r_A \sin \phi_A + r_B \sin \phi_B}{L} \tag{5}$$

where (r_A, ϕ_A) and (r_B, ϕ_B) represent the journal radii and friction angles at A and B. If, for example, we consider a link of length $L = 10$ in. with $\frac{1}{2}$-in.-diameter hinge pins at either end and the rather high friction coefficient of $\mu = 0.4$ ($\phi = 21.8°$), the angle of inclination of the force S is

$$\alpha = \sin^{-1}\left[\frac{2(0.25 \sin 21.8°}{10}\right] = 1.06° \tag{6}$$

Thus, we should not expect hinge friction to alter the directions of reactions in mechanisms by more than a degree or two under normal circumstances. This suggests that we can *usually* ignore hinge friction completely—which is correct, as will be illustrated in Example 1. However, this conclusion is unjustified if a small change in the *direction* of a force produces large changes in the *magnitudes* of related forces. This can happen in mechanisms which are close to a singular configuration, as will be illustrated in Example 2.

EXAMPLE 1. Typical Situation Where Effects of Hinge Friction are Negligible
    Let us reconsider the slider-crank mechanism of Fig. 11.61-3, allowing this time for friction in the hinge joints as well as in the slider. The conditions are as stated in Sec. 11.61, but we assume that the friction coefficient $\mu = \tan 20° = 0.364$ also acts on each of the hinges, which are assumed to have radii $r = 0.25$ (this is a realistic value for links OA and AB of lengths 3.90 and 10.8 units, respectively).
    The corresponding friction radius is therefore

$$\rho = r \sin \phi = 0.25 \sin 20° = 0.0855$$

The frictional torques acting at each joint are found from Eq. (2) to be

$$T_O = \rho R_O = 0.0855(X_0^2 + Y_0)^{1/2}$$
$$T_A = \rho R_A = 0.0855(X_A^2 + Y_A)^{1/2}$$
$$T_B = \rho R_B = 0.0855(X_B^2 + Y_B^2)^{1/2}$$

The reader should verify that these torques act on the links in the directions shown in Fig. 3 when the slider is moving to the right. In all other respects, Fig. 3 is a repeat of Fig. 11.61-3.
    From the free-body diagram of the slider [Fig. 3(d)], we see that

$$X_B = -(P_3 + \mu N_B) \tag{7a}$$
$$Y_B = N_B \tag{7b}$$
$$M_B = \mu N_B b_1 - T_B \tag{7c}$$

From Fig. 3(c), we see that the resultant moment about point A, on link 2, is

$$-P_2 h_2 + Y_B x_{BA} + X_B y_{AB} - T_A - T_B = 0 \tag{8}$$

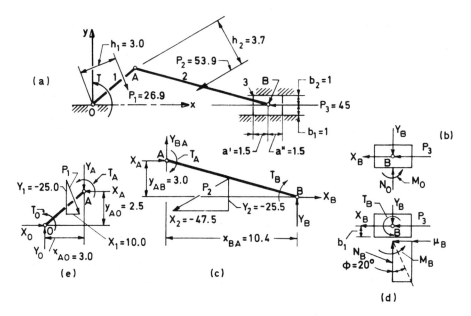

Figure 11.62-3. Example 1—Slider-crank with hinge and slider friction. Point B is moving to right; note directions of frictional hinge torques (T_A, T_B, T_0) required to resist relative rotations of links.

or

$$-P_2h_2 + N_Bx_{BA} - (P_3 + \mu N_B)y_{AB} - T_A - T_B = 0 \qquad (9)$$

where use has been made of Eqs. (7). Equation (9) may be solved to give

$$N_B = \frac{P_2h_2 + P_3y_{AB} + T_A + T_B}{x_{BA} - \mu y_{AB}} = 35.93 + \frac{T_A + T_B}{9.308} \qquad (10)$$

If N_B is known, X_B, Y_B, and M_B are given by Eqs. (7), and the remaining reactions are readily verified to be given by

$$R_B = (X_B^2 + Y_B^2)^{1/2} \qquad (11)$$

$$T_B = \rho R_B = 0.0855 R_B \qquad (12)$$

$$X_A = -X_B - X_2 = -X_B + 47.5 \qquad (13)$$

$$Y_A = -Y_2 - Y_B = 25.5 - Y_B \qquad (14)$$

$$R_A = (X_A^2 + Y_A^2)^{1/2} \qquad (15)$$

$$T_A = \rho R_A = 0.0855 R_A \qquad (16)$$

$$X_O = X_A - X_1 = X_A - 10 \qquad (17)$$

$$Y_O = Y_A - Y_1 = Y_A + 25 \qquad (18)$$

$$R_O = (X_O^2 + Y_O^2)^{1/2} \qquad (19)$$

$$T_O = \rho R_O = 0.0855 R_O \qquad (20)$$

Finally, the required torque T needed to equilibrate the crank OA is given by

$$T = P_1 h_1 + Y_A x_{AO} - X_A y_{AO} - T_O - T_A$$
$$= 80.7 + 3Y_A - 2.5X_A - T_O - T_A \qquad (21)$$

Unfortunately, we cannot initially use Eq. (10) to evaluate N_B, because the hinge torques T_A and T_B are unknown. Furthermore, the system of equations is nonlinear in the unknowns, because of Eqs. (15) and (19).

The recommended procedure is to first neglect hinge friction (i.e., assume $T_A = T_B = 0$) and use Eq. (10) to find a *first estimate* of N_B. This value may then be used in Eqs. (7) and Eqs. (11)–(16) to find estimates of $(X_B, Y_B, T_B, X_A, Y_A, T_A)$. Then Eq. (10) may be used to give a refined estimate of N_B, using the updated values of T_A and T_B. The procedure should be repeated as often as needed until N_B stabilizes.

The calculations are summarized in Table 1.

TABLE 11.62-1

Eq. Nos.	Eq.	Trial 1	Trial 2	Trial 3
10, 7b	$Y_B = N_B = 35.93 + \dfrac{T_A + T_B}{9.308}$	35.93	37.53	37.55
7a	$X_B = -(45 + 0.364N_B)$	−58.08	−58.66	−58.66
12	$T_B = 0.0855(X_B^2 + Y_B^2)^{1/2}$	5.839	5.954	5.954
13	$X_A = 47.5 - X_B$	105.58	106.15	106.16
14	$Y_A = 25.5 - Y_B$	−10.43	−12.03	−12.04
15, 16	$T_A = 0.0855(X_A^2 + Y_A^2)^{1/2}$	9.071	9.134	9.135

Note that the solution has stabilized by the third trial. In fact, the second trial gives completely satisfactory answers.

The remaining unknowns, calculated from Eqs. (17)–(21), are

$$X_O = 96.16, \qquad Y_O = 12.96, \qquad T_O = 8.296, \qquad T = -238.2$$

To complete the comparison, we note from Example 1 of Sec. **11.61** that *when hinge friction is neglected*

$$X_O = 95.58, \qquad Y_O = 14.57, \qquad T = -214.6, \qquad T_O = T_A = T_B = 0$$

Discussion: Example 1 illustrates the rapid convergence (usually by the second trial) of the recommended procedure for systems with hinge friction. It also suggests that the effects of hinge friction are not great in mechanisms which are not near singular configurations.

It should be mentioned that direct (noniterative) methods of treating special cases are frequently used as illustrative problems in books on dynamics of machinery. For example, one often sees—e.g., in Holowenko [1955, p. 173]—*direct* (graphical) solutions for a slider-crank mechanism using the concept of the friction circle. However, it is usually not pointed out that the "direct" solution given will only work in the special case when the active loads P_1 and P_2 (see Fig. 3) vanish. On the other hand, the analytical method given here is much more general and is usually rapidly convergent. When convergence difficulties arise, it is indicative that the system is near a singularity (as in Example 2) or that friction locking (as discussed in Sec. **11.61**) is likely to occur.

EXAMPLE 2. A Case Where Hinge Friction is Significant
Consider the *toggle mechanisms* shown in Fig. 4, where a small force P produces a large reaction S_2, which may be used to clamp a workpiece.

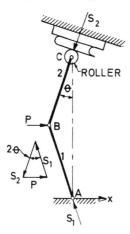

Figure 11.62-4. Toggle mechanism.

The inset force polygon (for ABC treated as a free-body) shows that, in the absence of friction, applied force $P = P_0$ produces a clamping force

$$S_2 = \frac{P_0/2}{\sin \theta} \tag{22}$$

For small values of angle θ, the force magnification can be immense. Because the smallness of angle θ is essential to the proper functioning of the mechanism, we should suspect that hinge friction will have a pronounced effect, even though it causes only small angular changes in the reactions.

To verify this conjecture, consider Fig. 5(a), which shows links AB and BC with friction circles (shown shaded with radii exaggerated) at the hinges A, B, C.

We suppose that $AB = BC = L$ and that the *friction radius*, $\rho = r \sin \phi$, is the same for all friction circles. With B moving to the right ($\dot{x}_B > 0$), the ground reaction S_1 on link 1 must resist the assumed rotation and act as shown in Fig. 5(a). To verify that the line of action of S_1 passes to the left of B, note that link 1 is rotating *clockwise relative to link 2*. Hence the reaction S_1' (on link 1, from link 2) must produce a (resisting) counterclockwise moment on link 1, as shown.

The reactions S_1 and S_2 are equal by symmetry and are inclined to the bar axes by the angle α, where

$$\sin \alpha = \frac{2\rho}{L} = \frac{2r}{L} \sin \phi \tag{23}$$

according to Eq. (5).

From the force polygon inset on Fig. 5(a), we see that the clamping force S_2 is given by

$$S_2 = \frac{P/2}{\sin (\theta + \alpha)} \tag{24}$$

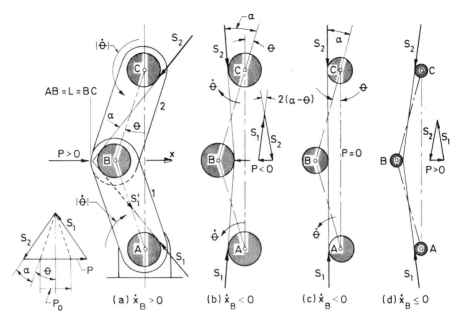

Figure 11.62-5. Effect of friction on toggle linkages: (a) Locking motion; (b) Unlocking motion requires negative P; (c) Unlocking motion requires zero P; (d) Unlocking motion requires positive P.

Comparing this result to Eq. (22), we see that friction causes a pronounced decrease in the clamping force. In particular, when $\theta \to 0$ the ideal clamping force predicted by Eq. (22) is *infinite*, but the presence of the nonvanishing term $\alpha = \sin^{-1}(2\rho/L)$ ($\doteq 2\mu r/L$) in Eq. (24) shows that the clamping force can never be infinite in the presence of friction. Equation (24) gives the applied force P necessary to build up any desired clamping force.

Once the force S_2 has been reached, due to some force P and angle θ, we may ask how much the clamping force will relax if P is reduced. To answer this question, we investigate the relation between P and S_2 when point B moves to the left ($\dot{x}_B < 0$) as shown in Fig. 5(b), where the angle θ and friction radii ρ are the same as in Fig. 5(a). When the rotation of the links impends in the directions shown in Fig. 5(b), the reactions S_1 and S_2 must act as shown to *resist* the relative rotations of the links. The corresponding force polygon [inset in Fig. 5(b)] shows that equilibrium requires a force P directed to the left. Therefore no such motion can impend if the original force P is reduced to zero, and we conclude that system is locked by friction in the position associated with the peak positive value of P. If the links are absolutely rigid, the force S_2 is determined by the load-deformation characteristics of the workpiece (see Fig. 4) and will not change when P is relaxed so long as $\dot{x}_B = 0$.

We next ask, what is the minimum friction radius that will keep the system locked after the original load is reduced to zero? The answer is given by Fig. 5(c), which illustrates the situation where the reactions S_1 and S_2 are self-equilibrating, i.e., where motion of point B impends to the left with $P = 0$. In this case it is seen that

$$\alpha = \theta$$

or, by Eqs. (3) and (23),

$$\rho \equiv r \sin \phi = \tfrac{1}{2} L \sin \theta$$

When $\alpha < \theta$ (or $\rho < \frac{1}{2}L \sin \theta$), as shown in Fig. 5(d), motion will impend to the left with a finite force P directed to the right, as shown in the inset force polygon.

We conclude that the toggle linkage is self-locking if the friction radius exceeds a minimum value

$$\rho_{min} = \tfrac{1}{2}L \sin \theta \tag{25}$$

or, equivalently, when

$$\sin \phi > \frac{1}{2}\frac{L}{r} \sin \theta \tag{26}$$

For a discussion of the related problem of *starting friction*, see Thumin [1951].

11.63 Dead Bands

The equilibrium *configuration* of a system with dry friction is not uniquely determined by the applied loads. To see this, consider the slider of Fig. 1(a) loaded by tangential force P and normal force N and restrained by a spring of stiffness k and free length L_f; the displacement of the spring measured from its unstrained position is denoted by x.

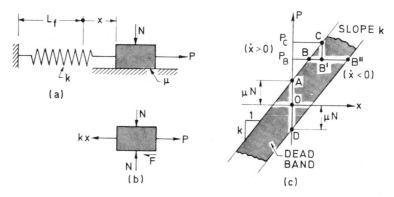

Figure 11.63-1. (a) Spring restrained slider with dry friction; (b) Free-body diagram; (c) Load-displacement diagram with dead band.

The free-body diagram of Fig. 1(b) indicates that motion to the right ($\dot{x} > 0$) is possible only if

$$P - kx \geq \mu N \tag{1}$$

and to the left ($\dot{x} < 0$) only if

$$kx - P \geq \mu N \tag{2}$$

Inequality (1) states that motion to the right is possible only for combinations of P and x which lie above the line AB in Fig. 1(c). Similarly, inequality (2) requires that P lies below line DB''. No motion is possible for the **dead band** included between these two parallel lines. For a specified load P_B, the displacement x may fall anywhere within the dead band along the line BB'', since $P = P_B = $ constant everywhere along this line. The actual position depends on the history of loading and can only be found from a complete dynamic analysis, taking into account the initial state of the system. To illustrate this history dependence, consider the special case of very slow loading (so that inertia forces are negligible) from $P = 0$ to $P = P_B$, starting when $x = 0$. If the load increases monotonically from 0 to P_B, the path followed in Fig. 1 is OAB, and the final state is represented by point B. Suppose, however, the load increases from 0 to P_C and then drops down to P_B. Since motion to the left is impossible at point C, the displacement remains fixed as the load drops from P_C to P_B, and the final displacement corresponds to point B' within the dead band.

Similarly, the dead band corresponding to the pendulum of Fig. 2(a) may be shown (Problem 11.64-19) to be given by the parallel curves of Fig. 2(b), which are described by the equations

$$\frac{P}{W} = \tan(\theta \pm \theta^*) \tag{3}$$

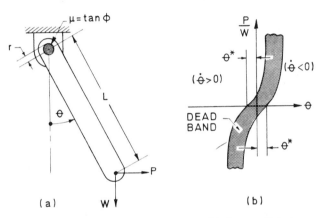

Figure 11.63-2. (a) Pendulum with friction coefficient μ; (b) Associated dead band.

where

$$\theta^* = \sin^{-1}\left(\frac{r \sin \phi}{L}\right) \doteq \frac{r}{L}\mu \tag{4}$$

The approximation in Eq. (4) is valid if $r\mu/L \ll 1$.

The presence of a dead band can be troublesome, especially in instrument applications. The dead band for a beam balance is found in Problem 11.64-21.

Although the configuration of a system is not uniquely specified by its loading, in the presence of dry friction, we have seen that it is possible to determine configurations of impending motion. For single-freedom mechanisms the dead band may be pictured in a load-displacement plane, but for multifreedom systems the "dead band" is a "region" enclosed by hypersurfaces of a multidimensional space [see Problem 11.64-20]. The determination of such multidimensional "dead regions" is a problem of *mathematical programming* which can be extremely cumbersome for nonlinear problems.

An alternative approach is to assume an arbitrary (nonequilibrium) initial configuration, compatible with the constraints of the system, and to solve the resulting differential equations of motion using the methods of dynamics developed in later chapters of this book. The equilibrium state will then be that for which all velocities vanish. If desired, some "artificial viscosity" (velocity-dependent damping) can be added to accelerate the approach to equilibrium.

11.64 Problems

General note: Unless otherwise indicated, assume a coefficient of static friction $\mu = \tan 20° = 0.364$ for all problems of this section.

1–5. Do Problems **11.53**-1–5 assuming that motion of the slider impends to the right $(\dot{x}_B > 0)$. Include friction on the *slider* but ignore friction in all *hinges*.

Note for Problems 6–17: Do the problem indicated, neglecting the effects of *slider* friction, but include friction in all *hinges*. Assume that each hinge pin has a radius of 0.1 in. and that motion impends with $\dot{x}_B > 0$.

6. Problem **11.53**-1.

7. Problem **11.53**-2.

8. Problem **11.53**-3.

9. Problem **11.53**-4.

10. Problem **11.53**-5.

11. Problem **11.53**-6.

12. Problem **11.53**-7.

13. Problem **11.53**-8.

14. Problem **11.53**-9.

15. Problem **11.53**-10.

16. Problem **11.53**-11.

17. Problem **11.53**-13.

18. Solve Example 1 of Sec. **11.61** if the slider motion impends to the left.

19. Show that the dead band for the pendulum of Fig. **11.63**-2(a) is bounded by the curves described by Eqs. (**11.63**-3) and (**11.63**-4).

20. In Fig. 1(a), L_1 and L_2 represent the free lengths of springs having stiffnesses k_1 and k_2. The sliders have coefficients of friction μ_1, μ_2 and are loaded by tangential forces (P_1, P_2) and normal forces (N_1, N_2).
  a. Show that the dead region in the four-space of variables (x_1, x_2, P_1, P_2) is described by the inequalities

$$-\mu_1 N_1 - P_1 \le k_2 x_2 - k_1 x_1 \le \mu_1 N_1 - P_1$$
$$P_2 - \mu_2 N_2 \le k_2 x_2 \le P_2 + \mu_2 N_2$$

  b. Show that for fixed values of the loads (P_1, P_2) the dead "region" is a parallelogram in the two-space of (x_1, x_2), as shown in Fig. 1(b). Express the intercepts (a_1, a_2, b_1, b_2), shown in the figure, in terms of the parameters $P_1, P_2, \mu_1, \mu_2, N_1, N_2$. What are the slopes of the sides of the parallelogram?

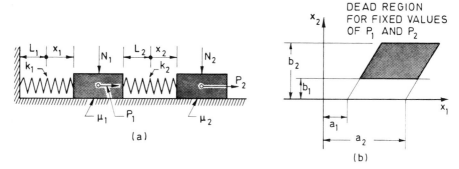

Figure 11.64-1. Prob. 20.

21. The beam AB shown in Fig. 2 carries weights P_1 and P_2 suspended from A and B by flexible strings. The pivot of the beam is located at C, a distance b from line AB, the coefficient of friction at C is $\mu = \tan\phi$, and the pivot radius at C is r.

  a. Derive the following expression for the ratio P_1/P_2

$$\frac{P_1}{P_2} = \frac{\cos\theta + (b/a)\sin\theta + (\rho'/a)}{\cos\theta - (b/a)\sin\theta - (\rho'/a)}$$

  where $\rho' = r\sin\phi \ \mathrm{sgn}\ \dot\theta$.

  b. Let $b/a = 0.2$, $r/a = 0.1$, and $\phi = 20°$. Plot P_1/P_2 versus θ (for impending motion with $\dot\theta > 0$) in the range $-10° < \theta < 10°$. Repeat for $\dot\theta < 0$.

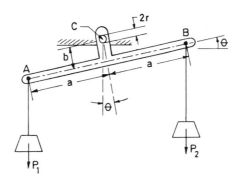

Figure 11.64-2. Prob. 21.

11.70 THE PRINCIPLE OF VIRTUAL WORK (PROOF)

It was mentioned in Sec. **11.30** that the principle of virtual work may be accepted as the fundamental axiom from which all the familiar results of *vector statics* follow. However, most students are familiar with that exposition of mechanics which is based on Newton's axioms and would rightfully question the validity of any other set of axioms, including the principle of virtual work. To relieve such anxieties we shall derive the principle from the familiar Newtonian laws.

Before reading the proofs given below it is advisable to review the summary of definitions given in Sec. **11.30**.

Proof 1

The first part of the principle states that if an ideal mechanical system is in equilibrium, the net virtual work of all the active forces vanishes for every set of virtual displacements.

To prove this statement, we note that the *given condition of equilibrium* implies that the resultant force $\mathbf{F}_i$ on any given particle P_i in the system must vanish; i.e.,

$$\mathbf{F}_i \equiv \mathbf{F}_i^a + \mathbf{F}_i^r = 0 \tag{1}$$

where $\mathbf{F}_i^a$ is the resultant *active* force applied to P_i and $\mathbf{F}_i^r$ is the resultant *reaction* on P_i due to the constraints of the system. The work of $\mathbf{F}_i$ on an *arbitrary virtual displacement* $\delta \mathbf{r}_i$ of particle P_i is given by

$$\delta W_i = \mathbf{F}_i \cdot \delta r_i = \mathbf{F}_i^a \cdot \delta r_i + \mathbf{F}_i^r \cdot \delta r_i \tag{2}$$

which vanishes because $\mathbf{F}_i = 0$, by hypothesis.

The net virtual work of all forces acting on the entire system is found by summation over all particles to be

$$\delta W = \sum_i \delta W_i \equiv \delta W^a + \delta W^r = 0 \tag{3}$$

where δW^a and δW^r denote the virtual work of the active forces and the virtual work of the reactions. The latter term is defined by the expression

$$\delta W^r \equiv \sum_i \mathbf{F}_i^r \cdot \delta r_i = 0 \tag{4}$$

and is zero, by definition, for an *ideal* mechanical system. From another point of view, we may regard Eq. (4) as a basic *postulate* which defines the meaning of the term **ideal mechanical system**. Postulate (4) brings us to the crux of the matter, as noted by Lanczos [1966, p. 76] in the following trenchant comment:

> This postulate is not restricted to the realm of statics. It applies equally to dynamics, when the principle of virtual work is suitably generalized by means of d'Alembert's principle. Since all of the fundamental variational principles of mechanics—the principles of Euler, Lagrange, Jacobi, Hamilton —are but alternative mathematical formulations of d'Alembert's principle, Postulate (4) is actually the *only* postulate of analytic mechanics, and is thus of fundamental importance.

Recall, from Eq. (**11.22**-9), that the virtual work of the active forces is given by

$$\delta W^a \equiv \sum_{i=1}^{F} \mathbf{F}_i^a \cdot \delta r_i \equiv Q_1 \delta q_1 + Q_2 \delta q_2 + \cdots + Q_F \delta Q_F \tag{5}$$

where the Q_i are generalized forces and the q_i are increments in generalized coordinates.

Taking note of Eqs. (4) and (5), we may reduce Eq. (3) to the final desired result:

$$\delta W^a = Q_1 \delta q_1 + Q_2 \delta q_2 + \cdots + Q_F \delta q_F = 0 \tag{6}$$

Equation (6) shows that the net virtual work of all the active forces must of necessity vanish for a system in equilibrium. Part 1 of the proof is now complete.

Proof 2

The second part of the principle states that if the net virtual work of all the active forces vanishes for every set of virtual displacements, an ideal mechanical system is in a state of generalized equilibrium.

The proof starts by accepting as given the equation

$$\delta W^a = Q_1 \delta q + Q_2 \delta q_2 + \cdots + Q_F \delta q_F = 0 \tag{7}$$

We may now visualize a set of virtual displacements in which all δq_i are made to vanish except for δq_1. Hence Eq. (7) requires that

$$Q_1 \delta q_1 = 0 \tag{8}$$

Because δq_1 is not zero, Eq. (8) tells us that $Q_1 = 0$. By repeating the argument with all δq_i set equal to zero except for δq_j, we deduce that

$$Q_j = 0 \qquad (j = 1, 2, \ldots, F) \tag{9}$$

If all generalized forces vanish as stipulated by Eq. (9), the system is, by *definition, in a state of generalized equilibrium.* The proof is now complete.

To show why we use the qualifier *generalized* before the word *equilibrium*, consider the case of a particle on a smooth horizontal tabletop with an inextensible string of length L tied between the particle and a fixed point O on the table, as shown in Fig. 1. A suitable generalized coordinate for the problem is the polar angle θ, measured as shown. The radial force F_r due to the tension in the string does no work during a virtual displacement $\delta\theta$; hence the virtual work of the active forces is

$$\delta W^a = F_\theta L \, \delta\theta \tag{10}$$

and the generalized force is therefore $Q = F_\theta L$. If we are told that δW^a vanishes, it follows that $F_\theta = 0$. Note, however, that we can say nothing about the reaction F_r. Indeed it is perfectly possible for the particle to move with constant speed v in a circle

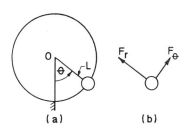

Figure 11.70-1. (a) Particle on a string; (b) Forces on the particle.

(a)   (b)

about point O, with $F_\theta = 0$. For such circular motion, the reaction F_r will not vanish but will have the finite value mv^2/L, where m is the mass of the particle. Thus, we see that *all* forces on the particle do not vanish and hence the particle is not in a state of "equilibrium" as usually defined. It is to describe such situations that we have coined the term *generalized equilibrium*.

If one stipulates the additional condition that all particles of the system are initially at rest (before the active forces are applied), it may be shown that the vanishing of δW^a assures that the system remains at rest and hence in a state of equilibrium. Hence no "centripetal" accelerations can develop and the qualifier *generalized* may be dropped from the expression *generalized equilibrium*. It is only for this restricted class of problems that one is justified in stating that the vanishing of δW^a is both necessary *and sufficient* for equilibrium.

EXAMPLE 1

A bead is free to slide without friction along a curved wire which lies in a horizontal plane. The only active force on the particle is the force of gravity. Is the particle in a state of equilibrium? Is it in a state of generalized equilibrium? Describe its state of motion.

Answers:

Not necessarily. Yes. It moves at constant speed (possibly zero speed) along the wire.

12

DYNAMICS OF SINGLE-FREEDOM SYSTEMS

12.10   INTRODUCTION

Given the active forces applied to a mechanical system and the initial state of the system, a detailed analysis of the subsequent motion entails three major steps:

1. Establishing the governing differential equations of motion
2. Integrating the equations of motion
3. Interpreting the results

Many of the simplest machines, such as the four-bar mechanism or reciprocating engine, are governed by nonlinear differential equations which cannot be solved in terms of well-known or "standard" functions. This means that it will usually be necessary to utilize a numerical or graphical method for step 2 of the process. Fortunately, the emergence of digital computers has taken the sting out of nonlinear differential equations, and we may consider the practical attack on step 2 to be a relatively straightforward and automatic procedure. To be sure, there are detailed questions of technique which require some attention, but these questions are best treated as part of the subject of numerical analysis.[1] For the present, we shall assume that step 2 can be treated in a uniform way by a standard subroutine for solving systems of differential equations with specified initial conditions.

Step 3 (interpretation) is too often dismissed as a trivial matter. However, in a complicated industrial setting, the interpretation of a mathematical solution can be the most difficult, and commercially important, phase of the problem. It is here that

---

[1]See Appendix **G**.

qualities of judgment, experience, and engineering insight are needed to evaluate the adequacy of the original model and to estimate the degree of validity of the predicted results. Because of the wide variety of technical aspects (e.g., manufacturing tolerances, stiffness, wear characteristics, etc.) and nontechnical aspects (e.g., marketing, man-machine interaction, safety, etc.) of the problem, no sustained attempt will be made to treat step 3 in this book.

Thus, the preeminent problem confronting us is step 1, namely to formulate the equations of motion in the most suitable way. What is considered to be "most suitable" depends on one's aims and point of view. For example, to a mathematician the most suitable formulation might be that which portrays the underlying mathematical structure of the problem; questions of existence, uniqueness, singularities, periodicity, and the relationship of solutions to known functions would be of interest and might best be examined from a formulation in terms of partial differential equations.[2] On the other hand, a physicist might wish to explore the relationship of classical mechanics to quantum mechanics and would therefore be most interested in expressing the equations of motion in terms which correspond to similar mathematical structures in quantum theory.[3]

Our point of view will be that of the engineer concerned with the performance, safety, and cost of a given machine—whether it be a large diesel engine, an orbiting satellite in the process of deploying large solar panels, or an intricate high-speed printer. For such systems it is necessary to have detailed knowledge of velocities for technical and economic performance estimates, knowledge of bearing and piston reactions for wear and lubrication analyses, and knowledge of accelerations for stress and safety analyses. Furthermore, we must be able to include a wide range of applied forces such as gas or hydraulic pressures in cylinders, aerodynamic and hydrodynamic forces, linear and nonlinear spring forces, and damping of mechanical or electromagnetic types. We must also be able to treat open and closed kinematic chains.

To successfully cope with this wide range of problems we shall use certain concepts from analytical mechanics. However, many practical machines and devices, such as internal combustion engines, flyball governors, punch presses, etc., may be idealized as systems with a single degree of freedom. Such systems are governed by a single differential equation of motion which is readily deduced from the *theorem of power balance*. This theorem, which is well known to students of elementary mechanics, states that the power supplied to an ideal mechanical system equals the rate of increase of the system's kinetic energy.

Accordingly, we shall develop the dynamics of single-freedom systems from considerations of power balance and shall call upon the more general methods of analytical mechanics only when we desire to investigate systems with two or more DOF.

12.20 NOTATION

In this chapter we shall focus our attention on ideal systems which, by definition, consist of an assemblage of resistant links joined together at frictionless joints. The effects of elasticity or damping in the system will be modeled by springs and dashpots

---

[2]See the discussion of Hamilton-Jacobi theory in Goldstein [1950, Chap. 2].
[3]See the discussion of Poisson brackets in Goldstein [1950, Chap. 8].

attached to appropriate points on the links. The methods to be presented in this chapter are applicable to all single-freedom systems, independently of the number of members or kinematic loops associated with the system.

The forces which act on the system may be prescribed as arbitrary functions of the system configuration and of the velocities of its various parts; they may also depend explicitly on time. For example, the gas pressure p acting on the piston of the engine in Fig. 1(a) is given by the *indicator diagram* of Fig. 1(b). The resisting torque shown in Fig. 1(c) could represent the load characteristic of an electric generator, a propeller, or other driven device.

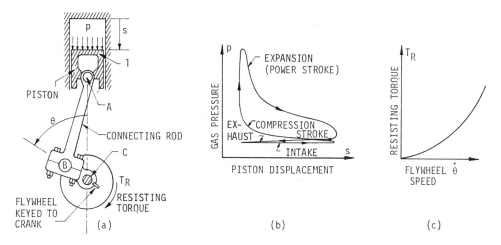

Figure 12.20-1. Four stroke reciprocating engine: (a) Engine cross-section; (b) Gas pressure vs piston displacement; (c) Load characteristic.

The inertial properties of a typical link are characterized by m_i, its mass, and J_i, its moment of inertia (**MOI**) with respect to its center of mass (**C.M.**); the subscript i represents the link identification number. The coordinates of the C.M. for link i will be specified by (x_i, y_i) measured along fixed orthogonal axes x and y.

Specified forces X_i and Y_i are assumed to act through the C.M. of link i (in the direction of the fixed axes x and y), and an active torque[4] M_i acts about the C.M. Any active forces which do not pass through the C.M. may be replaced by a set of *equipollent* forces acting on the C.M. together with a couple about the C.M. as explained in connection with Fig. **11.10**-2.

A massless spring may be replaced by a pair of equal but opposite collinear forces which act on the links at the terminal points of the spring (as in Fig. **11.41**-1). Note that there is no requirement for the springs to have linear force-displacement characteristics. Similarly, velocity-dependent dampers may be replaced by pairs of equal and opposite collinear forces which are dependent on the rate of relative extension of the line joining the terminals of the damper. Coulomb damping may be modeled as in Secs. **11.61** and **11.62**.

---

[4]The symbol M_i, signifying a moment, must not be confused with m_i, which denotes a mass.

Recall [Eq. (**11.22**-4a)] that the instantaneous power (rate of work) of all the active forces on link i is given by

$$P_i = X_i \dot{x}_i + Y_i \dot{y}_i + M_i \dot{\theta}_i \tag{1}$$

where $(\dot{x}_i, \dot{y}_i)$ are the velocities of the C.M. and $\dot{\theta}_i$ is the angular velocity of link i.

The net power of all active forces applied to the linkage system (including forces due to springs and dampers) is given by

$$P = \sum_i P_i = Q\dot{q} \tag{2}$$

where Q is the generalized force associated with the generalized coordinate q, as defined by Eq. (**11.22**-8b).

Alternatively, we can include the effect of elastic (i.e., reversible) springs by introducing the potential energy function V, which represents the energy stored in all such springs of the system.[5]

Then, as shown by Eq. (**11.42**-14), the springs contribute a term

$$Q_s = -\frac{\partial V}{\partial q} \tag{3}$$

to the generalized force, so that the power of the active forces is given by

$$P = \left[Q^{(e)} - \frac{\partial V}{\partial q} \right] \dot{q} \tag{4}$$

where $Q^{(e)}$ is the generalized force due to all active forces other than spring forces.

Equation (4) may be interpreted more generally by treating V as the potential energy of any conservative system of active forces and $Q^{(e)}$ as the generalized force associated with all other working forces.

12.30   KINETIC ENERGY AND GENERALIZED INERTIA

From elementary dynamics, we know that the kinetic energy of a typical link i is given by

$$T_i = \tfrac{1}{2} m_i (\dot{x}_i^2 + \dot{y}_i^2) + \tfrac{1}{2} J_i \dot{\theta}_i^2 \tag{1}$$

Recall from Eqs. (**10.50**-2) and (**10.50**-3) that

$$\dot{x}_i = u_i \dot{q}, \qquad \dot{y}_i = v_i \dot{q}, \qquad \dot{\theta}_i = \omega_j \dot{q} \tag{2}$$

where the *velocity coefficients* (u_i, v_i, ω_i) are functions of position only.

Accordingly, Eq. (1) can be recast in the form

$$T_i = \tfrac{1}{2} [m_i (u_i^2 + v_i^2) + J_i \omega_i^2] \dot{q}^2 \tag{3}$$

---

[5] Recall from Eqs. (**11.42**-20) and (**11.42**-22) that the potential of a linear spring is $V = ke^2/2$, where k is the spring's stiffness and e is its extension.

and the kinetic energy of the entire system of M moving links can be expressed as

$$T = \sum_{i=1}^{M} T_i = \tfrac{1}{2}\mathscr{I}\dot{q}^2 \tag{4}$$

where

$$\mathscr{I} \equiv \sum_{i=1}^{M} [m_i(u_i^2 + v_i^2) + J_i\omega_i^2] \tag{5}$$

From Eq. (4) we see that $\mathscr{I}$ plays the role of a **generalized inertia coefficient**.[6] For a translating rigid body, $\mathscr{I}$ would simply be the mass m. For general plane motion of a rigid body with angular velocity $\dot{q}$, $\mathscr{I}$ is the MOI with respect to the instantaneous center. It may be observed from Eq. (5) that $\mathscr{I}$ can never be negative so long as m_i and J_i are positive, as they must be in a real mechanism.

Because the concept of *generalized inertia* will play a central role in the dynamics of single-freedom systems, it is worthwhile to give a few examples of how it may be calculated.

EXAMPLE 1. Open-Loop System

In Fig. 1, a slider (2) of mass m_2 and MOI J_2 with respect to its C.M. at P glides smoothly along a pendulum (1) suspended from a fixed pivot A. The C.M. of the pendulum is at G, a distance b from A, and its MOI with respect to G is J_1. The slider is attached to a massless cable which is wrapped around a fixed disk, of radius a, which is concentric with A. The pendulum makes an angle θ with the downward vertical, and the C.M. of the slider is at distance r from A. Find the generalized inertia coefficient corresponding to the generalized coordinate θ.

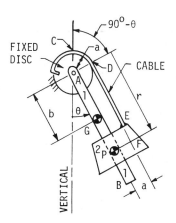

Figure 12.30-1. Example 1.

The kinetic energy of the pendulum is

$$T_1 = \tfrac{1}{2}J_1\dot{\theta}_1^2 + \tfrac{1}{2}m_1(b\dot{\theta})^2 = \tfrac{1}{2}J_{1A}\dot{\theta}^2$$

where $J_{1A} \equiv J_1 + m_1b^2$. The kinetic energy of the slider is

$$T_2 = \tfrac{1}{2}J_2\dot{\theta}^2 + \tfrac{1}{2}m_2[\dot{r}^2 + (r\dot{\theta})^2] \tag{6}$$

---

[6]Other names for $\mathscr{I}$ include *reduced mass* (Wittenbauer [1923, p. 710]), *reduced moment of inertia* (Timoshenko and Young [1948, p. 175]), and *effective moment of inertia* (Biezeno and Grammel [1954, p. 88]).

Noting that $CD + DF = \text{constant} = a[(\pi/2) - \theta] + r$, we can write

$$r = a\theta + \text{constant} = a\theta + r_0 \qquad (7)$$

where r_0 is the value of r when $\theta = 0$. From Eq. (7) we see that

$$\dot{r} = a\dot{\theta} \qquad (8)$$

When Eq. (8) is substituted into Eq. (6) we find the total kinetic energy:

$$T = T_1 + T_2 = \tfrac{1}{2}\mathcal{I}\dot{\theta}^2 = \tfrac{1}{2}[J_{1A} + J_2 + m_2(a^2 + r^2)]\dot{\theta}^2$$

The term in brackets is the generalized inertia. Utilizing Eq. (7), we can express $\mathcal{I}$ in the form

$$\mathcal{I} = J_{1A} + J_2 + m_2[a^2 + (r_0 + a\theta)^2] \qquad (9)$$

This example illustrates that for open-loop systems the generalized inertia is most conveniently found by a direct calculation of the kinetic energy.

EXAMPLE 2. Closed-Loop System

Find the generalized inertia for the four-bar mechanism of Fig. 2, where the C.M. of link i is located at $P_i (i = 1, 2, 3)$, and θ_3 is the generalized coordinate q.

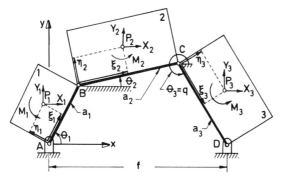

Figure 12.30-2. Example 2.

The kinetic energies of the side links are

$$T_1 = \tfrac{1}{2}J_{1A}\dot{\theta}_1^2 \qquad (10)$$

$$T_3 = \tfrac{1}{2}J_{3D}\dot{q}^2 \qquad (11)$$

where J_{1A} and J_{3D} are the moments of inertia of links 1 and 3 with respect to the fixed pivots A and D. The kinetic energy of the coupler link (2) is

$$T_2 = \tfrac{1}{2}J_2\dot{\theta}_2^2 + \tfrac{1}{2}m_2(\dot{x}_2^2 + \dot{y}_2^2) \qquad (12)$$

We now utilize the fact that

$$\dot{\theta}_1 = \omega_1\dot{q}, \qquad \dot{\theta}_2 = \omega_2\dot{q}, \qquad \dot{x}_2 = u_2\dot{q}, \qquad \dot{y}_2 = v_2\dot{q} \qquad (13)$$

where, from Eq. (10.51-5b), we see that

$$\omega_1 = \frac{\dot{\theta}_1}{\dot{\theta}_3} = \frac{a_3}{a_1} \frac{\sin(\theta_2 - q)}{\sin(\theta_1 - \theta_2)}; \tag{14}$$

$$\omega_2 = \frac{\dot{\theta}_2}{\dot{\theta}_3} = \frac{a_3}{a_2} \frac{\sin(q - \theta_1)}{\sin(\theta_1 - \theta_2)} \tag{15}$$

and, from Eq. (**10.51**-19), we see that

$$\begin{aligned} u_2 &= -y_{BA}\omega_1 - y_{2B}\omega_2 \\ v_2 &= x_{BA}\omega_1 + x_{2B}\omega_2 \end{aligned} \tag{16}$$

where

$$\begin{aligned} x_{BA} &= a_1 \cos\theta_1, \qquad y_{BA} = a_1 \sin\theta_1 \\ x_{2B} &= \xi_2 \cos\theta_2 - \eta_2 \sin\theta_2 \\ y_{2B} &= \xi_2 \sin\theta_2 + \eta_2 \cos\theta_2 \end{aligned} \tag{17}$$

Upon substitution of Eqs. (13) into Eqs. (10)–(12), we find

$$T = T_1 + T_2 + T_3 = \tfrac{1}{2}\mathcal{I}\dot{q}^2 \tag{18}$$

where

$$\mathcal{I} \equiv J_{1A}\omega_1^2 + J_{3D} + J_2\omega_2^2 + m_2(u_2^2 + v_2^2) \tag{19}$$

It may be verified that the above expression for $\mathcal{I}$ also follows directly from the general expression

$$\mathcal{I} = \sum_{i=1}^{3} [m_i(u_i^2 + v_i^2) + J_i\omega_i^2] \tag{20}$$

provided by Eq. (5) when the appropriate values of u_i and v_i are obtained from Eqs. (**10.51**-18)–(**10.51**-21).

12.31 *Dynamically Equivalent Links*

Although the generalized inertia may always be calculated from the defining equation (**12.30**-5), simplifications are possible in special cases. For example, suppose [as in Fig. 1(a)] that a link's mass center G lies on a line AB, fixed in the link, and that the speeds (v_A, v_B) of points A and B are known.

The speed v_G of the mass center is given by Eq. (**5.31**-12) in the form

$$v_G^2 = v_A^2 \frac{b}{L} + v_B^2 \frac{a}{L} - ab\dot{\phi}^2 \tag{1}$$

where a and b are the oriented distances AG and GB respectively, and $\dot{\phi}$ is the angular velocity of the link. If m is the link's mass and J_G its MOI (about G), its kinetic energy is therefore given by

$$\begin{aligned} T &= \frac{m}{2}v_G^2 + \frac{J_G}{2}\dot{\phi}^2 \\ &= \frac{1}{2}m\frac{b}{L}v_A^2 + \frac{1}{2}m\frac{a}{L}v_B^2 + \frac{1}{2}(J_G - mab)\dot{\phi}^2 \end{aligned} \tag{2}$$

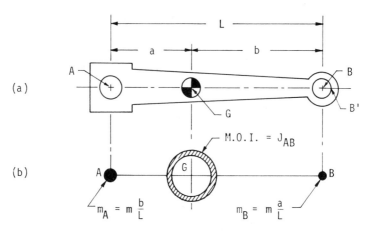

Figure 12.31-1. (a) Given link; (b) Dynamically equivalent replacement.

Upon introducing the notation

$$m_A = m\frac{b}{L}, \qquad m_B = m\frac{a}{L} \tag{3}$$

$$r_G^2 = \frac{J_G}{m}, \qquad J_{AB} = J_G - mab = m(r_G^2 - ab) \tag{4}$$

we see that the kinetic energy of the link is given by

$$T \equiv \tfrac{1}{2}\mathcal{I}_{AB}\dot{q}^2 = \tfrac{1}{2}(m_B v_B^2 + m_A v_A^2 + J_{AB}\dot{\phi}^2) \tag{5}$$

and the generalized inertia of the link is accordingly

$$\mathcal{I}_{AB} = m_B\left(\frac{v_B}{\dot{q}}\right)^2 + m_A\left(\frac{v_A}{\dot{q}}\right)^2 + J_{AB}\left(\frac{\dot{\phi}^2}{\dot{q}}\right)^2 \tag{6}$$

The general expression (5) suggests that any link may be replaced [as indicated in Fig. 1(b)] by an equivalent link consisting of a straight rigid massless rod AGB with point masses (m_A, m_B) at its ends (A, B) and a ring of negligible mass but finite MOI (J_{AB}) located anywhere along line AB.

Equations (3) show that the original link and the replacement system have the same total mass

$$m_A + m_B = m \tag{7}$$

and the same centroid, since

$$m_A a = m_B b \tag{8}$$

The centroidal moment of inertia for the substitute link is given by

$$J' = m_A a^2 + m_B b^2 + J_{AB} \tag{9}$$

Upon using the definitions of m_A, m_B, and J_{AB} provided by Eqs. (3) and (4), we find

$$J' = mba\left(\frac{b+a}{L}\right) + J_G - mab = J_G \qquad (10)$$

which shows that the substitute link has the proper MOI.

It is readily seen that the kinetic energy of this replacement link is given by Eq. (5), which was derived for the original link.

In summary, any link may be replaced (*conceptually*) by two lumped masses and a massless ring with finite MOI such that the original link and substitute link have the same

1. Mass
2. Centroidal location
3. Centroidal moment of inertia

Any replacement systems for which conditions (1) and (2) are satisfied [i.e., Eqs. (3) hold] are said to be **statically equivalent**. If, in addition, condition (3) is satisfied [i.e., Eq. (10) holds], the systems are said to be **dynamically equivalent**. This *conceptual* replacement may not be *physically* realizable, since the term J_{AB} will often be negative (see Problems **12.100**-31, 32, and 33). Nevertheless, Eq. (5) always gives the correct kinetic energy regardless of the sign of J_{AB}; it is not essential that any physical interpretation be attached to the term J_{AB}.

Several different schemes exist for generating **dynamically equivalent** links.[7] None of them are strictly necessary since the generalized inertia may always be calculated from the general equation (**12.30**-5) or from its special case, Eq. (6). The latter form is especially convenient when the speeds at points A and B are readily expressed in terms of the generalized velocity, as in the following example.

EXAMPLE 1. Kinetic Energy of a Four-bar Linkage

We consider the frequently occurring case where the mass center of the coupler link, in Fig. **12.30**-2, lies on the center line BC at distance ξ_2 from B. The kinetic energy of the coupler link is therefore

$$T_2 = \tfrac{1}{2}(m_B v_B^2 + m_C v_C^2 + J_{BC}\dot{\theta}_2^2) \qquad (11)$$

where

$$m_B = \frac{m_2(a_2 - \xi_2)}{a_2}$$

$$m_C = \frac{m_2\xi_2}{a_2} \qquad (12)$$

$$J_{BC} = J_2 - m_2\xi_2(a_2 - \xi_2)$$

Since

$$v_B^2 = (a_1\dot{\theta}_1)^2, \qquad v_C^2 = (a_3\dot{\theta}_3)^2$$

the coupler's kinetic energy can be expressed as

$$T_2 = \tfrac{1}{2}[m_B(a_1\dot{\theta}_1)^2 + m_C(a_3\dot{\theta}_3)^2 + J_{BC}\dot{\theta}_2^2] \qquad (13)$$

---

[7]See Hirschhorn [1962, pp. 152–156] and Beyer [1960, pp. 50–52].

Thus the total kinetic energy for the system is given by

$$T = T_1 + T_2 + T_3 = \tfrac{1}{2}J_{1A}\dot{\theta}_1^2 + T_2 + \tfrac{1}{2}J_{3D}\dot{\theta}_3^2 \tag{14}$$

Utilizing the standard notation

$$\dot{\theta}_i = \omega_i \dot{q}$$

in Eq. (13) and (14), we find that the kinetic energy of a four-bar mechanism (when the coupler centroid lies on the coupler axis) is

$$T = \tfrac{1}{2}\mathcal{I}\dot{q}^2$$

where

$$\mathcal{I} = (J_{1A} + m_B a_1^2)\omega_1^2 + J_{BC}\omega_2^2 + (J_{3D} + m_C a_3^2)\omega_3^2 \tag{15}$$

The generalized inertia may be expressed more compactly as

$$\mathcal{I} = J_1^*\omega_1^2 + J_2^*\omega_2^2 + J_3^*\omega_3^2 \tag{16}$$

where

$$J_1^* \equiv J_{1A} + m_B a_1^2 = J_1 + m_1(\xi_1^2 + \eta_1^2) + m_2\left(1 - \frac{\xi_2}{a_2}\right)a_1^2 \tag{17a}$$

$$J_2^* \equiv J_{BC} = J_2 - m_2\xi_2(a_2 - \xi_2) \tag{17b}$$

$$J_3^* \equiv J_{3D} + m_C a_3^2 = J_3 + m_3[(a_3 - \xi_3)^2 + \eta_3^2] + m_2\frac{\xi_2}{a_2}a_3^2 \tag{17c}$$

The appropriate values of ω_i to be used in Eq. (16) depend on which of the angles $(\theta_1, \theta_2, \theta_3)$ are to be chosen as generalized coordinates. If, for example, $\theta_3 = q$, then $\omega_3 = 1$, and ω_1, ω_2 are given by Eqs. (12.30-14), and (12.30-15).

12.40 THE GENERALIZED EQUATION OF MOTION

For an ideal mechanism, it follows from elementary dynamics, or from the first law of thermodynamics, that the power input of the active forces equals the rate of increase of the mechanism's kinetic energy; that is,

$$P = \frac{dT}{dt} \equiv \dot{T} \tag{1}$$

Recall that for a system with 1 DOF specified by the generalized coordinate q,

$$P = Q\dot{q} \quad \text{and} \quad T = \tfrac{1}{2}\mathcal{I}\dot{q}^2 \tag{2}$$

Hence the power balance equation (1) assumes the form

$$\frac{d}{dt}\left(\frac{1}{2}\mathcal{I}\dot{q}^2\right) = Q\dot{q} \tag{3}$$

or

$$\frac{1}{2}\mathcal{I}2\dot{q}\ddot{q} + \frac{1}{2}\frac{d\mathcal{I}}{dt}\dot{q}^2 = Q\dot{q} \tag{4}$$

Recognizing that $\mathcal{G}$ can be expressed as a function of q, we may write

$$\frac{d\mathcal{G}}{dt} = \frac{d\mathcal{G}(q)}{dq}\frac{dq}{dt} \tag{5}$$

Thus Eq. (4) assumes the final form

$$\boxed{\mathcal{G}\ddot{q} + \frac{1}{2}\frac{d\mathcal{G}}{dq}\dot{q}^2 = Q} \tag{6}$$

Equation (6) is the **generalized equation of motion**[8] for all ideal single-freedom systems. By including the effects of elastic restoring forces and frictional forces in the generalized force Q, we may apply Eq. (6) to the widest variety of practical problems.

Frequently we shall find it convenient to express the generalized equation (6) in the form

$$\mathcal{G}\ddot{q} + \mathcal{C}\dot{q}^2 = Q \tag{7}$$

where

$$\mathcal{C} \equiv \frac{1}{2}\frac{d\mathcal{G}}{dq} \tag{8}$$

A general expression for $\mathcal{C}$ follows upon differentiation of the general expression (**12.30**-5) for $\mathcal{G}$; i.e.,

$$\mathcal{G} = \sum_{i=1}^{M}[m_i(u_i^2 + v_i^2) + J_i\omega_i^2] \tag{9}$$

$$\mathcal{C} = \frac{1}{2}\frac{d\mathcal{G}}{dq} = \sum_{i=1}^{M}[m_i(u_iu_i' + v_iv_i') + J_i\omega_i\omega_i'] \tag{10}$$

where $(\ )' \equiv d(\ )/dq$.

A formal expression for the generalized force Q follows from the expression for the power of the active forces,

$$Q\dot{q} = \sum_{i=1}^{M}(X_i\dot{x}_i + Y_i\dot{y}_i + M_i\dot{\theta}_i)$$

which reduces to

$$Q = \sum_{i=1}^{M}(X_iu_i + Y_iv_i + M_i\omega_i) \tag{11}$$

upon utilization of the velocity ratios $u_i = \dot{x}_i/\dot{q}, \ldots .$

From the defining relationship $T = \frac{1}{2}\mathcal{G}\dot{q}^2$, we see that

$$|\dot{q}| = \sqrt{2\left(\frac{T}{\mathcal{G}}\right)} = \sqrt{2\tan\gamma} \tag{12}$$

---

[8]This equation plays a central role in a series of papers on the dynamical analysis of machines by Eksergian [1930–1931]. Wittenbauer [1923, p. 669] recognized that Eq. (6), which he called the *fundamental dynamic law*, is a generalization of Newton's second law. Curiously, Wittenbauer never really *applied* the equation anywhere in his 800-page treatise.

where

$$\gamma = \tan^{-1}\left(\frac{T}{\mathcal{I}}\right) \tag{13}$$

If the kinetic energy is plotted against the generalized inertia, as shown schematically in Fig. 1, we obtain a curve called the *Massen-Wucht* diagram by Wittenbauer [1923, p. 735]. From such a diagram, which we shall call the **energy-inertia diagram**, it is possible to graphically find the extreme values of the generalized speed:

$$\begin{aligned}
|\dot{q}|_{\max} &= \sqrt{2 \tan \gamma_{\max}} \\
|\dot{q}|_{\min} &= \sqrt{2 \tan \gamma_{\min}}
\end{aligned} \tag{14}$$

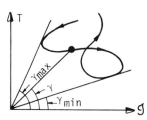

Figure 12.40-1. Energy-inertia diagram.

Extremal speeds are of interest in problems of flywheel design and in the study of speed fluctuations (see Sec. **12.93**).

Equation (6) may be recast into another useful form by noting that since $\mathcal{I}$ is a function of q, the kinetic energy may be expressed as a function of q and $\dot{q}$ of the form

$$T(q, \dot{q}) = \tfrac{1}{2}\mathcal{I}(q)\dot{q}^2 \tag{15}$$

Therefore

$$\frac{\partial T(q, \dot{q})}{\partial \dot{q}} = \mathcal{I}(q)\dot{q} \tag{16}$$

$$\frac{\partial T(q, \dot{q})}{\partial q} = \frac{1}{2}\frac{d\mathcal{I}(q)}{dq}\dot{q}^2 \tag{17}$$

Differentiation across Eq. (16) gives

$$\frac{d}{dt}\frac{\partial T(q, \dot{q})}{\partial \dot{q}} = \mathcal{I}(q)\ddot{q} + \frac{d\mathcal{I}(q)}{dq}\frac{dq}{dt}\dot{q} \tag{18}$$

Upon substituting $\mathcal{I}\ddot{q}$ from Eq. (18) into Eq. (6) and utilizing Eq. (17) to eliminate $d\mathcal{I}/dq$, we find

$$\frac{d}{dt}\frac{\partial T(q, \dot{q})}{\partial \dot{q}} - \frac{\partial T(q, \dot{q})}{\partial q} = Q \tag{19}$$

Equation (19) is a special case of **Lagrange's equations of motion**. The general form of Lagrange's equations for a system with F DOF is

$$\frac{d}{dt}\frac{\partial T}{\partial \dot{q}_r} - \frac{\partial T}{\partial q_r} = Q_r \qquad (r = 1, 2, \ldots, F) \tag{20}$$

Both $\mathcal{I}$ and $\mathcal{C}$ are functions of q, and Q is generally a function of both q and $\dot{q}$. Therefore Eq. (7) is a nonlinear second-order ordinary differential equation for $q(t)$.

Any such equation may be integrated numerically (e.g., by the Runge-Kutta algorithm) provided that the *initial values* of q and $\dot{q}$ are specified.

A simple example will be used to illustrate the application of the generalized equation of motion.

EXAMPLE 1. Oscillating Elliptic Trammel

A uniform rod AB of mass m_1, length L, and MOI $J_1 = m_1 L^2/12$ is attached to sliders of mass m_2 and m_3 at A and B, respectively. Point A slides along the downward vertical y axis, and point B slides along the horizontal x axis, as shown in Fig. 2, where θ (positive clockwise) is the angle between BA and the downward vertical. We seek the equation of motion under the influence of gravity forces.

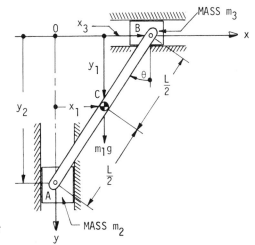

Figure 12.40-2. Elliptic trammel under gravity loading.

To evaluate the generalized inertia coefficient $\mathcal{I}$, we record the coordinates of the mass center of each body:

$$
\begin{aligned}
x_1 &= \tfrac{1}{2}L \sin \theta, & y_1 &= \tfrac{1}{2}L \cos \theta \\
x_2 &= 0, & y_2 &= L \cos \theta \\
x_3 &= L \sin \theta, & y_3 &= 0
\end{aligned}
\tag{21}
$$

The associated velocities are

$$
\begin{aligned}
\dot{x}_1 &= \tfrac{1}{2}L\dot{\theta} \cos \theta, & \dot{y}_1 &= -\tfrac{1}{2}L\dot{\theta} \sin \theta \\
\dot{x}_2 &= 0, & \dot{y}_2 &= -L\dot{\theta} \sin \theta \\
\dot{x}_3 &= L\dot{\theta} \cos \theta, & \dot{y}_3 &= 0
\end{aligned}
\tag{22}
$$

The kinetic energy is therefore

$$
T = \tfrac{1}{2}m_1(\dot{x}_1^2 + \dot{y}_1^2) + \tfrac{1}{2}J_1\dot{\theta}_1^2 + \tfrac{1}{2}m_2\dot{y}_2^2 + \tfrac{1}{2}m_3\dot{x}_3^2
\tag{23}
$$

Upon substituting Eqs. (22) into Eq. (23), we find

$$
T = \frac{1}{2}\left[m_1\left(\frac{L}{2}\right)^2 + J_1 + m_2 L^2 \sin^2 \theta + m_3 L^2 \cos^2 \theta \right]\dot{\theta}^2
\tag{24}
$$

If θ is taken as the generalized coordinate, the generalized inertia $\mathscr{I}$ is the quantity within brackets; i.e.,

$$\mathscr{I} = J_A + L^2(m_2 \sin^2 \theta + m_3 \cos^2 \theta) \tag{25}$$

where $J_A \equiv J_1 + m_1 L^2/4 = m_1 L^2/3$.

By definition,

$$\mathfrak{C} \equiv \frac{1}{2}\frac{d\mathscr{I}}{d\theta} = L^2(m_2 - m_3) \sin \theta \cos \theta \tag{26}$$

The power of the gravity forces is

$$P \equiv Q\dot{\theta} = m_1 g_1 \dot{y}_1 + m_2 g \dot{y}_2 + m_3 g \dot{y}_3 = -(\tfrac{1}{2}m_1 L \sin \theta + m_2 L \sin \theta)g\dot{\theta} \tag{27}$$

where use has been made of Eqs. (22). Thus the generalized force corresponding to θ is

$$Q = -gL(m_2 + \tfrac{1}{2}m_1) \sin \theta \tag{28}$$

The generalized equation of motion (7) is therefore

$$[\tfrac{1}{3}m_1 L^2 + L^2(m_2 \sin^2 \theta + m_3 \cos^2 \theta)]\ddot{\theta}$$
$$+ \tfrac{1}{2}L^2(m_2 - m_3)\dot{\theta}^2 \sin 2\theta + gL(m_2 + \tfrac{1}{2}m_1) \sin \theta = 0 \tag{29}$$

Equation (29) is a nonlinear differential equation which will, in general, have to be integrated numerically. However, for the special case $m_2 = m_3$, it takes the form

$$\ddot{\theta} + \frac{g}{L}\frac{m_2 + \tfrac{1}{2}m_1}{m_2 + \tfrac{1}{3}m_1} \sin \theta = 0 \tag{30}$$

Equation (30) is the equation of a mathematical pendulum of length

$$L_e \equiv L\frac{m_2 + \tfrac{1}{3}m_1}{m_2 + \tfrac{1}{2}m_1} \tag{31}$$

Therefore, the general character of the motion can be deduced from the known behavior of a pendulum. In particular, the motion approximates simple harmonic oscillation, with period $\tau = 2\pi\sqrt{L_e/g}$, provided that the angle θ remains small. For sufficiently large initial velocities, the pendulum (and hence the trammel) will go "over the top" and rotate continuously with $\dot{\theta}$ fluctuating but never changing sign.

12.50 THE ENERGY INTEGRAL METHOD FOR CONSERVATIVE SYSTEMS

12.51 General Results

Although Eq. (12.40-6) is the starting point for the solution of general problems, certain simplifications arise for the special case of *conservative* systems. Conservative systems, as defined in Sec. 11.42, are characterized by the fact that the work W of the generalized force Q is a unique function of the generalized coordinate q; i.e.,

$$W = \int_{q_0}^{q} Q \, dq = -[V(q) - V(q_0)] \tag{1}$$

where $V(q)$ is the potential energy and q_0 is an arbitrary reference position.

Note that the power balance equation, $\dot{T} = Q\dot{q}$, can be expressed in the differential form

$$dT = Q \, dq \tag{2}$$

If the system is conservative, Eq. (1) holds, and Eq. (2) may be integrated to give the kinetic energy explicitly as

$$T - T_0 = \int_{T_0}^{T} dT = \int_{q_0}^{q} Q \, dq = -[V - V_0]$$

or

$$T + V = T_0 + V_0 \equiv E = \text{constant} \tag{3}$$

where T_0 is the value of the kinetic energy when $q = q_0$, and $V_0 = V(q_0)$. Equation (3) states the well-known fact that the total energy E (kinetic plus potential) of a conservative system remains constant at all times.

Since $T \equiv \frac{1}{2}\mathcal{g}\dot{q}^2$, Eq. (3) can be written in the form

$$\tfrac{1}{2}\mathcal{g}\dot{q}^2 = E - V \tag{4}$$

Therefore, the velocity is given explicitly by

$$\frac{dq}{dt} = \dot{q} = \pm \left\{ \frac{2[E - V(q)]}{\mathcal{g}(q)} \right\}^{1/2} \tag{5}$$

where the choice of plus or minus sign depends on considerations of continuity, to be discussed in Sec. **12.52**.

Equation (5), which gives velocity as a function of position, is often called a **first integral** or **energy integral** of the governing second-order differential equation. To complete the solution, Eq. (5) is written in the form

$$dq \left\{ \frac{2[E - V(q)]}{\mathcal{g}(q)} \right\}^{-1/2} = \pm dt \tag{6}$$

Integrating on the left between q_0 and q and on the right between 0 and t, we find a **second integral** in the form

$$t = \pm \int_{q_0}^{q} \left\{ \frac{\mathcal{g}(q)}{2[E - V(q)]} \right\}^{1/2} dq \tag{7}$$

In general, the integral (7) will have to be evaluated by a numerical quadrature procedure (e.g., Simpson's rule), but it does give the required explicit numerical relationship between time and displacement. Thus Eq. (7) represents a general and complete solution for all single-freedom conservative systems.

12.52 Qualitative Discussion—Stability of Equilibrium

A qualitative picture of the system's motion can be obtained from a graph of the potential energy function $V(q)$, as shown by the curve in Fig. 1 for a hypothetical system.

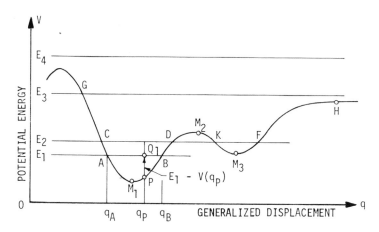

Figure 12.52-1. Potential energy function.

If we draw a horizontal line at a height equal to the total energy E of the system, we see that the directed distance PQ from a representative point P on the curve to the energy line gives the energy difference $E - V$. For example, when $E = E_1$, $PQ_1 = E_1 - V(q_P)$. Then Eq. (**12.51**-4) implies that the directed distance PQ equals $\frac{1}{2}\mathscr{I}\dot{q}^2$. Since $\mathscr{I}$ is an intrinsically positive quantity, we see that PQ must be positive. Thus, for a total energy corresponding to E_1, point P must remain below the line AB, and the motion would be restricted to the range $q_A \leq q \leq q_B$; that is, the system would oscillate between two well-defined configurations. As the representative point P travels from A to B, $\dot{q} > 0$, and the positive sign would have to be used in Eq. (**12.51**-5). When P reaches B, the velocity vanishes, and as P moves back from B toward A, the negative sign is required in Eq. (**12.51**-5). Thus we see that the signs to be used in Eqs. (**12.51**-5)–(**12.51**-7) are initially those which make $\dot{q}$ consistent with the given initial velocity; thereafter, a reversal of sign takes place whenever $\dot{q}$ becomes zero.

For an energy level E_2, the system could oscillate between points C and D or between points K and F, depending on the initial state. However, it is not possible for the representative point P to cross the **potential barrier** DM_2K.

For a higher total energy content E_3, P can move freely past the barrier at M_2. If, for example, the motion starts at q_P with $\dot{q} < 0$, P will move to the left until it comes to rest at G; then $\dot{q}$ will reverse sign, and P will move toward the right past M_1, M_2, M_3, and H. If the potential curve never again crosses the line E_3, q will increase indefinitely.

Finally, if the energy level is so high (e.g., E_4) that the potential curve never reaches the energy line, the point P will move toward $+\infty$ or $-\infty$ accordingly as $\dot{q}$ is initially positive or negative.

In general, we can get a reasonable qualitative picture of the motion by visualizing the potential curve as a roller-coaster track with P representing a car moving under the action of gravity. Although the height of P gives the correct gravitational potential, the generalized *mass* $\mathcal{G}(q)$ gradually changes with position. Therefore the roller-coaster analogy can only give a general idea of the motion, but it is qualitatively correct in all respects.

It is left as a problem to show that the time required for P (in Fig. 1) to move between A and B is finite when the energy of the system is E_1 as shown. It may also be verified that the time required to go from A to B is the same as that to go from B to A and that the motion is periodic with period

$$\tau = 2 \int_{q_A}^{q_B} \left\{ \frac{\mathcal{G}(q)}{2[E_1 - V(q)]} \right\}^{1/2} dq \tag{1}$$

The potential curve is also a useful guide to the study of equilibrium. We recall, from Eq. (**11.42**-14), that the generalized force Q is given by

$$Q = -\frac{dV}{dq} \tag{2}$$

Hence Q vanishes at each of the extremum points (such as M_1, M_2, M_3) of the potential function, and the system is therefore in a state of *generalized equilibrium* at such points.

If $Q = 0$ and simultaneously $\dot{q} = 0$, it follows from the generalized equation of motion that

$$\mathcal{G}\ddot{q} = Q - \mathcal{C}\dot{q}^2 = 0 \tag{3}$$

By successive differentiations of Eq. (3) it may be shown that $\ddot{q}$ and all higher derivatives vanish at the equilibrium point. Therefore, the system remains at rest forever at the equilibrium point. With $\dot{q} = 0$ and $dV/dq = 0$, the energy line E (in Fig. 2) will just touch the curve of the extremum point M.

Now suppose that the system is given a small impulse so that $\dot{q}$ suddenly acquires a small increment $\Delta\dot{q}$.[9] This increases the kinetic energy and raises the energy line by

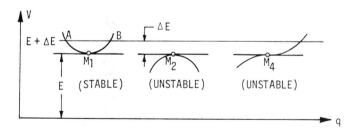

Figure 12.52-2. Illustrating stability of equilibrium.

---

[9]**Generalized impulse** is defined as $\int_0^t Q \, dt$. For impulses of vanishingly small duration (ϵ) it follows, from Eq. (3), that the jump in velocity $\Delta\dot{q}$ is given by $\mathcal{G}\Delta\dot{q} = \lim_{\epsilon \to 0} \int_0^\epsilon Q \, dt$, and $\Delta\dot{q}$ will be finite for finite impulse. The corresponding jump in displacement would then be $\Delta q = \dot{q}_{mean}\epsilon = (\dot{q} + \frac{1}{2}\Delta\dot{q})\epsilon$, which vanishes as $\epsilon \to 0$; i.e., a *finite impulse, of vanishingly small duration, produces a jump in velocity but not in displacement.*

a small amount ΔE. Our previous discussion of Fig. 1 (the roller-coaster analogy) shows that at a minimum of V, such as M_1, the subsequent motion is confined to the small neighborhood AB surrounding M_1. Therefore M_1 is called a point of **stable equilibrium**. The roller-coaster analogy tells us that at a *maximum* of V, such as at M_2, the representative point P will move away from M_2; hence we call M_2 an **unstable equilibrium** point. Similarly, we deduce that the representative point P will move away from a "horizontal" *inflection point* such as M_4, which is an *unstable* equilibrium point for all practical purposes.

In short, stable equilibrium exists only at those positions where the potential function $V(q)$ has a minimum.

12.53 *Examples*

EXAMPLE 1. Fig. 1

A uniform platform AB of length b and mass m (representing a trapdoor, tailgate, etc.) is pivoted at A and supported at B by a linear spring BD of stiffness k, which is tied to a fixed point D at a distance $\frac{3}{2}b$ vertically above A. The spring is unstrained when $\theta = 30°$. This is the same system examined in Example 3 of Sec. **11.42**, where the potential energy $V(\theta)$ was evaluated for the numerical values $b = 6$ ft, $k = 10$ lb/ft, and $mg = 50$ lb ($m = 1.553$ lb-sec²/ft).

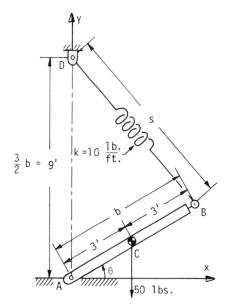

Figure 12.53-1. Example.

We seek to find the motion of the system if the platform is released without initial velocity after its outboard tip at B is hooked to the unstretched spring (with initial angle $\theta_0 = 30°$). The MOI of the uniform "rod" AB, with respect to A, is $J_A = mb^2/3$; hence the kinetic energy of the system is

$$T = \tfrac{1}{2}J_A\dot{\theta}^2 = \tfrac{1}{6}mb^2\dot{\theta}^2 \tag{1}$$

Note that the generalized inertia for this problem is simply J_A.

With V measured from θ_0, the total energy of the system vanishes since both T and V vanish in the initial state; i.e.,

$$T + V = E = 0 \tag{2}$$

Hence

$$T = \tfrac{1}{6}mb^2\dot\theta^2 = -V(\theta) \tag{3}$$

or

$$\dot\theta = \mp\left[\frac{-6V(\theta)}{mb^2}\right]^{1/2} = \mp 0.3276\sqrt{-V(\theta)} \tag{4}$$

where the minus sign is needed initially in Eq. (4) because, at $t = 0$, the trapdoor begins to fall and $\dot\theta$ begins to decrease; thereafter Eq. (4) is multiplied by -1 each time $\dot\theta$ becomes zero.

The potential energy $V(\theta)$ was shown in Example 3 of Sec. **11.42** to be given in foot-pounds by

$$V = 5[(117 - 108\sin\theta)^{1/2} - (63)^{1/2}]^2 + 150\sin\theta - 75 \tag{5}$$

The graph of $V(\theta)$, given in Fig. **11.42**-6, shows that a stable equilibrium position exists at point M_2, where $\theta = -2°$. For the initial condition given, the total energy $E = 0$; therefore the abscissa itself corresponds to the energy line labeled E in Fig. **12.52**-1. Accordingly, the motion is contained within the range of curve AM_2B (see the blowup of this region in Fig. 2); i.e., the platform oscillates between $\theta_A = -47.9537°$ and $\theta_B = 30°$. Values of $V(\theta)$ are calculated in column 4 of Table 1 for $5°$ intervals of θ.

Having found $V(\theta)$, the velocity $\dot\theta$ may be calculated from Eq. (4),[10] giving the

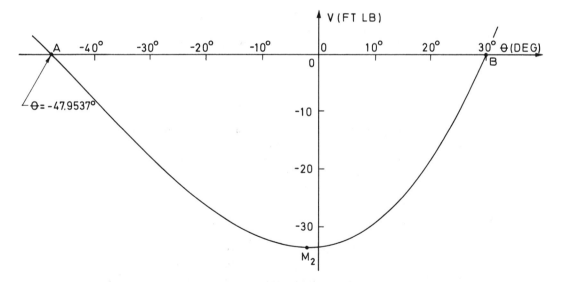

Figure 12.53-2. Potential energy V versus θ.

[10]The plus sign is to be used in Eq. (4) for motion from A to B, and the minus sign for motion from B to A, as indicated in Fig. 3.

TABLE 12.53-1 *Calculations for Example 1*

1	2	3	4 $V(\theta)$, Eq. (5) (ft-lb)	5 $\dot{\theta}_j$ Eq. (4) (rad/sec)	6 $\Delta t = t_j - t_{j-2}$, Eq. (7) (sec)	7 $t = t_j$ $= t_{j-2} + \Delta t$ (sec)
j	θ_j (deg)	θ_j (rad)				
1	−47.9537	−0.8369	0.000	0.0	—	0.0
2	−45	−0.7854	−2.956	0.5631		0.1831*
3	−40	−0.6981	−8.034	0.9286		
4	−35	−0.6109	−13.078	1.1848	0.2015	0.3846
5	−30	−0.5236	−17.930	1.3872		
6	−25	−0.4363	−22.428	1.5515	0.1271	0.5117
7	−20	−0.3491	−26.402	1.6834		
8	−15	−0.2618	−29.676	1.7847	0.1042	0.6159
9	−10	−0.1745	−32.075	1.8554		
10	−5	−0.0873	−33.422	1.8940	0.0944	0.7103
11	0	0	−33.545	1.8975		
12	5	0.0873	−32.276	1.8613	0.0923	0.8026
13	10	0.1745	−29.456	1.7781		
14	15	0.2618	−24.938	1.6360	0.0989	0.9016
15	20	0.3491	−18.592	1.4126		
16	25	0.4363	−10.306	1.0517	0.1278	1.0293
17	30	0.5236	0.0	0.0	—	1.195†

*From Eq. (11).
†From Eq. (12).

result shown in column 5 of Table 1 and in Fig. 3. Such a plot of velocity versus position is often called a *phase-plane* diagram.

The displacement-time relation follows from the definition $dt \equiv d\theta/\dot{\theta}$, which may be integrated to give

$$t - t_1 = \int_{\theta_1}^{\theta} \frac{d\theta}{\dot{\theta}} \tag{6}$$

where θ_1 is any convenient starting position and t_1 is the corresponding starting time, which we may set equal to zero for convenience.

Expression (6) is readily evaluated by means of Simpson's rule, which can be expressed in the form [see Appendix H, Eq. (H-12)]

$$t_{j+2} - t_j = \frac{\Delta\theta}{3}\left(\frac{1}{\dot{\theta}_j} + \frac{4}{\dot{\theta}_{j+1}} + \frac{1}{\dot{\theta}_{j+2}}\right) \tag{7}$$

where $\Delta\theta \; (\equiv \theta_{j+1} - \theta_j)$ is any convenient interval along the θ axis.

To integrate Eq. (6) in the range of increasing θ, we may start the process at point A in Fig. 2 and integrate forward toward B; i.e., we set $\theta_1 = -47.9537°$ (-0.83695 rad) and use the plus sign in Eq. (4).

In the attempt to use Simpson's rule in the neighborhood of the initial point, it is necessary to evaluate the integrand of Eq. (6) where $\dot{\theta}$ is zero; i.e., the integrand $1/\dot{\theta}$ is

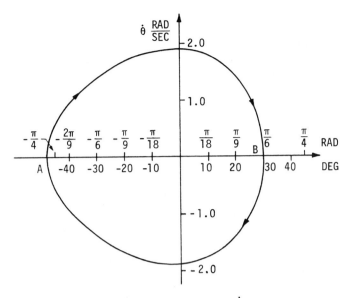

Figure 12.53-3. Angular velocity $\dot{\theta}$ versus θ.

infinite at the first step of integration. *This situation will always arise whenever one attempts to integrate across a point where the velocity vanishes.* A simple but effective remedy consists in using a power series expansion for $\theta(t)$ over a short time interval t_2 wherein θ passes from θ_1 to θ_2. Since $\dot{\theta}_1 = 0$, we may write

$$\theta = \theta_1 + \tfrac{1}{2}\ddot{\theta}_1 t^2 + \cdots \tag{8}$$

$$\dot{\theta} = \ddot{\theta}_1 t + \cdots \tag{9}$$

Neglecting all terms not explicitly shown in the above series is tantamount to assuming that the acceleration is essentially constant in the short time span t_2. This is a reasonable assumption if the applied force Q does not change too rapidly. Eliminating $\ddot{\theta}_1$, we find that in the *earliest stages of motion*

$$t = \frac{2(\theta - \theta_1)}{\dot{\theta}} \tag{10}$$

Accordingly, the time t_2, where $\theta = \theta_2$ and $\dot{\theta} = \dot{\theta}_2$, is given by[11]

$$t_2 = \frac{2(\theta_2 - \theta_1)}{\dot{\theta}_2} \tag{11}$$

Equation (11) gives a starting relationship which allows us to begin the numerical integration at any convenient point θ_2 where $\dot{\theta}_2$ is known and does not vanish. In this example we may choose $\theta_2 = -45°$, where $\dot{\theta}_2 = 0.5631$. Equation (11) then gives $t_2 = 0.1831$ sec.

By starting from θ_2 and using an interval of $\Delta\theta = 5°$ in Simpson's rule, Eq. (7)

---

[11]Equation (11) states that the time interval equals the displacement interval divided by the mean velocity $\tfrac{1}{2}(0 + \dot{\theta}_2)$.

was used to calculate[12] the values of Δt shown in column 6 of Table 1 for $\theta_4, \theta_6, \ldots, \theta_{16}$. All calculations were done on a small calculator.

Since $\dot{\theta}$ vanishes at the final point B where $\theta = \theta_n = \theta_{17} = 30°$, it is not possible to calculate t_{17} by means of Simpson's rule. The remedy, of course, is similar to that used for the starting relationship (11). By analogy with Eq. (11) one finds (see Problem **12.100**-12)

$$t_n - t_{n-1} = \frac{2(\theta_n - \theta_{n-1})}{\dot{\theta}_{n-1}} \tag{12}$$

Upon setting $n = 17$ and utilizing the previously calculated values $t_{16}, \dot{\theta}_{16}$ shown in Table 1, Eq. (12) gives

$$t_{17} = 1.195 \text{ sec}$$

The computed relation between θ and t is plotted in Fig. 4 in the range $\theta_A \equiv \theta_1 \le \theta \le \theta_{17} \equiv \theta_B$. The motion from B to C in Fig. 4 is the mirror image of that from B to A about the ordinate through $t = t_{17}$.

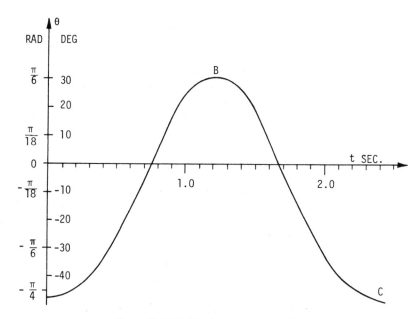

Figure 12.53-4. Displacement versus time.

Thereafter, the motion repeats itself periodically with period $\tau = 2t_{17} = 2.390$ sec. By shifting the curve of Fig. 4 to the left by $t_{17} = 1.195$ sec, the starting configuration will correspond to point B rather than A.

Note that Table 1 gives values of t_j only for even values of j (excluding the end points t_1 and t_{17}). Although enough values have been found to plot a satisfactory curve, we may (as described in Appendix H) find the values $t_5, t_7, \ldots, t_{15}$ by successive application of Eq. (7), starting with $j = 3$, provided that we know t_3. To find t_3 it is only necessary to find $\dot{\theta}$ at the point $\theta_{2.5} = -42.5°$, which is midway between θ_2 and θ_3.

---

[12]All angles should be expressed in radians for numerical work.

Calling this velocity $\dot{\theta}_{2.5}$, we may use Eq. (7) in the form

$$t_3 = t_2 + \frac{\Delta\theta/2}{3}\left(\frac{1}{\dot{\theta}_2} + \frac{4}{\dot{\theta}_{2.5}} + \frac{1}{\dot{\theta}_3}\right) \tag{13}$$

where it should be noted that the step size ($h = \Delta\theta/2$) is now 0.04363 rad (2.5°), i.e., half the step size previously used. It may be verified from Eqs. (4) and (5) that

$$\dot{\theta}_{2.5} = +0.3276\sqrt{-V(-42.5°)} = 0.7676 \text{ rad/sec}$$

Then Eq. (13) gives

$$t_3 = 0.1831 + \frac{0.04363}{3}\left(\frac{1}{0.5631} + \frac{4}{0.7676} + \frac{1}{0.9286}\right) = 0.3004 \text{ sec}$$

From this point onward, we may resume the use of Eq. (7), with a step size of $\Delta\theta = 0.08727$ rad (5°), to find $t_5, t_7, \ldots, t_{15}$; for example,

$$t_5 = t_3 + \frac{0.08727}{3}\left(\frac{1}{\dot{\theta}_3} + \frac{4}{\dot{\theta}_4} + \frac{1}{\dot{\theta}_5}\right) = 0.4509 \text{ sec}$$

EXAMPLE 2. Fig. 5

Most of our discussion has been concerned with systems containing a discrete number of rigid bodies. However, the general methods developed are equally valid for systems of continuous deformable bodies. As an example, consider a perfectly flexible but inextensible cable or chain, of length L, constrained to lie on a known curve in a vertical plane. Figure 5 shows such a case where the vertical ordinate y of the curve is given by a known function of arc length s in the form

$$y = f(s) \tag{14}$$

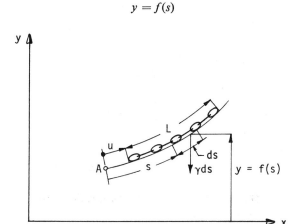

Figure 12.53-5. Chain on a curve.

where s is measured from any convenient point A on the curve. If one end of the cable moves through a displacement u (measured from A) along the curve, all points of the cable move through arc distance u, and the kinetic energy can be expressed as

$$T = \frac{m\dot{u}^2}{2} \tag{15}$$

where m is the total mass of the cable.

If the cable has uniform weight γ per unit length, the potential energy of a small segment of length ds at height y is $y\gamma \, ds$, and the potential energy of the entire cable is

$$V = \int_u^{u+L} y\gamma \, ds = \gamma \int_u^{u+L} f(s) \, ds \tag{16}$$

Since $f(s)$ is assumed to be known, the integral in Eq. (16) may be calculated for any desired value of s; i.e., we can tabulate

$$H(s) \equiv \int_0^s f(s) \, ds \tag{17}$$

Since the system is conservative, we can state that

$$T + V = \text{constant} \equiv E \tag{18}$$

Upon substituting Eqs. (15)–(17) into Eq. (18), we find

$$\frac{m\dot{u}^2}{2} + \gamma H(u + L) - \gamma H(u) = E \tag{19}$$

Equation (19) defines $\dot{u}^2$ as a known function of u for any given initial state, and we may proceed to integrate the expression $dt = du/\dot{u}$ in exactly the same manner as in Example 1.

Let us consider the particular case where the given curve is described by the simple expression

$$y = f(s) = \frac{s^2}{8R} \tag{20}$$

where R is a constant. It may be recalled from Eq. (**5.22**-6) that Eq. (20) describes a cycloid. From Eqs. (17) and (20) we see that

$$H(s) = \int_0^s f(s) \, ds = \frac{s^3}{24R} \tag{21}$$

Accordingly, Eq. (19) becomes

$$\frac{m\dot{u}^2}{2} + \frac{\gamma}{24R}[(u + L)^3 - u^3] = E \tag{22}$$

Upon differentiating across Eq. (22) and dividing through by $\dot{u}$, we find the second-order differential equation

$$\ddot{u} + \frac{\gamma L}{4mR}\left(u + \frac{L}{2}\right) = 0 \tag{23}$$

Recognizing that γL equals the total weight (mg) of the cable and calling $u +$

$(L/2) \equiv u_c$ (the distance from A to the midpoint of the cable), we can express Eq. (23) in the form

$$\ddot{u}_c + \frac{g}{4R}u_c = 0 \tag{24}$$

Equation (24) describes a simple harmonic motion with period

$$\tau = 2\pi\sqrt{\frac{4R}{g}} \tag{25}$$

This result, for a cable, is a generalization of Huygens' discovery of 1673 that a particle oscillates on a smooth cycloid with perfectly constant period independently of the amplitude of motion; i.e., the cycloid is an **isochrone**[13] for motion under gravity loading. Prior to this discovery, the best clocks (also invented by Huygens) used ordinary pendulums which gave the bob a circular motion that was only approximately isochronous. This defect in isochronism was appreciated by Huygens, who wrote the following[14]: "It is the oscillations of marine clocks that most noticeably become unequal, because of the ship's continual shaking. So that it is necessary to take care that oscillations of large and small amplitudes should be isochronous." Huygens' ingenious kinematic invention (cycloidal cheeks) to ensure cycloidal motion has already been described in connection with Fig. **5.22**-3.

12.60   INTEGRATING THE EQUATION OF MOTION

We now return to the general case where the active forces are not necessarily conservative. In this case, the generalized force Q may be an explicit function of velocities as well as configuration variables, and the generalized equation of motion, Eq. (**12.40**-7), is of the form

$$\mathcal{I}(q, \phi_i)\ddot{q} + \mathcal{C}(q, \phi_i)\dot{q}^2 = Q(q, \phi_i, \dot{q}, \dot{\phi}_i) \tag{1}$$

where ϕ_i represents a typical secondary position variable; $\mathcal{I}$ and $\mathcal{C}$ are known functions of position, defined by Eqs. (**12.40**-9) and (**12.40**-10); and Q is a function of position and velocity defined by Eq. (**12.40**-11).

The primary and secondary position variables are related, in general, by a set of N algebraic equations of the form [see Eq. (**9.10**-2)]

$$f_i(q, \phi_1, \phi_2, \ldots, \phi_N) = 0 \qquad (i = 1, 2, \ldots, N) \tag{2}$$

These equations of constraint come from the complete loop equations of the system and from whatever auxiliary equations exist due to the presence of gears and cams, as explained in Secs. **7.30**–**7.43**.

To integrate the second-order differential equation (1), we first convert it[15] into

---

[13]Isochrone = a curve of equal time.
[14]Quoted from Dugas [1955, p. 181].
[15]Following the procedure outlined in Appendix **G**.

two first-order equations with the aid of the auxiliary variables

$$w_1 \equiv q, \qquad w_2 \equiv \dot{q} \tag{3}$$

from which it follows that

$$\dot{w}_1 = \dot{q} = w_2 \tag{4}$$

$$\dot{w}_2 = \ddot{q} = \frac{1}{\mathcal{I}(q, \phi)}[Q(q, \phi, \dot{q}, \dot{\phi}) - \dot{q}^2 \mathcal{C}(q, \phi)] \tag{5}$$

As mentioned in Appendix **G**, the solution of a set of two first-order differential equations depends on two *initial values* which we shall assume are given values of q and $\dot{q}$ at time $t = 0$; in other words, we consider as known the initial values

$$w_1(0) \equiv q_0, \qquad w_2(0) \equiv \dot{q}_0 \tag{6}$$

The problem at hand is to solve the *initial value problem* represented by Eqs. (4)–(6) subject to the N algebraic equations (2). For future reference, we note that the latter equations may be represented in the *rate form*

$$\dot{f}_i = \frac{\partial f_i}{\partial q} \dot{q} + \sum_{j=1}^{N} \frac{\partial f_i}{\partial \phi_j} \dot{\phi}_j = 0 \tag{7}$$

Further progress may follow either of two paths, analogous to those described in Secs. **10.41** and **10.42**, respectively. The first approach is called the **method of minimal differential equations** and is described in Sec. **12.61**; the second approach, the method of **excess differential equations**, is described in Sec. **12.62**.

12.61   *The Method of Minimal Differential Equations*

In most cases, the algebraic equations (**12.60**-2) will be nonlinear, and we shall be unable to find explicit expressions for the variables ϕ_i in terms of q. However, as noted in Chapter 9, we can usually find numerical values of the ϕ_i for a given value of q by use of the Newton-Raphson method or by some other algorithm for the numerical solution of nonlinear algebraic equations. At the very worst, we can always make a large scale drawing and find the quantities ϕ_i graphically. Hence we shall assume that we can find the ϕ_i associated with any given value of q and shall symbolically represent this fact by the statement

$$\phi_i = \phi_i[q] \qquad (i = 1, 2, \ldots, N) \tag{1}$$

For a specified value of q, Eqs. (1) may be used to find the generalized inertia $\mathcal{I}(q, \phi)$ and the centripetal coefficient $\mathcal{C}(q, \phi)$, but we cannot, in general, evaluate the generalized force $Q(q, \phi, \dot{q}, \dot{\phi})$, which may depend on the secondary velocities. However, for given values of q and $\dot{q}$ we may utilize Eq. (**12.60**-7) in the matrix form

$$\mathbf{A}\dot{\boldsymbol{\phi}} = \mathbf{b}\dot{q} \tag{2}$$

where

$$\mathbf{A} \equiv \left[\frac{\partial f_i}{\partial \phi_j}\right], \qquad \mathbf{b} = -\left[\frac{\partial f_i}{\partial q}\right] \tag{3}$$

The solution of Eq. (2) can be expressed in the form

$$\dot{\phi}_i = k_i \dot{q} = k_i w_2 \tag{4}$$

where the velocity ratios $k_i \, (\equiv \dot{\phi}_i/\dot{q})$ are well-defined functions of position only. With the aid of Eqs. (1) and (4) we can calculate $Q(q, \phi, \dot{q}, \dot{\phi})$ for any given values of $q \equiv w_1$ and $\dot{q} \equiv w_2$. Thus the right-hand side of Eq. (12.60-5) can be evaluated; i.e., we have

$$\dot{w}_2 = \frac{[Q - \dot{q}^2 \mathfrak{C}]}{\mathfrak{g}} = \dot{w}_2[w_1, w_2] \tag{5}$$

In short, we have developed expressions of the form

$$\dot{w}_1 = F_1(w_1, w_2) \equiv w_2 \tag{6}$$
$$\dot{w}_2 = F_2(w_1, w_2) \tag{7}$$

which together with the initial conditions

$$w_1(0) = q_0, \qquad w_2(0) = \dot{q}_0 \tag{8}$$

represent an initial value problem in standard form. In other words, we may use a standard algorithm[16] for such problems to find w_1 and w_2 after a small time step $\Delta t = h$. Then the process is repeated to find w_1 and w_2 at $t = 2h, 3h$, etc. The process would usually be stopped after t reaches a specified value $t_{\max}$.

At any stage, one may find the secondary accelerations $\ddot{\phi}_i$ from the differentiated form of the velocity equations (2); i.e.,

$$\mathbf{A}\ddot{\boldsymbol{\phi}} = \mathbf{b}\ddot{q} + \dot{\mathbf{b}}\dot{q} - \dot{\mathbf{A}}\dot{\boldsymbol{\phi}} \equiv \mathbf{S} \tag{9}$$

as explained in Sec. 10.10.

If desired, velocities and accelerations, at points of interest, may be computed at any time from equations of the form

$$\dot{x} = u\dot{q}, \qquad \ddot{x} = u\ddot{q} + u'\dot{q}^2 \tag{10}$$
$$\dot{y} = v\dot{q}, \qquad \ddot{y} = v\ddot{q} + v'\dot{q}^2$$

as described in Sec. 10.51.

The entire procedure is summarized in the flowchart shown in Fig. 1.

---

[16]For example, the Euler method gives $w_1(h) \doteq w_1(0) + h\dot{w}_1(0) = q_0 + h\dot{q}_0$, $w_2(h) \doteq w_2(0) + h\dot{w}_2(0) = \dot{q}_0 + hF_2(q_0, \dot{q}_0)$. Other *Runge-Kutta* algorithms are described in Appendix G.

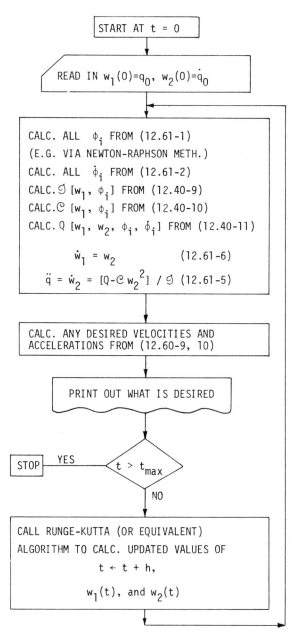

Figure 12.61-1. Flow chart for method of minimal differential equations.

12.62   The Method of Excess Differential Equations

We have seen that the motion of a system with 1 DOF may be determined from the solution of two first-order differential equations. However, in the process, it is necessary to continually update all of the N secondary position variables ϕ_i by means of Newton-Raphson iteration.

As pointed out by Paul and Krajcinovic [1970b], it is possible to eliminate the Newton-Raphson phase of the calculations by solving a system of $2 + N$ first-order differential equations. Since the method introduces N additional differential equations (beyond the minimal number of two dictated by the equation of motion), we call it **the method of excess differential equations**.

The equations to be solved are the equations of motion [Eqs. (**12.60**-4) and (**12.60**-5)]; i.e.,

$$\dot{w}_1 = \dot{q} = w_2 \tag{1}$$

$$\dot{w}_2 = \ddot{q} = \frac{Q - w_2^2 \mathcal{C}}{\mathcal{I}} \tag{2}$$

subject to the kinematic constraint Eq. (**12.61**-2); i.e.

$$\mathbf{A}\dot{\boldsymbol{\phi}} = \dot{q}\mathbf{b} \tag{3}$$

This set of linear equations is readily solved to find the velocity ratios

$$k_i = \frac{\dot{\phi}_i}{\dot{q}} \tag{4}$$

In this method we view the set of Eqs. (3) as *differential* equations in ϕ_i—not as *algebraic* equations for $\dot{\phi}_i$ as in Sec. **12.61**.

We now introduce the additional notation

$$\{w_3, w_4, \ldots, w_{N+2}\} \equiv \{\phi_1, \phi_2, \ldots, \phi_N\} \tag{5}$$

whereupon the Eqs. (1)–(3) may be expressed in the form

$$
\begin{aligned}
\dot{w}_1 &= w_2 \equiv F_1 \\
\dot{w}_2 &= \frac{Q - \mathcal{C}w_2^2}{\mathcal{I}} \equiv F_2 \\
\dot{w}_3 &\equiv \dot{\phi}_1 = k_1 w_2 \equiv F_3 \\
\dot{w}_4 &\equiv \dot{\phi}_2 = k_2 w_2 \equiv F_4 \\
&\quad \cdots \\
\dot{w}_{N+2} &\equiv \dot{\phi}_N = k_N w_2 \equiv F_{N+2}
\end{aligned}
\tag{6}
$$

Since $\mathcal{I}$ and $\mathcal{C}$ are explicit functions of position only, they are readily calculated from a knowledge of w_1. The generalized force Q may depend on the velocities $\dot{q}$ and $\dot{\phi}_i$, but these velocities are readily expressed via Eqs. (6) in terms of w_i. Therefore, the right-hand sides of Eqs. (6) may all be calculated from a knowledge of w_i, and we may think of $F_1, F_2, \ldots, F_{N+2}$ as known functions of the variables w_i.

In short, Eqs. (6) represent a set of ordinary differential equations in the standard form:

$$\dot{w}_i = F_i[w_i, t] \qquad (i = 1, 2, \ldots, N + 2) \tag{7}$$

If the initial conditions $(w_1, w_2, \ldots, w_{N+2})$ are specified at $t = 0$, Eqs. (7) may be solved by any of the standard numerical procedures (e.g., Runge-Kutta) for initial value problems.

In general, we shall know the initial values of the generalized displacement $q \equiv w_1$

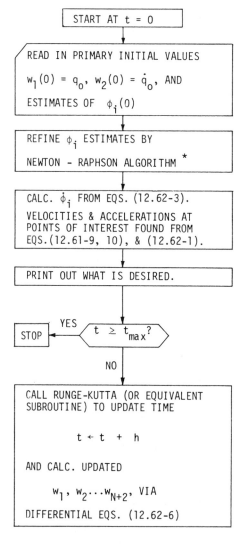

*For alternatives to this block, see text.

Figure 12.62-2. Flow chart for method of excess differential equations.

and generalized velocity $\dot{q} \equiv w_2$, but we shall not know (precisely) the initial values of the secondary position variables $\phi_i = w_{i+2}$. To find sufficiently accurate initial values of $(w_3, w_4, \ldots, w_{2+N})$, we can use one of the following methods:

1. Measure ϕ_i from a large scale drawing, with q set at its initial value.
2. Use the Newton-Raphson method (cf. Chapter 9) only to establish accurate *initial* values of ϕ_i.
3. Make small adjustments on the link lengths (or other system parameters) so that the assumed approximate values of $\phi_i(0)$ are sufficiently accurate for a "substitute mechanism" which is sufficiently "close" to the given mechanism.

The flowchart of Fig. 1 illustrates the logic of a computer program for the method of excess equations using method 2 to find the *initial* values of ϕ_i. For $t > 0$, no further use is made of the Newton-Raphson algorithm.

12.70 KINETOSTATICS

After the generalized differential equation of motion (**12.40**-7) has been solved for the generalized coordinate q as a function of time, it is possible to find the position, velocity, and acceleration of any point in a mechanism by the methods of Chapter 10. Thus Eqs. (**10.50**-2)–(**10.50**-4) give the acceleration components $(\ddot{x}_i, \ddot{y}_i)$ of the mass center of a typical link i and its angular acceleration $\ddot{\theta}_i$ in the form

$$\{\ddot{x}_i, \ddot{y}_i, \ddot{\theta}_i\} = \{u_i, v_i, \omega_i\}\ddot{q} + \{u_i', v_i', \omega_i'\}\dot{q}^2 \tag{1}$$

To find the reactions at the joints of the mechanism, one may write the three equations of motion for each link in the form

$$
\begin{aligned}
X_i - m_i\ddot{x}_i + X_i' &= 0 \\
Y_i - m_i\ddot{y}_i + Y_i' &= 0 \\
M_i - J_i\ddot{\theta}_i + M_i' &= 0
\end{aligned}
\tag{2}
$$

where (X_i, Y_i, M_i) is the active *force set* applied at the mass center of link i, (X_i', Y_i') represents the net reaction forces acting on all the joints of link i, and M_i' is the moment of the joint forces about the mass center. We may interpret Eqs. (2) to state that quantities $(-m_i\ddot{x}_i, -m\ddot{y}_i, -J_i\ddot{\theta}_i)$ play the role of *fictitious forces* (called **inertia forces**) which are in equilibrium with the real forces acting on the links—this is **d'Alembert's principle**. Thus Eqs. (2) may be viewed as equations of equilibrium (sometimes called **equations of dynamic equilibrium**), and the calculation of joint reactions may proceed exactly as in Chapter 11. Since the determination of joint reactions is essentially a problem of *statics* (provided that the inertia forces have been established from a prior dynamics analysis), this class of problems is sometimes referred to as **kinetostatics**.

Thus the discussion of reaction forces in Chapter 11 applies equally well to problems of dynamic equilibrium, provided that the active force sets (X_i, Y_i, M_i) used in the statics analysis are to be replaced by the **effective static force**[17] sets

$$\{X_i^*, Y_i^*, M_i^*\} = \{(X_i - m_i\ddot{x}_i), (Y_i - m_i\ddot{y}_i), (M_i - J_i\ddot{\theta})\} \tag{3}$$

In all other respects the analyses are identical and need not be discussed further here, except for a comment on the influence of friction.

Viscous friction enters the generalized equations of motion through velocity-dependent terms in the generalized force Q and causes no difficulties at all. On the other hand, Coulomb friction depends on the joint reactions (see Sec. **11.60** et seq.), and its contribution to Q cannot be determined until the joint reactions are known. But the joint reactions cannot be determined until the equations of motion are solved. This reasoning suggests an iterative procedure wherein the equations of motion are solved over a small time step h, with Coulomb friction ignored (or estimated). Then the joint reactions, including friction, may be found from a kinetostatic analysis, and the dynamics problem may be solved again using the updated estimate of frictional forces in the calculation of the generalized force Q. For the small time steps normally used in a dynamics analysis, it will be found that one cycle of iteration is usually adequate.

As a practical matter, due to inherent uncertainties in our knowledge of friction coefficients, we cannot model the effects of dry friction with great precision. It is therefore suggested that the iterative procedure just described is unnecessarily elaborate in most cases, and it is adequate to use the following simpler procedure:

1. At time $t = 0$, ignore the presence of Coulomb friction and solve the dynamics problem to find q, $\dot{q}$, $\ddot{q}$ at time $t + h$. Then find the inertia forces $(-m_i\ddot{x}_i, -m_i\ddot{y}_i, -J_i\ddot{\theta}_i)$.

2. With the inertia forces known (from step 1), at time $t + h$, solve the kinetostatic problem to find the current joint reactions, including frictional effects as in Sec. **11.60**.

3. Assume that the normal reactions (N) at sliders and the hinge forces (R) at hinges remain constant over the next time interval h. Then, for $h < t < 2h$, slider friction is given by $\mu_k N$, and hinge friction moment is given by $R\rho_k$, where μ_k is the coefficient of *kinetic* friction and ρ_k is the corresponding friction radius of a hinge (see Sec. **11.62**). *Note*: The direction of the frictional forces and moments must be chosen so as to oppose the relative motion of the pair elements. For practical purposes we may assume that the relative velocities of pair elements do not change sign during the time interval h, so that the direction of the frictional reactions to be assumed is that prevailing at the beginning of the time interval.

4. Calculate the updated inertia forces with t replaced by $t + h$ and repeat steps 2–4 as often as desired.

This procedure will usually give results well within the limits of accuracy possible when dry friction plays a moderate role. For extremely slow-moving and poorly

---

[17]This term is being used because the simpler expression *effective force* is frequently used in the literature for the negative of the inertia forces.

lubricated machines, the magnitude of the effective forces ($|X_i - m_i\ddot{x}_i|$, etc.) may not be large enough to exceed the limiting *static friction* ($\mu_s N$ or $\rho_s R$) at the joints. In such cases, one or more joints may "freeze" or "stick," thereby reducing the degree of freedom of the system during the next time interval. When the probability of such "sticking" is high, one should check at each time step whether the effective loads exceed the friction thresholds needed for the assumed motion. Well-designed machinery should have provision for sufficient lubrication to prevent sticking under normal operating conditions.

12.80   RECIPROCATING ENGINE DYNAMICS

As an *illustration* of the general methods developed in this chapter, we shall consider the response of reciprocating engines under the action of given forces. It should be borne in mind that the dynamic analysis of *any* single-freedom planar machine will parallel very closely that given here for the slider-crank (engine) mechanism. It is for this reason that we have gone into sufficient detail to generate working computer programs which enable us to simulate the complete response (transient and steady state) of this prototypic machine.

First we develop expressions and subroutines for the generalized inertia and centripetal coefficients $\mathscr{I}$ and $\mathscr{C}$ and for the generalized force Q. It is then shown how the generalized differential equation of motion may be integrated for the case of arbitrary loads and how certain "simplifications" may be introduced when the resisting forces do not depend on the velocities of the links. After the motion of each link has been explicitly determined, the reaction forces at the bearings are found.

Figure 1 shows the mechanism of one cylinder of a reciprocating engine. The

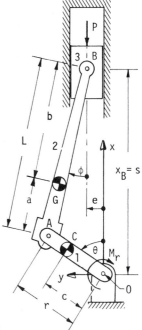

Figure 12.80-1. Reciprocating engine.

origin O of the coordinate axes (x, y) is placed at the fixed pivot of the crank, and the x axis is parallel to the cylinder axis.

The following terminology will be used:

L = length AB of connecting rod (abbreviated **conrod**)

r = length OA of crank

$$\lambda = \frac{r}{L}$$

a = distance from A to mass center G of conrod

b = distance from B to G

c = distance from O to mass center C of crank

e = offset of piston pin (B) from x axis

$W_1 = m_1 g$ = weight of crank assembly[18] (g = acceleration of gravity)

$W_2 = m_2 g$ = weight of conrod

$W_3 = m_3 g$ = weight of piston (including piston pin at B)

$J_1 = J_C = m_1 r_C^2$ = MOI about C of crank assembly (r_C is radius of gyration)

$J_2 = J_G = m_2 r_G^2$ = MOI of conrod about G (r_G is radius of gyration)

$J_O = J_1 + m_1 c^2$ = MOI of crank assembly about O

θ = crank angle (positive counterclockwise from x axis)

ϕ = connecting rod obliquity (positive clockwise as shown)

$s = x_B = x$ coordinate of piston pin

P = force due to gas pressure on piston (positive as shown)

M_r = resisting torque exerted by the driven device on the crankshaft (positive clockwise)

For simplicity, we restrict our attention to symmetric connecting rods (i.e., where A, G, B are collinear). For the case of nonsymmetric rods, see Problem **12.100**-39.

The following exact relations, recalled from Sec. **1.41**, will be needed in subsequent analyses[19]:

$$\phi = \sin^{-1}\left(\lambda \sin\theta - \frac{e}{L}\right) \tag{1}$$

$$s = r\left(\cos\theta + \frac{1}{\lambda}\cos\phi\right) \tag{2}$$

$$k_\phi \equiv \frac{\dot{\phi}}{\dot{\theta}} = \frac{\lambda \cos\theta}{\cos\phi} \tag{3}$$

---

[18]The **crank assembly** includes all masses rigidly fixed to the crank, e.g., crank cheeks OA, crank pin at A, crankshaft, flywheel, and balancing weights.

[19]The rightmost term in Eq. (4) follows from Eq. (**11.22**-13).

$$k_s \equiv \frac{\dot{s}}{\dot{\theta}} = -r\left(\sin\theta + \frac{1}{\lambda}k_\phi \sin\phi\right) = -(e + s\tan\phi) \qquad (4)$$

$$k'_\phi \equiv \frac{dk_\phi}{d\theta} = -\lambda\frac{\sin\theta}{\cos\phi} + k_\phi^2 \tan\phi \qquad (5)$$

$$k'_s = -r\left(\cos\theta + \frac{1}{\lambda}k'_\phi \sin\phi + \frac{1}{\lambda}k_\phi^2 \cos\phi\right) \qquad (6)$$

In addition, the following approximate relations are recalled from Sec. **1.41**:

$$k_\phi = \frac{\dot{\phi}}{\dot{\theta}} \doteq \lambda C_1 \cos\theta \qquad (7)$$

$$k_s = \frac{\dot{s}}{\dot{\theta}} \doteq -r\left(\sin\theta + \frac{A_2}{2}\sin 2\theta\right) \qquad (8)$$

where

$$C_1 \doteq 1 + \frac{\lambda^2}{8}, \qquad A_2 \doteq \lambda + \frac{\lambda^3}{4} \qquad (9)$$

For later reference we list here FORTRAN names (shown in brackets) which will be used in later sections:

L [L]	r [R]	λ [LAMDA]	a [A]	b [B]
c [C]	e [E]	m_1 [M1]	m_2 [M2]	m_3 [M3]
J_o [JO]	J_G [JG]	θ [THETA]	ϕ [PHI]	s [S]
P [P]	M_r [MR]	k_ϕ [KPHI]	k_s [KS]	k'_ϕ [KPHIPR]
k'_s [KSPR]				

12.81 *Generalized Inertia and Kinetic Energy of Engine*

The kinetic energy (K.E.) consists of three terms:

$$T = T_1 + T_2 + T_3 \qquad (1)$$

where

$$T_1 = \tfrac{1}{2}J_o\dot{\theta}^2 \qquad \text{(crank K.E.)} \qquad (2)$$
$$T_3 = \tfrac{1}{2}m_3\dot{s}^2 \qquad \text{(piston K.E.)} \qquad (3)$$
$$T_2 = \tfrac{1}{2}[m_B\dot{s}^2 + m_A(r\dot{\theta})^2 + J_{AB}\dot{\phi}^2] \qquad \text{(conrod K.E.)} \qquad (4)$$

where the last equation stems from Eq. (**12.31**-5) and the following parameters are defined by Eqs. (**12.31**-3) and (**12.31**-4):

$$m_A = \frac{m_2 b}{L}, \qquad m_B = \frac{m_2 a}{L} \qquad (5)$$

$$J_{AB} = J_G - m_2 ab = m_2(r_G^2 - ab) \qquad (6)$$

If we let θ represent the generalized coordinate, the kinetic energy can be expressed via Eqs. (1)–(4) in the form

$$T = \frac{1}{2}\mathcal{g}\dot{\theta}^2 = \frac{1}{2}\left[(J_O + m_A r^2) + (m_3 + m_B)\left(\frac{\dot{s}}{\dot{\theta}}\right)^2 + J_{AB}\left(\frac{\dot{\phi}}{\dot{\theta}}\right)^2\right]\dot{\theta}^2 \qquad (7)$$

Utilizing the velocity ratios defined by Eqs. (**12.80**-3) and (**12.80**-4), we find the final expression

$$\mathcal{g} = J_O + m_A r^2 + (m_3 + m_B)k_s^2 + J_{AB}k_\phi^2 \qquad (8)$$

Equation (8) is exact when k_ϕ and k_s are given by Eqs. (**12.80**-3) and (**12.80**-4). However, if one chooses to work with the leading terms of the approximate expres-

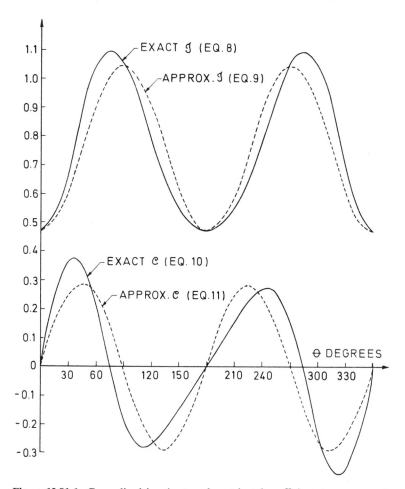

Figure 12.81-1. Generalized inertia $\mathcal{g}$, and centripetal coefficient $\mathcal{C}$ versus crank angle θ for engine with $r = 4''$, $L = 12''$, $a = 4''$, $e = 0$, $J_0 = 0.0129$ lb sec^2 in., $J_2 = 0.8290$ lb sec^2 in., $m_2 = 0.05181$ lb sec^2/in., $m_3 = 0.01295$ lb sec^2/in.

sions (**12.80**-7) through (**12.80**-9), one finds the *approximate* expression (for $e = 0$)

$$\mathcal{J} \doteq J_O + m_A r^2 + (m_3 + m_B)r^2 \sin^2 \theta + J_{AB}\lambda^2 \cos^2 \theta$$
$$= A - B \cos 2\theta \tag{9}$$

where

$$A = J_O + m_A r^2 + \tfrac{1}{2}[(m_3 + m_B)r^2 + J_{AB}\lambda^2] \tag{9a}$$

$$B = \tfrac{1}{2}[(m_3 + m_B)r^2 - J_{AB}\lambda^2] \tag{9b}$$

The upper solid curve in Fig. 1 shows the variations of $\mathcal{J}$ with θ for a particular engine, and the upper broken curve shows the approximation of Eq. (9).

The centripetal coefficient is

$$\mathcal{C} = \frac{1}{2}\frac{d\mathcal{J}}{d\theta} = (m_3 + m_B)k_s k_s' + J_{AB}k_\phi k_\phi' \tag{10}$$

where k_ϕ' and k_s' are defined by Eqs. (**12.80**-5) and (**12.80**-6). The approximation corresponding to Eq. (9) is given by

$$\mathcal{C} = \frac{1}{2}\frac{d\mathcal{J}}{d\theta} \doteq B \sin 2\theta \tag{11}$$

Equations (10) and (11) are shown in Fig. 1 by the lower solid and broken curves, respectively.

The functions $\mathcal{J}$ and $\mathcal{C}$ were computed by a small FORTRAN subroutine called GICC, which is listed below. FORTRAN names introduced in this subroutine are shown below (in brackets), next to their algebraic counterparts:

$$\mathcal{J}\,[\text{GI}] \qquad \mathcal{C}\,[\text{CC}] \qquad m_A\,[\text{MA}] \qquad m_B\,[\text{MB}] \qquad J_{AB}\,[\text{JAB}]$$

```
      SUBROUTINE GICC
C
C     THIS ROUTINE EVALUATES GENERALIZED INERTIA (GI), AND
C     CENTRIPETAL COEFFFICIENT (CC).
C
      IMPLICIT REAL*8 (A-Z)
      COMMON/BLK1/R,L,MA,MB,M1,M2,M3,JO,JG,JAB,LAMDA,KPHI,KS,KPHIPR,KSPR
     1,GI,CC,THETA,THETAF,PHI,S,D,E,A,C,ILEVEL,IP
      COMMON/BLK3/CTHETA,STHETA,CPHI,SPHI
      INTEGER ILEVEL,IP
      STHETA=DSIN(THETA)
      CTHETA=DCOS(THETA)
      PHI=DASIN(LAMDA*STHETA-E/L)
      SPHI=DSIN(PHI)
      CPHI=DCOS(PHI)
      S=R*(CTHETA+CPHI/LAMDA)
      KPHI=LAMDA*CTHETA/CPHI
      KS=-R*(STHETA+KPHI*SPHI/LAMDA)
      KPHIPR=-LAMDA*STHETA/CPHI+KPHI*KPHI*DTAN(PHI)
      KSPR=-R*(CTHETA+KPHIPR*SPHI/LAMDA+KPHI*KPHI*CPHI/LAMDA)
      GI=JO+MA*R*R+(M3+MB)*KS*KS+JAB*KPHI*KPHI
      CC=(M3+MB)*KS*KSPR+JAB*KPHI*KPHIPR
      RETURN
      END
```

12.82 *Generalized Force*

For the moment, we shall exclude the effects of gravity (which are relatively insignificant in high-speed engines) and of friction (which may be of interest for studies of engine efficiency and metal wear). For a discussion of these effects, see Biezeno and Grammel [1954], Taylor [1966, 1968], and Problems **12.100**-40 and 41.

From Fig. **12.80**-1 we see that the power supplied to the engine by the gas force P and resisting torque M_r is

$$\text{power} = -P\dot{s} - M_r\dot{\theta} = -\left[P\left(\frac{\dot{s}}{\dot{\theta}}\right) + M_r\right]\dot{\theta} \equiv Q\dot{\theta} \tag{1}$$

where Q is the generalized force corresponding to the generalized coordinate $q \equiv \theta$. It follows that the generalized force is

$$Q = -(Pk_s + M_r) = M_p - M_r \tag{2}$$

where

$$M_p \equiv -k_s P \tag{2a}$$

is the **turning moment** of the gas forces and $k_s \equiv s/\dot{\theta}$ is given by Eq. (**12.80**-4).

The gas pressure p is given by the **indicator diagram**—i.e., a plot of p versus piston position. Typical indicator diagrams are described in books on internal combustion engines, such as Taylor [1966, 1968]. For purposes of illustration[20] it suffices to consider the idealized **combination cycle** shown in Fig. 1. By proper choice of point 5 this cycle can be made to approximate the performance of either a spark ignition or diesel engine.

Since p is the absolute pressure, the net gas force is given by

$$P = (p - p_a)A_p \tag{3}$$

where A_p is the piston cross-sectional area and p_a is the ambient pressure which acts on the underside of the piston; for the most common case of *single-acting* engines, p_a is just the atmospheric pressure.

The distance d is the length of an idealized right circular cylinder of area A_p which has a volume equal to that of the gas trapped between the cylinder head and piston crown.

From Fig. 1 we see that

$$d = d_1 + s_{\max} - s \tag{4}$$

and the volume of the working gas is

$$V = A_p d \tag{5}$$

---

[20]We only illustrate the case for a four-stroke cycle. The alterations needed for a two-stroke cycle are straightforward.

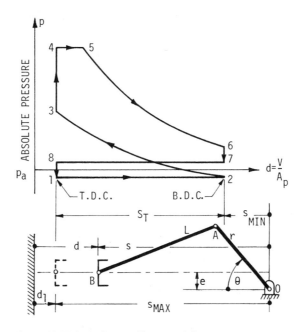

Figure 12.82-1. Indicator diagram giving pressure p versus piston displacement (d or s) for a four-stroke "combination" cycle. (T.D.C. = top dead center; B.D.C. = bottom dead center).

The cycle is summarized in Table 1, where the subscripted numbers correspond to the points $1, \ldots, 8$ in Fig. 1.

TABLE 12.82-1 *Summary of a Four-Stroke Cycle*

Path	θ (deg)	Process	p vs. V
$1 \longrightarrow 2$	$0 \longrightarrow 180$	Intake (suction)	$p = p_{\text{in}}$
$2 \longrightarrow 3$	$180 \longrightarrow 360$	Compression	$p = p_2(V_2/V)^{k_c}$
$3 \longrightarrow 4$	360	Combustion at const. vol.	$V = V_1$
$4 \longrightarrow 5$	$360 \longrightarrow \theta_5$	Combustion at const. p	$p = p_4 = p_5$
$5 \longrightarrow 6$	$\theta_5 \longrightarrow 540$	Expansion	$p = p_5(V_5/V)^{k_e}$
$6 \longrightarrow 7$	540	Const. vol.	$V = V_2$
$7 \longrightarrow 8$	$540 \longrightarrow 720$	Exhaust	$p = p_{\text{ex}}$
$8 \longrightarrow 1$	720	Const. vol.	$V = V_1$

The **polytropic exponents** k_c and k_e (which govern the compression and expansion processes) vary from 1.35 for low-speed engines to 1.39 for high-speed engines; for most purposes an average value $k_c = k_e = 1.37$ may be used.[21]

[21]Maleev [1945, p. 264].

The maximum and minimum values of s are given via Eqs. (**1.34**-7a) and (**1.34**-7b) in the form

$$s_{\text{max}} = [(L + r)^2 - e^2]^{1/2} \tag{6a}$$

$$s_{\text{min}} = [(L - r)^2 - e^2]^{1/2} \tag{6b}$$

and the total piston stroke S_T is given by

$$S_T = s_{\text{max}} - s_{\text{min}} \doteq 2r\left(1 + \frac{e^2}{2(L^2 - r^2)} + \cdots\right) \tag{6c}$$

Note that for the (usual) case, where $e = 0$, the total stroke is simply

$$S_T = 2r \quad \text{(when } e = 0\text{)} \tag{6d}$$

For the practical range of offsets used in engines, the introduction of offset changes the stroke by less than $\frac{1}{2}\%$ (see Problem **12.100**-46).

The minimum cylinder volume V_1 is called the **clearance volume**, and the corresponding distance $d_1 = V_1/A_p$ is called the **clearance distance**. Because of the generally complicated shape of the clearance space, d_1 is best expressed in terms of the **compression ratio**

$$CR \equiv \frac{V_2}{V_1} = \frac{V_1 + A_p S_T}{V_1} = 1 + \frac{A_p S_T}{A_p d_1} \tag{7a}$$

That is,

$$d_1 = \frac{S_T}{CR - 1} \tag{7b}$$

The pressure is a periodic function of crank angle, with period $N\pi$, where $N = 4(2)$ for a four-stroke (two-stroke) cycle. It therefore proves convenient to subtract a suitable multiple of $N\pi$ to give the **fundamental angle** θ_F in the range $0 \rightarrow N\pi$. In other terms,

$$\theta_F = \theta - N\pi \cdot \text{INT}\left(\frac{\theta}{N\pi}\right) \tag{8a}$$

where $\text{INT}(x)$ represents the integer part of x, i.e., what remains after dropping all digits to the right of the decimal point in the floating-point number x. This relation can also be expressed by use of the MOD function (a standard FORTRAN library function—called DMOD in the double-precision version):

$$\theta_F = \text{MOD}(\theta, N\pi) \tag{8b}$$

Now we may express the pressures throughout the combination cycle of Fig. 1 as follows:

$$p = p_1 = p_{\text{in}} \qquad \text{for } 0 \leq \theta_F < \pi \tag{9a}$$

$$p = p_{\text{in}}\left(\frac{d_2}{d}\right)^{k_c} \qquad \text{for } \pi \leq \theta_F < 2\pi \tag{9b}$$

$$p = \frac{p_3 + p_4}{2} \qquad \text{for } \theta_F = 2\pi \tag{9c}$$

$$p = p_4 = p_5 \qquad \text{for } 2\pi < \theta_F < \theta_5 \tag{9d}$$

$$p = p_5 \left(\frac{d_5}{d}\right)^{k_e} \qquad \text{for } \theta_5 < \theta_F < 3\pi \tag{9e}$$

$$p = \frac{p_6 + p_7}{2} \qquad \text{for } \theta_F = 3\pi \tag{9f}$$

$$p = p_7 = p_{ex} \qquad \text{for } 3\pi < \theta_F < 4\pi \tag{9g}$$

In practice, the pressure will not jump instantaneously from p_3 to p_4 or from p_6 to p_7; therefore it should be sufficiently accurate to use the corresponding mean pressures given by Eqs. (9c) and (9f).

For an Otto cycle, $d_5 = d_1$ and $\theta_5 = 2\pi$. For a typical diesel engine, θ_5 is about $375°$, and the corresponding value of d_5 is given by

$$d_5 = a_r d_1 \tag{10}$$

where the **relative admission** a_r is typically about 1.25 (Purday [1962, p. 4]). For $a_r = 1$, the combination cycle reduces to the Otto cycle.

For spark-ignition engines the peak pressure is in the range $300 \leq p_4$ (psi) ≤ 1000, and for Diesel engines, $500 \leq p_4$ (psi) ≤ 1500 (Lichty [1958]).

For future reference, we note that the work done by the gas forces during a polytropic process from state (p_i, V_i) to state (p_j, V_j) may be expressed[22] as

$$\Delta W_{ij} = \frac{p_i V_i - p_j V_j}{k - 1} \tag{11}$$

where k is the polytropic exponent. Therefore, it may be verified[22] that the work per cycle of the combination cycle illustrated in Fig. 1 is given by

$$\Delta W_p \equiv p_4(V_5 - V_4) + \frac{p_5 V_5 - p_6 V_6}{k_e - 1} + \frac{p_2 V_2 - p_3 V_3}{k_c - 1}$$
$$- (p_8 - p_1)(V_2 - V_1) \tag{12}$$

The resisting torque M_r depends on the characteristics of the driven device (e.g., generator, compressor, etc.). For example, a centrifugal compressor or a fan might generate a resisting torque proportional to the square of its shaft speed; i.e.,

$$M_r = C\dot{\theta}|\dot{\theta}| \tag{13}$$

where C is a constant. The effects of windage and friction torques in the engine itself may also be included in M_r.

The gas force P, the resisting torque M_r, and the generalized force Q may be calculated in a subroutine called FORCES, such as that listed below, for the combination cycle of Fig. 1 and the *velocity-squared* torque of Eq. (13).

In the following list of input parameters, the FORTRAN name is shown in brackets. The numerical data shown will be used for example problems in subsequent sections.

---

[22]See Problem **12.100**-47.

p_{in}: intake pressure, p_1 [PIN] $= 13.0$ psi

p_4: maximum pressure [PM] $= 650$ psi

p_{ex}: exhaust pressure, p_7 [PEX] $= 18$ psi

p_a: atmospheric pressure [PA] $= 14.7$ psi

C: coefficient for Eq. (13) [CD] $= 0.02$ lb.-in.-sec.²

CR: compression ratio [CR] $= 8.0$

a_r: admission ratio [AR] $= 1$ (Otto cycle)

k_e: polytropic exponent [KE] $= 1.37$

k_c: polytropic exponent [KC] $= 1.37$

A_p: piston area [AP] $= 7.06853$ in.²

Other FORTRAN names used and their algebraic counterparts are

d [D], d_1 [D1], d_2 [D2], d_5 [D5], θ_F [THETAF], p [P], M_r [MR]

```
      SUBROUTINE FORCES(T,DTHETA,Q)
C
C     THIS USER-SUPPLIED ROUTINE PROVIDES GAS FORCE P ASSOCIATED WITH
C     A FOUR-STROKE COMBINATION CYCLE, AND RESISTING MOMENT
C     MR DUE TO VELOCITY-SQUARED DAMPING
C
C-------------------------------------------------------------------
C     CONSTANTS USED
C     CN(1)=CR=COMPRESSION RATIO(V2/V1)
C     CN(2)=AR=ADMISSION RATIO(V5/V4). (AR=1 FOR OTTO CYCLE)
C     CN(3)=KC=POLYTROPIC EXPONENT FOR COMPRESSION
C     CN(4)=KE=POLYTROPIC EXPONENT FOR EXPANSION
C     CN(5)=PM=MAXIMUM PRESURE
C     CN(6)=PEX=EXHAUST PRESSURE
C     CN(7)=PIN=INTAKE PRESSURE
C     CN(8)=PA=ATMOSPHERIC PRESSURE
C     CN(9)=AP=AREA OF PISTON
C     CN(10)=CD=COEFFICIENT FOR VELOCITY-SQUARED RESISTING MOMENT
C-------------------------------------------------------------------
      IMPLICIT REAL*8 (A-Z)
      COMMON/BLK1/R,L,MA,MB,M1,M2,M3,J0,JG,JAB,LAMDA,KPHI,KS,KPHIPR,KSPR
     1,GI,CC,THETA,THETAF,PHI,S,D,E,A,C,ILEVEL,IP
      COMMON/BLK2/MP,MR,P
      COMMON/CONSTS/CR,AR,KC,KE,PM,PEX,PIN,PA,AP,CD,CN(14)
      INTEGER ILEVEL,IP,IFLAG
      DATA PI1,PI2,PI3,PI4,IFLAG/3.141592854,6.283185307,9.424777961,
     112.56637061,0/
      IF(IFLAG.NE.0)GO TO 100
C
C     ON FIRST CALL TO FORCES, CALCULATE PRELIMINARY RESULTS
      IFLAG=1
      D1=2.*R/(CR-1.)
      D5=AR*D1
      D2=D1+R+R
      P2=PIN
      P3=PIN*CR**KC
      P5=PM
      P6=PM*(D5/D2)**KE
      V2=AP*D2
      V3=AP*D1
      V5=AP*D5
```

```
          V6=V2
          V4=V3
          V1=V3
          DELWP=(P5*V5-P6*V6)/(KE-1.)+(P2*V2-P3*V3)/(KC-1.)+P5*(V5-V4)
         1+(PEX-PIN)*(V1-V2)
          MEP=DELWP/(V2-V1)
          THDRMS=DSQRT(DELWP/(CD*PI4))
          RMSRPM=THDRMS*9.549296586
          PRMS=DELWP*THDRMS/PI4
          WRITE(6,50)DELWP,MEP,THDRMS,RMSRPM,PRMS
    50    FORMAT(//5X,'PRELIMINARY CALCULATIONS FROM SUBROUTINE FORCES:'
         1/5X,'NOTE: UNITS CONSISTENT WITH INPUT DATA'//
         25X,'WORK OF GAS FORCE PER CYCLE   ',1PD14.6/
         35X,'MEAN EFFECTIVE PRESSURE       ',1PD14.6/
         45X,'STEADY STATE R.M.S. SPEED     ',1PD14.6,2X,'RAD/SEC,',1PD14.6,
         52X,'RPM'/5X,'R.M.S. POWER                 ',1PD14.6)
    100   CONTINUE
          D=D1+L+R-S
          IF(THETAF.GT.0.0.AND.THETAF.LE.PI1)P=PIN
          IF(THETAF.GT.PI1.AND.THETAF.LT.PI2)P=PIN*(D2/D)**KC
          IF(THETAF.EQ.PI2)P=(PI*CR**KC+PM)/2.
          IF(THETAF.GT.PI2.AND.D.LE.D5)P=PM
          IF(THETAF.GT.PI2.AND.D.GT.D5)P=PM*(D5/D)**KE
          IF(THETAF.EQ.PI3)P=(P+PEX)/2.
          IF(THETAF.GT.PI3)P=PEX
          IF(THETAF.EQ.0.0.OR.THETAF.EQ.PI4)P=(PIN+PEX)/2.
    C
    C     CONVERT PRESSURE TO FORCE
          P=(P-PA)*AP
          MP=-P*KS
          MR=CD*DTHETA*DABS(DTHETA)
          Q=MP-MR
          RETURN
          END
```

12.83 Transient Engine Performance—General Case

Having derived general expressions for s, $\mathcal{C}$, and Q, we are now in a position to integrate the differential equation of motion

$$s\ddot{\theta} = Q - \mathcal{C}\dot{\theta}^2 = M_p - M_r - \mathcal{C}\dot{\theta}^2 \tag{1}$$

starting from any given initial state (at $t = 0$) where

$$q(0) \equiv \theta_0, \qquad \dot{q}(0) \equiv \dot{\theta}_0 \tag{2}$$

Following the pattern of Sec. **12.61**, we define the auxiliary functions

$$w_1 \equiv q = \theta \tag{3}$$
$$w_2 \equiv \dot{q} = \dot{\theta} \tag{4}$$

Then the initial value problem posed by Eqs. (1) and (2) can be stated in the form

$$\dot{w}_1 = w_2 \equiv F_1(w_2) \tag{5}$$

$$\dot{w}_2 = \frac{M_p - M_r - \mathcal{C}w_2^2}{s} \equiv F_2(w_1, w_2, t) \tag{6}$$

$$w_1(0) = \theta_0 \tag{7}$$
$$w_2(0) = \dot{\theta}_0 \tag{8}$$

This initial value problem is readily solved by any standard program for integrating systems of differential equations. If, for example, we use the Runge-Kutta-based subroutine DRKGS (see Appendix **G**), it will be necessary to evaluate the functions F_1 and F_2 in a subroutine called FCT. The subroutine FCT can, in turn, call upon the subroutine GICC (Sec. **12.81**) to provide g and c, and it can call upon subroutine FORCES (Sec. **12.82**) to find the gas force P and resisting torque M_r which make up the generalized force Q. A standard subroutine FCT for all reciprocating engines can be written as follows:

```
      SUBROUTINE FCT(T,Y,DERY)
C
C     THIS ROUTINE EVALUATES THE RIGHT HAND SIDES--DERY(1),AND DERY(2) OF
C     THE TWO FIRST ORDER DIFFERENTIAL EQUATIONS.
C
      IMPLICIT REAL*8 (A-Z)
      DIMENSION Y(2),DERY(2)
      COMMON/BLK1/R,L,MA,MB,M1,M2,M3,J0,JG,JAB,LAMDA,KPHI,KS,KPHIPR,KSPR
     1,GI,CC,THETA,THETAF,PHI,S,D,E,A,C,ILEVEL,IP
      INTEGER ILEVEL,IP
      THETA=Y(1)
      DTHETA=Y(2)
      CALL GICC
      THETAF=DMOD(THETA,12.566370614)
      CALL FORCES(T,DTHETA,Q)
      DERY(1)=Y(2)
      DERY(2)=(Q-CC*Y(2)*Y(2))/GI
      RETURN
      END
```

Subroutine DRKGS also calls upon a subroutine OUTP which determines the arrangement of output and calls upon a subroutine REACTS[23] only if the user wants to find reaction forces at the bearing surfaces. Finally, a MAIN program, which we call DYREC (DYnamics of RECiprocating Machines), is needed to read input and call DRKGS. The organization of subprograms is shown in Fig. 1.

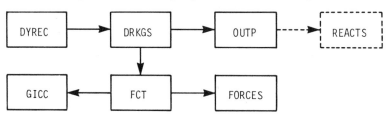

Figure 12.83-1. Relationship of main program, DYREC, to subroutines. Arrows point from calling programs to called programs.

The listings of DYREC and OUTP are given in Appendix **I**. Note that most of program DYREC consists of instructions for reading data and echoing them in printout. The essential logic is implemented in the 15 statements prior to (and including) the call to DRKGS.

Note on Steady-State Conditions

An engine is said to be in the **steady state** when all conditions repeat after every $N\pi$ rad of crankshaft rotation, where N is 4 (2) for a four-stroke (two-stroke) cycle.

---

[23]For details on the calculation of reaction forces, see Sec. **12.86**.

To determine the conditions necessary for a *steady state*, we note that during a crank rotation of $d\theta$ the kinetic energy increases by

$$dT = (M_p - M_r)d\theta \tag{9}$$

Integrating over a cycle of rotation, we find

$$T(\theta + N\pi) - T(\theta) = \int_{\theta}^{\theta+N\pi} (M_p - M_r)\, d\theta \tag{10}$$

But $T = \mathscr{s}\dot{\theta}^2/2$, and both $\mathscr{s}(\theta)$ and $\dot{\theta}$ have period $N\pi$, so that $T(\theta + N\pi) = T(\theta)$, and

$$\int_{\theta}^{\theta+N\pi} (M_p - M_r)\, d\theta = 0 \tag{11}$$

is a necessary condition for a steady state. In Sec. **12.84** it will be seen that Eq. (11) is also a sufficient condition for a steady state. For later use, we define the **inertia torque** M_{in} as

$$M_{\mathrm{in}} \equiv -(\mathscr{s}\ddot{\theta} + \mathscr{c}\dot{\theta}^2) = -(M_p - M_r) \tag{12}$$

where the last equation is merely a restatement of the fundamental equation of motion (1). It then follows from Eqs. (11) and (12) that the work of the inertia torque over a cycle vanishes; i.e.,

$$\int_{\theta}^{\theta+N\pi} M_{\mathrm{in}}\, d\theta = -\int_{\theta}^{\theta+N\pi} (\mathscr{s}\ddot{\theta} + \mathscr{c}\dot{\theta}^2)\, d\theta = 0 \tag{13}$$

If we are interested in the performance of the engine in its steady state, we can save computer time by simulating the motion with an initial speed near the mean speed in its steady state.

For example, if the resistance is of the velocity-squared type,

$$M_r = C\dot{\theta}^2$$

and steady state will be achieved when the work dissipated per cycle at the shaft equals

$$\int_0^{N\pi} M_r\, d\theta = \int_0^{N\pi} C\dot{\theta}^2\, d\theta = \Delta W_p \tag{14}$$

where ΔW_p is the work per cycle of the gas forces [cf. Eq. **(12.82-12)**]. Thus the root-mean-square (rms) speed is given by

$$\dot{\theta}_{\mathrm{rms}} \equiv \left[\frac{1}{N\pi} \int_0^{N\pi} \dot{\theta}^2\, d\theta \right]^{1/2} = \left[\frac{\Delta W_p}{CN\pi} \right]^{1/2} \tag{15}$$

If $\dot{\theta}$ does not fluctuate too greatly over a cycle, it will remain in the neighborhood of $\dot{\theta}_{\mathrm{rms}}$. Thus, if we simulate the behavior of the engine with initial velocity

$$\dot{\theta}(0) = \left(\frac{\Delta W_p}{CN\pi} \right)^{1/2} \tag{16}$$

we should expect to see steady-state behavior established after just a few cycles of motion for any choice of the initial angle $\theta(0)$.

If the resisting torque M_r is assumed to be constant, a steady state will be reached only if

$$\int_0^{N\pi} M_r \, d\theta = \Delta W_p$$

or

$$M_r = \frac{\Delta W_p}{N\pi} \tag{17}$$

and the mean speed achieved will depend on the initial conditions.

Simulation with Program DYREC

To illustrate the use of program DYREC, let us simulate the performance of the four-stroke engine whose dimensions and inertial properties are given in the caption of Fig. **12.81**-1 (except that $J_O = 2.5$ lb-sec²-in.). The resisting torque is assumed to be of the velocity-squared type, $M_r = C\dot\theta^2$. The gas forces follow the combination cycle of Fig. **12.82**-1, with the numerical data listed near the end of Sec. **12.82** .

For the given data, the work of the gas forces per cycle is found from Eq. (**12.82**-12) to be

$$\Delta W_p = 4703.6 \text{ in.-lb}$$

and the steady state rms velocity predicted by Eq. (15) is

$$\dot\theta_{\text{rms}} = \left(\frac{\Delta W_p}{CN\pi}\right)^{1/2} = 136.8 \text{ rad/sec}$$

We therefore set the initial conditions as $\theta_0 = 0$, $\dot\theta_0 = \dot\theta_{\text{rms}}$.

To compute about 20 points per cycle, we choose the time step Δt to be given approximately by

$$\dot\theta_{\text{rms}} \, \Delta t \doteq \frac{4\pi}{20}$$

or $\Delta t = 0.005$ sec.

The program was run for about 7 complete cycles on a UNIVAC 9070 computer for a total of 35 CPU seconds (including compilation). The total cost for the 140 time steps printed out was $2.45, or about 1.75 cents per calculated point.

As an example of the output data, we show in Fig. 2 a plot of the crankshaft speed $\dot\theta$ versus the fundamental crank angle θ_F for the first, fourth, and sixth cycles. Note that the system has, for all practical purposes, achieved its steady state by the fourth cycle. Results for the fifth cycle (not shown) are indistinguishable from the fourth and sixth cycles. From the graph we see that $\dot\theta_{\text{max}} = 145.5$ rad/sec, and $\dot\theta_{\text{min}} = 124.5$ rad/sec; therefore the arithmetic mean speed is $\dot\theta_a = 135$ rad/sec, which is extremely close to the predicted rms value of $\dot\theta_{\text{rms}} = 137$ rad/sec.

At each time step, the program automatically prints out displacements (θ, d, ϕ), velocities $(\dot\theta, \dot d, \dot\phi)$, accelerations $(\ddot\theta, \ddot d, \ddot\phi)$, inertia coefficients $(\mathcal{I}, \mathcal{C})$, and the quantities

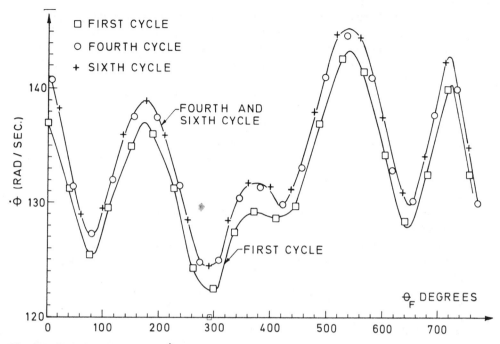

Figure 12.83-2. Crankshaft speed $\dot\theta$ versus fundamental crank angle θ_F, for one cylinder engine with four-stroke (Otto) cycle, and velocity-squared resisting torque. $J_0 = 2.5$ lb sec^2 in. Initial conditions were $\theta_0 = 0$, $\dot\theta_0 = 136.8$ rad/sec. All other data as in examples of sections **12.81** and **12.82**.

$$M_t = M_p - \mathcal{C}\dot\theta^2 \qquad \text{(turning moment)} \tag{18}$$

$$M_s = M_t - M_r \qquad \text{(surplus moment)} \tag{19}$$

The last two quantities are of interest in flywheel design (see Sec. **12.91**) and in studies of crankshaft torsional vibrations. From the printout, the user can easily reconstruct the additional quantities:

$$M_r = M_t - M_s \qquad \text{(resisting moment)} \tag{20}$$

$$M_p = M_t + \mathcal{C}\dot\theta^2 \qquad \text{(turning moment of gas pressure)} \tag{21}$$

$$P = \frac{\dot\theta}{d} M_p \qquad \text{(gas force)} \tag{22}$$

The program will also print out the reaction forces on each bearing surface (see Sec. **12.86**) if the user so desires.

12.84   Performance Under Velocity-Independent Loading

We shall now consider certain "simplifications" in the analysis which are possible when the resisting torque M_r is known to be independent of the velocity $\dot\theta$; i.e., $M_r = M_r(\theta)$. The most common example of this condition occurs when an engine is running in a test stand against a *constant* torque provided by a friction brake dynamometer. The resisting torque provided by an eddy current (or hydraulic) absorption dynamometer is velocity-dependent. However, when the engine has a large flywheel and is run under steady state conditions at high speeds, the fluctuations from the mean speed may be negligible. The term *steady state* (for an engine) implies not that the crank velocity $\dot{\theta}$ is constant but that it is a periodic function of θ with period $N\pi$, where N is the number of strokes per engine cycle.

Although it is common practice to assume, for design purposes, that an engine is loaded by a constant resisting torque M_r, it must be remembered that, even for steady-state operation, this is an idealization which is acceptable only under the special circumstances described in the previous paragraph.

When the torque M_r is velocity-independent, so too is the generalized force

$$Q = -(Pk_s + M_r) = M_p - M_r \qquad (12.82\text{-}2)$$

and the engine may be analyzed by the method of work and energy described in Sec. **12.50**.

For example, suppose that at time $t_0 = 0$, the initial conditions are $\theta = \theta_0, \dot{\theta} = \dot{\theta}_0$. Then the net work done on the engine as the shaft turns through the angle $\theta - \theta_0$ is

$$\int_{\theta_0}^{\theta} Q \, d\theta = \int_{\theta_0}^{\theta} (M_p - M_r) \, d\theta \equiv W(\theta) \qquad (1)$$

Since M_p and M_r are well-defined functions of θ, the quadrature in Eq. (1) can be done analytically or by a numerical procedure such as Simpson's rule (see Appendix **H**).

The change in kinetic energy is given by

$$T - T_0 = \tfrac{1}{2}(\mathcal{I}\dot{\theta}^2 - \mathcal{I}_0\dot{\theta}_0^2) = W(\theta) \qquad (2)$$

where

$$\mathcal{I}_0 = \mathcal{I}(\theta_0) \qquad (3)$$

Thus the velocity, at any stage, is given by

$$\dot{\theta} = \pm \left\{ \frac{1}{\mathcal{I}}[\mathcal{I}_0\dot{\theta}_0^2 + 2W(\theta)] \right\}^{1/2} \qquad (4)$$

where (as explained in Sec. **12.52**) the plus sign is used at $\theta = \theta_0$ if $\dot{\theta}_0 > 0$, and the sign is reversed each time $\dot{\theta}$ passes through zero. Usually, $\dot{\theta}$ is monotonic over the range of interest, and the plus sign should be used exclusively over that range; we shall therefore omit the negative sign in subsequent discussion of Eq. (4).

In Sec. **12.83** we saw that a necessary condition for a steady state to exist is that

$$W(\theta + N\pi) - W(\theta) = \int_{\theta}^{\theta + N\pi} (M_p - M_r) \, d\theta = 0 \qquad (12.83\text{-}11)$$

If this condition is known to be satisfied, it follows from Eq. (2) and the condition $\mathcal{I}(\theta + N\pi) = \mathcal{I}(\theta)$ that $\dot{\theta}$ has period $N\pi$; i.e.

$$\dot{\theta}(\theta + N\pi) - \dot{\theta}(\theta) = 0 \qquad (5)$$

In other words, Eq. (**12.83**-11) is *sufficient* (as well as *necessary*) to ensure that the engine is in a steady state.

Equation (4) gives the velocity for any crank angle θ. To find the corresponding time t we note that

$$t = \int_0^t dt = \int_{\theta_0}^{\theta} \frac{d\theta}{\dot{\theta}} \tag{6}$$

The quadrature indicated in Eq. (6) can also be performed numerically. Thus Eqs. (1), (4), and (6) comprise a complete solution, involving two successive quadratures, for any engine where the resisting force is *not* velocity-dependent.

It has been suggested (Givens and Wolford [1969]) that this method of analysis may also be used for systems with velocity-dependent resistance. The procedure advocated is to divide the range of integration into short intervals $\Delta\theta$ and to substitute for M_r in Eq. (1) that value which it had at the beginning of each interval. After the velocity is calculated at the end of the interval, the mean value of M_r for the interval is recalculated, and the procedure is repeated as often as necessary until the velocity at the end of the interval stabilizes. It should be recognized that this iterative process represents a relatively crude method of numerical integration which is less accurate, and more time-consuming, than a direct integration of the fundamental equation of motion, as in program DYREC.

It is not difficult to generate an efficient computer program for analysis of engines with *velocity-independent* loads based on the energy method just described. However, the *more general* program DYREC (see Sec. **12.83**) will accomplish the same end for *any* type of loading.

12.85 *Power Smoothing in Multicylinder Engines*

In an engine with several cylinders, the gas pressures, inertia forces, and kinetic energy associated with the individual cylinders must be added together in proper phase relationships.

For example, consider an **in-line** arrangement where all cylinder axes are parallel and lie in a common plane, as illustrated in Fig. 1 for a six-cylinder engine.

The angular arrangement of the cranks in Fig. 1 is described by the notation (0, 240, 120, 120, 240, 0), representing the angles, in degrees, of the successive cranks relative to an end crank.

The cylinders are usually numbered sequentially from one end, and the sequence in which they execute their power strokes—i.e., the **firing sequence** or **firing order**—is indicated by a string of cylinder numbers. For example, the cylinders in Fig. 1 are numbered 1, 2, 3, 4, 5, 6, starting from the left end, and the shaft rotates counter-clockwise (as viewed from the left end). Top dead center is reached simultaneously by cranks 1 and 6; after an additional 120° of rotation, cranks 2 and 5 reach top dead center; and after another 120° of rotation, cranks 3 and 4 reach top dead center. Firing may be assumed to occur at top dead center, so the only possible sequences of firing one cylinder after each 120° of rotation are as follows:

$$1 \longrightarrow \begin{cases} 2 \longrightarrow \begin{cases} 3 \longrightarrow 6 \longrightarrow 5 \longrightarrow 4 \\ 4 \longrightarrow 6 \longrightarrow 5 \longrightarrow 3 \end{cases} \\ 5 \longrightarrow \begin{cases} 3 \longrightarrow 6 \longrightarrow 2 \longrightarrow 4 \\ 4 \longrightarrow 6 \longrightarrow 2 \longrightarrow 3 \end{cases} \end{cases}$$

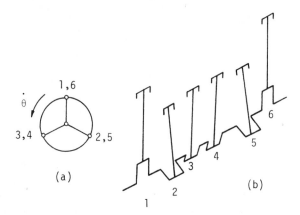

Figure 12.85-1. Six-cylinder in-line engine. A possible firing sequence is (1-5-3-6-2-4): (a) View from left end, in which shaft rotates counterclockwise; (b) Crankshaft layout.

Assuming that the third of these four possibilities is chosen, we would describe the firing sequence as (1–5–3–6–2–4). Since the firing sequence repeats indefinitely, all cyclic permutations of the above number–strings are equivalent; e.g., the chosen sequence could be represented as (5–3–6–2–4–1). Reversal of this order gives (1–4–2–6–3–5), which represents the same pattern of firing but with the shaft rotating clockwise (as viewed from the left end).

The angle of crankshaft rotation between successive firings is called the **firing interval**. For the particular firing sequence and crankshaft design shown, the firing interval is uniformly 120°. *Uniform* firing intervals are usually desirable, but nonuniform intervals may be desired for special applications. Uniform firing intervals tend to produce smaller fluctuations of engine speed and thereby contribute greatly to **power smoothing**. A table of appropriate firing sequences needed for uniform firing intervals of in-line engines is given by Maleev [1945, p. 527].

If α_j represents the angle by which the power stroke of cylinder j lags behind that of cylinder 1, the crank angle for cylinder j is given by

$$\theta_j = \theta_1 - \alpha_j \tag{1}$$

We shall call α_j the **firing lag** for cylinder j.

For the example of Fig. 1, with firing sequence (1–5–3–6–2–4) and firing interval 120°, we have $(\alpha_1, \alpha_5, \alpha_3, \alpha_6, \alpha_2, \alpha_4) \equiv (0, 120°, 240°, 360°, 480°, 600°)$.

All results that we have developed for a single-cylinder engine with crank angle θ can now be used for cylinder j if we replace θ by θ_j $(\equiv \theta_1 - \alpha_j)$. For example, suppose that all cylinders have the same dimensions and the same indicator diagrams; let $\mathcal{I}_1(\theta_1)$, $\mathcal{C}_1(\theta_1)$, $M_{p_1}(\theta_1)$ represent the generalized inertia[24], centripetal coefficient, and turning moment of the gas forces for cylinder 1. Then the corresponding quantities for a complete engine with N_c such cylinders are the following functions of θ_1:

---

[24]The constant J_O, which occurs in the definition of $\mathcal{I}$ [Eq. (12.81-8)], may be distributed among the individual slider-crank mechanisms in any way (e.g., uniformly) as long as the sum of the contributions equals the MOI of the complete crankshaft-flywheel assembly.

$$\mathcal{J}(\theta_1) \equiv \mathcal{J}_1(\theta_1) + \mathcal{J}_1(\theta_1 - \alpha_2) + \cdots + \mathcal{J}_1(\theta_1 - \alpha_{N_c}) \tag{2}$$

$$\mathcal{C}(\theta_1) \equiv \mathcal{C}_1(\theta_1) + \mathcal{C}_1(\theta_1 - \alpha_2) + \cdots + \mathcal{C}_1(\theta_1 - \alpha_{N_c}) \tag{3}$$

$$M_p(\theta_1) \equiv M_{p_1}(\theta_1) + M_{p_1}(\theta_1 - \alpha_2) + \cdots + M_{p_1}(\theta_1 - \alpha_{N_c}) \tag{4}$$

If M_r represents the total resisting moment applied to the crankshaft assembly, the equation of motion for the system is

$$\mathcal{J}(\theta)\ddot{\theta} + \mathcal{C}(\theta)\dot{\theta}^2 = M_p(\theta) - M_r \tag{5}$$

where we have defined the generalized coordinate θ by

$$\theta \equiv \theta_1 \tag{6}$$

Since all previous discussions of the system's response stemmed from Eq. (5), we see that all previous results for a single-cylinder engine have now been generalized by the introduction of Eqs. (1)–(6).

Similarly, the contribution to $\mathcal{J}$, $\mathcal{C}$, and M_p from each cylinder of a V-type or radial engine can be readily found.

12.86 *Bearing Reactions*

Once the fundamental equation of motion $\mathcal{J}\ddot{q} + \mathcal{C}\dot{q}^2 = Q$ has been solved for q ($q = \theta$) at a given time t, the position, velocity, and acceleration may be found for each point in the system. In particular, the acceleration components of the links' mass centers are given by Eqs. (**10.50**-3) and (**10.50**-4) in the form

$$\ddot{x}_i = u_i\ddot{q} + u_i'\dot{q}^2 \tag{1}$$

$$\ddot{y}_i = v_i\ddot{q} + v_i'\dot{q}^2 \tag{2}$$

and the angular accelerations of the links are given by Eq. (**10.50**-2) in the form

$$\ddot{\theta}_i = \omega_i\ddot{q} + \omega_i'\dot{q}^2 \tag{3}$$

The influence coefficients $(u_i, v_i, \omega_i, u_i', v_i', \omega_i')$ may be computed by the methods described in Sec. **10.51**.

Alternatively, the kinematic state of the mass centers can be found directly (see Fig. **12.80**-1) from the expressions

$$\begin{Bmatrix} x_1 \\ y_1 \end{Bmatrix} = c \begin{Bmatrix} \cos\theta \\ \sin\theta \end{Bmatrix} \tag{4a}$$

$$\begin{Bmatrix} x_3 \\ y_3 \end{Bmatrix} = \begin{Bmatrix} s \\ e \end{Bmatrix} \tag{4b}$$

$$\begin{Bmatrix} x_2 \\ y_2 \end{Bmatrix} = r \begin{Bmatrix} \cos\theta \\ \sin\theta \end{Bmatrix} + a \begin{Bmatrix} \cos\phi \\ -\sin\phi \end{Bmatrix} \tag{4c}$$

$$\begin{Bmatrix} \dot{x}_1 \\ \dot{y}_1 \end{Bmatrix} = c\dot{\theta} \begin{Bmatrix} -\sin\theta \\ \cos\theta \end{Bmatrix} \tag{5a}$$

$$\begin{Bmatrix} \dot{x}_3 \\ \dot{y}_3 \end{Bmatrix} = \begin{Bmatrix} \dot{s} \\ 0 \end{Bmatrix} \tag{5b}$$

$$\begin{Bmatrix} \dot{x}_2 \\ \dot{y}_2 \end{Bmatrix} = r\dot{\theta} \begin{Bmatrix} -\sin\theta \\ \cos\theta \end{Bmatrix} - a\dot{\phi} \begin{Bmatrix} \sin\phi \\ \cos\phi \end{Bmatrix} \tag{5c}$$

$$\begin{Bmatrix} \ddot{x}_1 \\ \ddot{y}_1 \end{Bmatrix} = c\ddot{\theta} \begin{Bmatrix} -\sin\theta \\ \cos\theta \end{Bmatrix} - c\dot{\theta}^2 \begin{Bmatrix} \cos\theta \\ \sin\theta \end{Bmatrix} \tag{6a}$$

$$\begin{Bmatrix} \ddot{x}_3 \\ \ddot{y}_3 \end{Bmatrix} = \begin{Bmatrix} \ddot{s} \\ 0 \end{Bmatrix} \tag{6b}$$

$$\begin{Bmatrix} \ddot{x}_2 \\ \ddot{y}_2 \end{Bmatrix} = r\ddot{\theta} \begin{Bmatrix} -\sin\theta \\ \cos\theta \end{Bmatrix} - a\ddot{\phi} \begin{Bmatrix} \sin\phi \\ \cos\phi \end{Bmatrix}$$
$$ - r\dot{\theta}^2 \begin{Bmatrix} \cos\theta \\ \sin\theta \end{Bmatrix} - a\dot{\phi}^2 \begin{Bmatrix} \cos\phi \\ -\sin\phi \end{Bmatrix} \tag{6c}$$

The rotational state of the links is given simply by

$$\{\theta_1, \theta_2, \theta_3\} = \{\theta, -\phi, 0\} \tag{7a}$$
$$\{\dot{\theta}_1, \dot{\theta}_2, \dot{\theta}_3\} = \{\dot{\theta}, -\dot{\phi}, 0\} \tag{7b}$$
$$\{\ddot{\theta}_1, \ddot{\theta}_2, \ddot{\theta}_3\} = \{\ddot{\theta}, -\ddot{\phi}, 0\} \tag{7c}$$

In the above equations, we need to calculate the terms ϕ, s, $\dot{\phi}$, $\dot{s}$, which may be found from Eqs. (**12.80**-1)–(**12.80**-4). We also need the terms

$$\ddot{\phi} = k_\phi \ddot{\theta} + k'_\phi \dot{\theta}^2 \tag{8}$$
$$\ddot{s} = k_s \ddot{\theta} + k'_s \dot{\theta}^2 \tag{9}$$

where k_ϕ, k_s, k'_ϕ, and k'_s may be found from Eqs. (**12.80**-3)–(**12.80**-6).

Having found numerical values for all three sets of accelerations ($\ddot{x}_i, \ddot{y}_i, \ddot{\theta}_i$), we can utilize d'Alembert's principle (see Sec. **12.70**), which states that the system is in a state of *dynamic equilibrium* under the combined effect of the *active forces* (X_i, Y_i, M_i) and the *inertia forces* ($-m_i\ddot{x}_i, -m_i\ddot{y}_i, -J_i\ddot{\theta}_i$).

Therefore, we may consider the system to be in a state of equilibrium under the influence of the *effective static force* sets

$$\{X_i^*, Y_i^*, M_i^*\} \equiv \{X_i - m_i\ddot{x}_i, Y_i - m_i\ddot{y}_i, M_i - J_i\ddot{\theta}_i\} \tag{10}$$

applied at the mass centers of each link.

To find the bearing reactions, we may utilize the static analysis of the mechanism depicted in Fig. **11.52**-1, provided that we replace the forces (X_i, Y_i, M_i) by (X_i^*, Y_i^*,

M_i^*) in Eqs. (**11.52**-1)–(**11.52**-9). These forces will act through the mass centers of the appropriate links provided that the parameters (ξ_i, η_i) used in Eqs. (**11.52**-11) are chosen as

$$(\xi_1, \eta_1) = (c, 0)$$
$$(\xi_2, \eta_2) = (a, 0) \qquad\qquad (11)$$
$$(\xi_3, \eta_3) = (0, 0)$$

The calculations for bearing reactions have been programmed in the FORTRAN subroutine REACTS. This subroutine, which is compatible with the main program DYREC (see Appendix **I**), is listed immediately below.

```
      SUBROUTINE REACTS
C
C     THIS ROUTINE COMPUTES REACTION FORCES AT THE JOINTS
C
C     NOTATION
C     (XA,YA): FORCE COMPONENTS ON CON. ROD AT HINGE A
C     (XB,YB): FORCE COMPONENTS ON CON. ROD AT HINGE B
C     (XO,YO): FORCE COMPONENTS ON CRANK AT HINGE O
C     (RA,RB,RO): RESULTANTS MAKING ANGLE (ANGA,ANGB,ANGO) WITH X AXIS
C     NB: REACTION FORCE OF FRAME ON SLIDER, THROUGH POINT B
C
      IMPLICIT REAL*8 (A-Z)
      COMMON/BLK1/R,L,MA,MB,M1,M2,M3,JO,JG,JAB,LAMDA,KPHI,KS,KPHIPR,KSPR
     1,GI,CC,THETA,THETAF,PHI,S,D,E,A,C,ILEVEL,IP
      COMMON/BLK2/MP,MR,P
      COMMON/BLK3/CTHETA,STHETA,CPHI,SPHI
      COMMON/BLK4/THETS(50),DTHETA(50),DPHI(50),DD(50),DDTHET(50),
     1DDPHI(50),DDD(50),RA(50),RB(50),RO(50),NB(50),ANGA(50),ANGB(50),
     2ANGO(50)
      INTEGER I,IP,ILEVEL
      X1=M1*C*(DTHETA(IP)*DTHETA(IP)*CTHETA+DDTHET(IP)*STHETA)
      Y1=M1*C*(DTHETA(IP)*DTHETA(IP)*STHETA-DDTHET(IP)*CTHETA)
      X2=M2*(R*DTHETA(IP)*DTHETA(IP)*CTHETA+R*DDTHET(IP)*STHETA
     1+A*DPHI(IP)*DPHI(IP)*CPHI+A*DDPHI(IP)*SPHI)
      Y2=M2*(R*DTHETA(IP)*DTHETA(IP)*STHETA-R*DDTHET(IP)*CTHETA
     1-A*DPHI(IP)*DPHI(IP)*SPHI+A*DDPHI(IP)*CPHI)
      X3=M3*DDD(IP)-P
      M2S=JG*DDPHI(IP)
      YAB=L*SPHI
      XBA=L*CPHI
      Y2A=-A*SPHI
      X2A=A*CPHI
      M2A=M2S+Y2*X2A-X2*Y2A
      XB=X3
      YB=-(M2A+X3*YAB)/XBA
      XA=-XB-X2
      YA=-Y2-YB
      XO=XA-X1
      YO=YA-Y1
      NB(IP)=YB
      RA(IP)=DSQRT(XA*XA+YA*YA)
      RB(IP)=DSQRT(XB*XB+YB*YB)
      RO(IP)=DSQRT(XO*XO+YO*YO)
      ANGA(IP)=DATAN2(YA,XA)*57.29577951
      ANGB(IP)=DATAN2(YB,XB)*57.29577951
      ANGO(IP)=DATAN2(YO,XO)*57.29577951
      RETURN
      END
```

If the user indicates that he wants to know the bearing reactions, program DYREC will print out, at each time step, the following information [output heading in brackets]:

R_A [RA]: Force exerted on connecting rod by crank pin (at A)

R_B [RB]: Force exerted on connecting rod by piston pin (at B)

R_O [RO]: Force exerted on crank by crankshaft journal (at O)

N_B [NB]: Force exerted on piston by cylinder wall

θ_A [ANGLE A]: Angle of force vector R_A with respect to x axis

θ_B [ANGLE B]: Angle of force vector R_B with respect to x axis

θ_O [ANGLE O]: Angle of force vector R_O with respect to x axis

12.90 FLYWHEEL CALCULATIONS

A **flywheel** is a heavy disk or wheel attached to a rotating shaft. Its purpose is to reduce the fluctuations in the shaft's speed $\dot{\theta}$ which arise, in reciprocating engines, from the periodic pressure pulses on the piston, the position-dependent generalized inertia $\mathcal{I}$, and the variations of resisting torque M_r.

The performance of a flywheel is expressed in terms of the **coefficient of fluctuation**

$$C_F \equiv \frac{1}{C_S} = \frac{\dot{\theta}_{max} - \dot{\theta}_{min}}{\dot{\theta}_a} \tag{1}$$

where the maximum, minimum, and average values of the shaft speed are represented by $\dot{\theta}_{max}$, $\dot{\theta}_{min}$, and

$$\dot{\theta}_a \equiv \tfrac{1}{2}(\dot{\theta}_{max} + \dot{\theta}_{min}) \tag{2}$$

The inverse of C_F is called the **coefficient of steadiness**.

Minimum recommended coefficients of steadiness for typical classes of driven devices are listed in Table 1. Note that discrepancies exist between the different sources quoted. It should also be noted[25] that flywheels are not necessary for two-stroke engines of more than six cylinders, for four-stroke engines of more than eight cylinders, or for radial airplane engines (since propellers make excellent flywheels).

The moment of inertia (I_w) of a flywheel may be related to the coefficient of fluctuation if we know the work ΔW which is done upon the flywheel as it accelerates from its minimum angular speed ($\dot{\theta}_{min}$) to its maximum speed ($\dot{\theta}_{max}$), for then the corresponding change in kinetic energy of the flywheel is given by

$$\Delta T = \tfrac{1}{2}I_w(\dot{\theta}_{max}^2 - \dot{\theta}_{min}^2) = \tfrac{1}{2}I_w(\dot{\theta}_{max} - \dot{\theta}_{min})(\dot{\theta}_{max} + \dot{\theta}_{min}) \tag{3}$$

Upon making use of Eqs. (1) and (2), Eq. (3) can be expressed in the form

$$\Delta T = C_F I_w \dot{\theta}_a^2 \tag{4}$$

---

[25]Maleev [1945, p. 555].

TABLE 12.90-1 *Suggested Coefficients of Steadiness C_s*

Driven Machine	$C_s = 1/C_F$	Source
Hammers, crushers	5	2
Concrete mixers	12	2
Pumps, shears	20–30	2
Boat propellers	20–40	2
Machine tools	35–40	1
Textile, flour, and paper mills	40	1, 2
Spinning mills	50–65	1, 2
dc generators, direct-connected	30–50	2, 3
dc generators, belt-connected	35	2
dc generators, belt-connected	100–120	3
ac generators, direct-connected	70–100	2, 3
ac generators, belt-connected	60	2
ac generators, belt-connected	125–150	3
ac generators	300	1
dc generators (lighting)	500	1
Automobile engine (idling)	5	2
Automobile engine (normal speed)	100	2

Sources: 1. Lessels et al. [1958, p. 8–96].
2. Maleev [1945, p. 554].
3. Lichty [1958, p. 9–168].

Upon setting $\Delta T = \Delta W$, we find the required relation in the form

$$I_w = \frac{\Delta T}{C_F \dot{\theta}_a^2} = \frac{\Delta W}{C_F \dot{\theta}_a^2} \tag{5}$$

Equation (5) tells us exactly what size flywheel is needed to provide a given coefficient of fluctuation at a given average speed *provided that ΔW is known.*

EXAMPLE 1. Flywheel for a Punch Press

An electric motor and its flywheel run at an average speed of $\dot{\theta}_a = 200$ rpm (20.9 rad/sec). Through a system of gears, the motor powers a punch press which punches one hole every 2 sec. If the actual punching operation takes place in 0.2 sec and absorbs 3000 ft-lb of energy for each hole punched, find the moment of inertia I_w of the flywheel needed to produce a coefficient of fluctuation of $C_F = 10\%$. It is conservative to assume that the inertia of all moving parts is negligible compared to that of the flywheel.

Solution:

The average power supplied by the motor is 3000 ft lb $\div$ 2 sec = 1500 ft-lb/sec. If the motor speed (and hence its torque output) is fairly constant, this power will be supplied at a fairly uniform rate. For most of the cycle, the motor just accelerates the flywheel (ignoring the inertia of the rest of the system) to its top speed. Top speed occurs just before a hole is punched, whereupon the flywheel speed is reduced within 0.2 sec to its minimum speed. During the actual punching, the motor will supply energy of (1500)(0.2) = 300 ft-lb to the workpiece, and the remaining energy of 2700 ft-lb (3000 − 300) must come from a sudden decrease in the kinetic energy of the flywheel. The motor gradually resupplies the 2700 ft-lb to the flywheel during the

1.8 sec between each impact. Hence $\Delta W = 2700$ ft-lb, and Eq. (5) provides the required MOI in the form

$$I_w = \frac{\Delta W}{C_F \dot{\theta}_a^2} = \frac{2700 \text{ ft-lb}}{(0.1)(20.9 \text{ rad/sec})^2} = 61.8 \text{ ft-lb-sec}^2$$

In Example 1, it was a simple matter to identify where in the cycle $\dot{\theta}_{min}$ and $\dot{\theta}_{max}$ occur and to find the work ΔW delivered by the flywheel as its speed changes from maximum to minimum. In machines where the work output is not confined to such a small segment of the cycle, we cannot always know a priori where in the cycle the extremes of speed occur, nor can we so easily identify the work done on the flywheel between these extremes. We therefore need to discuss more general methods of determining the work term ΔW.

In Sec. **12.91** we shall present a simple and efficient method of sizing flywheels by means of computer simulation. This method is valid for both steady-state and non-steady engine conditions and for large or small values of speed fluctuations.

In the literature,[26] one will find many schemes for sizing flywheels for *steady-state* operations. The most commonly advocated of these methods (the semigraphical energy loop method, which is valid only for small values of C_F) is described in Sec. **12.92**, where we have provided a fully automatic computer program that circumvents the tedium of the classical version.

The Wittenbauer method—which is useful only for relatively large values of C_F—is described in Sec. **12.93**. Both of these well-known methods are strictly valid only in the absence of velocity-dependent resisting torque M_r.

12.91   Flywheel Design by Computer Simulation

A simple method for sizing a flywheel is to use a computer program (such as DYREC) to generate steady-state performance curves, like those in Fig. **12.83**-2, for about four different values of J_O. From the curves (or directly from computer printout), one may find $\dot{\theta}_{max}$ and $\dot{\theta}_{min}$ for each value of J_O and plot the corresponding value of C_F given by Eq. (**12.90**-1). From the resulting curve of C_F versus J_O one may find the required value of J_O for any desired value of C_F.

Let us introduce the notation

$$I_w = \text{MOI, about } O, \text{ of the flywheel}$$

$$J_{OE} = \text{MOI about } O, \text{ all of mass (exclusive of flywheel)}$$
$$\text{attached to the rotating crankshaft}$$

Then

$$J_O = I_w + J_{OE} \tag{1}$$

Since J_{OE} is fixed for a given crank assembly, Eq. (1) gives I_w once J_O is known. To illustrate this procedure, the sample problem of Sec. **12.83** was recalculated

---

[26]See Biezeno and Grammel [1954, pp. 101–117] for analytical methods and Maleev [1945, pp. 552–555] for empirical methods used with internal combustion engines.

with the following values of J_O: 0, 2.5, 5, 10, 20 lb-in.-sec². From the computer print-out, the values of $\dot{\theta}_{max}$ and $\dot{\theta}_{min}$ were found for each case, and the corresponding value of C_F was found from Eq. (**12.90**-1). The results are plotted in Fig. 1.

The figure shows that for a coefficient of fluctuation of 10% we must have $J_O = 4.30$ lb-in.-sec², whereas $J_O = 10.0$ lb-in.-sec² is necessary to obtain $C_F = 5\%$. If $J_{OE} = 0.0129$ and we want $C_F = 10\%$, Eq. (1) shows that the required flywheel inertia is

$$I_w = J_O - J_{OE} = 4.30 - 0.01 = 4.29 \text{ lb-in.-sec}^2$$

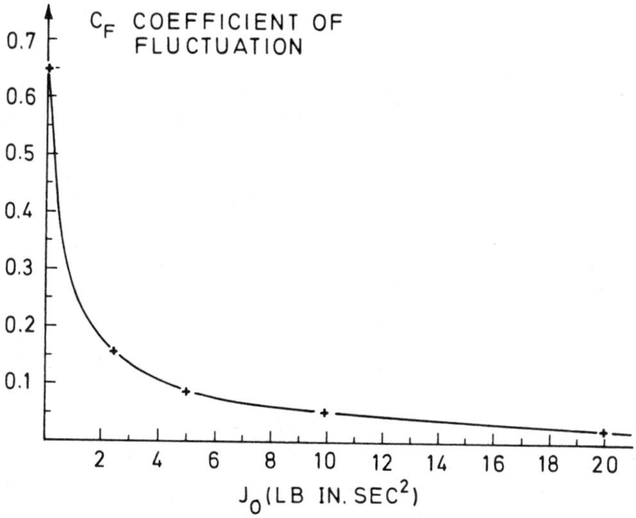

Figure 12.91-1. Coefficient of fluctuation C_F versus moment of inertia J_0 of crank assembly.

We emphasize that the proposed method is perfectly general for any value of C_F (large or small) and for any kind of torque-speed characteristic of the driven system (e.g., M_r may be a function of θ, $\dot{\theta}$, or both).

12.92 Flywheel Design by Energy Loop Method

In Sec. **12.90** we saw that the flywheel MOI required for *small* coefficients of fluctuation C_F is given by

$$I_w = \frac{\Delta W}{C_F \dot{\theta}_a^2} \qquad (1)$$

where

$$\dot{\theta}_a = \tfrac{1}{2}(\dot{\theta}_{max} + \dot{\theta}_{min}) \qquad (2)$$

and ΔW is the work done on the flywheel as its velocity $\dot{\theta}$ changes from $\dot{\theta}_{min}$ to $\dot{\theta}_{max}$.

To find ΔW, in the general case, we start by expressing the generalized equation of motion (**12.40**-7) in the form

$$(I_w + \mathcal{I}_E)\ddot{\theta} + \mathbb{C}\dot{\theta}^2 = Q \tag{3}$$

where

$$\mathcal{I}_E \equiv \mathcal{I} - I_w \tag{4}$$

is the generalized inertia of the system *excluding* the MOI of the flywheel. For small values of C_F, $\mathcal{I}_E$ will be small compared to I_w.

From Eq. (3) it is apparent that the torque acting on the flywheel must be

$$M_w \equiv I_w \ddot{\theta} \equiv Q - \mathbb{C}\dot{\theta}^2 - \mathcal{I}_E\ddot{\theta} \tag{5}$$

For the case of small speed fluctuations, we may replace $\dot{\theta}$ in Eq. (5) by its average value $\dot{\theta}_a$ and ignore the term $\mathcal{I}_E\ddot{\theta}$ in comparison with $\mathbb{C}\dot{\theta}^2$. Then we may approximate Eq. (5) by

$$I_w\ddot{\theta} = M_w \doteq Q - \mathbb{C}\dot{\theta}_a^2 \tag{6}$$

If $\dot{\theta}_{\min}$ and $\dot{\theta}_{\max}$ occur at angles θ_A and θ_B, respectively, the work done on the flywheel between its extremes of velocity is

$$W_{AB} = \int_{\theta_A}^{\theta_B} M_w \, d\theta = \int_{\theta_A}^{\theta_B} (Q - \mathbb{C}\dot{\theta}_a^2) \, d\theta \tag{7}$$

and Eq. (1) can be expressed in the form

$$I_w = \frac{W_{AB}}{C_F\dot{\theta}_a^2} \tag{8}$$

The key to flywheel design, for any machine, is knowledge of the work W_{AB}. To make the discussion concrete, we shall show how W_{AB} may be found in a reciprocating engine; the procedure would be similar for any other machine operating cyclically in a steady state.

We recall, from Eq. (**12.82**-2), that the generalized force for an engine is

$$Q \equiv M_p - M_r \tag{9}$$

where M_p is the torque due to the gas pressure and M_r is the resisting torque exerted on the engine by the driven device.

Accordingly, the torque on the flywheel is given by Eq. (6) in the form

$$M_w = M_p - M_r - \mathbb{C}\dot{\theta}_a^2 \tag{10}$$

The combination of terms

$$M_t \equiv M_p - \mathbb{C}\dot{\theta}_a^2 \tag{11}$$

represents the useful torque which is available at the engine shaft for distribution to the driven device or to the flywheel. This torque is referred to by engine designers as turning moment, tangential effort, or crank effort.[27] The difference between the available torque (M_t) and that actually delivered to the driven device (M_r) is called the surplus torque M_s (or deficit torque, if negative); i.e.,

$$M_s \equiv M_t - M_r = M_p - \mathcal{C}\dot{\theta}_a^2 - M_r \tag{12}$$

By comparison with Eq. (10) we see that the surplus torque is identical with the torque (M_w) acting on the flywheel. Thus, the work done on the flywheel between any two states θ_i and θ_j is

$$W_i^j = \int_{\theta_i}^{\theta_j} M_s \, d\theta = \int_{\theta_i}^{\theta_j} (M_p - \mathcal{C}\dot{\theta}_a^2 - M_r) \, d\theta \tag{13}$$

Both M_p and $\mathcal{C}$ are functions of θ alone. Hence, the integral in Eq. (13) may be evaluated, provided that M_r is velocity-independent. When M_r is a function of $\dot{\theta}$, it may be evaluated for the approximately constant speed $\dot{\theta}_a$. If M_r is a constant, its steady-state value must be given [see Eq. (12.83-17)] by

$$M_r = \frac{\Delta W_p}{N\pi} \tag{14}$$

where ΔW_p is the cyclic work of the gas pressure and N is the number of strokes per cycle (two or four). Equation (14) also represents the mean value of M_r over a cycle when M_r is any given function of θ.

If we plot the surplus torque M_s versus θ over a fundamental interval (of length $N\pi$), we get a curve like that shown in Fig. 1. This curve corresponds to the four-stroke engine which was analyzed in Secs. **12.83** and **12.91**.

Note that the surplus moment M_s vanishes at a number of **critical points** θ_j (labeled $\theta_0, \theta_1, \theta_2, \ldots, \theta_{10}$) in Fig. 1; these are listed in the second line of Table 1.

If we express Eq. (6) in the form

$$I_w \frac{d\dot{\theta}}{dt} = M_w = M_s \tag{15}$$

it follows that $\dot{\theta}$ achieves its extremum values at these critical points. Therefore θ_A (the point where $\dot{\theta}$ is minimum) occurs at one of these critical points, and so too must θ_B (the point where $\dot{\theta}$ is maximum).

The M_s curve is said to describe a **loop** between any two successive critical points, and the work represented by this loop is positive or negative accordingly as the loop is above or below the abscissa. The signed area of the loop to the left of point θ_j is

---

[27]Throughout much of the literature on engine dynamics the turning moment is defined not as in Eq. (11) but by the expression

$$M_t = M_p - m_{rec}\ddot{s}\left(\frac{\dot{s}}{\dot{\theta}}\right) = -(P + m_{rec}\ddot{s})\left(\frac{\dot{s}}{\dot{\theta}}\right) \tag{11a}$$

where $m_{rec} = m_3 + m_B$. It can be shown (Problem **12.100**-49) that Eq. (11) does, in fact, reduce to Eq. (11a) when the term J_{AB} is so small that the rotational inertia ($J_{AB}\ddot{\phi}$) associated with the conrod is negligible. In Sec. **13.21**, we shall show that it is feasible and desirable to make $J_{AB} = 0$ by suitably "balancing" the conrod; if this is done, the approximation (11a) becomes identical with Eq. (11).

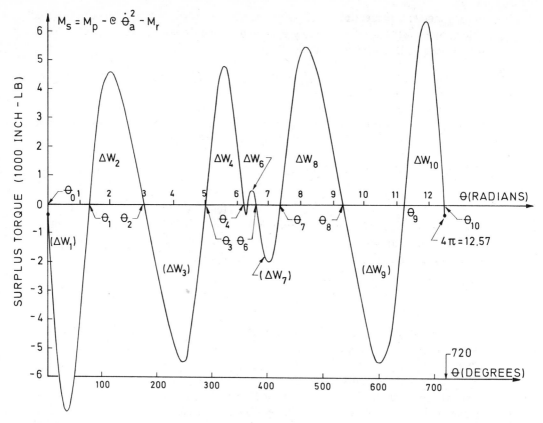

Figure 12.92-1. Surplus torque M_s versus θ. Areas ΔW_j of loops above abscissa are positive; those below (ΔW_j enclosed in parentheses) are negative.

TABLE 12.92-1 *(Refer to Fig. 1)*

Location (j):	1	2	3	4	5	6	7	8	9	10
θ_j (deg)	77	176	289	358	363	381	425	536	647	718
W_{j-1}^{j} (in.-lb)	−6213	4954	−6562	3581	−18	104	−984	6512	−6477	5106
$W_0^{j} = \sum \Delta W_j$	−6213	−1259	−7821	−4240	−4258	−4154	−5138	1374	−5103	3

denoted by

$$W_{j-1}^{j} \equiv \int_{\theta_{j-1}}^{\theta_j} M_s \, d\theta \tag{16}$$

These signed areas are listed in Table 1 under the corresponding value of θ_j. The numerical values may be found by graphical or numerical integration. Equating changes in kinetic energy to work done, we find

$$\tfrac{1}{2}I_w(\dot{\theta}_1^2 - \dot{\theta}_0^2) = W_0^1 = -6213$$

Similarly,

$$\tfrac{1}{2}I_w(\dot\theta_2^2 - \dot\theta_0^2) = W_0^2 \equiv W_0^1 + W_1^2 = -6213 + 4954 = -1259$$

and in general

$$\tfrac{1}{2}I_w(\dot\theta_j^2 - \dot\theta_0^2) = W_0^j = W_0^{j-1} + W_{j-1}^j = \sum_{k=1}^{j} W_{k-1}^k \qquad (17)$$

The energy increments per loop are shown in the third line of Table 1, and the cumulative sum of these increments is listed in the fourth line. From Eq. (17) we see that the minimum (maximum) value of $\dot\theta$ occurs where W_0^j is minimum (maximum). In Table 1, these values are $W_0^3 = -7821$ and $W_0^8 = 1374$. In other terms, $\dot\theta_{min}$ occurs at θ_3, and $\dot\theta_{max}$ occurs at θ_8. Hence, in this example

$$W_{AB} = \int_{\theta_3}^{\theta_8} M_s\, d\theta = W_3^8 = 1374 - (-7821) = 9195 \text{ in.-lb}$$

and the flywheel inertia I_w is given by Eq. (8) in appropriate units.

Suppose we desire a coefficient of fluctuation of 15.6% at a mean speed of $\dot\theta_a = 135$ rad/sec (so that we can compare with Fig. **12.83**-2). Then Eq. (8) predicts a flywheel moment of inertia

$$I_w = \frac{W_{AB}}{C_F \dot\theta_a^2} = \frac{9195}{(.155)(135)^2} = 3.26 \text{ lb-in.sec}^2$$

This value is within 31% of the value $(2.50 - 0.0129 = 2.49)$ predicted by the more accurate (simulation method) of Sec. **12.91** (see Fig. **12.91**-1) and errs on the high (safe) side, as it should.

Although Eq. (6) is sufficiently accurate (for small values of C_F), it does lead to an overestimate of the required value of I_w. To see this, we can rewrite Eq. (5) in the form

$$(I_w + \mathscr{I}_E)\ddot\theta = Q - \mathcal{C}\dot\theta^2 \qquad (18)$$

In finding I_w via Eq. (6) we have ignored the contribution of $\mathscr{I}_E$ to the system's inertia. We can take this neglected inertia into account, approximately, by replacing I_w, in all equations following Eq. (5), by $I_w + \bar{\mathscr{I}}_E$, where $\bar{\mathscr{I}}_E$ is a suitably defined average value of $\mathscr{I}_E$ over the cycle. Then Eq. (8) will assume the form

$$I_w = \frac{W_{AB}}{C_F \dot\theta_a^2} - \bar{\mathscr{I}}_E \qquad (19)$$

For engines without offset, we may use the approximation of Eq. (**12.81**-9) to find the arithmetic mean value of $\mathscr{I}_E$; i.e.

$$\bar{\mathscr{I}}_1 = \mathscr{I}_E^{av} \doteq A \equiv J_{OE} + m_A r^2 + \tfrac{1}{2}[(m_3 + m_B)r^2 + J_{AB}\lambda^2] \qquad (20a)$$

To insure that we do not overestimate $\bar{\mathscr{I}}$ (and thereby underestimate I_w) we may use the minimum value of ϑ_E predicted by Eq. (**12.81**-9); i.e,

$$\bar{\vartheta}_2 = \vartheta_E^{min} \equiv A - B = J_{OE} + m_A r^2 + J_{AB}\lambda^2 \tag{20b}$$

If we totally ignore the contribution of the conrod and piston to the generalized inertia, we get the most conservative estimate

$$\bar{\vartheta}_3 = J_{OE} \tag{20c}$$

For this example (with $J_{OE} = 0.0129$), Fig. **12.81**-1 shows that $\bar{\vartheta}_1 = 0.76$, $\bar{\vartheta}_2 = 0.47$. Accordingly, Eq. (19) gives

$$I_w = 3.26 - 0.76 = 2.50 \quad \text{(using Eq. (20a)]}$$

$$I_w = 3.26 - 0.47 = 2.79 \quad \text{[using Eq. (20b)]}$$

$$I_w = 3.26 - 0.01 = 3.25 \quad \text{[using Eq. (20c)]}$$

The first of these values agrees almost exactly with the value (2.49) predicted by the computer simulation method of Sec. 12.91; the second (conservative) value is only 11.5% higher; and the third value is essentially that predicted by the simplest (and most conservative) approximation of Eq. (1). For multicylinder engines, with customary low values of C_F, any one of the three approximations should be sufficiently accurate.

It is obvious from Fig. 1 that the calculations needed to generate the complicated function $M_s(\theta)$ are extremely time-consuming, as are the subsequent estimates of the loop areas.

The generation of the curve $M_s(\theta)$ may be readily programmed if we use the logic previously generated for finding M_p and $\mathcal{C}$. A FORTRAN listing for such a program, called FLYLOOP (FLYwheel design by energy LOOP method), is given in Appendix J. The main program calls subroutine CENTCO to calculate the centripetal coefficient $\mathcal{C}$ and calls subroutine FORCES (essentially the same subroutine described in Sec. **12.82**) to calculate the gas force P associated with the combination cycle of Fig. 12.82.1. On the first call to FORCES, the mean resisting moment M_r [MR] is calculated via Eqs. (14) and (**12.82**-12). The running integral $W_0^j = \int_0^{\theta_j} M_s \, d\theta$ is calculated at user-selected intervals $\Delta\theta$, by means of Simpson's rule—Eq. (**H**-14).

The arrangement of output is shown in Fig. 2 for the example illustrated in Fig. 1. In the printout, flywheels "1", "2", and "3" correspond to the successively more conservative estimates of Eqs. (20a), (20b), and (20c). When using program FLYLOOP, there is no need to make a plot such as Fig. 1 or a table such as Table 1, since the printed output gives all the results which can be gleaned from such plots and tables. However, the program does print out the surplus torque M_s[MS] as well as the cumulative work w_0^j[WS] for background information.

The complete calculation for the example problem (using 720 points at 1° spacing) was performed on a UNIVAC 9070 computer for under one dollar.

It should be noted that the output of program FLYLOOP can be utilized for values of $\dot{\theta}_a$ and C_F other than those read in as data. In fact, any combination of $\dot{\theta}_a$,

```
PROGRAM FLYLOOP

FOUR-STROKE IC ENGINE

INPUT PARAMETERS

CRANK RADIUS.................................... R=        4.000000D 00
CON.ROD LENGTH................................. L=        1.200000D 01
DISTANCE BETWEEN CRANKPIN & C.G. OF CON.ROD A=            4.000000D 00
PISTON OFFSET DISTANCE......................... E=        0.000000D 00
MOMENT OF INERTIA ABOUT CRANKSHAFT AXIS....JO=           1.290000D-02
MOMENT OF INERTIA ABOUT CG. OF CON.ROD.....JG=           8.290000D-01
MASS OF CON.ROD...............................M2=        5.181000D-02
MASS OF PISTON...............................M3=         1.295000D-02
PERIOD OF MS VS. THETA CURVE...........PERIOD=           7.200000D 02
STEP SIZE (DELTA THETA), DEGREES.............H=          1.000000D 00
COEFFICIENT OF FLUCTUATION..................CF=          1.555000D-01
AVERAGE CRANK SPEED (RAD/SEC).........VELAVR=            1.350000D 02




OPTIONAL CONSTANTS

CN( 1)= 8.000000D 00
CN( 2)= 1.000000D 00
CN( 3)= 1.370000D 00
CN( 4)= 1.370000D 00
CN( 5)= 6.500000D 02
CN( 6)= 1.800000D 01
CN( 7)= 1.300000D 01
CN( 8)= 1.470000D 01
CN( 9)= 7.068583D 00


-------------------------------------------------------------------


RESULTS

MEAN RESISTING MOMENT              MR=   3.742982D 02
NET(SURPLUS-DEFICIT) WORK          WAB=  9.195713D 03
FLYWHEEL (1) MOMENT OF INERTIA     IW1=  2.483548D 00
FLYWHEEL (2) MOMENT OF INERTIA     IW2=  2.771359D 00
FLYWHEEL (3) MOMENT OF INERTIA     IW3=  3.231897D 00

MINIMUM SPEED= 1.245038D 02  RAD/SEC, AT THETA= 289.00  DEG.
MAXIMUM SPEED= 1.454963D 02  RAD/SEC, AT THETA= 536.00  DEG.
```

INDEX	CRANK ANGLE THETA(DEG)	SURPLUS TORQUE MS	CUMULATIVE WORK WS
1	0.00	-3.742982D 02	0.000000D 00
2	1.00	-6.747860D 02	-9.155406D 00
3	2.00	-9.746971D 02	-2.355112D 01
4	3.00	-1.273456D 03	-4.317205D 01
5	4.00	-1.570490D 03	-6.799308D 01
6	5.00	-1.865228D 03	-9.797913D 01
	6.00		-1.330852D 02

Figure 12.92-2. Sample of output from Program Flyloop.

SECTION 12.92 FLYWHEEL DESIGN BY ENERGY LOOP METHOD

C_F, and I_w satisfying Eq. (1) in the form

$$C_F I_w \dot{\theta}_a^2 = W_{AB} = 9195 \text{ in.-lb} \tag{23}$$

will be valid for the particular machine analyzed. Furthermore, the extreme speeds are, by definition of C_F,

$$\dot{\theta}_{\min} = (1 - \tfrac{1}{2}C_F)\dot{\theta}_a$$
$$\dot{\theta}_{\max} = (1 + \tfrac{1}{2}C_F)\dot{\theta}_a \tag{24}$$

for any value of $\dot{\theta}_a$.

As an example, we may ask for the values of C_F, $\dot{\theta}_{\max}$, and $\dot{\theta}_{\min}$ if the machine just analyzed has a flywheel with $I_w = 50$ in.-lb-sec^2 and runs at an average speed of $\dot{\theta}_a = 1000$ rpm (104.7 rad/sec).

From Eq. (23) we find

$$C_F = \frac{W_{AB}}{I_w \dot{\theta}_a^2} = \frac{9195 \text{ in.-lb}}{50 \text{ in.-lb-sec}^2 \ (104.7)^2 \text{ sec}^{-2}} = 0.0168$$

and from Eqs. (24)

$$\dot{\theta}_{\min} = (1 - 0.0168/2)1000 = 991.6 \text{ rpm}$$
$$\dot{\theta}_{\max} = (1 + 0.0168/2)1000 = 1008 \text{ rpm}$$

12.93 Wittenbauer's Semigraphical Method[28]

If $\mathscr{I}_E$ is the generalized inertia of the engine exclusive of the flywheel inertia I_w, the change in kinetic energy as θ changes from 0 to some arbitrary value is given by

$$\tfrac{1}{2}(\mathscr{I}_E + I_w)\dot{\theta}^2 - \tfrac{1}{2}(\mathscr{I}_{E,0} + I_w)\dot{\theta}_0^2 = \int_0^\theta (M_p - M_r)\, d\theta \tag{1}$$

where $\mathscr{I}_{E,0}$ is the known value of $\mathscr{I}_E$ when $\theta = 0$ and $\dot{\theta}_0$ is the unknown crank velocity at the same point. Let us introduce the nondimensional ratios

$$z \equiv \left(\frac{\dot{\theta}}{\dot{\theta}_a}\right)^2, \qquad z_0 = \left(\frac{\dot{\theta}_0}{\dot{\theta}_a}\right)^2 \tag{2}$$

where

$$\dot{\theta}_a \equiv \tfrac{1}{2}(\dot{\theta}_{\max} + \dot{\theta}_{\min}) \tag{3}$$

Then the energy equation (1) can be expressed in the form

$$z \equiv \frac{w + (\mathscr{I}_{E,0} + I_w)z_0}{\mathscr{I}_E + I_w} \tag{4}$$

where

$$w \equiv \frac{2}{\dot{\theta}_a^2} \int_0^\theta (M_p - M_r)\, d\theta \tag{5}$$

is proportional to the work done by the generalized force $M_p - M_r$.

---

[28]Wittenbauer [1923, pp. 759, 775].

If the resisting moment M_r is velocity-dependent, we may approximate M_r by its mean value, so that w becomes a known function of θ. For any desired value of $\dot{\theta}_a$, we may then evaluate w and $\mathcal{I}_E$ for any value of θ and plot a curve of w versus $\mathcal{I}_E$ over a complete cycle of motion. This curve, which is necessarily closed, will look somewhat as in Fig. 1.

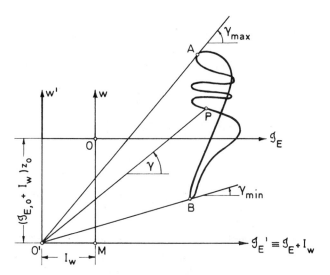

Figure 12.93-1. Wittenbauer's construction.

If we knew the values of I_w and z_0, we could locate the point O', as shown in the figure, and erect axes $\mathcal{I}'_E$ and w', through O', parallel to the axes of $\mathcal{I}_E$ and w. Then the coordinates $\mathcal{I}'_E$ and w' of a point P on the curve would define the angle γ such that

$$\tan \gamma = \frac{w'}{\mathcal{I}'_E} = \frac{(\mathcal{I}_{E,0} + I_w)z_0 + w}{\mathcal{I}_E + I_w} \tag{6}$$

Comparing Eqs. (6) and (4), we conclude that

$$\left(\frac{\dot{\theta}}{\dot{\theta}_a}\right)^2 = z = \tan \gamma \tag{7}$$

Thus, if we knew values of I_w and $\dot{\theta}_0$ (and hence could locate point O'), Eq. (7) would give the speed $\dot{\theta}$ for any point on the curve. To find that I_w needed to provide a given coefficient of fluctuation

$$C_F \equiv \frac{\dot{\theta}_{max} - \dot{\theta}_{min}}{\dot{\theta}_{av}} \tag{8}$$

we combine Eqs. (3) and (8) to give

$$\frac{\dot{\theta}_{max}}{\dot{\theta}_a} = 1 + \frac{1}{2}C_F \tag{9a}$$

$$\frac{\dot{\theta}_{min}}{\dot{\theta}_a} = 1 - \frac{1}{2}C_F \tag{9b}$$

Then Eq. (7) shows that

$$\tan \gamma_{\max} = \left(\frac{1 + C_F}{2}\right)^2 \tag{10a}$$

$$\tan \gamma_{\min} = \left(\frac{1 - C_F}{2}\right)^2 \tag{10b}$$

We can now draw straight lines with these slopes that are tangent to the known curve (at points A and B) as shown in Fig. 1; these straight lines intersect at O'. With O' now known, we can read the desired value I_w from the curve; i.e.,

$$I_w = O'M \tag{11}$$

Then, if desired, $z_0 = (\dot{\theta}_0/\dot{\theta}_a)^2$ may be calculated from the measured distance

$$OM = (\mathcal{I}_{E,0} + I_w)z_0 \tag{12}$$

When C_F is extremely small, $\tan \gamma_{\min}$ and $\tan \gamma_{\max}$ are practically equal, and it is difficult to locate point O' graphically. In such cases, the method of Sec. **12.92** (valid only for small values of C_F) is applicable.

Note that this **method of Wittenbauer** essentially constitutes a solution for $\dot{\theta}$ by the first energy integral [see Eq. (**12.84**-4)], with the effects of velocity-dependent resistance M_r averaged out. In addition, the tedious computations involved invite the use of a computer (or calculator) program. It is therefore the opinion of the author that the more general method of computer simulation, described in Sec. **12.91**, is preferable to Wittenbauer's method of sizing flywheels when C_F is large. As previously mentioned, Wittenbauer's method is inaccurate when C_F is small; in such a case, one should use either the energy loop method of Sec. **12.92** or the computer simulation method of Sec. **12.91**.

12.100 PROBLEMS

1. Prove that the generalized inertia $\mathcal{I}$ for a rigid lamina is the moment of inertia of the lamina with respect to the instantaneous center.

2. Starting from the general definition

$$\mathcal{I} = \sum_{i=1}^{M} [m_i(u_i^2 + v_i^2) + J_i\omega_i^2]$$

show that the generalized inertia coefficient for the four-bar linkage of Fig. **12.30**-2, with $\theta_3 = q$, is given by

$$\mathcal{I} = m_2(u_2^2 + v_2^2) + J_{1A}k_1^2 + J_2k_2^2 + J_{3D}$$

where the MOI of link 1 (3) with respect to point A (D) is denoted by J_{1A} (J_{3D}) and k_1, k_2 are the velocity ratios $\dot{\theta}_1/\dot{\theta}_3$ and $\dot{\theta}_2/\dot{\theta}_3$.

3. Using θ as generalized coordinate, find the generalized inertia $\mathscr{J}$ of the ladder AB which slides on frictionless orthogonal guides OA and OB (Fig. 1). AB is a uniform rod of mass m.

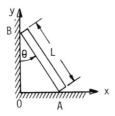

Figure 12.100-1. (Prob. 3).

4. A cylinder of radius a, mass m, and MOI J(with respect to its axis) rolls inside a larger hollow fixed cylinder of radius b (Fig. 2). The axes of the cylinders are parallel and horizontal. Let θ, the angle between the downward vertical direction and the line joining the centers of the cylinders, be the generalized coordinate. Find the generalized inertia coefficient and derive the equation of motion:

$$\ddot{\theta} + \frac{g}{b-a}\left[\frac{1}{1+J/(ma^2)}\right]\sin\theta = 0$$

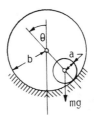

Figure 12.100-2. (Prob. 4).

5. If a system is in a *state of equilibrium*, defined by the instantaneous conditions $Q = 0$, $\dot{q} = 0$, show that it must follow that all higher derivatives dq^n/dt^n $(n = 2, 3, \ldots)$ must vanish. Then prove that the system remains at rest forever. Assume that all derivatives of q and Q are finite and continuous.

6. Using Eq. (**12.51**-7), show that point P in Fig. **12.52**-1 will reach point A in a finite time if the total energy of the system is E_1. Also show that the time to move from A to B equals the time required to move from B to A.

7. Using the graph of potential energy (Fig. **11.42**-6), explain in qualitative terms the behavior of the platform in Fig. **12.53**-1 if it is dropped from an initial angle $\theta = 60°$ with initial velocity $\dot{\theta} = 0$. Neglect any possible buckling of the spring, which is strain-free at $\theta = 30°$.

8. A car of constant mass m moves along an undulating roller-coaster track in a vertical plane. If the horizontal displacement x of the car is taken as the generalized coordinate, show that the generalized inertia coefficient is given by $\mathscr{J} = m[1 + (dy/dx)^2]$, where y is the vertical displacement.

9. Using θ as generalized coordinate, find an expression for the generalized inertia $\mathscr{I}$ for the shaper mechanism of Fig. **10.60**-3.

10. Find the differential equation of motion for the shaper of Fig. **10.60**-3 using θ as generalized coordinate. Express all coefficients needed in terms of position variables. A horizontal cutting force X_4 on the ram and the driving torque M_{1B} on link AB are assumed to be known functions of position and velocity.

11. Figure 3 shows a pulley (1) of radius a which is pivoted at A and has MOI about A of J_1. A pin P is located at radius $AP = r$ in the pulley and drives the vertically slotted rod (2), of total mass m_2, along the horizontal x direction. A string tied to and wrapped around the pulley hangs vertically and supports a weight W. If the angle ϕ is chosen as generalized coordinate, find
    a. The generalized inertia coefficient.
    b. The equation of motion.
    c. The initial angular acceleration $\ddot{\phi}$ if the motion starts from a state of rest.

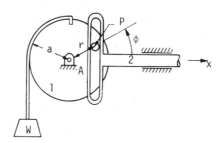

Figure **12.100**-3. (Prob. 11).

12. If it is known that $\dot{\theta}_n = 0$, show that a Taylor series expansion about the point t_{n-1} leads to Eq. (**12.53**-12).

13. In Fig. 4, a uniform rod AB of length a is pivoted at a fixed point A. Point B slides in a straight slot of a member 2 which is constrained to slide along a straight line which is perpendicular to the slot and makes an angle δ with the downward vertical. The MOI of AB about point A is J_{1A}. The mass of member 2 is m_2.
    a. Find the generalized inertia coefficient corresponding to the angle θ.
    b. Find an expression for the velocity $\dot{\theta}$ as a function of θ.

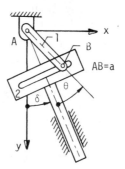

Figure **12.100**-4. (Prob. 13). Scotch yoke mechanism.

c. Find the period of the motion for small angles ($\theta \ll 1$).

d. Find and plot the displacement θ over a complete period if the system is initially at rest with $\theta = 45°$. Use the following numerical data: $AB = 1$ ft, $J_{1A} = 0.5$ ft-lb-sec^2, $m_2 = 1$ lb-sec^2/ft, $\delta = 30°$.

14. A circular disc cam (1) of radius R rotates about a fixed pivot A and drives a flat-faced translating follower (2) along the vertical y axis (Fig. 5). A drum of radius r is centered at A and keyed to the cam. A load (3) is suspended vertically from a weightless cable which is wrapped about the drum. The center of the cam C is at a distance e from A, which lies on the y axis. The weights of cam, follower, and load are W_1, W_2, and W_3, and the moment of inertia of the cam and drum with respect to A is J_{1A}.

   a. Find the generalized inertia coefficient of the system corresponding to the generalized coordinate θ shown.

   b. Find an expression for $\ddot{\theta}$ as a function of θ.

   c. If the system starts from rest with $\theta = 0$, find the displacement versus time during the first revolution of the cam. Use the following numerical data[29]: $W_1 = 3.49$ lb, $W_2 = 3.49$ lb, $W_3 = 10$ lb, $R = 4$ in., $J_{1A} = 0.10844$ in.-lb-sec^2, $e = r = 2$ in.

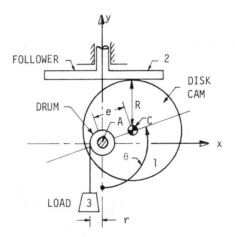

Figure 12.100-5. (Prob. 14).

15. A bell crank (1) of weight W consists of two orthogonal uniform rods AB and AC, each of length a (Fig. 6). The bell crank is pivoted at the fixed point A and drives identical scotch yokes (2 and 3) of weight W along the (downward vertical) x axis and along the (horizontal) y axis. A horizontal spring of stiffness k is fixed between fixed point D and E, the midpoint of the slot in member 3. The free length of the spring is a.

   a. Find the generalized inertia coefficient of the system corresponding to the generalized coordinate θ, the angle measured from the downward vertical to AB.

   b. Find an expression for $\ddot{\theta}$ as a function of θ.

   c. Find the displacement θ versus time over a complete period if the initial velocity is $\dot{\theta} = 10$ rad/sec when $\theta = 0$. Use the following numerical data: $a = 10$ in., $W = 20$ lb, $k = 4$ lb/in.

[29]Quinn [1949].

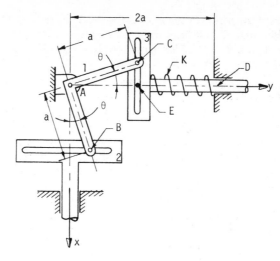

Figure 12.100-6. (Prob. 15).

16. A frictionless fluid oscillates in the thin bent tube shown in Fig. 7. If the tube is of uniform cross section and the fluid column has length L, show that the period of oscillation is

$$T = 2\pi \left[\frac{L}{g(\sin \alpha + \sin \beta)} \right]^{1/2}$$

for small amplitudes.

Figure 12.100-7. (Prob. 16).

17. A liquid column of length L and density ρ is trapped within a bent tube of uniform cross section (Fig. 8). Air is trapped in the space between each meniscus and the closed ends

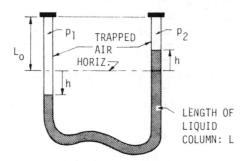

Figure 12.100-8. (Prob. 17).

CHAPTER 12 DYNAMICS OF SINGLE-FREEDOM SYSTEMS

of the tube. When the fluid is at rest, the pressure p_0 and the volume V_0 are the same in each air space. When the liquid columns rise and fall through a distance h, the pressure p and volume V within the air spaces are related by $pV^n = p_0V_0^n$.
a. Find the equation of motion for the oscillating liquid column.
b. Show that the frequency of oscillation for small amplitudes of the variable h is

$$f = \frac{1}{2\pi}\left[\frac{2g}{L}\left(1 + \frac{np_0}{\rho g L_0}\right)\right]^{1/2}$$

18. A small bead P slides smoothly on an inextensible string of length $2a$ which is suspended between fixed points A and B (Fig. 9). The line AB is of length $2c$ and makes an angle β with the horizontal.
a. Find the equilibrium position of the particle.
b. Find the period of small oscillations about the position of equilibrium.

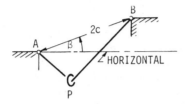

Figure 12.100-9. (Prob. 18).

19. In Fig. 10, gear 1 is fixed to the pulley of radius a, from which the mass m is suspended at the end of a vertical massless cable. J_1 is the MOI, with respect to the fixed center A, of the gear and pulley. J_2 is the MOI, with respect to fixed center B, of the mating gear 2. N_1 and N_2 are the numbers of teeth in gears 1 and 2, respectively. Show that the angular acceleration of gear 1 is $(\ddot{\theta}_1/g) = ma/[ma^2 + J_1 + J_2(N_1/N_2)^2]$.

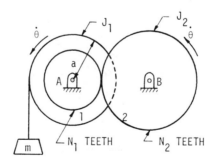

Figure 12.100-10. (Prob. 19).

20. For the epicyclic gear train of Fig. 11, M_2 and M_3 are the external torques applied to gear 2 and arm 3; J_2 is the MOI of gear 2 about B; J_{3A} is MOI of arm 3 about fixed center A; m_2 is the mass of gear 2; N_2 and N_1 are the number of teeth in planet gear 2 and sun gear 1, respectively; $a = AB$. Show that the angular acceleration $\ddot{\theta}_3$ of the arm is

$$\ddot{\theta}_3 = \frac{M_3 + k_{23}M_2}{J_{3A} + m_2a^2 + J_2k_{23}^2}$$

where $k_{23} \equiv 1 + (N_1/N_2)$.

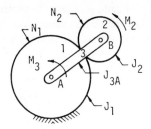

Figure 12.100-11. (Prob. 20).

21. For the system described in Problem 20, find
    a. The tension in link AB.
    b. The force transmitted to the ground.
    c. The moment (with respect to A) transmitted to ground.

22. In a gear train of N gears, each gear rotates about a fixed center. Let M_i be the external torque applied to gear i, and let J_i be the MOI of gear i about its fixed axis. If $\dot{\theta}_i$ is the angular velocity of gear i and the velocity ratios $k_i = \dot{\theta}_i/\dot{\theta}_1$ are all known, show that
    a. The generalized inertia corresponding to θ_1 is $\mathcal{I} = \sum_{i=1}^{N} J_i k_i^2$.
    b. The generalized force corresponding to θ_1 is $Q = \sum_{i=1}^{N} M_i k_i$.
    c. The angular acceleration $\ddot{\theta}_1 = Q/\mathcal{I}$.

23. A thin uniform rod of mass m and length L is initially at rest and centered on the top of a fixed cylinder of diameter L (Fig. 12). If the cylinder axis is horizontal and the rod rolls without slipping over the cylinder, find
    a. The generalized inertia coefficient corresponding to angular motion of the rod.
    b. The generalized differential equation of motion.
    c. The natural frequency of small rocking motions.

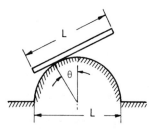

Figure 12.100-12. (Prob. 23).

24. If the mass center of the rod in Problem 23 is located at a small distance e from the geometric center of the rod, show that small stable rocking motions are possible about an equilibrium position where $\theta = 2(e/L)$.

25. An ideal plane pendulum consists of a 2-ft-long massless rod, hinged at a fixed pivot and carrying a point mass weighing 10 lb at the free end.
    a. Sketch a plot of potential energy of the system versus a suitable generalized coordinate.
    b. Use the sketch of part a to qualitatively describe the behavior of the system if the

point mass swings through its lowermost position with a speed of 12 ft/sec. Find the maximum angular displacement of the pendulum from the vertical.

c. Repeat part b but change the speed at the lowermost point to 17 ft/sec.

26. a. Show that the period of free vibration for a pendulum of length L can be expressed in the form $T = C2\pi(L/g)^{1/2}$, where g = acceleration of gravity and C is an amplitude-dependent constant, which represents a certain definite integral.

b. Use numerical integration (e.g., Simpson's rule) to evaluate C for an amplitude of $140°$.

27. The crank (link 2) of the shaper mechanism shown in Fig. **10.60**-5 is driven through a speed reduction unit by means of a motor which produces a torque versus speed characteristic as shown in Fig. 13. Write a computer program to find the motion of the sys-

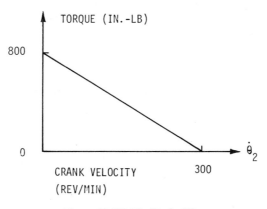

Figure **12.100-13.** (Prob. 27).

tem over the first three cycles of the crank rotation, starting from a state of rest with $\theta_2 = 90°$ at time $t = 0$. Assume that the cutting force is constant at $F = 100$ lb during the cutting stroke ($\dot{x} > 0$) and then drops to zero during the return stroke. Link 2 (although pictured as a rod) is a flywheel with moment of inertia about its centroid C of 6.00 lb-sec^2-in. The other links are uniform bars with mass properties indicated below:

Link No.	Weight (lb)	Mass (lb-sec^2/in.)	Centroidal MOI (lb-sec^2-in.)
1	35	0.09067	0.1209
—	—	—	—
3	3	0.00777	0.00583
4	10	0.02590	0.95185
5	4	0.01036	0.01381

Print out θ_2, $\dot{\theta}_2$, y_D, $\dot{y}_D$, $\ddot{y}_D$ versus time over a complete revolution of the flywheel, starting from $\theta_2 = 90°$ at $t = 0$. Print out between 10 and 20 times.

28. A crank OA (link 2) drives a reciprocating cutting tool (link 4), as shown in Fig. 14. The crank is driven through a speed reduction unit by means of a motor which produces a torque versus speed characteristic as shown. Prepare a computer program to find the motion of the system over the first three cycles of the crank rotation, starting from a state of rest with $\theta_2 = 0$ at time $= 0$. Note: e' is the negative of e defined in Fig. **1.41**-1.

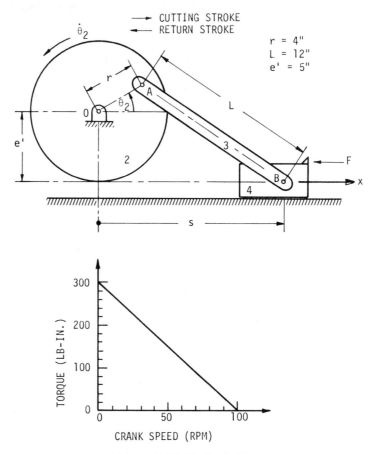

Figure 12.100-14. (Prob. 28).

Assume that the cutting force is constant at $F = 50$ lb during the cutting stroke and then drops to zero during the return stroke. Link 2 is a flywheel centered at O, and it may be idealized as a circular disc of radius 7 in. and weight 21.8 lb. Links 3 and 4 are uniform steel bars weighing 12 and 20 lb, respectively. Using units of inches, seconds, and radians,

a. Print out θ_2, $\dot{\theta}_2$, s, $\dot{s}$ versus time over the first three revolutions of the flywheel, starting from $\theta_2 = 0°$ at $t = 0$. You should print results between 10 and 20 times for each revolution of the flywheel.

b. Plot $\dot{\theta}_2$ versus t.

c. Calculate the coefficient of fluctuation C_F of the flywheel for the last cycle computed, where C_F is defined by Eq. (**12.90**-1).

29. A pendulum consists of a weightless rod (OA) suspended from pivot O and a concentrated mass (m_1) at A, where $OA = L$. A concentrated mass (m_2) slides freely along OA and is simultaneously restrained to the horizontal plane at distance h beneath O (Fig. 15). Use the angle θ between OA and the downward vertical as generalized coordinate, and find expressions for $\mathcal{T}$, $\mathcal{C}$, and Q in terms of θ. Write the differential equation of motion. Find the frequency of small vibrations.

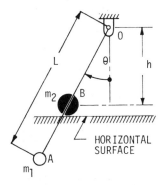

Figure 12.100-15. (Prob. 29).

30. In the Scotch yoke shown in Fig. 16, $OA = R$, the piston has mass m, and the flywheel MOI is I_w. The weight P is suspended from a string wrapped around the flywheel rim (of radius R). and the piston acts against a return spring of stiffness K. Find expressions for $\mathcal{T}$, $\mathcal{C}$, and Q in terms of the generalized coordinate θ. Assume that the spring is unstrained when $\theta = 0$.

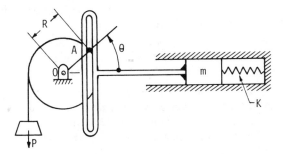

Figure 12.100-16. (Prob. 30).

31. If the link shown in Fig. **12.31-1** is a homogeneous uniform rod, show that the term J_{AB} defined by Eq. (**12.31-4**) is given by $J_{AB} = -2J_G$, where the notation is as defined in Sec. **12.31**. *Note the negative sign.*

32. Suppose that the link shown in Fig. **12.31-1** is a homogeneous circular cone with axis AB, vertex at B, and diameter D of the transverse cross section at A. Show that

$$J_{AB} \equiv J_G - mab = J_G\left[1 - \frac{5}{1 + (D/L)^2}\right]$$

and hence that $J_{AB} < 0$ when $0 < D/L < 2$.

33. Figure **12.31**-1(a) represents the connecting rod of an automotive engine with the following data: $a = 3$ in., $b = 8$ in., weight $(W) = 3$ lb, and centroidal radius of gyration $(r_G) = 4.4$ in. Find the tip masses m_A, m_B and the MOI (J_{AB}) of the dynamically equivalent rod shown in Fig. **12.31**-1(b).

34. The link of Fig. **12.31**-1 is suspended from B as a pendulum and is observed to have a vibration period τ_B. Its mass m is measured, and the distance $AG = a$ to its mass center is found by balancing the link on a horizontal knife edge.
    a. Find an expression for τ_B as a function of m, a, and the centroidal MOI J_G.
    b. If the link weighs 125 lb, $L = 40$ in., $a = 14.5$ in., and the time for 60 complete oscillations is 100 sec, find J_G.

35. When the link of Fig. **12.31**-1 is suspended as a pendulum from A, its period of vibration is observed to be τ_A. When suspended from B, its period is τ_B. If its mass m has been measured but the location of the centroid is not known, show that J_G, the centroidal MOI, can be found from the data (m, L, τ_A, τ_B). (Hint: Express the periods in terms of the given data and the unknowns a, b. Then use the relation $L = a + b$ to solve for the three unknowns a, b, J_G.) Find J_G for $mg = 125$ lb, $L = 40$ in., $\tau_A = 1.667$ sec, and $\tau_B = 1.818$ sec.

36. The link in Fig. **12.31**-1(a) weighs 3.69 lb and has dimensions $a = 3.25$ in. and $L = 11$ in. When suspended as a pendulum with a knife edge support at B', its period for small vibrations was observed to be 1.023 sec. Find J_G and J_{AB} for this link if the distance $BB' = \frac{7}{16}$ in.

37. For the problem defined in Example 1 of Sec. **12.53**,
    a. Find expressions for $\mathcal{I}$, $\mathcal{C}$, Q, and give the differential equation of motion.
    b. Numerically integrate the generalized differential equation of motion using the data given in Sec. **12.53**. Compare your results with those in Figs. **12.53**-3 and 4.

38. Show the influence of engine offset on $\mathcal{I}$ and $\mathcal{C}$ by replotting the curves of Fig. **12.81**-1 when e is changed from zero to $e = 2$ in.

39. If the mass center G of a connecting rod does not lie on the line of centers AB,
    a. What expressions would you use to calculate $\mathcal{I}$ and $\mathcal{C}$?
    b. Modify the subroutine GICC of Sec. **12.81** for this general case.

40. *Turning Moment Due to Gravity.* Suppose that the x axis in Fig. **12.80**-1 is inclined from the upward vertical direction by an angle δ, measured positive clockwise.
    a. Using the terminology of Secs. **12.80** and **12.81**, show that the turning moment (generalized force) due to gravity is given by (with $m^1{}_B \equiv m_B + m_3$):

$$M_g = g[(m_1 c + m_A r + m'_B r) \sin (\theta - \delta) + m'_B L k_\phi \sin (\phi + \delta)] \qquad \text{(i)}$$

    b. Show that for a horizontal engine $(\delta = 90°)$ the expression for M_g given in Eq. (i) reduces to

$$M_g = -g(m_1 c + m_A r) \cos \theta \qquad \text{(ii)}$$

    The same result occurs for $\delta = 90°$ when one simply drops the terms multiplied by M_B in Eq. (i). Why?
    c. Show that Eq. (i) can also be expressed in the form

$$M_g = (cW_1 + rW_2 + rW_3) \sin (\theta - \delta) + (aW_2 + LW_3) k_\phi \sin (\phi + \delta) \qquad \text{(iii)}$$

41. Express M_g, defined by Eq. (i) in Problem 40, as a Fourier series in θ, retaining terms up to and including (a) second harmonics or (b) fourth harmonics.

42. Suppose that there exists viscous damping at the joints of the slider-crank mechanism shown in Fig. **12.80**-1. Let the viscous resisting moments on joints O, A, B be given, respectively, by $C_O\dot{\theta}$, $C_A(\dot{\theta} + \dot{\phi})$, $C_B\dot{\phi}$, and let the resisting force on the slider be given by $C_s\dot{s}$, where C_O, C_A, C_B, C_s are all constants. Show that this viscous joint damping contributes to the overall resisting moment M_r an additional term $M_{rv} = [C_O + C_A(1 + k_\phi)^2 + C_B k_\phi^2 + C_s k_s^2]\dot{\theta}$.

43. The angular arrangement of cranks on the crankshaft of a four-cylinder, four-stroke cycle, in-line engine is (0, 180°, 180°, 0). Sketch the crankshaft, and find all possible firing sequences if the firing interval is to be uniform. Specify whether the sequences given correspond to counterclockwise or clockwise rotation as seen from the left end of your sketch.

44. Solve Problem 28 using Program DYREC.

45. Verify that the approximation in Eq. (**12.82**-6c) is valid for small ratios of e/r.

46. For typical reciprocating internal combustion engines, the ratio r/L lies in the range $\frac{1}{6}$ to $\frac{2}{3}$. When offset e exists, the ratio e/r is most effective in the range $\frac{1}{8}$ to $\frac{1}{4}$. Find the percentage change in the total stroke S_T due to the introduction of offset using
   a. Average values of r/L and e/L in the given range.
   b. Extreme values of r/L and e/L in the given range.

47. a. Verify Eq. (**12.82**-11) for the case of a polytropic process where pressure p and volume V are related by the expression $pV^k = $ constant and k is also constant.
   b. Use Eq. (**12.82**-11) to verify Eq. (**12.82**-12).

48. a. Show that the relative admission a_r defined in Eq. (**12.82**-10) is related to compression ratio CR, the angle θ_5 (see Fig. **12.82**-1), and $\lambda = r/L$ by the expression

$$a_r = 1 + \frac{CR - 1}{2}\left[1 - \cos\theta_5 + \frac{1}{\lambda} - \left(\frac{1}{\lambda^2} - \sin^2\theta_5\right)^{1/2}\right]$$

   b. Find the compression ratio for an engine in which $a_r = 1.25$, $\theta_5 = 15°$, and $\lambda = 1/3.5$.

49. Show that the neglect of the connecting rod's angular velocity $(\dot{\phi})$ in the expression for kinetic energy is equivalent to neglecting the terms containing J_{AB} in the expression for $\mathcal{g}$ and $\mathcal{C}$. If in addition it is assumed that the crank speed is a constant $(\dot{\theta}_a)$, show that the turning moment defined by Eq. (**12.92**-11) can be approximated as $M_t = M_p - \mathcal{C}\dot{\theta}_a^2$ $\doteq M_p - m_{rec}\ddot{s}k_s = -(P + m_{rec}\ddot{s})k_s$, where $m_{rec} = m_3 + m_B$. [*Hint:* Show that $k_s' = (d/d\theta)(\dot{s}/\dot{\theta}) \doteq \ddot{s}/\dot{\theta}_a^2$, and use Eq. (**12.81**-10) to find $\mathcal{C}$].

50. The turning moment $M_t = M_p + M_{in}$ for a single-cylinder, four-stroke engine is approximated as shown in Fig. 17. For a mean steady-state speed of 1000 rpm and a constant resisting moment M_r, find
   a. The resisting moment.
   b. The horsepower of the engine.
   c. The flywheel moment of inertia I_w needed to assure a coefficient of fluctuation of $C_F = 0.05$.

51. Repeat Problem 50, but this time assume that two identical cylinders fire at a uniform firing interval of 360°. Draw the appropriate curve of M_t for $0° \leq \theta \leq 360°$.

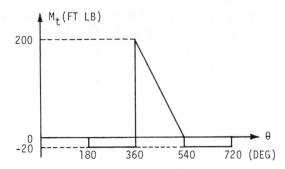

Figure 12.100-17. (Prob. 50, 51, 52).

52. Repeat Problem 50, but this time assume that four identical cylinders fire at a uniform firing interval of 180°. Draw the appropriate curve of M_t versus θ for $0 \le \theta \le 180°$.

53. A single-cylinder, four-stroke engine, running at 1000 rpm, has the following characteristics (notation of Sec. **12.80** et seq.): $r = 2$ in., $L = 8$ in., $W_3 = 1.5$ lb, $A_p = 10$ in.^{2}, $W_2 = 1.25$ lb, $a = 2.5$ in., and $J_G = 0.044527$ lb-sec^2-in. The variation of pressure (p) with crank angle (θ) may be approximated as follows: p rises linearly with s from 0 at $s = 9.6$ in. to a peak of 600 psi at top dead center where $s = 10$ in., and then falls linearly to zero at $s = 8$ in.; p is zero everywhere else.
   a. Find the mean effective pressure.
   b. Find the indicated horsepower.
   c. Plot the turning moment M_t. For this purpose, you may use the approximation (**12.92**-11a) given in footnote 27.
   d. Find I_w, the flywheel MOI, needed for a coefficient of fluctuation of 10%.

54. Repeat Problem 53 but assume that six identical cylinders act at a uniform firing interval of 120°.

13

BALANCING OF MACHINERY

13.00 INTRODUCTION

It is commonly known that lead weights are placed around the rims of automobile wheels to prevent excessive vibrations of the wheels at highway speeds. Similarly, a washing machine can vibrate excessively if the material being washed all piles up on one side of the spinning drum. The vibration will be abated if the wash is redistributed uniformly around the drum axis. This is rotor balancing in its most primitive form.

In this chapter, we shall examine the general nature of the problem of machinery balance. We shall start with the simplest case of rigid rotor systems and then shall consider the well-understood procedures for balancing slider-crank mechanisms (e.g., piston engines). Finally, we shall discuss briefly the less well-understood problem of balancing general planar mechanisms.

13.10 BALANCE OF RIGID ROTORS— GENERAL RESULTS

A rigid body constrained to rotate about a fixed axis is called a **rotor**. We shall use d'Alembert's principle (Sec. **12.70**) to find the forces exerted by the rotor on its bearings.

Figure 1 shows a rotor of arbitrary shape rotating about the z axis of a fixed orthogonal coordinate system (x, y, z), with bearings indicated at C and D.

Orthogonal axes (ξ, η, ζ) are fixed in the rotor, with the ζ and z axes coinciding. The angle between the ξ and x axes is denoted by ϕ, so that $\dot{\phi}$ and $\ddot{\phi}$ represent the angular velocity and acceleration of the rotor.

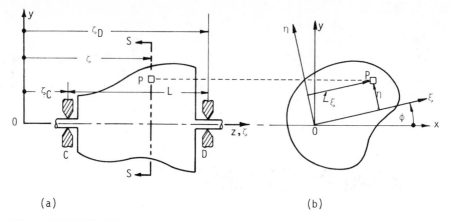

Figure 13.10-1. Rigid body (rotor) rotating about z axis: (a) Side view; (b) Cross-section at distance ζ from 0.

The acceleration components of an elemental mass dm located at a point P with coordinates (ξ, η, ζ) is given [cf. Eqs. **(5.32-15)**] by

$$
\begin{aligned}
a_\xi &= -\xi\dot{\phi}^2 - \eta\ddot{\phi} \\
a_\eta &= -\eta\dot{\phi}^2 + \xi\ddot{\phi} \\
a_\zeta &= 0
\end{aligned}
\tag{1}
$$

Hence the corresponding components of inertia force are

$$
\begin{aligned}
dF_\xi &= -a_\xi \, dm = \dot{\phi}^2\xi \, dm + \ddot{\phi}\eta \, dm \\
dF_\eta &= -a_\eta \, dm = \dot{\phi}^2\eta \, dm - \ddot{\phi}\xi \, dm \\
dF_\zeta &= -a_\zeta \, dm = 0
\end{aligned}
\tag{2}
$$

Summing these forces over the whole body, we find the components of the resultant inertia force in the form

$$
F_\xi = \int dF_\xi = \dot{\phi}^2\xi_G m + \ddot{\phi}\eta_G m
$$

$$
F_\eta = \int dF_\eta = \dot{\phi}^2\eta_G m - \ddot{\phi}\xi_G m
\tag{3}
$$

$$
F_\zeta = 0
$$

where m is the total mass of the rotor and

$$
\xi_G \equiv \frac{1}{m}\int \xi \, dm, \qquad \eta_G \equiv \frac{1}{m}\int \eta \, dm
\tag{4}
$$

are the coordinates of the *mass center G*.

Since the only active forces on the rotor are the lateral bearing reactions, the resultant bearing forces (R_ξ, R_η) are equal and opposite to the inertia forces (F_ξ, F_η).

These forces, which clearly represent the net force transmitted to the foundation by the rotor, are often called **shaking forces**.

For the special case of constant angular speed ($\dot{\phi} = \omega$, $\ddot{\phi} = 0$), the components (F_ξ, F_η) of the shaking force are constant; the resultant shaking force has magnitude

$$F = me\omega^2 \tag{5}$$

and makes an angle β with the rotating ξ axis, where the **eccentricity** e and **location angle** β are

$$e = (\xi_G^2 + \eta_G^2)^{1/2} \tag{6a}$$

$$\beta = \arctan_2(\eta_G, \xi_G) \tag{6b}$$

The vector (F_ξ, F_η) turns with the rotor and produces harmonically varying components along the fixed x, y axes, given by

$$\begin{aligned} F_x &= me\omega^2 \cos(\omega t + \beta) \\ F_y &= me\omega^2 \sin(\omega t + \beta) \end{aligned} \tag{7}$$

From Eqs. (3) and (7) we see that the shaking force vanishes when ξ_G and η_G vanish, i.e., when the mass center lies on the axis of rotation. Any axis passing through the mass center of a body is called a **central axis**; hence we may state that the *resultant shaking force vanishes when the axis of rotation is a central axis of the rotor*. Under this condition, the rotor is said to be in **static balance**.

As we shall see, it is relatively easy to bring a rotor into static balance. However, a statically balanced rotor may still exert a significant *couple* on the foundation, e.g., by producing equal but oppositely directed reaction forces at each of two bearings on the spin axis. To investigate this possibility, consider the moments produced by the elemental inertial forces (dF_ξ, dF_η, dF_ζ) about the body axes (ξ, η, ζ). With the help of Fig. 1 and Eqs. (2), it is readily seen that these **inertia moments** are given by[1]

$$\begin{aligned} dM_\xi &= -\zeta\, dF_\eta = -\dot{\phi}^2\zeta\eta\, dm + \ddot{\phi}\zeta\xi\, dm \\ dM_\eta &= +\zeta\, dF_\xi = +\dot{\phi}^2\zeta\xi\, dm + \ddot{\phi}\zeta\eta\, dm \\ dM_\zeta &= -\eta\, dF_\xi + \xi\, dF_\eta = -\ddot{\phi}(\eta^2 + \xi^2)\, dm \end{aligned} \tag{8}$$

The resultant inertia moments are therefore given by[2]

$$M_\xi = \int dM_\xi = \dot{\phi}^2 I_{\zeta\eta} - \ddot{\phi}I_{\zeta\xi}$$

$$M_\eta = \int dM_\eta = -\dot{\phi}^2 I_{\zeta\xi} - \ddot{\phi}I_{\zeta\eta} \tag{9}$$

$$M_\zeta = \int dM_\zeta = -\ddot{\phi}I_\zeta$$

where the **products of inertia** ($I_{\zeta\eta}$, $I_{\zeta\xi}$) and the **moment of inertia** (I_ζ) are defined as

---

[1]The same result follows from the definition of the moment of the force as the vector cross product

$$(dM_\xi, dM_\eta, dM_\zeta) = (\xi, \eta, \zeta) \times (dF_\xi, dF_\eta, dF_\zeta)$$

[2]Equations (9) are a special case of Euler's equations for rotation of a body about a point (cf. Timoshenko and Young [1948, p. 333]).

$$I_{\zeta\eta} \equiv -\int \zeta\eta \, dm, \qquad I_{\zeta\xi} \equiv -\int \zeta\xi \, dm, \qquad I_\zeta = \int (\xi^2 + \eta^2) \, dm \qquad (10)$$

The resultant inertia moments (M_ξ, M_η) are transmitted to the foundation of the rotor and are referred to as **shaking moments**. The moment $-M_\zeta$ is the **driving moment** supplied by a motor; when the rotor speed $\dot\phi$ is constant, M_ζ vanishes.

From Eqs. (9) it is apparent that when $\dot\phi = $ constant $= \omega$ the shaking moment is a rotating vector (fixed in the rotor) of magnitude M at an angle γ to the ξ axis. By analogy to Eqs. (5)–(7), the components of the shaking couple on the fixed x, y axes are readily seen to be

$$M_x = M \cos(\omega t + \gamma)$$
$$M_y = M \sin(\omega t + \gamma) \qquad (11)$$

where

$$M = \omega^2 (I_{\zeta\eta}^2 + I_{\zeta\xi}^2)^{1/2} \qquad (12)$$

$$\gamma = \arctan_2 (I_{\zeta\xi}, I_{\zeta\eta}) \qquad (13)$$

From Eqs. (11), it is apparent that the shaking moments about the coordinate axes will vanish if, and only if, the products of inertia $I_{\zeta\xi}$ and $I_{\zeta\eta}$ both vanish; if such is the case, the ζ axis is called a **principal axis of inertia**.

If the spin axis is a *central* axis of the rotor, the rotor is *statically balanced* and cannot exert a resultant shaking *force*. At most, such a rotor can exert a *couple* on its bearings. If, in addition, the rotor axis is a *principal axis of inertia*, the moment of the couple also vanishes.

In short, *a body which rotates about a principal central axis exerts no resultant force or couple on its foundation*. Such a body is said to be **dynamically balanced**.

Note that a *rigid* body which is statically or dynamically balanced at one speed will be so balanced at all speeds, since the conditions for balance depend only on mass distribution. Therefore, in subsequent discussions of *balancing*, we shall often assume for simplicity that the body is rotating at a constant speed $\dot\phi = \omega$. Of course the *magnitudes* of the shaking forces and moments, produced by an unbalanced rotor, depend on speed $\dot\phi$ and acceleration $\ddot\phi$.

If the rotor is *simply supported*[3] at two bearings C and D, located along the spin axis at ζ_C and ζ_D as in Fig. 1, we may easily find the reaction forces (R_ξ^C, R_η^C) at support C in the coordinate directions (ξ, η) and similar reactions (R_ξ^D, R_η^D) at support D.

Figure 2 shows the force set (F_η, M_ξ) at O, which is equipollent to all inertia forces in the direction of the η axis at any given instant. Also shown in Fig. 2 are the η components of the bearing reactions at C and D. By summing moments about D and then about C, we find

$$R_\eta^C = -\frac{\zeta_D F_\eta + M_\xi}{L}$$

$$R_\eta^D = \frac{\zeta_C F_\eta + M_\xi}{L} \qquad (14)$$

---

[3]The term *simply supported* implies that the bearings are free to pivot about points C and D; this is depicted symbolically in Fig. 1 by *knife edge* supports.

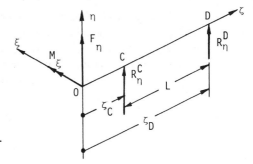

Figure 13.10-2. Reactions (in η direction) at supports C and D.

where

$$L \equiv \zeta_D - \zeta_C \tag{15}$$

is the distance between supports.

Similarly, the reactions in the direction of the ξ axis are readily shown to be

$$R_\xi^C = \frac{-\zeta_D F_\xi + M_\eta}{L}$$

$$R_\xi^D = \frac{\zeta_C F_\xi - M_\eta}{L} \tag{16}$$

13.11 Rotors with Known Unbalance[4]

We shall now consider rotors which may be modeled as an assemblage of particles of weight W_i attached to a weightless rigid shaft at stations denoted by $i = 1, 2, \ldots$, N, with eccentricities (radii) r_i and location angles β_i, at a number of known axial locations z_i along the shaft. The corresponding Cartesian coordinates of these particles are therefore

$$(\xi_i, \eta_i) = r_i(\cos \beta_i, \sin \beta_i) \tag{1}$$

and the inertia force produced by each weight has components given via Eq. **(13.10-1)** in the form

$$F_{\xi i} = -\frac{W_i}{g} a_{\xi i} = \frac{W_i}{g} \xi_i \dot{\phi}^2 + \frac{W_i}{g} \eta_i \ddot{\phi} = f_{\xi i} \frac{\dot{\phi}^2}{g} + f_{\eta i} \frac{\ddot{\phi}}{g}$$

$$F_{\eta i} = -\frac{W_i}{g} a_{\eta i} = \frac{W_i}{g} \eta_i \dot{\phi}^2 - \frac{W_i}{g} \xi_i \ddot{\phi} = f_{\eta i} \frac{\dot{\phi}^2}{g} - f_{\xi i} \frac{\ddot{\phi}}{g} \tag{2}$$

where the quantities

$$f_{\xi i} \equiv W_i \xi_i = W_i r_i \cos \beta_i$$

$$f_{\eta i} \equiv W_i \eta_i = W_i r_i \sin \beta_i \tag{3}$$

serve as *influence coefficients* and represent the inertia forces $(F_{\xi i}, F_{\eta i})$ when

$$\dot{\phi}^2 = g \quad \text{and} \quad \ddot{\phi} = 0$$

---

[4]A computer program—ROBAL (ROtor BALancing)—which performs the calculations described in this section is available from the author.

We shall refer to this state of rotation as the **standard state** and to the "forces" defined by Eq. (3) as the **standard forces**.

Since a rotor which is balanced at one constant speed remains balanced at all speeds, we shall assume, until further notice, that the rotor speed has the constant standard value ($\dot{\phi} = \sqrt{g}$). Then the shaking force components for the rotor as a whole are given by

$$f_\xi = \sum f_{\xi i} = \sum W_i r_i \cos \beta_i$$
$$f_\eta = \sum f_{\eta i} = \sum W_i r_i \sin \beta_i \tag{4}$$

where summations are made over all particles W_i. The resultant shaking force has magnitude

$$f = (f_\xi^2 + f_\eta^2)^{1/2} \tag{5a}$$

and makes an angle

$$\beta = \arctan_2 (f_\eta, f_\xi) \tag{5b}$$

with the ξ axis.

Accordingly, the shaking force can be nullified by adding (anywhere along the shaft) a balancing weight W' at radius r' and location angle

$$\beta' = \beta + 180° \tag{6a}$$

where

$$W'r' = f \tag{6b}$$

To verify this statement, we note that the inertia force produced by this balancing mass has components

$$f'_\xi \equiv W'r' \cos \beta' = -f \cos \beta = -f_\xi$$
$$f'_\eta \equiv W'r' \sin \beta' = -f \sin \beta = -f_\eta$$

which null out the forces (f_ξ, f_η) produced by the aggregate of given weights. In other terms, the addition of the single balancing weight described by Eqs. (6) results in a *statically balanced* rotor.

To achieve dynamic balance, we shall first show that any rotor is dynamically equivalent to a system of two weights (W_A, W_B) located by polar coordinates (r_A, β_A) and (r_B, β_B) in arbitrarily chosen *transverse* planes (A and B) as shown in Fig. 1.

Let the *oriented* axial distance from a given transverse plane A to particle W_i be denoted by a_i; the *oriented* axial distance from W_i to plane B will be denoted by b_i. Note that a_i and b_i are both positive when W_i lies between planes A and B, as in Fig. 1. However, a_i (b_i) will be negative if W_i is outboard of plane A (B); for example, particle W_j in Fig. 1 is outboard of plane A, making $a_j < 0$.

Due to particle W_i, the inertia force components

$$f_{\xi i} = W_i r_i \cos \beta_i$$
$$f_{\eta i} = W_i r_i \sin \beta$$

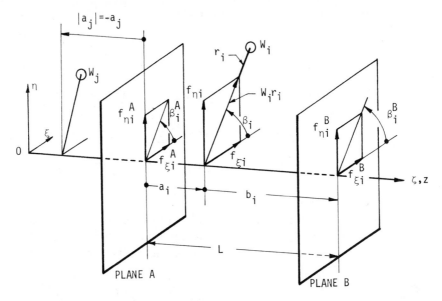

Figure 13.11-1. Unbalance vector $(W_i r_i)$ resolved into equipollent vectors in planes A and B.

act through station i on the shaft. The force $f_{\xi i}$ is equipollent to a *pair* of parallel forces $(f_{\xi i}^A, f_{\xi i}^B)$, acting in planes A and B, respectively; these equipollent forces are[5]

$$f_{\xi i}^A = \frac{f_{\xi i} b_i}{L}, \qquad f_{\xi i}^B = \frac{f_{\xi i} a_i}{L} \tag{7a}$$

where L is the distance between planes A and B. Similarly, the component $f_{\eta i}$ acting at station i is equipollent to a pair of parallel forces in planes A and B given, respectively, by

$$f_{\eta i}^A = \frac{f_{\eta i} b_i}{L}, \qquad f_{\eta i}^B = \frac{f_{\eta i} a_i}{L} \tag{7b}$$

Similar results exist for any mass particle in the rotor, so the resultant of all inertia forces is equipollent to a set of forces, in planes A and B, with components

$$f_{\xi}^A = \sum f_{\xi i}^A = \frac{1}{L} \sum W_i r_i b_i \cos \beta_i \tag{8a}$$

$$f_{\eta}^A = \sum f_{\eta i}^A = \frac{1}{L} \sum W_i r_i b_i \sin \beta_i \tag{8b}$$

$$f_{\xi}^B = \sum f_{\xi i}^B = \frac{1}{L} \sum W_i r_i a_i \cos \beta_i \tag{8c}$$

$$f_{\eta}^B = \sum f_{\eta i}^B = \frac{1}{L} \sum W_i r_i a_i \sin \beta_i \tag{8d}$$

---

[5] A doubtful reader should verify that the equipollent system has the same force resultant and moment (about any axis) as the original system.

The net shaking force has components

$$f_\xi = f_\xi^A + f_\xi^B$$
$$f_\eta = f_\eta^A + f_\eta^B \tag{9}$$

Upon recalling that $b_i + a_i = L$, it follows from Eqs. (9) that

$$f_\xi = \sum W_i r_i \cos \beta_i \frac{b_i + a_i}{L} = \sum W_i r_i \cos \beta_i$$

$$f_\eta = \sum W_i r_i \sin \beta_i \frac{b_i + a_i}{L} = \sum W_i r_i \sin \beta_i \tag{10}$$

which agrees with Eqs. (4).

Rather than using the four equations (8) to find the force coefficients $(f_\xi^A, f_\eta^A, f_\xi^B, f_\eta^B)$, we may use only two of them—say the last two—to find f_ξ^B and f_η^B. Then Eqs. (10) provide the shaking forces (f_ξ, f_η), and from Eqs. (9) we find

$$f_\xi^A = f_\xi - f_\xi^B$$
$$f_\eta^A = f_\eta - f_\eta^B \tag{11}$$

It is strongly recommended (especially when calculations are being done "by hand") that Eqs. (9) and (10) be used as a check on the predictions of Eqs. (8), as illustrated in Example 2, below.

Having calculated the *standard forces* (f_ξ^A, f_η^A) and (f_ξ^B, f_η^B), we may find the unbalances $(W_A r_A$ and $W_B r_B)$, located in planes A and B, which will produce these forces; namely,

$$W_A r_A = [(f_\xi^A)^2 + (f_\eta^A)^2]^{1/2}$$
$$W_B r_B = [(f_\xi^B)^2 + (f_\eta^B)^2]^{1/2} \tag{12}$$

The location angles of these equivalent unbalances are clearly

$$\beta_A = \arctan_2 (f_\eta^A, f_\xi^A)$$
$$\beta_B = \arctan_2 (f_\eta^B, f_\xi^B) \tag{13}$$

The explicit formulas derived for the two-mass replacement model verify that *any rotor is dynamically equivalent to a point mass in each of two arbitrarily selected transverse planes.*

Thus, we see that once the planes A and B are selected, *the equivalent two-mass rotor is uniquely determined for any given rotor.* Moreover, Eqs. (12) and (13) suggest that we can balance a rotor with known unbalance by placing a weight W_A' (W_B') at a location angle β_A' (β_B') which is diametrically opposite W_A (W_B), in plane A (B), at a radius r_A' (r_B'); i.e.,

$$W_A' r_A' = W_A r_A, \qquad \beta_A' = \beta_A + 180°$$
$$W_B' r_B' = W_B r_B, \qquad \beta_B' = \beta_B + 180° \tag{14}$$

The net effect of adding these balancing weights is to nullify completely the centrifugal force of both replacement masses (W_A and W_B) of the two-mass equivalent rotor. In short, the addition of the balancing weights W'_A and W'_B in the positions indicated results in *complete dynamic balancing*.

In practice, the **balancing planes** (A and B) are chosen at accessible points (usually at or near the ends of the rotor), and the balancing weights are placed at as large a radius as is practicable to minimize the amount of added mass. Instead of adding mass diametrically opposite the calculated unbalance locations, it is possible to drill a hole of appropriate size and shape on the same side of the spin axis as the calculated unbalanced masses, i.e., to "add negative mass."

Finding Bearing Reactions

Once the standard forces ($f_\xi^A, f_\eta^A, f_\xi^B, f_\eta^B$) are known for a given rotor, it is a simple matter to find the reactions at bearings C and D along the rotor axis. For example, Fig. 2 shows forces (f_ξ^A, f_ξ^B) acting through points A and B. The **standard reactions** (r_ξ^C, r_ξ^D) at bearings C and D represent the bearing reactions, in the direction of the ξ axis, for the standard rotational state ($\dot{\phi}^2 = g, \ddot{\phi} = 0$).

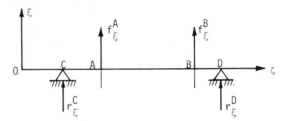

Figure 13.11-2. Shaking forces and bearing reactions in ξ direction.

The reactions at C and D are readily found by summation of moments about point D and C, respectively:

$$r_\xi^C = -\frac{f_\xi^A AD + f_\xi^B BD}{CD} \tag{15a}$$

$$r_\xi^D = -\frac{f_\xi^A CA + f_\xi^B CB}{CD} \tag{15b}$$

Similarly,

$$r_\eta^C = -\frac{f_\eta^A AD + f_\eta^B BD}{CD} \tag{15c}$$

$$r_\eta^D = -\frac{f_\eta^A CA + f_\eta^B CB}{CD} \tag{15d}$$

Note that these equations are universally valid, regardless of the relative position of points A, B, C, D, provided that all distances indicated are considered to be oriented in sense of CD. For example, if A were to the left of C in Fig. 1, we would treat $CA = -AC$ as a negative number.

Having found the standard reactions, we merely multiply them by $\dot{\phi}^2/g$ to find the reactions at any constant speed. More generally, we can find the reactions for any combination of $\dot{\phi}$ and $\ddot{\phi}$ by analogy with Eqs. (2); i.e., the bearing reactions at any speed and acceleration are

$$R_\xi^C = r_\xi^C \left(\frac{\dot\phi^2}{g} \right) + r_\eta^C \left(\frac{\ddot\phi}{g} \right) \tag{16a}$$

$$R_\xi^D = r_\xi^D \left(\frac{\dot\phi^2}{g} \right) + r_\eta^D \left(\frac{\ddot\phi}{g} \right) \tag{16b}$$

$$R_\eta^C = r_\eta^C \left(\frac{\dot\phi^2}{g} \right) - r_\xi^C \left(\frac{\ddot\phi}{g} \right) \tag{16c}$$

$$R_\eta^D = r_\eta^D \left(\frac{\dot\phi^2}{g} \right) - r_\eta^D \left(\frac{\ddot\phi}{g} \right) \tag{16d}$$

Vector Methods

All results of this section can be expressed compactly in complex vector notation, which lends itself readily to graphical methods of analysis. For example, the inertia force due to a weight W_k—with cylindrical coordinates given as usual by (r_k, β_k, z_k)—is described by the complex vector

$$
\begin{aligned}
F_k &= \frac{\dot\phi^2}{g} W_k r_k e^{i\beta_k} - i \frac{\ddot\phi}{g} W_k r_k e^{i\beta} \\
&= \left(\frac{\dot\phi^2}{g} - i \frac{\ddot\phi}{g} \right) W_k r_k e^{i\beta_k} \equiv C f_k
\end{aligned}
\tag{17}
$$

where

$$C \equiv \frac{\dot\phi^2 - i\ddot\phi}{g} \tag{18}$$

will be called the complex **acceleration ratio**,

$$f_k \equiv W_k r_k e^{i\beta_k} \tag{19}$$

will be called the **standard unbalance force** or **standard unbalance vector**, and $i = \sqrt{-1}$.

From Eq. (17) we see that the resultant inertia force for the rotor is given, for any rotational state, by

$$F = C \sum f_k = C \sum W_k r_k e^{i\beta_k} \tag{20}$$

Hence, the resultant inertia force will vanish (i.e., static balance is achieved) when the vector sum in Eq. (20) is null; i.e.,

$$\sum f_k = \sum W_k r_k e^{i\beta_k} = 0 \tag{21}$$

is a necessary and sufficient condition for *static* balance for any values of $\dot\phi$ and $\ddot\phi$.

Now consider the moment M_k of the inertia force F_k with respect to the fixed point O on the spin axis. This moment is a vector represented by the cross product $\mathbf{M}_k = z_k \hat{\mathbf{z}} \times \mathbf{F}_k$ (where $\hat{\mathbf{z}}$ is a unit vector along the z axis). It is easily seen that $\mathbf{M}_k$ has magnitude $z_k |\mathbf{F}_k|$, lies in the ξ, η plane, and leads $\mathbf{F}_k$ by 90° (in the same sense that $\hat{\boldsymbol{\eta}}$ leads $\hat{\boldsymbol{\xi}}$). Thus, the complex vector representation of this moment vector is

$$M_k = i z_k F_k = i C z_k f_k \tag{22}$$

Accordingly, the resultant moment about O of all inertia forces is

$$M = \sum M_k = iC \sum z_k f_k \tag{23}$$

For this moment to vanish, it is necessary and sufficient that

$$\sum z_k f_k \equiv \sum z_k W_k r_k e^{i\beta_k} = 0 \tag{24}$$

regardless of the values of $\dot{\phi}$ and $\ddot{\phi}$. Thus, *dynamic balance* is achieved when Eqs. (21) and (24) are simultaneously satisfied.

Equation (21) can be interpreted to state that the standard forces $f_k = W_k r_k e^{i\beta_k}$ form a closed vector polygon, and Eq. (24) states that the vectors $z_k f_k = z_k W_k e^{i\beta_k}$ form a second closed polygon. In short, *static balance is achieved when the standard forces f_k form a closed vector polygon, and dynamic balance is achieved if (in addition) the standard moments $z_k f_k$ form a closed vector polygon.*

These vector methods form the basis of the following semigraphical method for finding the balance vectors ($W'_A r'_A e^{i\beta_{A'}}$ and $W'_B r'_B e^{i\beta_{B'}}$) needed in preselected balancing planes A and B to achieve dynamic balance.

First Eq. (24) is expressed in the form

$$\sum_{k=1}^{N} z_k W_k r_k e^{i\beta_k} + z_A W'_A r'_A e^{i\beta_{A'}} + z_B W'_B r'_B e^{i\beta_{B'}} = 0 \tag{25}$$

where N is the number of unbalanced weights. If the origin O of the z axis is placed at point A so that $z_A = 0$, it follows from Eq. (25) that the required balance vector in the known plane B is given explicitly by

$$W'_B r'_B e^{i\beta_{B'}} = -\frac{1}{z_B} \sum_{K=1}^{N} z_k W_k r_k e^{i\beta_k} \tag{26}$$

The vector represented by the right-hand side of Eq. (26) is readily constructed by graphical addition; however, it is more accurately found (and with no more effort) by means of an electronic calculator.

The required balance vector in plane A follows from Eq. (21) in the explicit form

$$W'_A r'_A e^{i\beta_{A'}} = -\sum_{k=1}^{N} W_k r_k e^{i\beta_k} - W'_B r'_B e^{i\beta_{B'}} \tag{27}$$

where the vector sum in Eq. (27) may be found graphically or numerically.

Numerous illustrations of the graphical approach are given by Wilcox [1967].

Special Case (Disk-like Rotors)

For the special case where all weights W_i lie in a single plane at $z_k = \text{constant} = z_0$ (e.g., when all weight is attached to a thin orthogonally mounted disc), the condition (24) for *dynamic* balance can be expressed in the form

$$z_0 \sum f_k = 0 \quad \text{or} \quad \sum f_k = \sum W_k r_j e^{i\beta_k} = 0 \tag{28}$$

In other terms, the condition of dynamic balance (28) has degenerated into the condition (21) for static balance. In short, *if a disk-like rotor* (*mounted normal to the shaft axis*) *is statically balanced in a single balancing plane through its C.G., it will automatically be dynamically balanced.*

This result explains why automobile wheels (which are reasonably disk-like) can usually be effectively balanced by merely attaching weights to the rim, which move the C.G. to the wheel axis. Of course, it is desirable that the weights be distributed symmetrically about the midplane of the wheel, so that these "balancing" weights do not aggravate the residual *dynamic* unbalance by increasing the products of inertia $(I_{z\eta}, I_{z\zeta})$ about the geometric center of the wheel.

Influence of Gravity

So far, we have excluded the force of gravity from our discussion. The effect of gravity is maximum for horizontal shafts, where the peak vertical shaking force $[(W/g)\,e\omega^2]$ adds to the rotor weight W and produces a peak load on the bearings of

$$F_y = W + \frac{We\omega^2}{g}$$

in the unbalanced state (where $\ddot{\phi} = 0, \dot{\phi} = \omega$).

After adding a balancing weight $W' = We/r'$ at a radius r', the shaking force disappears, but the static vertical load due to gravity alone becomes

$$F'_y = W + W' = W + \frac{We}{r'}$$

Balancing will result in a reduction of net bearing load only if the unbalanced load F_y exceeds the dead weight F'_y, i.e., if

$$\frac{\omega^2}{g} > \frac{1}{r'}$$

If the rotor speed (in revolutions per minute) is denoted by N_m and r' is expressed in inches, the condition for balancing to be effective (in reducing reactions) is

$$N_m \frac{2\pi}{60} > \sqrt{\frac{386}{r'}} \quad \text{or} \quad N_m > \frac{188}{\sqrt{r'}}$$

Thus, for a balancing radius $r' = 6$ in., balancing reduces the maximum reaction when $N_m > 76.6$ rpm (about the speed of an old-fashioned phonograph record).

Influence of Rotor Flexibility

Up to now we have assumed that the rotor is perfectly rigid. However, the unbalance forces will cause the centerline of the rotor to undergo a nonuniform lateral displacement. This elastic displacement may either increase or decrease the eccentricity at any cross section along the rotor. Since the initial eccentricities of a well-balanced rotor are measured in thousandths of an inch, it does not require a great deal of elastic deformation to radically alter the distribution of unbalances along the shaft. In particular, it is shown in the theory of vibrations (see Den Hartog [1956]) that the lateral displacements of a rotating shaft become extremely great when the rotational speed ω approaches one of the (infinitely many) *natural frequencies* of lateral vibration for the rotor. Therefore, near these so-called **critical speeds** of rotation the elastic displacements dominate the initial eccentricities, and the conditions for balance become totally different from those of the *rigid* rotor. In particular, different balancing masses are required at different speeds, the balancing planes cannot be chosen arbitrarily (they must not pass through nodes of the vibration mode at a given critical speed), and the number of required balancing planes will exceed two if the rotor is to be in balance at more than one speed.

Compliance in the bearings and in the support structure (e.g., bearing pedestals) has the same qualitative effect as compliance in the rotor itself. That is, the critical speeds occur at the natural frequencies of the *rotor-bearing-support* system.

A rotor may be assumed to be *rigid* only if its speed lies below about *half* of the lowest natural frequency of the rotor-bearing-support system.

EXAMPLE 1.[6] Single-Plane Balancing

A uniform thin disk is mounted on a shaft without eccentricity or obliquity. It has 14 equal holes spaced as shown in Fig. 3. Holes 1, 2, 3, and 4 contain bolts weighing 1 oz each, and holes 12 and 14 contain bolts weighing 2 oz each. We desire to balance the disk by adding suitable weights in any of the unfilled holes.

We first recall that static balance of such a disk is sufficient to ensure dynamic

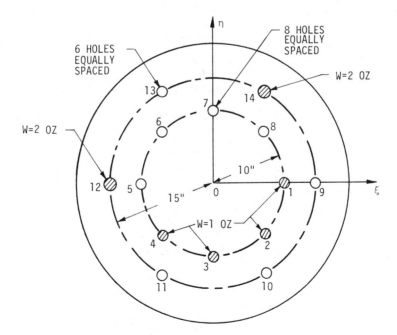

Figure 13.11-3. Disk with holes and bolts.

---

[6]This problem has been solved by a graphical method in Wilcox [1967, pp. 37–39].

balance. Then we find the unbalance vector (due to the bolts) which has components

$$(f_\xi, f_\eta) = \sum_{k=1}^{14} W_k r_k (\cos \beta_k, \sin \beta_k) \tag{29}$$

This unbalance can be nullified by adding weights W'_j and W'_m in holes j and m (to be selected later) such that

$$\begin{aligned}
W'_j r'_j \cos \beta'_j + W'_m r'_m \cos \beta'_m &= -f_\xi \\
W_j r'_j \sin \beta_k + W'_m r'_m \sin \beta'_m &= -f_\eta
\end{aligned} \tag{30}$$

The calculations are best arranged in the form of Table 1. From the sum of the last two rows we see that

TABLE 13.11-1

				i			
	1	2	3	4	12	14	
W (oz)	1	1	1	1	2	2	
r (in.)	10	10	10	10	15	15	
β (deg)	0	-45	-90	-135	180	60	Row Sums
$Wr \cos \beta$ (in.-oz)	10	7.0711	0	-7.0711	-30	15	$-5 \quad \equiv f_\xi$
$Wr \sin \beta$ (in.-oz)	0	-7.0711	-10	-7.0711	0	25.981	$1.838 \equiv f_\eta$

$$f_\xi = -5 \text{ in.-oz,} \qquad f_\eta = 1.838 \text{ in.-oz}$$

The location angle of this unbalance vector is

$$\beta = \arctan_2 (1.838, -5) = 159.8°$$

We therefore seek to add balancing weights whose resultant centrifugal force has location angle

$$\beta' = 159.8° + 180° = 339.8° \qquad (\text{or } \beta' = -20.2°)$$

The ray with angle β' falls between holes 9 and 10, which are located (see Fig. 3) by angles $\beta_9 = 0$, $\beta_{10} = -60°$; note that $r_9 = r_{10} = 15$ in. Letting $j = 9$ and $m = 10$, we can express Eqs. (30) in the form

$$\begin{aligned}
W'_9 r_9 \cos \beta_9 + W'_{10} r_{10} \cos \beta_{10} &= W'_9(15)(1) + W'_{10}(15) \cos(-60°) = 5 \\
W'_9 r_9 \sin \beta_9 + W'_{10} r_{10} \sin \beta_{10} &= W'_9(15)(0) + W'_{10}(15) \sin(-60°) = -1.838
\end{aligned}$$

Hence

$$W'_{10} = \frac{1.836}{15 \sin 60°} = 0.141 \text{ oz}$$

$$W'_9 = \frac{5 - W'_{10}(15) \cos 60°}{15} = 0.263 \text{ oz}$$

are the balancing weights to be placed in holes 9 and 10.

EXAMPLE 2. General Case of Known Lumped Unbalance

The rotor shown in Fig. 4 has an instantaneous angular speed of $\dot{\phi} = 31.46$ rad/sec and an acceleration of $\ddot{\phi} = 62.83$ rad/sec². We wish to find

a. The balancing weights W_A and W_B which are needed at a radius of $e'_A = e'_B = 10$ in. in balancing planes A and B and the corresponding angles (β'_A, β'_B) of these weights for dynamic balance of the rotor

b. The reaction forces (R_ξ^C, R_η^C, R_ξ^D, R_η^D), at the specified speed and acceleration, of bearings at C and D

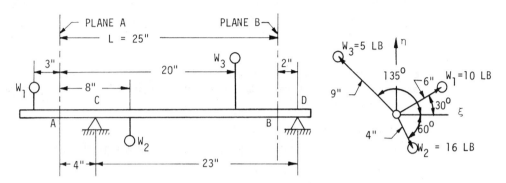

Figure 13.11-4. Rotor of example No. 2.

Solution:

The work is best arranged in tabular form, as in Table 2, where the first six rows represent the input data. We are working in units of inches, pounds, seconds, and degrees. Rows 7 and 8 are not strictly necessary but will provide a very useful check against arithmetic blunders.

TABLE 13.11-2

Row	Data and Calculations for Example 2			
1	i	1	2	3
2	W (lb)	10	16	5
3	r (in.)	6	4	9
4	β (deg)	30	300	135
5	b (in.)	28	17	5
6	a (in.)	−3	8	20	Row Sums
7	$Wr \cos \beta$	51.961	32.000	−31.820	52.14
8	$Wr \sin \beta$	30.000	−55.426	31.820	6.394
9	$bWr \cos \beta$	1454.9	544.00	−159.10	1839.8
10	$bWr \sin \beta$	840.00	−942.24	159.10	56.86
11	$aWr \cos \beta$	−155.88	256.00	−636.4	−536.3
12	$aWr \sin \beta$	−90.000	−443.41	636.4	103.0

Noting that $L = 25$ in. and making use of the row sums in rows 9–12 of the table, we see from Eqs. (8) that the standard forces in planes A and B are

$$f_\xi^A = \frac{1}{L} \sum (Wrb \cos \beta)_i = \frac{1}{25}(1840) = 73.59$$

$$f_\eta^A = \frac{1}{L} \sum (Wrb \sin \beta)_i = \frac{1}{25}(56.86) = 2.274$$

$$f_\xi^B = \frac{1}{L} \sum (Wra \cos \beta)_j = \frac{1}{25}(-536.3) = -21.45$$

$$f_\eta^B = \frac{1}{L} \sum (Wra \sin \beta)_i = \frac{1}{25}(103.0) = 4.120$$

As a check on these force coefficients, we may now use Eqs. (9) and (10) in the form

$$f_\xi = f_\xi^A + f_\xi^B = 52.14 = \sum (Wr \cos \beta)_i$$
$$f_\eta = f_\eta^A + f_\eta^B = 6.394 = \sum (Wr \sin \beta)_i$$

which is confirmed by the row sums in rows 7 and 8 of the table.

Part a. To find the equivalent unbalances in planes A and B and their respective location angles we use Eqs. (12) and (13), which yield

$$W_A r_A = [(f_\xi^A)^2 + (f_\eta^A)^2]^{1/2} = (73.59^2 + 2.27^2)^{1/2} = 73.62 \text{ in.-lb}$$
$$W_B r_B = [(f_\xi^B)^2 + (f_\eta^B)^2]^{1/2} = [(-21.45)^2 + 4.12^2]^{1/2} = 21.84 \text{ in.-lb}$$
$$\beta_A = \arctan_2 (2.274, 73.59) = 1.770°$$
$$\beta_B = \arctan_2 (4.120, -21.45) = 169.1°$$

Since the balance weights are to be installed at a radius of $r_A' = r_B' = 10$ in., it follows from Eq. (14) that

$$W_A' = \frac{W_A r_A}{r_A'} = \frac{73.62}{10} = 7.362 \text{ lb}$$

$$W_B' = \frac{W_B r_B}{r_B'} = \frac{21.84}{10} = 2.184 \text{ lb}$$

$$\beta_A' = 180 + \beta_A = 180 + 1.77 = 181.8°$$
$$\beta_B' = 180 + \beta_B = 180 + 169.1 = 349.1°$$

This completes part a of the problem, which is all that is needed to achieve dynamic balance.

Part b. To find the bearing reactions at operating conditions, we first find them, for the standard state, by using the results of part a in Eqs. (15), together with the observation (see Fig. 4) that $CD = 23$, $AD = 27$, $BD = 2$, $CA = -AC = -4$, $CB = 21$; i.e.,

$$r_\xi^C = -\frac{f_\xi^A AD + f_\xi^B BD}{CD} = -\frac{(73.59)(27) + (-21.45)(2)}{23} = -84.52$$

$$r_\xi^D = -\frac{f_\xi^A CA + f_\xi^B CB}{CD} = -\frac{(73.59)(-4) + (-21.45)(21)}{23} = 32.38$$

$$r_\eta^C = -\frac{f_\eta^A AD + f_\eta^B BD}{CD} = -\frac{(2.274)(27) + (4.120)(2)}{23} = -3.028$$

$$r_\eta^D = -\frac{f_\eta^A CA + f_\eta^B CB}{CD} = -\frac{(2.274)(-4) + (4.120)(21)}{23} = -3.366$$

At this stage it is worth checking that the resultant reactions are equal and opposite to the resultant shaking forces (for the standard state), i.e.,

$$r_\xi^C + r_\xi^D = -52.14 = -(f_\xi^A + f_\xi^B)$$
$$r_\eta^C + r_\eta^D = -6.394 = -(f_\eta^A + f_\eta^B)$$

Finally, the bearing reactions at the operating state ($\dot\phi = 31.46$ rad/sec, $\ddot\phi = 62.83$ rad/sec^2) are given by Eqs. (16) in the form

$$R_\xi^C = r_\xi^C\left(\frac{\dot\phi^2}{g}\right) + r_\eta^C\left(\frac{\ddot\phi}{g}\right) = \frac{(-84.52)(31.46)^2 + (-3.028)(62.83)}{386.1} = -217.2 \text{ lb}$$

$$R_\eta^C = r_\eta^C\left(\frac{\dot\phi^2}{g}\right) - r_\xi^C\left(\frac{\ddot\phi}{g}\right) = \frac{(-3.028)(31.46)^2 - (-84.52)(62.83)}{386.1} = 5.992 \text{ lb}$$

$$R_\xi^D = r_\xi^D\left(\frac{\dot\phi^2}{g}\right) + r_\eta^D\left(\frac{\ddot\phi}{g}\right) = \frac{(32.28)(31.46)^2 - (-3.366)(62.83)}{386.1} = 82.45 \text{ lb}$$

$$R_\eta^D = r_\eta^D\left(\frac{\dot\phi^2}{g}\right) - r_\xi^D\left(\frac{\ddot\phi}{g}\right) = \frac{(-3.366)(31.46)^2 - (32.28)(62.83)}{386.1} = -13.90 \text{ lb}$$

This completes part b of the problem.

13.12 Balancing Machines and Instrumentation

We have seen in the previous section how to balance any rotor when the unbalances are known a priori. Unfortunately, unbalance which arises from manufacturing tolerances, assembly operations, thermal distortions, and other unpredictable sources is not usually known with sufficient precision to be counterbalanced by purely computational procedures. Therefore, experimental procedures and equipment are necessary to define the unbalance in a given rotor.

Static Balance: For purposes of static balance, one may place the rotor axis across two smooth parallel horizontal rails and observe the lowest point on the rotor. The C.G. will be between the rotor axis and the low point, so a trial weight (e.g., a lump of putty) can be attached to the high side of the rotor. The trial weight may be varied until the rotor is in neutral equilibrium (it does not roll along the rails from its initial location). This determines the amount and location of balancing weight required.

For disk-like rotors (e.g., automobile wheels) the center O of the rotor may be supported on a spherical pivot with the rotor axis approximately vertical. A spherical *spirit level*[7] may be used to indicate the *direction* (in the rotor's plane) of the eccentricity vector $\overrightarrow{OG}$ to the mass center G. Trial weights can be placed at the "high points" indicated by the spirit level.

Dynamic Balance: We have seen in Sec. **13.11** that any dynamically unbalanced rotor is equivalent to a rotor consisting of a lumped mass located in each of two transverse balancing planes. It is the function of a **dynamic balancing machine** to locate these two equivalent masses for any set of user-selected balancing planes.

---

[7]This is a flat disc capped by a transparent shallow spherical shell, concave inward. The interior is filled with a liquid, leaving a small bubble of trapped air. Because of its buoyancy, the bubble always seeks the highest point on the spherical surface. Therefore, the bubble acts like an arrowhead pointing along the direction of greatest height gradient.

A wide variety of such machines has evolved over the years, and a detailed description of them would be beyond the scope of this book. For descriptions of such machines see McQuery [1973], Wilcox [1967], Muster and Stadelbauer [1976], and the technical literature of manufacturers, such as Avery [1964], IRD [1977], and Schenck Trebel [1973]. Some background material on older forms of balancing machines is given in Den Hartog [1965] and Kroon [1943, 1944].

Most balancing machines consist of a framework which carries two bearings C and D into which the rotor journals are placed, usually with the rotor axis horizontal. These bearings are permitted to oscillate in one direction (usually horizontal) normal to the rotor axis.

For example, in Fig. 1 each bearing is mounted in a rigid horizontal beam attached to flexible pedestals that permit motion of the bearing in the horizontal (x) direction.

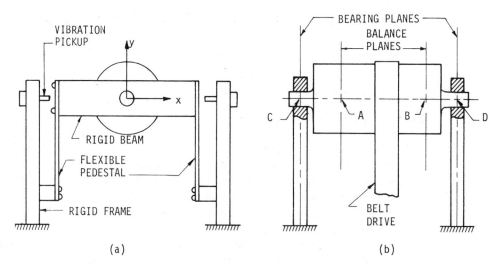

Figure 13.12-1. Schematic of typical balancing machine. (a) End view (b) Axial cross-section.

Such a machine is called a **soft bearing** (or **flexible bearing) machine**. If the pedestals shown in Fig. 1 are made rigid, the machine is called a **hard bearing** (or **rigid bearing) machine.** The rotor is brought up to a desired angular speed ω by means of a belt drive, or through other means (e.g., a stream of compressed air or a universal joint driving at one end). The resulting lateral vibrations of the bearings at C and D are sensed by **vibration pickups** (e.g., moving-coil-type transducers), and the signals are processed through electronic readout devices. The machine is calibrated so that the horizontal components of the bearing *reactions*

$$R_x^C = \frac{\omega^2}{g} r^C \cos(\omega t + \beta_C)$$

$$R_x^D = \frac{\omega^2}{g} r^D \cos(\omega t + \beta_D)$$

(1)

are known once the horizontal motions at the bearings are measured. The terms r^C, r^D

are the magnitudes of the reactions at C and D for the standard state $(\omega^2 = g, \dot{\omega} = 0)$, and β_C, β_D are the angles (relative to the rotating ξ axis) of the resultant reactions.

In a *hard* bearing machine, the reaction forces in Eqs. (1) are sensed directly by force transducers (e.g., of strain gage or piezoelectric type). In any case, the circuitry is designed to identify the standard reaction magnitudes r^C, r^D and the location angles β_C, β_D. From this information, the reaction components along the body fixed (ξ, η) axes are readily calculated to be

$$r_\xi^C = r^C \cos \beta_C, \qquad r_\eta^C = r^C \sin \beta_C$$
$$r_\xi^D = r^D \cos \beta_D, \qquad r_\eta^D = r^D \sin \beta_D \tag{2}$$

These four values may be placed into the four equations (**13.11**-15), which are then readily solved explicitly for the standard force components in any two user-selected balancing planes A and B. The solution (in matrix form) is

$$\begin{bmatrix} f_\xi^A & f_\eta^A \\ f_\xi^B & f_\eta^B \end{bmatrix} = \frac{1}{AB} \begin{bmatrix} BC & BD \\ CA & DA \end{bmatrix} \begin{bmatrix} r_\xi^C & r_\eta^C \\ r_\xi^D & r_\eta^D \end{bmatrix} \tag{3}$$

where $AB, BC, \ldots$ are the *oriented* line segments shown in Fig. **13.11**-2. These same results may, of course, be derived by summing moments about points A and B in Fig. **13.11**-2 and in an analogous figure wherein η replaces ξ everywhere.

Finally, the desired balancing weights (W_A, W_B) and their location angles (β_A, β_B) in the desired balancing planes are given by Eqs. (**13.11**-12) and (**13.11**-13); i.e.,

$$W_A r_A = [(f_\xi^A)^2 + (f_\eta^A)^2]^{1/2}, \qquad \beta_A = \arctan_2 (f_\eta^A, f_\xi^A)$$
$$W_B r_B = [(f_\xi^B)^2 + (f_\eta^B)^2]^{1/2}, \qquad \beta_B = \arctan_2 (f_\eta^B, f_\xi^B) \tag{4}$$

Equations (2)–(4) may be compactly expressed in the complex vector notation:

$$\begin{bmatrix} W_A r_A e^{i\beta_A} \\ W_B r_B e^{i\beta_B} \end{bmatrix} = \frac{1}{AB} \begin{bmatrix} BC & BD \\ CA & DA \end{bmatrix} \begin{bmatrix} r^C e^{i\beta_C} \\ r^D e^{i\beta_D} \end{bmatrix} \tag{5}$$

Modern balancing machines perform all of the required calculations automatically by *compensating networks* after the user "dials in" the appropriate lengths between balancing planes (A, B) and bearing planes (C, D).

Standards for Tolerable Unbalance

Because the resolution of any measurement instrument is limited, we cannot expect absolutely perfect balancing, even for hypothetically rigid rotors. The amount of residual eccentricity that can be tolerated depends on the nature of the machinery being balanced and the judgment of the end user.

Committees of the International Standards Organization (ISO) and the American National Standards Institute (ANSI) have formulated recommendations for the allowable **quality grade**

$$G = e\omega \tag{6}$$

for different classes of machinery, where

$$e = \frac{\sum Wr}{W} \text{ is eccentricity in } \textit{millimeters}$$

$$\omega = \text{angular speed in radians per second}$$

Typical recommended[8] values of G are

$G = e\omega$ (mm/sec)	Rotor Types
4000	Crankshaft assembly of slow marine diesel engines with uneven number of cylinders
1600	Crankshaft assembly of large two-cycle engines
630	Crankshaft assembly of large four-cycle engines
100	Complete engines for cars, trucks, locomotives
40	Car wheels, drive shafts, crankshaft assembly for engines of cars, trucks, locomotives
16	Individual engine components
6.3	Fans, flywheels, machine tools
2.5	Gas and steam turbines, small electric armatures
1	Tape recorder and phonograph drives, grinding machine driver
0.4	Gyroscopes, precision grinder components

For machines which are not mentioned specifically in the ISO standard, one may use the following recommendations of Wilcox [1967, p. 33] for allowable eccentricity:

Speed (rpm):	300	1000	3000	10,000	30,000
Eccentricity e (0.001 in.):	10	2	0.5	0.1	0.02

Wilcox further recommends that his tabulated values of e be divided by 4 for "precision machines."

13.20 BALANCE OF RECIPROCATING (SLIDER-CRANK) MACHINES

We shall now consider the inertia forces generated by the moving links of the slider-crank mechanism. In particular, we seek to find those forces and moments which are transmitted to the frame of the machine and to show how the severe vibrational effects of these forces on the supporting substructure can be alleviated.

There are essentially three ways of controlling such vibrations:

1. Vibration isolation
2. Active balancing
3. Passive balancing

---

[8]*Extracted* from *ISO Standard DIS 1940*, reported in Schenck Trebel [1973], and in Muster and Stadelbauer [1976].

Vibration isolation is achieved by "tuning" the supporting structure so that the frequencies of the forces transmitted from the machine are well removed from the important natural frequencies of the support system. This is sometimes difficult to achieve, especially when the *structure-borne* vibrations are propagated through the ground or other solid media to excite resonance in components hundreds of feet from the vibration source. A common way to isolate machine vibration is by means of suitable *vibration mounts* of rubber or other compliant materials. The topic of vibration isolation is discussed in texts on the theory of vibrations (e.g., Crede [1951]) and will not be pursued here.

By **active balancing** we refer to the introduction of dummy pistons, geared revolving counterweights, and other means of intentionally producing inertia forces designed to annul the shaking forces and moments which are due to the accelerating masses of a given machine. We shall briefly consider some forms of active balancing; however, it is generally an expensive and cumbersome remedy which should be avoided if possible.

We shall use the term **passive balancing** to describe the alleviation of shaking forces and moments by the addition or removal of mass from various portions of the moving links. Passive balancing is by far the simplest and least expensive solution to the problem, and it should be used whenever possible. We shall therefore concentrate on this mode of inertia balancing.

There is a substantial body of literature on this topic and a seemingly infinite number of specific configurations that have been discussed in the past (including in-line, V, W, H, radial, rotary, and other arrangements of cylinders, with various cylinder spacings in multiple banks or rows). Extensive discussions of the subject may be found in the specialized literature on engine dynamics, such as Root [1932] and Judge [1947], and in books on dynamics of machinery, such as Ham and Crane [1948] and Holowenko [1955]. A comprehensive mathematical analysis of engine balancing is given by Biezeno and Grammel [1954], and a compact but incisive discussion of the topic is given by Taylor [1966, 1968].

In our discussion, we intend to present only the fundamental principles of inertia balancing but in such a way that readers should be able to solve most of the balancing problems they may encounter and be well equipped to consult the literature on more difficult problems.

13.21 *Shaking Forces and Torque of Single-Cylinder Engines*

Figure 1(a) shows a single cylinder engine model where the connecting rod has been replaced, as in Sec. **12.81**, by equivalent tip masses (m_A, m_B) and a massless ring of MOI J_{AB}. Recall[9] that the nomenclature used previously was

m_1, m_2, m_3:   masses of crank, conrod, and piston

J_C, J_G: centroidal MOI of crank and conrod (also J_1, J_2)

$m_A = \dfrac{m_2 b}{L}, m_B = \dfrac{m_2 a}{L}$:   equivalent conrod tip masses

$J_{AB} = J_G - m_2 ab$:   MOI associated with equivalent conrod

---

[9]See Secs. **12.80**–**12.82** for a review of terminology.

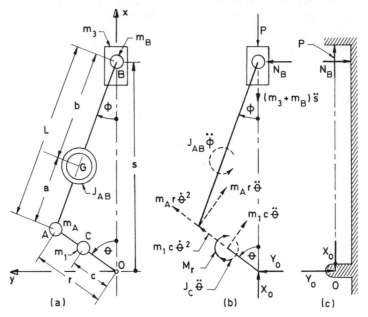

Figure 13.21-1. (a) Single cylinder engine; (b) Free-body diagram of engine mechanism; (c) Forces on frame.

Figure 1(b) is a free-body diagram of the slider-crank mechanism in which the inertia forces (and moments) are indicated by broken-line arrows. The noninertial forces, indicated by solid-line arrows, are

P: gas force

M_r: resisting moment on the crank

N_B: normal force exerted on the piston by the frame

(X_0, Y_0): reaction forces exerted on the crank by the frame

From Fig. 1(c) we see that the net force components (X_f, Y_f) exerted by the engine on the frame and the torque (M_f) on the frame (about point O) are

$$X_f = P - X_0 \tag{1}$$

$$Y_f = -Y_0 - N_B \tag{2}$$

$$M_f = -N_B s \tag{3}$$

To find the reaction forces (X_0, Y_0, N_B) we note that for the free body of Fig. 1(b) to be in *dynamic equilibrium* it is necessary that

$$\sum X = X_0 + (m_1 c + m_A r)\dot{\theta}^2 \cos\theta - (m_3 + m_B)\ddot{s} \tag{4}$$
$$+ (m_1 c + m_A r)\ddot{\theta} \sin\theta - P = 0$$

$$\sum Y = Y_0 + (m_1 c + m_A r)\dot{\theta}^2 \sin\theta - (m_1 c + m_A r)\ddot{\theta}\cos\theta + N_B = 0 \tag{5}$$

$$\sum M_0 = N_B s + J_{AB}\ddot{\phi} - (m_A r^2 + m_1 c^2 + J_c)\ddot{\theta} - M_r = 0 \tag{6}$$

We now introduce the notation

$$m_{\text{rec}} \equiv m_3 + m_B = m_3 + m_2 \frac{a}{L}$$

$$m_{\text{rot}} \equiv m_1 \frac{c}{r} + m_A = m_1 \frac{c}{r} + m_2 \frac{b}{L} \tag{7}$$

for the **reciprocating mass** and **rotating mass**. Then the forces and torques transmitted from engine to frame are given by Eqs. (1)–(7) in the simple form

$$X_f = P - X_0 = m_{\text{rot}} r \dot{\theta}^2 \cos\theta + m_{\text{rot}} r \ddot{\theta} \sin\theta - m_{\text{rec}}\ddot{s} \tag{8}$$

$$Y_f = -Y_0 - N_B = m_{\text{rot}} r \dot{\theta}^2 \sin\theta - m_{\text{rot}} r \ddot{\theta} \cos\theta \tag{9}$$

$$M_f = -N_B s = J_{AB}\ddot{\phi} - (m_A r^2 + m_1 c^2 + J_c)\ddot{\theta} - M_r \tag{10}$$

It is worth noting that the *force* on the frame equals the resultant *inertia force* of the moving parts. However, the resultant **inertia torque** of the moving parts equals $M_f + M_r$; i.e., the inertia torque is resisted by both the frame and by the driven device. This is the most common situation. However, a driven device (e.g., an electric generator) may be mounted on the frame in such a way as to transmit the torque M_r to the frame. In such a case, we say that there is **torque return** to the frame, and torque $+M_r$ cancels the last term in Eq. (10). Thus we see that the torque on the frame equals the inertia torque of the moving parts when a torque return path exists.

To make further progress, we now introduce into Eqs. (8) and (10) the Fourier expansions for $\ddot{s}$ and $\ddot{\phi}$ given by Eqs. (**1.41**-23) and (**1.41**-26), thereby finding

$$\begin{aligned}
X_f = {}& (m_{\text{rot}} + m_{\text{rec}})r\dot{\theta}^2 \cos\theta \\
&+ m_{\text{rec}} r\dot{\theta}^2(A_2 \cos 2\theta - A_4 \cos 4\theta + A_6 \cos 6\theta - \cdots) \\
&+ (m_{\text{rot}} + m_{\text{rec}})r\ddot{\theta} \sin\theta \\
&+ m_{\text{rec}} r\ddot{\theta}\left(\frac{A_2}{2} \sin 2\theta - \frac{A_4}{4} \sin 4\theta + \frac{A_6}{6} \sin 6\theta - \cdots\right)
\end{aligned} \tag{11a}$$

$$Y_f = m_{\text{rot}} r\dot{\theta}^2 \sin\theta - m_{\text{rot}} r\ddot{\theta} \cos\theta \tag{11b}$$

$$\begin{aligned}
M_f + M_r = {}& -J_{AB}\lambda\dot{\theta}^2(C_1 \sin\theta - C_3 \sin 3\theta + C_5 \sin 5\theta - \cdots) \\
&+ J_{AB}\lambda\ddot{\theta}\left(C_1 \cos\theta - \frac{C_3}{3} \cos 3\theta + \frac{C_5}{5} \cos 5\theta - \cdots\right) \\
&- (m_A r^2 + m_1 c^2 + J_c)\ddot{\theta}
\end{aligned} \tag{11c}$$

where $\lambda = r/L$, and the Fourier amplitudes are approximated by

$$A_2 \doteq \lambda, \qquad A_4 \doteq \frac{\lambda^3}{4}, \qquad A_6 \doteq \frac{9\lambda^5}{128}$$

$$C_1 \doteq 1, \qquad C_3 \doteq \frac{3\lambda^2}{8}, \qquad C_5 \doteq \frac{15\lambda^4}{128} \tag{12}$$

More exact expressions for A_i and C_i are given in Sec. **1.41**, but Eqs. (12) demonstrate the rapidly decreasing significance of the higher-order harmonics. For example, when $\lambda = 1/3.5$, $A_4/A_2 \doteq 0.020$, and $A_6/A_2 \doteq 0.00047$. Usually, it is sufficient to consider only the first two harmonics for practical engines.

Force Balancing

We note from Eq. (11b) that the horizontal[10] force (Y_f) may be eliminated entirely (even if $\ddot{\theta} \neq 0$) merely by making $m_{\text{rot}} = 0$, i.e., by enforcing the condition

$$rm_{\text{rot}} = m_1 c + m_A r = 0 \tag{13}$$

This is easily achieved by clamping a temporary mass $m_A = m_2 b/L$ around the crankpin axis and then *statically balancing* the entire crank assembly (see Sec. **13.11**). The counterweights needed to make c negative in Eq. (13) can be placed on extensions of the crank cheeks, as shown in Fig. 2, or they can be placed on flywheels attached to the crankshaft.

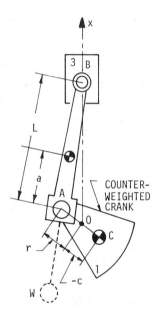

Figure 13.21-2. Illustrating counterbalancing of crank for elimination of Y_f, and theoretical counterbalancing of connecting rod by weight W.

From Eq. (11a) we see that the vertical force (X_f) will vanish (at all speeds) only if Eq. (13) is enforced and in addition $m_{\text{rec}} = 0$, i.e., if

$$Lm_{\text{rec}} = m_3 L + m_2 a = 0 \tag{14}$$

In principle, we can satisfy Eq. (14) by placing a sufficiently large counterweight W on an extension (shown dashed in Fig. 2) of the crankshaft. Such an awkward space-consuming arrangement has apparently never been used in practice. *Therefore,*

---

[10]For convenience, we assume that the x axis is vertical; this purely semantic convention may be modified as required.

the vertical inertia force in a single-cylinder engine (with conventional connecting rod) cannot be passively balanced.

However, the *first-order* vertical shaking force can be balanced passively. To see this, we use Eqs. (11a) and (11b) to express the first-order force components in the form

$$X_1 \equiv (m_{\text{rot}} + m_{\text{rec}})r\omega^2 \cos\theta$$

$$Y_1 \equiv m_{\text{rot}}r\omega^2 \sin\theta$$

where we assume, for the moment, that $\dot{\theta} = \omega$ is constant.

By adding to the crank of Fig. 2 a countermass

$$m' = m_{\text{rot}} + km_{\text{rec}}$$

at a distance r from O, diametrically opposite point A, the first-order forces are changed to

$$X_1 = (m_{\text{rot}} - m' + m_{\text{rec}})r\omega^2 \cos\theta = (1 - k)m_{\text{rec}}r\omega^2 \cos\theta$$

$$Y_1 = (m_{\text{rot}} - m')r\omega^2 \sin\theta = -km_{\text{rec}}r\omega^2 \sin\theta$$

By setting k equal to 1 or 0, we can eliminate X_1 or Y_1, respectively, but there is no value of k which results in the simultaneous elimination of both X_1 and Y_1.

If we set $k = \frac{1}{2}$, the peak shaking force components become

$$|X_1|_{\text{max}} = |Y_1|_{\text{max}} = \tfrac{1}{2}m_{\text{rec}}r\omega^2$$

For any value of k other than $\frac{1}{2}$, either $|X_1|_{\text{max}}$ or $|Y_1|_{\text{max}}$ will exceed $\frac{1}{2}m_{\text{rec}}r\omega^2$. In this sense, we may consider the countermass

$$m' = m_{\text{rot}} + \tfrac{1}{2}m_{\text{rec}}$$

(at radius r) to be the *best compromise* for simultaneous reduction of both $|X_1|$ and $|Y_1|$.

A similar discussion applies to the *nonsteady* first-order shaking forces (due to the existence of a nonzero $\ddot{\theta}$), but these forces are usually of less importance than the second-order *steady* forces [terms in Eqs. (11) multiplied by $A_2\dot{\theta}^2$], which cannot be *passively* balanced in engines with conventional conrod design.

However, at constant speed, it is possible to use *active* balancing methods to eliminate any of the harmonics which occur in Eq. (11a). For example, we can introduce *pairs* of counterweights, rotating in opposite directions in the plane of the engine, such that the horizontal forces they induce cancel out, while their vertical forces add together. To cancel the vertical harmonic $m_{\text{rec}}r\dot{\theta}^2 A_n \cos n\theta$, two counterweights must rotate at speeds $n\dot{\theta}$ and $-n\dot{\theta}$, respectively; this can be achieved by a system of gears or pulleys.[11] Alternatively, it is possible to arrange a system of dummy pistons[11] which produce inertia forces that completely cancel the inertia forces of a given engine.

---

[11]For various arrangements of geared counterweights or dummy pistons, see Biezeno and Grammel [1954, p. 38], Judge [1947, Chap. 9], and Holowenko [1955, Chap. 19].

These *active* methods of balancing are effective but expensive, and they require proper lubrication of the associated gears, pistons, and other surfaces subject to wear.

Torque Balancing

The right-hand side of Eq. (11c) defines the *inertia torque*, which is seen to vanish only if the following two conditions are simultaneously satisfied:

$$J_{AB} = J_G - m_2 a(L - a) = 0 \tag{15a}$$

$$m_A r^2 + m_1 c^2 + J_C = 0 \tag{15b}$$

These conditions are *mutually incompatible* since Eq. (15a) requires that $L - a > 0$, and Eq. (15b) requires that $m_A \equiv m_2 (L - a)/L < 0$. Thus, the *inertia torque cannot be passively eliminated for nonsteady operation.* However, the terms which are multiplied by $\ddot{\theta}$ in Eq. (11c) are *relatively* insignificant in a modern high-speed engine running with $\dot{\theta}$ nearly constant. Thus, only the terms multiplied by $\dot{\theta}^2$ in Eq. (11c) need to be considered, and we may conclude that an engine is *torque balanced* (at constant speed) when $J_{AB} = 0$.

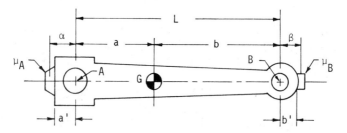

Figure 13.21-3. Connecting rod with tip masses μ_A and μ_B added to produce $J_{AB} = 0$.

It is possible to enforce the condition $J_{AB} = 0$ by simply adding tabs of mass μ_A and μ_B at the two ends of the rod, as shown in Fig. 3.

If the distance between the centroid of the mass μ_A (μ_B) and point A (B) is α (β), the required tab masses must satisfy the relationship (Kožešnik [1962, p. 125])

$$\alpha(L + \alpha)\mu_A + \beta(L + \beta)\mu_B = -J_{AB} \tag{16}$$

where J_{AB} represents the value of $J_G - mab$ prior to the addition of the tabs.

Since J_{AB} will be negative for any realistic form of connecting rod, Eq. (16) has real solutions which increase the original weight of the rod (typically) by 10–20% (see Problems **12.40**-27 and 28).

It is therefore possible to passively balance the inertia torque transmitted to the frame of a single-cylinder engine operating at constant speed.

An alternative representation of the shaking torque on the frame follows from the equilibrium of moments about point A in Fig. 1(b) when considering the conrod and piston as a single free body; i.e.,

$$N_B L \cos \phi - (P + m_{\text{rec}} \ddot{s})L \sin \phi + J_{AB}\ddot{\phi} = 0 \tag{17}$$

Now, using Eq. (3), we see that the shaking torque on the frame can be expressed in the form

$$M_f = -N_B s = -(P + m_{rec}\ddot{s})s \tan \phi + J_{AB}\ddot{\phi}\left(\frac{s}{L}\right)\sec \phi \qquad (18)$$

This representation for the shaking moment explicitly shows the influence of the gas force P. To express M_f as a Fourier series in θ it is necessary to expand the gas force turning moment[12] into a Fourier series of the form

$$M_p \equiv Ps \tan \phi = b_0 + b_1 \cos \psi + b_2 \cos 2\psi + \cdots + b_n \cos n\psi + \cdots$$
$$+ a_1 \sin \psi + a_2 \sin 2\psi + \cdots + a_n \sin n\psi + \cdots \qquad (19)$$

where

$$\psi = \theta \qquad \text{(for two-stroke cycles)}$$
$$\psi = \frac{\theta}{2} \qquad \text{(for four-stroke cycles)} \qquad (20)$$

The harmonic coefficients (a_i, b_i) have been tabulated by Porter [1943] for a variety of typical engine types. To complete the Fourier series for M_f, the terms $\ddot{s}s \tan \phi$ and $\ddot{\phi}s \sec \phi$ which appear in Eq. (18) must also be expressed as Fourier series in a manner analogous to that used in Sec. **1.41** for the expansions of $\cos \phi$, s, and $\ddot{s}$. The details will be omitted here, but interested readers will find the results given (for $\ddot{\theta} = 0$) in Crandall [1962].

13.22 *Inertia Balancing of Multicylinder Engines*

In-Line Engines

We start by considering in-line engines with N_c cylinders, such as that shown in Fig. 1, where θ_j represents the crank angle of cylinder j, and z_j is the axial distance from the origin O to the axis of cylinder j. We shall assume, for simplicity, that each cylinder has identical[13] construction and that $\dot{\theta}_j = \omega$ is constant. Then Eqs. (**13.21**-11) show that the resultant shaking forces are

$$X = \sum_j X_j \qquad (1a)$$

$$Y = \sum_j Y_j \qquad (1b)$$

where j ranges from 1 to N_c and

$$X_j = m_{rot}r\omega^2 \cos \theta_j + m_{rec}r\omega^2(\cos \theta_j + A_2 \cos 2\theta_j$$
$$- A_4 \cos 4\theta_j + A_6 \cos 6\theta_j - \cdots) \qquad (1c)$$

$$Y_j = m_{rot}r\omega^2 \sin \theta_j \qquad (1d)$$

---

[12]Recall the definition of M_p, Eqs. (**12.82**-2a) and (**12.80**-4).
[13]This is the usual situation for internal combustion engines. The discussion is readily generalized for nonuniform construction.

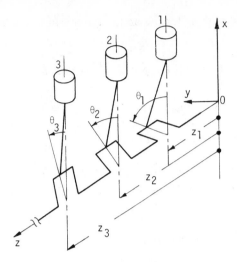

Figure 13.22-1. In-line arrangement of cylinders.

The shaking forces of the individual cylinders contribute to a **yawing moment**[14] (M_x) and a **pitching moment**[14] (M_y) about the x and y axes, given by

$$M_x = -\sum_j z_j Y_j \tag{2a}$$

$$M_y = \sum_j z_i X_j \tag{2b}$$

Finally, the **rolling moment**[14] M_z of the *inertia* forces is found, from Eq. (**13.21**-11c), by summing up the inertia torque of each cylinder; i.e.,

$$M_z = -J_{AB}\lambda\omega^2 \sum_j (C_1 \sin \theta_j - C_3 \sin 3\theta_j + C_5 \sin 5\theta_j - \cdots) \tag{3}$$

Henceforth, we shall assume that the connecting rods have been balanced, as in Fig. **13.21**-3, so that $J_{AB} = 0$, and the rolling moment (M_z) is fully balanced.

Next let us consider the contribution of the terms multiplied by m_{rot}; i.e.,

$$X = \omega^2 \sum_j rm_{rot} \cos \theta_j + \cdots \tag{4a}$$

$$Y = \omega^2 \sum_j rm_{rot} \sin \theta_j + \cdots \tag{4b}$$

$$M_x = -\omega^2 \sum_j rm_{rot} z_j \cos \theta_j + \cdots \tag{4c}$$

$$M_y = \omega^2 \sum_j rm_{rot} z_j \sin \theta_j + \cdots \tag{4b}$$

If each crank is counterbalanced, as in Fig. **13.21**-2, so that $m_{rot} = 0$ for each cylinder, all the forces and moments in Eqs. (4) vanish.

It should also be noted [see Eqs. (**13.11**-10), (**13.11**-8c) and (**13.11**-8d)] that the vanishing of each sum in Eqs. (4) is equivalent to *dynamically balancing* a rotor which consists of a set of lumped masses $m_{rot} = m_A + m_1 c/r$ located at coordinates (r, θ_j,

---

[14]These are literally correct terms if the engine is mounted in a vehicle with the z axis pointing forward and the y axis pointing athwart.

z_j). Thus, the elimination of the sums in Eqs. (4) can also be achieved by placing balancing weights in only two transverse planes (e.g., at the ends of the crankshaft). The appropriate balancing weights may be calculated by the methods of Sec. **13.11**, or they may be found by clamping temporary masses m_A to each crankpin and then balancing the crankshaft on a dynamic balancing machine.

Henceforth, we shall assume that the crankshafts have been dynamically balanced as described, so that all terms involving m_{rot} in Eqs. (1) and (2), vanish. This means that the horizontal force

$$Y = 0 \tag{5a}$$

and the yawing moment

$$M_x = 0 \tag{5b}$$

Under the stipulated conditions, the only nonvanishing shaking force is the *vertical force X*, and the only surviving shaking moment is the *pitching moment M_y*. Upon setting $m_{rot} = 0$ in Eqs. (1) and (2) we find

$$X = F_o(\sum_j \cos \theta_j + A_2 \sum_j \cos 2\theta_j - A_4 \sum_j \cos 4\theta_j + A_6 \sum \cos 6\theta_j - \cdots) \tag{6}$$

$$M_y = F_o(\sum_j z_j \cos \theta_j + A_2 \sum z_j \cos 2\theta_j - A_4 \sum z_j \cos 4\theta_j + \cdots) \tag{7}$$

where

$$F_o \equiv m_{rec} r\omega^2 = \left[m_3 + m_2 \left(\frac{a}{L} \right) \right] r\omega^2 \tag{8}$$

The terms multiplied by $\cos n\theta_j$ are said to be **forces and moments of order** n. From Eqs. (6) and (7) we see that the conditions for balance are

$$\textit{for force balance of order } n: \quad \sum_j \cos n\theta_j = 0 \tag{9a}$$

$$\textit{for moment balance of order } n: \quad \sum_j z_j \cos n\theta_j = 0 \tag{9b}$$

where $n = 1, 2, 4, 6, \ldots$.

Suppose crank j leads crank 1 by an angle δ_j; then Eqs. (9) may also be expressed in the form

$$\sum_j \cos n(\theta_1 + \delta_j) = \cos n\theta_1 \sum_j \cos n\delta_j - \sin n\theta_1 \sum_j \sin n\delta_j = 0 \tag{10a}$$

$$\sum_j z_j \cos n(\theta_1 + \delta_j) = \cos n\theta_1 \sum_j z_j \cos n\delta_j - \sin n\theta_1 \sum_j z_j \sin n\delta_j = 0 \tag{10b}$$

where $\delta_1 \equiv 0$ and $n = 1, 2, 4, \ldots$.

These equations in turn will be satisfied if the crankshaft conforms to the following criteria:

$$\textit{for force balance of order } n: \quad \sum_j \cos n\delta_j = 0, \qquad \sum \sin n\delta_j = 0 \tag{11a}$$

$$\textit{for moment balance of order } n: \quad \sum z_j \cos n\delta_j = 0, \qquad \sum z_j \sin n\delta_j = 0 \tag{11b}$$

To find the magnitude of the unbalanced force and moment of order n, let

$$C_n = \sum_j \cos n\delta_j, \qquad S_n = \sum_j \sin n\delta_j \qquad (12)$$

$$B_n = (C_n^2 + S_n^2)^{1/2}, \qquad \beta_n \equiv \arctan_2(S_n, C_n) \qquad (13)$$

Then, from Eq. (10a), we see that

$$\sum_j \cos n(\theta_1 + \delta_j) = C_n \cos n\theta_1 - S_n \sin n\theta_1$$

$$= B_n \cos(n\theta_1 + \beta_n) \qquad (14)$$

The last term in Eq. (14) follows from the geometry of Fig. 2, and it may be interpreted as the projection on the x axis of a vector of length B_n which makes an angle $n\theta_1 + \beta_n$ with the x axis; since $n\theta_1 = n\omega t$, this vector is rotating at n times the crankshaft speed.

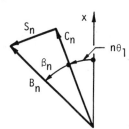

Figure 13.22-2. Rotating vector interpretation of nth order harmonic.

Upon substituting Eq. (14) into Eq. (6), we may express the shaking force in the form

$$X = X_1 \cos(\theta_1 + \beta_1) + X_2 \cos(2\theta_1 + \beta_2) - X_4 \cos(4\theta_1 + \beta_4) + \cdots \qquad (15)$$

where the **magnitudes of the nth-order shaking forces** are defined by

$$\begin{aligned}
X_1 &= F_o B_1 = m_{\text{rec}} r\omega^2 B_1 \\
X_2 &= F_o B_2 A_2 = m_{\text{rec}} r\omega^2 B_2 (\lambda + \tfrac{1}{4}\lambda^3 + \tfrac{15}{128}\lambda^5 + \cdots) \\
X_4 &= F_o B_4 A_4 = m_{\text{rec}} r\omega^2 B_4 (\tfrac{1}{4}\lambda^3 + \tfrac{3}{16}\lambda^5 + \cdots) \\
X_6 &= F_o B_6 A_6 = m_{\text{rec}} r\omega^2 B_6 (\tfrac{9}{128}\lambda^5 + \cdots)
\end{aligned} \qquad (16)$$

To express the shaking moment in an analogous fashion, we introduce the notation

$$C_n' = \sum_j z_j \cos n\delta_j, \qquad S_n' = \sum_j z_j \sin n\delta_j \qquad (17a)$$

$$B_n' = [(C_n')^2 + (S_n')^2]^{1/2}, \qquad \beta_n' = \arctan_2(S_n', C_n') \qquad (17b)$$

whereupon we can express Eq. (7) in the form

$$M_y = M_1 \cos(\theta_1 + \beta_1') + M_2 \cos(2\theta_1 + \beta_2') + M_4 \cos(4\theta_1 + \beta_4') + \cdots \qquad (18)$$

where the **magnitudes of the nth-order shaking moments** are defined as in Eq. (16), with B_n replaced by B'_n; i.e.,

$$M_n = F_o B'_n A_n \equiv \frac{X_n B'_n}{B_n} \tag{19}$$

where $A_1 = 1$.

Thus Eqs. (15) and (18) can be expressed in the alternative forms

$$\left(\frac{X}{F_o}\right) \doteq B_1 \cos(\theta_1 + \beta_1) + \left(\frac{r}{L}\right) B_2 \cos(2\theta_1 + \beta_2)$$
$$- \frac{1}{4}\left(\frac{r}{L}\right)^3 B_4 \cos(4\theta_1 + \beta_4) + \cdots \tag{20a}$$

$$\left(\frac{M_y}{F_o}\right) \doteq B'_1 \cos(\theta_1 + \beta'_1) + \left(\frac{r}{L}\right) B'_2 \cos(2\theta_1 + \beta'_2)$$
$$- \frac{1}{4}\left(\frac{r}{L}\right)^3 B'_4 \cos(4\theta_1 + \beta'_4) + \cdots \tag{20b}$$

Before applying these results to specific engines, let us consider briefly some restrictions on the forms of crankshaft which are likely to be encountered. First, we shall assume that firing occurs at regular intervals. This requires that the cranks are spaced at a uniform angle α, where, for N_c cylinders,

$$\alpha = \frac{360°}{N_c} \qquad \text{for two-stroke cycles} \tag{21a}$$

$$\alpha = \frac{720°}{N_c} \qquad \text{for four-stroke cycles} \tag{21b}$$

The end views of various crankshafts which conform to Eqs. (21) are shown in Fig. 3.

The number of **crank positions** shown in Fig. 3 for any given engine is designated by N_{cp}. For four-stroke engines with an even number of cylinders, N_{cp} is half the

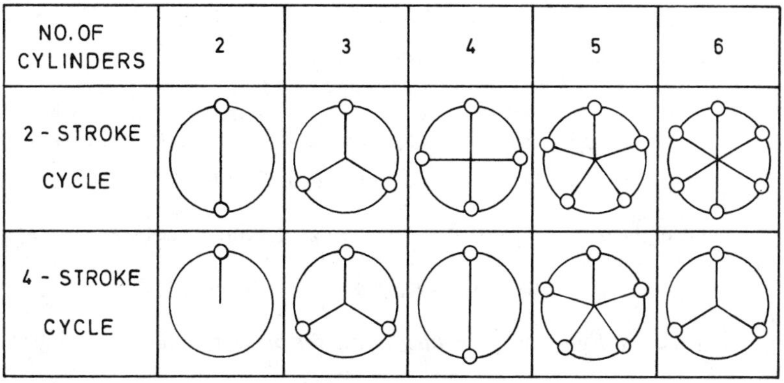

NO. OF CYLINDERS	2	3	4	5	6
2 - STROKE CYCLE					
4 - STROKE CYCLE					

Figure 13.22-3. End views of crankshafts for engines with uniform firing intervals.

number of cylinders; for all other engines N_{cp} is equal to the number of cylinders. The smallest (nonzero) angular distance between any two cranks is designated as α_p, and by definition of N_{cp} is (see Fig. 3)

$$\alpha_p = \frac{360°}{N_{cp}} \tag{21c}$$

We shall call α_p the **crank interval**.

In all subsequent discussions, we shall locate the origin O of the z axis midway between the two end cylinders, and we shall assume that the axial distance between any two adjacent cylinders is a constant d.

EXAMPLE 1. Two-Cylinder Engines

a. *Two-Stroke Cycles*. From Fig. 3 we see that the crankshaft must have two opposing cranks, as shown in Fig. 4(a). Hence, $\delta_1 = 0$, $\delta_2 = \pi$, $z_1 = -(d/2)$, and $z_2 = (d/2)$.

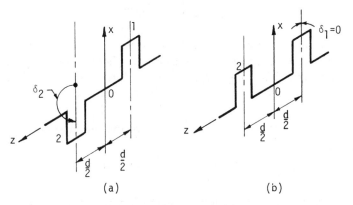

Figure 13.22-4. Crankshafts for 2 cylinder engines: (a) Two-stroke engine; (b) Four-stroke engine.

The first-order harmonics are given by Eqs. (12) and (13) as

$$C_1 = \cos 0 + \cos \pi = 0, \qquad S_1 = \sin 0 + \sin \pi = 0, \qquad B_1 = 0$$

Hence the *first-order shaking force vanishes*.

For $n = 2, 4, 6, \ldots$, Eqs. (12) and (13) give

$$C_n = \cos 0 + \cos n\pi = 2, \qquad S_n = \sin 0 + \sin n\pi = 0$$
$$B_n = 2, \qquad \beta_n = \arctan_2 (0, 2) = 0$$

Hence, the magnitudes of the higher-order shaking forces are given by Eqs. (16) as

$$X_n = m_{rec} r\omega^2 2 A_n \qquad (n = 2, 4, 6, \ldots)$$

That is, *all higher-order shaking forces are present*, and Eq. (20a) gives

$$X \doteq 2m_{rec} r\omega^2 \left[\lambda \cos 2\theta_1 - \left(\frac{\lambda^3}{4}\right) \cos 4\theta_1 + \left(\frac{9\lambda^5}{128}\right) \cos 6\theta_1 - \cdots \right]$$

Similarly, we find from Eqs. (17) that (for $n = 1$)

$$C'_1 = -\frac{d}{2}\cos 0 + \frac{d}{2}\cos \pi = -d, \quad S'_1 = -\frac{d}{2}\sin 0 + \frac{d}{2}\sin 0 = 0$$

$$B'_1 = d, \quad \beta'_1 = \arctan_2 (0, -d) = \pi$$

Thus, *the magnitude of the primary shaking moment is given* by Eq. (19) as

$$M_1 = m_{rec} r \omega^2 d$$

For $n = 2, 4, 6, \ldots$, Eqs. (17) give

$$C'_n = -\frac{d}{2}\cos 0 + \frac{d}{2}\cos n\pi = 0, \quad S'_n = -\frac{d}{2}\sin 0 + \frac{d}{2}\sin n\pi = 0$$

Hence $M_n = 0$; i.e., *all higher-order shaking moments are balanced*, and the complete pitching moment is given by Eq. (20b) in the form

$$M_y = m_{rec} r \omega^2 d \cos (\theta_1 + \pi)$$

b. *Four-Stroke Engine.* From Fig. 3, we see that the crankshaft must have its two cranks aligned as shown in Fig. 4(b). Therefore, the shaking forces are precisely those of a single-cylinder engine (Sec. **13.21**) with all masses (e.g., m_{rec}) doubled. By symmetry the *shaking moments about point O are fully balanced for all orders.*

Example 1 illustrates how Eqs. (16) and (19) may be used to find the shaking forces and moments of any order for any in-line engine. These equations (and Fig. 2) suggest the use of vector diagrams or **star diagrams**, as shown in Fig. 5 for a three-cylinder engine.

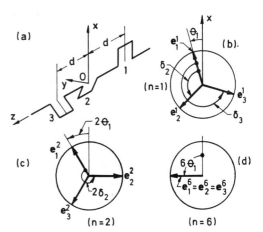

Figure 13.22-5. (a) Three-throw crank-shaft ($\delta_1 = 0, \delta_2 = 120°, \delta_3 = 240°$; $z_1 = -d, z_2 = 0, z_3 = d$). (b, c, d) Star diagrams for 1st, 2nd and 6th order forces.

In Fig. 5(b), crank 1 is represented by a vector e^1_1 which makes an angle θ_1 with the x axis. Similarly, cranks 2, 3, ... are represented by vectors $e^1_2, e^1_3, \ldots$, which lead e^1_1 by the fixed angles $\delta_2, \delta_3, \ldots$. The length of vector e^1_j should be proportional to $F_j \equiv m_{rec} r \omega^2$ for crank j. In the usual case (where $F_j = F_0$ is the same for all cranks), all vectors e_j^1 have the same length. The resultant vector

$$\mathbf{e}^1 = \mathbf{e}_1^1 + \mathbf{e}_2^1 + \mathbf{e}_3^1 + \cdots \tag{22}$$

has a projection on the x axis given by

$$(\mathbf{e}^1)_x = \cos \theta_1 + \cos (\theta_1 + \delta_2) + \cos (\theta_1 + \delta_2) + \cdots \tag{23}$$

By comparison with Eq. (6) we see that

$$F_0(\mathbf{e}^1)_x = F_0 \sum_j \cos \theta_j \tag{24}$$

is the *first-order* vertical shaking force on the frame.

Thus, we see that the necessary and sufficient condition for vanishing of the first-order vertical shaking force is that the vector $\mathbf{e}^1$ should vanish. By resolving the component vectors $\mathbf{e}_j^1$ along and perpendicular to $\mathbf{e}_1^1$, we see that the vanishing of $\mathbf{e}^1$ is equivalent to the two scalar equations

$$\sum_j \cos \delta_j = 0, \qquad \sum_j \sin \delta_j = 0 \tag{25}$$

which agree with Eq. (11a), as they should, for the case of $n = 1$.

To analyze the second-order shaking force, we draw a star diagram, as in Fig. 5(c), where $\mathbf{e}_1^2$ is located at angle $2\theta_1$ from the x axis, and $\mathbf{e}_j^2$ leads $\mathbf{e}_1^2$ by the angle $2\delta_j$. The resultant vector

$$F_o A_2 \mathbf{e}^2 \equiv F_o A_2 \sum \mathbf{e}_j^2 \tag{26}$$

has a projection on the x axis of magnitude

$$F_o A_2 (\mathbf{e}^2)_x \equiv F_o A_2 \sum \cos 2\theta_j \tag{27}$$

which clearly represents the second-order vertical shaking force in Eq. (6).

Thus the condition for balance of second-order vertical forces is

$$\mathbf{e}^2 \equiv \mathbf{e}_1^2 + \mathbf{e}_2^2 + \mathbf{e}_3^2 + \cdots \tag{28}$$

or, equivalently,

$$\sum_j \cos 2\delta_j = 0, \qquad \sum_j \sin 2\delta_j = 0$$

In general, the nth-order vertical shaking force is given by the x component of

$$F_o A_n \mathbf{e}^n \equiv F_o A_n \sum_j \mathbf{e}_j^n \qquad (n = 1, 2, 4, 6, \ldots) \tag{29}$$

(where $A_1 = 1$), and *balance of the nth-order forces is assured if the resultant $(\mathbf{e}^n)$ of the nth-order star diagram vanishes.*

For the crankshaft with three evenly distributed cranks, shown in Fig. 5(a), it is apparent from symmetry that $\mathbf{e}^1 = 0$ and $\mathbf{e}^2 = 0$. The reader may verify that, for this engine, $\mathbf{e}^3 = \mathbf{e}^4 = \mathbf{e}^5 = 0$. That is, the forces, up to the fifth order, are balanced. However, since $6\delta_2 = 720° = 4\pi$ rad and $6\delta_3 = 1440° = 8\pi$ rad, the three vectors (e_1^6, e_2^6, e_3^6) are identical and add together to produce an unbalanced shaking force of magnitude

$$3F_0A_6 = 3m_{rec}r\omega^2\,\frac{9\lambda^5}{128} \qquad (30)$$

Because λ is usually smaller than $\frac{1}{3}$, this unbalanced shaking force is usually small.[15]

This example illustrates the following general rule for in-line engines with equal crank spacing α_p: *The resultant shaking force vanishes for all orders except for those that are even multiples of N_{cp} (the number of crank positions), in which case the force vectors for each cylinder coincide.*

To prove this rule, we number the cranks so that e_1, e_2, e_3, ... follow in sequence around the crankshaft. We now note from any star diagram (e.g., any in Fig. 5) that e_2^n will coincide with e_1^n when

$$n\alpha_p = K(360°) \qquad (31a)$$

where α_p is the crank interval and K is any integer. When e_1^n coincides with e_2^n, then e_3^n will coincide with e_2^n, and so forth, because of the uniform crank spacing assumed. From Eq. (21c) we recall that $\alpha_p = 360/N_{cp}$; hence Eq. (31a) can be expressed in the form

$$n = KN_{cp} \qquad (31b)$$

where N_{cp} is the number of crank positions.

According to Eq. (1c), n must be 1 or an *even* number. Therefore Eq. (31b) predicts the results shown in the last column of Table 1, as K ranges over those integer values which make n even. The values of N_{cp} shown in Table 1 follow directly from Fig. 3.

TABLE 13.22-1

No. of Cylinders	Strokes Per Cycle	N_{cp}	Orders (n) Which Add
2	4	1	1, 2, 4, 6, ...
2	2	2	2, 4, 6, 8, ...
3	2 or 4	3	6, 12, 18, 24, ...
4	4	2	2, 4, 6, 8, ...
4	2	4	4, 8, 12, 16, ...
5	2 or 4	5	10, 20, 30, 40, ...
6	4	3	6, 12, 18, 24, ...
6	2	6	6, 12, 18, 24, ...

If we ignore all harmonics above the second, it is apparent from the table that, with *one exception*, all in-line engines with three or more cylinders and uniform firing interval are well balanced with respect to inertia *forces*. The exception is the four-cylinder,

---

[15]For example, when $\lambda = r/L = \frac{1}{3}$, $27\lambda^5/128 = 0.000868$. Then, if $m_{rec} = \frac{5}{386}$ lb-sec^2 in.$^{-1}$, $r = 4$ in. and $\omega = 1000$ rpm $= 104.7$ rad/sec, the shaking force predicted by Eq. (30) is only 0.493 lb.

four-stroke engine, which experiences a second-order force of magnitude

$$X_2 = m_{rec}r\omega^2 B_2 A_2 \doteq m_{rec}r\omega^2 4\lambda$$

For the analysis of shaking *moments* of various orders, we may use the graphical approach just described if we multiply each of the vectors e_j^n, shown in Fig. 5, by the appropriate moment arm (z_j). Then, according to Eq. (7), the shaking moment of the nth order is the projection on the x axis of the vector

$$F_o A_n \mathbf{m}^n \equiv F_o A_n (z_1 e_1^n + z_2 e_2^n + z_3 e_3^n + \cdots) \qquad (32)$$

This vector relationship may be represented graphically, or the analytical procedure embodied in Eqs. (17)–(19) may be used. For example, in the case of Fig. 5, these equations give

$$C_1' = \sum z_j \cos \delta_j = -d \cos 0 + 0 + d \cos 240° = -1.5\,d$$
$$S_1' = \sum z_j \sin \delta_j = -d \sin 0 + 0 + d \sin 240° = -\tfrac{1}{2}\sqrt{3}\,d$$
$$B_1' = [(C_1')^2 + (S_1')^2]^{1/2} = \sqrt{3}\,d; \qquad M_1 = \sqrt{3}\,dF_o$$
$$\beta_1' = \arctan_2 (S_1', C_1') = \arctan_2 (-\tfrac{1}{2}\sqrt{3}, -1.5) = 210°$$

Hence, the first-order shaking moment is

$$M_y = M_1 \cos (\theta_1 + \beta_1') = \sqrt{3}\,dM_{rec}r\omega^2 \cos (\theta_1 + 210°)$$

Similarly, it may be shown that the second-order shaking moment is

$$M_y = M_2 \cos (2\theta_1 + \beta_2') \doteq \sqrt{3}\,d\left(\frac{r}{L}\right)m_{rec}r\omega^2 \cos (2\theta_1 + 150°)$$

Note on Crankshaft Symmetry

From Fig. 3, we infer that four-stroke engines with 4, 6, 8, ... cylinders always have two cranks passing through top dead center simultaneously (to maintain a uniform firing interval). The crankshafts for these engines may be designed with mirror symmetry[16] about the x, y plane, as in Fig. 6(a). For such symmetric shafts, the pitching moment produced by an unbalanced force at any crank is balanced by that produced at its mirror image crank. Therefore such crankshafts will result in complete elimination of pitching moments, and they are commonly used for in-line four-stroke engines with four, six, or eight cylinders. From Fig. 3, we see that such crankshafts would not be used for engines with an odd number of cylinders if a *uniform* firing interval is desired.

Extensive tables of crankshaft arrangements for in-line, opposed, V, and radial engines, together with the associated balance characteristics, are given by Taylor [1966, 1968, Chap. 8].

---

[16]That is, the end cranks have the same position angle; those second from the ends have the same angle; etc.

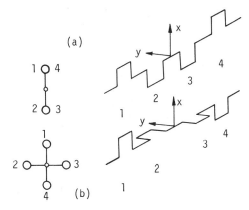

Figure 13.22-6. Four throw crankshafts:
(a) Symmetric; (b) Antisymmetric.

EXAMPLE 2

To illustrate the systematic analysis of an in-line engine, we shall find the shaking force X and pitching moment M_y for the four-cylinder in-line engine whose crankshaft is shown in Fig. 7. The required quantities are found from Eqs. (15)–(20), and the calculations are most conveniently arranged in a table similar to Table 2. We shall carry out the calculations up to the fourth harmonic ($n = 4$), but it should be clear from the table how to extend the calculations for any number of harmonics desired.

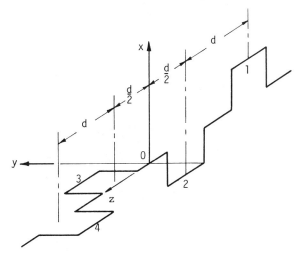

Figure 13.22-7. Crankshaft for example 2. $(\delta_1, \delta_2, \delta_3, \delta_4) =$ $(0, 180°, 90°, 270°)$.

The next to last column consists of the sums of the respective rows. The last column contains the amplitudes and phase angles

$$B_n = (C_n^2 + S_n^2)^{1/2}$$

$$\beta_n = \arctan_2 (S_n, C_n)$$

$$B_n' = [(C_n')^2 + (S_n')^2]^{1/2}$$

$$\beta_n' = \arctan_2 (S_n', C_n')$$

TABLE 13.22-2

Cyl. No.	4	3	2	1	Row Sum $\begin{bmatrix} C_n \\ S_n \\ C'_n \\ S'_n \end{bmatrix}$	$\begin{bmatrix} B_n \\ \beta_n \\ B'_n \\ \beta'_n \end{bmatrix}$
z	1.5d	0.5d	−0.5d	−1.5d		
δ	270	90	180	0		
$\cos \delta$	0	0	−1	1	$0 = C_1$	$0 = B_1$
$n = 1$ $\quad \sin \delta$	−1	1	0	0	$0 = S_1$	
$z \cos \delta$	0	0	0.5d	−1.5d	$-d = C'_1$	$\sqrt{2}\,d = B'_1$
$z \sin \delta$	−1.5d	0.5d	0	0	$-d = S'_1$	$225° = \beta'_1$
2δ	540(0)	180	360(0)	0		
$\cos 2\delta$	−1	−1	1	1	$0 = C_2$	$0 = B_2$
$n = 2$ $\quad \sin 2\delta$	0	0	0	0	$0 = S_2$	
$z \cos 2\delta$	−1.5d	−0.5d	−0.5d	−1.5d	$-4d = C'_2$	$4d = B'_2$
$z \sin 2\delta$	0	0	0	0	$0 = S'_2$	$180° = \beta'_2$
4δ	1080(0)	360(0)	720(0)	0		
$\cos 4\delta$	1	1	1	1	$4 = C_4$	$4 = B_4$
$n = 4$ $\quad \sin 4\delta$	0	0	0	0	$0 = S_4$	$0 = \beta_4$
$z \cos 4\delta$	1.5d	0.5d	−0.5d	−1.5d	$0 = C'_4$	$0 = B'_4$
$z \sin 4\delta$	0	0	0	0	$0 = S'_4$	

associated with the nth-order force and moment. The phase angles need not be entered in the table where B_n or B'_n vanish. Thus, for this example, Eqs. (20) give (up to fourth-order terms)

$$\frac{X}{m_{rec} r \omega^2} = 0 + 0 - \frac{1}{4}\left(\frac{r}{L}\right)^3 B_4 \cos (4\theta_1 + \beta_4) = -\left(\frac{r}{L}\right)^3 \cos 4\theta_1$$

$$\frac{M_y}{M_{rec} r \omega^2} = B'_1 \cos (\theta_1 + \beta'_1) + \frac{r}{L} B'_2 \cos (2\theta_1 + \beta'_2) + 0$$

$$= \sqrt{2}\, d \cos (\theta_1 + 225°) + 4d \frac{r}{L} \cos (2\theta_1 + 180°)$$

V Engines

If the inertia forces and moments of any order are balanced in each individual bank of a V engine, they will be balanced for the engine as a whole. However, in many cases, a V engine may be extremely well balanced even if there is a residual unbalance in each of its banks considered as an individual in-line engine. To analyze a V engine, we start by considering a single *row* consisting of a left and right cylinder acting on a common crank, as in Fig. 8(a).

We shall be interested in the components of the inertia forces (X, Y) along the axes (x, y), where the x axis is the bisector of the V angle, β. The axes (x_L, y_L) are directed along and transverse to the left cylinder axis, and the corresponding inertia forces for the left cylinder are denoted by X_L, Y_L. For the right cylinders a similar notation is used with R replacing L. From Fig. 8(b), we see that

$$X = (X_L + X_R) \cos \frac{\beta}{2} - (Y_L - Y_R) \sin \frac{\beta}{2} \qquad (33a)$$

$$Y = (X_L - X_R) \sin \frac{\beta}{2} + (Y_L + Y_R) \cos \frac{\beta}{2} \qquad (33b)$$

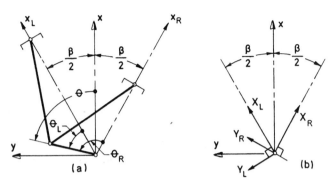

Figure 13.22-8. One row of a V engine: (a) Kinematic skeleton; (b) Forces (X_L, Y_L) due to left cylinder, and forces (X_R, Y_R) due to right cylinder.

The individual cylinder forces are given by Eqs. (1) with subscript j replaced by R or L. Hence Eqs. (33) become

$$X = \cos \frac{\beta}{2} \{(m'_{rot} + m_{rec})r\omega^2(\cos \theta_L + \cos \theta_R)$$

$$+ m_{rec}r\omega^2[A_2(\cos 2\theta_L + \cos 2\theta_R) - A_4(\cos 4\theta_2 + \cos 4\theta_R) + \cdots]\}$$

$$- \sin \frac{\beta}{2} m'_{rot}r\omega^2(\sin \theta_L - \sin \theta_R) \qquad (34a)$$

$$Y = \sin \frac{\beta}{2} \{(m'_{rot} + m_{rec})r\omega^2(\cos \theta_L - \cos \theta_R)$$

$$+ m_{rec}r\omega^2[A_2(\cos \theta_L - \cos \theta_R) - A_4(\cos 4\theta_L - \cos 4\theta_R) + \cdots]\}$$

$$+ \cos \frac{\beta}{2} m'_{rot}r\omega^2(\sin \theta_L + \sin \theta_R) \qquad (34b)$$

In these equations, we define m_{rec} as usual:

$$m_{rec} = m_2 \frac{a}{L} + m_3 \qquad (35a)$$

But only half the crank mass m_1 must be assigned to each cylinder, so that the *rotating mass*, for each cylinder, is

$$m'_{rot} = \frac{1}{2} m_1 \frac{c}{r} + m_2 \frac{b}{L} \qquad (35b)$$

rather than m_{rot} as defined in Eq. (**13.21-7**).

We now express the individual crank angles in the symmetric form

$$\theta_L = \theta - \frac{\beta}{2}$$

$$\theta_R = \theta + \frac{\beta}{2} \tag{36}$$

where θ is the crank angle measured from the x axes shown in Fig. 7. Accordingly,

$$\cos n\theta_L \pm \cos n\theta_R = \cos \frac{n\beta}{2}(\cos n\theta \pm \cos n\theta) + \sin \frac{n\beta}{2}(\sin n\theta \mp \sin n\theta)$$

$$\sin \theta_L \pm \sin \theta_R = \cos \frac{\beta}{2}\sin \theta \pm \sin \theta) - \sin \frac{\beta}{2}(\cos \theta \mp \cos \theta) \tag{37}$$

and Eqs. (34) assume (after algebraic simplification) the form

$$X = 2\left(m'_{rot} + \cos^2 \frac{\beta}{2} m_{rec}\right) r\omega^2 \cos \theta$$

$$+ 2m_{rec}r\omega^2 \cos \frac{\beta}{2}(A_2 \cos \beta \cos 2\theta - A_4 \cos 2\beta \cos 4\theta + \cdots) \tag{38a}$$

$$Y = 2\left(m'_{rot} + \sin^2 \frac{\beta}{2} m_{rec}\right) r\omega^2 \sin \theta$$

$$+ 2m_{rec}r\omega^2 \sin \frac{\beta}{2}(A_2 \sin \beta \sin 2\theta - A_4 \sin 2\beta \sin 4\theta + \cdots) \tag{38b}$$

For the *primary* forces to vanish, it is necessary that

$$m'_{rot} + \cos^2 \frac{\beta}{2} m_{rec} = 0 \tag{39a}$$

$$m'_{rot} + \sin^2 \frac{\beta}{2} m_{rec} = 0 \tag{39b}$$

or, equivalently [by addition and subtraction of Eqs. (39)],

$$2m'_{rot} + m_{rec} = m_1 \frac{c}{r} + 2m_2 \frac{b}{L} + m_2 \frac{a}{L} + m_3 = 0 \tag{40a}$$

$$\cos^2 \frac{\beta}{2} - \sin^2 \frac{\beta}{2} = 0 \tag{40b}$$

Equation (40a) can be satisfied by placing a counterweight opposite the crank, and Eq. (40b) can be satisfied by letting $\beta = 90°$.

When $\beta = 90°$, the second-order component of the (vertical) force X vanishes, but the second-order component of the (horizontal) Y force is present and given by

$$Y_2 \doteq m_{rec}r\omega^2\sqrt{2}\,\frac{r}{L}\sin 2\theta \equiv F_2 \sin 2\theta \tag{41}$$

where

$$F_2 = m_{rec} r \omega^2 \sqrt{2} \, \frac{r}{L} \tag{42}$$

In short, the *two-cylinder V engine may be passively balanced for first-order forces, but the second-order horizontal force remains unbalanced.*

One may proceed in a similar fashion to find the forces (X_j, Y_j) at all axial stations z_j in any *multirow* V engine, where the crank angle θ_j for row j is given by

$$\theta_1 = \theta, \qquad \theta_2 = \theta + \delta_2, \qquad \theta_3 = \theta + \delta_3, \qquad \ldots \tag{43}$$

where δ_j is the angle by which crank j leads crank 1.

If we concentrate on 90 V engines and assume that the first-order forces are eliminated by counterbalancing, their moments will also be absent. However, the unbalanced secondary horizontal force at station j produces a (yawing) moment about the x axis of amount

$$-z_j F_2 \sin 2\theta_j$$

where use has been made of Eq. (42). Hence, the total yawing moment for all cylinders is given by

$$M_x = -F_2 \sum_j z_j \sin 2\theta_j = -F_2 \sum_j z_j \sin 2(\theta + \delta_j) \tag{44}$$

where j ranges over the number of rows (i.e., the number of cranks). Note that

$$\sum z_j \sin (2\theta + \delta_j) = \sin 2\theta \sum_j z_j \cos 2\delta_j + \cos 2\theta \sum_j z_j \sin 2\delta_j$$
$$= B_2' \sin (2\theta + \beta_2') \tag{45}$$

where B_2' and β_2' are defined as in Eqs. (17a) and (17b), with j ranging over the number of cranks. Thus the second-order yawing moment in a multirow V engine with properly counterbalanced crankshaft is

$$M_x^2 = -m_{rec} r \omega^2 \sqrt{2} \, \frac{r}{L} B_2' \sin (2\theta + \beta_2') \tag{46}$$

As an example, consider the *antisymmetric*[17] crankshaft shown in Fig. 6(b), where

$$(\delta_1, \delta_2, \delta_3, \delta_4) = (0, 90°, 270°, 180°)$$
$$(z_1, z_2, z_3, z_4) = (\tfrac{3}{2}d, \tfrac{1}{2}d, -\tfrac{1}{2}d, -\tfrac{3}{2}d)$$

Then

$$C_2' = \sum_j z_j \cos 2\delta_j = \tfrac{3}{2}d - \tfrac{1}{2}d + \tfrac{1}{2}d - \tfrac{3}{2}d = 0$$
$$S_2' = \sum_j z_j \sin 2\delta_j = 0 + 0 + 0 + 0 = 0$$
$$B_2' = [(C_2')^2 + (S_2')^2]^{1/2} = 0$$

---

[17]In an **antisymmetric** crankshaft, the last crank is opposite the first crank, the second from last is opposite the second crank, etc.

Hence $M_x^2 = 0$, and we conclude that *the V8 engine with antisymmetric crankshaft is fully balanced for forces and moments of first and second order.*

In fact all V engines with antisymmetric crankshaft may be so balanced.

Finally, we note that the analysis for radial engines is quite similar to that given for V engines. For a complete discussion of the topic of inertia balancing for a wide variety of reciprocating engine types, see Biezeno and Grammel [1954].

13.30 BALANCING OF PLANAR LINKAGES

Although the theory of balancing is well developed for slider-crank mechanisms, methods for balancing more general linkages are not so thoroughly understood. Much of the pertinent research in this area has been surveyed by Lowen and Berkof [1968] and more recently by Berkof et al. [1977].[18]

In general, it will be found that it is often possible to *passively* balance the inertia *forces* in a plane linkage, but an unbalanced shaking *couple* in the plane of the mechanism will persist. For example, we shall see in Sec. **13.31** that four-bar linkages fall into this category. It has been shown, by Tepper and Lowen [1972, 1973], that the shaking forces in any planar mechanism without axisymmetric link groupings can be passively balanced if, and only if, there is a path from the ground to each link by way of revolutes only. Furthermore, they have shown that such a mechanism, with N links, can be force-balanced by no more than $N/2$ counterweights. A procedure for force balancing a general planar *linkage* has been given by Walker and Oldham [1978].

Although force balancing alone is acceptable in many situations, it should be recognized that the addition of the necessary balancing weights will tend to increase the shaking moment, bearing forces, and the required input torque. Such increases were found to be of the order of 50% in a study by Lowen et al. [1974][19] of 39 families of four-bar linkages with in-line couplers and side links (i.e., where the C.G. of a link lies on the link's bar axis).

To avoid an undue increase in bearing reactions, it may be desirable to reduce the size of the counterweights, thereby achieving only **partial force balance**. Tepper and Lowen [1975] have worked out a technique of partial balancing which ensures that the individual ground reactions at each of the two fixed pivots never exceed specified values, which are lower than those associated with full force balancing. Looking at the problem of residual shaking moment from a different point of view, Wiederrich and Roth [1976] have shown how to balance out a substantial part of the residual angular momentum fluctuation (remember that the shaking moment equals the rate of change of angular momentum).

Because of the many design variables and conflicting objectives involved in linkage balancing problems, such problems can be profitably studied from the point of view of *optimization theory*. Promising studies along such lines have been made by Sadler and Mayne [1973] and their associates Conte et al. [1975]. Of course, it is always possible to introduce *active balancing* procedures which depend on the use of geared rotating

---

[18]The author is indebted to Dr. Berkof for providing a prepublication copy of this reference.
[19]This reference contains charts which enable a user to thoroughly analyze the state of balance of four-bar linkages having a certain "standard configuration."

counterweights or dummy links. Publications on such techniques are cited in the survey papers of Berkof, Lowen, and Tepper, referred to above.

In Sec. **13.31**, we shall show how it is always possible to force-balance four-bar linkages, and in Sec. **13.32** we shall indicate how to find the joint reactions for balanced, or unbalanced, four-bar mechanisms.

13.31 *Inertia Balancing of Four-Bar Linkages*

Active Balancing

Figure 1 shows the kinematic skeleton of a four-bar linkage $ABCD$. It is our objective to find ways of *passively* balancing this linkage. However, let us first examine the possibility of *actively* balancing the mechanism by adjoining to it a dummy mechanism $(AB'C'D)$ which is its mirror image about the "horizontal" frame axis AD. To maintain the mirror symmetry shown, it is necessary that the two angles (θ, θ') be kept equal at all times by gearing, belting, or similar means. Since the "vertical" inertia force on any mass element is canceled out by that on its symmetric counterpart, the two linkages together produce no net vertical shaking force and no shaking moment about any point on line AD. The resultant of all elemental inertia forces is therefore statically equipollent to a "horizontal" shaking force acting along the line AD. This residual horizontal force can in turn be annulled by adjoining the two dummy linkages, shown dashed, which are formed by reflecting the solid-line mechanism about a vertical axis.

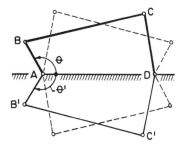

Figure 13.31-1. Active balancing of a four-bar linkage.

This method of *active* balancing is complicated, costly, and mainly of theoretical interest. We therefore turn now to strictly *passive* methods of balancing.

First, we note that the side links can always be individually counterbalanced such that their individual mass centers coincide with their respective permanent pivots. Since the frame experiences no forces or couples from such counterbalanced side links, we need only consider the inertia forces produced by the coupler link.

In-line Couplers

In the special case (Fig. 2) where the mass center G of the coupler is on the coupler axis, we can, as suggested by Crossley [1954, p. 373], replace link BC by a dynamically equivalent representation, e.g., as in Fig. **12.31-2**.

This replacement link has concentrated masses fixed to points B and C of magnitude

$$m_B = m_2 \frac{a_2 - g_2}{a_2}, \qquad m_C = \frac{m_2 g_2}{a_2} \tag{1}$$

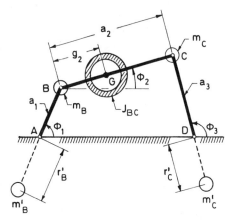

Figure 13.31-2. Inline coupler link replaced by equivalent link.

where m_2 is the mass of the coupler link. The replacement link also carries a massless ring with finite MOI J_{BC}. The inertia couple $J_{BC}\ddot{\phi}_2$ has no influence on the shaking *force*, but it will influence the shaking *moment*. Fortunately, J_{BC} can be eliminated by balancing tabs of the type shown in Fig. **13.21**-3. Since the two replacement masses (m_B and m_C) are fixed to the side links, their associated inertia *forces* may be eliminated by attaching countermasses (m'_B, m'_C) to the side links at radii (r'_B, r'_C) opposite the fixed pivots such that

$$m'_B r'_B = m_B a_1, \qquad m'_C r'_C = m_C a_3 \qquad (2)$$

This method of counterbalancing eliminates the resultant inertia force but leaves a shaking couple proportional to J_{BC}.

General Four-Bar Linkages

We now consider the method of Berkof and Lowen [1969] for force-balancing a four-bar linkage such as that shown in Fig. 3.

The lengths of the moving links are denoted, as usual, by a_i and the frame length by f. To conform with the nomenclature of Berkof and Lowen, we use the symbols ϕ_i

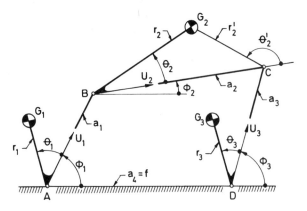

Figure 13.31-3. General four-bar linkage.

(defined as shown) to denote the time-varying angles of the various bars. The mass centers (G_i) of the links are located by time-invariant polar coordinates (r_i, θ_i), and the masses of the links are denoted by m_i. Note that G_2 can also be located by the alternative polar coordinates (r_2', θ_2').

It proves convenient to work with the complex vectors

$$Z_k = \overrightarrow{AG_k} = \text{position vector of mass center } G_k \ (k = 1, 2, 3)$$
$$U_k = e^{i\phi_k} = \text{unit vector along axis of moving link } k$$
$$\zeta_k = r_k e^{i\theta_k} = \text{position of } G_k \text{ relative to link } k$$
$$\zeta_2' = r_2' e^{i\theta_2'} = \text{alternative representation for defining } G_2$$

Since the net inertia force is

$$F = -(m_1 \ddot{Z}_1 + m_2 \ddot{Z}_2 + m_3 \ddot{Z}_3) = -\frac{d^2}{dt^2}(m_1 Z_1 + m_2 Z_2 + m_3 Z_3)$$

the shaking force will vanish if

$$mZ_G \equiv m_1 Z_1 + m_2 Z_2 + m_3 Z_3 = \text{constant} \tag{3}$$

where

$$m \equiv m_1 + m_2 + m_3 \tag{4}$$

That is, the *mass center G (of all moving links) must remain fixed in space.*

From Fig. 3, we observe that

$$Z_1 = r_1 e^{i(\theta_1 + \phi_1)} = r_1 e^{i\theta_1} e^{i\phi_1} = \zeta_1 U_1 \tag{5a}$$
$$Z_2 = a_1 U_1 + \zeta_2 U_2 \tag{5b}$$
$$Z_3 = f + \zeta_3 U_3 \tag{5c}$$

Therefore, Eq. (3) can be expressed as

$$mZ_G = (m_1 \zeta_1 + m_2 a_1)U_1 + m_2 \zeta_2 U_2 + m_3 \zeta_3 U_3 + m_3 f = \text{constant} \tag{6}$$

However, the unit vectors are related by the loop closure condition:

$$a_1 U_1 + a_2 U_2 = f + a_3 U_3 \tag{7}$$

Upon solving Eq. (7) for U_2 and substituting into Eq. (6), we find

$$mZ_G = C_1 U_1 + C_2 U_3 + C_0 = \text{constant} \tag{8}$$

where

$$a_2 C_1 \equiv a_2 m_1 \zeta_1 + a_2 a_1 m_2 - a_1 m_2 \zeta_2 = a_2 m_1 \zeta_1 - a_1 m_2 (\zeta_2 - a_2) \tag{9a}$$
$$a_2 C_2 \equiv a_3 m_2 \zeta_2 + a_2 m_3 \zeta_3 \tag{9b}$$
$$a_2 C_0 \equiv f(a_2 m_3 + m_2 \zeta_2) \tag{9c}$$

are complex constants (time-invariant).

Since U_1 and U_3 are time-varying quantities, Z_G can be a constant only if the coefficients C_1 and C_2 vanish. If we note, from Fig. 3, that

$$\zeta_2' = \zeta_2 - a_2 \tag{10}$$

the requirements for *force balance* are given by Eqs. (9a) and (9b) as[20]

$$a_2 C_1 = a_2 m_1 \zeta_1 - a_1 m_2 \zeta_2' = 0 \tag{11a}$$

$$a_2 C_2 = a_3 m_2 \zeta_2 + a_2 m_3 \zeta_3 = 0 \tag{12a}$$

In real terms, these two vector relations express the four conditions

$$a_2 m_1 r_1 = a_1 m_2 r_2' \tag{11b}$$

$$\theta_1 = \arg \zeta_1 = \arg \zeta_2' = \theta_2' \tag{11c}$$

$$a_3 m_2 r_2 = a_2 m_3 r_3 \tag{12b}$$

$$\theta_2 = \arg \zeta_2 = \arg \zeta_3 \pm \pi = \theta_3 \pm \pi \tag{12c}$$

From these equations, we see that if the mass distribution of one link is given, the *mass moments* $(m_i r_i)$ and the angular locations of the C.G. (θ_i) for the other links are determined. Therefore, these equations can be used to design a force-balanced linkage.

More frequently, the equations will be used to determine the size and location of counterweights which must be added to a *given* unbalanced mechanism in order to eliminate the shaking force. For this purpose, it is convenient to introduce the notation

$$m_k^\circ, r_k^\circ, \theta_k^\circ, \zeta_k^\circ: \quad \text{parameters of the unbalanced linkage}$$

$$m_k^*, r_k^*, \theta_k^*, \zeta_k^*: \quad \text{parameters associated with counterweights alone}$$

It then follows that the parameters for the counterweighted linkage are expressed in the form

$$m_k r_k e^{i\theta_k} = m_k^\circ r_k^\circ e^{i\theta_k^\circ} + m_k^* r_k^* e^{i\theta_k^*} \tag{13a}$$

or

$$m_k \zeta_k = m_k^\circ \zeta_k^\circ + m_k^* \zeta_k^* \tag{13b}$$

Of course, the total mass of link k is

$$m_k = m_k^\circ + m_k^* \tag{14}$$

To balance a given mechanism by means of counterweights on the side links (1 and 3) only, we let $(\ )_2 = (\ )_2^\circ$ for any parameter with subscript 2; then we use Eqs. (11b) and (11c) to find

$$m_1 r_1 = \frac{a_1}{a_2} m_2^\circ r_2'^\circ \tag{15a}$$

$$\theta_1 = \theta_2'^\circ \tag{15b}$$

---

[20]For an ingenious interpretation of Eqs. (11a) and (12a), due to Freudenstein [1973], refer to Problem **13.40-47**.

and we use Eqs. (12b) and (12c) to find

$$m_3 r_3 = \frac{a_3}{a_2} m_2^\circ r_2^\circ \tag{16a}$$

$$\theta_3 = \theta_2^\circ \mp \pi \tag{16b}$$

These equations uniquely define the complex vectors

$$m_k \zeta_k \equiv m_k r_k e^{i\theta_k} \qquad (k = 1, 3) \tag{17}$$

Finally, we may use Eqs. (13) to find the desired balance vectors (for $k = 1, 3$):

$$m_k^* r_k^* e^{i\theta_k^*} \equiv m_k^* \zeta_k^* = m_k \zeta_k - m_k^\circ \zeta_k^\circ \tag{18}$$

The *magnitudes* of the unbalance vectors are the mass–moments ($m_k^* r_k^*$) needed on each side link, and the *arguments* of the unbalance vectors are the location angles θ_k^* (relative to vectors U_1 and U_3) for the balancing masses m_k^*.

It is left as an exercise to show that magnitude and argument of the required balancing vectors are given explicitly (for $k = 1, 3$) by

$$m_k^* r_k^* = [(m_k r_k)^2 + (m_k^\circ r_k^\circ)^2 - 2(m_k r_k)(m_k^\circ r_k^\circ)\cos(\theta_k - \theta_k^\circ)]^{1/2} \tag{19a}$$

$$\theta_k^* = \arctan_2\left[(m_k r_k \sin\theta_k - m^\circ r^\circ \sin\theta_k^\circ), (m_k r_k \cos\theta_k - m_k^\circ r_k^\circ \cos\theta_k)\right] \tag{19b}$$

The following example, from Berkof and Lowen [1969], illustrates the numerical calculations involved. In the reference cited, it is also shown how a six-link mechanism can be balanced by a similar procedure.

EXAMPLE 1.

The following data are given for the four-bar mechanism of Fig. 3:

k	1	2	3
a_k (in.)	2	6	3
r_k° (in.)	1	3.190	1.500
θ_k° (deg)	0	16°	0
m_k (lb[21])	0.1020	0.2740	0.1200

We are also given the frame length $f = 5.5$ in. and the alternative parameters for the coupler link: $r_2'^\circ = 3.063$ in., $\theta_2'^\circ = 163.3°$.

From Eqs. (15) and (16) we find

$$m_1 r_1 = \frac{a_1}{a_2} m_2^\circ r_2'^\circ = \frac{2}{6}(0.274)(3.063) = 0.2797 \text{ lb-in.}$$

$$\theta_1 = \theta_2'^\circ = 163.3°$$

$$m_3 r_3 = \frac{a_3}{a_2} m_2^\circ r_2^\circ = \frac{3}{6}(0.274)(3.19) = 0.4370 \text{ lb-in.}$$

$$\theta_3 = \theta_2^\circ + \pi = 16° + 180° = 196°$$

---

[21]Note: We may interpret the symbol m as weight (in this example) since the gravitational acceleration g would cancel out of all calculations involving the mass $m = W/g$, where W is the weight.

Therefore,

$$m_1\zeta_1 \equiv m_1r_1e^{i\theta_1} = 0.2797(\cos 163.3° + i\sin 163.3°) = -0.2679 + i\,0.08037$$
$$m_3\zeta_3 \equiv m_3r_3e^{i\theta_1} = 0.4370(\cos 196° + i\sin 196°) = -0.4201 - i\,0.1204$$

From the given data we have

$$m_1^°\zeta_1^° \equiv m_1^°r_1^°e^{i\theta_1°} = (0.102)(1)(\cos 0 + i\sin 0) = 0.102$$
$$m_3^°\zeta_3^° \equiv m_3^°r_3^°e^{i\theta_3°} = (0.120)(1.5)(\cos 0 + i\sin 0) = 0.180$$

Hence, Eq. (18) gives the desired *balance vectors* as

$$m_1^*r_1^*e^{i\theta_1^*} = m_1\zeta_1 - m_1^°\zeta_1^° = -0.3699 + i\,0.08037$$
$$m_3^*r_3^*e^{i\theta_3^*} = m_3\zeta_3 - m_3^°\zeta_3^° = -0.6001 - i\,0.1204$$

The magnitudes of these vectors are the desired mass–moments:

$$m_1^*r_1^* = [(0.3699)^2 + (0.08037)^2]^{1/2} = 0.3785 \text{ lb-in.}$$
$$m_3^*r_3^* = [(0.6001)^2 + (0.1204)^2]^{1/2} = 0.6120 \text{ lb-in.}$$

The arguments of the balance vectors are the location angles of the balance weights (relative to the unit vectors U_1 and U_2):

$$\theta_1^* = \arctan_2 (0.08037, -0.3699) = 167.7°$$
$$\theta_3^* = \arctan_2 (-0.1204, -0.6001) = 191.3°$$

These last four results could also have been found from Eqs. (19).

Note that the actual masses (m_1^*, m_2^*) to be used will depend on the designer's choice for the radii (r_1^*, r_3^*), since only the mass moments $m_k^* r_k^*$ are determined by the calculations.

13.32 *Joint Reactions Due to Inertia Forces*

We have just seen how the *resultant* shaking force on the frame of a four-bar linkage may be eliminated by passive balancing. Nevertheless, the forces transmitted to each of the fixed pivots do not vanish by virtue of the balancing procedure used, nor do the reactions on the moving hinges. The calculation of these reactions is a problem of *kinetostatics*. As explained in Sec. **12.70**, the problem is reduced to a statics problem if the applied forces and moments (X_i, Y_i, M_i) are replaced by the *effective static force sets*

$$\{X_i^*, Y_i^*, M_i^*\} = \{(X_i - m_i\ddot{x}_i), (Y_i - m_i\ddot{y}_i), (M_i - J_i\ddot{\theta}_i)\} \tag{1}$$

Note that here $\ddot{\theta}_i$ represents the angular acceleration of the links. The acceleration terms needed in Eq. (1) are given by Eqs. (**10.51**-31) and (**10.51**-42).

If no real forces (X_i, Y_i, M_i) are acting, the mechanism will be *coasting*,[22] and the

---

[22]If a mechanism is brought up to speed and then all power input is discontinued, the motion will continue with a cyclic exchange of kinetic and potential energy ($T + V = E$), in the absence of friction. This kind of motion we shall call **coasting**. Ultimately, of course, friction would dissipate the motion.

effective static forces of Eq. (1) are simply the inertia forces ($-m\ddot{x}_i$, etc.). Frequently, however, we shall be interested in the condition where a **driving torque** T_A acts on the crank (link 1) and maintains a specified state of motion (e.g., $\dot{\theta}_1 = \text{constant} = \Omega$). In such a case, the required driving torque may be found from the principle of virtual work, expressed in the form

$$T_A \dot{\theta}_1 + \sum_{i=1}^{3} (X_i^* \dot{x}_i + Y_i^* \dot{y}_i - J_i \ddot{\theta}_i \dot{\theta}_i) = 0 \tag{2}$$

Upon making use of the velocity coefficients

$$\{u_i, v_i, \omega_i\} \equiv \{\dot{x}_i, \dot{y}_i, \dot{\theta}_i\}/\dot{\theta}_1$$

defined by Eqs. (**10.50**-2)–(**10.50**-4), we may solve Eq. (2) to give

$$T_A = -\sum_{i=1}^{3} (X_i^* u_i + Y_i^* v_i - J_i \ddot{\theta}_i \omega_i)$$

$$= \sum_{i=1}^{3} (m_i \ddot{x}_i u_i + m_i \ddot{y}_i v_i + J_i \ddot{\theta}_i \omega_i) \tag{3}$$

where $\ddot{x}_i$, $\ddot{y}_i$, $\ddot{\theta}_i$ may be found from Eqs. (**10.51**-31), (**10.51**-42), and the associated equations of Sec. **10.51**.

With the driving torque T_A calculated from Eq. (3), we can evaluate the effective static couples

$$M_1^* = T_A - J_1 \ddot{\theta}_1, \qquad M_2^* = -J_2 \ddot{\theta}_2, \qquad M_3^* = -J_3 \ddot{\theta}_3 \tag{4}$$

The following procedure may now be followed to find the joint reactions (X_A, Y_A, X_B, . . .) which are defined in Fig. **11.52**-2(b):

1. Find (X_i^*, Y_i^*, M_i^*) as defined by Eq. (1).
2. Find reactions (X_A, Y_A, . . .) from Eqs. (**11.52**-14)–(**11.52**-20).

After finding the reactions at all four hinges, we may calculate the components of the shaking force on the frame [see Fig. **11.52**-2(b)]:

$$X_f = X_A - X_D \tag{5}$$
$$Y_f = Y_A - Y_D \tag{6}$$

The shaking moment (on the frame) with respect to point A is

$$M_f = -Y_D a_4 \tag{7}$$

where a_4 is the frame length AD.

We have tacitly assumed that there is no *torque return* to the frame from the driving device. If, however, the driving motor is rigidly fixed to the frame link, it may be seen, from a free-body diagram of the frame, that the shaking moment is given by

$$M_f = -Y_D a_4 - T_A \tag{8}$$

where T_A is the driving torque given by Eq. (3).

13.40 PROBLEMS

Note: In all problems, a right-handed orthogonal Cartesian coordinate system (ξ, η, ζ) is embedded in the rotor with ζ along the spin axis z. Point masses $(m_i = W_i/g)$ are located by cylindrical polar coordinates (r_i, β_i, z_i) as in Fig. **13.11**-1.

1. If Fig. 1 represents a thin uniform rectangular flat plate of mass m rotating about the diagonal CD with constant angular speed ω, show that the reactions at the bearings C and D are given by

$$R_\eta^D = -R_\eta^C = \frac{m\omega^2 hd(h^2 - d^2)}{12(h^2 + d^2)^{3/2}}$$

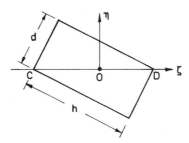

Figure 13.40.1. (Probs. 1, 2).

2. If Fig. 1 represents a uniform right cylinder of diameter $d = 2R$ and mass m rotating about diagonal CD with constant angular speed ω, show that the reactions at the bearings C and D are given by

$$R_\eta^D = -R_\eta^C = \frac{m\omega^2 Rh(h^2 - 3R^2)}{6(h^2 + 4R^2)^{3/2}}$$

3. If, in Problem 1, the angular acceleration is α and the angular velocity is ω, find the components of the reactions at C and D in the directions of ξ and η.

4. The axis of a weightless shaft passes through the center of a uniform square plate whose plane makes an angle of 70° with the shaft axis. The plate weighs 20 lb and has a side length of 10 in. Bearings support the shaft at 15 in. from either side of the plate center.
    a. Find the bearing reactions when the shaft rotates at a constant angular speed of 1200 rpm.
    b. Where, along the edges of the plate, would you locate balancing masses? How much mass would you use at each such location to provide dynamic balancing?

5. A shaft passes through the centroid of a uniform steel circular cylinder of diameter 16 in. and length 10 in. The shaft and cylinder axes intersect at an angle $\psi = 15°$. If the rotation speed is $\dot{\phi} = 100$ rpm and $\ddot{\phi} = 50$ rad/sec², find
    a. The shaking moments (M_ξ, M_η) and the driving moment $(-M_\zeta)$.
    b. The magnitudes and location angles of the reactions at bearings C and D, which are located 8 in. to the left, and 12 in. to the right of the cylinder's center.

6. Weights of 5, 10, 15, and 20 lb are located on a circle along the rim of a 48-in.-diameter wheel at location angles $\beta = 0°$, 90°, 180°, and 270°, respectively. Where on the rim should a single weight be placed, and how much weight should be used to provide dynamic balance?

7. Thin sheet metal discs used as rotors in automobile speedometers have been balanced by the following procedure (Senger [1961]): Let O be the disc center, and G be the center of gravity (C.G.). The direction OG is determined statically (e.g., by suspending the disc from a horizontal axis through O). Then a hole H_1 of weight H is punched out at a radius R located 90° clockwise from radius OG. Then the direction OG' of G', the new C.G., is measured (statically) and found to be at the angle α, counterclockwise from OG. Finally, two equal holes H_2 and H_3 of weight H are punched out, at equal radii R, making equal angles β on either side of the known radius OG'. Show that the angle β required for static balance (and hence dynamic balance) of the disc is given by $\beta = \cos^{-1}\left(\frac{1}{2}\csc\alpha\right)$.

8. Bolts weighing 4 oz each are added to a thin uniform disc at angles $\beta_1 = 0°$, $\beta_2 = 45°$ along a circle of radius 5 in. Holes, drilled at a radius of 8 in., deduct weights of 2 oz each at locations $\beta_3 = 110°$ and $\beta_4 = 190°$. What weight should be drilled out at a radius $e' = 10$ in. in order to statically balance the disc? What is the location angle for this *balancing hole*?

9. Suppose that the *unbalanced* disc of Problem 8 is mounted on a shaft through the center O of the disc, which is also the origin of the ξ, η, ζ coordinate axes. The plane of the disc makes an angle of $\psi = 40°$ with the η axis, and the disc spins about the ζ axis with constant angular speed $\dot{\phi} = 1000$ rpm. The location angles β_i are all measured in the plane of the disc from the ξ axis.
   a. Find the unbalanced shaking forces (F_ξ, F_η) and the shaking moments M_ξ, M_η about point O.
   b. Find the reaction forces $(R_{\xi C}, R_{\eta C})$ and $(R_{\xi D}, R_{\eta D})$ at knife edge bearings located at points C and D on the shaft with coordinates $\zeta_C = -6$ in. and $\zeta_D = 6$ in.

10. Assume that the disc of Problem 9 has been *statically* balanced, leaving a residual unbalanced couple of $M_\xi = 10$ in.-oz.
    a. What is the minimum number of holes that you can drill to produce dynamic balance, and where would you suggest locating these holes?
    b. What weight would you drill out at each hole to achieve dynamic balance?

11. A rotor is characterized by the following distribution of lumped weights:

Plane	W (lb)	r (in.)	β (deg)	z (in.)
1	25	3	30	0
2	30	2	80	8
3	35	1	150	11
A	W'_A	3	β'_A	4
B	W'_B	2	β'_B	16

   a. Find $W'_A, W'_B, \beta'_A, \beta'_B$ required for dynamic balance.
   b. For the unbalanced rotor, running at a speed of 1000 rpm, find the magnitude and location angle of the reactions on bearings located axially at points C and D which have axial coordinates $z_C = -4$ in. and $z_D = 20$ in. Assume constant speed.
   c. Do part b, but assume an angular speed of 600 rpm and an angular acceleration of $\ddot{\phi} = -2000$ rad/sec² at a given instant.

12. A rotor has the following distribution of lumped weight:

Plane	W (lb)	r (in.)	β (deg)	z (in.)
1	5	2	0	0
2	5	2	270	5
3	5	2	180	10
4	5	2	90	15
A	W'_A	2	β'_A	2
B	W'_B	2	β'_B	12

a. Find the balancing weights (W'_A, W'_B) and their angular location (β'_A, β'_B) required for dynamic balance.

b. Find the magnitudes and location angles of reactions at bearings C, D located at axial coordinates $z_C = -2$, $z_D = 18$ for the original unbalanced rotor when the angular speed is $\dot{\phi} = 200$ rpm and the angular acceleration is $\ddot{\phi} = 400$ rad/sec².

13. A rotor has three unbalanced weights with the following characteristics:

i	W_i (oz.)	r_i (in.)	β_i (deg)	z_i (in.)
1	40	6	0	48
2	50	6	230	108
3	60	6	115	156

a. Find the balancing weights (W'_A, W'_B) and location angles (β'_A, β'_B) if the weights are attached at a radius of 5 in. in balancing planes (A, B) having axial coordinates $z_A = 0$ and $z_B = 189$ in.

b. Find the reactions (magnitudes and location angles) at bearings C and D located at $z_C = 12$ in. and $z_D = 14$ in. when the unbalanced rotor has a uniform speed of $\dot{\phi} = 200$ rpm.

c. Repeat part b if the rotor has an acceleration of $\ddot{\phi} = 500$ rad/sec² and instantaneous speed $\dot{\phi} = 200$ rpm.

14. The mass center G of a rotor weighing 450 lb is known to be in the plane $z = 0$. Balancing weights, in planes A and B, bring the rotor into *static balance*, where $W_A = 20$ lb, $r_A = 15$ in., $\beta_A = 0°$, $z_A = -20$ in., $W_B = 24$ lb, $r_B = 18$ in., $\beta_B = 90°$, and $z_B = 16$ in.

a. Find the *initial* eccentricity (r_G) of the rotor and its location angle (β_G) relative to weight W_A.

b. For a constant rotor speed of 50 rpm, find the magnitudes and location angles (relative to weight W_A) of bearing reactions at bearings C and D, located 46 in. apart (after the rotor has been statically balanced).

15. Five discs are attached to a shaft in the manner shown in the accompanying table. Note that disc 4 is in balance, by itself, and that the location (both axial and angular) of disc 5 is unspecified. The system is to be brought into dynamic balance by drilling a hole in disc 4 at a 4-in. radius and by properly adjusting the position of disc 5. Find

CHAPTER 13 BALANCING OF MACHINERY

i	W_i (lb)	r_i (in.)	β_i (deg)	z_i (in.)
1	1.8	1	0	-8
2	3.0	1	190	7
3	1.4	1	270	25
4	—	0	—	0
5	1.5	1	?	?

a. The correct location angle β_5 and axial position (z_5) of disc 5.

b. The weight (W'_4) to be drilled out of disc 4 and the location angle β'_4 of the drilled hole.

16. The rotor shown in Fig. 2 was temporarily balanced, in a balancing machine, by placing 0.15-lb trial weights at two angular locations (A_1, A_2) in plane A and at two angular locations (B_1, B_2) in plane B. Find the weights (W'_C, W'_D) to be drilled out of the rotor at a radius of 6 in. in planes C and D for dynamic balance with the trial weights removed. Also find the angular locations β'_C and β'_D corresponding to the drilled holes.

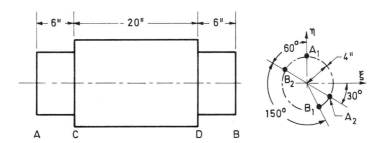

Figure 13.40-2. (Prob. 16).

17. Verify Eqs. (**13.12**-3) by solving Eqs. (**13.11**-15).

18. Verify Eqs. (**13.12**-3) by using equations of moment equilibrium in Fig. **13.11**-2 and in an analogous figure.

19. Verify by algebra that Eq. (**13.12**-5) follows from Eqs. (**13.12**-2)–(**13.12**-4).

20. $F_A \equiv r_a W_A e^{i\beta_A}$ and $F_B \equiv r_B W_B e^{i\beta_B}$ are complex vectors of unbalance in planes A and B, and $R_C \equiv r^C e^{i\beta_C}$ and $R_D \equiv r^D e^{i\beta_D}$ are the complex bearing reaction vectors at C and D (for $\omega^2 = g$, $\dot{\omega} = 0$). Verify Eqs. (**13.12**-5) directly by using two equations of moment equilibrium about suitable points in Fig. **13.12**-1.

21. The rotor of Fig. **13.12**-1 has the following axial dimensions: $CA = 6$ in., $AB = 18$ in., $BD = 4$ in. When rotating at a constant speed of 1000 rpm in a balancing machine, the reactions at bearings C and D were observed to have magnitudes of 20 and 30 lb, respectively, and location angles of $\beta_C = 20°$, $\beta_D = 135°$. What balancing weights (W'_A, W'_B) should be located in planes A and B at a radius of 6 in., and what should be the location angles (β'_A, β'_B) of these weights?

22. Conditions are as in Problem 21, but $CA = -6$ in. Sketch the rotor and solve as in Problem 21.

23. Conditions are as in Problem 21, but $BD = -4$ in. What weights (W'_A, W'_B) should be *drilled out* in planes A and B at a radius of 6 in., and what should be the location angles (β'_A, β'_B) of the drill holes?

24. A uniform circular disc (center O) is modified by drilling five holes along a 10-in.-diameter circle (center O) at angles $(0, \pm 45°, \pm 90°)$ from a fixed radius in the disc. The weight removed at each hole is 2 oz. The holes at 45° and 90° are filled with bolts weighing 5 oz each. Find the weight and its location along the 10-in.-diameter circle needed to statically balance the disc.

25. Sketch an arrangement of gears and rotating counterweights which is suitable to eliminate the first two harmonics of vertical shaking force (X_f) in a single-cylinder engine. Indicate the amount of unbalance (Wr) required on each gear. Note that your counterweights should not produce horizontal shaking forces (Y_f) on the frame.

26. Sketch an arrangement of dummy pistons operating off the crankshaft of a single-cylinder engine such that the vertical shaking force X_f of the engine is balanced out without introducing any additional horizontal shaking force Y_f. Give information on the required masses and lengths of the parts you add to the system.

27. Verify that the addition of tab masses (μ_A, μ_B) which satisfy Eq. (13.21-16) will make J'_{AB} (the value of J_{AB} for the modified connecting rod) vanish. [Hint: First show that, for the original rod, $J_{AB} = J_B - m_2 bL$, where J_B is the MOI of the rod with respect to point B; then find the modified values (m'_2, b', J'_B) of (m_2, b, J_B), and set $J'_{AB} = J'_B - m'_2 b'L = 0$].

28. The connecting rod of Fig. 13.21-3 has weight $W_2 = 3$ lb, centroidal radius of gyration $r_G = 4.4$ in., $a = 3$ in., and $b = 8$ in. If $a' = 2$ in. and $b' = 1$ in., reasonable locations for the end tabs are $\alpha = 2.5$ and $\beta = 1.25$ in. Using the above data, find the required tab masses (μ_A, μ_B) and the volume of steel (specific weight, 0.284 lb/in.³) required for each tab if (a) $\mu_B = 0$ and (b) $\mu_B = \mu_A$.

29. The following data are given for the single-cylinder engine of Fig. 13.21-1: $m_1 g = 5$ lb, $m_2 g = 3.69$ lb, $m_3 g = 3.84$ lb, $r = 2.5$ in., $L = 11$ in., $a = 3.25$ in., $c = 0$ (crankshaft statically balanced), $\dot{\theta} = 500$ rpm, $\ddot{\theta} = 0$. Find the maximum values of shaking forces (X_f, Y_f) transmitted to the engine frame.

30. The connecting rod of the engine in Problem 29 has a radius of gyration $r_G = 3.5$ in. about its centroid G, and the crank assembly has radius of gyration $r_C = 5$ in. about its centroid C. In the absence of gas force (P) and resisting torque (M_r), find the maximum value of the shaking torque (M_f) transmitted to the engine frame when $\dot{\theta} = 500$ rpm, $\ddot{\theta} = 0$.

31. The crankshaft of the engine described in Problem 29 has been statically balanced $(c = 0)$. What counterweight W' must be added at the radius $OC = 2$ in. (see Fig. 13.21-2) in order to balance out the horizontal shaking force Y_f?

32. The connecting rod of the engine of Problem 29 is suspended as a pendulum with a knife edge pivot located 7/16 in. vertically above point B (the center of the piston pin bearing). The period of pendulum vibration is observed to be 1.023 sec. Calculate the maximum shaking torque M_f (due to inertia only) when the engine runs at a constant speed of 500 rpm.

33. For a three-cylinder engine with crankshaft as in Fig. 13.22-5, show that the second-order shaking moment is $M_y = \sqrt{3}\ d(r/L) m_{rec} r\omega^2 \cos(2\theta_1 + 150°)$.

34–41. Find the vertical shaking force X_f and shaking moment M_y for an in-line engine with N_c cylinders spaced at equal distances d along the z axis. The cylinders are numbered sequentially $1, 2, \ldots, N_c$, as in Fig. 13.22-1, and δ_j is the angle (in degrees) between crank 1 and crank j. Unless otherwise specified, consider all orders up to and including the *fourth.* Express your answer in terms of $F_o = m_{rec}r\omega^2$, $\lambda = r/L$, and the crank angle θ of the first crank. Assume that the crankshafts have been balanced to eliminate horizontal shaking forces (Y), yawing moment (M_x), and rolling moment (M_z).

Problem No.	N_c	$(\delta_1, \delta_2, \ldots, \delta_{N_c})$
34	4	(0, 90°, 180°, 270°)
35	4	(0, 90°, 270°, 180°)
36	4	(0, 180°, 90°, 270°)
37	4	(0, 180°, 180°, 0°)
38	4	(0, 90°, 270°, 180°)
39	6	(0, 120°, 240°, 240°, 120°, 0°)
40	6	(0, 240°, 120°, 120°, 240°, 0°)
41	6	(0, 120°, 240°, 60°, 300°, 180°)

42. A two-cylinder *opposed* engine can be considered a V engine with V–angle $\beta = 180°$. Show that the crankshaft of such an engine can be balanced in such a way that either of the first-order shaking forces (X or Y) may be eliminated but not both of them. What can you say about the forces of higher order?

43. Verify that Eqs. (**13.31**-19a) and (**13.31**-19b) follow from Eq. (**13.31**-18).

44–46. The given data refer to Fig. **13.31**-3. In each case, find the balancing weights to be attached to the side links (1 and 3) at a radius of 2 in. in order to guarantee *force balance.* Also find the angular location (θ_1^*, θ_3^*) of the balance weights, and make a sketch showing the mechanism with balance weights attached.

Problem No.	Link (k)	a_k (in.)	r_k^o (in.)	θ_k^o (deg)	m_k^o (lb)
44	1	1	0.5	90	0.2
44	2	2.5	1.25	45	1.0
44	3	2.0	1.0	30	0.6
44	4	3	—	—	—
45	1	2	1	30	0.2
45	2	3.5	3.5	60	1.5
45	3	4	2	−30	0.5
45	4	1	—	—	—
46	1	1	0.5	90	0.2
46	2	2.5	1.25	−45	1.0
46	3	2.0	1.0	30	0.6
46	4	3.0	—	—	—

47. Using the result a of Problem **2.50**-2, express the acceleration of point G_2 in Fig. **13.31**-3 in the form of a complex vector:

$$\ddot{Z}_2 = (1 - \lambda)\ddot{Z}_B + \lambda\ddot{Z}_C \qquad \text{(i)}$$

where λ is a complex constant. Then show that

$$\ddot{Z}_B = -a_1 U_1 (\dot{\phi}_1^2 - i \ddot{\phi}_1) \tag{ii}$$

$$\ddot{Z}_C = -a_3 U_3 (\dot{\phi}_3^2 - i \ddot{\phi}_3) \tag{iii}$$

$$\ddot{Z}_k = -r_k e^{i\theta_k} U_k (\dot{\phi}_k^2 - i \ddot{\phi}_k) \qquad (k = 1, 3) \tag{iv}$$

where Z_1 and Z_3 represent vectors $\overrightarrow{OG}_1$ and $\overrightarrow{OG}_3$, from an arbitrary origin O to points G_1 and G_3 in Fig. **13.31**-3. If the condition of vanishing shaking force is expressed in the form

$$m_1 \ddot{Z}_1 + m_2 \ddot{Z}_2 + m_3 \ddot{Z}_3 = 0 \tag{v}$$

show that the substitution of Eqs. (i)–(iv) into Eq. (v) leads directly to the Berkof-Lowen balancing criteria, Eqs. (11a) and (12a) of Sec. **13.31**. Freudenstein [1973] interpreted the terms $1 - \lambda$ and λ in Eq. (i) as *complex masses*, which revolve with points B and C; he then reasoned that the inertia force $\lambda \ddot{Z}_C$ could be annulled by a compensating complex balancing mass rotating with link 3, and similarly for force $(1 - \lambda) \ddot{Z}_B$ and link 1.

48. Sketch all the forces acting on the frame of the four-bar linkage of Fig. **11.52**-2. With the aid of this sketch, show that the moment transmitted to the frame is given by Eq. (**13.32**-8) or (**13.32**-7) accordingly as the driving motor is or is not mounted on the frame of the mechanism.

49. For the four-bar linkage of Problem 44,
    a. Find all joint reactions $(X_A, Y_A, \ldots)$ defined in Fig. **11.52**-2 before the linkage is balanced.
    b. Find the shaking forces and shaking moment predicted by Eqs. (**13.32**-5)–(**13.32**-7) before any balancing is done.
    c. Do a complete force balance of the linkage (i.e., solve Problem 44), and repeat parts a and b for the balanced linkage.
    d. What is the percentage change in the maximum resultant force at each joint, in the maximum unbalanced shaking couple, and in the total weight of the linkage due to the force balancing.

50. Repeat Problem 49, but for the mechanism of Problem 45.

51. Repeat Problem 49, but for the mechanism of Problem 46.

52. Write a computer program that will produce the following output for any given four-bar mechanism with constant crank speed: (a) all joint reactions, (b) driving torque, (c) shaking forces and couples, and (d) magnitude and locations of balancing weights needed on the side links to produce complete force balancing. Items (a), (b), and (c) should be printed out at uniform user-specified intervals of crank rotation.

14

INTRODUCTION TO DYNAMICS OF MULTIFREEDOM MECHANISMS AND GENERAL-PURPOSE COMPUTER PROGRAMS

14.00   INTRODUCTION

In this final chapter, we shall generalize the previously derived results for single-freedom systems to include the most general multifreedom planar mechanical system. Multiple degrees of freedom are common in open-loop systems (e.g., multiple pendulums and human body models used in vehicle crash simulations). They occur less frequently in idealized models of linkage machinery. However, when realistic conditions (such as joint tolerances, member flexibility, cam-follower jump, etc.) are to be analyzed, multiple degrees of freedom must be included in the modeling of all machinery.

In Sec. **14.10** we shall show how the inclusion of inertia forces in the principle of virtual work leads to Lagrange's form of d'Alembert's principle. This fundamental principle permits us to establish the governing differential equations of motion for multifreedom systems in a form which is well suited for solution by numerical integration. The methods of numerical integration used for single-freedom systems in Chapter 12 are generalized in Sec. **14.20** to the multifreedom case. Although the procedures given for formulating and solving the governing differential and algebraic equations are straightforward, it would be extremely tedious and time-consuming to carry out such a solution without the aid of digital computers. Fortunately, a number of general-purpose computer programs for the dynamic analysis of multifreedom mechanisms are now available to the general public. A list is given in Sec. **14.30** of such programs known to the author at the time of writing. From the references cited, the reader should be able to decide which of these programs is best suited to his needs.

Finally, we shall present an introduction to the characteristics, and capabilities of one such general-purpose program—DYMAC. The logical structure of this program is based on the analytical and numerical procedures described in this book. In Sec. **14.40** it is shown how such mechanism characteristics as loop topology, applied forces, link geometry, and inertial properties are communicated to the program in digitized form.

14.10 LAGRANGE'S FORM OF D'ALEMBERT'S PRINCIPLE

We recall from Sec. **12.70** that *d'Alembert's principle* permits us to treat a typical member (i) of a mechanism as though it were in a state of equilibrium, provided that the active force set (X_i, Y_i, M_i) applied to the mass center of the link is replaced by the **effective static force** set

$$\{X_i^*, Y_i^*, M_i^*\} = \{(X_i - m_i\ddot{x}_i), (Y_i - m_i\ddot{y}_i), (M_i - J_i\ddot{\theta}_i)\} \tag{1}$$

where we use the notation of Chapter 12.

The rate at which these forces do work on an ideal mechanical system with M moving members is given by

$$P = \sum_{i=1}^{M} [(X_i - m_i\ddot{x})\dot{x}_i + (Y_i - m_i\ddot{y}_i)\dot{y}_i + (M_i - J_i\ddot{\theta})\dot{\theta}_i] \tag{2}$$

If the velocities ($\dot{x}_i, \dot{y}_i, \dot{\theta}_i$) are compatible with the kinematic constraints imposed on the system, they represent *virtual velocities*, and Eq. (2) gives the *virtual power* (the rate of virtual work). According to the statement of the *principle of virtual work* given in Sec. **11.30**, the system will be in a state of generalized equilibrium if, and only if, the virtual power vanishes for every set of virtual velocities. Hence Eq. (2) leads to the conclusion that

$$\sum_{i=1}^{M} [(X_i - m_i\ddot{x})\dot{x}_i + (Y_i - m_i\ddot{y}_i)\dot{y}_i + (M_i - J_i\ddot{\theta}_i)\dot{\theta}_i] = 0 \tag{3}$$

for any set of velocities ($\dot{x}_i, \dot{y}_i, \dot{\theta}_i$) that is compatible with the kinematic constraints imposed on the system.

We must view Eq. (3) as the embodiment of a *principle* rather than as a single scalar *equation*—a principle which Lagrange made the foundation of analytic mechanics. We therefore refer to the principle embodied in Eq. (3) as **Lagrange's form of d'Alembert's principle** (or as the **d'Alembert-Lagrange principle**), although it is sometimes referred to in the literature simply as *d'Alembert's principle*.[1] Other names used for Eq. (3) include *indeterminate* or *variational equation of motion*,[2] *generalized d'Alembert principle*,[3] *first form of the fundamental equation*,[4] and *general equation of dynamics*.[5]

---

[1]Goldstein [1950, p. 16].
[2]Kelvin and Tait [1879, Vol. I, p. 269].
[3]Meirovitch [1970].
[4]Pars [1968].
[5]Gantmacher [1970, p. 20].

By drawing upon kinematic results developed earlier, we may utilize the d'Alembert-Lagrange principle to quickly derive the *set* of differential equations which govern the motion of a system with F degrees of freedom.

To ensure that the velocities $(\dot{x}_i, \dot{y}_i, \dot{\theta}_i)$ are indeed compatible with the constraints, we express them in terms of the generalized coordinates $(q_1, q_2, \ldots, q_F)$, where F is the degree of freedom of the system. The desired kinematic relations are given by Eqs. (10.52-3)–(10.52-6) in the form

$$\{\dot{x}_i, \dot{y}_i, \dot{\theta}_i\} \equiv \sum_{r=1}^{F} \{U_{ir}, V_{ir}, \Omega_{ir}\}\dot{q}_r \tag{4}$$

$$\{\ddot{x}_i, \ddot{y}_i, \ddot{\theta}_i\} = \{U_i, V_i, \Omega_i\} + \sum_{j=1}^{F} \{U_{ij}, V_{ij}, \Omega_{ij}\}\ddot{q}_j \tag{5}$$

where the velocity coefficients $(U_{ij}, \ldots)$ depend only on the displacements and the acceleration terms $(U_i, \ldots)$ depend on both velocities and displacements.

If Eqs. (4) and (5) are substituted into Eq. (3), the latter becomes

$$\sum_{i=1}^{M} \sum_{r=1}^{F} \left[(X_i U_{ir} + Y_i V_{ir} + M_i \Omega_{ir}) - (m_i U_i U_{ir} + m_i V_i V_{ir} + J_i \Omega_i \Omega_{ir}) \right.$$
$$\left. - \sum_{j=1}^{F} (m_i U_{ij} U_{ir} + m_i V_{ij} V_{ir} + J_i \Omega_{ij} \Omega_{ir})\ddot{q}_j \right]\dot{q}_r = 0 \tag{6}$$

If we first sum over the index i and then over index r, Eq. (6) can be expressed in the form

$$\sum_{r=1}^{F} \left(Q_r - C_r - \sum_{j=1}^{F} I_{rj}\ddot{q}_j \right)\dot{q}_r = 0 \tag{7}$$

where

$$Q_r \equiv \sum_{i=1}^{M} (X_i U_{ir} + Y_i V_{ir} + M_i \Omega_{ir}) \tag{8}$$

$$C_r \equiv \sum_{i=1}^{M} (m_i U_i U_{ir} + m_i V_i V_{ir} + J_i \Omega_i \Omega_{ir}) \tag{9}$$

$$I_{rj} \equiv \sum_{i=1}^{M} (m_i U_{ij} U_{ir} + m_i V_{ij} V_{ir} + J_i \Omega_{ij} \Omega_{ir}) \tag{10}$$

Note that the quantities Q_r defined by Eq. (8) are precisely the *generalized forces* defined previously by Eq. (11.22-8a). By reference to Eqs. (10.52-32) and (10.52-35), it may be verified that the quantities C_r defined by Eqs. (9) can be expressed as quadratic forms in the *generalized velocities* $\dot{q}_j$; i.e., they represent inertia forces due to centripetal and Coriolis accelerations.

The coefficients I_{rj} defined by Eqs. (10) satisfy the **symmetry relations**

$$I_{rj} = I_{jr} \tag{11}$$

and form the elements of a symmetric square matrix which we shall call the **generalized inertia matrix**. We refer to the elements I_{rj} as **inertia coefficients**.

Since Eq. (7) must be valid for every possible combination of the generalized velocities $\dot{q}_r$, it must hold for the particular set

$$\dot{q}_1 = 1, \qquad \dot{q}_r = 0 \qquad (r \neq 1) \tag{12}$$

When Eqs. (12) are substituted into Eq. (7), the only surviving terms are

$$\sum_{j=1}^{F} I_{1j}\ddot{q}_j = Q_1 - C_1 \tag{13}$$

Similarly, if we let $\dot{q}_s = 1$, $\dot{q}_r = 0$ $(r \neq s)$, where s is any integer in the range $1, 2, \ldots, F$, Eq. (7) reduces to the form

$$\boxed{\sum_{j=1}^{F} I_{sj}\ddot{q}_j = Q_s - C_s} \qquad (s = 1, 2, \ldots, F) \tag{14}$$

Equations (14) represent a set of F *differential equations*, each of second order in the generalized coordinates q_j. These **generalized equations of motion** were developed in Paul and Krajcinovic [1970b] and form the basis of the author's general-purpose computer program DYMAC, described in Sec. **14.40**. These equations are closely related to the so-called *explicit form of Lagrange's Equations*, but as explained in Paul [1975], they are in a form which, for closed-loop systems, is much better suited for numerical integration.

It may easily be shown that the generalized equations of motion (14) reduce to the single differential equation (**12.40**-7) for systems with only a single degree of freedom.

For *open-loop systems* (e.g., a double pendulum) there are no explicit equations of constraint relating the F generalized coordinates q_j, and the F equations (14) form a *complete* set of differential equations, which may be numerically integrated (see Appendix **G**) to find the F Lagrangian variables $q_j(t)$ associated with any given set of initial postions, $q_j(0)$, and initial velocities, $\dot{q}_j(0)$.

However, closed-loop systems are much more prevalent in linkage machinery. For such systems, the differential equations of motion must be supplemented by the algebraic equations of constraint

$$f_i(\phi_1, \phi_2, \ldots, \phi_N, q_1, \ldots, q_F) = 0 \qquad (i = 1, 2, \ldots, N) \tag{15}$$

which represent explicit relationships between the secondary variables ϕ_i and the generalized coordinates.[6] Examples of such equations are *loop closure equations*, e.g., Eqs. (**9.22**-4), or auxiliary equations for gears and cams, e.g., Eqs. (**7.30**-19) and (**7.43**-5).

Before the equations of motion can be integrated, it is necessary to establish appropriate initial conditions for all of the Lagrangian position variables ψ_i. Starting with given values of the independent variables q_i, the secondary variables ϕ_i may be found from the constraint equations (15) using a numerical technique, e.g., the Newton-

---

[6]Recall from Sec. **8.13** that the set of Lagrangian variables ψ_i is the union of the set of independent variables q_r and the set of secondary variables ϕ_j.

Raphson method, as described in Chapter 9. With the initial values of all ψ_i known, one may proceed to find the initial values for all secondary velocities $\dot{\phi}_i$ by first differentiating Eqs. (15) to find the matrix equation

$$\mathbf{A}\dot{\phi} \equiv \left[\frac{\partial f_i}{\partial \phi_j}\right]\{\dot{\phi}_j\} = \left[\frac{-\partial f_i}{\partial q_j}\right]\{\dot{q}_j\} \equiv \mathbf{B}\dot{q} \tag{16}$$

This linear system of N equations may be solved[7] for the values of $\dot{\phi}_j$ associated with any given set of initial generalized velocities $\dot{q}_j$; i.e.,

$$\dot{\phi} = \mathbf{A}^{-1}\mathbf{B}\dot{q} \equiv \mathbf{K}\dot{q} \tag{17}$$

With all initial displacements and velocities known, the differential equations of motion may be integrated.

14.20 NUMERICAL INTEGRATION OF THE EQUATIONS OF MOTION

The initial value problem just formulated can be solved by numerical integration in a manner analogous to that used in Sec. **12.60** for single-freedom dynamic systems. We begin by introducing the variables w_i such that

$$\{q_1, q_2, \ldots, q_F\} \equiv \{w_1, w_2, \ldots, w_F\} \tag{1}$$

$$\{\dot{q}_1, \dot{q}_2, \ldots, \dot{q}_F\} \equiv \{w_{F+1}, w_{F+2}, \ldots, w_{2F}\} \tag{2}$$

Then the equations of motion (**14.10**-14) can be expressed in the form

$$\sum_{j=1}^{F} I_{sj}\dot{w}_{F+j} = Q_s - C_s \qquad (s = 1, 2, \ldots, F) \tag{3}$$

With the initial values of all ψ_i and $\dot{\psi}_i$ known, we can evaluate the terms I_{sj}, Q_s, and C_s, which appear in Eq. (3). Hence, we can solve the linear equations (3) for the initial values of $\dot{w}_{F+j}$. Moreover, we can repeat this procedure at any stage of the solution where displacements (ψ_i) and velocities ($\dot{\psi}_i$) are known. In short, we may write Eqs. (3) in the symbolic form

$$\dot{w}_{F+p} = F_p[w_1, \ldots, w_{2F}] \qquad (p = 1, \ldots, F) \tag{4}$$

to express the fact that $\dot{w}_{F+p}$ can be calculated from $w_1, \ldots, w_{2F}$. If Eq. (2) is rewritten in the form

$$\dot{w}_i = \ddot{q}_i = w_{F+i} \qquad (i = 1, \ldots, F) \tag{5}$$

we see that $\dot{w}_1, \dot{w}_2, \ldots, \dot{w}_{2F}$ may all be found, from Eqs. (4) and (5), for known values

---

[7]We assume that the system starts from a configuration in which matrix **A** is not singular. If the problem posed is physically meaningful, an appropriate choice of generalized coordinates will ensure that **A** is not singular (see Hud [1976]).

of $w_1 \ldots w_{2F}$. Hence, Eqs. (4) and (5) represent a system of differential equations of order $2F$ in the standard form described in Appendix **G**.

Therefore, a Runge-Kutta (or equivalent) routine can be called to solve for the updated values $(w_1, \ldots, w_{2F})$ at time $t + \Delta t$. The values of the secondary coordinates ϕ_i at time $t + \Delta t$ can be calculated by means of a Newton-Raphson procedure, and the updated values of the secondary velocities $\dot{\phi}_i$ can then be found directly from Eq. (**14.10**-17). Thus, all of the state variables (ψ_i and $\dot{\psi}_i$) become known at time $t + \Delta t$. We may now regard $t + \Delta t$ as a new *initial time* and repeat every step of the above procedure to find $w_1, \ldots, w_{2F}$ at $t + 2\Delta t$. The process may be repeated over the entire range of interest $(0 \leq t \leq t_{max})$.

This method of solution entails the integration of a system of $2F$ first-order equations. It is not possible to express the equations of motion for nondegenerate[8] mechanical systems by a system of differential equations of order less than $2F$. This procedure is an obvious extension of the *method of minimal differential equations* described in Sec. **12.61**.

For computations, it is efficient to work with more than the minimal number of differential equations, i.e., to use an *excess* number of differential equations, as in Sec. **12.62**. Toward this end, we supplement the differential equations of motion by the velocity constraint equations (**14.10**-17), expressed in the form

$$\dot{\phi}_i = \sum_{j=1}^{F} K_{ij} \dot{q}_j \equiv \sum_{j=1}^{F} K_{ij} w_{F+j} \qquad (i = 1, 2, \ldots, N) \tag{6}$$

If we supplement our new notation by the definitions

$$\{\phi_1, \phi_2, \ldots, \phi_N\} \equiv \{w_{2F+1}, w_{2F+2}, \ldots, w_{2F+N}\} \tag{7}$$

Eqs. (6) assume the form

$$\dot{w}_{2F+i} = \sum_{j=1}^{N} K_{ij} w_{F+j} \qquad (i = 1, 2, \ldots, N) \tag{8}$$

Equations (8) together with Eqs. (4) and (5) constitute a system of order $2F + N$. For a regular[9] linkage with L loops, $N = 2L$, and the order of the system of differential equations is therefore $2F + 2L$. Thus, for example, if $F = 1$ and $L = 2$, this method is of order 6 as opposed to the minimal order 2. Nonetheless, numerical experiments by the author indicate that, with the use of program DYMAC, this *method of excess differential equations* typically requires about 40% of the computer time required by the method of minimal equations. These experiments confirm the conjecture (cf. Sec. **10.42**) made by Paul and Krajcinovic [1970-b] regarding the relative efficiency of the two methods.

---

[8]So-called *degenerate systems* of differential equations (having order less than $2F$) may arise when some inertia forces are assumed to be negligible in comparison to the active forces. For examples, see Panovko and Gubanova [1965] or Paul and Soler [1972]. When disproportionately small inertia coefficients occur in matrix $[I_{ij}]$, the system is nondegenerate but is governed by "stiff differential equations" which can introduce numerical difficulties (Gear [1971]).

[9]Recall from Sec. **8.21** that a *regular* linkage has only lower pairs (i.e., hinges or sliders) and contains no loops along which all joints are sliding pairs.

14.30 SURVEY OF GENERAL-PURPOSE COMPUTER PROGRAMS FOR DYNAMICS OF MACHINERY

Several general-purpose computer programs for the dynamic analysis of mechanisms have been described in the literature. These programs, which automatically generate and numerically integrate the equations of motion are based on different, but related, analytical and numerical principles. The paper by Paul [1975] reviews the various principles and techniques available for formulating the equations of motion, for integrating them numerically, and for solving the associated kinetostatic problem for the determination of bearing reactions. The relative advantages of vector methods, d'Alembert's principle, Lagrange's equations with and without multipliers, Hamilton's equations, virtual work, and energy methods are discussed. Particular emphasis is placed on how well suited the various methods are to the automatic generation of the equations of motion and to the form and order of the corresponding systems of differential equations. It is shown how velocity ratios, influence coefficients, centripetal coefficients, generalized inertia coefficients, and Christoffel symbols interrelate the various methods and tie them to classical results of analytical dynamics such as the *explicit* equations of motion and the power balance principle. Methods for solving both the general dynamics problem and the kinetostatic problem are reviewed, and the particular methods of implementation used in some available general-purpose computer programs and in other recent literature are described.

The major general-purpose programs available at the time of writing—and sources of information on each—are given in the following list (also see the lists of Kaufman [1975] and Dix [1978]).

IMP (Integrated Mechanisms Program) is described by Sheth and Uicker [1972], Sheth [1972], and Uicker [1973]. A version, called **IMP-UM**, was developed by Sheth at the University of Michigan (3D).[10]

DRAM (Dynamic Response of Articulated Machinery), described in general terms by Chace and Sheth [1973], is an updated version of **DAMN**, reported upon by Chace and Smith [1971] and Smith et al. [1972] (2D).

MEDUSA (Machine Dynamics Universal System Analyzer) is described by Dix and Lehman [1972] (2D).

DYMAC (Dynamics of Machinery), developed by B. Paul, G. Hud, and A. Amin at the University of Pennsylvania, is based on the work of Paul and Krajcinovic [1970b]. Also see Paul [1977] and Paul and Hud [1976] (2D).

VECNET (Vector Network) has been described by Andrews and Kesavan [1975] as "a research tool" for mass particles and spherical joints. It has spawned a second-generation program, **PLANET**, for plane motion of rigid bodies; see Rogers and Andrews [1975, 1977] (2D).

ADAMS, developed at the University of Michigan by Orlandea et al. [1977], is described in Orlandea and Chace [1977] (3D).

The above-mentioned programs are all applicable to the closed-loop systems, with sliding and rotating pairs, that typify linkage machinery. They can also be used for special cases where sliding pairs are not present or only open loops occur (e.g., in

---

[10]"3D" indicates the capability of analyzing three-dimensional mechanisms. "2D" denotes a restriction to planar mechanisms.

modeling human or animal bodies). Special-purpose programs suitable for such cases include the following.

AAPD (Automated Assembly Program—Displacement Analysis), a program for three-dimensional mechanisms, is described by Langrana and Bartel [1975]. Only revolute and spherical pairs are allowed (3D).

UCIN, developed by Huston et al. [1977] at the University of Cincinnati, is for strictly open-loop (chain-like) systems, e.g., the human body (3D).

With the exception of DYMAC, all of the general-purpose programs mentioned above require the user to communicate with the program by means of a *problem-oriented language* that is unique to the individual program. Through such a language the user indicates that certain links are interconnected by means of certain types of kinematic pairs (e.g., revolutes or sliders) and that certain types of forces (e.g., spring or gravity) act at certain points on certain members. This procedure relieves the user of the need to write any FORTRAN code, but it does require him to learn the vocabulary and conventions of a particular problem-oriented language.

Program DYMAC, on the other hand, requires that the user provide simple FORTRAN subroutines which describe the forces and imposed motions applied to the mechanism. This approach has the advantage of permitting complete generality in the nature of the forces and constraints and obviates the need to learn a special problem-oriented language. Of course, it presupposes that the user has a rudimentary knowledge of FORTRAN. It will be seen below that the level of FORTRAN coding required to use DYMAC is quite elementary.

14.40 INTRODUCTION TO PROGRAM DYMAC

The goal of this section is to indicate how the methods of analytical mechanics may be incorporated into a general-purpose digital computer program for the dynamic analysis of an arbitrary planar mechanism. The program DYMAC (**DY**namics of **MAC**hinery) is chosen for this purpose, since its logical structure most closely parallels the analytical treatment developed in this book. A more complete description of the program and examples of its capabilities will be found in the user's manual (Paul and Hud [1976]). Updated versions of the program and its documentation are available from the author.

14.41 Program Characteristics

The program is written entirely in FORTRAN IV and has been tested on IBM and UNIVAC computers. With double precision used throughout, the program requires 165,000 bytes of storage. This amount of storage accommodates problems involving up to 20 links, 10 degrees of freedom, 10 independent loops, 50 points of interest, and 30 edges (line segments in kinematic network). The source deck contains approximately 1850 cards of which 500 are comment cards that may be deleted at will.

Program DYMAC enables users to do a complete dynamic analysis of multidegree of freedom, arbitrary planar linkages under the influence of user-specified applied forces and moments which may be arbitrary functions of displacement and velocity.

In addition, the user may specify a wide variety (virtually unlimited) of relationships between the gross motions of various links or specified points on any of the links. Such *control constraints* may be used, for example, to ensure that a given link moves with a specified angular velocity or to have points in a mechanism move along user-specified paths. The control constraints may also be used to include the effects of gears and cams. The mechanism may contain closed loops or open loops.

The user provides only the minimum information needed to describe the geometric properties and mass distribution of the system (e.g., link lengths, mass, moments of inertia, initial positions and velocities, etc.) The program internally formulates the appropriate dynamic equations of motion, automatically integrates them, and makes whatever internal checks are necessary to maintain accuracy.

The *principal results* displayed are the position, velocity, and acceleration associated with each link at user-specified time intervals as well as the forces and moments applied at the mass centers of all links. In addition, displacement, velocity, and acceleration components of user-selected *points of interest*, anywhere in the system, are displayed at user-specified time intervals. Joint reactions will be printed, if desired.

The program is so organized that the user can easily request a wide variety of postprocessing procedures, such as plotting, card punching, tape generation, or any further calculations on output data that he considers desirable.

14.42 Modeling of Mechanism

To communicate the necessary topological, geometrical, and inertial characteristics of the mechanism to the program, the required information must be suitably digitized and stored. This is accomplished by suitable program logic in conjunction with the following modeling procedure.

Define Variables

After establishing a fixed set of global (x, y) axes and identifying a complete set of Lagrangian variables $(\psi_1, \psi_2, \ldots, \psi_M)$, as illustrated in Fig. 1(a), a kinematic network should be drawn, as in Fig. 1(b).

Identify Oriented Edges

The motion of each link is defined by that of a **characteristic edge** embedded in the link. These characteristic edges are numbered sequentially from 1 to M, as illustrated in Fig. 1(b) by the numbers enclosed within small squares. The remaining edges are numbered from $M + 1$ onward, as shown within the small semicircles in Fig. 1(b). The user decides upon an (arbitrary) orientation for each edge in the network, as indicated by the arrowheads on the edges in Fig. 1(b).

Identify Independent Loops

After selecting a complete set of independent loops (see Sec. **8.22**), the loop topology is communicated to the program by means of a **complete loop table.** This table specifies which edges are contained in each loop and how they are oriented relative to the loop direction.

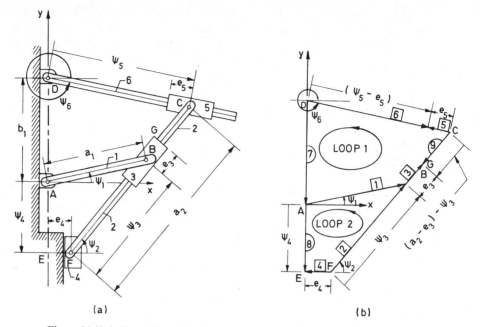

Figure 14.42-1. Example mechanism: (a) Given mechanism; (b) Kinematic network.

Ground Path Table

We next construct a ground path table, used to define the origin of the characteristic edge of any given link, relative to the ground. This table specifies a vector path from the global origin to the tail of the characteristic edge vector of each link.

Link Data

To describe each link completely, it is necessary to specify its mass m_i, moment of inertia I_i about its center of mass, and the location of its mass center. The units are arbitrary but must be consistent.

The location of the mass center is given in local polar coordinates (r_c, α_c) measured with respect to the characteristic edge vector on each link, as shown in Fig. 2 for a typical member called link 2. This information is recorded in a **link data table**.

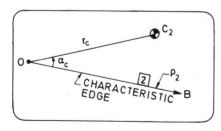

Figure 14.42-2. Polar coordinates (r_c, α_c) defining mass center C_2 on link 2.

Edge Data

Every edge i in a mechanism possesses a length ρ_i inclined at an angle α_i to the positive x axis (measured positive counterclockwise). In turn, each ρ_i and α_i is composed of a constant part and a variable part. For example, edge 9 in Fig. 1 has

$$\rho_9 = (a_2 - e_3) - \psi_3 \tag{1a}$$

$$\alpha_9 = \pi + \psi_2 \tag{1b}$$

More generally, for edge i

$$\rho_i = CL_i \pm \psi_{il} \tag{2a}$$

$$\alpha_i = CA_i \pm \psi_{ia} \tag{2b}$$

where CL_i and CA_i are the constant parts of ρ_i and α_i, il is the index of the length variable, and ia is the index of the angle variable.

The edge data are recorded as shown in Table 1 for the example of Fig. 1. The $+$ or $-$ should precede the entries for il and ia accordingly as a $+$ or $-$ appears in Eq. (2a) or (2b).

TABLE 14.42-1 *Edge Data*

Edge (i)	$\pm il$	CL_i	$\pm ia$	CA_i
1		a_1	1	
2	3		2	
3		e_3	2	
4		e_4		π
5		e_5	6	π
6	5	$-e_5$	6	
7		b_1		$3\pi/2$
8	4			$3\pi/2$
9	-3	$a_2 - e_3$	2	π

Initial Conditions

Recall that in a mechanism with F degrees of freedom, F of the variables ψ_i are designated as primary variables (or generalized coordinates) and the remainder are said to be secondary variables. The physics of the problem requires that initial displacement and velocity conditions, corresponding to *primary* ψ_i and $\dot{\psi}_i$, must be prescribed precisely. However, merely approximate estimates are required for the initial values of secondary position variables ψ_i because their values are refined by a Newton-Raphson iteration. *Secondary* initial velocities $\dot{\psi}_i$ are *not* to be specified because they are implied by the primary data and are calculated internally. The *user* identifies which ψ_i are *initially*[11] to be considered as primary, which as secondary, and which as *controlled*. The concept of a controlled variable is explained in Sec. **14.44**.

---

[11]During the course of calculation, the program will automatically decide which variables are primary in such a way as to minimize roundoff errors.

Points of Interest

If it is desired to obtain the positions, velocities, or accelerations of selected points, a *points of interest table* is formed. The locations of points of interest are described in the manner shown in Fig. 3 for a link (4) with two points of interest, P_5 and P_6. The angular location of a point is given by α_j and the radial location by r_j, where j is the identification index of the point ($j \leq 50$).

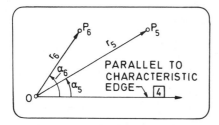

Figure 14.42-3. Typical link (No. 4) with two points of interest (Nos. 5 and 6).

14.43 *Modeling of Active Forces, Springs, and Dampers*

The active forces and moments may be functions of time, displacements, and velocities. It should be noted that the weights of links are treated as active forces. To permit the solution of a wide range of problems, the user is required to write FORTRAN expressions for all nonzero values of active forces and torques and to include them in an appropriate location in a subroutine called FORCES.

Specified resultant external force components X_i, Y_i act through the mass center of link i, together with a resultant external torque T_i *acting* about the mass center. The forces and torques may be functions of position, velocity, and time.

Any external forces or torques which do not act through the center of mass may readily be transferred into statically equipollent systems. Internal forces, such as those due to dampers and springs, can be treated as pairs of external forces acting on the members to which they are attached.

For example, Fig. 1 shows a spring and viscous damper which act in parallel between the two points D and E. If the damper is linear, it produces a compressive force F_v which is proportional to the rate of contraction of line DE; i.e., $F_v = -C_d \dot{\psi}_s$, where C_d is a given damping coefficient. Similarly, a linear spring of stiffness K_s will

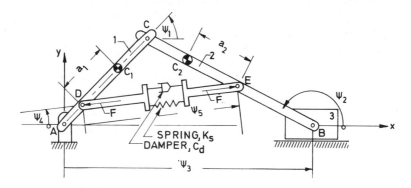

Figure 14.43-1. Illustrating method for modelling springs and dampers.

produce an additional compression $F_s = -K_s(\psi_5 - \psi_{5,0})$, where $\psi_{5,0}$ is the length of DE when the spring is unstrained. In short, the compressive force F exerted on link 1 at D and on link 2 at E is

$$F = -C_d\dot{\psi}_5 - K_s(\psi_5 - \psi_{5,0}) \tag{1}$$

The force F acting on point D of member 1 may be resolved into its Cartesian components

$$X_1 = -F\cos\psi_4 \tag{2a}$$

$$Y_1 = -F\sin\psi_4 \tag{2b}$$

The moment of these force components about the mass center (C_1) of link 1 is seen to be

$$T_1 = X_1 a_1 \sin\psi_1 - Y_1 a_1 \cos\psi_1 \tag{2c}$$

Equations (2) describe a system of forces statically equipollent to the action, on link 1, of the spring and damper.

We now focus attention on the compressive damping force acting at point E on member 2. Since this force is equal and opposite to that acting at point D on member 1, we may write

$$X_2 = -X_1 \tag{3a}$$

$$Y_2 = -Y_1 \tag{3b}$$

and the moment of these two force components about the mass center (C_2) of link 2 is seen to be

$$T_2 = X_2 a_2 \sin\psi_2 - Y_2 a_2 \cos\psi_2 \tag{3c}$$

The action of the spring and damper has now been replaced, as required, by an appropriate pair of statically equipollent active force sets.

DYMAC requires the user to provide FORTRAN expressions for all nonzero applied forces in an otherwise standard subroutine called FORCES. For example, to model the forces of Fig. 1 we would place the following statements in subroutine FORCES:

```
C*****USER-SUPPLIED EXPRESSIONS FOLLOW***
      A1=2.0D0
      A2=1.5D0
      CD=2.5D0
      SK=5.0D0
      PSI50=6.0D0
      F=-CD*PSID(5)-SK*(PSI(5)-PSI50)
      FX(1)=-F*DCOS(PSI(4))
      FY(1)=-F*DSIN(PSI(4))
      TOR(1)=FX(1)*A1*DSIN(PSI(1))-FY(1)
              -FY(1)*A1*DCOS(PSI(1))
      FX(2)=-FX(1)
      FY(2)=-FY(1)
      TOR(2)=FX(2)*A2*DSIN(PSI(2))
              -FY(2)*A1*DCOS(PSI(1))
```

The FORTRAN statements parallel Eqs. (1)–(3) exactly. Note that the FORTRAN names for X_i, Y_i, T_i are FX(I), FY(I), TOR(I), respectively. The FORTRAN names for position variables ψ_i, velocities $\dot{\psi}_i$, and time t are PSI(I), PSID(I), and T, respectively. This is a standard DYMAC convention for all problems. However, the FORTRAN names for the constants $a_1, a_2, C_d, K_s, \psi_{5,0}$, and force F were chosen as A1, A2, CD, SK, PSI50, and F for this particular problem. A more efficient and convenient method of specifying constants is described in the user's manual.

14.44 Modeling Motion Generator Constraints

In addition to permitting virtually arbitrary applied forces on a linkage, DYMAC gives the user the opportunity to specify relationships of his choosing between the variables ψ_i. These relationships are called **motion generator constraints** or **control constraints**.[12] These relationships are of the general form

$$g_i(\psi_j, t) = 0, \qquad (i = 1, 2, \ldots, N_g) \tag{1}$$

where t is time and N_g is the number of such relationships to be specified. DYMAC requires that FORTRAN statements equivalent to Eqs. (1) be placed in an otherwise standard subroutine called MOTGEN. If no such conditions are being specified, the user need not touch the given (dummy) subroutine MOTGEN. DYMAC also requires the expressions for the nonzero partial derivatives of g_i of the form $\partial g_i/\partial \psi_j$ and $\partial g_i/\partial t$ and for the total derivatives $(d/dt)(\partial g_i/\partial \psi_j)$ and $(d/dt)(\partial g_i/\partial t)$. The FORTRAN names to be used for g_i and its various derivatives are

g_i	$\dfrac{\partial g_i}{\partial \psi_j}$	$\dfrac{d}{dt}\left(\dfrac{\partial g_i}{\partial \psi_j}\right)$	$\dfrac{\partial g_i}{\partial t}$	$\dfrac{d}{dt}\left(\dfrac{\partial g_i}{\partial t}\right)$
G(I)	DG(I, J)	DTDG(I, J)	DTG(I)	DTDTG(I)

For each control equation introduced, the user should designate one variable (other than a primary variable) that appears in the equation as a **controlled variable**. As an example where the motion generation capability is useful, suppose that a user wishes to specify that link 4 of a mechanism rotates with constant angular velocity $\dot{\psi}_4 = C$. In other terms,

$$g_1 = \psi_4 - Ct - \psi_{4,0} \tag{2}$$

where $\psi_{4,0}$ is the initial value of ψ_4 at $t = 0$. The only nonvanishing derivatives of g_1 are

$$\frac{\partial g_1}{\partial \psi_4} = 1, \quad \frac{\partial g_1}{\partial t} = -C \tag{3}$$

---

[12]The theory of control constraints is given in Paul and Amin [1978b].

The corresponding FORTRAN statements to be placed in subroutine MOTGEN are

$$G(1)=PSI(4)-C*T-P40$$
$$DG(1,4)=1.0$$
$$DTG(1)=-C$$

where C and P40 are constants that the user chooses. A more sophisticated example of motion generation constraints follows.

14.45 *Example*

Let us consider the pitching and heaving motions of an automobile as it travels over a rough road. For purposes of illustration we shall assume that the rear wheels move with constant horizontal speed V_o while the wheel axles C and D move along a curve which is parallel to a user-specified road profile $y = f(x)$ as shown in Fig. 1. We shall also assume that the axes of the suspension springs AD and BC always remain normal to the fixed line AB in the rigid chassis.

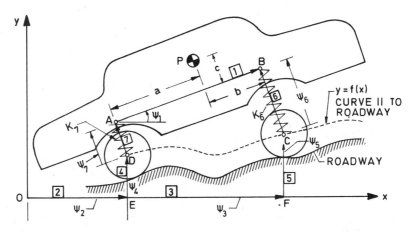

Figure 14.45-1. Spring suspended vehicle on rough road. Dampers (not shown) act in parallel with springs.

We are to find the displacements, velocities, and accelerations $(x, y, \dot{x}, \dot{y}, \ddot{x}, \ddot{y})$ versus time for the three points of interest A, B, P. At time $t = 0$, the vehicle is moving forward at speed V_o on level ground with the front wheels just about to pass onto the curve $y = f(x)$, as shown in Fig. 2.

Upon introducing the seven Lagrangian coordinates $(\psi_1, \ldots, \psi_7)$ shown in Fig. 1 and recognizing the single[13] independent loop $ADEFCBA$, we may formulate all input data needed by DYMAC in the format described in Sec. **14.42**.

If we name the spring compressive forces F6 and F7, the FORTRAN expressions

---

[13]This example problem has been modeled as a single closed-loop system. It could have been modeled as an open-loop system, with different motion generation constraints. Examples of multiloop, closed, and open mechanisms are given in the DYMAC user's manual.

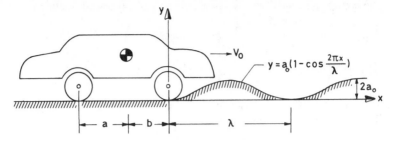

Figure 14.45-2. Initial state of vehicle of example 3.

needed by subroutine FORCES for the forces and torque applied to body 1 are

$$
\begin{aligned}
&\text{F6}=-\text{K6}*(\text{PSI}(6)-\text{PSI6F})-\text{CD}*\text{PSID}(6)\\
&\text{F7}=-\text{K7}*(\text{PSI}(7)-\text{PSI7F})-\text{CD}*\text{PSID}(7)\\
&\text{FX(1)}=-(\text{F6}+\text{F7})*\text{DSIN(PSI(1))}\\
&\text{FY(1)}=\text{F6}+\text{F7})*\text{DCOS(PSI(1))}-\text{WEIGHT}\\
&\text{TOR(1)}=\text{B}*\text{F6}-\text{A}*\text{F7}
\end{aligned}
\tag{1}
$$

The control constraints which keep the wheels on the arbitrary road profile $y = f(x)$ are

$$g_1 = \psi_4 - f(\psi_2) = 0 \tag{2}$$

$$g_2 = \psi_5 - f(\psi_2 + \psi_3) = 0 \tag{3}$$

and the constraint which keeps the rear wheels moving at horizontal speed V_0 is

$$g_3 = \psi_2 - V_o t + (a + b) = 0 \tag{4}$$

where $-(a + b)$ is the value of ψ_2 at $t = 0$.

In the user's manual for DYMAC, and in Paul [1977], it is illustrated how to incorporate these *motion generation constraints* into a small subroutine MOTGEN. No other user-supplied coding is needed for this example.

The standard arrangement for printing the points-of-interest output is indicated in Table 1. The postprocessing routine was utilized to provide the printer-plot of the path (y versus x) of the mass center, shown in Fig. 3. Such plots come off the printer exactly as shown except for the curve, which is easily penciled through the printed crosses. Similar plots and accurate listings of velocities and accelerations of all points of interest are easily obtained.

The total computer time for 173 time steps was 22 sec on an IBM 370/168 computer. The total cost, including extensive printout and four printer-plots, was $9.00. When the same problem was modeled as a strictly open-loop system, the computing cost was reduced by 30%, but considerably more user effort was needed to prepare subroutine MOTGEN.

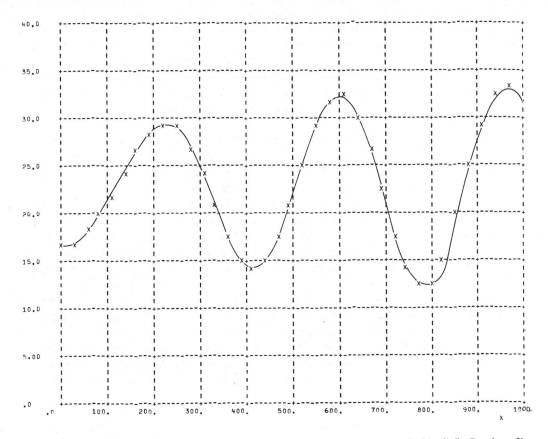

Figure 14.45-3. Path of mass center for automobile of Fig. 14.45-1. Speed $v_0 = 31.25$ mile/h. Road profile: $y = 6 (1 - \cos 2\pi x/360)$ in. Printer-plot is shown.

TABLE 14.45-1 *Partial Output Arrangement of Points of Interest*

Time	Point	Position		Velocity		Acceleration	
		X	Y	XDOT	YDOT	XDDOT	YDDOT
0.0	1	−108	16.2	550	0	.3 E-10	.5 E-5
	2	−54	16.2	550	0	.3 E-10	.5 E-5
	3	0	16.2	550	0	.3 E-10	.5 E-5
0.05	1	−80.5	16.1	549	−0.52	−27	−22.5
	2	−26.5	16.2	549	1.51	−27.1	68
	3	27.5	16.2	549	3.56	−27.2	158

Summary

Program DYMAC is a general-purpose computer program for solving problems of dynamics of planar machinery. It has been described with sufficient detail to suggest to the reader how the system must be modeled, how the input data must be organized, and the level of effort required to write the user-supplied subroutines which specify the forces and motion constraints acting on the system. A more complete description of program usage is given in the user's manual (Paul and Hud [1976].)

Although DYMAC requires that users be capable of writing simple FORTRAN subroutines (which never exceed the level of difficulty of the example described), it has the great advantage of permitting virtually unlimited generality in the nature of the forces and constraints.

The standard printed output of DYMAC consists of the position, velocity, and acceleration associated with each link and the applied force and moment components on each link at user-specified time intervals. The standard output also contains the x and y components of displacement, velocity, and acceleration of user-selected points of interest at the desired time intervals. If desired, joint reactions will be printed out. Provision is made for the user to postprocess the output data in order to obtain plots, punched cards, tapes, additional calculations (e.g., root-mean-square displacements), etc. The presence of gears and cams can be accommodated in the current version of DYMAC by appropriate statements in the user-supplied subroutine MOTGEN.

Experience shows that DYMAC is a versatile, efficient, and easily used program which is capable of solving virtually any problem of dynamics of planar machinery. DYMAC may be used to solve problems of *statics* or *kinematics*, but for such simpler classes of problems it is more efficient to use related programs such as STATMAC (STATics of MAChinery) or KINMAC (KINematics of MAChinery). Information on the availability of all these programs will be supplied by the author upon request.

APPENDIX A

GRASHOF'S THEOREM

As in Sec. **1.31**, let s, l, p, q denote the lengths of the shortest, longest, and two intermediate bars of a four-bar mechanisn. It is often stated, without proof, that the Grashof criterion

$$s + l \le p + q \tag{1}$$

is a necessary and sufficient condition for one or more bars to be capable of *revolving* through 360° relative to some other bar. Grashof [1883] proved this theorem by systematically investigating all possible cases that can arise. A more compact proof, together with a list of related references on this topic, will be found in Paul [1979], where the classification scheme of Table **1.31**-1 is verified. It is also shown there that the equality form of Grashof's criterion (i.e., $s + l = p + q$) is a necessary and sufficient condition for the existence of a *change-point mechanism* (as defined in Sec. **1.33**).

These and other aspects of Grashof's criterion are also discussed in Chapter 10 of Dijksman [1976] where it is shown that condition (1) must be satisfied for any four-bar mechanism which is known to be a crank,[1] i.e., that Grashof's criterion is *necessary* for a crank to exist. A concise proof of the *sufficiency* of Grashof's criterion [i.e., that any four bars which satisfy condition (1) can be assembled into a four-bar mechanism with a crank] is given in Paul [1979].

---

[1]Recall that a *crank* is defined as a side link which can undergo a complete revolution relative to the fixed link.

APPENDIX B

COMPLEX NUMBERS AS ROTATION OPERATORS

Euler's formula[1]

$$e^{i\theta} = \cos\theta + i\sin\theta \tag{1}$$

is usually derived by expanding the exponential and trigonometric functions in Taylor series. An alternative approach, described by Paul [1973], can be summarized as follows.

If a vector is represented by plane polar coordinates (r, θ), we may express it in the symbolic form

$$Z \equiv r\langle\theta\rangle \tag{2}$$

where r is the *magnitude* and $\langle\theta\rangle$ is the *angle* (or *argument*) of the vector. We now introduce the **turning operator** τ, which rotates a vector 90° counterclockwise; i.e.,

$$\tau Z \equiv r\langle\theta + 90°\rangle \tag{3}$$

Hence

$$\tau(\tau Z) \equiv \tau^2 Z = r\langle\theta + 180°\rangle = -r\langle\theta\rangle \tag{4}$$

From Eq. (4) we see that the operation $\tau\tau = \tau^2$ is equivalent to multiplication by -1, and we can write

$$\tau^2 = -1 \tag{5}$$

---

[1]"Gentlemen, that is surely true; it is absolutely paradoxical; we cannot understand it and we don't know what it means, but we have proved it, and therefore we know it must be the truth."—Benjamin Peirce, quoted by Kasner and Newman [1940].

Using τ, we can express an arbitrary vector in the *rectangular form*

$$Z = r\langle\theta\rangle = r\cos\theta\langle0\rangle + r\sin\theta\langle90°\rangle = r\cos\theta + \tau r\sin\theta \qquad (6)$$

where we have utilized the natural notation

$$\langle0\rangle = 1\langle0\rangle = 1 \qquad (7)$$

When $r = 1$, Eq. (6) is reduced to the form

$$\langle\theta\rangle = \cos\theta + \tau\sin\theta \qquad (8)$$

which is analogous to Eq. (1).

If we represent the three vectors Z_1, Z_2, Z_3 by the notation

$$Z_k = r_k(\cos\theta_k + \tau\sin\theta_k) \qquad (9)$$

the formal result of multiplication is

$$\begin{aligned}
Z_3 = Z_1Z_2 &= r_1(\cos\theta_1 + \tau\sin\theta_1)r_2(\cos\theta_2 + \tau\sin\theta_2) \\
&= r_1r_2[(\cos\theta_1\cos\theta_2 - \sin\theta_1\sin\theta_2) \\
&\quad + \tau(\sin\theta_1\cos\theta_2 + \sin\theta_2\cos\theta_1)] \\
&= r_1r_2[\cos(\theta_1 + \theta_2) + \tau\sin(\theta_1 + \theta_2)] = r_1r_2\langle\theta_1 + \theta_2\rangle \qquad (10)
\end{aligned}$$

That is, *the product of Z_1 and Z_2 is a vector whose magnitude is the product of the magnitudes of Z_1 and Z_2 and whose angle is the sum of the angles of Z_1 and Z_2.* To find the rule for division, we let

$$r_3\langle\theta_3\rangle \equiv Z_3 = \frac{Z_1}{Z_2} \quad \text{or} \quad Z_1 = Z_2Z_3 \qquad (11)$$

Hence

$$Z_1 \equiv r_1\langle\theta_1\rangle = r_2r_3\langle\theta_2 + \theta_3\rangle = Z_2Z_3 \qquad (12)$$

Now multiply by the vector $(1/r_2)\langle-\theta_2\rangle$, to give

$$r_1\langle\theta_1\rangle\frac{1}{r_2}\langle-\theta_2\rangle = r_2r_3\langle\theta_2 + \theta_3\rangle\frac{1}{r_2}\langle-\theta_2\rangle$$
$$\frac{r_1}{r_2}\langle\theta_1 - \theta_2\rangle = \frac{r_2r_3}{r_2}\langle\theta_2 + \theta_3 - \theta_2\rangle = r_3\langle\theta_3\rangle \qquad (13)$$

Upon comparing Eqs. (13) and (11) we see that the magnitude of the quotient of two vectors is the ratio of the magnitudes of numerator and denominator and that the angle of the quotient vector is the difference in angles of the numerator and denominator.

In summary, we see that, insofar as vector operations are concerned, τ is equivalent to the imaginary unit i, $\langle\theta\rangle$ is equivalent to $e^{i\theta}$, and the rules for multiplication and division of complex vectors are justified without recourse to series expansions or other processes involving the concept of a limit.

TWO-ARGUMENT INVERSE TANGENT FUNCTION

Suppose that we wish to find the angle θ from given values (x, y) which are known to be proportional to $\cos \theta$ and $\sin \theta$, respectively. That is, we are given x and y and wish to find the value of θ which satisfies

$$x = a \cos \theta \tag{1}$$

$$y = a \sin \theta \tag{2}$$

such that

$$-\pi \leq \theta \leq \pi \tag{3}$$

Equations (1) and (2) imply that point (x, y) lies on the circle of radius a, as shown in Fig. 1, where

$$a = (x^2 + y^2)^{1/2} \tag{4}$$

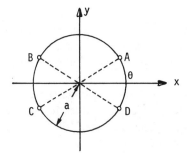

Figure C-1. Possibilities for θ, given x and y.

The standard definition of the **principal value** of the inverse tangent function is

$$\theta_1 = \tan^{-1}\left(\frac{y}{x}\right) \tag{5}$$

$$-\frac{\pi}{2} \le \theta_1 \le \frac{\pi}{2} \tag{6}$$

The range limitation (3) requires that $\theta = \theta_1$ at points such as A and D in Fig. 1 (where $x \ge 0$). But at points such as B and C (where $x < 0$), $\theta = \theta_1 + \pi$ or $\theta = \theta_1 - \pi$ accordingly as y is positive or negative.

In short, we can write

$$\theta = \arctan_2\,(y, x) \tag{7}$$

where the **two-argument inverse tangent function** is defined by

$$
\begin{aligned}
\arctan_2\,(y, x) &\equiv \theta_1 && \text{if } x > 0 \\
&\equiv \theta_1 + \pi\,\text{sgn}\,(y) && \text{if } x < 0 \\
&\equiv \frac{\pi}{2}\,\text{sgn}\,(y) && \text{if } x = 0
\end{aligned}
\tag{8}
$$

$$
\text{sgn}\,(y) =
\begin{cases}
1 & \text{if } y \ge 0 \\
-1 & \text{if } y < 0
\end{cases}
\tag{9}
$$

Note that $\arctan_2\,(0, 0)$, like $\tan^{-1}\,(0/0)$, is undefined.

For digital computer applications, the two-argument inverse tangent function is a standard FORTRAN-supplied subprogram, called ATAN2(Y, X); its double-precision version is called DATAN2(Y, X).

Since $\arctan_2\,(y, x)$ differs from $\tan^{-1}\,(y/x)$ by a constant at most, it follows that

$$\frac{d}{dt}\arctan_2\,(y, x) = \frac{d}{dt}\tan^{-1}\left(\frac{y}{x}\right) \tag{10}$$

where x and y are functions of the variable t.

CONCEPTS FROM THEORY OF GRAPHS (NETWORK TOPOLOGY)

This appendix provides a review of some basic definitions and concepts in graph theory (the topology of networks) which will enable us to state important results in kinematics with precision and brevity. It will also provide the background needed to follow the developing literature which utilizes graph theory in kinematics and dynamics of machinery. Since the introduction by Paul [1960] of some rudimentary topological concepts in a discussion of mobility criteria, there have been numerous more sophisticated applications to kinematics analysis and synthesis, e.g., Crossley [1965], Dobrjansky and Freudenstein [1967], and Woo [1967]. More recently, graph theory has played a major role in the development of computer programs for the dynamic analysis of machinery. Typical programs of this type are IMP (Sheth and Uicker [1972]), DRAM (Chace and Smith [1971]), VECNET (Andrews and Kesavan [1975]), DYMAC (Paul [1977]), and KINMAC (Paul and Amin [1978]).

For introductory references, the author highly recommends the two small books by Ore [1963] and Lietzmann [1955]. More formal references are Seshu and Reed [1961], Branin [1962], Roe [1966], and Koenig et al. [1967].

The terminology of graph theory given below is illustrated by Fig. 1.

D.1 TERMINOLOGY

vertex or **node:** a point. In Fig. 1 the nodes are indicated by small circles and are given identification numbers ($i = 1, 2, \ldots, 8$) enclosed in circles.

edge: a line segment. In Fig. 1(a) edge identification numbers ($j = 1, 2, \ldots, 10$) are placed close to the edges.

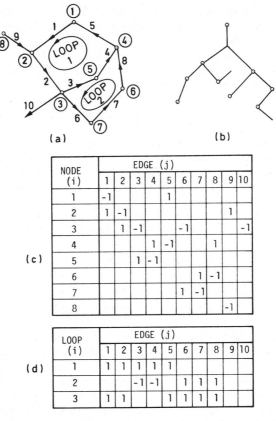

(a)

(b)

(c)

NODE (i)	EDGE (j) 1	2	3	4	5	6	7	8	9	10
1	-1				1					
2	1	-1							1	
3		1	-1			-1				-1
4				1	-1			1		
5			1	-1						
6							1	-1		
7						1	-1			
8									-1	

(d)

LOOP (i)	EDGE (j) 1	2	3	4	5	6	7	8	9	10
1	1	1	1	1	1					
2			-1	-1		1	1	1		
3	1	1			1	1	1	1		

Figure D-1. Typical graphs (networks), and associated matrices. Nodes are indicated by small circles. Node identification numbers are encircled; edge numbers are not: (a) An oriented graph; (b) A tree; (c) Network incidence matrix; (d) Oriented loop matrix. Loop 3 consists of edges 1, 2, 6, 7, 8, 5.

end point: a node lying on only one edge (e.g., node 8 in Fig. 1).

end line: an edge attached to only one node (e.g., edge 10 in Fig. 1).

connected: A system of edges and nodes is *connected* if any two nodes can be joined by a sequence of edges (a path) of the system. The system of Fig. 1(a) is connected, and the system of Fig. 1(b) is connected, but the combined system is not connected.

graph[1] or **network:** a *connected* system of edges and nodes. We shall only consider *graphs in the restricted sense*, where no edge begins and ends at the same node without passing through another node. Figures 1(a) and (b) represent two different graphs.

planar graph: a graph which can be drawn in a plane such that the edges have no intersections except at the vertices. Figure 1 shows planar graphs. Figure 5 shows nonplanar graphs.

closed graph: a graph without end points or end lines [e.g., Fig. 2].

oriented or **directed edge:** an edge with an arrow to indicate a positive direction.

oriented graph: a graph in which every edge is oriented, as in Fig. 1(a).

---

[1]Frequently called *linear graph* in the older literature.

incidence: An edge is incident on a node if the node is at an end of the edge. An oriented edge is **positively (negatively) incident** on the node the arrow points toward (from); this convention is usually reversed for electrical networks. In Fig. 1(a), edge 5 is positively incident on node 1 and negatively incident on node 4.

subgraph: a graph which is a subset of the edges and nodes of another graph.

circuit or **closed loop**: a subgraph such that on every node there are incident exactly two edges. In Fig. 1, the edges (1, 2, 3, 4, 5) form a closed loop (loop 1).

simple circuit or **simple loop**: a closed loop which encloses exactly one polygon with non-intersecting sides. In Fig. 1, loops 1 and 2 are simple circuits, but the loop consisting of edges (1, 2, 6, 7, 8, 5) is not a simple circuit because it encloses two polygons (loops 1 and 2). The polygons enclosed by simple circuits are sometimes called **faces**.

polygonal net: a closed planar graph; e.g., Fig. 2.

tree: a graph without closed loops, e.g., Fig. 1(b). The edges in a tree are sometimes called **branches**.

chords: a set of edges which if removed from a graph would convert the graph into a tree. In Fig. 1, edges 1 and 3 are a set of chords. Many other choices of chords are possible (e.g., edges 2 and 8).

incidence matrix: a matrix $[m_{ij}]$ such that

$$m_{ij} = +1 \ (-1) \text{ if edge } j \text{ is positively (negatively) incident on node } i.$$

$$= 0 \text{ if edge } j \text{ is not incident on node } i$$

Figure 1(c) shows $[m_{ij}]$ for the graph of Fig. 1(a); blank entries represent zeros. Note that column 10 has only one entry because there is no node of the graph at the leading end of edge 10.

oriented loop matrix: Let the closed loops of a graph be numbered $(i = 1, 2, \ldots, L)$. As each loop is traversed in a definite sense (say counterclockwise), a set of edges will be traversed in either a positive or negative sense. The matrix $[a_{ij}]$ is defined by

$$a_{ij} = +1 \ (-1) \text{ if edge } j \text{ is traversed positively (negatively) along loop } i$$

$$= 0 \text{ if edge } j \text{ is not part of loop } i$$

Figure 1(d) shows a loop matrix corresponding to Fig. 1(a).

isomorphism: Two graphs are said to be isomorphs if they have the same incidence properties (same incidence matrix). Figure 4 shows three isomorphic graphs.

D.2 LOOP INDEPENDENCE

Let us represent the oriented edges $j = 1, 2, \ldots, E$ by vectors $\mathbf{e}_1, \mathbf{e}_2, \ldots, \mathbf{e}_E$. Then the closed vector polygons representing loops 1 and 2 in Fig. 1(a) can be expressed as vector equations of the form

$$\mathbf{e}_1 + \mathbf{e}_2 + \mathbf{e}_3 + \mathbf{e}_4 + \mathbf{e}_5 + 0 + 0 + 0 = 0 \tag{1}$$

$$0 + 0 - \mathbf{e}_3 - \mathbf{e}_4 + 0 + \mathbf{e}_6 + \mathbf{e}_7 + \mathbf{e}_8 = 0 \tag{2}$$

A third loop can be formed from edges (1, 2, 6, 7, 8, 5), giving rise to the equation

$$\mathbf{e}_1 + \mathbf{e}_2 + 0 + 0 + \mathbf{e}_5 + \mathbf{e}_6 + \mathbf{e}_7 + \mathbf{e}_8 = 0 \tag{3}$$

Note that Eq. (3) results from merely adding Eqs. (1) and (2). This fact is reflected in the loop matrix [Fig. 1(d)] where row 3 results from adding rows 1 and 2.

A set of vector equations (or rows of a matrix) are said to be linearly independent when no equation (or row) is a linear combination of the others (see Appendix **E.6**). We shall call a set of loops in a network *independent* (or *dependent*) accordingly as the vector equations associated with them are linearly independent (or dependent).

The number L of independent loops in a *planar graph* equals the number L_s of *simple* circuits in the *graph*. That L cannot be smaller than L_s follows from the fact that we may order the vector equations of the simple circuits such that each equation involves a new edge and is therefore independent of prior equations. For example, starting with the central loop (1) in Fig. 2 and successively writing vector loop equations for loops 2, 3, and 4, we see that each successive loop contains at least one edge that did not appear in a previous loop.

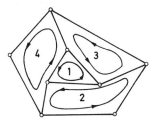

Figure D-2. The simple circuits of a typical planar graph.

To show that L cannot exceed L_s we note that any nonsimple circuit can be synthesized by superposing all the simple circuits enclosed by it, as in the case of Eqs. (1)–(3).

Thus the *number of independent loops in a planar network equals the number of simple circuits in the network*. In other words, the simple circuits provide a **complete** set of independent loops for a planar graph.

Because the network incidence matrix contains all relevant information on the topology of an oriented network, it should be possible to find a loop matrix with independent rows (i.e., to define a complete set of independent circuits) by operating on the information in the incidence matrix. Branin [1962] gives such an algorithm, which has been utilized by Smith [1971] and Chace et al. in program DRAM; Sheth [1972] describes another such algorithm used in program IMP. The author's programs, DYMAC and KINMAC require the user to specify which independent loops he prefers.[2]

D.3 EULER'S THEOREM

Euler showed[3] that in a polygonal net with V vertices, E edges, and L simple loops,

$$V - E + L = 1 \qquad (4)$$

---

[2]See Paul and Hud [1976], Paul and Amin [1978a].
[3]Euler published the result in 1752 without proof and in 1758 with proof. In 1860 it was discovered that Descartes (1596–1650) had enunciated the same theorem. There is speculation that Archimedes was aware of the theorem (Lietzmann [1955, p. 90]).

To prove this, let us examine the so-called **Euler invariant**

$$I \equiv V - E + L$$

For a simple polygon with n vertices, n edges, and one loop, $I = n - n + 1 = 1$. Now, any polygonal net can be built up by starting with a simple polygon and adding a chain of edges to its outside. For example, starting with polygon $(1, 2, 3, 4)$ in Fig. 3 (for which $I = 1$), we can add the *chain* of three vertices $(5, 6, 7)$ and four links $(2, 5), (5, 6), (6, 7), (7, 4)$, thereby adding the polygon (ii).

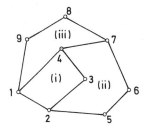

Figure D-3. Building up a polygonal net.

This modification causes a change in I of

$$\Delta I = \Delta V - \Delta E + \Delta L = 3 - 4 + 1 = 0$$

Next we add the chain $(7, 8), (8, 9), (9, 1)$ defining polygon (iii); this time $\Delta I = \Delta V - \Delta E + \Delta L = 2 - 3 + 1 = 0$.

Continuing in this fashion, we see that I is always of the form

$$\Delta I = \Delta V - (\Delta V + 1) + 1 = 0$$

Hence Eq. (4) is valid for a general polygonal net.

D.4 CRITERIA FOR PLANAR GRAPHS

The discussion of the previous two sections was restricted to planar graphs. We need a means of recognizing whether a graph is planar or not. Usually, this can be determined by applying a few simple techniques. For example, if four vertices $(1, 2, 3, 4)$ lie on a circuit, as in Fig. 4(a) and are joined by edges $(1, 3)$ and $(2, 4)$, it is always possible to place one of these added edges (either one) outside the circuit, as in Fig. 4(b).

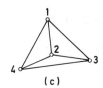

Figure D-4. Isomorphs of a planar graph (The complete square graph).

However, suppose six vertices (1, 2, . . . , 6) follow in order around a circuit and all pairs of opposite vertices, (1, 4), (2, 5), (3, 6), are joined to form a so-called **hexagonal graph**, as in Fig. 5(a). If all three added edges (call them *diagonals*) lie inside the circuit, as in Fig. 5(a), they must cross. If one diagonal lies outside the circuit, as in Fig. 5(b), the inside two diagonals cross. If only one diagonal remains inside, the outer two cross, as in Fig. 5(c). In every case, at least two edges must cross, and the hexagonal graph is clearly not planar.

Similar reasoning applies to the **pentagonal graph** of Fig. 6(a), which is likewise not planar.

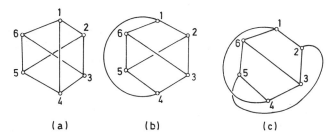

Figure D-5. Isomorphs of the hexagonal graph.

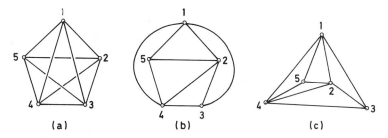

Figure D-6. (a) Pentagonal graph; (b) Pentagonal graph with edge (5, 3) deleted; (c) Isomorph of Fig. (b).

Any graph which has a hexagonal or pentagonal subgraph embedded in it cannot be planar. To sharpen the idea of an embedded subgraph we need the following definition: A graph is said to be **contracted** when a vertex joining only two edges is removed and **expanded** when a vertex is added somewhere along an edge.

We can now state **Kuratowski's theorem**[4] for determining whether or not a graph is planar: *A graph is planar if and only if it does not contain within it any graph which can be contracted or expanded to the hexagonal graph (Fig. 5) or pentagonal graph [Fig. 6(a)].*

As an example of Kuratowski's theorem, consider the graph of Fig. 7(a), which is a subgraph of the crossed mechanism of Fig. **8.22**-3. By contracting nodes *A* and *D*, the subgraph of Fig. 7(b) is obtained. This is precisely the hexagonal graph, proving that the graph of Fig. **8.22**-3 is not planar.

---

[4]Ore [1963, p. 96].

Finally, we note that though we have freely distorted edges into curves whenever convenient, it is always possible to draw any planar graph as a system of noncrossing straight lines; Figs. 4(c) and 6(c) illustrate the procedure.

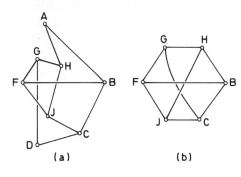

(a)

(b)

Figure D-7. (a) Subgraph of Fig. **8.22**-3; (b) Same with vertices *A* and *D* contracted.

APPENDIX E

MATRICES AND SYSTEMS OF LINEAR ALGEBRAIC EQUATIONS

It has been estimated[1] that linear systems of equations occur in about 75% of all scientific problems. Indeed, we have found in this book that such equations arise in the calculation of link displacements, velocities, accelerations, reaction forces, and in solving the differential equations of motion for general mechanisms.

The purpose of this appendix is to provide a compact compilation of the major definitions and results from linear (and computational) algebra that are used freely within the text. For the most part, proofs are omitted, but appropriate references are given. For additional background in this general area, the reader may consult Kreyszig [1975], Bocher [1907], Fox [1965], Hamming [1971], Ralston [1965], and Forsythe et al. [1977].

E.1 MATRIX NOTATION

A rectangular array of m rows and n columns of numbers in the form

$$\begin{bmatrix} a_{11} & a_{12} & \cdots & a_{1n} \\ a_{21} & a_{22} & \cdots & a_{2n} \\ \cdot & & & \\ \cdot & & & \\ \cdot & & & \\ a_{m1} & a_{n2} & \cdots & a_{mn} \end{bmatrix} = \mathbf{A}$$

---

[1] Dahlquist and Björck [1974, p. 137].

is said to be a **matrix,** and the numbers a_{ij} are said to be its **elements.** Matrices will be represented by boldface type, e.g., **A.** When $m = n$ the matrix is said to be **square,** and of **order** n. When $n = 1$ the matrix is called a **column matrix** or **column vector;** similarly, when $m = 1$, the matrix is said to be a **row matrix** or **row vector.**[2] Occasionally the word **vector** without qualifier is used to represent a column vector.

To specifically indicate the **dimensions** (i.e., numbers of rows and columns of a matrix) we shall occasionally write

$$\mathop{\mathbf{A}}_{m \times n}$$

The special matrices shown below occur sufficiently frequently that they have been given the names indicated:

$$\begin{bmatrix} 0 & 0 & 0 & \cdots \\ 0 & 0 & \cdots & \\ \cdot & & & \\ \cdot & & & \\ \cdot & & m \times n \end{bmatrix} \qquad \begin{bmatrix} 1 & 0 & 0 & \cdots & 0 \\ 0 & 1 & 0 & \cdots & 0 \\ 0 & 0 & 1 & \cdots & 0 \\ \cdots & & & \cdots & \\ \cdots & & & \cdots & 1 \end{bmatrix}$$

null matrix   **identity matrix**
(all zeros)   or **unit matrix**

$$\begin{bmatrix} a_{11} & 0 & 0 & \cdots \\ 0 & a_{22} & 0 & \cdots \\ 0 & 0 & a_{33} & \cdots \\ \cdots & & & a_{nn} \end{bmatrix} \qquad \begin{bmatrix} a_{11} & a_{12} & a_{13} & \cdots & a_{1n} \\ 0 & a_{22} & a_{23} & \cdots & a_{2n} \\ 0 & 0 & a_{33} & \cdots & a_{3n} \\ \cdots & & & & \\ 0 & 0 & 0 & \cdots & a_{nn} \end{bmatrix}$$

diagonal matrix   **triangular matrix**
(upper)

We shall designate that the elements of a matrix **B** are b_{ij} (i = row number, j = column number) by the notation

$$\mathbf{B} = [b_{ij}]$$

Two matrices of equal dimensions are **equal** when their corresponding elements are equal, i.e.,

$$\mathbf{B} = \mathbf{A}$$

if, and only if, $b_{ij} = a_{ij}$ for all i and j. The following definitions will also be used:

---

[2]To emphasize that a matrix is of column form, its elements may be enclosed in *braces,* thus: $\{x_i\}$. To save space, we shall sometimes denote the elements of a column matrix in the form $\mathbf{x} = \{x_1, x_2, \ldots, x_n\}$ and the elements of a row matrix by $\mathbf{x} = \lfloor x_1, x_2, \ldots, x_n \rfloor$.

addition or **subtraction:** $\quad \mathbf{A} \pm \mathbf{B} \equiv [a_{ij} \pm b_{ij}]$

multiplication by a scalar c: $\quad c\mathbf{A} = [ca_{ij}]$

transposition (interchange of rows and columns): $\quad$ if $\mathbf{A} = [a_{ij}]$, $\quad$ then $\mathbf{A}^T \equiv [a_{ji}]$

Note that if $\mathbf{x} = [x_1, x_2, \ldots, x_n]$ is a row vector, then $\mathbf{x}^T = [x_1, x_2, \ldots, x_n]^T$ is a column vector, which may be alternatively expressed as $\mathbf{x}^T = \{x_1, x_2, \ldots, x_n\}$.

The above definitions lead to the following rules of matrix algebra:

$$\mathbf{A} \pm \mathbf{B} = \mathbf{B} \pm \mathbf{A}$$

$$(\mathbf{A} + \mathbf{B}) + \mathbf{C} = \mathbf{A} + (\mathbf{B} + \mathbf{C}) = \mathbf{A} + \mathbf{B} + \mathbf{C}$$

$$(a + b)\mathbf{A} = a\mathbf{A} + b\mathbf{A}$$

$$a(b\mathbf{A}) = b(a\mathbf{A}) = ab\mathbf{A}$$

E.2 MATRIX MULTIPLICATION

If the number of columns in a matrix $\mathbf{A}_{m \times p}$ equals the number of rows in another matrix $\mathbf{B}_{p \times n}$, the two matrices are said to be **conformable**. The **product** of two conformable matrices is a third matrix $\mathbf{P}_{m \times n}$, defined by the rule

$$[P_{ij}] \equiv \underset{m \times n}{\mathbf{P}} \equiv \underset{m \times p}{\mathbf{A}} \underset{p \times n}{\mathbf{B}} \equiv \left[\sum_{k=1}^{p} A_{ik} B_{kj} \right] \tag{1}$$

This definition leads to the conclusion that any matrix is unaltered when multiplied "from the left" or "from the right" by an identity matrix of suitable order; i.e.,

$$\underset{m \times m}{\mathbf{I}} \underset{m \times n}{\mathbf{A}} = \underset{m \times n}{\mathbf{A}} = \underset{m \times n}{\mathbf{A}} \underset{n \times n}{\mathbf{I}} \tag{2}$$

Using Eq. (1), we may prove that

$$(\mathbf{A}\mathbf{B})^T = \mathbf{B}^T \mathbf{A}^T \tag{3}$$

That is, the transpose of a product is the product, *in reverse order*, of the transpose of each factor.

If the elements of $\mathbf{A}$ and $\mathbf{B}$ are functions of time t, it is readily shown from Eq.(1) that the **derivative** of the product matrix $\mathbf{A}\mathbf{B}$ is given by the conventional **chain rule**

$$\dot{\mathbf{P}} = \frac{d}{dt}(\mathbf{A}\mathbf{B}) = \mathbf{A}\dot{\mathbf{B}} + \mathbf{B}\dot{\mathbf{A}} \tag{4}$$

where

$$\dot{\mathbf{A}} = [\dot{a}_{ij}], \qquad \dot{\mathbf{B}} \equiv [\dot{b}_{ij}]$$

E.3 PARTITIONING OF MATRICES

When two matrices are multiplied together we may subdivide the work into separate compartments by drawing horizontal and vertical lines called **partitions** through the matrices, as shown in the example below:

$$
\underset{5\times 6 \; 6\times 3}{\mathbf{A} \; \mathbf{B}} = \left[\begin{array}{cc|cccc} a_{11} & a_{12} & a_{13} & a_{14} & a_{15} & a_{16} \\ a_{21} & a_{22} & a_{23} & a_{24} & a_{25} & a_{26} \\ \hline a_{31} & a_{32} & a_{33} & a_{34} & a_{35} & a_{36} \\ a_{41} & a_{42} & a_{43} & a_{44} & a_{45} & a_{46} \\ a_{51} & a_{52} & a_{53} & a_{54} & a_{55} & a_{56} \end{array}\right] \left[\begin{array}{ccc} b_{11} & b_{12} & b_{13} \\ b_{21} & b_{22} & b_{23} \\ \hline b_{31} & b_{32} & b_{33} \\ b_{41} & b_{42} & b_{43} \\ b_{51} & b_{52} & b_{53} \end{array}\right]
$$

$$
= \begin{bmatrix} \mathbf{A}_1 & \mathbf{A}_2 \\ \mathbf{A}_3 & \mathbf{A}_4 \end{bmatrix} \begin{bmatrix} \mathbf{B}_1 \\ \mathbf{B}_2 \end{bmatrix} \tag{1}
$$

where

$$
\mathbf{A}_1 = [a_{ij}] \qquad (i = 1, 2; j = 1, 2)
$$
$$
\mathbf{A}_2 = [a_{ij}] \qquad (i = 1, 2; j = 3, 4, 5, 6)
$$
$$
\cdots, \text{etc.}
$$

The matrices $\mathbf{A}_1$, etc., are called **submatrices**, and a matrix whose *elements* are matrices is called a **supermatrix**. By applying the usual rules of matrix multiplication to the supermatrices of Eq. (1) we find that

$$
\mathbf{AB} = \begin{bmatrix} \mathbf{A}_1\mathbf{B}_1 + \mathbf{A}_2\mathbf{B}_2 \\ \mathbf{A}_3\mathbf{B}_1 + \mathbf{A}_4\mathbf{B}_2 \end{bmatrix} \tag{2}
$$

By expansion of Eq. (1), it will be found that Eq. (2) is correct. In general, we may partition matrices in any way desired as long as all submatrices which are to be multiplied together are conformable, e.g., $\underset{3\times 2 \; 2\times 3}{\mathbf{A}_3 \; \mathbf{B}_1}$.

E.4 GAUSSIAN ELIMINATION METHOD FOR SOLVING LINEAR EQUATIONS

Given n linear equations, $\mathbf{Ax} = \mathbf{c}$, in n unknowns x_i, i.e.,

$$
a_{11}x_1 + a_{12}x_2 + \cdots + a_{1n}x_n = c_1
$$
$$
a_{21}x_1 + a_{22}x_2 + \cdots + a_{2n}x_n = c_2 \tag{1}
$$
$$
\cdots
$$
$$
a_{n1}x_1 + a_{n2}x_2 + \cdots + a_{nn}x_n = c_n
$$

we may begin the solution by seeking to eliminate the variable x_1 between some selected equation (called the **pivot equation**) and all the other equations. If, for example, $a_{11} \neq 0$, we may use the first equation as pivot. If we multiply the first equation throughout by a_{21}/a_{11} and subtract the resulting equation from the second equation, we obtain a new equation in which x_1 does not appear.

We now repeat this type of **row operation** on each of the succeeding equations using a_{i1}/a_{11} as the **multiplier** for the ith equation. The net result is a new set of $(n-1)$ **modified equations** of the form

$$\sum_{j=2}^{n} a_{ij}^2 x_j = c_i^2 \qquad (i = 2, 3, \ldots, n) \qquad (2)$$

in which x_1 does not appear.

If x_1 does not appear in the first equation (i.e., $a_{11} = 0$), we cannot calculate the needed multipliers a_{i1}/a_{11}. In such a case, we can interchange the names of the variable x_1 and some other variable, say x_s, which does appear in the first equation; this is equivalent[3] to interchanging the columns 1 and s in the matrix $\mathbf{A}$. Elements used as divisors (e.g., a_{11}) are called **pivot elements**.

We now perform a similar row operation on each of the modified equations (2), interchanging columns or rows, if necessary, to find a new set of $n-2$ modified equations in which x_2 does not appear. Suppose that the process may be repeated $n-1$ times. Then the first equation of each modified set forms the set of equations

$$\begin{bmatrix} a_{11} & a_{12} & a_{13} & \cdots & a_{1n} \\ 0 & a_{22}^2 & a_{23}^2 & \cdots & a_{2n}^2 \\ 0 & 0 & a_{33}^3 & \cdots & a_{3n}^3 \\ 0 & 0 & 0 & \cdots & a_{4n}^4 \\ \cdot & & & & \\ \cdot & & & & \\ \cdot & & & & \\ 0 & 0 & 0 & \cdots & a_{nn}^n \end{bmatrix} \begin{bmatrix} x_1' \\ x_2' \\ \cdot \\ \cdot \\ \cdot \\ x_n' \end{bmatrix} = \begin{bmatrix} c_1' \\ c_2' \\ \cdot \\ \cdot \\ \cdot \\ c_n' \end{bmatrix} \equiv \mathbf{A}'\mathbf{x}' = \mathbf{c}' \qquad (3)$$

where the *primes* on x_i' reflect the fact that some variable names may have been interchanged (e.g., $x_2' \leftarrow x_3$, $x_3' \leftarrow x_2$) and $\mathbf{A}'$, $\mathbf{c}'$ are called **modified matrices**.

The triangular nature of the matrix $\mathbf{A}'$ enables us to solve by the **back-substitution** process:

$$x_n' = \frac{c_n'}{a_{nn}'}$$

$$x_{n-1}' = \frac{c_{n-1}' - a_{n-1,n}' x_n'}{a_{n-1,n-1}'}$$

$$\cdots$$

$$x_{n-j}' = \frac{c_{n-j}' - \sum_{k=n-j+1}^{n} a_{n-j,k}' x_k'}{a_{n-j,n-j}'} \qquad (4)$$

Note that the solution depends only on the two matrices $\mathbf{A}'$ and $\mathbf{c}'$ which can be

---

[3]Alternatively, we can interchange the pivot equation with some other equation that does contain the variable being eliminated. This procedure is equivalent to a **row exchange** and does not require the renaming of variables.

combined into the single partitioned matrix [**A′ c′**]. The latter may be generated by performing the elementary row operations directly upon the so called **augmented matrix** of the system [**A c**], formed by adjoining the given column **c** to the given matrix **A**.

Suppose that, at some stage of the process, we find that all of the coefficients $a^s_{s,j}$ in the row s of modified matrix **A′** vanish. When this happens, the matrix **A** is called **singular**. For singular matrices we cannot find any pivotal element, and the method needs revision. We shall see shortly that the vanishing row is due to the fact that not all of the original equations were *independent*. For such cases, an infinite number of solutions may exist, or no solutions may exist (see Sec. **E.6**).

It will be noted that we have used the diagonal elements as pivots (divisors) in our row operations. If a diagonal element has a small absolute value, it could lead to appreciable roundoff errors. For this reason, it is advisable to interchange columns (i.e., temporarily rename variables) so that the diagonal (pivotal) element has the greatest numerical value in its row. This technique is called **partial pivoting**. Some computer programs go further and also interchange equations at each step so that the pivotal element has the greatest numerical value of all elements in all the rows awaiting processing. This method is called **complete pivoting**. Some investigators (e.g., Hamming [1971, p. 110]) claim that partial pivoting is essential, but complete pivoting is not usually worth the effort.

A listing is given in Sec. **E.8** for a FORTRAN subroutine, GELG, which utilizes the Gauss elimination method with complete pivoting. For a program with partial pivoting, see Forsythe et al. [1977].

E.5 INVERSE MATRIX

The **inverse** of a square matrix **A** is a matrix $\mathbf{A}^{-1}$ such that $\mathbf{A}^{-1}\mathbf{A} = \mathbf{I}$. Not every square matrix has an inverse. If no inverse exists, the original matrix is *singular*. A matrix and its inverse can be shown to be commutative; i.e.,

$$\mathbf{A}\mathbf{A}^{-1} = \mathbf{A}^{-1}\mathbf{A} = \mathbf{I} \tag{1}$$

The inverse is useful in solving systems of linear equations of the type $\mathbf{A}\mathbf{x} = \mathbf{c}$, since $\mathbf{A}^{-1}\mathbf{A}\mathbf{x} = \mathbf{A}^{-1}\mathbf{c}$ implies $\mathbf{x} = \mathbf{A}^{-1}\mathbf{c}$. Ordinarily, it is not computationally efficient to solve such systems by use of the inverse. Suppose, however, that one has to solve several *sets* of equations with the same **A** matrix and different right-hand side vectors, $\mathbf{c}_1, \mathbf{c}_2, \ldots, \mathbf{c}_n$. Then it is efficient to find the required solutions, $\mathbf{x}_1, \mathbf{x}_2, \ldots, \mathbf{x}_n$, of the equations

$$\mathbf{A}\mathbf{X} \equiv \mathbf{A}[\mathbf{x}_1\mathbf{x}_2 \cdots \mathbf{x}_n] = [\mathbf{c}_1, \mathbf{c}_2, \ldots, \mathbf{c}_n] \equiv \mathbf{C} \tag{2}$$

in the form

$$\mathbf{X} = \mathbf{A}^{-1}\mathbf{C} \tag{3}$$

If, in Eq. (3), we let **C** be the identity matrix **I**, it follows that the square matrix **X** is the inverse of matrix **A**. In other words, the inverse of a matrix **A** is given by the solution $\mathbf{X}$ of

$$\mathbf{AX} = \mathbf{I} \tag{4}$$

Therefore, if one ever needs to calculate the inverse of a matrix $\mathbf{A}$, the recommended procedure is to solve Eq. (4) by the Gauss elimination procedure and set

$$\mathbf{A}^{-1} = \mathbf{X} \tag{5}$$

Note on Efficiency of Inversion Versus Direct Solution

When we have a single right-hand side vector $\mathbf{c}$, we can solve the equations $\mathbf{Ax} = \mathbf{c}$ by a *direct* application of the Gauss elimination method, or we can first calculate the *inverse* $\mathbf{A}^{-1}$ and then find $\mathbf{x} = \mathbf{A}^{-1}\mathbf{c}$. Clearly, the direct solution is more efficient than the solution by inversion. However, when one has to solve for several (say k) different right-hand side vectors, the solution by inversion may be economical. To decide how large k must be before inversion becomes competitive with the direct solution, one may count the arithmetic operations in the two methods. For example, Fox [1965, pp. 176–178] shows that, for large n (the order of the matrix $\mathbf{A}$), inversion requires approximately three times the computational effort as a direct solution. This would imply that inversion should be used only if $k \geq 3$.

However, numerical experiments by the author, with program DGELG (see Sec. E.8) on the UNIVAC 7090 computer, showed an increase in computer (CPU) time required for inversion over that required for a direct solution of only 4%, 25%, and 58% for $n = 10, 20$, and 30, respectively. In other words, inversion becomes economical for $k \geq 2$ for the program and computer cited.

E.6   LINEAR DEPENDENCE AND RANK

Let us designate by $\mathbf{V}_1, \mathbf{V}_2, \ldots, \mathbf{V}_F$ a set of (row or column) vectors of the same dimension and by $\mu_1, \mu_2, \ldots, \mu_F$ a set of scalars, called **weights**. The **weighted sum**

$$\mathbf{L} = \mu_1\mathbf{V}_1 + \mu_2\mathbf{V}_2 + \cdots + \mu_F\mathbf{V}_F \tag{1}$$

is called a **linear combination** of the vectors $\mathbf{V}_i$. By varying the weights, we can find any number of such linear combinations, and it is easily shown that any linear combination of these linear combinations is itself a linear combination of the vectors $\mathbf{V}_i$. If it is possible to find any set of weights (which are not all zero) such that $\mathbf{L} = \mathbf{0}$ in Eq. (1), the vectors $\mathbf{V}_i$ are said to be **linearly dependent**; otherwise they are called **linearly independent**.

The **rank** r of a matrix is defined as the maximum number of linearly independent rows or columns in the matrix. For a matrix of dimensions $m \times n$ the maximum possible rank is the lesser of m and n. A square matrix of order n is **singular** if $r < n$.

Given the set of m equations in n unknowns,

$$\mathbf{Ax} = \mathbf{c} \tag{2}$$

it may happen that the equations are *inconsistent* and no solution exists, a unique solution may exist, or an infinite number of solutions may exist—depending on the nature of the *augmented matrix*

$$[\mathbf{Ac}] \equiv [a_{ij}, c_j] \tag{3}$$

The following three theorems[4] summarize the most important cases.

Theorem 1

A necessary and sufficient condition for a system of linear equations to be consistent is that the matrix of the system has the same rank as the augmented matrix.

Theorem 2

If, in a system of n linear equations, the matrix of the system and the augmented matrix have the same rank r, the values of $n - r$ of the unknowns may be assigned at pleasure and the others will then be uniquely determined. The $n - r$ unknowns, whose values may be assigned at pleasure, may be chosen in any way, provided that the matrix of the coefficients of the remaining unknowns is of rank r.

Theorem 3

The n **homogeneous equations** $\mathbf{Ax} = \mathbf{0}$, in n unknowns, have a nontrivial solution if, and only if, $\mathbf{A}$ is singular.

E.7 CRAMER'S RULE

If $\mathbf{A}$ is a nonsingular square matrix of order n, the solution of the n equations, in n unknowns,

$$\mathbf{Ax} = \mathbf{c} \tag{1}$$

is given by

$$x_1 = \frac{D_1}{D}, \quad x_2 = \frac{D_2}{D}, \quad \dots, \quad x_n = \frac{D_n}{D} \tag{2}$$

where

$$D = \det \mathbf{A} \tag{3}$$

and D_i is the *determinant* of the matrix formed by replacing column i of $\mathbf{A}$ by the column vector $\mathbf{c}$.

For $n = 2$, these rules give

$$x_1 = \frac{c_1 a_{22} - c_2 a_{12}}{D}, \qquad x_2 = \frac{a_{11} c_2 - a_{21} c_1}{D}, \qquad D = a_{11} a_{22} - a_{21} a_{12} \tag{4}$$

Cramer's rule is convenient for small systems ($n \le 3$) but cannot compete, in efficiency, with the Gauss elimination method for larger systems.[5]

---

[4]Bocher [1907, pp. 46, 47], or see any work on linear algebra, e.g., Kreyszig [1975].
[5]To solve a set of 20 equations by Cramer's rule would require about 3×10^8 years on a "fast modern computer," according to Forsythe et al. [1977, p. 30]. On the other hand, it may be shown that the determinant of any square matrix $\mathbf{A}$ is given by the product of the diagonal elements of the modified matrix $\mathbf{A}'$ which results from the Gaussian elimination scheme of Eq. (E.4-3).

E.8 FORTRAN LISTING FOR
A LINEAR EQUATION SOLVER

Numerous library programs exist for solving systems of linear algebraic equations. One such FORTRAN program is IBM's DGELG (GELG, in single precision), which is based on the Gauss elimination algorithm with complete pivoting. This subroutine has been embedded in several of the computer programs developed elsewhere in this book (e.g., NEWTR in Appendix F and KINAL in Sec. **10.43**) and is therefore listed, for those who wish to use the programs, as is. The information supplied here should enable those who wish to use other subroutines to alter the calling instructions appropriately.

The program listing is almost self-documented, but the following points are worth emphasizing:

1. The subroutine solves the matrix equation

$$\mathbf{Ax} = \mathbf{r} \tag{1}$$

for a given $M \times M$ matrix $\mathbf{A}$ and right-hand side vector $\mathbf{r}$ of M components. The matrix $\mathbf{A}$ is destroyed in the process, and the solution vector $\mathbf{x}$ is returned in the locations which originally held $\mathbf{r}$.

2. To simultaneously solve several equations of the form (1) with the same $\mathbf{A}$ matrix but different $\mathbf{r}$ vectors, e.g.,

$$\mathbf{Ax}_1 = \mathbf{r}_1, \quad \mathbf{Ax}_2 = \mathbf{r}_2, \quad \ldots, \quad \mathbf{Ax}_N = \mathbf{r}_N \tag{2}$$

the problem is posed in the form

$$\mathbf{AX} \equiv \mathbf{A}[\mathbf{x}_1 \mathbf{x}_2 \cdots \mathbf{x}_N] = [\mathbf{r}_1 \mathbf{r}_2 \cdots \mathbf{r}_N] \equiv \mathbf{R} \tag{3}$$

The user supplies matrix $\mathbf{A}$ and $M \times N$ matrix $\mathbf{R}$ whose columns consist of the vectors $\mathbf{r}_i$. The subroutine returns the N solutions $\mathbf{x}_1, \mathbf{x}_2, \ldots, \mathbf{x}_N$ in the locations originally occupied by the columns of $\mathbf{R}$.

3. It is a requirement of the subroutine that the given matrix $\mathbf{A} = [A_{ij}]$ must be packed columnwise into the linear array[6]

$$A(I) = \{A_{11}, A_{21}, \ldots, A_{M1}, A_{12}, A_{22}, \ldots, A_{M2}, \ldots, A_{1M}, A_{2M}, \ldots, A_{MM}\}$$

prior to calling DGELG. Similarly, the matrix $\mathbf{R}$ must be packed columnwise into a linear array

$$R(I) = \{R_{11}, R_{21}, \ldots, R_{M1}, \ldots, R_{1N}, R_{2N}, \ldots, R_{MN}\}$$

4. The determinant of $\mathbf{A}$ (called DET) is considered to be *numerically zero* when its magnitude falls below $\text{EPS} \times \text{MAX}[\text{ABS}(A_{ij})]$. Experience shows that a value of $\text{EPS} = 10^{-10}$ works well in most problems of kinematics. $|\text{DET}|$ is returned by

---

[6]This is readily accomplished, e.g., by the six FORTRAN statements in subroutine NEWTR (Sec. **F.2**) following the comment "Repack Matrix. . . ."

DGELG, along with an error indicator, IER, which is always zero if |DET| is greater than the specified numerical tolerance. When IER exceeds 0, it represents the apparent rank of an apparently singular matrix.

The following listing[7] is a double-precision version capable of handling 20 simultaneous equations with a single right-hand side. In general, for M equations and N right-hand side vectors, the dimensions of $A(I)$ and $R(I)$ should be M^2 and MN, respectively.

```
C           SUBROUTINE DGELG
C
C           PURPOSE
C               TO SOLVE A GENERAL SYSTEM OF SIMULTANEOUS LINEAR EQUATIONS.
C
C           USAGE
C               CALL DGELG (R,A,M,N,EPS,IER,DET)
C
C           DESCRIPTION OF PARAMETERS
C               R        - DOUBLE PRECISION M BY N RIGHT HAND SIDE MATRIX
C                          (DESTROYED). ON RETURN R CONTAINS THE SOLUTIONS
C                          OF THE EQUATIONS.
C               A        - DOUBLE PRECISION M BY M COEFFICIENT MATRIX
C                          (DESTROYED).
C               M        - THE NUMBER OF EQUATIONS IN THE SYSTEM.
C               N        - THE NUMBER OF RIGHT HAND SIDE VECTORS.
C               EPS      - SINGLE PRECISION INPUT CONSTANT WHICH IS USED AS
C                          RELATIVE TOLERANCE FOR TEST ON LOSS OF
C                          SIGNIFICANCE.
C               IER      - RESULTING ERROR PARAMETER CODED AS FOLLOWS
C                          IER=0  - NO ERROR,
C                          IER=-1 - NO RESULT BECAUSE OF M LESS THAN 1 OR
C                                   PIVOT ELEMENT AT ANY ELIMINATION STEP
C                                   EQUAL TO 0,
C                          IER=K  - WARNING DUE TO POSSIBLE LOSS OF SIGNIFI-
C                                   CANCE INDICATED AT ELIMINATION STEP K+1,
C                                   WHERE PIVOT ELEMENT WAS LESS THAN OR
C                                   EQUAL TO THE INTERNAL TOLERANCE EPS TIMES
C                                   ABSOLUTELY GREATEST ELEMENT OF MATRIX A.
C               DET      - DETERMINANT OF MATRIX A.
C
C           REMARKS
C               INPUT MATRICES R AND A ARE ASSUMED TO BE STORED COLUMNWISE
C               IN M*N RESP. M*M SUCCESSIVE STORAGE LOCATIONS. ON RETURN
C               SOLUTION MATRIX R IS STORED COLUMNWISE TOO.
C               THE PROCEDURE GIVES RESULTS IF THE NUMBER OF EQUATIONS M IS
C               GREATER THAN 0 AND PIVOT ELEMENTS AT ALL ELIMINATION STEPS
C               ARE DIFFERENT FROM 0. HOWEVER WARNING IER=K - IF GIVEN -
C               INDICATES POSSIBLE LOSS OF SIGNIFICANCE. IN CASE OF A WELL
C               SCALED MATRIX A AND APPROPRIATE TOLERANCE EPS, IER=K MAY BE
C               INTERPRETED THAT MATRIX A HAS THE RANK K. NO WARNING IS
C               GIVEN IN CASE M=1.
C
C           SUBROUTINES AND FUNCTION SUBPROGRAMS REQUIRED
C               NONE
C
C           METHOD
C               SOLUTION IS DONE BY MEANS OF GAUSS-ELIMINATION WITH
C               COMPLETE PIVOTING.
C
C           ..............................................................
C
```

[7]Reprinted by permission from *System/360 Scientific Subroutine Package, Version III, Programmer's Manual—GH20-0205-4*, p. 121, © 1970 by International Business Machines Corporation, White Plains, N.Y.

```
       SUBROUTINE DGELG (R,A,M,N,EPS,IER,DET)
C
C
       DOUBLE PRECISION R,A,PIV,TB,TOL,PIVI,DABS,EPS,DET
       DIMENSION A(400),R(20)
       IF(M)23,23,1
C
C      SEARCH FOR GREATEST ELEMENT IN MATRIX A
     1 IER=0
       DET=1.0D0
       PIV=0.D0
       MM=M*M
       NM=N*M
       DO 3 L=1,MM
       TB=DABS(A(L))
       IF(TB-PIV)3,3,2
     2 PIV=TB
       I=L
     3 CONTINUE
       TOL=EPS*PIV
C      A(I) IS PIVOT ELEMENT. PIV CONTAINS THE ABSOLUTE VALUE OF A(I).
C
C
C      START ELIMINATION LOOP
       LST=1
       DO 17 K=1,M
       DET=DET*PIV
C
C      TEST ON SINGULARITY
       IF(PIV)23,23,4
     4 IF(IER)7,5,7
     5 IF(PIV-TOL)6,6,7
     6 IER=K-1
     7 PIVI=1.D0/A(I)
       J=(I-1)/M
       I=I-J*M-K
       J=J+1-K
C      I+K IS ROW-INDEX, J+K COLUMN-INDEX OF PIVOT ELEMENT
C
C      PIVOT ROW REDUCTION AND ROW INTERCHANGE IN RIGHT HAND SIDE R
       DO 8 L=K,NM,M
       LL=L+I
       TB=PIVI*R(LL)
       R(LL)=R(L)
     8 R(L)=TB
C
C      IS ELIMINATION TERMINATED
       IF(K-M)9,18,18
C
C      COLUMN INTERCHANGE IN MATRIX A
     9 LEND=LST+M-K
       IF(J)12,12,10
    10 II=J*M
       DO 11 L=LST,LEND
       TB=A(L)
       LL=L+II
       A(L)=A(LL)
    11 A(LL)=TB
C
C      ROW INTERCHANGE AND PIVOT ROW REDUCTION IN MATRIX A
    12 DO 13 L=LST,MM,M
       LL=L+I
       TB=PIVI*A(LL)
       A(LL)=A(L)
    13 A(L)=TB
C
```

```
C        SAVE COLUMN INTERCHANGE INFORMATION
         A(LST)=J
C
C        ELEMENT REDUCTION AND NEXT PIVOT SEARCH
         PIV=0.D0
         LST=LST+1
         J=0
         DO 16 II=LST,LEND
         PIVI=-A(II)
         IST=II+M
         J=J+1
         DO 15 L=IST,MM,M
         LL=L-J
         A(L)=A(L)+PIVI*A(LL)
         T3=DABS(A(L))
         IF(T3-PIV)15,15,14
      14 PIV=T3
         I=L
      15 CONTINUE
         DO 16 L=K,NM,M
         LL=L+J
      16 R(LL)=R(LL)+PIVI*R(L)
      17 LST=LST+M
C        END OF ELIMINATION LOOP
C
C
C        BACK SUBSTITUTION AND BACK INTERCHANGE
      18 IF(M-1)23,22,19
      19 IST=MM+M
         LST=M+1
         DO 21 I=2,M
         II=LST-I
         IST=IST-LST
         L=IST-M
         L=A(L)+.5D0
         DO 21 J=II,NM,M
         T3=R(J)
         LL=J
         DO 20 K=IST,MM,M
         LL=LL+1
      20 T3=T3-A(K)*R(LL)
         K=J+L
         R(J)=R(K)
      21 R(K)=T3
C
      22 RETURN
C
C
C        ERROR RETURN
      23 IER=-1
         RETURN
         END
```

APPENDIX F

THE NEWTON-RAPHSON ALGORITHM

F.1 ANALYSIS

Given N algebraic equations in N unknowns,

$$f_1(x_j) \equiv f_1(x_1, x_2, \ldots, x_N) = 0$$
$$f_2(x_j) \equiv f_2(x_1, x_2, \ldots, x_N) = 0$$
$$\cdots \tag{1}$$
$$f_N(x_j) \equiv f_N(x_1, x_2, \ldots, x_N) = 0$$

we seek to find that solution vector

$$\mathbf{x} \equiv \{x_1, x_2, \ldots, x_N\} = \{x_1^*, x_2^*, \ldots, x_N^*\} \tag{2}$$

which will satisfy Eqs. (1).

In the neighborhood of an arbitrary point $x_j^{(r)}$ we may approximate Eqs. (1) by the linear terms of a Taylor series:

$$f_i(x_j) \doteq f_i^{(r)} + \left(\frac{\partial f_i}{\partial x_1}\right)^{(r)} \Delta x_1 + \left(\frac{\partial f_i}{\partial x_2}\right)^{(r)} \Delta x_2 + \cdots$$
$$+ \left(\frac{\partial f_i}{\partial x_N}\right)^{(r)} \Delta x_N \qquad (i = 1, 2, \ldots, N) \tag{3}$$

where $f_i^{(r)}$ and $(\partial f_i / \partial x_j)^{(r)}$ represent the functions f_i and their partial derivatives, evaluated at the point $x_j^{(r)}$ and

$$\Delta x_j \equiv x_j^{(r+1)} - x_j^{(r)} \tag{4}$$

is the correction required for the next trial solution $x_j^{(r+1)}$. The functions $f_i(x_j)$ in Eqs. (3) will be approximately[1] zero if all the right-hand sides of Eqs. (3) are set equal to zero, giving the matrix equation

$$\mathbf{A}^{(r)} \, \Delta \mathbf{x} = -\mathbf{f}^{(r)} \tag{5}$$

where

$$\mathbf{f}^{(r)} = \{f_1, f_2, \ldots, f_N\}^{(r)} \tag{6}$$

and the **Jacobian matrix A** is given by

$$\mathbf{A} = \left[\frac{\partial f_i}{\partial x_j} \right] = \begin{bmatrix} \partial f_1/\partial x_1 & \partial f_1/\partial x_2 & \cdots & \partial f_1/\partial x_N \\ \partial f_2/\partial x_1 & \partial f_2/\partial x_2 & \cdots & \partial f_2/\partial x_N \\ \cdots & & & \\ \partial f_N/\partial x_1 & \partial f_N/\partial x_2 & \cdots & \partial f_N/\partial x_N \end{bmatrix} \tag{7}$$

The superscript (r) denotes that $\mathbf{A}$ and $\mathbf{f}$ are evaluated at the point $x_j \equiv x_j^{(r)}$.

If we start with an initial approximate solution $x_j^{(1)}$, we may then evaluate all of the terms $f_i(x_j^{(1)}) \equiv f_i^{(1)}$; these **residuals** will not vanish because $x_j^{(1)}$ is not an exact solution. However, by solving Eq. (5) we get a set of **corrections** Δx_j which will (if the method converges) lead to a better estimate of the true solution in the form

$$x_j^{(2)} = x_j^{(1)} + \Delta x_j$$

The process is now repeated, with $x_j^{(2)}$ playing the role of $x_j^{(1)}$, to find the new approximation $x_j^{(3)}$, and so forth. The general rule is

$$x_j^{(r+1)} = x_j^{(r)} + \Delta x_j \tag{8}$$

The process should be continued until one of the following conditions is satisfied:

$$|\Delta x_j| \le \text{XTOL}_i \qquad (i = 1, 2, \ldots, N) \tag{9a}$$
$$|f_i| \le \text{FTOL}_i \qquad (i = 1, 2, \ldots, N) \tag{9b}$$
$$r = \text{ITMAX} \tag{9c}$$

where XTOL_i and FTOL_i are user-specified tolerances and ITMAX is a user-supplied bound on the number of iterations permitted.

---

[1] We must say "approximately" because the Taylor series was truncated after the linear terms.

Equations (5)–(9) constitute the **Newton-Raphson algorithm**. It is an *iterative* procedure which—when it works—will ultimately double the number of correct decimal places after each iteration (this is called **quadratic convergence**).

If the initial estimate $\mathbf{x}^{(1)}$ deviates too far from the desired solution, the algorithm may fail to converge, or it may converge to a solution other than that desired (e.g., to a lagging mode four-bar mechanism when a leading mode configuration is desired). However, experience shows that, in problems of kinematics, initial estimates of a position variable can be in error by more than 20% of the variable's total range without causing difficulties (see Sec. **9.21**).

To avoid the analytical, and programming, effort associated with determining the many derivatives needed for the Jacobian $\mathbf{A}$, one may replace the derivatives by *finite difference* approximations such as

$$\frac{\partial f_i}{\partial x_j} = \frac{f_i(\mathbf{x} + \boldsymbol{\epsilon}) - f_i(\mathbf{x} - \boldsymbol{\epsilon})}{2\epsilon_j} \tag{10a}$$

or

$$\frac{\partial f_i}{\partial x_j} = \frac{f_i(\mathbf{x} + \boldsymbol{\epsilon}) - f_i(\mathbf{x})}{\epsilon_j} \tag{10b}$$

where ϵ_j is a small increment in x_j and all elements of $\boldsymbol{\epsilon}$ are null except for the jth component; i.e.,

$$\boldsymbol{\epsilon} = \{0, 0, \ldots, 0, \epsilon_j, 0, \ldots, 0\}$$

The term ϵ_j may be chosen initially as a small fraction (e.g., 0.0001) of the initial guess at x_j (for $x_j \neq 0$); at later stages one can choose ϵ_j as a fraction of the latest calculated value of Δx_j.

The *central difference* formula (10a) is more accurate than the *forward difference* formula (10b), but use of the latter will require fewer function evaluations [since the terms $f_i(\mathbf{x})$ computed for Eq. (10b) will be utilized in Eq. (5) too].

If the matrix $\mathbf{A}$ is nearly singular, large errors can occur in the solution of Eq. (5), and convergence may not occur. This "failure" is in fact a welcome occurrence because it signals that the mechanism is at, or near, a singular configuration (see Sec. **8.23**), e.g., a critical form, change point, etc., where the given equations (1) do not uniquely determine the motion.

Other aspects of the question of convergence are discussed by Hamming [1971, p. 50] and McCracken and Dorn [1964, p. 133].

For a single equation, $f(x) = 0$, the computational procedure reduces to the formula

$$x^{(r+1)} = \left[x - \frac{f(x)}{df/dx} \right]^{(r)} \tag{11}$$

For two equations, the matrix equation (5) is easily solved by Cramer's rule as in the example of Sec. **9.21**.

For more than two equations, it is preferable to solve Eqs. (5) by means of a linear equation solving routine such as the Gauss elimination method.

F.2 COMPUTER IMPLEMENTATION

The flowchart of a FORTRAN subroutine NEWTR for implementing the Newton-Raphson procedure is shown in Fig. 1. For an alternative program, see Carnahan et al. [1969, pp. 319–329].

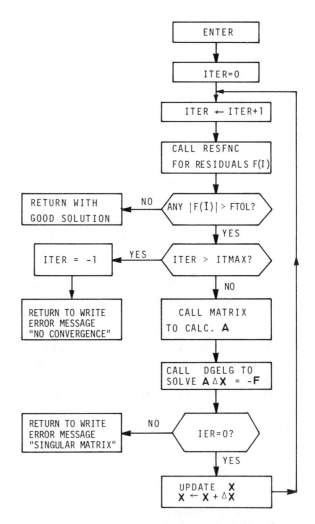

Figure F-1. Flow chart for subroutine NEWTR.

The following notation is used in the flowchart:

ITER: iteration counter; corresponds to r in Eq. (5)

ITMAX: maximum number of iterations to be allowed

FTOL: tolerance permitted on residuals f_i

RESFNC: a user-supplied subroutine[2] which calculates the residual functions $f_i(x_j)$

MATRIX: a user-supplied subroutine[2] which calculates the Jacobian matrix **A**

DGELG: subroutine from the IBM Scientific Subroutine Package[3] which solves systems of linear algebraic equations by the Gauss elimination method

IER: normally zero, but nonzero values are returned by DGELG to indicate various errors, including apparent singularity of **A**

Note that NEWTR will return either a valid solution or one of two error indications. A typical *main* program which calls upon NEWTR is given in Sec. **9.23**, where it is shown how the error returns IER and ITER are processed.

A listing of NEWTR follows. Note that ITMAX has been set at 10. This is a practical choice for most problems, but it is easily changed if the user so desires. Perhaps the only essential point not covered by the many comment statements is that DGELG requires that the columns of matrix A be reassembled into a *vector* array A1. The repacking of A(I,J) into A1(K) is done in NEWTR right after the statement CALL MATRIX.

Background information on the usage of the linear equation subroutine DGELG is provided in Appendix **E**.

```
C          SUBROUTINE NEWTR
C          BY B. PAUL, 1976
C          PURPOSE
C             TO SOLVE A SYSTEM OF NONLINEAR ALGEBRAIC EQUATIONS
C             OF FORM   F(X)=0
C
C          USAGE
C             CALL NEWTR(X,A,NEQS,FTOL,ITER,IER,F)
C
C          DESCRIPTION OF PARAMETERS
C             X(I)    - DOUBLE PRECISION VECTOR OF INITIAL ESTIMATES.
C                       ON RETURN X CONTAINS THE SOLUTIONS OF THE
C                       EQUATIONS.
C             A(I,J)  - DOUBLE PRECISION JACOBIAN MATRIX SUPPLIED BY
C                       SUBROUTINE MATRIX
C             NEQS    - THE NUMBER OF EQUATIONS IN THE SYSTEM.
C             FTOL    - DOUBLE PRECISION INPUT CONSTANT WHICH IS USED
C                       AS TOLERANCE FOR RESIDUAL TEST.
C             ITER    - AN OUTPUT VALUE, WHICH SPECIFIES THE NUMBER OF
C                       ITERATIONS PERFORMED. IF ITER GETS GREATER THAN
```

---

[2]For an example, see Sec. **9.23**.
[3]See Appendix **E.8** for listing.

```
C                      ITMAX(=10),SUBROUTINE NEWTR RETURNS WITH ERROR
C                      MESSAGE ITER =-1 INTO MAIN PROGRAM
C                      TO CHANGE VALUE OF ITMAX CHANGE STATEMENT 1
C           IER      - RESULTING ERROR PARAMETER FROM SUBROUTINE DGELG.
C                      DESCRIBED IN THE INTRODUCTION TO SUBROUTINE DGELG.
C                      IF IER IS NOT EQUAL TO ZERO, SUBROUTINE NEWTR
C                      RETURNS TO MAIN PROGRAM WITH THIS VALUE OF IER.
C           F(I)     - RESIDUAL FUNCTIONS RETURNED FROM SUBROUTINE RESFNC
C        SUBROUTINES AND FUNCTION SUBPROGRAMS REQUIRED:
C           1. SUBROUTINE DGELG (R,A,M,N,EPS,IER,DET), A DOUBLE PRECI-
C              SION PROGRAM FROM IBM'S SSP LIBRARY USED TO SOLVE A
C              GENERAL SYSTEM OF LINEAR ALGEBRAIC EQUATIONS.
C           2. SUBROUTINE MATRIX, JACOBIAN MATRIX
C           3. SUBROUTINE RESFNC, RESIDUAL FUNCTIONS, F(I)
C
C        METHOD
C           SOLUTION IS DONE BY MEANS OF A NEWTON - RAPHSON ITERATION.
C        ....................................................................
      SUBROUTINE NEWTR(X,A,NEQS,FTOL,ITER,IER,F)
      DOUBLE PRECISION X,A,FTOL,F,DABS,A1,DET,EPS
C
C        DIMENSION SET FOR 20 EQS. AND 20 VALUES OF X
C         USER MAY ALTER IF DESIRED
      DIMENSION X(20),A(20,20),F(20),A1(400)
C
C        SET EPS=.1D-13 FOR NUMERICAL ZERO OF DETERMINANT IN SUBROUTINE
C        DGELG. USER CAN CHANGE VALUE IF NECESSARY.
      EPS=.1D-13
C
C          SET MAXIMUM ITERATIONS TO BE PERFORMED
    1 ITMAX=10
C
C        BEGIN ITERATIONS AND COMPUTE RESIDUALS
      ITER=0
    5 ITER=ITER+1
      CALL RESFNC
C
C        SCAN RESIDUALS FOR EXCESSIVE TOLERANCE
      DO 10 I=1,NEQS
      IF (DABS(F(I)).GT.FTOL) GO TO 15
   10 CONTINUE
C
C        ALL RESIDUALS SMALL ENOUGH. RETURN
      ITER=ITER-1
      RETURN
C
C        CHECK WHETHER NUMBER OF ITERATIONS .EQ. ITMAX
   15 IF(ITER.EQ.ITMAX)GO TO 40
C
C        FORM JACOBIAN MATRIX. CONTINUE ITERATING
      CALL MATRIX
C
C          REPACK MATRIX A BY COLUMNS TO FORM VECTOR A1
      K=1
      DO 25 I=1,NEQS
      DO 25 J=1,NEQS
      A1(K)=A(J,I)
      K=K+1
   25 CONTINUE
C
C        FIND INCREMENTS IN EQUATION VARIABLES
      DO 30 I=1,NEQS
   30 F(I)=-F(I)
      CALL DGELG (F,A1,NEQS,1,EPS,IER,DET)
      IF (IER.NE.0) RETURN
```

```
C
C      UPDATE ALL EQUATION VARIABLES. CONTINUE ITERATING
       DO 35 I=1,NEQS
 35    X(I)=X(I)+F(I)
       GO TO 5
C
C       RETURNS WITH ERROR MESSAGE
  40 ITER=-1
     RETURN
     END
```

APPENDIX G

NUMERICAL SOLUTION OF DIFFERENTIAL EQUATIONS

G.1 GENERAL

A set of n first-order differential equations of the form

$$y'_1 = \frac{dy_1}{dx} = f_1(x, y_1, y_2, \ldots, y_n)$$
$$\ldots$$
$$y'_n = \frac{dy_n}{dx} = f_n(x, y_1, y_2, \ldots, y_n)$$

(1a)

is said to be in **standard form**. These equations may be compactly expressed in the subscripted form

$$y'_i = \frac{dy_i}{dx} = f_i(x, y_j) \qquad (i = 1, 2, \ldots, n)$$

(1b)

or in the matrix-vector form

$$\mathbf{y}' = \frac{d\mathbf{y}}{dx} = \mathbf{f}(x, \mathbf{y})$$

(1c)

If the known **initial values** $y_{10}, y_{20}, \ldots, y_{n0}$ are specified at the point x_0, Eqs. (1) constitute an **initial value problem** for the determination of a set of functions $y_1(x)$, $y_2(x), \ldots, y_n(x)$ which may be viewed as n curves drawn in the x, y plane. These curves will exist and be unique[1] provided that certain (rather weak) smoothness conditions are satisfied by the functions $f_i(x, y_j)$.

---

[1]For existence and uniqueness theorems on *systems* of ODE's, see Ince [1956, Chap. 3].

A **numerical solution** of the nth-order initial value problem consists of a set of numerical values for each of the n functions $y_i(x)$ at a discrete number of predetermined values $x_2, x_3, \ldots, x_j, \ldots$ of the independent variable x. We shall use capital letters Y_i to distinguish the (approximate) calculated values from y_i, the (exact) theoretical values.

Numerical methods[2] for initial value problems of ODE's fall into two broad categories:

1. **One-step** or **direct methods** (e.g., Runge-Kutta methods)
2. **Multistep** methods; also called **predictor-corrector** or **iterative** methods (e.g., methods of Adams, Moulton, Hamming, etc.)

In the first category, the desired values $Y_i(x_j)$ are calculated by a definite number of calculations utilizing information about the values Y_i and their derivatives at the previous point x_{j-1}. On the other hand, most[3] predictor-corrector methods can only be used after the solution has been started by a one-step method and tend to require a constant step size $h \equiv x_{j+1} - x_j$.

Well-tested and efficient computer programs exist for both one-step and multistep methods, and both have been applied with great success to problems of kinematics and dynamics. However, since a one-step program must be available to "start up" a multistep program and because of the flexibility they permit in choice of step size, it is sufficient to focus our attention on the one-step methods.

G.2 EULER'S METHOD AND EULER'S IMPROVED METHOD

Given a single equation in one variable

$$y' = f(x, y) \tag{1}$$

with initial condition

$$y(x_1) \equiv y_1 \tag{2}$$

we seek to find the integral curve $y(x)$ passing through the point $P_1(x_1, y_1)$ in the x, y plane.

More precisely, we seek numerical approximations Y_2, Y_3, etc., corresponding to $y(x)$ at the points $x_2, x_3, \ldots$ which are spaced at given intervals

$$h_j \equiv x_{j+1} - x_j \tag{3}$$

along the x axis. The quantity h_j is called the **step size**. The value Y_1, of course, is not approximate but equals the given initial value y_1.

---

[2]The literature on the topic is vast. A few selected references, in (roughly) increasing order of difficulty are Hamming [1971], James et al. [1968], McCracken and Dorn [1964], Carnahan et al. [1969], Ralston [1965], and Gear [1971].

[3]Starting difficulties for multistep methods can be circumvented by the use of *variable-order* integration rules, described by Gear [1971].

A first approximation to Y_2 is given by the straight-line extrapolation formula

$$Y_2 = Y_1 + k_1 h \tag{4}$$

where h is the local step size, and

$$k_1 \equiv \left(\frac{dy}{dx}\right)_{P_1} \equiv f(x_1, Y_1) \tag{5}$$

is the initial slope of the integral curve.

The point $P_2(x_2, Y_2)$ may be considered a new initial point, and the process just described can be repeated to find additional points $(P_3, P_4, \ldots)$ on the approximate integral curve. The general formulas are

$$k_1 = f(x_i, Y_i) \tag{6}$$
$$Y_{i+1} = Y_i + k_1 h \tag{7}$$

Equations (6) and (7) constitute Euler's method. Comparing Eqs. (7) against the Taylor series

$$Y_{i+1} = Y_i + hf(x_i, Y_i) + \tfrac{1}{2}h'^2 f'(x_i, Y_i) + \cdots \tag{7a}$$

we see that the first term neglected or *truncated* in the Taylor series is that involving h^2. Thus the method is said to have **second-order truncation error** (or **first-order accuracy**).

Obviously the calculated solution will drift away from the true solution as x increases. Of course, reducing h to an extremely small value will postpone the inevitable drift but will increase the amount of labor significantly if step-size control is the only weapon used in the battle for accuracy.

A more sophisticated method of error control is utilized in the so-called improved Euler method. In this method, it is recognized that the slope k_1, used in determining Y_2, should really have been replaced by the mean slope $\bar{k}$ of the solution curve in the interval $[x_1, x_2]$. A good approximation to the mean slope is given by the average value k of the slopes at P_1 and P_2; i.e.,

$$\bar{k} \doteq \tfrac{1}{2}[k_1 + f(x_2, y_2)] \tag{8}$$

Unfortunately, the true value y_2 is unknown, but Eq. (4) provides a good approximation to it, and we may use the approximation

$$f(x_2, y_2) \doteq f(x_2, Y_1 + k_1 h_1) \equiv k_2 \tag{9}$$

Thus, the mean slope is given to a good approximation by

$$\bar{k} \equiv \tfrac{1}{2}(k_1 + k_2) \tag{10}$$

and we may replace Eq. (4) by the better approximation

$$Y_2 = Y_1 + \bar{k}h \tag{11}$$

Utilizing the same idea to generate $Y_3, Y_4, \ldots$, we formulate the algorithm

$$k_1 = f(x_i, Y_i) \tag{12}$$

$$k_2 = f(x_i + h, Y_i + k_1 h) \tag{13}$$

$$\bar{k} = \tfrac{1}{2}(k_1 + k_2) \tag{14}$$

$$Y_{i+1} = Y_i + \bar{k}h \tag{15}$$

Note that the subscripts in k_1 and k_2 are not related to the subscript of point x_i. Equations (12)–(15) constitute an algorithm known as **Euler's improved method**, or **Heun's method**.

To estimate the truncation error in Euler's improved method, we write Eq. (15) in the slightly rearranged form

$$Y_{i+1} = Y_i + hf(x_i, Y_i) + \frac{1}{2}h^2 \left[\frac{f(x_i + h, Y_i + k_1 h) - f(x_i, Y_i)}{h} \right] \tag{16}$$

In the limit, as $h \to 0$, the term in brackets approaches $df/dx = d^2y/dx^2$, evaluated at point x_i. Hence Eq. (16) represents a Taylor series truncated after the term containing h^2, and Euler's improved method is said to have **second-order accuracy** (or **third-order truncation error**).

G.3 RUNGE-KUTTA METHODS

The Euler methods belong to a wider class of methods called **Runge-Kutta methods**. All Runge-Kutta methods have the following characteristics:

1. They are self-starting methods which do not require calculation of any derivatives of $f(x, y)$.
2. They calculate succeeding Y values by an algorithm of the type

$$Y_{i+1} = Y_i + \bar{k}h \tag{1}$$

where $\bar{k}$ is an *average* slope calculated from a formula of type

$$\bar{k} = w_1 k_1 + w_2 k_2 + \cdots + w_m k_m \tag{2}$$

where

$$k_j = f(x + \alpha_j h, \ Y_i + \beta_1 k_1 + \beta_2 k_2 + \cdots + \beta_{j-1} k_{j-1}) \tag{3}$$

and w_j, α_j, β_j are all fixed constants.
3. They agree with Taylor series expansions up to terms including h^p, where p is called the **order** of the method.

Euler's method is therefore a first-order Runge-Kutta method, and Euler's improved method is a second-order Runge-Kutta method. The most commonly used

Runge-Kutta methods are fourth-order methods. The general algorithm for one such method (using Runge's choice of coefficient) is

$$k_1 = f(x_i, Y_i) \tag{4}$$

$$k_2 = f(x_i + \tfrac{1}{2}h, Y_i + \tfrac{1}{2}hk_1) \tag{5}$$

$$k_3 = f(x_i + \tfrac{1}{2}h, Y_i + \tfrac{1}{2}hk_2) \tag{6}$$

$$k_4 = f(x_i + h, Y_i + hk_3) \tag{7}$$

$$\bar{k} = \tfrac{1}{6}(k_1 + 2k_2 + 2k_3 + k_4) \tag{8}$$

$$Y_{i+1} = Y_i + \bar{k}h \tag{9}$$

Other choices of constants in Eqs. (2) and (3) result in other forms of Runge-Kutta formulas. Examples are given by Romanelli [1960], who includes a set of formulas devised by Gill in 1950 in order to cut down on computer storage requirements and to control the growth of roundoff error. Gill's coefficients are used in the IBM subroutine DRKGS, described below.

G.4 ACCURACY CONTROL

To judge the accuracy of the Runge-Kutta method, it is a common practice to compute two solutions simultaneously. The first solution is designated by $Y^{(1)}$ at a typical point x (we can drop the subscripts on Y_i and x_i for simplicity at this point) and is computed for two successive steps of length h to find $Y^1(x + 2h)$. The second solution is designated by $Y^{(2)}$ (at the same point x) and is computed using a step size of $2h$ to find $Y^{(2)}(x + 2h)$.

If the method used has accuracy of order p, it may be shown, by a method known as *Richardson extrapolation*,[4] that the truncation error in $Y^{(1)}$ is approximately

$$\delta = y - Y^{(1)} = \frac{Y^{(1)} - Y^{(2)}}{2^p - 1} \tag{1}$$

For a fourth-order Runge-Kutta method, we have $p = 4$, and

$$\delta = y - Y^{(1)} = \tfrac{1}{15}(Y^{(1)} - y^{(2)}) \tag{2}$$

Therefore an effective strategy for controlling error is to compute the error term δ at each point x_j. If $|\delta|$ is greater than some user-supplied tolerance[5] ϵ_2, then h is too large, and it should be cut in half. If $|\delta|$ is smaller than $\epsilon_2/50$, the h value is unnecessarily small, and it may be doubled.[6] However, care should be taken that h is never permitted to exceed the maximum value required by the user. When $|\delta|$ falls within the desired limits, Eq. (2) should be approximated in the form

$$y = Y^{(1)} + \tfrac{1}{15}(Y^{(1)} - Y^{(2)}) \tag{3}$$

---

[4]Hildebrand [1956, pp. 78, 238].
[5]Typically, choose ϵ_2 to be about 10^{-3} or 10^{-4} times a representative expected value of y.
[6]This is the strategy followed in IBM subroutine DRKGS.

G.5   EXTENSION TO SYSTEMS OF EQUATIONS

Given a system of ordinary differential equations in the standard form

$$\frac{dy_i}{dx} = f_i(x, y_1, y_2, \ldots, y_n) \qquad (i = 1, 2, \ldots, n) \tag{1}$$

with initial values

$$y_i(x_1) \equiv y_{i1} \tag{2}$$

we may readily extend the Runge-Kutta algorithm, defined by Eqs. (4)–(9) of Sec. **G.3**, to the case in hand. First we introduce the notation

$$Y_{ij} = Y_i(x_j) \tag{3}$$

Then the intermediate slopes defined by Eqs. (4)–(7) of Sec. **G.3** must be generalized to give four such slopes for each of the n different functions $y_i(x)$ being developed, and we must take into account the fact that f_i depends not only on x_j but also on each of the n functions $y_1(x), y_2(x), \ldots, y_n(x)$. Hence the formulas for the four intermediate slopes associated with Runge's coefficients are

$$k_{1i} = f_i(x_j, Y_{1j}, Y_{2j}, \ldots, Y_{nj}) \tag{4a}$$
$$k_{2i} = f_i(x_j + \tfrac{1}{2}h, Y_{1j} + \tfrac{1}{2}hk_{11}, Y_{2j} + \tfrac{1}{2}hk_{12}, \ldots, Y_{nj} + \tfrac{1}{2}hk_{1n}) \tag{4b}$$
$$k_{3i} = f_i(x_j + \tfrac{1}{2}h, Y_{1j} + \tfrac{1}{2}hk_{21}, Y_{2j} + \tfrac{1}{2}hk_{22}, \ldots, Y_{nj} + \tfrac{1}{2}hk_{2n}) \tag{4c}$$
$$k_{4i} = f_i(x_j + h, Y_{1j} + hk_{31}, Y_{2j} + hk_{32}, \ldots, Y_{nj} + hk_{3n}) \tag{4d}$$

The mean slope, given by Eq. (8) of Sec. **G.3**, generalizes to

$$\bar{k}_i = \tfrac{1}{6}(k_{1i} + 2k_{2i} + 2k_{3i} + k_{4i}) \tag{5}$$

and the final updated values of Y_i are

$$Y_i(x_j + h) \equiv Y_{i, j+1} = Y_{ij} + \bar{k}_i h \tag{6}$$

In a similar manner, the **error monitor** δ, defined by Eq. (**G.4**-2), must be generalized to read

$$\delta = \tfrac{1}{15} \sum [a_1|Y_1^{(1)} - Y_1^{(2)}| + a_2|Y_2^{(1)} - Y_2^{(2)}| + \cdots + a_n|Y_n^{(1)} - Y_n^{(2)}|] \tag{7}$$

where $Y_i^{(1)}$ is the value of $Y_i(x_j + 2h)$ calculated with a step size of h and $Y_i^{(2)}$ is the same, calculated with a step size $2h$.

The terms $a_1, a_2, \ldots, a_n$ are user-supplied weighting factors which are introduced to help scale down (or up) any variables Y_i which may be disproportionately large (or small). The only requirement on the **error weights** a_i is that they must all be *positive* and *their sum must be 1*. When no a priori information is available on the relative size of the (as yet unknown) functions $y_i(x)$, it is usually safe to set all a_i equal, i.e., to set

$$a_i = \frac{1}{n} \qquad (8)$$

To illustrate other choices of a_i, suppose that we are seeking three functions (y_1, y_2, y_3) where $y_1 =$ order of $10y_3$ and $y_2 =$ order of $100y_3$. If we set

$$a_1 = 0.1, \qquad a_2 = 0.01, \qquad a_3 = 1 - 0.1 - 0.01 = 0.89$$

then $\sum a_i = 1$, and all terms in Eq. (7) would contribute in roughly the same order to the error monitor δ.

If we wish to keep each of the terms $a_i | Y_i^{(2)} - Y_i^{(1)} |$ less than 10^{-4}, it is necessary, in this example, that

$$\delta < \tfrac{1}{15}(10^{-4} + 10^{-4} + 10^{-4}) = 2 \times 10^{-5}$$

and we would select the tolerance on δ to be

$$\epsilon_2 = 2 \times 10^{-5}$$

G.6 COMPUTER PROGRAM DRKGS

The fourth-order Runge-Kutta algorithm with Gill's coefficients has been implemented in the IBM subroutines RKGS (single precision) and DRKGS (double precision). Full documentation on these FORTRAN subroutines will be found in IBM [1970]. However, the program source listing is almost self-documented, as may be seen below in the listing of DRKGS.

The user must supply a MAIN PROGRAM which supplies all input data and calls upon DRKGS to solve the nth-order initial value problem:

$$\left. \begin{aligned} \frac{dy_i}{dx} &= f_i(y_1, y_2, \ldots, y_n; x) \\ y_i(x_0) &= y_{i0} \end{aligned} \right\} \quad (i = 1, 2, \ldots, n) \qquad \begin{matrix} (1) \\ \\ (2) \end{matrix}$$

Within DRKGS, x is called X, and the values of y_i are stored in the array Y(I).

The user must provide a subroutine called FCT(X, Y, DERY), wherein the right-hand sides of Eqs. (1) are calculated and stored in the array DERY (I).

In addition, the user must provide a subroutine called OUTP (X, Y, DERY, IHLF, NDIM, PRMT), wherein

IHLF: an output value which specifies the number of times that the user-supplied initial time step (h) gets bisected

NDIM: n (number of equations)

PRMT(1): x_0 (initial x)

PRMT(2): x_{final} (user-desired final value of x)

PRMT(3): h (user-supplied initial step length)

PRMT(4): ϵ_2 (tolerance on δ; see Sec. **G.5**)

PRMT(5): this is initially set at zero by DRKGS and is changed to a nonzero value when the user desires to exit from DRKGS

PRMT(6), ... PRMT(M): optional constants which the user may introduce for his own purposes

On first calling DRKGS, the user must read the initial values of y_{i0} into the array Y(I), and the error weights a_i [see Eq. (**G.5**-7)] must be read into the array DERY(I). Typically, the user-supplied main program will start with the statements

```
IMPLICIT  REAL * 8 (A-H,O-Z)
EXTERNAL  OUTP,FCT
DIMENSION  PRMT(99),DERY(50),AUX(8,50).
```

The IMPLICIT statement is a way of specifying double precision for all floating-point numbers. The EXTERNAL statement is needed to specify that the subprogram names OUTP and FCT appear in the *argument list* of the call to DRKGS by the main program, i.e.,

```
CALL  DRKGS(PRMT,Y,DERY,NDIM,IHLF,PRMT,FCT,OUTP, AUX)
```

The DIMENSION statement in this example makes provision for up to 99 elements in the array PRMT and for up to 50 differential equations ($n = 50$).

The array AUX is used internally by DRKGS and must always have the dimensions $(8, n)$.

For an example of how to prepare subroutines FCT, see Sec. **12.83**. An example of how to prepare subroutine OUTP is given in Appendix **I**, where a main program (DYREC) calls upon DRKGS, which in turn calls upon both OUTP (listed after DYREC in Appendix **I**) and FCT.

The source listing[7] follows.

---

[7]Reprinted by permission from *System/360 Scientific Subroutine Package, Version III, Programmer's Manual—GH20-0205-4*, p. 333, © 1970 by International Business Machines Corporation, White Plains, N.Y.

```
C          SUBROUTINE DRKGS
C
C          PURPOSE
C             TO SOLVE A SYSTEM OF FIRST ORDER ORDINARY DIFFERENTIAL
C             EQUATIONS WITH GIVEN INITIAL VALUES.
C
C          USAGE
C             CALL DRKGS (PRMT,Y,DERY,NDIM,IHLF,FCT,OUTP,AUX)
C             PARAMETERS FCT AND OUTP REQUIRE AN EXTERNAL STATEMENT.
C
C          DESCRIPTION OF PARAMETERS
C             PRMT    - DOUBLE PRECISION INPUT AND OUTPUT VECTOR WITH
C                       DIMENSION GREATER THAN OR EQUAL TO 5, WHICH
C                       SPECIFIES THE PARAMETERS OF THE INTERVAL AND OF
C                       ACCURACY AND WHICH SERVES FOR COMMUNICATION BETWEEN
C                       OUTPUT SUBROUTINE (FURNISHED BY THE USER) AND
C                       SUBROUTINE DRKGS. EXCEPT PRMT(5) THE COMPONENTS
C                       ARE NOT DESTROYED BY SUBROUTINE DRKGS AND THEY ARE
C             PRMT(1)- LOWER BOUND OF THE INTERVAL (INPUT),
C             PRMT(2)- UPPER BOUND OF THE INTERVAL (INPUT),
C             PRMT(3)- INITIAL INCREMENT OF THE INDEPENDENT VARIABLE
C                       (INPUT),
C             PRMT(4)- UPPER ERROR BOUND (INPUT). IF ABSOLUTE ERROR IS
C                       GREATER THAN PRMT(4), INCREMENT GETS HALVED.
C                       IF INCREMENT IS LESS THAN PRMT(3) AND ABSOLUTE
C                       ERROR LESS THAN PRMT(4)/50, INCREMENT GETS DOUBLED.
C                       THE USER MAY CHANGE PRMT(4) BY MEANS OF HIS
C                       OUTPUT SUBROUTINE.
C             PRMT(5)- NO INPUT PARAMETER. SUBROUTINE DRKGS INITIALIZES
C                       PRMT(5)=0. IF THE USER WANTS TO TERMINATE
C                       SUBROUTINE DRKGS AT ANY OUTPUT POINT, HE HAS TO
C                       CHANGE PRMT(5) TO NON-ZERO BY MEANS OF SUBROUTINE
C                       OUTP. FURTHER COMPONENTS OF VECTOR PRMT ARE
C                       FEASIBLE IF ITS DIMENSION IS DEFINED GREATER
C                       THAN 5. HOWEVER SUBROUTINE DRKGS DOES NOT REQUIRE
C                       AND CHANGE THEM. NEVERTHELESS THEY MAY BE USEFUL
C                       FOR HANDING RESULT VALUES TO THE MAIN PROGRAM
C                       (CALLING DRKGS) WHICH ARE OBTAINED BY SPECIAL
C                       MANIPULATIONS WITH OUTPUT DATA IN SUBROUTINE OUTP.
C             Y       - DOUBLE PRECISION INPUT VECTOR OF INITIAL VALUES
C                       (DESTROYED). LATERON Y IS THE RESULTING VECTOR OF
C                       DEPENDENT VARIABLES COMPUTED AT INTERMEDIATE
C                       POINTS X.
C             DERY    - DOUBLE PRECISION INPUT VECTOR OF ERROR WEIGHTS
C                       (DESTROYED). THE SUM OF ITS COMPONENTS MUST BE
C                       EQUAL TO 1. LATERON DERY IS THE VECTOR OF
C                       DERIVATIVES, WHICH BELONG TO FUNCTION VALUES Y AT
C                       INTERMEDIATE POINTS X.
C             NDIM    - AN INPUT VALUE, WHICH SPECIFIES THE NUMBER OF
C                       EQUATIONS IN THE SYSTEM.
C             IHLF    - AN OUTPUT VALUE, WHICH SPECIFIES THE NUMBER OF
C                       BISECTIONS OF THE INITIAL INCREMENT. IF IHLF GETS
C                       GREATER THAN 20, SUBROUTINE DRKGS RETURNS WITH
C                       ERROR MESSAGE IHLF=21 INTO MAIN PROGRAM. ERROR
C                       MESSAGE IHLF=112 OR IHLF=113 APPEARS IN CASE
C                       PRMT(3)=0 OR IN CASE SIGN(PRMT(3)).NE.SIGN(PRMT(2)-
C                       PRMT(1)) RESPECTIVELY.
C             FCT     - THE NAME OF AN EXTERNAL SUBROUTINE USED. THIS
C                       SUBROUTINE COMPUTES THE RIGHT HAND SIDES DERY OF
C                       THE SYSTEM TO GIVEN VALUES X AND Y. ITS PARAMETER
C                       LIST MUST BE X,Y,DERY. SUBROUTINE FCT SHOULD
C                       NOT DESTROY X AND Y.
C             OUTP    - THE NAME OF AN EXTERNAL OUTPUT SUBROUTINE USED.
C                       ITS PARAMETER LIST MUST BE X,Y,DERY,IHLF,NDIM,PRMT.
C                       NONE OF THESE PARAMETERS (EXCEPT, IF NECESSARY,
C                       PRMT(4),PRMT(5),...) SHOULD BE CHANGED BY
```

```
C                         SUBROUTINE OUTP. IF PRMT(5) IS CHANGED TO NON-ZERO,
C                         SUBROUTINE DRKGS IS TERMINATED.
C            AUX     - DOUBLE PRECISION AUXILIARY STORAGE ARRAY WITH 8
C                      ROWS AND NDIM COLUMNS.
C
C       REMARKS
C           THE PROCEDURE TERMINATES AND RETURNS TO CALLING PROGRAM, IF
C           (1) MORE THAN 10 BISECTIONS OF THE INITIAL INCREMENT ARE
C               NECESSARY TO GET SATISFACTORY ACCURACY (ERROR MESSAGE
C               IHLF=11),
C           (2) INITIAL INCREMENT IS EQUAL TO 0 OR HAS WRONG SIGN
C               (ERROR MESSAGES IHLF=12 OR IHLF=13),
C           (3) THE WHOLE INTEGRATION INTERVAL IS WORKED THROUGH,
C           (4) SUBROUTINE OUTP HAS CHANGED PRMT(5) TO NON-ZERO.
C
C       SUBROUTINES AND FUNCTION SUBPROGRAMS REQUIRED
C           THE EXTERNAL SUBROUTINES FCT(X,Y,DERY) AND
C           OUTP(X,Y,DERY,IHLF,NDIM,PRMT) MUST BE FURNISHED BY THE USER.
C
C       METHOD
C           EVALUATION IS DONE BY MEANS OF FOURTH ORDER RUNGE-KUTTA
C           FORMULAE IN THE MODIFICATION DUE TO GILL. ACCURACY IS
C           TESTED COMPARING THE RESULTS OF THE PROCEDURE WITH SINGLE
C           AND DOUBLE INCREMENT.
C           SUBROUTINE DRKGS AUTOMATICALLY ADJUSTS THE INCREMENT DURING
C           THE WHOLE COMPUTATION BY HALVING OR DOUBLING. IF MORE THAN
C           10 BISECTIONS OF THE INCREMENT ARE NECESSARY TO GET
C           SATISFACTORY ACCURACY, THE SUBROUTINE RETURNS WITH
C           ERROR MESSAGE IHLF=11 INTO MAIN PROGRAM.
C           TO GET FULL FLEXIBILITY IN OUTPUT, AN OUTPUT SUBROUTINE
C           MUST BE FURNISHED BY THE USER.
C           FOR REFERENCE, SEE
C           RALSTON/WILF, MATHEMATICAL METHODS FOR DIGITAL COMPUTERS,
C           WILEY, NEW YORK/LONDON, 1960, PP.110-120.
C
C   .............................................................
C
      SUBROUTINE DRKGS(PRMT,Y,DERY,NDIM,IHLF,FCT,OUTP,AUX)
C
C
      DIMENSION Y(2),DERY(NDIM),AUX(8,NDIM),A(4),B(4),C(4),PRMT(10)
      DOUBLE PRECISION PRMT,Y,DERY,AUX,A,B,C,X,XEND,H,AJ,BJ,CJ,R1,R2,
     1DELT    ,DABS
      DO 1 I=1,NDIM
    1 AUX(8,I)=.06666666666666667D0*DERY(I)
      X=PRMT(1)
      XEND=PRMT(2)
      H=PRMT(3)
      PRMT(5)=0.D0
      CALL FCT(X,Y,DERY)
C
C       ERROR TEST
      IF(H*(XEND-X))38,37,2
C
C       PREPARATIONS FOR RUNGE-KUTTA METHOD
    2 A(1)=.5D0
      A(2)=.2928932188134524800
      A(3)=1.707106781186547500
      A(4)=.16666666666666667D0
      B(1)=2.D0
      B(2)=1.D0
      B(3)=1.D0
      B(4)=2.D0
      C(1)=.5D0
      C(2)=.2928932188134524800
```

```
        C(3)=1.7071067811865475D0
        C(4)=.5D0
C
C       PREPARATIONS OF FIRST RUNGE-KUTTA STEP
        DO 3 I=1,NDIM
        AUX(1,I)=Y(I)
        AUX(2,I)=DERY(I)
        AUX(3,I)=0.D0
     3  AUX(6,I)=0.D0
        IREC=0
        H=H+H
        IHLF=-1
        ISTEP=0
        IEND=0
C
C
C       START OF A RUNGE-KUTTA STEP
     4  IF((X+H-XEND)*H)7,6,5
     5  H=XEND-X
     6  IEND=1
C
C       RECORDING OF INITIAL VALUES OF THIS STEP
     7  CALL OUTP(X,Y,DERY,IREC,NDIM,PRMT)
        IF(PRMT(5))40,8,40
     8  ITEST=0
     9  ISTEP=ISTEP+1
C
C
C       START OF INNERMOST RUNGE-KUTTA LOOP
        J=1
    10  AJ=A(J)
        BJ=B(J)
        CJ=C(J)
        DO 11 I=1,NDIM
        R1=H*DERY(I)
        R2=AJ*(R1-BJ*AUX(6,I))
        Y(I)=Y(I)+R2
        R2=R2+R2+R2
    11  AUX(6,I)=AUX(6,I)+R2-CJ*R1
        IF(J-4)12,15,15
    12  J=J+1
        IF(J-3)13,14,13
    13  X=X+.5D0*H
    14  CALL FCT(X,Y,DERY)
        GOTO 10
C       END OF INNERMOST RUNGE-KUTTA LOOP
C
C
C       TEST OF ACCURACY
    15  IF(ITEST)16,15,20
C
C       IN CASE ITEST=0 THERE IS NO POSSIBILITY FOR TESTING OF ACCURACY
    16  DO 17 I=1,NDIM
    17  AUX(4,I)=Y(I)
        ITEST=1
        ISTEP=ISTEP+ISTEP-2
    18  IHLF=IHLF+1
        X=X-H
        H=.5D0*H
        DO 19 I=1,NDIM
        Y(I)=AUX(1,I)
        DERY(I)=AUX(2,I)
    19  AUX(6,I)=AUX(3,I)
        GOTO 9
C
```

```
C       IN CASE ITEST=1 TESTING OF ACCURACY IS POSSIBLE
   20 IMOD=ISTEP/2
      IF(ISTEP-IMOD-IMOD)21,23,21
   21 CALL FCT(X,Y,DERY)
      DO 22 I=1,NDIM
      AUX(5,I)=Y(I)
   22 AUX(7,I)=DERY(I)
      GOTO 9
C
C       COMPUTATION OF TEST VALUE DELT
   23 DELT=0.D0
      DO 24 I=1,NDIM
   24 DELT=DELT+AUX(8,I)*DABS(AUX(4,I)-Y(I))
      IF(DELT-PRMT(4))28,28,25
C
C       ERROR IS TOO GREAT
   25 IF(IHLF-20)26,36,36
   26 DO 27 I=1,NDIM
   27 AUX(4,I)=AUX(5,I)
      ISTEP=ISTEP+ISTEP-4
      X=X-H
      IEND=0
      GOTO 18
C
C       RESULT VALUES ARE GOOD
   28 CALL FCT(X,Y,DERY)
      DO 29 I=1,NDIM
      AUX(1,I)=Y(I)
      AUX(2,I)=DERY(I)
      AUX(3,I)=AUX(6,I)
      Y(I)=AUX(5,I)
   29 DERY(I)=AUX(7,I)
      CALL OUTP(X-H,Y,DERY,IHLF,NDIM,PRMT)
      IF(PRMT(5))40,30,40
   30 DO 31 I=1,NDIM
      Y(I)=AUX(1,I)
   31 DERY(I)=AUX(2,I)
      IREC=IHLF
      IF(IEND)32,32,39
C
C       INCREMENT GETS DOUBLED
   32 IHLF=IHLF-1
      ISTEP=ISTEP/2
      H=H+H
      IF(IHLF)4,33,33
   33 IMOD=ISTEP/2
      IF(ISTEP-IMOD-IMOD)4,34,4
   34 IF(DELT-.02D0*PRMT(4))35,35,4
   35 IHLF=IHLF-1
      ISTEP=ISTEP/2
      H=H+H
      GOTO 4
C
C
C       RETURNS TO CALLING PROGRAM
   36 IHLF=21
      CALL FCT(X,Y,DERY)
      GOTO 39
   37 IHLF=112
      GOTO 39
   38 IHLF=113
   39 CALL OUTP(X,Y,DERY,IHLF,NDIM,PRMT)
   40 RETURN
      END
```

APPENDIX H

NUMERICAL QUADRATURE

Suppose that we wish to calculate a **running integral** (i.e., *indefinite* integral)

$$I(x) = \int_{x_1}^{x} y(x)\, dx \tag{1}$$

where the integrand $y(x)$ is known at points $x_1, x_2, \ldots, x_n$, evenly spaced at interval h along the abscissa; i.e.,

$$x_{i+1} = x_i + h \tag{2}$$

Figure 1 shows a typical curve $y(x)$. Since the integral $I(x)$ is the area under the curve $y(x)$, we may find $I(x)$ by summing the areas of the crosshatched vertical strips shown in the figure.

The simplest approximation to the area under the curve segment AB is found by replacing arc AB with an approximating straight line. Then the area of the trapezoidal

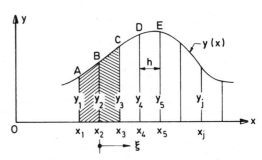

Figure H-1. Illustrating numerical quadrature.

strip under AB is given by

$$\int_{x_1}^{x_2} y \, dx = \tfrac{1}{2}(y_1 + y_2)h \tag{3}$$

Similarly,

$$\int_{x_j}^{x_{j+1}} y \, dx = \tfrac{1}{2}(y_j + y_{j+1})h \tag{4}$$

for any other strip. Equation (4) is called the **trapezoidal rule**.

Considerably greater accuracy will result if the arc ABC is approximated by a parabola passing through points A, B, C rather than by a broken straight line. The equation of the approximating parabola can be expressed in the form

$$y(\xi) = y_2 + \alpha\xi + \tfrac{1}{2}\beta\xi^2 \tag{5}$$

where ξ is measured from x_2 as shown in Fig. 1. From Eq. (5) we find that

$$y_3 \equiv y(h) = y_2 + \alpha h + \tfrac{1}{2}\beta h^2 \tag{6}$$
$$y_1 \equiv y(-h) = y_2 - \alpha h + \tfrac{1}{2}\beta h^2 \tag{7}$$

Upon "adding" Eqs. (6) and (7) we find

$$\beta = \frac{1}{h^2}(y_1 - 2y_2 + y_3) \tag{8}$$

The area under ABC is given by

$$I(x_3) = \int_{-h}^{h} (y_2 + \alpha\xi + \tfrac{1}{2}\beta\xi^2) \, d\xi = \frac{h}{3}(6y_2 + \beta h^2) \tag{9}$$

When β, from Eq. (8), is placed in Eq. (9) we find

$$I(x_3) = \int_{x_1}^{x_3} y \, dx = \frac{h}{3}(y_1 + 4y_2 + y_3) \tag{10}$$

Similarly, the area under arc CDE may be approximated by

$$\int_{x_3}^{x_5} y \, dx = \frac{h}{3}(y_3 + 4y_4 + y_5) \tag{11}$$

In general we have the result

$$\int_{x_j}^{x_{j+2}} y(x) \, dx = \frac{h}{3}(y_j + 4y_{j+1} + y_{j+2}) \tag{12}$$

which is known as **Simpson's rule**. The increase of accuracy when using Simpson's rule rather than the trapezoidal rule more than compensates for the very small increase in computational labor required. It may be shown (Carnahan et al. [1969, p. 173]) that the truncation error (per step) in Simpson's rule is of the order h^5, whereas that in the trapezoidal rule is of order h^3.

To find the running integral $I(x) \equiv \int_{x_1}^{x} y\, dx$, we note that $I(x_1) = 0$ by definition, $I(x_3)$ is given by Eq. (10), and Eq. (12) gives

$$I(x_{j+2}) = I(x_j) + \frac{h}{3}(y_j + 4y_{j+1} + y_{j+2}) \tag{13}$$

With $I(x_3)$ known, Eq. (13) permits calculation of the desired running integral for $x = x_5, x_7, \ldots, x_n$, where n is any odd integer.

If we could find $I(x_2)$, Eq. (13) would also provide the running integral for the points $(x = x_4, x_6, x_8, \ldots, x_{n-2})$ with even subscripts.

One way to find the starting value $I(x_2)$ is to calculate the integrand $y_{1.5} \equiv y[(x_1 + x_2)/2]$ and then use Eq. (12) in the form

$$I(x_2) = \int_{x_1}^{x_2} y\, dx = \frac{h/2}{3}(y_1 + 4y_{1.5} + y_2) \tag{14}$$

From this point on, Eq. (13) may be used to find $I(x_4)$, $I(x_6)$, ... Other procedures could be devised to avoid calculation of the additional ordinate $y_{1.5}^*$, but the use of Eq. (14), as recommended, preserves the high accuracy (fifth-order truncation error) of Simpson's rule.

If the abscissa is divided up into an even number of intervals, we can find the total integral up to any odd-numbered abscissa x_n as follows:

$$\int_{x_1}^{x_n} y\, dx = \frac{h}{3}(y_1 + 4y_2 + y_3) + \frac{h}{3}(y_3 + 4y_4 + y_5)$$

$$+ \frac{h}{3}(y_5 + 4y_6 + y_7) + \cdots + \frac{h}{3}(y_{n-2} + 4y_{n-1} + y_n)$$

$$= \frac{h}{3}[y_1 + 4(y_2 + y_4 + \cdots + y_{n-1}) + 2(y_3 + y_5 + \cdots + y_{n-2}) + y_n] \tag{15}$$

Equation (15) is the form of Simpson's rule best suited to finding a *definite integral* between the limits x_1 and x_n when n is an odd number. When n is even, one may use Eq. (15) to find $\int_{x_1}^{x_{n-1}} y\, dx$ and then compute the area $\int_{x_{n-1}}^{x_n} y\, dx$ of the final strip by a formula analogous to Eq. (14); i.e.,

$$\int_{x_{n-1}}^{x_n} y\, dx = \frac{h/2}{3}(y_{n-1} + 4y_{n-\frac{1}{2}} + y_n) \tag{16}$$

APPENDIX I

PROGRAM DYREC FOR DYNAMICS OF RECIPROCATING MACHINES[1]

```
C      PROGRAM DYREC
C
C      BY B.PAUL AND A.AMIN
C
C      PURPOSE.....
C          DYNAMIC ANALYSIS OF RECIPROCATING MACHINES (SLIDER-CRANK TYPE)
C
C      PROGRAM DESCRIPTION.....
C          THIS PROGRAM COMPUTES TRANSIENT RESPONSE OF SLIDER-CRANK
C      MECHANISM TO USER-SPECIFIED FORCES
C
C      DESCRIPTION OF INPUT VARIABLES.....
C      (ANY CONSISTENT SYSTEM OF UNITS MAY BE USED WITH TIME IN SECONDS
C        AND ANGLES IN RADIANS)
C        TITLE    ANY TITLE DESCRIBING PROBLEM (UP TO 80 CHARACTERS)
C        TZERO    INITIAL TIME
C        TFINAL   FINAL TIME
C        TDELTA   TIME INCREMENT
C        ILEVEL   INDEX FOR LEVEL OF ANALYSIS DESIRED
C                   =0 FOR DISPLACEMENT ANALYSIS ONLY
C                   =1 FOR DISPLACEMENT, VELOCITY, & ACCELERATION ANALYSIS
C                   =2 FOR DISP., VEL., ACCEL., AND REACTION FORCES
C        NCONST   NUMBER OF USER-SUPPLIED CONSTANTS (UP TO 24)
C        R        CRANK RADIUS
C        L        CONNECTING ROD LENGTH
C        A        DISTANCE BETWEEN CRANKPIN & C.G. OF CON. ROD
C        C        DISTANCE FROM CRANK AXIS TO C.G. OF CRANK
C        E        PISTON OFFSET DISTANCE
C        JO       M.O.I. OF CRANK ASSEMBLY ABOUT CRANKSHAFT AXIS
C        JG       M.O.I. OF CON. ROD ABOUT ITS C.G.
C        M1       MASS OF CRANK ASSEMBLY
C        M2       MASS OF CON. ROD
C        M3       MASS OF PISTON
C        THETA    INITIAL VALUE OF CRANK ANGLE IN RADIANS
C        DTHETA   INITIAL VALUE OF CRANK VELOCITY IN RAD/SEC
C        CN(I)    OPTIONAL CONSTANTS TO BE USED IN SUBROUTINE FORCES
C
C      DESCRIPTION OF MAJOR VARIABLES USED INTERNALLY.....
C        B        (L-A)
C        LAMDA    R/L
C        THETA    CRANK ANGLE
C        PHI      CON. ROD ANGLE
```

---

[1]Program version DYREC-MC, for *multicylinder* machines is available from the author upon request.

631

```
C         S         PISTON DISTANCE FROM CRANK AXIS
C         D         PISTON DISTANCE FROM CYLINDER HEAD (D=D1+L+R-S)
C         D1        CLEARANCE DISTANCE (D AT TOP DEAD CENTER)
C         T         TIME
C         DTHETA    (D/DT)THETA
C         DPHI      (D/DT)PHI
C         DD        (D/DT)D=-(D/DT)S
C         DDTHET    (D/DT)DTHETA
C         DDPHI     (D/DT)DPHI
C         DDD       (D/DT)DD
C         GI        GENERALIZED INERTIA
C         CC        CENTRIPETAL COEFFICIENT
C         Q         GENERALIZED FORCE
C         P         GAS FORCE
C         MR        RESISTING MOMENT ABOUT CRANK SHAFT
C         MP        TURNING MOMENT DUE TO PRESSURE ONLY
C         MT        TOTAL TURNING MOMENT (DUE TO PRESSURE AND INERTIA)
C
C         NOTE:     FOR REACTION FORCES SEE NOTATION DESCRIBED UNDER
C                   SUBROUTINE REACTS
C
C         USER-SUPPLIED SUBPROGRAM.....
C            SUBROUTINE FORCES(T,DTHETA,Q) RETURNING Q
C
C         INPUT DATA ARRANGEMENT.....
C                   CARD NO.        FORMAT          VARIABLES
C
C                   1               (20A4)     TITLE
C
C                   2               (3D10.0,2I5) TZERO,TFINAL,
C                                                TDELTA,ILEVEL,NCONST
C                   3               (5D10.0)    R,L,A,C,E
C                   4               (5D10.0)    J0,JG,M1,M2,M3
C                   5               (2D10.0)    THETA,DTHETA
C
C                   6 ONLY IF NCONST.GT.0
C                                   (8F10.0)    CN(I) I=1,2,...
C
C                   7 ONLY IF NCONST.GT.8
C                                   (8F10.0)    CN(I) I=9,10,...
C
C                   8 ONLY IF NCONST.GT.16
C                                   (8F10.0)    CN(I) I=17,18,...
C
      IMPLICIT REAL*8 (A-Z)
      EXTERNAL FCT,OUTP
      DIMENSION Y(2),DERY(2),PRMT(10),AUX(8,2),TITLE(20)
      COMMON/BLK1/R,L,MA,MB,M1,M2,M3,J0,JG,JAB,LAMDA,KPHI,KS,KPHIPR,KSPR
     1,GI,CC,THETA,THETAF,PHI,S,D,E,A,C,ILEVEL,IP
      COMMON/CONSTS/CN(24)
      INTEGER ILEVEL,NCONST,IP,I
   10 READ(5,100,END=99)TITLE
      READ(5,110)TZERO,TFINAL,TDELTA,ILEVEL,NCONST
      READ(5,120)R,L,A,C,E
      READ(5,120)J0,JG,M1,M2,M3
      READ(5,130)THETA,DTHETA
      IF(NCONST.NE.0)READ(5,140)(CN(I),I=1,NCONST)
      WRITE(6,150)TITLE,R,L,A,C,E,J0,JG,M1,M2,M3
      WRITE(6,160)TZERO,TFINAL,TDELTA,THETA,DTHETA
      IF(NCONST.EQ.0)GO TO 20
      WRITE(6,170)
      WRITE(6,180)(I,CN(I),I=1,NCONST)
   20 WRITE(6,190)
      TFINAL=TFINAL+TDELTA-DMOD(TFINAL,TDELTA)
      PRMT(1)=TZERO
      PRMT(2)=TFINAL
```

```
               PRMT(3)=TDELTA
               PRMT(4)=.1D-4*L
               Y(1)=THETA
               Y(2)=DTHETA
               DERY(1)=.5
               DERY(2)=.5
               IP=0
               B=L-A
               LAMDA=R/L
               MA=M2*B/L
               MB=M2*A/L
               JAB=JG-M2*A*B
               CALL DRKGS(PRMT,Y,DERY,2,IHLF,FCT,OUTP,AUX)
               WRITE(6,200)
               GO TO 10
99             WRITE(6,210)
               STOP
100            FORMAT(20A4)
110            FORMAT(3D10.0,2I5)
120            FORMAT(5D10.0)
130            FORMAT(2D10.0)
140            FORMAT(8F10.0)
150            FORMAT(1H1,4X,'PROGRAM DYREC'//5X,20A4///
              15X,'INPUT PARAMETERS'//
              15X,'CRANK RADIUS.................................. R=',1PD14.6/
              25X,'CON.ROD LENGTH............................... L=',1PD14.6/
              35X,'DISTANCE BETWEEN CRANKPIN & C.G. OF CON.ROD  A=',1PD14.6/
              35X,'DISTANCE FROM CRANK AXIS TO C.G. OF CRANK..  C=',1PD14.6/
              45X,'PISTON OFFSET DISTANCE...................... E=',1PD14.6/
              55X,'MOMENT OF INERTIA ABOUT CRANKSHAFT AXIS.... JO=',1PD14.6/
              65X,'MOMENT OF INERTIA ABOUT CG. OF CON.ROD..... JG=',1PD14.6/
              65X,'MASS OF CRANK.............................. M1=',1PD14.6/
              75X,'MASS OF CON.ROD............................ M2=',1PD14.6/
              85X,'MASS OF PISTON............................. M3=',1PD14.6//)
160            FORMAT
              1(5X,'INITIAL TIME.........................TZERO=',1PD14.6/
              15X,'FINAL TIME..........................TFINAL=',1PD14.6/
              25X,'TIME INCREMENT.....................TDELTA=',1PD14.6////
              35X,'INITIAL CONDITIONS: THETA=',0PF7.2,2X,'DTHETA=',1PD16.6)
170            FORMAT(////5X,'OPTIONAL CONSTANTS'//)
180            FORMAT(5X,'CN(',I2,')=',1PD13.6)
190            FORMAT(////5X,'NOTE: IN OUTPUT, ALL ANGULAR RATES ARE GIVEN IN '
              1,'UNITS OF RADIANS AND SECONDS.')
200            FORMAT(//10X,'ANALYSIS COMPLETED')
210            FORMAT(//10X,'LAST CASE COMPLETED')
               END
C
C......................................................................


               SUBROUTINE OUTP(T,Y,DERY,IHLF,NDIM,PRMT)
C
C       THIS ROUTINE CONTROLS ARRANGEMENT OF OUTPUT
C
               IMPLICIT REAL*8 (A-Z)
               DIMENSION Y(2),DERY(2),PRMT(10)
               COMMON/BLK1/R,L,MA,MB,M1,M2,M3,JO,JG,JAB,LAMDA,KPHI,KS,KPHIPR,KSPR
              1,GI,CC,THETA,THETAF,PHI,S,D,E,A,C,ILEVEL,IP
               COMMON/BLK2/MP,MR,P
               COMMON/BLK4/THETS(50),DTHETA(50),DPHI(50),DD(50),DDTHET(50),
              1DDPHI(50),DDD(50),RA(50),RB(50),RO(50),NB(50),ANGA(50),ANGB(50),
              2ANGO(50)
               INTEGER I,IP,NP,JP,ILEVEL
               DATA JP/0/
               TP=DMOD(T,PRMT(3))
               IF(PRMT(3)*.98.GT.TP.AND.TP.NE.0.0)RETURN
```

```
              CALL FCT(T,Y,DERY)
              IF(IP.EQ.0)WRITE(6,5)
              IP=IP+1
              JP=JP+1
              THETAF=THETAF*57.295779.51
              PHI=PHI*57.29577951
              MT=MP-CC*DERY(1)*DERY(1)
              MS=MT-MR
              WRITE(6,10)JP,T,THETAF,D,PHI,GI,CC,MT,MS
              IF(ILEVEL.EQ.0)GO TO 35
              THETS(IP)=THETAF
              DTHETA(IP)=DERY(1)
              DPHI(IP)=DTHETA(IP)*KPHI
              DD(IP)=-DTHETA(IP)*KS
              DDTHET(IP)=DERY(2)
              DDPHI(IP)=DDTHET(IP)*KPHI+DTHETA(IP)*DTHETA(IP)*KPHIPR
              DDD(IP)=-DDTHET(IP)*KS-DTHETA(IP)*DTHETA(IP)*KSPR
              IF(ILEVEL.EQ.1)GO TO 35
              CALL REACTS
      35      IF(IP.EQ.50)GO TO 15
              IF(T.LE.PRMT(2)-.1D-5)RETURN
      15      NP=IP
              IP=0
              IF(T.GE.PRMT(2)-.1D-5)PRMT(5)=1.
              IF(ILEVEL.EQ.0)RETURN
              WRITE(6,20)
              JP=JP-NP
              DO 50 I=1,NP
              JP=JP+1
              WRITE(6,25)JP,THETS(I),DTHETA(I),DD(I),DPHI(I),DDTHET(I),DDD(I)
             1,DDPHI(I)
      50      CONTINUE
              IF(ILEVEL.EQ.1)RETURN
              WRITE(6,30)
              JP=JP-NP
              DO 60 I=1,NP
              JP=JP+1
              WRITE(6,40)JP,RA(I),ANGA(I),RB(I),ANGB(I),RO(I),ANGO(I),NB(I)
      60      CONTINUE
              RETURN
      5       FORMAT(1H1//38X,´DISPLACEMENTS´,20X,´GEN.INERTIA´5X,´CEN.COEFF.´,5
             1X,´TRNG. MOMENT´,5X,´SURPLUS MOMENT´//´ INDEX´,3X,´TIME´,13X,
             2´THETA(DEG)´,8X,´D´,14X,´PHI(DEG)´,11X,´I´,15X,´C´,14X,´MT´,16X,
             3´MS´//)
      25      FORMAT(1X,I3,5X,0PF7.2,4X,1P6D18.6)
      30      FORMAT(1H1//55X,´REACTION FORCES´//32X,´R=RESULTANT OF X AND Y´,15
             1X,´ANGLE=ANGLE BETWEEN R AND X AXIS (DEG)´//
             25X,´INDEX´,11X,´RA´,11X,´ANGLE A´,11X,´RB´,11X,´ANGLE B´,11X,´RO´,
             311X,´ANGLE O´,11X,´NB´//)
      40      FORMAT(5X,I3,4X,1PD18.6,4X,0PF7.2,3X,1PD18.6,4X,0PF7.2,
             11PD18.6,4X,0PF7.2,2X,1PD18.6)
      10      FORMAT(1X,I3,1PD16.6,5X,0PF7.2,4X,1PD16.6,5X,0PF7.2,4X,1P4D16.6)
      20      FORMAT(1H1//50X,´VELOCITIES AND ACCELERATIONS´//´ INDEX´,4X,
             1´THETA(DEG)´,8X,´DTHETA´,12X,´DD´,16X,´DPHI´,14X,´DDTHET´,14X,´DDD
             2´,12X,´DDPHI´//)
              END
```

APPENDIX J

PROGRAM FLYLOOP FOR SIZING FLYWHEELS[1]

```
C      PROGRAM FLYLOOP
C
C      BY B.PAUL AND A.AMIN
C
C      PURPOSE.....
C          TO FIND M.O.I. OF FLYWHEEL FOR GIVEN LEVEL OF SPEED
C          FLUCTUATION
C
C      METHOD.....
C          ENERGY LOOP METHOD.
C          FINDS KINETIC ENERGY DUE TO SURPLUS  TORQUE
C
C      DESCRIPTION OF INPUT VARIABLES.....
C          TITLE     ANY TITLE DESCRIBING PROBLEM (UP TO 80 CHARACTERS)
C          R         CRANK RADIUS
C          L         CONNECTING ROD LENGTH
C          A         DISTANCE BETWEEN CRANKPIN &C.G. OF CON. ROD
C          E         PISTON OFFSET DISTANCE
C          JO        M.O.I. OF CRANK ASSEMBLY ABOUT CRANKSHAFT AXIS
C          JG        M.O.I. OF CON. ROD ABOUT ITS C.G.
C          M2        MASS OF CON. ROD
C          M3        MASS OF PISTON
C          CN(I)     OPTIONAL CONSTANTS TO BE USED IN SUBROUTINE FORCES
C          PERIOD    PERIOD OF MS VS. THETA CURVE, IN DEGREES
C          H         STEP SIZE (DELTA THETA) IN DEGREES
C          CF        COEFFICIENT OF FLUCTUATION
C          VELAVR    AVERAGE CRANK SPEED IN RAD/SEC
C          NCONST    NUMBER OF USER-SUPPLIED CONSTANTS (UP TO 24)
C
C      DESCRIPTION OF MAJOR VARIABLES USED INTERNALLY.....
C          B         (L-A)
C          LAMDA     R/L
C          THETA     CRANK ANGLE
C          S         PISTON DISTANCE FROM CRANK AXIS
C          D         PISTON DISTANCE FROM CYLINDER HEAD (D=D1+L+R-S)
C          D1        CLEARANCE DISTANCE (D AT TOP DEAD CENTER)
C          CC        CENTRIPETAL COEFFICIENT
C          P         GAS FORCE
C          MR        RESISTING MOMENT ABOUT CRANK SHAFT
C          MP        TURNING MOMENT DUE TO GAS FORCE
C          MS        SURPLUS TURNING MOMENT
C          KS        (KS=-MP/P)
```

---

[1]Program version FLYLOOP-MC, for *multicylinder* machines is available from the author upon request.

```
C
C          USER-SUPPLIED SUBPROGRAM.....
C             SUBROUTINE FORCES(P,MR) RETURNING P AND MR
C             NOTE: STANDARD SUBROUTINE FORCES FOR COMBINATION CYCLE IS
C             SUPPLIED
C
C          INPUT DATA ARRANGEMENT.....
C                    CARD NO.          FORMAT          VARIABLES
C
C                       1              (20A4)          TITLE
C                       2              (4D10.0)        R,L,A,E
C                       3              (4D10.0)        J0,J6,M2,M3
C                       4              (2D10.0)        PERIOD,H
C                       5              (2D10.0,I10)    CF,VELAVR,NCONST
C                    6 ONLY IF NCONST.GT.0
C                                      (8F10.0)        CN(I) I=1,2,...
C                    7 ONLY IF NCONST.GT.8
C                                      (8F10.0)        CN(I) I=9,16,..
C                    8 ONLY IF NCONST.GT.16
C                                      (8F10.0)        CN(I) I=17,18,...
C
      IMPLICIT REAL*8 (A-Z)
      DIMENSION MS(201),W(201),THETAS(201),TITLE(20)
      COMMON/BLK1/R,L,A,E,M3,MB,LAMDA,S,D,CC,KS,THETA
      COMMON/CONSTS/CN(24)
      INTEGER NPTS,NCONST,NREV,I
      DATA PI/3.141592654/
C
C.....READ AND PRINT INPUT.....
1     READ(5,5,END=99)TITLE
      READ(5,20)R,L,A,E
      READ(5,20)J0,J6,M2,M3
      READ(5,10)PERIOD,H
      READ(5,25)CF,VELAVR,NCONST
      IF(NCONST.NE.0)READ(5,15)(CN(I),I=1,NCONST)
      WRITE(6,30)TITLE,R,L,A,E,J0,J6,M2,M3,PERIOD,H,CF,VELAVR
      IF(NCONST.EQ.0)GO TO 2
      WRITE(6,43)
      WRITE(6,44)(I,CN(I),I=1,NCONST)
C
C.....INITIALIZE VARIABLES.....
2     WMAX=0.
      WMIN=0.
      W(1)=0.
      THETA=0.
      NPTS=PERIOD/H+1.
      H=H*PI/180.
      HD3=H/3.
      B=L-A
      LAMDA=R/L
      MA=M2*B/L
      MB=M2*A/L
      JAB=J6-M2*A*B
C
C.....COMPUTE SURPLUS TORQUE AT EACH INTERVAL OVER THE CYLCLE.....
      DO 100 I=1,NPTS
      CALL CENTCO
      CALL FORCES(P,MR)
      MP=-KS*P
      MS(I)=MP-MR-CC*VELAVR*VELAVR
      THETAS(I)=THETA*57.29577951
      THETA=THETA+H
100   CONTINUE
C
C.....COMPUTE SURPLUS TORQUE AT HALF- POINT ON THE FIRST INTERVAL
C        INTERVAL.....
      THETA=H/2.
```

```
        CALL CENTCO
        CALL FORCES(P,MR)
        MSH2=-KS*P-MR-CC*VELAVR*VELAVR
C
C.....COMPUTE SURPLUS K.E. OVER THE CYCLE BY SIMPSON'S RULE.....
        W(2)=HD3*(MS(1)+4.*MSH2+MS(2))/2.
        DO 150 I=3,NPTS
        W(I)=W(I-2)+HD3*(MS(I-2)+4.*MS(I-1)+MS(I))
C
C.....SEARCH FOR MAXIMUM & MINIMUM K.E. VALUE.....
        IF(WMAX.GE.W(I))GO TO 110
        WMAX=W(I)
        XVMAX=THETAS(I)
        GO TO 150
110     IF(WMIN.LE.W(I))GO TO 150
        WMIN=W(I)
        XVMIN=THETAS(I)
150     CONTINUE
C
C.....COMPUTE M.O.I. OF FLYWHEEL.....
        WAB=WMAX-WMIN
        IW0=WAB/(CF*VELAVR*VELAVR)
        GIMIN=J0+MA*R*R+JAB*LAMDA*LAMDA
        GIAV=J0+MA*R*R+.5*((M3+MB)*R*R+JAB*LAMDA*LAMDA)
        IW1=IW0-GIAV
        IW2=IW0-GIMIN
        IW3=IW0-J0
        VMIN=VELAVR*(1-CF/2.)
        VMAX=VELAVR*(1.+CF/2.)
C
C.....PRINT THE RESULTS.....
        WRITE(6,50)MR,WAB,IW1,IW2,IW3,VMIN,XVMIN,VMAX,XVMAX.
        WRITE(6,60)
        WRITE(6,70)(I,THETAS(I),MS(I),W(I),I=1,NPTS)
        WRITE(6,80)
        GO TO 1
99      WRITE(6,90)
        STOP
5       FORMAT(20A4)
10      FORMAT(2F10.0)
15      FORMAT(8F10.0)
25      FORMAT(2D10.0,I10)
20      FORMAT(4D10.0)
30      FORMAT(1H1,4X,'PROGRAM FLYLOOP'//5X,20A4///
       15X,'INPUT PARAMETERS'//
       15X,'CRANK RADIUS............................... R=',1PD16.6/
       25X,'CON.ROD LENGTH............................. L=',1PD16.6/
       35X,'DISTANCE BETWEEN CRANKPIN & C.G. OF CON.ROD A=',1PD16.6/
       45X,'PISTON OFFSET DISTANCE..................... E=',1PD16.6/
       55X,'MOMENT OF INERTIA ABOUT CRANKSHAFT AXIS....J0=',1PD16.6/
       65X,'MOMENT OF INERTIA ABOUT CG. OF CON.ROD.....JG=',1PD16.6/
       75X,'MASS OF CON.ROD............................M2=',1PD16.6/
       85X,'MASS OF PISTON.............................M3=',1PD16.6/
       95X,'PERIOD OF MS VS. THETA CURVE..........PERIOD=',1PD16.6/
       15X,'STEP SIZE (DELTA THETA), DEGREES...........H=',1PD16.6/
       25X,'COEFFICIENT OF FLUCTUATION................CF=',1PD16.6/
       35X,'AVERAGE CRANK SPEED (RAD/SEC).........VELAVR=',1PD16.6//)
43      FORMAT(//5X,'OPTIONAL CONSTANTS'/)
44      FORMAT(5X,'CN(',I2,')=',1PD13.6)
50      FORMAT(/5X,70('-')//5X,'RESULTS'//
       15X,'MEAN RESISTING MOMENT                  MR=',1PD14.6/
       15X,'NET(SURPLUS-DEFICIT) WORK             WAB=',1PD14.6/
       25X,'FLYWHEEL (1) MOMENT OF INERTIA        IW1=',1PD14.6/
       35X,'FLYWHEEL (2) MOMENT OF INERTIA        IW2=',1PD14.6/
       35X,'FLYWHEEL (3) MOMENT OF INERTIA        IW3=',1PD14.6//
       45X,'MINIMUM SPEED=',1PD14.6,' RAD/SEC, AT THETA=',0PF7.2,
       52X,'DEG.'/5X,'MAXIMUM SPEED=',1PD14.6,' RAD/SEC, AT THETA=',
       60PF7.2,2X,'DEG.')
60      FORMAT(1H1//5X,'INDEX',5X,'CRANK ANGLE',3X,'SURPLUS TORQUE',5X
       1,'CUMULATIVE WORK'/15X,'THETA(DEG)',10X,'MS',16X,'WS'//)
```

```
70     FORMAT(5X,I3,7X,0PF7.2,4X,1PD16.6,2X,D16.6)
80     FORMAT(//10X,'ANALYSIS COMPLETED')
90     FORMAT(//10X,'LAST CASE COMPLETED')
       END
C
C.............................................................
C
C
       SUBROUTINE CENTCO
C
C     THIS ROUTINE EVALUATES CENTRIPETAL COEFFICIENT (CC).
C
       IMPLICIT REAL*8 (A-Z)
       COMMON/BLK1/R,L,A,E,M3,MB,LAMDA,S,D,CC,KS,THETA
       STHETA=DSIN(THETA)
       CTHETA=DCOS(THETA)
       PHI=DASIN(LAMDA*STHETA-E/L)
       SPHI=DSIN(PHI)
       CPHI=DCOS(PHI)
       S=R*(CTHETA+CPHI/LAMDA)
       KPHI=LAMDA*CTHETA/CPHI
       KS=-R*(STHETA+KPHI*SPHI/LAMDA)
       KPHIPR=-LAMDA*STHETA/CPHI+KPHI*KPHI*DTAN(PHI)
       KSPR=-R*(CTHETA+KPHIPR*SPHI/LAMDA+KPHI*KPHI*CPHI/LAMDA)
       CC=(M3+MB)*KS*KSPR+JAB*KPHI*KPHIPR
       RETURN
       END
C
C.............................................................
C
C
       SUBROUTINE FORCES(P,MR)
C
C     THIS ROUTINE PROVIDES GAS FORCE P ASSOCIATED WITH
C     A FOUR-STROKE COMBINATION CYCLE, AND AVERAGE MOMENT MR
C     DUE TO WORK DONE BY THE ENGINE.
C
C-------------------------------------------------------------
C     CONSTANTS USED
C     CN(1)=CR=COMPRESSION RATIO(V2/V1)
C     CN(2)=AR=ADMISSION RATIO(V5/V4). (AR=1 FOR OTTO CYCLE)
C     CN(3)=KC=POLYTROPIC EXPONENT FOR COMPRESSION
C     CN(4)=KE=POLYTROPIC EXPONENT FOR EXPANSION
C     CN(5)=PM=MAXIMUM PRESURE
C     CN(6)=PEX=EXHAUST PRESSURE
C     CN(7)=PIN=INTAKE PRESSURE
C     CN(8)=PA=ATMOSPHERIC PRESSURE
C     CN(9)=AP=AREA OF PISTON
C-------------------------------------------------------------
       IMPLICIT REAL*8 (A-Z)
       COMMON/BLK1/R,L,A,E,M3,MB,LAMDA,S,D,CC,KS,THETA
       COMMON/CONSTS/CR,AR,KC,KE,PM,PEX,PIN,PA,AP,CN(15)
       INTEGER IFLAG
       DATA PI1,PI2,PI3,PI4,IFLAG/3.141592654,6.283185307,9.424777961,
      112.56637061,0/
       IF(IFLAG.NE.0)GO TO 100
C
C...... ON FIRST CALL TO FORCES, CALCULATE MEAN RESISTING MOMENT MR.....
       IFLAG=1
       D1=2.*R/(CR-1.)
       D5=AR*D1
       D2=D1+R+R
       P2=PIN
       P3=PIN*CR**KC
       P5=PM
       P6=PM*(D5/D2)**KE
       V2=AP*D2
```

```
        V3=AP*D1
        V5=AP*D5
        V6=V2
        V4=V3
        V1=V3
        DELWP=(P5*V5-P6*V6)/(KE-1.)+(P2*V2-P3*V3)/(KC-1.)+P5*(V5-V4)+
       1(PEX-PIN)*(V1-V2)
        MR=DELWP/PI4
C
C..... CALCULATE PRESSURES AND GAS FORCE ON PISTON .....
100     CONTINUE
        D=D1+L+R-S
        IF(THETA.GT.0.0.AND.THETA.LE.PI1)P=PIN
        IF(THETA.GT.PI1.AND.THETA.LT.PI2)P=PIN*(D2/D)**KC
        IF(THETA.EQ.PI2)P=(PI*CR**KC+PM)/2.
        IF(THETA.GT.PI2.AND.D.LE.D5)P=PM
        IF(THETA.GT.PI2.AND.D.GT.D5)P=PM*(D5/D)**KE
        IF(THETA.EQ.PI3)P=(P+PEX)/2.
        IF(THETA.GT.PI3)P=PEX
        IF(THETA.EQ.0.0.OR.THETA.EQ.PI4)P=(PIN+PEX)/2.
C
C.....CONVERT PRESSURE TO FORCE.....
        P=(P-PA)*AP
        RETURN
        END
```

REFERENCES*

ALBERT, C. D. and F. S. ROGERS [1938]. *Kinematics of Machinery*, Wiley, New York. [**3.10, 3.41**].

ANDREWS, G. C. and H. K. KESAVAN [1975]. "The Vector Network Model: A New Approach to Vector Dynamics," *Mech. Mach. Theory, 10*, 57–75. [**14.30, D**].

ANSDALE, R. F. [1968]. *The Wankel RC Engine*, Iliffe, London. [**5.22**].

ARTOBOLEVSKII, I. I. [1964]. *Mechanisms for the Generation of Plane Curves* (trans. by R. D. Wills), Macmillan, New York. [**1.55**].

ARTOBOLEVSKII, I. I. [1975]. *Mechanisms in Modern Engineering Design*, Vol. 1, Lever Mechanisms (trans. from Russian), Mir Publishers, Moscow. Also see *Mekhanizmih*, Acad. of Sci. of U.S.S.R. Inst. of Machine Science, Moscow and Leningrad (1948). [**8.31**].

AVERY, W. & T. [1964]. "Symposium on Dynamic Balancing," Univ. of Birmingham, sponsored by W. & T. Avery Ltd., Birmingham, England. [**13.12**].

BAGCI, C. [1971]. "Degrees of Freedom of Motion in Mechanisms," *J. Eng. Ind. Trans. ASME Ser. B*, 93, 140–148. [**8.21**].

BARBER, T. W. [1934]. *Engineers Sketch Book*, 7th ed., Chemical Publishing, New York. [**1.55**].

BEDINI, S. A. and F. R. MADDISON [1966]. "Mechanical Universe, the Astrarium of Giovanni de' Dondi," *Trans. Am. Philos. Soc., New Series*, 56, Part 5, 1-69. [**3.10**].

BEER, F. P. and E. R. JOHNSTON, JR., [1972]. *Vector Mechanics for Engineers*, McGraw-Hill, New York. [**11.60**].

---

*Boldface characters indicate Sections where a work is cited. An alphabetical list of second and subsequent authors will be found after the main list.

BEGGS, J. S. [1955]. *Mechanism*, McGraw-Hill, New York. **[1.23, 1.55]**.

BELL, E. T. [1965]. *Men of Mathematics*, Simon and Schuster, New York. **[Intro. to Part 3]**.

BERKOF, R. S. and G. G. LOWEN [1969]. "A New Method for Completely Force Balancing Simple Linkages," *J. Eng. Ind. Trans. ASME Ser. B*, 91, 21–26. **[13.31]**.

BERKOF, R. S., G. LOWEN, and F. R. TEPPER [1977]. "Balancing of Linkages," *Shock Vibration Dig.*, 9, no. 6, 3–10. **[13.30]**.

BERRY, A. [1898]. *A Short History of Astronomy*, reprinted by Dover, New York (1961). **[5.14]**.

BEYER, R. [1960]. *Kinematisch-getriebedyamisches Praktikum*, Springer, Berlin. **[12.31]**.

BEYER, R. [1963]. *The Kinematic Synthesis of Mechanisms* (trans. from the German by H. Kuenzel), Chapman and Hall, London. **[5.12]**.

BICKFORD, J. H. [1972]. *Mechanisms for Intermittent Motion*, Industrial Press, New York. **[1.53, 1.55]**.

BIEZENO, C. B. and R. GRAMMEL [1954]. *Engineering Dynamics*, Vol. IV Internal-Combustion Engines (trans. by M. P. White), Blackie, London. **[1.41, 12.30, 12.82, 12.90, 13.20, 13.21, 13.22]**.

BILLINGS, J. H. [1943]. *Applied Kinematics*, Van Nostrand Reinhold, New York. **[3.42]**.

BOCHER, M. [1907]. *Introduction to Higher Algebra*, Macmillan, New York. **[D, E]**.

BOTTEMA, O. [1953]. "On Gruebler's Formulae for Mechanisms," *Appl. Sci. Res.*, A2, 162–164. **[8.21]**.

BOTTEMA, O. [1961]. "Some Remarks on Theoretical Kinematics," in *Proc. Int. Conf. for Teachers of Kinematics*, Shoe String Press, Hamden, Conn., pp. 157–167. **[5.42]**.

BOWDEN, F. P. and D. TABOR [1954]. *The Friction and Lubrication of Solids*, Clarendon Press, Oxford (1954, Part 1; 1964, Part 2). **[11.60]**.

BOWDEN, F. P. and D. TABOR [1956]. *Friction and Lubrication*, Methuen, London. **[11.60]**.

BRANIN, F. H., JR. [1962]. "D.C. and Transient Analysis of Networks Using a Digital Computer" *IRE International Convention Record*, Part 2, 236–256. **[D]**.

BROWN, E. D., R. S. OWENS, and E. R. BOOSER [1969]. "Friction of Dry Surfaces," in Ling et al. [1969]. **[11.60]**.

BROYDEN, C. G. [1970]. "Recent Developments in Solving Nonlinear Algebraic Systems," in *Numerical Methods for Nonlinear Algebraic Equations*, ed. by P. Rabinowitz, Gordon & Breach, New York, pp. 61–73. **[9.30]**.

BUCHSBAUM, F. and F. FREUDENSTEIN [1970]. "Synthesis of Kinematic Structure of Geared Kinematic Chains and Other Mechanisms," *J. Mech.*, 5, 357–397. **[7.30]**.

BUCKINGHAM, E. [1963]. *Analytical Mechanics of Gears*, Dover, New York. **[3.10, 3.41, 3.42]**.

BURSTALL, A. F. [1965]. *A History of Mechanical Engineering*, M.I.T. Press, Cambridge, Mass. **[1.20, 3.10]**.

BURSTALL, A. F. [1969]. *Simple Working Models of Historic Machines*, M.I.T. Press, Cambridge, Mass., or Edward Arnold & Co., London (1968). **[3.10]**.

CAMERON, A. [1966]. *The Principles of Lubrication*, Longmans, Green, London, and Wiley, New York. **[11.60]**.

CAMERON, A. [1971]. *Basic Lubrication Theory*, Halsted Press Div. of J. Wiley, New York. **[11.60]**.

CANDEE, A. H. [1961]. *Introduction to the Kinematic Geometry of Gear Teeth*, Chilton, Philadelphia. **[3.44]**.

CARNAHAN, B., H. A. LUTHER, and J. O. WILKES [1969]. *Applied Numerical Methods*, Wiley, New York. **[F, G, H]**.

CHACE, M. A. and P. N. SHETH [1973]. "Adaptation of Computer Techniques to the Design of Mechanical Dynamic Machinery," *ASME Paper 73-DET-58*, ASME, New York. **[14.30]**.

CHACE, M. A. and D. A. SMITH [1971]. "DAMN—Digital Computer Program for the Dynamic Analysis of Generalized Mechanical Systems," *SAE Trans.*, *80*, 969–983. **[14.30, D]**.

CHAKRABORTY, J. and S. G. DHANDE [1977]. *Kinematics and Geometry of Planar and Spatial Cam Mechanisms*, Wiley, New York. **[4.10]**.

CHEN, F. Y. [1977]. "A Survey of the State of the Art of Cam System Dynamics," *Mech. Mach. Theory*, *12*, 201–224. **[4.10]**.

CHIRONIS, N. P. (ed.) [1965]. *Mechanisms, Linkages, and Mechanical Controls*, McGraw-Hill, New York. **[1.55]**.

CHIRONIS, N. P. (ed.) [1966]. *Machine Devices and Instrumentation*, McGraw-Hill, New York. **[1.55, 9.30]**.

CHIRONIS, N. P. (ed.) [1967]. *Gear Design and Application*, McGraw-Hill, New York. **[1.55, 3.10, 3.20, 3.30]**.

CLARK, R. H. [1963]. "The Steam Engine in Industry and Road Transport," in *Engineering Heritage*, Vol. I, Institution of Mechanical Engineers, London, p. 64. **[5.22]**.

CLARK, W. M. [1943]. *A Manual of Mechanical Movements*, Doubleday, New York [This edition is based on *Five Hundred and Seven Mechanical Movements*, New York (1868), ed. by H. T. Brown, from a series in *The American Artisan*, started in 1864.) **[1.55]**.

COHEN, M. R. and I. E. DRABKIN (eds.) [1948]. *A Source Book in Greek Science*, McGraw-Hill, New York, and Harvard Univ. Press, Cambridge, Mass. **[3.34]**.

CONTE, F. L., G. R. GEORGE, R. W. MAYNE, and J. P. SADLER [1975]. "Optimum Mechanism Design Combining Kinematic and Dynamic-Force Considerations," *J. Eng. Ind. Trans. ASME Ser. B*, *97*, 662–670. **[13.30]**.

COOLIDGE, J. L. [1963]. *A History of Geometrical Methods*, Dover, New York. **[Intro. to Part 1]**.

CRANDALL, S. H. [1962]. "Rotating and Reciprocating Machines," in *Handbook of Engineering Mechanics*, ed. by W. Flügge, McGraw-Hill, New York, Chap. 58. **[13.21]**.

CREDE, C. E. [1951]. *Vibration and Shock Isolation*, Wiley, New York. **[13.20]**.

CROSSLEY, F. R. E. [1954]. *Dynamics in Machines*, Ronald, New York. **[13.31]**.

CROSSLEY, F. R. E. [1964]. "A Contribution to Gruebler's Theory in the Number Synthesis of Plane Mechanisms," *J. Eng. Ind. Trans. ASME Ser. B.*, *86*, 1–8. **[8.21]**.

CROSSLEY, F. R. E. [1965]. "The Permutations of Kinematic Chains of Eight Members or Less from the Graph-Theoretic Viewpoint," *Developments in Theoretical and Applied Mechanics*, Vol. 2 (*Proc. 2nd Southeastern Conf.*, Atlanta, Ga.), 467–486, Pergamon Press, Oxford. **[D]**.

DAHLQUIST, G. and BJÖRCK [1974]. *Numerical Methods*, Prentice-Hall, Englewood Cliffs, N.J. **[E]**.

DAVIES, T. H. [1968]. "An Extension of Manolescu's Classification of Planar Kinematic Chains and Mechanisms of $M \geq 1$, Using Graph Theory," *J. Mech.*, *3*, 87–100. **[8.21]**.

DAVIES, T. H. and F. R. E. CROSSLEY [1966]. "Structural Analysis of Plane Linkages by Fraenke's Condensed Notation," *J. Mech.*, *1*, 171–183. **[8.21]**.

DAVISON, C. ST. C. B. [1963]. "Geared Power Transmission," in *Engineering Heritage*, Vol. I, Institution of Mechanical Engineers, London, 118–123. **[3.10]**.

DEAN, P. M., JR. [1962]. "Gear-Tooth Proportions," in *Gear Handbook*, ed. by D. W. Dudley, McGraw-Hill, New York, Chap. 5. **[3.41, 3.42, 3.43]**.

DE CAMP, L. S. [1974]. *The Ancient Engineers*, Ballantine Books, Division of Random House, New York. **[3.10]**.

DEERE, J. & CO. [1969]. *Fundamentals of Service, Power Trains*, J. Deere & Co., Moline, Ill. **[3.31]**.

DE GARMO, E. P. [1969]. *Materials and Processes in Manufacturing*, Collier-Macmillan, Toronto. **[3.44]**.

DE JONGE, A. E. R. [1943]. "A Brief Account of Modern Kinematics," *Trans. ASME*, *65*, 663–683. **[5.42]**.

DEN HARTOG, J. P. [1956]. *Mechanical Vibrations*, 4th ed., McGraw-Hill, New York. **[2.20, 13.11, 13.12]**.

DE SOLLA PRICE, D.: *See* under Price.

DIJKSMAN, E. A. [1976]. *Motion Geometry of Mechanisms*, Cambridge University Press, Cambridge. **[Intro. to Part 1, 1.33, 2.10, A]**.

DIMAROGONAS, A. D., G. N. SANDOR, and A. G. ERDMAN [1971]. "Synthesis of a Geared N-Bar Linkage," *J. Eng. Ind. Trans. ASME Ser. B*, *93*, 157–164. **[7.30]**.

DIX, R. [1978]. "Dynamic Analysis for Rigid-Link Mechanisms," *Shock and Vibration Digest*, *10*, No. 1, 31–33. **[14.30]**.

DIX, R. C. and LEHMAN, T. J. [1972]. "Simulation of the Dynamics of Machinery," *J. Eng. Ind. Trans. ASME Ser. B*, *94*, 433–438. **[14.30]**.

DOBRJANSKYJ, L. and F. FREUDENSTEIN [1967]. "Some Applications of Graph Theory to the Structural Analysis of Mechanisms," *J. Eng. Ind. Trans. ASME Ser. B*, *89*, 153–158. **[D]**.

DOWSON, D. [1978]. "Men of Tribology," *J. Lubrication Technology. Trans. ASME*, *100*, 148–155. **[11.60]**.

DRACHMANN, A. G. [1963]. *The Mechanical Technology of Greek and Roman Antiquity*, Munksgaard, Copenhagen. **[3.10]**.

DUDLEY, D. W. [1954]. *Practical Gear Design*, McGraw-Hill, New York. **[3.10]**.

DUDLEY, D. W. (ed.) [1962a]. *Gear Handbook*, McGraw-Hill, New York. **[3.10]**.

DUDLEY, D. W. [1962b]. "Gear Arrangements," in *Gear Handbook*, McGraw-Hill, New York, pp. 3.1–3.44. **[3.30, 3.33]**.

DUDLEY, D. W. [1969]. *The Evolution of the Gear Art*, American Gear Manufacturers Association, Washington, D.C. **[3.10]**.

Dugas, R. [1955]. *A History of Mechanics*, Editions du Griffon, Neuchatel-Switzerland. **[Intro. to Part 2, 2.42, 11.30, 12.53]**.

Edwards, J. [1892]. *Calculus*, Macmillan, New York. **[5.21]**.

Ehrich, F. F. [1967]. "The Influence of Trapped Fluids on High Speed Rotor Vibrations," *J. Eng. Ind., Ser. B, 89*, 804. **[2.20]**.

Eksergian, R. [1930–1931]. "Dynamical Analysis of Machines," a series of 15 installments, appearing in *J. Franklin Inst., 209, 210, 211*. **[12.40]**.

Ferguson, E. S. [1962]. "Kinematics of Mechanisms from the Time of Watt," *U.S. Nat. Museum Bull. 228, Paper 27*, contributions from the Museum of History and Technology, 185–230 (available from Supt. Doc., U.S. Govt. Printing Office, Washington, D.C.). **[Intro. to Part 1, 1.20, 1.55, 5.22]**.

Forsythe, G. E., M. A. Malcom, and C. B. Moler [1977]. *Computer Methods for Mathematical Computations*, Prentice-Hall, Englewood Cliffs, N.J. **[D,E]**.

Fox, L. [1965]. *An Introduction to Numerical Linear Algebra*, Oxford Univ. Press, New York. **[D, E]**.

Fox, R. L. [1971]. *Optimization Methods for Engineering Design*, Addison-Wesley, Reading, Mass. **[9.30]**.

Franke, R. [1958]. *Vom Aufbau der Getriebe*, 3rd ed., VDI-Verlag, Dusseldorf. **[8.21]**.

Freudenstein, F. [1955]. "Approximate Synthesis of Four-Bar Linkages," *Trans. ASME, 77*, 853–861. **[9.21]**.

Freudenstein, F. [1962]. "On the Variety of Motions Generated by Mechanisms," *J. Eng. Ind. Trans. ASME Ser. B, 84*, 156–160. **[8.23]**.

Freudenstein, F. [1973]. "Quasi Lumped-Parameter Analysis of Dynamical Systems," Paper 27, *Proc. 3rd Appl. Mech. Conf.*, Oklahoma State Univ., Stillwater, Okla. **[13.31]**.

Freudenstein, F. and E. J. F. Primrose [1963]. "Geared Five-Bar Motion (Parts I and II), *J. Appl. Mech., 30, Trans. ASME Ser. E, 85*, 161–169, 170–175. **[7.30]**.

Freudenstein, F. and B. Roth [1963]. "Numerical Solutions of Systems of Nonlinear Equations," *J. Assoc. Comp. Mach., 10*, 550–556. **[9.30]**.

Freudenstein, F. and G. N. Sandor [1961]. "On the Burmester Points of a Plane," *J. Appl. Mech. 28, Trans. ASME Ser. E, 83*, 41–49. (Discussion 473–475). **[5.12]**.

Freudenstein, F. and G. N. Sandor [1964]. "Kinematics of Mechanisms," in *Mechanical Design and Systems Handbook*, ed. by H. Rothbart, McGraw-Hill, New York, pp. 4.1–4.67. **[1.55]**.

Fuller, D. D. [1958]. "Friction," in *Mechanical Engineers' Handbook*, McGraw-Hill, New York, ed. by T. Baumeister, pp. 3–39 through 3–52. **[11.60]**.

Gantmacher, F. [1970]. *Lectures in Analytical Mechanics*, MIR Publishers, Moscow. **[14.10]**.

Gear, C. W. [1971]. *Numerical Initial Value Problems in Ordinary Differential Equations*, Prentice-Hall, Englewood Cliffs, N.J. **[14.20, G]**.

Gillmor, C. S. [1971]. *Coulomb and the Evolution of Physics and Engineering in Eighteenth-Century France*, Princeton Univ. Press, Princeton, N.J. **[11.60]**.

Givens, E. J. and J. C. Wolford [1969]. "Dynamic Characteristics of Spatial Mechanisms," *J. Eng. Ind. Trans. ASME Ser. B, 91*, 228–234. **[12.84]**.

GLOVER, J. H. [1969]. "How To Construct Ratio and Efficiency Formulas for Planetary Gear Trains," *AGMA Paper* 279.01, Am. Gear Manufacturers Assoc., Washington, D.C. **[3.33]**.

GOLDSTEIN, H. [1950]. *Classical Mechanics*, Addison-Wesley, Reading, Mass. **[12.10, 14.10]**.

GOODMAN, L. E. and W. H. WARNER [1961]. *Statics and Dynamics*, Wadsworth, Belmont, Calif. **[11.60]**.

GOODMAN, T. P. [1958]. "An Indirect Method for Determining Accelerations in Complex Mechanisms," *Trans. ASME, 81*, 1676–1682. **[10.50]**.

GRAFSTEIN, P. and O. B. SCHWARZ [1971]. *Pictorial Handbook of Technical Devices*, Chemical Publications, New York. **[1.55]**.

GRASHOF, F. [1883]. *Theoretische Maschinenlehre*, Verlag L. Voss, Leipzig, pp. 113–118. **[1.31, A]**.

GREEN, W. G. [1956]. *Theory of Machines*, Blackie, London. **[11.60]**.

GREENWOOD, D. T. [1965]. *Principles of Dynamics*, Prentice-Hall, Englewood Cliffs, N.J. **[8.13]**.

GUPTA, K. C., A. BANERJEE, and R. L. FOX [1972]. "Automated Kinematic Analysis of Planar Mechanisms," *ASME Paper* 72-Mech-90, ASME, New York. **[9.30]**.

HAIN, K. [1967]. *Applied Kinematics* (trans. from German ed. of 1961), ed. by D. P. Adams and T. P. Goodman, McGraw-Hill, New York. **[1.55, 5.12, 6.10, 8.21]**.

HALL, A. S., JR. [1958]. "Inflection Circle and Polode Curvature," in *Trans. 5th Conf. Mech.*, Penton Publ. Co., Cleveland, pp. 207–231. **[5.42]**.

HALL, A. S., JR. [1966]. *Kinematics and Linkage Design*, Balt Publishers, West Lafayette, Ind., p. 29. **[1.31, 1.34, 5.12, 8.24, 9.30]**.

HALL, A. S. and E. S. AULT [1943]. "Auxiliary Points Aid Acceleration Analysis," *Mach. Des., 15*, no. 11, 120–123, 234. **[6.11]**.

HAM, C. W. and E. J. CRANE [1948]. *Mechanics of Machinery*, McGraw-Hill, New York. **[3.44, 13.20]**.

HAMMING, R. H. **[1971]**. *Introduction to Applied Numerical Analysis*, McGraw-Hill, New York. **[9.30, D, E, G]**.

HARRISBERGER, L. [1961]. *Mechanization of Motion*, Wiley, New York. **[5.12]**.

HARTENBERG, R. S. and J. DENAVIT [1964]. *Kinematic Synthesis of Linkages*, McGraw-Hill, New York. **[1.20, 1.23, 1.34, 1.42, 1.55, 2.10, 5.12, 5.42, 8.21]**.

HERTZ, H. [1900]. *The Principles of Mechanics*, reprinted in 1956 by Dover, New York. **[8.14]**.

HILDEBRAND, F. B. [1956]. *Introduction to Numerical Analysis*, McGraw-Hill, New York. **[G]**.

HINKLE, R. T. [1960]. *Kinematics of Machines*, 2nd ed., Prentice-Hall, Englewood Cliffs, N.J. **[8.21]**.

HIRSCHHORN, J. [1962]. *Kinematics and Dynamics of Plane Mechanisms*, McGraw-Hill, New York. **[6.22, 12.31]**.

HOLOWENKO, A. R. [1955]. *Dynamics of Machinery*, Wiley, New York. **[11.62, 13.20, 13.21]**.

HRONES, J. A. and G. L. NELSON [1951]. *Analysis of the Four-Bar Linkage*, Technology Press of M.I.T., Cambridge, Mass., and Wiley, New York. **[1.31]**.

HUD, G. C. [1976]. "Dynamics of Inertia Variant Machinery," Ph.D. dissertation, Univ. of Pennsylvania, Philadelphia. [**10.42, 14.10, 14.30**].

HUNT, K. H. [1959]. *Mechanisms and Motion*, Wiley, New York. [**8.21**].

HUSTON, R. L., C. E. PASSERELLO, and M. W. HARLOW [1977]. *UCIN Vehicle-Occupant/Crash Victim Simulation Model*, Structural Mechanics Software Series, Univ. Press of Virginia, Charlottesville, Va. [**14.30.**]

IBM [1970]. *System/360 Scientific Subroutine Package, Version III, Programmer's Manual, Doc. GH20-0205-4*, IBM Technical Publ. Dept., White Plains, New York. [**E, G**].

INCE, E. L. [1956]. *Ordinary Differential Equations*, ed. of 1926 reprinted by Dover, New York. [**8.14, G**].

IRD [1977]. "Dynamic Balancing," Application Rept. No. 111, IRD Mechanalysis, Inc., Columbus, Ohio. [**13.12**].

JAMES, M. L., G. M. SMITH, and J. C. WOLFORD [1968]. *Applied Numerical Methods for Digital Computation, with FORTRAN*, International Textbook Co., Scranton, Pa. [**9.20, 9.21, G**].

JENSEN, P. W. [1965]. *Cam Design and Manufacture*, Industrial Press, New York. [**4.10, 4.20**].

JENSEN, P. W. [1966]. "How To Design Cycloid Gear Mechanisms," in Chironis [1968, pp. 88–91]. [**5.22**].

JONES, F. D. (ed.) [1936–1967]. *Ingenious Mechanisms for Designers and Inventors*, Vol. 1 (1948), Vol. 2 (1936), Vol. 3 (1951), Vol. 4 (1967) Industrial Press, New York. [**1.55**].

JUDGE, A. W. [1947]. *Automobile and Aircraft Engines*, Vol. 1, The Mechanics of Petrol and Diesel Engines, Pitman, New York. [**13.20, 13.21**].

KANE, T. R. [1968]. *Dynamics*, Holt, Rinehart and Winston, New York, pp. 205–206. [**9.30**].

KASNER, E. and J. NEWMAN [1940]. *Mathematics and the Imagination*, Simon and Schuster, New York. [**B**].

KAUFMAN, R. E. [1975]. "Kinematic and Dynamic Design of Mechanisms," in *Shock and Vibration Computer Programs Reviews and Summaries*, ed. by W. & B. Pilkey, Shock and Vibration Information Center, Naval Research Lab., Code 8404, Washington, D.C. [**14.30**].

LORD KELVIN and P. G. TAIT [1879]. *Treatise on Natural Philosophy* (Parts I and II in two vols.), Cambridge Univ. Press, Cambridge; reprinted numerous times through 1923. [**5.42, 8.13, 14.10**].

KHOL, R. (ed.) [1976]. "Mechanical Drives Reference Issue," *Mach. Des.*, 48, no. 13, 12. [**3.42**].

KILMISTER, C. W. and J. E. REEVE [1966]. *Rational Mechanics*, American Elsevier, New York. [**8.11, 8.14**].

KLEIN, A. W. [1917]. *Kinematics of Machinery*, McGraw-Hill, New York. [**1.55**].

KLEVEN, P. S. [1976]. "Curing Slider Hang-ups," *Mach. Des.*, 45, no. 15, 105–109. [**11.61**].

KOENIG, H. E., Y. TOKAD, and H. K. KESAVAN [1967]. *Analysis of Discrete Physical Systems*, McGraw-Hill, New York. [**D**].

KOSTER, M. P. [1974]. *Vibrations of Cam Mechanisms*, Macmillan, London. [**4.10**].

KOŽEŠNÍK, J. [1962]. *Dynamics of Machines*, Noordhoff, Groningen, Netherlands. [**13.21**].

KREYSZIG, E. [1975]. *Advanced Engineering Mathematics*, Wiley, New York. [**D, E**].

KROON, R. P. [1943]. "Balancing of Rotating Apparatus—I," *J. Appl. Mech.*, *10*, *Trans. ASME*, *65*, A225–A228. **[13.12]**.

KROON, R. P. [1944]. "Balancing of Rotating Apparatus—II," *J. Appl. Mech.*, *11*, *Trans. ASME*, *66*, A47–A50. **[13.12]**.

LAMB, H. [1928]. *Statics*, Cambridge Univ. Press, Cambridge, England. **[7.12]**.

LANCZOS, C. [1966]. *The Variational Principles of Mechanics*, 3rd ed., Univ. of Toronto Press, Toronto. **[Intro. to Part 3, 8.13, 11.30]**.

LANGRANA, N. A. and D. L. BARTEL [1975]. "An Automated Method for Dynamic Analysis of Spatial Linkages for Biomechanical Applications," *J. Eng. Ind. Trans. ASME Ser. B*, *97*, 556–574. **[14.30]**.

LESSELLS, J. M., R. A. BAREISS, and A. M. WAHL [1958]. "Machine Elements," in *Mechanical Engineers' Handbook*, ed. by T. Baumeister, McGraw-Hill, New York, pp. 8–11 through 8–98. **[12.90]**.

LEVAI, Z. [1969]. *Bibliography of Planetary Mechanisms*, Budapest Univ. of Technology, Automobile Engineering Department, Budapest. **[3.10]**.

LICHTY, L. [1958]. "Internal Combustion Engines," in *Mechanical Engineer's Handbook*, ed. by T. Baumeister, McGraw-Hill, New York, pp. 9.104, 9.105, 9.158. **[12.90]**.

LIETZMANN, W. [1955]. *Visual Topology*, American Elsevier, New York. [Translated by M. Bruckheimer, from *Anschauliche Topologie*, Oldenbourg-Verlag, Munich (1965).] **[D]**.

LING, F. F., E. E. KLAUS, and R. S. FEIN (eds.) [1969]. *Boundary Lubrication, An Appraisal of World Literature*, ASME, New York. **[11.60]**.

LOWEN, G. G. and R. S. BERKOF [1968]. "Survey of Investigations into the Balancing of Linkages," *J. Mech.*, *3*, 221–231. (This issue contains 11 translations of articles from the Russian and German.) **[13.30]**.

LOWEN, G. G., F. R. TEPPER, and R. S. BERKOF [1974]. "The Quantitative Influence of Complete Force Balancing on the Forces and Moments of Certain Families of Four-Bar Linkages," *Mech. Mach. Theory*, *9*, 299–323. For *Erratum*, see *10*, 361–363 (1975). **[13.30]**.

MABIE, H. H. and F. W. OCVIRK [1975]. *Mechanisms and Dynamics of Machinery*, Wiley, New York. **[4.10]**.

MACH, E. [1960]. *The Science of Mechanics, A Critical and Historical Account of its Development* (trans. by J. McCormack), Open Court, LaSalle, Ill. **[Intro. to Part 3, 11.30]**.

MALEEV, V. L. [1945]. *Internal-Combustion Engines*, McGraw-Hill, New York. **[12.82, 12.85, 12.90]**.

MAXWELL, J. CLERK [1849]. "On the Theory of Rolling Curves," *Trans. Roy. Soc. Edinburgh*, *XVI*, Part V. [In *Collected Papers of J. C. Maxwell*, Vol. 1, Dover, New York (1965), pp. 4–29. **[5.21]**.

MAXWELL, R. L. [1960]. *Kinematics and Dynamics of Machinery*, Prentice-Hall, Englewood Cliffs, N.J. **[4.30, 4.31, 4.32, 4.33, 4.34]**.

McCRACKEN, D. D. and W. S. DORN [1964]. *Numerical Methods and FORTRAN Programming*, Wiley, New York. **[G]**.

McQUERY, D. E. [1973]. "Understanding Balancing Machines," *Am. Mach. Spec. Rept. No. 656*. (Avail. from Schenk Trebel Corp., Deer Park, N.Y. **[13.12]**.

MEIROVITCH, L. [1970]. *Methods of Analytical Mechanics*, McGraw-Hill, New York. **[14.10]**.

MERRITT, H. E. [1971]. *Gear Engineering*, Halsted Press, Div. of Wiley, New York. [**3.20, 3.30, 3.33, 3.40, 3.44**].

MEWES, E. [1968]. "Unbalanced Inertia Forces in Slider-Crank Mechanisms of Large Eccentricity," *J. Appl. Mech., 25, Trans. ASME Ser. E, 90*, 225–232. [**1.41**].

MICHALEC, G. W. [1966]. *Precision Gearing: Theory and Practice*, Wiley, New York. [**3.44**].

MODREY, J. [1959]. "Analysis of Complex Kinematic Chains with Influence Coefficients," *J. Appl. Mech., 26, Trans. ASME Ser. E*, 184–188. Also see discussion by T. J. Goodman, *J. Appl. Mech., 27*, 215–216 (1960). [**6.11, 10.50**].

MOHAN RAO, A. V. and G. N. SANDOR [1971]. "Extension of Freudenstein's Equation to Geared Linkages," *J. Eng. Ind. Trans. ASME Ser. B, 93*, 201-210. [**7.30**].

MOLIAN, S. [1968a]. "Solution of Kinematics Equations by Newton's Method," *J. Mech. Eng. Sci., 10*, 360–362. [**9.20, 9.30**].

MOLIAN, S. [1968b]. *The Design of Cam Mechanisms and Linkages*, American Elsevier, New York. [**1.55, 4.10, 4.20, 4.40, 5.15**].

MOORE, D. F. [1975]. *Principles and Applications of Tribology*, Pergamon, Elmsford, N.Y. [**11.60**].

MOSKALENKO, V. A. [1964]. *Mechanisms* (trans. from Russian ed. of 1963, Iliffe, London), Hayden, New York. [**1.55**].

MURRAY, S. (ed.) [1974]. *Basic Clutches and Transmissions*, Peterson Publishing Co., Los Angeles. [**3.34**].

MUSTER, D. and D. G. STADELBAUER [1976]. "Balancing of Rotating Machinery," in *Shock and Vibration Handbook*, 2nd ed., by C. M. Harris and C. E. Crede, McGraw-Hill, New York, Chap. 39. [**13.12**].

NEEDHAM, J. [1965]. *Science and Civilisation in China*, Vol. 4, Cambridge Univ. Press, Cambridge, pp. 286–303. [**3.10**].

NEKLUTIN, C. N. [1969]. *Mechanisms and Cams for Automatic Machines*, American Elsevier, New York. [**4.10, 7.60**].

NOLLE, H. [1974a]. "Linkage Coupler Curve Synthesis: A Historical Review—I. Development up to 1875," *Mech. Mach. Theory, 9*, 147–168. [**1.31**].

NOLLE, H. [1974b]. "Linkage Coupler Curve Synthesis: A Historical Review—II. Developments After 1875," *Mech. Mach. Theory, 9*, 325–348. [**1.31**].

NOLLE, H. [1975]. "Linkage Coupler Curve Synthesis: A Historical Review—III. Spatial Synthesis and Optimization," *Mech. Mach. Theory, 10*, 41–55. [**1.31**].

NORBYE, J. P. [1971]. *The Wankel Engine*, Chilton, Philadelphia. [**5.22**].

O'CONNOR, J. J. (ed.) [1968]. *Standard Handbook of Lubrication Engineering*, McGraw-Hill, New York. [**11.60**].

OLEKSA, S. A. and D. TESAR [1971]. "Multiply Separated Position Design of the Geared Five-Bar Generator," *J. Eng. Ind. Trans. ASME Ser. B, 93*, 74-84. [**7.30, 8.24**].

OLSSON, U. [1953]. *Non-circular Cylindrical Gears*, Mechanical Engineering Series, Vol. 2, no. 10, Acta Polytechnica, Stockholm (book in English). [**3.20**].

ORAVAS, G. Æ. and L. MCLEAN [1966]. "Historical Development of Energetical Principles in Elastomechanics," *Appl. Mech. Rev.*, Parts I and II, *8*, 647–658, 919–933. [**11.30**].

ORE, O. [1963]. *Graphs and Their Uses*, Mathematical Association of America, Washington. [**D**].

ORLANDEA, N. and M. A. CHACE [1977]. "Simulation of a Vehicle Suspension with the ADAMS Computer Program," *SAE Paper No. 770053*, Soc. of Automotive Engineers, Warrendale, Pa. [**14.30**].

ORLANDEA, N., M. A. CHACE, and D. A. CALAHAN [1977]. "A Sparsity-Oriented Approach to the Dynamic Analysis and Design of Mechanical Systems—Parts I and II," *J. Eng. Ind. Trans. ASME Ser. B*, 99, 733–779, 780–784 [**14.30**].

ORTEGA, J. M. and W. C. RHEINBOLDT [1970]. *Iterative Solution of Nonlinear Equations in Several Variables*, Academic Press, New York. [**9.30**].

PANOVKO, Y. G. and I. I. GUBANOVA [1965]. *Stability and Oscillations of Elastic Systems* (trans. by C. V. Larrick), Consultants Bureau, New York, pp. 127–130. [**14.20**].

PARS, L. A. [1968]. *A Treatise on Analytical Dynamics*, Wiley, New York. [**8.11, 8.13, 11.00, 14.10**].

PAUL, B. [1960]. "A Unified Criterion for the Degree of Constraint of Plane Kinematic Chains," *J. Appl. Mech.*, 27, *Trans. ASME Ser. E*, 82, 196–200 (also see discussion, 751–752). [**Intro. to Part 2, 8.22, D**].

PAUL, B. [1963]. "Planar Librations of an Extensible Dumbbell Satellite," *AIAA J.*, 2, 411–418. [**11.42**].

PAUL, B. [1973]. "Complex Vectors in Kinematics," *Mech. Eng. News*, 10, no. 3, 31–33. Also see *11*, no. 6, for errata. [**B**].

PAUL, B. [1975]. "Analytical Dynamics of Mechanisms—A Computer-Oriented Overview," *Mech. Mach. Theory*, 10, 481–507. [**11.50, 14.10, D**].

PAUL, B. [1977]. "Dynamic Analysis of Machinery Via Program DYMAC," *SAE Paper 770049*, Soc. of Automotive Engineers, Warrendale, Pa. [**10.43, 14.30, D**].

PAUL, B. [1979]. "A Reassessment of Grashof's Criterion," *J. Mechanical Design, Trans. ASME* (to be published). [**1.33, A**].

PAUL, B. and A. AMIN [1978a]. "User's Manual for Program KINMAC (KINematics of MAChinery)," MEAM Rept. Dept. of Mech. Eng. and Appl. Mech., Univ. of Pennsylvania, Philadelphia. [**10.43, D**].

PAUL, B. and A. AMIN [1978b]. "Automated Dynamic Analysis of Machinery with Arbitrary Equations of Constraint," MEAM Report, Dept. of Mech. Eng. and Appl. Mech., Univ. of Pennsylvania, Philadelphia. [**14.44**].

PAUL, B. and G. HUD [1975]. "User's Manual for Program DKINAL, *MEAM Rept.* 74–2 (rev. 1975), Dept. of Mech. Eng. and Appl. Mech., Univ. of Pennsylvania, Philadelphia. [**9.23, 10.43**].

PAUL, B. and G. HUD [1976]. "User's Manual for Program DYMAC-L2 (DYnamics of MAChinery-Lower Pairs Version 2), *MEAM Report 76–5*, Dept. of Mech. Eng. and Appl. Mech., Univ. of Pennsylvania, Philadelphia. [**10.43, 14.30, 14.40, 14.45, D**].

PAUL, B. and D. KRAJCINOVIC [1970a]. "Computer Analysis of Machines with Planar Motion—Part 1: Kinematics," *J. Appl. Mech.*, 37, *Trans. ASME Ser. E*, 92, 697–702. [**Intro. to Part 2, 10.20, 10.42**].

PAUL, B. and D. KRAJCINOVIC [1970b]. "Computer Analysis of Machines with Planar Motion—Part 2: Dynamics," *J. Appl. Mech.*, 37, *Trans. ASME Ser. E*, 92, 703–712. [**Intro. to Part 2, 10.62, 12.62, 14.10, 14.20**].

PAUL, B. and A. SOLER [1972]. "Cable Dynamics and Optimum Towing Strategies for Tethered Submersibles," *J. Marine Tech.*, *6*, no. 2, 34–42. [**14.20**].

PAUL, B., J. W. WEST, and E. Y. YU [1963]. "A Passive Gravitational Attitude Control System for Satellites," *Bell Sys. Tech. J.*, *42*, 2195–2238. [**11.42**].

PELECUDI, C. [1967]. *Bazele analizei mecanismelor* (in Rumanian), Acad. Republicii Socialiste Romania, Bucarest. [**8.21**].

PETERSON, R. E. [1958]. "Gearing," in *Mechanical Engineers' Handbook*, ed. by T. Baumeister, McGraw-Hill, New York, p. 8–102. [**3.42**].

POLLITT, E. P. [1960]. "Some Applications of the Cycloid in Machine Design," *J. Eng. Ind.*, *Trans. ASME Ser. B*, *82*, 407–414. [**5.22**].

PORTER, F. P. [1943]. "Harmonic Coefficients of Engine Torque Curves," *J. Appl. Mech.*, *10*, *Trans. ASME*, *65*, 33–48. [**13.21**].

PRICE, D. DE SOLLA [1974]. "Gears from the Greeks, the Antikythera Mechanism—A Calendar Computer from ca. 80 B.C.," *Trans. Am. Philos. Soc. New Ser.*, *64*, Part 7, 1–70. [**3.10**].

PRICE, D. DE SOLLA [1975]. *Science since Babylon*, University Press, New Haven. [**3.10**].

PROCTOR, R. A. [1878]. *A Treatise on the Cycloid*, London. [**3.10**].

PURDAY, H. F. P. [1962]. *Diesel Engine Designing*, Van Nostrand Reinhold, New York. [**12.82**].

QUINN, B. E. [1949]. "Energy Method for Determining Dynamic Characteristics of Mechanisms," *J. Appl. Mech.*, *16*, *Trans. ASME*, *71*, 283–288. [**12.100**].

RABINOWICZ, E. [1965]. *Friction and Wear of Materials*, Wiley, New York. [**11.60**].

RABINOWITZ, P. [1970]. "A Short Bibliography on Solutions of Systems of Nonlinear Algebraic Equations" in *Numerical Methods for Non-Linear Algebraic Equations*, ed. by P. Rabinowitz, Gordon & Breach, New York, pp. 195–199. [**9.30**].

RAIMONDI, A. A. and J. BOYD [1958]. "A Solution for the Finite Journal Bearing and its Application to Analysis and Design: Parts I, II, III," *Trans. ASLE*, *1*, 159–209. [**11.60**].

RALSTON, A. [1965]. *A First Course in Numerical Analysis*, McGraw-Hill, New York. [**D, G**].

RAMOUS, A. J. [1972]. *Applied Kinematics*, Prentice-Hall, Englewood Cliffs, N.J. [**3.31**].

RAVEN, F. H. [1958]. "Velocity and Acceleration Analysis of Plane and Space Mechanisms by Means of Independent-Position Equations," *J. Appl. Mech.*, *25*, *Trans. ASME*, *Ser. E*, *80*, 1–6. [**Intro. to Part. 2**].

RAVEN, F. H. [1959]. "Analytical Design of Disk Cams and Three-Dimensional Cams by Independent Position Equations," *J. Appl. Mech.*, *26*, *Trans. ASME Ser. E*, *81*, 18–24. [**4.41, 4.43**].

REULEAUX, F. [1876]. *The Kinematics of Machinery* (trans. and annotated by A. B. W. Kennedy), reprinted by Dover, New York (1963). [**1.20, 1.21, 1.22, 1.33, 1.55, 3.10, 5.13, 5.21, 5.22, 8.14**].

ROE, P. H. [1966]. *Networks and Systems*, Addison-Wesley, Reading, Mass. [**D**].

ROGERS, R. J. and G. C. ANDREWS [1975]. "Simulating Planar Systems Using a Simplified Vector-Network Method," *Mech. Mach. Theory*, *10*, 509–519. [**14.30**].

ROGERS, R. J. and G. C. ANDREWS [1977]. "Dynamic Simulation of Planar Mechanical

Systems with Lubricated Bearing Clearances Using Vector Network Methods," *J. Eng. Ind. Trans. ASME Ser. B, 99*, 131–137. [**14.30**].

ROMANELLI, M. J. [1960]. "Runge-Kutta Methods for the Solution of Ordinary Differential Equations," in *Mathematical Methods for Digital Computers*, Vol. 1, ed. by A. Ralston and H. S. Wilf, Wiley, New York, pp. 110–120. [**G**].

ROOT, R. E. [1932]. *Dynamics of Engine and Shaft*, Wiley, New York. [**13.20**].

ROSENAUER, N. and A. H. WILLIS [1953]. *Kinematics of Mechanisms*, Associated General Publications, Sydney, Australia. [**6.10, 6.22, 8.21**].

ROTHBART, H. A. [1956]. *Cams, Design, Dynamics and Accuracy*, Wiley, New York. [**4.10, 4.20, 4.30**].

ROUTH, E. J. [1868]. *Dynamics of a System of Rigid Bodies*, Part I, Macmillan, London (5th ed., 1891). [**8.13**].

SADLER, J. P. and R. W. MAYNE [1973]. "Balancing of Mechanisms by Nonlinear Programming," in *Proc. 3rd Appl. Mech. Conf.*, Oklahoma State Univ., Stillwater, Okla., Paper No. 29. [**13.30**].

SANDOR, G. N. [1959]. "A General Complex Number Method for Plane Kinematic Synthesis with Applications," Dr. of Eng. Sci. thesis, Columbia Univ., New York. [**5.12**].

SANDOR, G. N, R. E. KAUFMAN, A. G. ERDMAN, T. J. FOSTER, J. P. SADLER, C. Z. SMITH, and T. N. KERSHAW [1970]. "Kinematic Synthesis of Geared Linkages," *J. Mech., 5*, 59–87. [**7.30**].

SAWYER, R. T. [1972]. "Gas Turbine Locomotives," *Mech. Eng., 94* Aug. issue, 26. [**1.33**].

SCHENCK TREBEL [1973]. "Theory of Balancing," Schenck Trebel Corp., Deer Park, N.Y. [**13.12**].

SCHWAMB, P., A. L. MERRILL, W. H. JAMES, and V. L. DOUGHTIE [1947]. *Elements of Mechanism*, 6th ed., Wiley, New York. (Note: The 1st ed. of 1904 was based on Prof. Schwamb's lecture notes of 1885.) [**1.55, 3.10**].

SCOTT, G. L. [1962]. "Elements of Gears and Basic Formulas," in *Gear Handbook*, ed. by D. W. Dudley, McGraw-Hill, New York, Chap. 4. [**3.40**].

SENGER, W. I. [1961]. "Balancing of Rotating Machinery, Part II. Practice of Balancing," in *Shock and Vibration Handbook*, Vol. 3, ed. by C. H. Harris and C. E. Crede, McGraw-Hill, New York, pp. 39.23–39.41. [**13.40**].

SESHU, S. and M. B. REED [1961]. *Linear Graphs and Electric Networks*, Addison-Wesley, Reading, Mass. [**D**].

SHETH, P. N. [1972]. "A Digital Computer Based Simulation Procedure for Multiple Degree of Freedom Mechanical Systems with Geometric Constraints," Ph.D. thesis, Univ. of Wisconsin, Madison, Wisc. [**14.30, D**].

SHETH, P. N. and J. J. UICKER, JR. [1972]. "IMP (Integrated Mechanisms Program), A Computer-Aided Design Analysis System for Mechanisms and Linkage," *J. Eng. Ind. Trans. ASME Ser. B, 94*, 454–464. [**14.30, D**].

SINCLAIR, A. [1907]. *Development of the Locomotive Engine*, Angus Sinclair Publishing Co., New York. Reprinted with annotation of J. H. White, Jr., M.I.T. Press, Cambridge, Mass. (1970). [**1.55**].

SINGER, C. S. [1959]. *A Short History of Scientific Ideas to 1900*, Oxford Univ. Press, New York. [**5.22**].

SMITH, D. A. [1971]. "Reaction Forces and Impact in Generalized Two-Dimensional Mechanical Dynamic Systems," Ph.D. dissertation, Mech. Eng., Univ. of Michigan, Ann Arbor, Mich. [**D**].

SMITH, D. A., M. A. CHACE, and A. C. RUBENS [1972]. "The Automatic Generation of a Mathematical Model for Machinery Systems," *ASME Paper 72-Mech-31*, ASME, New York. [**14.30**].

SMITH, D. E. [1958]. *History of Mathematics* (2 vols.), Dover, New York. [**3.34**].

SOMMERFELD, A. [1964]. *Mechanics*, Academic Press, New York. [**5.22**].

SONI, A. H. [1971]. "Structural Analysis of Two General Constraint Kinematic Chains and Their Practical Applications," *J. Eng. Ind. Trans. ASME Ser. B*, *93*, 231–238. [**8.21**].

SONI, A. H. [1974]. *Mechanism Synthesis and Analysis*, McGraw-Hill, New York. [**5.12**].

SPOTTS, M. F. [1967]. "Formula for Epicyclic Gear Train of Large Reduction," in Chironis [1967, p. 114]. [**3.34**].

STONG, C. L. [1975]. "The Amateur Scientist," *Sci. Am.*, *233*, no. 6, 120–123.

SUH, C. H. and C. W. RADCLIFFE [1978]. *Kinematics and Mechanisms Design*, Wiley, New York. [**5.12**].

SVOBODA, A. [1948]. *Computing Mechanisms and Linkages*, McGraw-Hill, New York (Reprinted by Dover, New York, 1965). [**1.55**].

SYNGE, J. L. [1960]. "Classical Dynamics," in *Encyclopedia of Physics*, Vol. III/1, ed. by S. Flugge, Springer-Verlag, Berlin, pp. 1–225. [**8.13**].

TAO, D. C. [1964]. *Applied Linkage Synthesis*, Addison-Wesley, Reading, Mass. [**5.12**].

TAO, D. C. [1967]. *Fundamentals of Applied Kinematics*, Addison-Wesley, Reading, Mass. [**3.10, 3.31**].

TAYLOR, C. F. [1966, 1968]. *The Internal Combustion Engine in Theory and Practice*, Vol. I (1966), Vol. II (1968), M.I.T. Press, Cambridge, Mass. [**1.41, 12.82, 13.20, 13.22**].

TEPPER, F. R. and G. G. LOWEN [1972]. "General Theorems Concerning Full Force Balancing of Planar Linkages by Internal Mass Redistribution," *J. Eng. Ind. Trans. ASME Ser. B*, *94*, 789–796. [**13.30**].

TEPPER, F. R. and G. G. LOWEN [1973]. "Two General Rules for Full Force Balancing of Planar Linkages," in *Proc. 3rd Appl. Mech. Conf.*, Oklahoma State Univ., Stillwater, Okla. Paper No. 10. [**13.30**].

TEPPER, F. R. and G. G. LOWEN [1975]. "Shaking Force Optimization of Four-Bar Linkages with Adjustable Constraints on Ground Bearing Forces," *J. Eng. Ind. Trans. ASME Ser. B*, *97*, 643–651. [**13.30**].

TESAR, D. and G. K. MATTHEW [1976]. *The Dynamic Synthesis, Analysis, and Design of Modeled Cam Systems*, Lexington Books, D. C. Heath, Lexington, Mass. [**4.10**].

THUMIM, C. [1951]. "Starting Friction in Mechanical Linkages," *Prod. Eng.*, 151–155. [**11.62**].

TIMOSHENKO, S. P. [1953]. *History of Strength of Materials*, McGraw-Hill, New York. [**Intro. to Part 3**].

TIMOSHENKO, S. P., and D. H. YOUNG [1948]. *Advanced Dynamics*, McGraw-Hill, New York. [**8.13, 12.30, 13.10**].

TOWNES, H. W. and D. O. BLACKKETTER [1971]. "Cam of Arbitrary Shape," *Mech. Eng. News*, *8*, no. 4, Am. Soc. Eng. Educ., 34–39. [**7.42**].

TURNBULL, H. W. [1951]. *The Great Mathematicians*, Methuen, London. [**Intro. to Part 2**].

UICKER, J. J., JR. [1973]. "Users' Guide for IMP (Integrated Mechanisms Program). A Problem-Oriented Language for the Computer-Aided Design and Analysis of Mechanisms," *NSF Rept.*, Res. Grant GK-4552, Univ. of Wisconsin, Madison, Wisc. [**14.30**].

UICKER, J. J., JR., J. DENAVIT, and R. S. HARTENBERG [1964]. "An Iterative Method for the Displacement Analysis of Spatial Mechanisms," *J. Appl. Mech., 31, Trans. ASME Ser. E, 86*, 309–314. [**9.20, 9.30**].

USHER, A. P. [1929]. *A History of Mechanical Inventions*, Harvard Univ. Press, reprinted by Beacon Press, Boston (1959). [**1.20, 1.21, 3.10**].

USPENSKII, V. A. [1961]. *Some Applications of Mechanics to Mathematics*, Ginn/Blaisdell, Waltham, Mass. [**11.42**].

VAN DYKE, C. and B. PAUL [1979]. "User's Manual for Program CAMKIN (CAM KINematics)," MEAM Report, Dept. of Mech. Eng. and Appl. Mech., University of Pennsylvania, Philadelphia. [**4.40**].

WALKER, M. J. and K. OLDHAM [1978]. *Mechanism and Machine Theory, 13*, 175–185. [**13.30**].

WEBSTER, A. G. [1912]. *Dynamics of Particles and of Rigid, Elastic and Fluid Bodies*, reprinted by Dover, New York (1959). [**8.13**].

WEINBERG, G. D., V. R. GRACE, G. R. EDWARDS, H. S. ROBINSON, P. THROCKMORTON, and E. K. RALPH [1965]. "The Antikythera Shipwreck Reconsidered," *Trans. Am. Philos. Soc. New Ser., 55*, Part 3, 1–48. [**3.10**].

WIEDERRICH, J. L. and B. ROTH [1976]. "Momentum Balancing of Four-Bar Linkages," *J. Eng. Ind. Trans. ASME Ser. B, 98*, 1289–1295. [**13.30**].

WILCOX, J. B. [1967]. *Dynamic Balancing of Rotating Machinery*, Pitman, London. [**13.11, 13.12**].

WILDE, D. J. and C. S. BEIGHTLER [1967]. *Foundations of Optimization*, Prentice-Hall, Englewood Cliffs, N.J. [**9.30**].

WILLIS, R. W. [1870]. *Principles of Mechanisms*, Longmans, Green, London. [**1.10, 1.55, 3.10, 5.15**].

WITTENBAUER, F. [1923]. *Graphische Dynamik*, Springer, Berlin. [**1.55, 8.24, 12.30, 12.40, 12.93**].

WOO, L. S. [1967]. "Type Synthesis of Plane Linkages," *J. of Eng. Ind. Trans. ASME Ser. B, 89*, 159–172. [**D**].

WOODBURY, R. S. [1959]. *History of the Gear-Cutting Machine*, M.I.T. Press, Cambridge, Mass. [**3.10, 3.44**].

YATES, R. C. [1959]. *A Handbook on Curves and Their Properties*, J. W. Edwards, Publ., Ann Arbor, Mich. [**5.21**].

ZIEGLER, H. [1968]. *Principles of Structural Stability*, Ginn/Blaisdell, Waltham, Mass. [**11.42, 11.43**].

ZWIKKER, C. [1963]. *The Advanced Geometry of Plane Curves and Their Applications*, Dover, New York. [**2.10**].

LISTING OF REFERENCE COAUTHORS*

Andrews, G. C.—*Rogers, R. J.*

Amin, A.—*Paul, B.*

Ault, E. S.—*Hall, A. S.*

Banerjee, A.—*Gupta, K. C.*

Bareiss, R. A.—*Lessells, J. M.*

Bartel, D. L.—*Langrana, N. A.*

Beightler, C. S.—*Wilde, D. J.*

Berkof, R. S.—*Lowen, G. G.*

Björck,—*Dahlquist, G.*

Blackketter, D. O.—*Townes, H. W.*

Booser, E. R.—*Brown, E. D.*

Boyd, J.—*Raimondi, A. A.*

Calahan, D. A.—*Orlandea, N.*

Chace, M. A.—*Orlandea, N.*
  Smith, D. A.

Crane, E. J.—*Ham, C. W.*

Crossley, F. R. E.—*Davies, T. H.*

Denavit, J.—*Hartenberg, R. S.*
  Uicker, J. J., Jr.

Dhande, S. G.—*Chakraborty, J.*

Dorn, W. S.—*McCracken, D. D.*

Doughtie, V. L.—*Schwamb, P.*

Drabkin, J. E.—*Cohen, M. R.*

Edwards, G. R.—*Weinberg, G. D.*

Erdman, A. G.—*Dimerogonas, A. D.*
  Sandor, G. N.

Fein, R. S.—*Ling, F. F.*

Foster, F. J.—*Sandor, G. N.*

Fox, R. L.—*Gupta, K. C.*

Freudenstein, F.—*Buschbaum, F.*
  Dobrjansky, L.

George, G. R.—*Conte, F. L.*

Grace, V. R.—*Weinberg, D. D.*

Grammel, R.—*Biezeno, C. B.*

Gubanova, I. I.—*Panovoko, Y. G.*

Harlow, M. W.—*Huston, R. L.*

Hartenberg, R. S.—*Uicker, J. J., Jr.*

Hud, G.—*Paul, B.*

James, W. H.—*Schwamb, P.*

Johnston, E. R.—*Beer, F. P.*

Kaufman, R. E.—*Sandor, G. N.*

Kershaw, T. N.—*Sandor, G. N.*

Kesavan, H. K.—*Andrews, G. C.*
  Koenig, H. E.

Klaus, E. E.—*Ling, F. F.*

Krajcinovic, D.—*Paul, B.*

Lehman, T. J.—*Dix, R. C.*

Lowen, G. G.—*Berkof, R. S.*
  Tepper, F. R.

Luther, H. A.—*Carnahan, B.*

Maddison, F. R.—*Bedini, S. A.*

Malcom, M. A.—*Forsythe, G. E.*

Matthew, G. K.—*Tesar, D.*

Mayne, R. W.—*Conte, F. L.*
  Sadler, J. P.

McClean, L.—*Oravas, G. E.*

Merrill, A. L.—*Schwamb, P.*

Moler, C. B.—*Forsythe, G. E.*

Newman, J.—*Kasner, E.*

Ocvirk, F. W.—*Mabie, H. H.*

Oldham, K.—*Walker, M. J.*

Owens, R. S.—*Brown, E. D.*

Passerello, C. E.—*Huston, R. L.*

Primrose, E. J. F.—*Freudenstein, F.*

Radcliffe, C. W.—*Suh, C. H.*

Ralph, E. K.—*Weinberg, G. D.*

Reed, M. B.—*Seshu, S.*

---

*For those coauthors whose names do not appear alphabetically in the main list of references, look up the publications of the coauthors indicated here in *italics*.

Reeve, J. E.—*Kilmister, C. W.*

Rheinboldt, W. C.—*Ortega, J. M.*

Robinson, H. S.—*Weinberg, G. D.*

Rogers, F. S.—*Albert, C. D.*

Roth, B.—*Freudenstein, F.*
Wiederrich, J. L.

Rubens, A. C.—*Smith, D. A.*

Sadler, J. P.—*Conte, F. L.*
Sandor, G. N.

Sandor, G. N.—*Dimerogonas, A. D.*
Freudenstein, F.
Mohan Rao, A. V.

Schwarz, O. B.—*Grafstein, P.*

Sheth, P. N.—*Chace, M. A.*

Smith, C. Z.—*Sandor, G. N.*

Smith, D. A.—*Chace, M. A.*

Smith, G. M.—*James, M. L.*

Soler, A.—*Paul., B.*

Stadelbauer, D. G.—*Muster, D.*

Tabor, D.—*Bowden, F. P.*

Tait, P. G.—*Kelvin (Lord)*

Tepper, F. R.—*Berkof, R. S.*
Lowen, G. G.

Tesar, D.—*Benedict, C. E.*
Oleksa, S. A.

Throckmorton, P.—*Weinberg, G. D.*

Tokad, Y.—*Koenig, H. E.*

Trummel, J. M.—*Carson, W. F.*

Uicker, J. J., Jr.—*Sheth, P. N.*

Wahl, A. M.—*Lessells, J. M.*

Warner, W. H.—*Goodman, L. E.*

West, J. W.—*Paul, B.*

Wilkes, J. O.—*Carnahan, B.*

Willis, A. H.—*Rosenauer, N.*

Wolford, J. C.—*Givens, E. J.*
James, M. C.

Young, D. H.—*Timoshenko, S. P.*

Yu, E. Y.—*Paul, B.*

ANSWERS TO
SELECTED PROBLEMS

1.37-2 True, False, True, False, True, True   **1.37-3** Structure cannot undergo finite displacement   **1.37-6** Continuously revolve   **1.37-8** $c = 12.04$ cm   **1.37-11** It is a change-point mechanism which can function as a double-crank   **1.37-13** Statement is consistent if change points are considered

1.37-21

Setting	ψ_{min}	θ	ψ_{max}	θ
Full	$-50.6°$	$18.42°$	$42.8°$	$207.8°$
Left	$-52.0°$	$41.8°$	$3.0°$	$244.7°$
Right	$0.0°$	$41.8°$	$55.0°$	$244.7°$
Center	$-22.4°$	$64.2°$	$23.0°$	$285.6°$

1.43-3 $\phi = \lambda\left(C_1 \sin\theta - \frac{1}{9} C_3 \sin 3\theta + \frac{1}{25} C_5 \sin 5\theta + \ldots\right)$ **1.43-4** For detailed results see references given in footnote 20 of Section **1.41**.

1.43-5

Case:	1	2	3	4	5	6
TR	1.11	1.26	1.47	1.18	1.43	1.92

1.43-6 $TR = 1.48$

2.50-4 $r = r_0 \cosh(\omega t/\sqrt{3}) + \sqrt{3}(\dot{r}_0/\omega) \sinh(\omega t/\sqrt{3})$ **2.50-6** $\ddot{Z} = -\omega^2\,\overrightarrow{OP}$ **2.50-7** $\ddot{Z} = -4\omega^2\,\overrightarrow{CP}$, $|\ddot{Z}| = 4\omega^2 r$ **2.50-11** $a_\varepsilon = -a\omega^2 + a[\omega + (V_r/a)]^2 \cos\theta + a\dot{\omega}$ $\sin\theta$ $a_\eta = a[\omega + (V_r/a)]^2 \sin\theta - a\dot{\omega}(\cos\theta - 1)$ **2.50-12** $a_x = 486,200$ in./sec^2, $a_y = 98,140$ in./sec^2 **2.50-22** $\ddot{Z}_Q = (a\dot{\theta}^2 + s\ddot{\theta})ie^{i\theta} - s\dot{\theta}^2 e^{i\theta}$ $\ddot{Z}_N = \ddot{Z}_Q - ib\dot{\theta}^2 e^{i\theta}$ $- b\ddot{\theta}e^{i\theta}$ **2.50-23** $Z = \frac{1}{2}[Z_0 + (\dot{Z}_0/p)]e^{pt} + \frac{1}{2}[Z_0 - (\dot{Z}_0/p)]e^{-pt}$ Trajectory asymptotically approaches straight line   **2.50-24** $Z = \frac{1}{2}[Z_0 + i(\dot{Z}_0/p)]e^{pt} + \frac{1}{2}[Z_0 - i(\dot{Z}_0/$

$p)]e^{-pt}$ Trajectory is ellipse **2.50-40** $x = (h \sin \theta + \xi) \cos \theta - \eta \sin \theta$ $y = (h \sin \theta + \xi) \sin \theta + \eta \cos \theta$ where (ξ, η) are rectangular coordinates of the point fixed in link 4, and θ is rotation angle of link 3.

3.34-3

Gear:	1	2	3	Reverse
Ratio:	3.240	1.636	1	-4.050

3.34-9

Fixed Gear	Ratio
A	$\omega_S/\omega_C = -3.5$
C	$\omega_S/\omega_A = 4.5$
S	$\omega_C/\omega_A = 9/7 = 1.286$

3.34-10 $\omega_6 = (\omega_5 - \bar{R}_{51}\omega_1)/(1 - \bar{R}_{51})$ $\bar{R}_{51} = N_1 N_3/(N_2 N_5) = 0.11494$ $\omega_6 = 0.0909$ rpm **3.34-11** $\omega_1/\omega_6 = -7.700$ **3.34-12** $\omega_5/\omega_6 = 0.885$ **3.34-13** $\omega_5/\omega_1 = 0.1149$ **3.34-14** $\omega_5 = 0.11494$ rpm **3.34-15** $\omega_D/\omega_C = 1 - N_A N_F/(N_B N_D)$ $\omega_D/\omega_C = 1/10{,}000$ **3.34-16** Hint: See footnote 14 in Sec. **3.31** **3.34-17** $\omega_C = 1508.2$ rpm **3.34-18** (a) $\dfrac{\omega_1}{\omega_7} = \dfrac{N_2 N_3 (N_6 + N_4)}{N_1 (N_2 N_6 - N_3 N_4)} = -39{,}406$ (b) See "Hint" in problem 16. Example: Change N_2 to 52. Then $\omega_1/\omega_7 = +558$ **3.34-19** $\omega_5/\omega_4 = 30.75$ **3.34-20** $\omega_5/\omega_4 = -29.75$ **3.34-21** $\omega_9/\omega_2 = -1.76$ **3.34-22** $\omega_4 = (N_2 \omega_2 + N_1 \omega_1)/(N_1 + N_2)$ **3.34-23** $\dfrac{\omega_5}{\omega_1} = \dfrac{1 - N_3 N_4/(N_2 N_5)}{1 + (N_3/N_1)} = \dfrac{1}{16}$ **3.34-24** (a) $\omega_1/\omega_4 = 20/3$; (b) load rises; (c) 40/3 **3.34-25** High: $\omega_4/\omega_1 = 1$ Low: $\omega_4/\omega_1 = 0.2826$ Neutral: ω_4 is independent of ω_1

3.34-26

Gear:	Low	Med.	High	Reverse
Ratio:	2.469	1.469	1.0	-2.130

3.34-27 (a) $\omega_7/\omega_6 = 1 + (N_1 N_4)/(N_3 N_5)$ (b) $N_1 = N_3 = 30$; $N_2 = 20$; $N_4 = N_5 = 50$, $N_8 = 30$ (Example) **3.34-29** (a) For one Rev. of: Dial 1, Dial 2, Dial 3 Distance (cubits): 2400, 2.16×10^7, 6.48×10^8 (b) All dials rotate clockwise. Worms F and K are left-handed. **3.34-30** $\dfrac{\omega_C}{\omega_S} = \dfrac{1 + \epsilon(N_A/N_S)(N_H/N_J)(N_D/N_E)}{1 + (N_A/N_S)}$ **3.34-31** $\omega_3 = (N_1/N_2)(\omega_9 - \omega_1)$; $\omega_4 = -(N_1/N_2)(\omega_9 + \omega_1)$ **3.34-32** (a) $\omega_Y/\omega_Z = 6250$; (b) $V_Z = 1/200$ in./sec (c) $V_Z = 1/2000$ in./sec; (d) clockwise **3.34-33** $\dfrac{\omega_3}{\omega_6} = -\dfrac{1}{49}$; $\dfrac{\omega_4}{\omega_6} = 0$; $\dfrac{\omega_5}{\omega_6} = \dfrac{1}{50}$ **3.45-7** $C = 3.625$ in. **3.45-8** $C = 4.775$ in. **3.45-9** $C = 0.875$ in. **3.45-10** $C = 1.592$ in. **3.45-12** (a) 32; (b) 1/4 in.; (c) 0.03925 in.; (d) 7.4215 in.; (e) 8.5 in.; (f) 7.7452 in. **3.45-13** (a) 32; (b) 1/4 in.; (c) 0.0625 in.; (d) 7.375 in.; (e) 8.5 in.; (f) 7.5175 in. **3.45-14** (a) 20; (b) 1/5 in.; (c) 0.314 in.; (d) 3.4215 in.; (e) 4.4 in.; (f) 3.8726 in. **3.45-15** $D_1 = 12$ in., $D_2 = 8$ in., $N_1 = 48$, $N_2 = 32$, $p = 0.7854$ in. **3.45-16** (a) $N_2 = 45$; (b) $c = 12.5$ in.; (c) $p = 1.0472$ in.; (d) $A = 0.3333$ in.; (e) 1.5; (f) $D_{b1} = 9.681$ in. **3.45-19** 1.442; **3.45-20** 1.755; **3.45-21** 1.843; **3.45-22** 2.299

Prob. **4.50**-	6	7	8	9	10	11	12	13	14
Max $\lvert \dot{y} \rvert =$	41.88	53.33	53.33	26.67	40.00	31.42	30.00	36.0	36.0
Max $\lvert \ddot{y} \rvert =$	1755	2233	1422	711.1	1600	1974	1131	864	1872

Prob. **4.50**-	15,30	16,31	17,32	18,33	19,34	20,35	21,36
min $\lvert \rho \rvert$	0.790	0.790	0.859	0.854	0.746	1.067	0.875
max $\lvert$press. $\angle \rvert$	36.9°	45.4°	39.0°	42.2°	49.2°	41.0°	—

5.16-5 (b) $x = \dfrac{u}{2} - \dfrac{v}{2}\dfrac{\sin\phi}{1-\cos\phi}$, $y = \dfrac{v}{2} + \dfrac{u}{2}\dfrac{\sin\phi}{1-\cos\phi}$ **5.16-6** (b) $x + iy = (u +$

$iv)/\left(1 - \dfrac{1}{2}e^{i\phi}\right)$ **5.16-19** P traces an ellipse. Both centrodes are circles. **5.16-**

23 Let $\psi = \sin^{-1}[(a/l)\sin\theta]$ Parametric Eqs. of fixed centrode are: $x_1 = a\cos\theta + l$ $\cos\psi$; $y_1 = x_1\tan\theta$. Parametric Eqs. of moving centrode are: $\xi_1 = y_1\cos\psi$; $\eta_1 = y_1\sin$ ψ (with y_1 as above). **5.16-25** For pulley A: the pulley is the moving centrode. Fixed centrode is a vertical line tangent to pulley A. For pulley B: Moving centrode is a circle of radius 1/3 that of pulley B. Fixed centrode is vertical line tangent to moving centrode. **5.16-26** Moving centrode is circle of radius $r = AB$, centered at B. Fixed centrode is circle of radius $2r$, centered at A. **5.16-27** Fixed centrode is circle of radius r, centered at A. Moving centrode is circle of radius $2r$, centered at B. **5.16-32** Fixed centrode is straight line. Moving centrode is circle of radius v/ω **5.16-33** Fixed centrode: $(x/b)^2 = 1 + (y/b)^2$ Moving centrode: $(\eta/b)^2 = (\xi/b)^4 - (\xi/b)^2$ **5.43-2** $\theta_e = 7.16°$; stable **5.43-8** Radius of curvature $= 0.441$ in. **5.43-10 and 5.43-11** Using notation of Eq. (5.20-4) and Fig. **5.42-1** $CP = 4\rho_m\sin\theta/(2 - |D/\rho_m|)$

6.14-3 (d) see Fig. **6.23**-1. **6.14-4** see Ans. to **6.24**-9. **6.14-8** 0.660 in./sec **6.24-9** (a) 5.27 ft/sec; (b) 1.09 rad/sec (cw); (c) 8.20 ft/sec; (d) -6.030 rad/sec^2

7.60-1 $\dot{x}_3 = (\xi_3'S_3 - \eta_3C_3)\dot{\theta}_3$, $\dot{y}_3 = -(\eta_3S_3 + \xi_3'C_3)\dot{\theta}_3$ $\ddot{x}_3 = (\xi_3'S_3 - \eta_3C_3)\ddot{\theta}_3 +$ $(\xi_3'C_3 + \eta_3S_3)\dot{\theta}_3^2$ $\ddot{y}_3 = -(\eta_3S_3 + \xi_3'C_3)\ddot{\theta}_3 - (\eta_3C_3 - \xi_3'S_3)\dot{\theta}_3^2$ where $\xi_3' \equiv a_3 - \xi_3$ **7.60-6** (a) $\theta = \pm\cos^{-1}[(a^2 + d^2 - r^2)/(2ad)]$ $\phi = \arctan_2[a\sin\theta, (d - a\cos\theta)]$ (b) $\dot{\theta} = \dfrac{\dot{r}}{a\sin(\theta+\phi)}$; $\dot{\phi} = \dfrac{r\cos(\theta+\phi)}{r\sin(\theta+\phi)}$ **7.60-9** $\dot{\theta}_1 = \dfrac{-a\dot{\theta}_2\cos\theta_2}{b\cos\theta_1}$; $V_c = -(b\dot{\theta}_1$ $\sin\theta_1 + a\dot{\theta}_2\sin\theta_2)$ **7.60-10** $\dot{s} = (ak_1\cos\theta_1 + bk_2\cos\theta_2)u$ where $k_2 = (d\sin\theta_2 - g\cos\theta_2 + b\sin\theta_2)^{-1}$ $k_1 = -k_2b\sin\theta_2/(a\sin\theta_1)$ **7.60-11** $\dot{\theta} = -(V_A/f)\cos^2\theta$ $\ddot{\theta} = -(V_A/f)\dot{\theta}\sin 2\theta + (\dot{V}_A/f)\cos^2\theta$ **7.60-12** $\dfrac{\dot{\phi}}{\dot{\psi}} = -\dfrac{a}{R}\dfrac{\cos(\theta_2-\psi)}{\cos(\theta_2-\phi)}$ where θ_2 = angle between $\overrightarrow{CP}$ and $\overrightarrow{AB}$ **7.60-13** $k_{51} = \left(1 + \dfrac{N_1}{N_5}\right)k_{21} - \dfrac{N_1}{N_5}$, where $k_{21} = \dfrac{a_1}{a_2}$ $\dfrac{\sin(\theta_3-\theta_1)}{\sin(\theta_2-\theta_3)}$, $k_{61} = \left(1 + \dfrac{N_5}{N_6}\right)k_{31} - \dfrac{N_5}{N_6}k_{51}$ where $k_{31} = \dfrac{a_1}{a_3}\dfrac{\sin(\theta_1-\theta_2)}{\sin(\theta_2-\theta_3)}$ **7.60-14** $\dot{\theta}_3 = a_1\dot{\theta}_1\dfrac{\sin(\theta_1+\phi)}{C_1\cos\phi + C_2\sin\phi}$ where $C_1 = e\sin\theta_3 + \rho_3\sin\theta_5$ $C_2 = e\cos\theta_3 + \rho_3\cos\theta_5$ **7.60-15** (a) $\dot{r}_3 = a\dot{\theta}_2\sin(\theta_4 - \theta_2)$; $r_3\dot{\theta}_4 = a\dot{\theta}_2\cos(\theta_2 - \theta_4)$ (c) Show that $i\dot{\theta}_2ae^{i\theta_2} - i\dot{\theta}_4r_3e^{i\theta_4} - \dot{r}_3e^{i\theta_4} = 0$ and that this complex equation prescribes velocity construction. Then show that derivative of this equation prescribes acceleration construction. **7.60-17** (a) $r = [a^2 + f^2 - b^2 - 2af\cos\theta]^{1/2}$; $\phi = \sin^{-1}\left(\dfrac{a\sin\theta}{c} - \beta\right)$ where $c = (b^2 + r^2)^{1/2}$; $\beta = \arctan_2(b, r)$ **7.60-18** $\omega_{\text{gear}} = [(a/b)\sin(\phi+\theta) + (a/r)\cos(\phi+\theta)]\dot{\theta}$ **7.60-20** $\dot{x}_c/\dot{\theta} = (b-s)\cos(\alpha-\theta) + h\cos\theta\csc\alpha\sin(\alpha-\theta)$, $\dot{y}_c/\dot{\theta}$

---

* With base circle radius of 2.63.

$= (b - s) \sin (\alpha - \theta) - h \cos \theta \csc \alpha \cos (\alpha - \theta)$ **7.60-21** $\dot{\theta} = \dfrac{\dot{s}}{a}$

$\dfrac{\sin \beta - \sin (\phi - \beta)}{\cos \theta + \cos (\phi + \theta)};$ $\dot{\phi} = \dfrac{\dot{s}}{b} \dfrac{-\cos (\theta + \beta)}{\cos \theta + \cos (\phi + \theta)}$ **7.60-27** $\dfrac{\omega_2}{\omega_1} =$

$1 - \dfrac{(e/a) \sin \theta_1}{[1 - (e/a)^2 \cos^2 \theta_1]^{1/2}}$ **7.60-28** $\theta = -$ arc tan $G(\phi)$; $G(\phi) \equiv$

$\dfrac{F'(\phi) \sin \phi + F(\phi) \cos \phi}{F'(\phi) \cos \phi - F(\phi) \sin \phi};$ $h = F(\phi) \sin (\theta + \phi);$ $\dot{\phi}/\dot{\theta} = -\sec^2 \theta/G'(\phi)$ $\dot{h}/\dot{\theta} = F \cos (\theta + \phi) + (\dot{\phi}/\dot{\theta})[F'(\phi) \sin (\theta + \phi) + F(\phi) \cos (\theta + \phi)]$ **7.60-33** $\nu = \arctan_2 (e \sin \phi, 1 - e \cos \phi)$

8.15-4   (a) $q_1 = \psi_1$, $q_2 = \psi_2$; (b) $b^2 = a^2 + d^2 - 2ad \cos (\psi_3 - \psi_2)$, $a^2 = b^2 + d^2 - 2bd \cos (\psi_4 - \psi_3)$; (c) Let ξ_5 be distance from B along link BC. Then $x = c \cos \psi_1 + a \cos \psi_2 + \xi_5 \cos \psi$; etc.; (d) $(x_B - x_A)^2 + (y_B - y_A)^2 = a^2$, etc.   **8.15-5**   Hint: $X_2 - X_1 = a \cos \theta$, etc.   **8.24-2**   (b) $\mathbf{C} = \begin{bmatrix} a_1 \sin \theta_1 & a_2 \sin \theta_2 \\ a_1 \cos \theta_1 & -a_2 \cos \theta_2 \end{bmatrix}$

8.24-16 through 19. $F = 1$.   **8.24-21** $F = 3$.   **8.24-22** $F = 1$.   **8.24-25, 26, 27** critical forms of various types.   **8.24-32** $F = 0$.   **8.24-33** $F = 1$.   **8.24-34** $F = 2$.   **8.24-36** $F = 1$.   **8.24-39** $F = 2$.   **8.24-40** $F = 2$.   **8.24-41** $F = 1$.   **8.24-42** $F = 2$.   **8.24-43, 44, 45** $F = 1$.

9.40-4 $\psi_1 = 0$, $\psi_3 = 262.1°$; $\psi_1 = 30°$, $\psi_3 = 246.3°$

9.40-6

θ_2:	$-90°$	$-60°$	$-30°$	0	30°	60°	90°
θ_4:	90°	75.960°	68.360°	69.844°	75.536°	82.559	90
r_3:	8	8.2445	9.3934	11.6086	13.8691	15.4426	16
r_5:	0	4.3666	6.6381	6.2023	4.4959	2.3312	0
r_1:	12	11.4623	10.7313	10.8977	11.4295	11.8484	12

9.40-7   (a) $x_P = 1.5 \cos \theta_1 + 10.5 \cos \theta_2 + 0.5 \sin \theta_2$

(b)

θ_1(Deg)	θ_2(Deg)	θ_3(Deg)	x_P(mm)	y_P(mm)
0	-8.263	229.50	11.819	-2.004
20	-12.743	233.45	11.541	-2.291
40	-15.616	240.78	11.127	-2.341

9.40-8

θ_1(Deg):	0	30	60
θ_4(Deg):	76.96	78.36	70.15
θ_5(Deg):	148.28	152.37	148.86

9.40-9

θ_1(Deg)	x_P(in.)	y_P(in.)	x_Q(in.)	y_Q(in.)
0	25.52	33.71	26.29	28.79
20	27.96	33.21	28.34	28.72
40	29.94	32.20	30.00	28.55

9.40-10   (a) $\theta = \cos^{-1} [(a^2 + d^2 - b^2)/(2ad)] + \arctan_2 (h, s)$ $\phi = \cos^{-1} [(b^2 + d^2 - a^2)/(2bd)] - \arctan_2 (h, s)$ (b) $\theta = 86.10°$, $\phi = 24.62°$ when $s = 11$.   **9.40-11**   (a) $\Delta s = (-f_2 \sin \phi + f_1 \cos \phi)/[\cos \phi - h'(s) \sin \phi]$ where $f_1 = a \cos \theta + b \cos \phi - s$, $f_2 = a \sin \theta - b \sin \phi - h$

(b)

θ(deg):	60	90
ϕ(deg):	12.26	24.60
s in.:	13.73	10.91

9.40-12 (a) $F(\phi) = b \sin \phi + f[b \cos \phi + a \cos \theta] - a \sin \theta = 0$ (b) See answer to problem **9.40-11**   **9.40-13** (a) $F(s) = s^2 + [f(s)]^2 - 2a[s \cos \theta + f(s) \sin \theta] + a^2 - b^2 = 0$ (b) See answer to problem **9.40-11**   **9.40-14** $r_2^2 = a_1^2 + f^2 + 2a_1 f \cos q$; $\theta_3 = \arctan_2 [-(a_1 \sin q + f), -a_1 \cos q]$ $\theta_4 = \pi + \sin^{-1}[(a_3/a_4) \sin \theta_3]$; $r_5 = a_3 \cos \theta_3 + a_4 \cos \theta_4$

9.40-15

θ	β	γ	ϕ
30°	32.09°	44.26°	9.59°
60°	35.99°	61.20°	16.78°

9.40-16 $\theta_1 = 85.17°$, $\theta_2 = 20.00°$, $r_1 = 6.142$ in., $r_4 = 3.857$ in.   **9.40-17** $r_1 = 6.142$ **9.40-19** $\phi = 19.107°$, $r = 6.614$ in., $x_P = -3.291$ in.   **9.40-20** $\phi_1 = 30.00°$, $\phi_2 = 4.330$ in., $\phi_3 = 68.43$ in., $\phi_4 = 10.08$ in.

10.60-3 (a) strong   **10.60-4** (a) -0.2175 rad/sec; (b) $(\dot{x}_F, \dot{y}_F) = (-1.406, 0.8118)$ ft/sec   **10.60-5** (a) $(\ddot{x}_C, \ddot{y}_C) = -(1.464, 1.899)$ ft/sec²; (b) $\alpha_{CB} = 1.0117$ rad/sec²   **10.60-6** (a) $\ddot{y}_D = -1.951$ ft/sec²; (b) $(\ddot{x}_F, \ddot{y}_F) = (-6.908, 0.801)$ **10.60-7** (a) $(\dot{x}_E, \dot{y}_E) = -(1.184, 1.888)$ in./sec; (b) $(\ddot{x}_E, \ddot{y}_E) = -(55.45, 85.31)$ in./sec²   **10.60-8** (a) $(\dot{x}_D, \dot{y}_D) = (2.086, 3.613)$ in./sec (b) $(\ddot{x}_D, \ddot{y}_D) = (11.80, -14.36)$ in./sec²   **10.60-9** (a) $\dot{x}_P = -(b + c)\dot{\phi} \sec^2 \phi$, $\dot{r} = -AB(b/r)\dot{\theta} \sin \theta$ (d) $\dot{x}_P = -9.550$ in./sec, $\dot{r} = -5.142$ in./sec (e) $\ddot{x}_P = 5.197$ in./sec², $\ddot{r} = -13.32$ in./sec²   **10.60-10** (b) $\dot{\phi}_1 = (\dot{q}/\phi_2) \cos (q - \phi_1)$; $\dot{\phi}_2 = \dot{q} \sin (\phi_1 - q)$

i:	1	2	3	4
(e) $\dot{\phi}_i$:	0.00	-12.50	0.00	0.00
(f) $\ddot{\phi}_i$:	-14.43	0.00	11.64	-86.82

(units are inches, radians, seconds)

10.60-11 (d) $\dot{r}_1 = -82.31$ in./sec   **10.60-12** (b) $\dot{\theta}_4 = 0.498$ rad/sec, $(\dot{r}_1, \dot{r}_3, \dot{r}_5) = (0.362, 1.653, -1.451)$ in./sec.   **10.60-13** (b) $-(\dot{r}_1/\dot{r}_4) = -2.332$ **10.60-18** (a) See Sec. **1.41**   **10.60-24** strongly, weakly;   **10.60-26** weakly, strongly   **10.60-28** 55,800 in./sec² at $\theta = 54°$

11.24-1 $Q = M + Pa \sin (\phi + \psi)/\cos \psi$ [$\psi = \angle$ of Conrod obliquity]   **11.24-2** $Q = -P - M \cos \psi/a \sin (\phi + \psi)$ [$\psi = \angle$ of Conrod obliquity]   **11.24-3** $Q = a[2F \cos \phi + S \sin (\theta + \phi)/\cos \theta]$ **11.24-4** $Q = -S - (2F \cos \phi \cos \theta)/\sin (\phi + \theta)$ **11.24-5** $Q_1 = a(H \cos \phi_1 - 3V \sin \phi_1)$, $Q_3 = a(H \cos \phi_3 - V \sin \phi_3)$ **11.24-6** $Q_1 = V(2 \tan \phi_2 - 3 \tan \phi_1)$, $Q_2 = V(\tan \phi_3 - 2 \tan \phi_2)$, $Q_3 = H - V \tan \phi_3$ **11.24-7** $Q_\phi = -2a(P + V) \sin \phi \cos \alpha$, $Q_\alpha = -2a(P + V) \cos \phi \sin \alpha$ **11.24-8** $Q = p$.   **11.24-9** $Q_1 = abc\sigma_y$, $Q_2 = abc\sigma_x$ **11.24-10** $P - 4F$.   **11.24-11** $P - 5F$.   **11.24-12** $P - 8F$.   **11.24-13** $P - 7F$.   **11.24-14** $P - 26F$.   **11.24-15** $Q = M - \frac{1}{4}P(D - d)$.   **11.24-16** $Q = M - \frac{1}{4}P(D - d) \sec \phi$ **11.24-17** $Q = \sum_{i=1}^{e} (M_i \omega_i + X_i u_i + Y_i v_i)$; ω_i from Eq. (**10.51**-5b), u_i and v_i from Eqs. (**10.51**-18, 19, 20).   **11.31-1, 2, 3** Refer to answers for Sec. **11.24**.   **11.31-5** $H = -13.29$ lb   **11.31-6** 4,773 in.lb (clockwise).   **11.31-8** 70.29 lb.   **11.31-9** 1,937 in.lb (clockwise).   **11.31-10 through 16** Refer to answers for Sec. **11.24**.   **11.31-17** (b) $\psi = 45°$, $\theta = 26.6°$.   **11.31-18** $\theta_1 = 57.0$, $\theta_2 = 17.1°$ **11.31-20** $R = \frac{1}{2}\sqrt{3} P$, $F = 2.5\sqrt{3} P$ **11.43-1** $\theta = 60°$.   **11.43-2** θ

= 26.3°.   **11.43-3** $\theta = 31.1°$.   **11.43-4** $\theta = 46.5°$.   **11.43-6** $\theta_1 = -14.66°$, $\theta_2 = 10.02°$, $\theta_3 = -1.73°$, $s = 6.88$ in.   **11.43-7** $\theta_1 = 71.57°$, $\theta_2 = 7.18°$, $\theta_3 = 302.79°$, $s = 1.2772$ **11.53-1** $M_1 = -813.4$ in.lb, $X_O = X_A = -X_B = 400$ lb $Y_O = Y_A = -Y_B = -N_B = -120.6$ lb, $M_B = 0$ **11.53-2** $X_3 = -98.35$ lb, $X_O = X_A = -X_B = 98.35$ lb $Y_O = Y_A = -Y_B = -N_B = -29.65$ lb, $M_B = 0$ **11.53-3** $M_1 = -700$ in.lb, $X_O = X_A = 300$ lb, $Y_O = Y_A = -180.5$ lb $X_B = -400$ lb, $Y_B = N_B = 30.5$ lb, $M_B = 0$ **11.53-4** $X_3 = 203.2$ lb, $X_O = X_A = -103.2$ lb, $X_B = 203.2$ lb $Y_O = Y_A = 121.2$ lb, $Y_B = N_B = 28.8$ lb, $M_B = 0$ **11.53-5** $F_1 = 215.9$ lb, $X_O = -87.0$ lb, $Y_O = -42.7$ lb, $X_A = 100$ lb $Y_A = -150.7$ lb, $X_B = -200$ lb, $Y_B = N_B = 0.7$ lb, $M_B = 0$

12.100-3 $\mathcal{I} = mL^2/3$ **12.100-11** (a) $\mathcal{I} = J_1 + Wa^2/g + m_2 r^2 \sin^2 \phi$; (b) $\mathcal{I}\ddot{\phi} + \dfrac{\dot{\phi}^2}{2} m_2 r^2 \sin 2\phi = Wa$ **12.100-13** $-72.73° < \theta \le 45°$ **12.100-14** $\mathcal{I} = J_{1A} + (W_2/g)e^2 \sin^2 \theta + (W_3/g)r^2$ $\dot{\theta} = \{2[E + r\theta W_3 + e(W_1 + W_2)(\cos \theta - 1)]/\mathcal{I}\}^{1/2}$ $\theta = 720°$ at $t = 0.606$ sec   **12.100-15** -0.910 rad $\le \theta \le 0.636$ rad, period $= 0.715$ sec   **12.100-17** $\ddot{h} + (2g/L)h = (p_1 - p_2)/\rho L$ **12.100-23** $(I + ma^2\theta)\ddot{\theta} + ma^2\theta\dot{\theta}^2 + mga\theta \cos \theta = 0$; $f = \sqrt{3ag}/(\pi L)$ **12.100-25** (b) $96.8°$; (c) pendulum revolves continuously.   **12.100-28** $3.8 \le \dot{\theta}_2$ (rad/sec) ≤ 12.2; $C_F = 1.05$ **12.100-29** $(m_1 L^2 + m_2 h^2 \sec^4 \theta)\ddot{\theta} + 2m_2 h^2 \sec^4 \theta \tan \theta\dot{\theta}^2 + W_1 L \sin \theta = 0$ **12.100-46** (a) $1/10\%$; (b) $1/2\%$ **12.100-48** 12.45 **12.100-53** (a) $MEP^* = 120$ psi; (b) $IHP = 6.06$ hp; (c) $I_w = 4.86$ in.lb sec^2 **12.100-54** (a) $MEP = 119.8$ psi; (b) $IHP = 36.35 H_p$; (c) $I_w = 2.674$ in.lb sec^2

13.40-6 14.14 lb at $45°$.   **13.40-8** Deduct 5.54 oz. at $3.03°$ **13.40-11** (a) $W'_A = 42.03$ lb, $\beta'_A = 230.1°$, $W'_B = 19.93$ lb, $\beta'_B = 334.1°$ (b) $R_C = 2345$ lb, $\beta_C = -125.5°$, $R_D = 1328$ lb, $\beta_D = -86.4°$.   **13.40-12** (a) $W'_A = 7.07$ lb, $\beta'_A = 135°$, $W'_B = 7.07$ lb, $\beta'_B = 315°$ (b) $R_C = 10.86$ lb, $\beta_C = 92.64°$, $R_D = 10.86$ lb, $\beta_D = -87.36°$ **13.40-13** (a) $W'_A = 16.25$ oz, $\beta'_A = 149.3°$, $W'_B = 44.54$ oz, $\beta'_B = 321.7°$ (b) $R_C = 22{,}751$ oz, $\beta_C = 141.93°$, $R_D = 22{,}912$ oz, $\beta_D = -38.10°$ (c) $R_C = 34{,}498$ oz, $\beta_C = 93.19°$, $R_D = 34{,}742$ oz, $\beta_D = -86.84°$ **13.40-14** (a) $r_G = 1.17$ in., $\beta_G = 235.2°$ (b) $W_C = 14.1$ lb, $\beta_C = 131.0°$, $W_D = 14.1$ lb, $\beta_D = -49.0°$ **13.40-15** (a) $\beta_5 = 47.77°$, $z_5 = 34.80$ in., (b) $W'_4 = 0.206$ lb, $\beta'_4 = -100.23°$ **13.40-16** Drill $W'_C = 0.145$ lb at $\beta'_C = 211.5°$, $W'_D = 0.096$ lb at $\beta'_D = 40.39°$ **13.40-24** 8.63 oz at $-81.4°$ **13.40-28** (a) $\mu_B = 0$, $\mu_A = 0.417$ lb (1.47 in.^{3}) (b) $\mu_A = \mu_B = 0.287$ lb (1.01 in.^{3})   **13.40-29** $|X_f| = 150.2$ lb, $|Y_f| = 46.2$ lb;   **13.40-31** 1.625 lb   **13.40-36** $|X_f| = -m_{\text{rec}} r\omega^2 4(\lambda/4)^3 \cos 4\theta_1$ $|M_y| = m_{\text{rec}} r\omega^2 [\sqrt{2}d \cos (\theta_1 + 45°) - 4d\lambda \cos 2\theta_1]$

---

* Mean effective pressure (MEP) = (Work per cycle)/$2rA_p$.

INDEX

INDEX

---

*Boldface numbers indicate pages where terms are defined.

Engine balance *(cont.):*
  torque balancing, 536
  V engines, 548-52
Engine dynamics, 465-85, 489-98
  bearing forces, 483-85
  equation of motion, 475
  indicator diagram, 436, **470**-75
  inertia coefficients, 468-69, 483
  nomenclature, 466-67
  resisting moment, 473, 478-79, 509
  steady state, **476**-81
  surplus moment, 479
  turning moment, **370, 470,** 479, 491
Epicycle and epicyclic, 81, **164**
Equilibrium, **376**
  equations of, **377**
  generalized, **376,** 403
  stability criteria, 388, 392
Equipollent, **361**
Equivalent links, **441,** 553
Equivalent mechanisms, 9, **34**-37, 212-13, 223
Equivalent rotor, 518
Euler, 430
Euler invariant, 594
Euler Method, 459, 617-619
Euler-Savary equation, **200-201,** 203
Euler's formula, **40,** 586
Euler's theorem, 284, 291, 593
Evolute, **183**
Excess differential equations, 572

Ferguson's paradox, 100
Finite difference, 613
Firing interval, lag, sequence, **481-82,** 546
Flexibility of rotors, 522
Flexion and extension, **14**
Floating (cam, hinge, link), 250, 252-53
Fluctuation, coefficient of, **486-98**
Follower (cam):
  flat-faced, 138-44, 146
  roller, diameter, 136
  roller, oscillating, 135, 148-51, 213
  roller, translating, 133-34, 144, 213
Force set, **366**
Four-bar mechanism, 8, 12-25
  balance of, 553-59
  basic types, 12
  coupler-point curve, **20**
  displacement analysis, 18
  geared, 214
  kinetic energy of, 441
  lagging or leading form, **19**
  velocity and acceleration, 20, 208, 219, 233-37
  velocity and centripetal coefficients, **21**

Frame, representation as tree, 281
Freudenstein's equation, 304
Friction:
  angle of, **413**
  coefficients of, **412**
  couple, circle and radius, **420**
  in hinge joints, 419-28
  laws of, 411-13
  lock, **418**-19
  in moving links, 464-65
  in sliding pairs, 414-19
  static and kinetic, **412**
  in toggle mechanism, 424
Friction wheels, **66, 67**
Fundamental angle, **472**
Fundamental auxiliary equation, cams, **250**
Fundamental equation of epicyclic trains, **82-83**
Fundamental law of gearing, 66, **104,** 168

Galileo, 229, 375
Galloway mechanism, **17**
Gauss elimination, 302, 600-08
Gear, 63, 67
  annular, internal, or ring, 69, 83
  bevel, 66, **70**-71, 87
  compound or cluster, **76**
  crown, **71**
  helical, 67-**68,** 71
  herringbone (double helical), **68**
  hyperboloidal and hypoid, **71**
  lantern, **64**
  manufacturing, 115-21
  miter, **72**
  nomenclature, 101-02
  noncircular, 67
  spur, 64, **66-68**
  sun and planet, **81,** 83
  worm, 64, **72**
Gear tooth (*see also* involute teeth):
  conjugate, **104, 106**
  cycloidal (trochoidal), 66, **106**-07
  nomenclature, 101-02
  profile, **101**-21
  standard proportions, **114**
  trace, **67**
Gear train:
  carrier or arm, **81,** 83
  center distance, **78, 105**-06, 119
  compound, **76,** 81, 89
  epicyclic, **81**
  fixed carrier ratio, **82,** 85
  Humpage's, 95
  planetary, **83,** 88
  principal members, **83**